TAKE A QUICK LOOK AT WHAT'S AHEAD

W9-BRN-227

FIND THESE USEFUL
APPENDICES ON THE WEB SITE, LOCATED AT
WWW.COURSE.COM/MIS/DBS8

DATABASE SYSTEMS

DESIGN, IMPLEMENTATION, AND MANAGEMENT

PETER ROB • CARLOS CORONEL

COURSE TECHNOLOGY
CENGAGE Learning™

Australia • Brazil • Japan • Korea • Mexico • Singapore • Spain • United Kingdom • United States

COURSE TECHNOLOGY
CENGAGE Learning™

Database Systems: Design, Implementation, and Management
by Peter Rob and Carlos Coronel

Product Manager: Kate Hennessy

Developmental Editor: Deb Kaufmann

Editorial Assistant: Patrick Frank

Marketing Manager: Bryant Chrzan

Marketing Coordinator: Victoria Ortiz

Content Project Manager: Jill Braiewa

Compositor: GEX Publishing Services

Print Buyer: Justin Palmeiro

© 2009 Course Technology, Cengage Learning

ALL RIGHTS RESERVED. No part of this work covered by the copyright herein may be reproduced, transmitted, stored or used in any form or by any means graphic, electronic, or mechanical, including but not limited to photocopying, recording, scanning, digitizing, taping, Web distribution, information networks, or information storage and retrieval systems, except as permitted under Section 107 or 108 of the 1976 United States Copyright Act, without the prior written permission of the publisher.

For product information and technology assistance, contact us at
Cengage Learning Customer & Sales Support, 1-800-354-9706

For permission to use material from this text or product, submit all requests online at **www.cengage.com/permissions**
Further permissions questions can be emailed to
permissionrequest@cengage.com

ISBN-13: 978-1-4239-0201-0

ISBN-10: 1-4239-0201-7

Course Technology
25 Thomson Place
Boston, Massachusetts 02210
USA

Cengage Learning is a leading provider of customized learning solutions with office locations around the globe, including Singapore, the United Kingdom, Australia, Mexico, Brazil, and Japan. Locate your local office at:
international.cengage.com/region

Cengage Learning products are represented in Canada by Nelson Education, Ltd.

To learn more about Course Technology, visit **www.cengage.com/coursetechnology**

To learn more about Cengage Learning, visit **www.cengage.com.**

Purchase any of our products at your local bookstore or at our preferred online store
www.ichapters.com

Microsoft, Windows 95, Windows 98, Windows 2000, and Windows XP are registered trademarks of Microsoft® Corporation. Some of the product names and company names used in this book have been used for identification purposes only and may be trademarks or registered trademarks of their manufacturers and sellers. SAP, R/3, and other SAP product/services referenced herein are trademarks of SAP Aktiengesellschaft, Systems, Applications and Products in Data Processing, Neurottstasse 16, 69190 Walldorf, Germany. The publisher gratefully acknowledges SAP's kind permission to use these trademarks in this publication. SAP AG is not the publisher of this book and is not responsible for it under any aspect of press law.

Printed in the United States of America
2 3 4 5 6 7 12 11 10 09 08

Dedication

To Anne, who remains my best friend after 46 years of marriage. To our son, Peter William, who turned out to be the man we hoped he would be and who proved his wisdom by making Sheena our treasured daughter-in-law. To Sheena, who stole our hearts so many years ago. To our grandsons, Adam Lee and Alan Henri, who are growing up to be the fine human beings their parents are. To my mother-in-law, Nini Fontein, and to the memory of my father-in-law, Henri Fontein—their life experiences in Europe and Southeast Asia would fill a history book and they enriched my life by entrusting me with their daughter's future. To the memory of my parents, Hendrik and Hermine Rob, who rebuilt their lives after World War II's horrors, who did it again after a failed insurgency in Indonesia, and who finally found their promised land in these United States. And to the memory of Heinz, who taught me daily lessons about loyalty, uncritical acceptance, and boundless understanding. I dedicate this book to you, with love.

Peter Rob

To Victoria, my beautiful and wonderful wife of 18 years, who does the hard work of keeping up with me and is a living example of beauty and sweetness. Thanks for being so caring. To Carlos Anthony, my son, who is his father's pride, always teaching me new things and growing up to be an intelligent and talented gentleman. To Gabriela Victoria, my daughter and the princess of the house, who is growing like a rose and becoming a gracious and beautiful angel. To Christian Javier, our little bundle of joy, who is learning and growing, always with so much energy and happiness and is like his father in more ways than one. To all my children, thanks for your laughs, your sweet voices, beautiful smiles, and frequent hugs. I love you; you are my Divine treasure. To my parents for their sacrifice and encouragement. And to Smokey, the laziest one in the family, with no cares, no worries, and with all the time in the world. To all, I dedicate the fruits of many long days and nights. Thanks for your support and understanding.

Carlos Coronel

DEDICATION

BRIEF CONTENTS

The following appendixes and answers to selected questions and problems are included in the Student Online Companion for this text, found at **oc.course.com/mis/dbs8**. CoursePort registration and login are required, using the keycode provided with this book.

TABLE OF CONTENTS

PART I DATABASE CONCEPTS

PART II   DESIGN CONCEPTS

TABLE OF CONTENTS

PART III ADVANCED DESIGN AND IMPLEMENTATION

TABLE OF CONTENTS

PART IV   ADVANCED DATABASE CONCEPTS

TABLE OF CONTENTS

TABLE OF CONTENTS

PART V   DATABASES AND THE INTERNET

PART VI   DATABASE ADMINISTRATION

IN THE STUDENT ONLINE COMPANION

The Student Online Companion can be found at oc.course.com/mis/dbs8. CoursePort registration and login are required, using the keycode provided with this book.

APPENDIX A   DESIGNING DATABASES WITH VISIO PROFESSIONAL: A TUTORIAL

APPENDIX B   THE UNIVERSITY LAB: CONCEPTUAL DESIGN

APPENDIX C   THE UNIVERSITY LAB: CONCEPTUAL DESIGN VERIFICATION, LOGICAL DESIGN, AND IMPLEMENTATION

APPENDIX D   CONVERTING AN ER MODEL INTO A DATABASE STRUCTURE

APPENDIX E   COMPARISON OF ER MODEL NOTATIONS

APPENDIX F   CLIENT/SERVER SYSTEMS

APPENDIX G   OBJECT-ORIENTED DATABASES

APPENDIX H   UNIFIED MODELING LANGUAGE (UML)

APPENDIX I   DATABASES IN ELECTRONIC COMMERCE

APPENDIX J   WEB DATABASE DEVELOPMENT WITH COLDFUSION

APPENDIX K   THE HIERARCHICAL DATABASE MODEL

APPENDIX L   THE NETWORK DATABASE MODEL

ANSWERS TO SELECTED QUESTIONS AND PROBLEMS

For many reasons, few books survive to reach their eighth edition. Authors and publishers who let the success of their earlier work produce the comfort of complacency usually pay the price of watching the marketplace dismantle their creations. This database systems book has been successful through seven editions because we—the authors, editors, and the publisher—paid attention to the impact of technology and to adopter questions and suggestions. We believe that this eighth edition successfully reflects the same attention to such stimuli.

In many respects, rewriting a book is more difficult than writing it the first time. If the book is successful, as this one is, a major concern is that the updates, inserts, and deletions will adversely affect writing style and continuity of coverage. (The medical profession's guiding principle comes to mind: First, do no harm.) Naturally, our own experience is a good starting point, but we also know that authors can develop a proprietary attitude that prevents them from spotting weaknesses or opportunities for improvement. Fortunately, the efforts of a combination of superb reviewers and editors, in addition to a wealth of feedback from adopters and students of the previous editions, helped provide the proper guidance and evaluation for rewriting. We believe that we have incorporated new material while maintaining the flow, integrity, and writing style that made the previous seven editions successful.

CHANGES TO THE EIGHTH EDITION

In this eighth edition, we have added some new features and we have reorganized some of the coverage to provide a better flow of material. Aside from enhancing the already strong database design coverage, we have made other improvements in the topical coverage. Here are a few of the highlights:

- New and updated Business Vignettes showing the impact of database technologies in the real world

- Additional UML (Unified Modeling Language) examples

- Expanded coverage of SQL Server functions

- Additional coverage on types of indexes used by DBMS

- New business intelligence coverage

- Added coverage on Java Database Connectivity (JDBC)

- Additional data security coverage including security vulnerabilities and measures

This eighth edition continues to provide a solid and practical foundation for the design, implementation, and management of database systems. This foundation is built on the notion that, while databases are very practical things, their successful creation depends on understanding the important concepts that define them. It's not easy to come up with the proper mix of theory and practice, but we are grateful that the previously mentioned feedback suggests that we largely succeeded in our quest to maintain the proper balance.

THE APPROACH: A CONTINUED EMPHASIS ON DESIGN

As the title suggests, *Database Systems: Design, Implementation, and Management* covers three broad aspects of database systems. However, for several important reasons, special attention is given to database design.

- The availability of excellent database software enables even database-inexperienced people to create databases and database applications. Unfortunately, the "create without design" approach usually paves the road to any number of database disasters. In our experience, many, if not most, database system failures are traceable to poor design and cannot be solved with the help of even the best programmers and managers. Nor is better DBMS software likely to overcome problems created or magnified by poor design. Using an analogy, even the best bricklayers and carpenters can't create a good building from a bad blueprint.

- Most of the vexing database system management problems seem to be triggered by poorly designed databases. It hardly seems worthwhile to use scarce resources to develop excellent and extensive database system management skills in order to exercise them on crises induced by poorly designed databases.

- Design provides an excellent means of communication. Clients are more likely to get what they need when database system design is approached carefully and thoughtfully. In fact, clients may discover how their organizations really function once a good database design is completed.

- Familiarity with database design techniques promotes one's understanding of current database technologies. For example, because data warehouses derive much of their data from operational databases, data warehouse concepts, structures, and procedures make more sense when the operational database's structure and implementation are understood.

Because the practical aspects of database design are stressed, we have covered design concepts and procedures in detail, making sure that the numerous end-of-chapter problems are sufficiently challenging so students can develop real and useful design skills. We also make sure that students understand the potential and actual conflicts between database design elegance, information requirements, and transaction processing speed. For example, it makes little sense to design databases that meet design elegance standards while they fail to meet end-user information requirements. Therefore, we explore the use of carefully defined trade-offs to ensure that the databases are capable of meeting end-user requirements while conforming to high design standards.

The Systems View

The book's title begins with *Database Systems*. Therefore, we examine the database and design concepts covered in Chapters 1–6 as part of a larger whole by placing them within the systems analysis framework of Chapter 9. We believe that database designers who fail to understand that the database is part of a larger system are likely to overlook important database design requirements. In fact, Chapter 9, Database Design, provides the map for the advanced database design developed in Appendixes B and C. Within the larger systems framework, we can also explore issues such as transaction management and concurrency control (Chapter 10), distributed database management systems (Chapter 12), business intelligence and data warehouses (Chapter 13), database connectivity and Web technologies (Chapter 14), and database administration and security (Chapter 15).

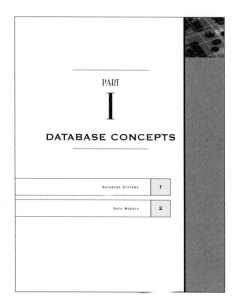

Database Design

The first item in the book's subtitle is *Design*, and our examination of database design is comprehensive. For example, Chapters 1 and 2 examine the development of databases and data models and illustrate the need for design. Chapter 3 examines the details of the relational database model; Chapter 4 provides extensive, in-depth, and practical database design coverage; and Chapter 5 is devoted to critical normalization issues that affect database efficiency and effectiveness. Chapter 6 explores advanced database design topics. Chapters 7 and 8 examine database implementation issues and how the data are accessed through Structured Query Language (SQL). Chapter 9 examines database design within the systems framework and maps the activities required to successfully design and implement the complex real-world database developed in Appendixes B and C.

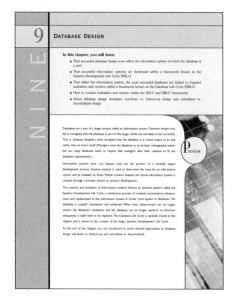

Because database design is affected by real-world transactions, the way data are distributed, and ever-increasing information requirements, we examine major database features that must be supported in current-generation databases and models. For example, Chapter 10, Transaction Management and Concurrency Control, focuses on the characteristics of database transactions and how they affect database integrity and consistency. Chapter 11, Database Performance Tuning and Query Optimization, illustrates the need for query efficiency in a real world that routinely generates and uses terabyte-sized databases and tables with millions of records. Chapter 12, Distributed Database Management Systems, focuses on data distribution, replication, and allocation. In Chapter 13, Business Intelligence and Data Warehouses, we explore the characteristics of the databases that are used in decision support and online analytical processing. Chapter 14, Database Connectivity and Web Technologies, covers the basic database connectivity issues encountered in a Web-based data world, and it shows the development of Web-based database front ends.

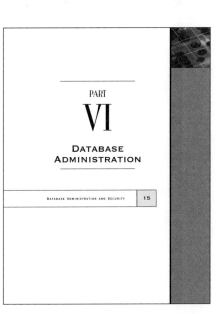

Implementation

The second portion of the subtitle is *Implementation*. We use Structured Query Language (SQL) in Chapters 7 and 8 to show how databases are implemented and managed. Appendixes B and C demonstrate the design of a database that was fully implemented and they illustrate a wide range of implementation issues. We had to deal with conflicting design goals: design elegance, information requirements, and operational speed. Therefore, we carefully audited the initial design (Appendix B) to check its ability to meet end-user needs and to establish appropriate implementation protocols. The result of this audit yielded the final, implementable design developed in Appendix C. The special issues encountered in an Internet database environment are addressed in Chapter 14, Database Connectivity and Web Technologies, and in Appendix J, Web Database Development with ColdFusion.

Management

The final portion of the subtitle is *Management*. We deal with database management issues in Chapter 10, Transaction Management and Concurrency Control; Chapter 12, Distributed Database Management Systems; and Chapter 15, Database Administration and Security. Chapter 11, Database Performance Tuning and Query Optimization, is a valuable resource that illustrates how a DBMS manages the data retrieval operations.

TEACHING DATABASE: A MATTER OF FOCUS

Given the wealth of detailed coverage, instructors can "mix and match" chapters to produce the desired coverage. Depending on where database courses fit into the curriculum, instructors may choose to emphasize database design or database management. (See Figure 1.)

The hands-on nature of database design lends itself particularly well to class projects for which students use instructor-selected software to prototype a student-designed system for the end user. Several of the end-of-chapter problems are sufficiently complex to serve as projects, or an instructor may work with local businesses to give students hands-on experience. Note that some elements of the database design track are also found in the database management track. This is because it is difficult to manage database technologies that are not understood.

FIGURE 1

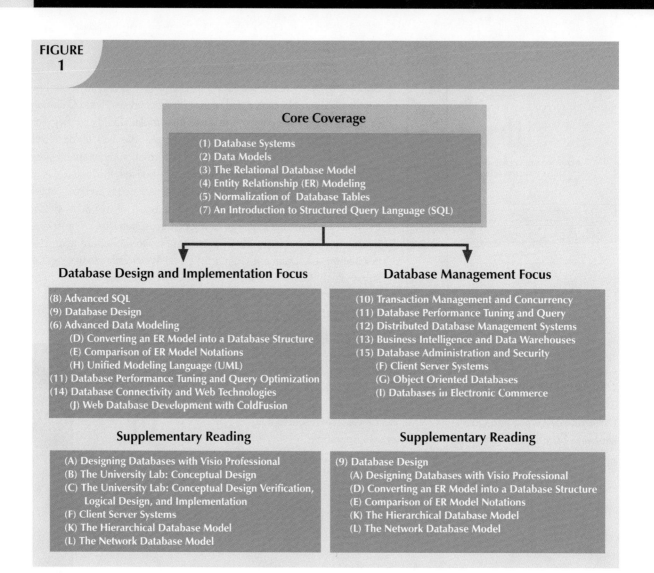

Core Coverage

(1) Database Systems
(2) Data Models
(3) The Relational Database Model
(4) Entity Relationship (ER) Modeling
(5) Normalization of Database Tables
(7) An Introduction to Structured Query Language (SQL)

Database Design and Implementation Focus

(8) Advanced SQL
(9) Database Design
(6) Advanced Data Modeling
   (D) Converting an ER Model into a Database Structure
   (E) Comparison of ER Model Notations
   (H) Unified Modeling Language (UML)
(11) Database Performance Tuning and Query Optimization
(14) Database Connectivity and Web Technologies
   (J) Web Database Development with ColdFusion

Database Management Focus

(10) Transaction Management and Concurrency
(11) Database Performance Tuning and Query
(12) Distributed Database Management Systems
(13) Business Intelligence and Data Warehouses
(15) Database Administration and Security
   (F) Client Server Systems
   (G) Object Oriented Databases
   (I) Databases in Electronic Commerce

Supplementary Reading

(A) Designing Databases with Visio Professional
(B) The University Lab: Conceptual Design
(C) The University Lab: Conceptual Design Verification, Logical Design, and Implementation
(F) Client Server Systems
(K) The Hierarchical Database Model
(L) The Network Database Model

Supplementary Reading

(9) Database Design
   (A) Designing Databases with Visio Professional
   (D) Converting an ER Model into a Database Structure
   (E) Comparison of ER Model Notations
   (K) The Hierarchical Database Model
   (L) The Network Database Model

The options shown in Figure 1 serve only as a starting point. Naturally, instructors will tailor their coverage based on their specific course requirements. For example, an instructor may decide to make Appendix I an outside reading assignment and Appendix A a self-taught tutorial, then use that time to cover client/server systems or object-oriented databases. The latter choice would serve as a gateway to UML coverage.

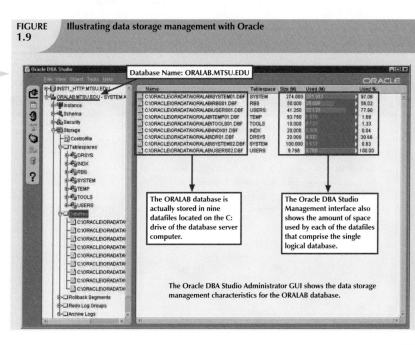

THE RELATIONAL REVOLUTION

Today, we take for granted the benefits brought to us by relational databases: the ability to store, access, and change data quickly and easily on low-cost computers. Yet, until the late 1970s, databases stored large amounts of data in a hierarchical structure that was difficult to navigate and inflexible. Programmers needed to know what clients wanted to do with the data before the database was designed. Adding or changing the way the data was analyzed was a time-consuming and expensive process. As a result, you searched through huge card catalogs to find a library book, you used road maps that didn't show changes made in the last year, and you had to buy a newspaper to find information on stock prices.

In 1970, Edgar "Ted" Codd, a mathematician employed by IBM, wrote an article that would change all that. At the time, nobody realized that Codd's obscure theories would

Business Vignettes highlight part topics in a real-life setting.

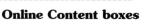

ONLINE CONTENT

Appendixes A through L are available in the Student Online Companion for this book.

Online Content boxes draw attention to material in the Student Online Companion for this text and provide ideas for incorporating this content into the course.

NOTE

No naming convention can fit all requirements for all systems. Some words or phrases are reserved for the DBMS's internal use. For example, the name ORDER generates an error in some DBMSs. Similarly, your DBMS might interpret a hyphen (-) as a command to subtract. Therefore, the field CUS-NAME would be interpreted as a command to subtract the NAME field from the CUS field. Because neither field exists, you would get an error message. On the other hand, CUS_NAME would work fine because it uses an underscore.

Notes highlight important facts about the concepts introduced in the chapter.

FIGURE 1.9   **Illustrating data storage management with Oracle**

A variety of **four-color figures**, including ER models and implementations, tables, and illustrations, clearly illustrate difficult concepts.

The ORALAB database is actually stored in nine datafiles located on the C: drive of the database server computer.

The Oracle DBA Studio Management interface also shows the amount of space used by each of the datafiles that comprise the single logical database.

The Oracle DBA Studio Administrator GUI shows the data storage management characteristics for the ORALAB database.

TEXT FEATURES

SUMMARY

▶ The ERM uses ERDs to represent the conceptual database as viewed by the end user. The ERM's main components are entities, relationships, and attributes. The ERD also includes connectivity and cardinality notations. An ERD can also show relationship strength, relationship participation (optional or mandatory), and degree of relationship (unary, binary, ternary, etc.).

▶ Connectivity describes the relationship classification (1:1, 1:M, or M:N). Cardinality expresses the specific number of entity occurrences associated with an occurrence of a related entity. Connectivities and cardinalities are usually based on business rules.

▶ In the ERM, a M:N relationship is valid at the conceptual level. However, when implementing the ERM in a relational database, the M:N relationship must be mapped to a set of 1:M relationships through a composite entity.

A robust **Summary** at the end of each chapter ties together the major concepts and serves as a quick review for students.

KEY TERMS

binary relationship, 120	identifying relationship, 115	relationship degree, 120
cardinality, 111	iterative process, 127	required attribute, 105
composite attribute, 108	mandatory participation, 118	simple attribute, 108
composite identifier, 107	multivalued attribute, 108	single-valued attribute, 108
connectivity, 111	non-identifying relationship, 113	strong relationship, 115
derived attribute, 110	optional attribute, 106	ternary relationship, 120
existence-dependent, 113	optional participation, 118	unary relationship, 120
existence-independent, 113	participants, 111	weak entity, 113
identifiers, 106	recursive relationship, 120	weak relationship, 113

An alphabetic list of **Key Terms** points to the pages where terms are first explained.

REVIEW QUESTIONS

1. What two conditions must be met before an entity can be classified as a weak entity? Give an example of a weak entity.
2. What is a strong (or identifying) relationship, and how is it depicted in a Crow's Foot ERD?
3. Given the business rule "an employee may have many degrees," discuss its effect on attributes, entities, and relationships. (*Hint:* Remember what a multivalued attribute is and how it might be implemented.)
4. What is a composite entity, and when is it used?
5. Suppose you are working within the framework of the conceptual model in Figure Q4.5.

Review Questions challenge students to apply the skills learned in each chapter.

PROBLEMS

1. Given the following business rules, create the appropriate Crow's Foot ERD.
   a. A company operates many departments.
   b. Each department employs one or more employees.
   c. Each of the employees may or may not have one or more dependents.
   d. Each employee may or may not have an employment history.
2. The Hudson Engineering Group (HEG) has contacted you to create a conceptual model whose application will meet the expected database requirements for the company's training program. The HEG administrator gives you

Problems become progressively more complex as students draw on the lessons learned from the completion of preceding problems.

CoursePort provides a central location from which you can access Course Technology's online learning solutions with convenience and flexibility. You can:

- Gain access to online resources including robust Student Online Companion Web sites.

- Simplify your coursework by reducing human error and the need to keep track of multiple passwords.

- Take advantage of CoursePort's tailored services including personalized homepages.

See the insert card at the back of this book for instructions on how to access this text's CoursePort site.

This Web resource, which you will find referenced throughout the book in the Online Content boxes, includes the following features:

APPENDIXES

Twelve appendixes provide additional material on a variety of important areas, such as using Microsoft® Visio®, ER model notations, UML, object-oriented databases, databases and electronic commerce, and Adobe® ColdFusion®.

ANSWERS TO SELECTED QUESTIONS AND PROBLEMS

The authors have provided answers to selected Review Questions and Problems from each chapter to help students check their comprehension of chapter content and database skills.

DATABASE, SQL SCRIPT, AND COLDFUSION FILES

The Student Online Companion includes all of the database structures and table contents used in the text. For students using Oracle® and Microsoft SQL Server™, the SQL scripts to create and load all tables used in the SQL chapters (7 and 8) are included. In addition, all ColdFusion scripts used to develop the Web interfaces shown Appendix J are included.

VIDEO TUTORIALS

Custom-made video tutorials by Peter Rob and Carlos Coronel, exclusive to this textbook, provide clear explanations of the essential concepts presented in the book. These unique tutorials will help the user gain a better understanding of topics such as SQL, Oracle, ERDs, and ColdFusion.

TEST YOURSELF ON DATABASE SYSTEMS

Brand new quizzes, created specifically for this site, allow users to test themselves on the content of each chapter and immediately see what answers they answered right and wrong. For each question answered incorrectly, users are provided with the correct answer and the page in the text where that information is covered. Special testing software randomly compiles a selection of questions from a large database, so students can take quizzes multiple times on a given chapter, with some new questions each time.

MICROSOFT POWERPOINT® SLIDES

Direct access is offered to the book's PowerPoint presentations that cover the key points from each chapter. These presentations are a useful study tool.

USEFUL WEB LINKS

Students can access a chapter-by-chapter repository of helpful and relevant links for further research.

GLOSSARY OF KEY TERMS

Students can view a PDF file of the glossary from the book. They can also search for keywords and terms in this file; it's quick and easy to use!

Q & A

Helpful question-and-answer documents are available for download. Here you will find supporting material in areas such as Data Dependency/Structural Dependency and Weak Entities/Strong Relationships.

Database Systems: Design, Implementation, and Management, Eighth Edition, includes teaching tools to support instructors in the classroom. The ancillaries that accompany the textbook are listed below. Most of the teaching tools available with this book are provided to the instructor on a single CD-ROM; they are also available on the Web at *www.course.com.*

INSTRUCTOR'S MANUAL

The authors have created this manual to help instructors make their classes informative and interesting. Because the authors tackle so many problems in depth, instructors will find the *Instructor's Manual* especially useful. The details of the design solution process are shown in detail in the *Instructor's Manual* as well as notes about alternative approaches that may be used to solve a particular problem. Finally, the book's questions and problems together with their answers and solutions are included.

SQL SCRIPT FILES FOR INSTRUCTORS

The authors have provided teacher's SQL script files to create and delete users. They have also provided SQL scripts to let instructors cut and paste the SQL code into the SQL windows. (Scripts are provided for Oracle as well as for MS SQL Server.) The SQL scripts, which have all been tested by Course Technology, are a major convenience for instructors. You won't have to type in the SQL commands and the use of the scripts eliminates errors due to "typos" that are sometimes difficult to trace.

COLDFUSION FILES FOR INSTRUCTORS

The ColdFusion Web development solutions are provided. Instructors have access to a menu-driven system that lets teachers show the code as well as the execution of that code.

DATABASES

Microsoft Access™ Instructor databases are available for many chapters that include features not found in the student databases. For example, the databases that accompany Chapters 7 and 8 include many of the queries that produce the problem solutions. Other Access databases, such as the ones accompanying Chapters 3, 4, and 5, include the implementation of the design problem solutions to let instructors illustrate the effect of design decisions. All the MS Access files are in the original 2000 format so that students can use them regardless of what version they have installed on their computers. In addition, instructors have access to all the script files for both Oracle and MS SQL Server so that all the databases and their tables can be converted easily and precisely.

SOLUTIONS

Instructors are provided with solutions to all Review Questions and Problems. Intermediate solution steps for the more complex problems are shown to make the instructor's job easier. Solutions may also be found on the Course Technology Web site at *www.course.com*. The solutions are password-protected.

EXAMVIEW®

This objective-based test generator lets the instructor create paper, LAN, or Web-based tests from test banks designed specifically for this Course Technology text. Instructors can use the QuickTest Wizard to create tests in fewer than five minutes by taking advantage of Course Technology's question banks, or instructors can create customized exams.

POWERPOINT PRESENTATIONS

Microsoft PowerPoint slides are included for each chapter. Instructors can use the slides in a variety of ways; for example, as teaching aids during classroom presentations or as printed handouts for classroom distribution. Instructors can modify the slides provided or include slides of their own for additional topics introduced to the class.

FIGURE FILES

Figure files for solutions presented in the Instructor's Manual allow instructors to create their own presentations. Instructors can also manipulate these files to meet their particular needs.

DISTANCE LEARNING CONTENT

Course Technology, the premier innovator in management information systems publishing, is proud to present online courses in Blackboard and WebCT.

- *Blackboard and WebCT Level 1 Online Content.* If you use Blackboard or WebCT, the test bank for this textbook is available at no cost in a simple, ready-to-use format. Go to *www.course.com* and search for this textbook to download the test bank.
- *Blackboard Level 2 and WebCT Level 2 Online Content.* Blackboard Level 2 and WebCT Level 2 are also available for *Database Systems: Design, Implementation, and Management.* Level 2 offers course management and access to a Web site that is fully populated with content for this book.

ACKNOWLEDGMENTS

Regardless of how many editions of this book are published, they will always rest on the solid foundation created by the first edition. We remain convinced that our work has become successful because that first edition was guided by Frank Ruggirello, a former Wadsworth senior editor and publisher. Aside from guiding the book's development, Frank also managed to solicit the great Peter Keen's evaluation (thankfully favorable) and subsequently convinced PK to write the foreword for the first edition. Although we sometimes found Frank to be an especially demanding taskmaster, we also found him to be a superb professional and a fine friend. We suspect Frank will still see his fingerprints all over our current work. Many thanks.

A difficult task in rewriting a book is deciding what new approaches, topical coverage, and depth of coverage changes can or cannot fit into a book that has successfully weathered the test of the marketplace. The comments and suggestions made by the book's adopters, students, and reviewers play a major role in deciding what coverage is desirable and how that coverage is to be treated.

Some adopters became extraordinary reviewers, providing incredibly detailed and well-reasoned critiques even as they praised the book's coverage and style. Dr. David Hatherly, a superb database professional who is a senior lecturer in the School of Information Technology, Charles Sturt University–Mitchell, Bathhurst, Australia, made sure that we knew precisely what issues led to his critiques. Even better for us, he provided the suggestions that made it much easier for us to improve the topical coverage in earlier editions. Dr. Hatherly's recommendations continue to be reflected in this eighth edition. All of his help was given freely and without prompting on our part. His efforts are much appreciated, and our thanks are heartfelt.

We also owe a debt of gratitude to Professor Emil T. Cipolla, who teaches at St. Mary College. Professor Cipolla's wealth of IBM experience turned out to be a valuable resource when we tackled the embedded SQL coverage in Chapter 8.

Every technical book receives careful scrutiny by several groups of reviewers selected by the publisher. We were fortunate to face the scrutiny of reviewers who were superbly qualified to offer their critiques, comments, and suggestions—many of which were used to strengthen this edition. While holding them blameless for any remaining shortcomings, we owe these reviewers many thanks for their contributions:

Amita G. Chin, Virginia Commonwealth University

Samuel Conn, Regis University

Bill Hochstettler, Franklin University

Larry Molloy, Oakland Community College

Kevin Scheibe, Iowa State University

G. Shankar, Boston University

Because this eighth edition is build solidly on the foundation of the previous editions, we would also like to thank the following reviewers for their efforts in helping to make the previous editions successful: Dr. Reza Barkhi, Pamplin College of Business, Virginia Polytechnic Institute and State University; Dr. Vance Cooney, Xavier University; Harpal S. Dillion, Southwestern Oklahoma State University; Janusz Szczypula, Carnegie Mellon University; Dr. Ahmad Abuhejleh, University of Wisconsin, River Falls; Dr. Terence M. Baron, University of Toledo; Dr. Juan Estava, Eastern Michigan University; Dr. Kevin Gorman, University of North Carolina, Charlotte; Dr. Jeff Hedrington, University of Wisconsin, Eau Claire; Dr. Herman P. Hoplin, Syracuse University; Dr. Sophie Lee, University of Massachusetts, Boston; Dr. Michael Mannino, University of Washington; Dr. Carol Chrisman, Illinois State University; Dr. Timothy Heintz, Marquette University; Dr. Herman Hoplin, Syracuse University; Dr. Dean James, Embry-Riddle University; Dr. Constance Knapp,

Pace University; Dr. Mary Ann Robbert, Bentley College; Dr. Francis J. Van Wetering, University of Nebraska; Dr. Joseph Walls, University of Southern California; Dr. Stephen C. Solosky, Nassau Community College; Dr. Robert Chiang, Syracuse University; Dr. Crist Costa, Rhode Island College; Dr. Sudesh M. Duggal, Northern Kentucky University; Dr. Chang Koh, University of North Carolina, Greensboro; Paul A. Seibert, North Greenville College; Neil Dunlop, Vista Community College; Ylber Ramadani, George Brown College; Samuel Sambasivam, Azusa Pacific University; Arjan Sadhwani, San Jose State University; Genard Catalano, Columbia College; Craig Shaw, Central Community College; Lei-da Chen, Creighton University; Linda K. Lau, Longwood University; Anita Lee-Post, University of Kentucky; Lenore Horowitz, Schenectady County Community College; Dr. Scott L. Schneberger, Georgia State University; Tony Pollard, University of Western Sydney; Lejla Vrazalic, University of Wollongong; and David Witzany, Parkland College.

In some respects, writing books resembles building construction: When 90 percent of the work seems done, 90 percent of the work remains to be done. Fortunately for us, we had a great team on our side.

- How can we possibly pay sufficient homage to Deb Kaufmann's many contributions? Even our best superlatives don't begin to paint a proper picture of our professional relationship with Deb Kaufmann, our developmental editor since the fifth edition. Deb has that magic combination of good judgment, intelligence, technical skill, and the rare ability to organize and sharpen an author's writing without affecting its intent or its flow. And she does it all with style, class, and humor. She is the best of the best.

- After writing so many books and eight editions of *this* book, we know just how difficult it can be to transform the authors' work into an attractive book. Jill Braiewa, the content project manager, made it look easy. Jill is one of those wonderful can-do people who moves the proverbial publishing mountains. The words *unflappable* and *professional* come to mind when we think about her.

- We also owe Kate Hennessy, our product manager, special thanks for her ability to guide this book to a successful conclusion. Kate's work touched all of the publication bases, and her managerial skills protected us from those publishing gremlins that might have become a major nuisance. Not to mention the fact that her skills in dealing with occasionally cranky authors far exceed those of any diplomat we can think of. And did we mention that Kate is, quite simply, a delightful person?

- Many thanks to Mary Kemper, our copyeditor. Given her ability to spot even the smallest discrepancies, we suspect that her middle name is "Thorough." We can only imagine the level of mental discipline required to perform her job and we salute her.

We also thank our students for their comments and suggestions. They are the reason for writing this book in the first place. One comment stands out in particular: "I majored in systems for four years, and I finally discovered why when I took your course." And one of our favorite comments by a former student was triggered by a question about the challenges created by a real-world information systems job: "Doc, it's just like class, only easier. You really prepared me well. Thanks!"

Last, and certainly not least, we thank our families for the solid home support. They graciously accepted the fact that during more than a year's worth of rewriting, there would be no free weekends, rare free nights, and even rarer free days. We owe you much, and the dedication we wrote to you is but a small reflection of the important space you occupy in our hearts.

Peter Rob

Carlos M. Coronel

PART

I

DATABASE CONCEPTS

THE RELATIONAL REVOLUTION

Business Vignette

Today, we take for granted the benefits brought to us by relational databases: the ability to store, access, and change data quickly and easily on low-cost computers. Yet, until the late 1970s, databases stored large amounts of data in a hierarchical structure that was difficult to navigate and inflexible. Programmers needed to know what clients wanted to do with the data before the database was designed. Adding or changing the way the data was analyzed was a time-consuming and expensive process. As a result, you searched through huge card catalogs to find a library book, you used road maps that didn't show changes made in the last year, and you had to buy a newspaper to find information on stock prices.

In 1970, Edgar "Ted" Codd, a mathematician employed by IBM, wrote an article that would change all that. At the time, nobody realized that Codd's obscure theories would spark a technological revolution on par with the development of personal computers and the Internet. Don Chamberlin, coinventor of SQL, the most popular computer language used by database systems today, explains, "There was this guy Ted Codd who had some kind of strange mathematical notation, but nobody took it very seriously." Then Ted Codd organized a symposium, and Chamberlin listened as Codd reduced complicated five-page programs to one line. "And I said, 'Wow,'" Chamberlin recalls.

The symposium convinced IBM to fund System R, a research project that built a prototype of a relational database and that would eventually lead to the creation of SQL and DB2. IBM, however, kept System R on the back burner for a number of crucial years. The company had a vested interest in IMS, a reliable, high-end database system that had come out in 1968. Unaware of the market potential of this research, IBM allowed its staff to publish these papers publicly.

Among those reading these papers was Larry Ellison, who had just founded a small company. Recruiting programmers from System R and the University of California, Ellison was able to market the first SQL-based relational database in 1979, well before IBM. By 1983, the company had released a portable version of the database, grossed over $5,000,000 annually and changed its name to Oracle. Spurred on by competition, IBM finally released SQL/DS, its first relational database, in 1980.

By 2007, global sales of database management systems topped $15 billion with Oracle capturing roughly half of the market share and IBM trailing at under a quarter. Microsoft's SQL Server market share grew faster than its competitors, climbing to 14%.

1

DATABASE SYSTEMS

ONE

In this chapter, you will learn:

- The difference between data and information

- What a database is, what the various types of databases are, and why they are valuable assets for decision making

- The importance of database design

- How modern databases evolved from file systems

- About flaws in file system data management

- What the database system's main components are and how a database system differs from a file system

- The main functions of a database management system (DBMS)

Preview

Good decisions require good information that is derived from raw facts. These raw facts are known as data. Data are likely to be managed most efficiently when they are stored in a database. In this chapter, you learn what a database is, what it does, and why it yields better results than other data management methods. You also learn about various types of databases and why database design is so important.

Databases evolved from computer file systems. Although file system data management is now largely outmoded, understanding the characteristics of file systems is important because file systems are the source of serious data management limitations. In this chapter, you also learn how the database system approach helps eliminate most of the shortcomings of file system data management.

1.1 DATA VS. INFORMATION

To understand what drives database design, you must understand the difference between data and information. **Data** are raw facts. The word *raw* indicates that the facts have not yet been processed to reveal their meaning. For example, suppose that you want to know what the users of a computer lab think of its services. Typically, you would begin by surveying users to assess the computer lab's performance. Figure 1.1, Panel A, shows the Web survey form that enables users to respond to your questions. When the survey form has been completed, the form's raw data are saved to a data repository, such as the one shown in Figure 1.1, Panel B. Although you now have the facts in hand, they are not particularly useful in this format—reading page after page of zeros and ones is not likely to provide much insight. Therefore, you transform the raw data into a data summary like the one shown in Figure 1.1, Panel C. Now it's possible to get quick answers to questions such as "What is the composition of our lab's customer base?" In this case, you can quickly determine that most of your customers are juniors (24.59%) and seniors (53.01%). Because graphics can enhance your ability to quickly extract meaning from data, you show the data summary bar graph in Figure 1.1, Panel D.

FIGURE 1.1   Transforming raw data into information

a) Initial Survey Screen

b) Raw Data

c) Information in Summary Format

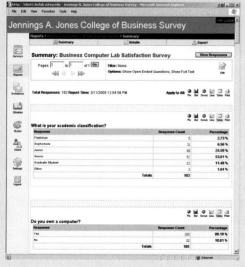

d) Information in Graphic Format

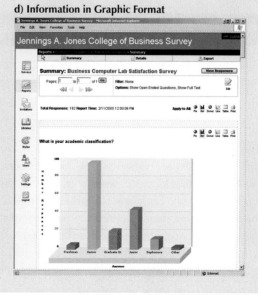

Information is the result of processing raw data to reveal its meaning. Data processing can be as simple as organizing data to reveal patterns or as complex as making forecasts or drawing inferences using statistical modeling. To reveal meaning, information requires *context*. For example, an average temperature reading of 105 degrees does not mean much unless you also know its context: Is this in degrees Fahrenheit or Celsius? Is this a machine temperature, a body temperature, or an outside air temperature? Information can be used as the foundation for decision making. For example, the data summary for each question on the survey form can point out the lab's strengths and weaknesses, helping you to make informed decisions to better meet the needs of lab customers.

Keep in mind that raw data must be properly *formatted* for storage, processing, and presentation. For example, in Panel C of Figure 1.1, the student classification is formatted to show the results based on the classifications Freshman, Sophomore, Junior, Senior, and Graduate Student. The respondents' yes/no responses might need to be converted to a Y/N format for data storage. More complex formatting is required when working with complex data types, such as sounds, videos, or images.

In this "information age," production of accurate, relevant, and timely information is the key to good decision making. In turn, good decision making is the key to business survival in a global market. We are now said to be entering the "knowledge age."[1] Data are the foundation of information, which is the bedrock of **knowledge**—that is, the body of information and facts about a specific subject. Knowledge implies familiarity, awareness, and understanding of information as it applies to an environment. A key characteristic of knowledge is that "new" knowledge can be derived from "old" knowledge.

Let's summarize some key points:

- Data constitute the building blocks of information.
- Information is produced by processing data.
- Information is used to reveal the meaning of data.
- Accurate, relevant, and timely information is the key to good decision making.
- Good decision making is the key to organizational survival in a global environment.

Timely and useful information requires accurate data. Such data must be generated properly, and it must be stored in a format that is easy to access and process. And, like any basic resource, the data environment must be managed carefully. **Data management** is a discipline that focuses on the proper generation, storage, and retrieval of data. Given the crucial role that data plays, it should not surprise you that data management is a core activity for any business, government agency, service organization, or charity.

1.2 INTRODUCING THE DATABASE AND THE DBMS

Efficient data management typically requires the use of a computer database. A **database** is a shared, integrated computer structure that stores a collection of:

- End-user data, that is, raw facts of interest to the end user.
- **Metadata**, or data about data, through which the end-user data are integrated and managed.

The metadata provide a description of the data characteristics and the set of relationships that link the data found within the database. For example, the metadata component stores information such as the name of each data element, the type of values (numeric, dates or text) stored on each data element, whether or not the data element can be left empty, and so on. Therefore, the metadata provide information that complements and expands the value and use of the data. In short, metadata present a more complete picture of the data in the database. Given the characteristics of metadata, you might hear a database described as a "collection of *self-describing* data."

---

[1] Peter Drucker coined the phrase "knowledge worker" in 1959 in his book *Landmarks of Tomorrow*. In 1994, Ms. Esther Dyson, Mr. George Gilder, Dr. George Keyworth, and Dr. Alvin Toffler introduced the concept of the "knowledge age."

A **database management system (DBMS)** is a collection of programs that manages the database structure and controls access to the data stored in the database. In a sense, a database resembles a very well-organized electronic filing cabinet in which powerful software, known as a *database management system*, helps manage the cabinet's contents.

1.2.1 ROLE AND ADVANTAGES OF THE DBMS

The DBMS serves as the intermediary between the user and the database. The database structure itself is stored as a collection of files, and the only way to access the data in those files is through the DBMS. Figure 1.2 emphasizes the point that the DBMS presents the end user (or application program) with a single, integrated view of the data in the database. The DBMS receives all application requests and translates them into the complex operations required to fulfill those requests. The DBMS hides much of the database's internal complexity from the application programs and users. The application program might be written by a programmer using a programming language such as Visual Basic.NET, Java, or C++, or it might be created through a DBMS utility program.

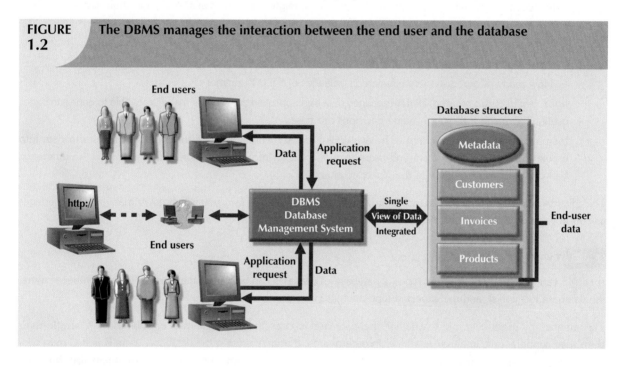

FIGURE 1.2   The DBMS manages the interaction between the end user and the database

Having a DBMS between the end user's applications and the database offers some important advantages. First, the DBMS enables the data in the database *to be shared* among multiple applications or users. Second, the DBMS *integrates* the many different users' views of the data into a single all-encompassing data repository.

Because data are the crucial raw material from which information is derived, you must have a good method to manage such data. As you will discover in this book, the DBMS helps make data management more efficient and effective. In particular, a DBMS provides advantages such as:

* *Improved data sharing.* The DBMS helps create an environment in which end users have better access to more data and better-managed data. Such access makes it possible for end users to respond quickly to changes in their environment.
* *Improved data security.* The more users access the data, the greater the risks of data security breaches. Corporations invest considerable amounts of time, effort, and money to ensure that corporate data are used properly. A DBMS provides a framework for better enforcement of data privacy and security policies.

- *Better data integration*. Wider access to well-managed data promotes an integrated view of the organization's operations and a clearer view of the big picture. It becomes much easier to see how actions in one segment of the company affect other segments.

- *Minimized data inconsistency*. **Data inconsistency** exists when different versions of the same data appear in different places. For example, data inconsistency exists when a company's sales department stores a sales representative's name as "Bill Brown" and the company's personnel department stores that same person's name as "William G. Brown" or when the company's regional sales office shows the price of a product as $45.95 and its national sales office shows the same product's price as $43.95. The probability of data inconsistency is greatly reduced in a properly designed database.

- *Improved data access*. The DBMS makes it possible to produce quick answers to ad hoc queries. From a database perspective, a **query** is a specific request issued to the DBMS for data manipulation—for example, to read or update the data. Simply put, a query is a question, and an **ad hoc query** is a spur-of-the-moment question. The DBMS sends back an answer (called the **query result set**) to the application. For example, end users, when dealing with large amounts of sales data, might want quick answers to questions (ad hoc queries) such as:

  - What was the dollar volume of sales by product during the past six months?
  - What is the sales bonus figure for each of our salespeople during the past three months?
  - How many of our customers have credit balances of $3,000 or more?

- *Improved decision making*. Better-managed data and improved data access make it possible to generate better quality information, on which better decisions are based.

- *Increased end-user productivity*. The availability of data, combined with the tools that transform data into usable information, empowers end users to make quick, informed decisions that can make the difference between success and failure in the global economy.

The advantages of using a DBMS are not limited to the few just listed. In fact, you will discover many more advantages as you learn more about the technical details of databases and their proper design.

1.2.2 TYPES OF DATABASES

A DBMS can support many different types of databases. Databases can be classified according to the number of users, the database location(s), and the expected type and extent of use.

The number of users determines whether the database is classified as single-user or multiuser. A **single-user database** supports only one user at a time. In other words, if user A is using the database, users B and C must wait until user A is done. A single-user database that runs on a personal computer is called a **desktop database**. In contrast, a **multiuser database** supports multiple users at the same time. When the multiuser database supports a relatively small number of users (usually fewer than 50) or a specific department within an organization, it is called a **workgroup database**. When the database is used by the entire organization and supports many users (more than 50, usually hundreds) across many departments, the database is known as an **enterprise database**.

Location might also be used to classify the database. For example, a database that supports data located at a single site is called a **centralized database**. A database that supports data distributed across several different sites is called a **distributed database**. The extent to which a database can be distributed and the way in which such distribution is managed is addressed in detail in Chapter 12, Distributed Database Management Systems.

The most popular way of classifying databases today, however, is based on how they will be used and on the time sensitivity of the information gathered from them. For example, transactions such as product or service sales, payments, and supply purchases reflect critical day-to-day operations. Such transactions must be recorded accurately and immediately. A database that is designed primarily to support a company's day-to-day operations is classified as an **operational database** (sometimes referred to as a **transactional** or **production database**). In contrast, a **data warehouse** focuses primarily on storing data used to generate information required to make tactical or strategic decisions. Such decisions typically require extensive "data massaging" (data manipulation) to extract information to formulate pricing decisions, sales forecasts, market positioning, and so on. Most decision-support data are based on historical data obtained from operational databases. Additionally, the data warehouse can store data derived from many sources. To make it easier to retrieve such data, the data warehouse structure is quite different from that of an operational or transactional database. The design, implementation, and use of data warehouses are covered in detail in Chapter 13, Business Intelligence and Data Warehouses.

Databases can also be classified to reflect the degree to which the data are structured. **Unstructured data** are data that exist in their original (raw) state, that is, in the format in which they were collected. Therefore, unstructured data exist in a format that does not lend itself to the processing that yields information. **Structured data** are the result of taking unstructured data and formatting (structuring) such data to facilitate storage, use, and the generation of information. You apply structure (format) based on the type of processing that you intend to perform on the data. Some data might be not ready (unstructured) for some types of processing, but they might be ready (structured) for other types of processing. For example, the data value 37890 might refer to a zip code, a sales value, or a product code. If this value represents a zip code or a product code and is stored as text, you cannot perform mathematical computations with it. On the other hand, if this value represents a sales transaction, it is necessary to format it as numeric.

To further illustrate the structure concept, imagine a stack of printed paper invoices. If you want to merely store these invoices as images for future retrieval and display, you can scan them and save them in a graphic format. On the other hand, if you want to derive information such as monthly totals and average sales, such graphic storage would not be useful. Instead, you could store the invoice data in a (structured) spreadsheet format so that you can perform the requisite computations. Actually, most data you encounter is best classified as semistructured. **Semistructured data** are data that have already been processed to some extent. For example, if you look at a typical Web page, the data are presented to you in a prearranged format to convey some information.

The database types mentioned thus far focus on the storage and management of highly structured data. However, corporations are not limited to the use of structured data. They also use semistructured and unstructured data. Just think of the very valuable information that can be found on company e-mails, memos, documents such as procedures and rules, Web page contents, and so on. Unstructured and semistructured data storage and management needs are being addressed through a new generation of databases known as XML databases. **Extensible Markup Language (XML)** is a special language used to represent and manipulate data elements in a textual format. An **XML database** supports the storage and management of semistructured XML data.

Table 1.1 compares features of several well-known database management systems.

TABLE 1.1	Types of Databases							
PRODUCT	**NUMBER OF USERS**			**DATA LOCATION**		**DATA USAGE**		**XML**
	SINGLE USER	MULTIUSER		CENTRALIZED	DISTRIBUTED	OPERATIONAL	DATA WAREHOUSE	
		WORK-GROUP	ENTER-PRISE					
MS Access	X	X		X		X		
MS SQL Server	X^2	X	X	X	X	X	X	X
IBM DB2	X^2	X	X	X	X	X	X	X
MySQL	X	X	X	X	X	X	X	X*
Oracle RDBMS	X^2	X	X	X	X	X	X	X
* Supports XML functions only. XML data is stored in large text objects.								

---

[2] Vendor offers single-user/personal DBMS version.

NOTE

Most of the database design, implementation, and management issues addressed in this book are based on production (transaction) databases. The focus on production databases is based on two considerations. First, production databases are the databases most frequently encountered in common activities such as enrolling in a class, registering a car, buying a product, or making a bank deposit or withdrawal. Second, data warehouse databases derive most of their data from production databases, and if production databases are poorly designed, the data warehouse databases based on them will lose their reliability and value as well.

1.3 WHY DATABASE DESIGN IS IMPORTANT

Database design refers to the activities that focus on the design of the database structure that will be used to store and manage end-user data. A database that meets all user requirements does not just happen; its structure must be designed carefully. In fact, database design is such a crucial aspect of working with databases that most of this book is dedicated to the development of good database design techniques. Even a good DBMS will perform poorly with a badly designed database.

Proper database design requires the designer to identify precisely the database's expected use. Designing a transactional database emphasizes accurate and consistent data and operational speed. The design of a data warehouse database recognizes the use of historical and aggregated data. Designing a database to be used in a centralized, single-user environment requires a different approach from that used in the design of a distributed, multiuser database. This book emphasizes the design of transactional, centralized, single-user, and multiuser databases. Chapters 12 and 13 also examine critical issues confronting the designer of distributed and data warehouse databases.

A well-designed database facilitates data management and generates accurate and valuable information. A poorly designed database is likely to become a breeding ground for difficult-to-trace errors that may lead to bad decision making—and bad decision making can lead to the failure of an organization. Database design is simply too important to be left to luck. That's why college students study database design, why organizations of all types and sizes send personnel to database design seminars, and why database design consultants often make an excellent living.

1.4 HISTORICAL ROOTS: FILES AND FILE SYSTEMS

Although managing data through the use of file systems is now largely obsolete, there are several good reasons for studying them in some detail:

- An understanding of the relatively simple characteristics of file systems makes the complexity of database design easier to understand.
- An awareness of the problems that plagued file systems can help you avoid those same pitfalls with DBMS software.
- If you intend to convert an obsolete file system to a database system, knowledge of the file system's basic limitations will be useful.

In the recent past, a manager of almost any small organization was (and sometimes still is) able to keep track of necessary data by using a manual file system. Such a file system was traditionally composed of a collection of file folders, each properly tagged and kept in a filing cabinet. Organization of the data within the file folders was determined by the data's expected use. Ideally, the contents of each file folder were logically related. For example, a file folder in a doctor's office might contain patient data, one file folder for each patient. All of the data in that file folder would describe only that particular patient's medical history. Similarly, a personnel manager might organize personnel data by category of employment (for example, clerical, technical, sales, and administrative). Therefore, a file folder labeled "Technical" would contain data pertaining to only those people whose duties were properly classified as technical.

As long as a data collection was relatively small and an organization's managers had few reporting requirements, the manual system served its role well as a data repository. However, as organizations grew and as reporting requirements became more complex, keeping track of data in a manual file system became more difficult. In fact, finding and using data in growing collections of file folders turned into such a time-consuming and cumbersome task that it became unlikely that such data could generate useful information. Consider just these few questions to which a retail business owner might want answers:

- What products sold well during the past week, month, quarter, or year?
- What is the current daily, weekly, monthly, quarterly, or yearly sales dollar volume?
- How do the current period's sales compare to those of last week, last month, or last year?
- Did the various cost categories increase, decrease, or remain stable during the past week, month, quarter, or year?
- Did sales show trends that could change the inventory requirements?

The list of questions such as these tends to be long and to increase in number as an organization grows.

Unfortunately, generating reports from a manual file system can be slow and cumbersome. In fact, some business managers faced government-imposed reporting requirements that required weeks of intensive effort each quarter, even when a well-designed manual system was used. Consequently, necessity called for the design of a computer-based system that would track data and produce required reports.

The conversion from a manual file system to a matching computer file system could be technically complex. (Because people are accustomed to today's relatively user-friendly computer interfaces, they have forgotten how painfully hostile computers used to be!) Consequently, a new kind of professional, known as a **data processing (DP) specialist**, had to be hired or "grown" from the current staff. The DP specialist created the necessary computer file structures, often wrote the software that managed the data within those structures, and designed the application programs that produced reports based on the file data. Thus, numerous homegrown computerized file systems were born.

Initially, the computer files within the file system were similar to the manual files. A simple example of a customer data file for a small insurance company is shown in Figure 1.3. (You will discover later that the file structure shown in Figure 1.3, although typically found in early file systems, is unsatisfactory for a database.)

FIGURE 1.3   Contents of the CUSTOMER file

C_NAME	C_PHONE	C_ADDRESS	C_ZIP	A_NAME	A_PHONE	TP	AMT	REN
Alfred A. Ramas	615-844-2573	218 Fork Rd., Babs, TN	36123	Leah F. Hahn	615-882-1244	T1	100.00	05-Apr-2008
Leona K. Dunne	713-894-1238	Box 12A, Fox, KY	25246	Alex B. Alby	713-228-1249	T1	250.00	16-Jun-2008
Kathy W. Smith	615-894-2285	125 Oak Ln, Babs, TN	36123	Leah F. Hahn	615-882-2144	S2	150.00	29-Jan-2009
Paul F. Olowski	615-894-2180	217 Lee Ln., Babs, TN	36123	Leah F. Hahn	615-882-1244	S1	300.00	14-Oct-2008
Myron Orlando	615-222-1672	Box 111, New, TN	36155	Alex B. Alby	713-228-1249	T1	100.00	28-Dec-2008
Amy B. O'Brian	713-442-3381	387 Troll Dr., Fox, KY	25246	John T. Okon	615-123-5589	T2	850.00	22-Sep-2008
James G. Brown	615-297-1228	21 Tye Rd., Nash, TN	37118	Leah F. Hahn	615-882-1244	S1	120.00	25-Mar-2009
George Williams	615-290-2556	155 Maple, Nash, TN	37119	John T. Okon	615-123-5589	S1	250.00	17-Jul-2008
Anne G. Farriss	713-382-7185	2119 Elm, Crew, KY	25432	Alex B. Alby	713-228-1249	T2	100.00	03-Dec-2008
Olette K. Smith	615-297-3809	2782 Main, Nash, TN	37118	John T. Okon	615-123-5589	S2	500.00	14-Mar-2009

C_NAME = Customer name
C_PHONE = Customer phone
C_ADDRESS = Customer address
C_ZIP = Customer zip code

A_NAME = Agent name
A_PHONE = Agent phone
TP = Insurance type
AMT = Insurance policy amount, in thousands of $
REN = Insurance renewal date

ONLINE CONTENT

The databases used in each chapter are available in the Student Online Companion for this book. Throughout the book, Online Content boxes highlight material related to chapter content located in the Student Online Companion.

The description of computer files requires a specialized vocabulary. Every discipline develops its own jargon to enable its practitioners to communicate clearly. The basic file vocabulary shown in Table 1.2 will help you understand subsequent discussions more easily.

TABLE 1.2	Basic File Terminology
TERM	**DEFINITION**
Data	"Raw" facts, such as a telephone number, a birth date, a customer name, and a year-to-date (YTD) sales value. Data have little meaning unless they have been organized in some logical manner. The smallest piece of data that can be "recognized" by the computer is a single character, such as the letter A, the number 5, or a symbol such as /. A single character requires 1 byte of computer storage.
Field	A character or group of characters (alphabetic or numeric) that has a specific meaning. A field is used to define and store data.
Record	A logically connected set of one or more fields that describes a person, place, or thing. For example, the fields that constitute a record for a customer named J. D. Rudd might consist of J. D. Rudd's name, address, phone number, date of birth, credit limit, and unpaid balance.
File	A collection of related records. For example, a file might contain data about vendors of ROBCOR Company, or a file might contain the records for the students currently enrolled at Gigantic University.

Using the proper file terminology given in Table 1.2, you can identify the file components shown in Figure 1.3. The CUSTOMER file shown in Figure 1.3 contains 10 records. Each record is composed of nine fields: C_NAME, C_PHONE, C_ADDRESS, C_ZIP, A_NAME, A_PHONE, TP, AMT, and REN. The 10 records are stored in a named file. Because the file in Figure 1.3 contains customer data for the insurance company, its filename is CUSTOMER.

Using the CUSTOMER file's contents, the DP specialist wrote programs that produced very useful reports for the insurance company's sales department:

- Monthly summaries that showed the types and amounts of insurance sold by each agent. (Such reports might be used to analyze each agent's productivity.)
- Monthly checks to determine which customers must be contacted for renewal.
- Reports that analyzed the ratios of insurance types sold by each agent.
- Periodic customer contact letters designed to summarize coverage and to provide various customer relations bonuses.

As time went on, the insurance company needed additional programs to produce new reports. Although it took some time to specify the report contents and to write the programs that produced the reports, the sales department manager did not miss the old manual system—using the computer saved much time and effort. The reports were impressive, and the ability to perform complex data searches yielded the information needed to make sound decisions.

Then the sales department at the insurance company created a file named SALES, which helped track daily sales efforts. Additional files were created as needed to produce even more useful reports. In fact, the sales department's success was so obvious that the personnel department manager demanded access to the DP specialist to automate payroll processing and other personnel functions. Consequently, the DP specialist was asked to create the AGENT file shown in Figure 1.4. The data in the AGENT file were used to write checks, keep track of taxes paid, and summarize insurance coverage, among other tasks.

FIGURE 1.4 Contents of the AGENT file

A_NAME	A_PHONE	A_ADDRESS	ZIP	HIRED	YTD_PAY	YTD_FIT	YTD_FICA	YTD_SLS	DEP
Alex B. Alby	713-228-1249	123 Toll, Nash, TN	37119	01-Nov-2000	26566.24	6641.56	2125.30	132737.75	3
Leah F. Hahn	615-882-1244	334 Main, Fox, KY	25246	23-May-1986	32213.78	8053.44	2577.10	138967.35	0
John T. Okon	615-123-5589	452 Elm, New, TN	36155	15-Jun-2005	23198.29	5799.57	1855.86	127093.45	2

A_NAME	= Agent name	YTD_PAY	= Year-to-date pay
A_PHONE	= Agent phone	YTD_FIT	= Year-to-date federal income tax paid
A_ADDRESS	= Agent address	YTD_FICA	= Year-to-date Social Security taxes paid
ZIP	= Agent zip code	YTD_SLS	= Year-to-date sales
HIRED	= Agent date of hire	DEP	= Number of dependents

As the number of files increased, a small file system, like the one shown in Figure 1.5, evolved. Each file in the system used its own application program to store, retrieve, and modify data. And each file was owned by the individual or the department that commissioned its creation.

FIGURE 1.5 A simple file system

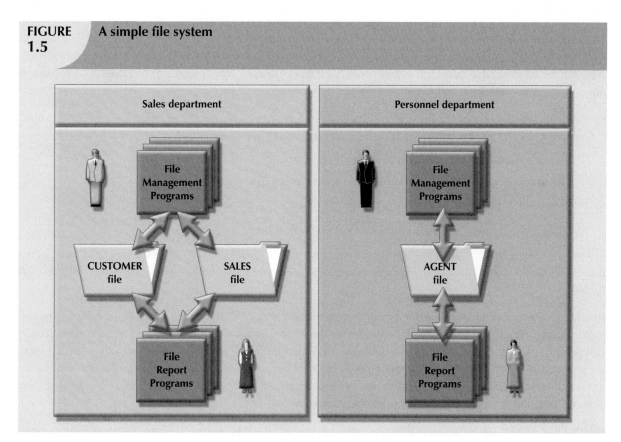

As the insurance company's file system grew, the demand for the DP specialist's programming skills grew even faster, and the DP specialist was authorized to hire additional programmers. The size of the file system also required a larger, more complex computer. The new computer and the additional programming staff caused the DP specialist to spend less time programming and more time managing technical and human resources. Therefore, the DP specialist's job evolved into that of a **data processing (DP) manager**, who supervised a new DP department. In spite of these organizational changes, however, the DP department's primary activity remained programming, and the DP manager inevitably spent much time as a supervising senior programmer and program troubleshooter.

1.5 PROBLEMS WITH FILE SYSTEM DATA MANAGEMENT

The file system method of organizing and managing data was a definite improvement over a manual system, and the file system served a useful purpose in data management for over two decades—a very long time in the computer era. Nonetheless, many problems and limitations became evident in this approach. A critique of the file system method serves two major purposes:

- Understanding the shortcomings of the file system enables you to understand the development of modern databases.

- Many of the problems are not unique to file systems. Failure to understand such problems is likely to lead to their duplication in a database environment, even though database technology makes it easy to avoid them.

The first and most glaring problem with the file system approach is that even the simplest data-retrieval task requires extensive programming. With the older file systems, programmers had to specify what must be done and how it was to be done. As you will learn in upcoming chapters, modern databases use a nonprocedural data manipulation language that allows the user to specify what must be done without specifying how it must be done. Typically, this nonprocedural language is used for data retrieval (such as query by example and report generator tools), is much faster, and can work with different DBMSs.

The need to write programs to produce even the simplest reports makes ad hoc queries impossible. Harried DP specialists and DP managers who work with mature file systems often receive numerous requests for new reports. They are often forced to say that the report will be ready "next week" or even "next month." If you need the information now, getting it next week or next month will not serve your information needs.

Furthermore, making changes in an existing structure can be difficult in a file system environment. For example, changing just one field in the original CUSTOMER file would require a program that:

1. Reads a record from the original file.

2. Transforms the original data to conform to the new structure's storage requirements.

3. Writes the transformed data into the new file structure.

4. Repeats steps 2 to 4 for each record in the original file.

In fact, any change to a file structure, no matter how minor, forces modifications in all of the programs that use the data in that file. Modifications are likely to produce errors (bugs), and additional time is spent using a debugging process to find those errors.

Another problem related to the need for extensive programming is that as the number of files in the system expands, system administration becomes more difficult. Even a simple file system with a few files requires the creation and maintenance of several file management programs (each file must have its own file management programs that allow the user to add, modify, and delete records, to list the file contents, and to generate reports). Because ad hoc queries are not possible, the file reporting programs can multiply quickly. The problem is compounded by the fact that each department in the organization "owns" its data by creating its own files.

Another fault of a file system database is that security features are difficult to program and are, therefore, often omitted in a file system environment. Such features include effective password protection, the ability to lock out parts of files or parts of the system itself, and other measures designed to safeguard data confidentiality. Even when an attempt is made to improve system and data security, the security devices tend to be limited in scope and effectiveness.

To summarize the limitations of file system data management so far:

- It requires extensive programming.

- It can not perform ad hoc queries.

- System administration can be complex and difficult.

- It is difficult to make changes to existing structures.
- Security features are likely to be inadequate.

Those limitations, in turn, lead to problems of structural and data dependency.

1.5.1 STRUCTURAL AND DATA DEPENDENCE

A file system exhibits **structural dependence,** which means that access to a file is dependent on its structure. For example, adding a customer date-of-birth field to the CUSTOMER file shown in Figure 1.3 would require the four steps described in the previous section. Given this change, none of the previous programs will work with the new CUSTOMER file structure. Therefore, all of the file system programs must be modified to conform to the new file structure. In short, because the file system application programs are affected by change in the file structure, they exhibit structural dependence. Conversely, **structural independence** exists when it is possible to make changes in the file structure without affecting the application program's ability to access the data.

Even changes in the characteristics of data, such as changing a field from integer to decimal, require changes in all the programs that access the file. Because all data access programs are subject to change when any of the file's data storage characteristics change (that is, changing the data type), the file system is said to exhibit **data dependence.** Conversely, **data independence** exists when it is possible to make changes in the data storage characteristics without affecting the application program's ability to access the data.

The practical significance of data dependence is the difference between the **logical data format** (how the human being views the data) and the **physical data format** (how the computer must work with the data). Any program that accesses a file system's file must tell the computer not only what to do, but also how to do it. Consequently, each program must contain lines that specify the opening of a specific file type, its record specification, and its field definitions. Data dependence makes the file system extremely cumbersome from the point of view of a programmer and database manager.

1.5.2 FIELD DEFINITIONS AND NAMING CONVENTIONS

At first glance, the CUSTOMER file shown in Figure 1.3 appears to have served its purpose well: requested reports usually could be generated. But suppose you want to create a customer phone directory based on the data stored in the CUSTOMER file. Storing the customer name as a single field turns out to be a liability because the directory must break up the field contents to list the last names, first names, and initials in alphabetical order.

Similarly, producing a listing of customers by city is a more difficult task than is necessary. From the user's point of view, a much better (more flexible) record definition would be one that anticipates reporting requirements by breaking up fields into their component parts. Thus, the revised customer file fields might be listed as shown in Table 1.3. (Note that the revised file is named CUSTOMER_V2 to indicate that this is the second version of the CUSTOMER file.)

TABLE 1.3   Sample Fields in the CUSTOMER_V2 File

FIELD	CONTENTS	SAMPLE ENTRY
CUS_LNAME	Customer last name	Ramas
CUS_FNAME	Customer first name	Alfred
CUS_INITIAL	Customer initial	A
CUS_AREACODE	Customer area code	615
CUS_PHONE	Customer phone	234-5678
CUS_ADDRESS	Customer street address or box number	123 Green Meadow Lane
CUS_CITY	Customer city	Murfreesboro
CUS_STATE	Customer state	TN
CUS_ZIP	Customer zip code	37130
AGENT_CODE	Agent code	502

Selecting proper field names is also important. For example, make sure that the field names are reasonably descriptive. In examining the file structure shown in Figure 1.3, it is not obvious that the field name REN represents the customer's insurance renewal date. Using the field name CUS_RENEW_DATE would be better for two reasons. First, the prefix CUS can be used as an indicator of the field's origin, which is the CUSTOMER_V2 file. Therefore, you know that the field in question yields a customer property. Second, the RENEW_DATE portion of the field name is more descriptive of the field's contents. With proper naming conventions, the file structure becomes *self-documenting*. That is, by simply looking at a field name, you can determine which file the field belongs to and what information the field is likely to contain.

> **NOTE**
>
> You might have noticed the addition of the AGENT_CODE field in Table 1.3. Clearly, you must know what agent represents each customer, so the customer file must include agent data. You will learn in Section 1.5.3 that storing the agent *name*, as was done in the original CUSTOMER file shown in Figure 1.3, will yield some major problems that are eliminated by using a unique code that is assigned to each agent. And you will learn in Chapter 2, Data Models, what other benefits are obtained from storing such a code in the (revised) customer table. In any case, because the agent code is an *agent* characteristic, its prefix is AGENT.

Some software packages place restrictions on the length of field names, so it is wise to be as descriptive as possible within those restrictions. In addition, very long field names make it difficult to fit more than a few fields on a page, thus making output spacing a problem. For example, the field name CUSTOMER_INSURANCE_RENEWAL_DATE, while being self-documenting, is less desirable than CUS_RENEW_DATE.

Another problem in Figure 1.3's CUSTOMER file is the difficulty of finding desired data efficiently. The CUSTOMER file currently does not have a unique record identifier. For example, it is possible to have several customers named John B. Smith. Consequently, the addition of a CUS_ACCOUNT field that contains a unique customer account number would be appropriate.

The criticisms of field definitions and naming conventions shown in the file structure of Figure 1.3 are not unique to file systems. Because such conventions will prove to be important later, they are introduced early. You will revisit field definitions and naming conventions when you learn about database design in Chapter 4, Entity Relationship (ER) Modeling and in Chapter 6, Advanced Data Modeling; when you learn about database implementation issues in Chapter 9, Database Design; and when you see an actual database design implemented in Appendixes B and C (The University Lab design and implementation). Regardless of the data environment, the design—whether it involves a file system or a database—must always reflect the designer's documentation needs and the end user's reporting and processing requirements. Both types of needs are best served by adhering to proper field definitions and naming conventions.

ONLINE CONTENT

Appendixes A through L are available in the Student Online Companion for this book.

NOTE

No naming convention can fit all requirements for all systems. Some words or phrases are reserved for the DBMSs internal use. For example, the name ORDER generates an error in some DBMSs. Similarly, your DBMS might interpret a hyphen (-) as a command to subtract. Therefore, the field CUS-NAME would be interpreted as a command to subtract the NAME field from the CUS field. Because neither field exists, you would get an error message. On the other hand, CUS_NAME would work fine because it uses an underscore.

1.5.3 DATA REDUNDANCY

The file system's structure makes it difficult to combine data from multiple sources and its lack of security renders the file system vulnerable to security breaches. The organizational structure promotes the storage of the same basic data in different locations. (Database professionals use the term **islands of information** for such scattered data locations.) Because it is unlikely that data stored in different locations will always be updated consistently, the islands of information often contain different versions of the same data. For example, in Figures 1.3 and 1.4, the agent names and phone numbers occur in both the CUSTOMER and the AGENT files. You need only one correct copy of the agent names and phone numbers. Having them occur in more than one place produces data redundancy. **Data redundancy** exists when the same data are stored unnecessarily at different places.

Uncontrolled data redundancy sets the stage for:

- *Data inconsistency*. Data inconsistency exists when different and conflicting versions of the same data appear in different places. For example, suppose you change an agent's phone number or address in the AGENT file. If you forget to make corresponding changes in the CUSTOMER file, the files contain different data for the same agent. Reports will yield inconsistent results depending on which version of the data is used.

NOTE

Data that display data inconsistency are also referred to as data that lack data integrity. **Data integrity** is defined as the condition in which all of the data in the database are consistent with the real-world events and conditions. In other words, data integrity means that:

- Data are *accurate*—there are no data inconsistencies
- Data are *verifiable*—the data will always yield consistent results.

Data entry errors are more likely to occur when complex entries (such as 10-digit phone numbers) are made in several different files and/or recur frequently in one or more files. In fact, the CUSTOMER file shown in Figure 1.3 contains just such an entry error: the third record in the CUSTOMER file has a transposed digit in the agent's phone number (615-882-2144 rather than 615-882-1244).

It is possible to enter a nonexistent sales agent's name and phone number into the CUSTOMER file, but customers are not likely to be impressed if the insurance agency supplies the name and phone number of an agent who does not exist. And should the personnel manager allow a nonexistent agent to accrue bonuses and benefits? In fact, a data entry error such as an incorrectly spelled name or an incorrect phone number yields the same kind of data integrity problems.

- *Data anomalies*. The dictionary defines *anomaly* as "an abnormality." Ideally, a field value change should be made in only a single place. Data redundancy, however, fosters an abnormal condition by forcing field value changes in many different locations. Look at the CUSTOMER file in Figure 1.3. If agent Leah F. Hahn decides to get married and move, the agent name, address, and phone are likely to change. Instead of making just a single name and/or phone/address change in a single file (AGENT), you also must make the change each time that agent's name, phone number, and address occur in the CUSTOMER file. You could be faced with the prospect of making hundreds of corrections, one for each of the customers served by that agent! The same problem occurs when an agent decides to quit. Each customer served by that agent must be assigned a new agent. Any change in any field value must be correctly made in many places to maintain data integrity. A **data anomaly** develops when all of the required changes in the redundant data are not made successfully. The data anomalies found in Figure 1.3 are commonly defined as follows:

- *Update anomalies*. If agent Leah F. Hahn has a new phone number, that number must be entered in each of the CUSTOMER file records in which Ms. Hahn's phone number is shown. In this case, only three changes must be made. In a large file system, such changes might occur in hundreds or even thousands of records. Clearly, the potential for data inconsistencies is great.

- *Insertion anomalies*. If only the CUSTOMER file existed, to add a new agent, you would also add a dummy customer data entry to reflect the new agent's addition. Again, the potential for creating data inconsistencies would be great.

- *Deletion anomalies*. If you delete the customers Amy B. O'Brian, George Williams, and Olette K. Smith, you will also delete John T. Okon's agent data. Clearly, this is not desirable.

1.6 DATABASE SYSTEMS

The problems inherent in file systems make using a database system very desirable. Unlike the file system, with its many separate and unrelated files, the database system consists of logically related data stored in a single logical data repository. (The "logical" label reflects the fact that, although the data repository appears to be a single unit to the end user, its contents may actually be physically distributed among multiple data storage facilities and/or locations.) Because the database's data repository is a single logical unit, the database represents a major change in the way end-user data are stored, accessed, and managed. The database's DBMS, shown in Figure 1.6, provides numerous advantages over file system management, shown in Figure 1.5, by making it possible to eliminate most of the file system's data inconsistency, data anomaly, data dependency, and structural dependency problems. Better yet, the current generation of DBMS software stores not only the data structures, but also the relationships between those structures and the access paths to those structures—all in a central location. The current generation of DBMS software also takes care of defining, storing, and managing all required access paths to those components.

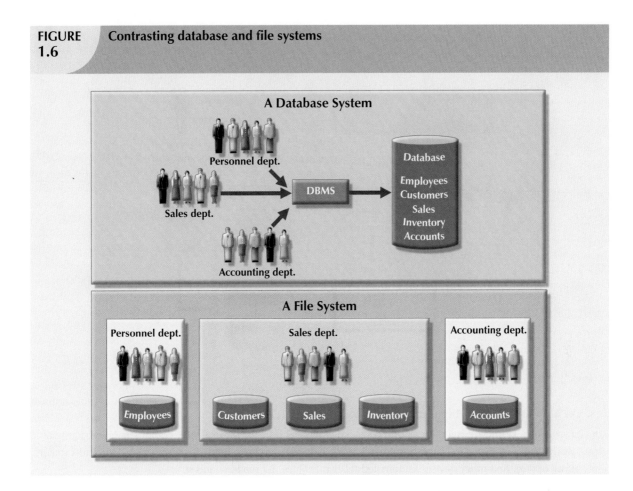

FIGURE 1.6 Contrasting database and file systems

Remember that the DBMS is just one of several crucial components of a database system. The DBMS may even be referred to as the database system's heart. However, just as it takes more than a heart to make a human being function, it takes more than a DBMS to make a database system function. In the sections that follow, you'll learn what a database system is, what its components are, and how the DBMS fits into the database system picture.

1.6.1 THE DATABASE SYSTEM ENVIRONMENT

The term **database system** refers to an organization of components that define and regulate the collection, storage, management, and use of data within a database environment. From a general management point of view, the database system is composed of the five major parts shown in Figure 1.7: hardware, software, people, procedures, and data.

FIGURE 1.7   The database system environment

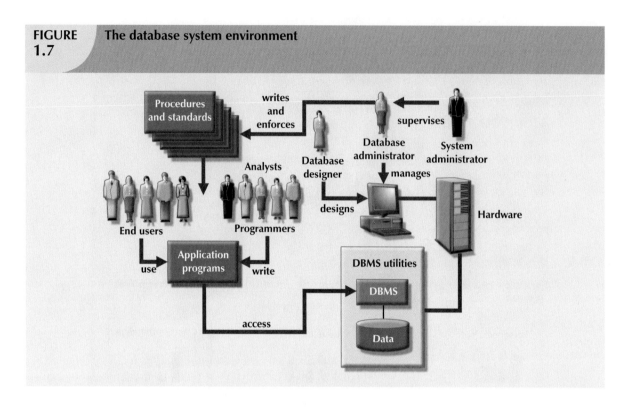

Let's take a closer look at the five components shown in Figure 1.7:

- *Hardware*. Hardware refers to all of the system's physical devices; for example, computers (microcomputers, workstations, servers, and supercomputers), storage devices, printers, network devices (hubs, switches, routers, fiber optics), and other devices (automated teller machines, ID readers, and so on).

- *Software*. Although the most readily identified software is the DBMS itself, to make the database system function fully, three types of software are needed: operating system software, DBMS software, and application programs and utilities.

    - *Operating system software* manages all hardware components and makes it possible for all other software to run on the computers. Examples of operating system software include Microsoft Windows, Linux, Mac OS, UNIX, and MVS.

    - *DBMS software* manages the database within the database system. Some examples of DBMS software include Microsoft SQL Server, Oracle Corporation's Oracle, MySQL AB's MySQL and IBM's DB2.

    - *Application programs and utility software* are used to access and manipulate data in the DBMS and to manage the computer environment in which data access and manipulation take place. Application programs are most commonly used to access data found within the database to generate reports, tabulations, and other information to facilitate decision making. Utilities are the software tools used to help manage the database system's computer components. For example, all of the major DBMS vendors now provide graphical user interfaces (GUIs) to help create database structures, control database access, and monitor database operations.

- *People*. This component includes all users of the database system. On the basis of primary job functions, five types of users can be identified in a database system: systems administrators, database administrators, database designers, systems analysts and programmers, and end users. Each user type, described below, performs both unique and complementary functions.

    - *System administrators* oversee the database system's general operations.

    - *Database administrators*, also known as DBAs, manage the DBMS and ensure that the database is functioning properly. The DBA's role is sufficiently important to warrant a detailed exploration in Chapter 15, Database Administration and Security.

- *Database designers* design the database structure. They are, in effect, the database architects. If the database design is poor, even the best application programmers and the most dedicated DBAs cannot produce a useful database environment. Because organizations strive to optimize their data resources, the database designer's job description has expanded to cover new dimensions and growing responsibilities.

- *Systems analysts and programmers* design and implement the application programs. They design and create the data entry screens, reports, and procedures through which end users access and manipulate the database's data.

- *End users* are the people who use the application programs to run the organization's daily operations. For example, salesclerks, supervisors, managers, and directors are all classified as end users. High-level end users employ the information obtained from the database to make tactical and strategic business decisions.

- *Procedures.* Procedures are the instructions and rules that govern the design and use of the database system. Procedures are a critical, although occasionally forgotten, component of the system. Procedures play an important role in a company because they enforce the standards by which business is conducted within the organization and with customers. Procedures also are used to ensure that there is an organized way to monitor and audit both the data that enter the database and the information that is generated through the use of that data.

- *Data.* The word *data* covers the collection of facts stored in the database. Because data are the raw material from which information is generated, the determination of what data are to be entered into the database and how that data are to be organized is a vital part of the database designer's job.

A database system adds a new dimension to an organization's management structure. Just how complex this managerial structure is depends on the organization's size, its functions, and its corporate culture. Therefore, database systems can be created and managed at different levels of complexity and with varying adherence to precise standards. For example, compare a local movie rental system with a national insurance claims system. The movie rental system may be managed by two people, the hardware used is probably a single microcomputer, the procedures are probably simple, and the data volume tends to be low. The national insurance claims system is likely to have at least one systems administrator, several full-time DBAs, and many designers and programmers; the hardware probably includes several servers at multiple locations throughout the United States; the procedures are likely to be numerous, complex, and rigorous; and the data volume tends to be high.

In addition to the different levels of database system complexity, managers must also take another important fact into account: database solutions must be cost-effective as well as tactically and strategically effective. Producing a million-dollar solution to a thousand-dollar problem is hardly an example of good database system selection or of good database design and management. Finally, the database technology already in use is likely to affect the selection of a database system.

1.6.2 DBMS FUNCTIONS

A DBMS performs several important functions that guarantee the integrity and consistency of the data in the database. Most of those functions are transparent to end users, and most can be achieved only through the use of a DBMS. They include data dictionary management, data storage management, data transformation and presentation, security management, multiuser access control, backup and recovery management, data integrity management, database access languages and application programming interfaces, and database communication interfaces. Each of these functions is explained below.

- *Data dictionary management.* The DBMS stores definitions of the data elements and their relationships (metadata) in a **data dictionary**. In turn, all programs that access the data in the database work through the DBMS. The DBMS uses the data dictionary to look up the required data component structures and relationships, thus relieving you from having to code such complex relationships in each program. Additionally, any changes made in a database structure are automatically recorded in the data dictionary, thereby freeing you from having to modify all of the programs that access the changed structure. In other words, the DBMS provides data abstraction, and it removes structural and data dependency from the system. For example, Figure 1.8 shows how Microsoft SQL Server Express presents the data definition for the CUSTOMER table.

FIGURE 1.8   Illustrating metadata with Microsoft SQL Server Express

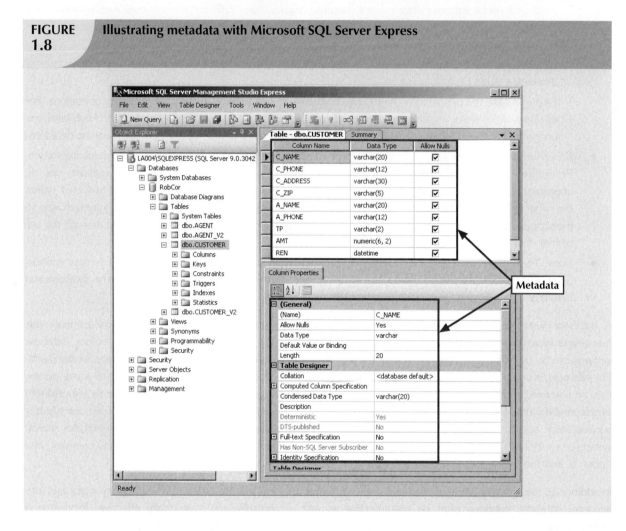

- *Data storage management*. The DBMS creates and manages the complex structures required for data storage, thus relieving you from the difficult task of defining and programming the physical data characteristics. A modern DBMS provides storage not only for the data, but also for related data entry forms or screen definitions, report definitions, data validation rules, procedural code, structures to handle video and picture formats, and so on. Data storage management is also important for database performance tuning. **Performance tuning** relates to the activities that make the database perform more efficiently in terms of storage and access speed. Although the user sees the database as a single data storage unit, the DBMS actually stores the database in multiple physical data files. (See Figure 1.9.) Such data files may even be stored on different storage media. Therefore, the DBMS doesn't have to wait for one disk request to finish before the next one starts. In other words, the DBMS can fulfill database requests concurrently. Data storage management and performance tuning issues are addressed in Chapter 11, Database Performance Tuning and Query Optimization.

FIGURE 1.9   Illustrating data storage management with Oracle

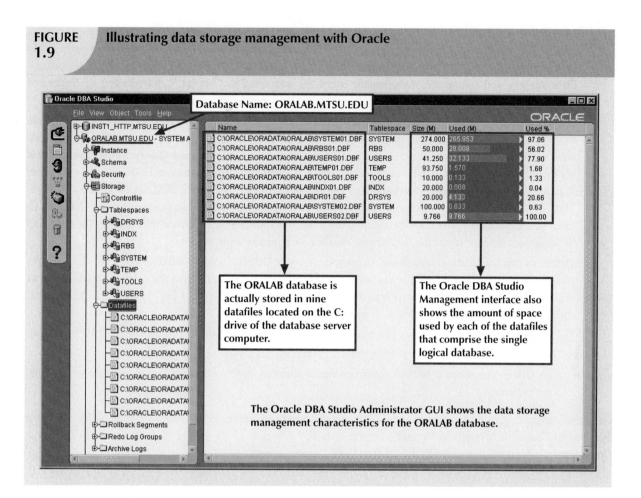

The Oracle DBA Studio Administrator GUI shows the data storage management characteristics for the ORALAB database.

- *Data transformation and presentation.* The DBMS transforms entered data to conform to required data structures. The DBMS relieves you of the chore of making a distinction between the logical data format and the physical data format. That is, the DBMS formats the physically retrieved data to make it conform to the user's logical expectations. For example, imagine an enterprise database used by a multinational company. An end user in England would expect to enter data such as July 11, 2008 as "11/07/2008." In contrast, the same date would be entered in the United States as "07/11/2008." Regardless of the data presentation format, the DBMS must manage the date in the proper format for each country.

- *Security management.* The DBMS creates a security system that enforces user security and data privacy. Security rules determine which users can access the database, which data items each user can access, and which data operations (read, add, delete, or modify) the user can perform. This is especially important in multiuser database systems. Chapter 15, Database Administration and Security, examines data security and privacy issues in greater detail. All database users may be authenticated to the DBMS through a username and password or through biometric authentication such as a fingerprint scan. The DBMS uses this information to assign access privileges to various database components such as queries and reports.

- *Multiuser access control.* To provide data integrity and data consistency, the DBMS uses sophisticated algorithms to ensure that multiple users can access the database concurrently without compromising the integrity of the database. Chapter 10, Transaction Management and Concurrency Control, covers the details of the multiuser access control.

- *Backup and recovery management.* The DBMS provides backup and data recovery to ensure data safety and integrity. Current DBMS systems provide special utilities that allow the DBA to perform routine and special backup and restore procedures. Recovery management deals with the recovery of the database after a failure, such as a bad sector in the disk or a power failure. Such capability is critical to preserving the database's integrity. Chapter 15 covers backup and recovery issues.

- *Data integrity management*. The DBMS promotes and enforces integrity rules, thus minimizing data redundancy and maximizing data consistency. The data relationships stored in the data dictionary are used to enforce data integrity. Ensuring data integrity is especially important in transaction-oriented database systems. Data integrity and transaction management issues are addressed in Chapter 7, Introduction to Structured Query Language (SQL), and Chapter 10, Transaction Management and Concurrency Control.

- *Database access languages and application programming interfaces*. The DBMS provides data access through a query language. A **query language** is a nonprocedural language—one that lets the user specify what must be done without having to specify how it is to be done. **Structured Query Language (SQL)** is the de facto query language and data access standard supported by the majority of DBMS vendors. Chapter 7, Introduction to Structured Query Language (SQL), and Chapter 8, Advanced SQL, address the use of SQL. The DBMS also provides application programming interfaces to procedural languages such as COBOL, C, Java, Visual Basic.NET, and C++. In addition, the DBMS provides administrative utilities used by the DBA and the database designer to create, implement, monitor, and maintain the database.

- *Database communication interfaces*. Current-generation DBMSs accept end-user requests via multiple, different network environments. For example, the DBMS might provide access to the database via the Internet through the use of Web browsers such as Mozilla Firefox or Microsoft Internet Explorer. In this environment, communications can be accomplished in several ways:

  - End users can generate answers to queries by filling in screen forms through their preferred Web browser.
  - The DBMS can automatically publish predefined reports on a Web site.
  - The DBMS can connect to third-party systems to distribute information via e-mail or other productivity applications.

Database communication interfaces are examined in greater detail in Chapter 12, Distributed Database Management Systems, in Chapter 14, Database Connectivity and Web Technologies, and in Appendix I, Databases in Electronic Commerce. (Appendixes are found in the Student Online Companion.)

1.6.3 MANAGING THE DATABASE SYSTEM: A SHIFT IN FOCUS

The introduction of a database system over the file system provides a framework in which strict procedures and standards can be enforced. Consequently, the role of the human component changes from an emphasis on programming (in the file system) to a focus on the broader aspects of managing the organization's data resources and on the administration of the complex database software itself.

The database system makes it possible to tackle far more sophisticated uses of the data resources as long as the database is designed to make use of that available power. The kinds of data structures created within the database and the extent of the relationships among them play a powerful role in determining the effectiveness of the database system.

Although the database system yields considerable advantages over previous data management approaches, database systems do carry significant disadvantages. For example:

- *Increased costs*. Database systems require sophisticated hardware and software and highly skilled personnel. The cost of maintaining the hardware, software, and personnel required to operate and manage a database system can be substantial. Training, licensing, and regulation compliance costs are often overlooked when database systems are implemented.

- *Management complexity*. Database systems interface with many different technologies and have a significant impact on a company's resources and culture. The changes introduced by the adoption of a database system must be properly managed to ensure that they help advance the company's objectives. Given the fact that databases systems hold crucial company data that are accessed from multiple sources, security issues must be assessed constantly.

- *Maintaining currency*. To maximize the efficiency of the database system, you must keep your system current. Therefore, you must perform frequent updates and apply the latest patches and security measures to all components. Because database technology advances rapidly, personnel training costs tend to be significant.

- *Vendor dependence*. Given the heavy investment in technology and personnel training, companies might be reluctant to change database vendors. As a consequence, vendors are less likely to offer pricing point advantages to existing customers, and those customers might be limited in their choice of database system components.

- *Frequent upgrade/replacement cycles*. DBMS vendors frequently upgrade their products by adding new functionality. Such new features often come bundled in new upgrade versions of the software. Some of these versions require hardware upgrades. Not only do the upgrades themselves cost money, but it also costs money to train database users and administrators to properly use and manage the new features.

SUMMARY

- Data are raw facts. Information is the result of processing data to reveal its meaning. Accurate, relevant, and timely information is the key to good decision making, and good decision making is the key to organizational survival in a global environment.

- Data are usually stored in a database. To implement a database and to manage its contents, you need a database management system (DBMS). The DBMS serves as the intermediary between the user and the database. The database contains the data you have collected and "data about data," known as metadata.

- Database design defines the database structure. A well-designed database facilitates data management and generates accurate and valuable information. A poorly designed database can lead to bad decision making, and bad decision making can lead to the failure of an organization.

- Databases evolved from manual and then computerized file systems. In a file system, data are stored in independent files, each requiring its own data management programs. Although this method of data management is largely outmoded, understanding its characteristics makes database design easier to understand. Awareness of the problems of file systems can help you avoid similar problems with DBMSs.

- Some limitations of file system data management are that it requires extensive programming, system administration can be complex and difficult, making changes to existing structures is difficult, and security features are likely to be inadequate. Also, independent files tend to contain redundant data, leading to problems of structural and data dependency.

- Database management systems were developed to address the file system's inherent weaknesses. Rather than depositing data in independent files, a DBMS presents the database to the end user as a single data repository. This arrangement promotes data sharing, thus eliminating the potential problem of islands of information. In addition, the DBMS enforces data integrity, eliminates redundancy, and promotes data security.

KEY TERMS

ad hoc query, 8

centralized database, 8

data, 8

data anomaly, 18

data dependence, 15

data dictionary, 21

data inconsistency, 8

data independence, 15

data integrity, 17

data management, 6

data redundancy, 17

data warehouse, 9

database, 6

database design, 10

database management system (DBMS), 7

database system, 19

desktop database, 8

distributed database, 8

enterprise database, 8

extensible markup language (XML), 9

field, 12

file, 12

information, 6

islands of information, 17

knowledge, 17

logical data format, 15

metadata, 15

multiuser database, 8

operational database, 8

performance tuning, 22

physical data format, 15

production database, 8

query, 9

query language, 8

query result set, 8

record, 8

single-user database, 8

structural dependence, 15

structural independence, 15

structured data, 9

Structured Query Language (SQL), 9

transactional database, 8

unstructured data, 9

workgroup database, 8

XML, 9

XML database, 9

ONLINE CONTENT

Answers to selected Review Questions and Problems for this chapter are contained in the Student OnlineCompanion for this book.

REVIEW QUESTIONS

1. Discuss each of the following terms:
    a. data
    b. field
    c. record
    d. file
2. What is data redundancy, and which characteristics of the file system can lead to it?
3. What is data independence, and why is it lacking in file systems?
4. What is a DBMS, and what are its functions?
5. What is structural independence, and why is it important?
6. Explain the difference between data and information.
7. What is the role of a DBMS, and what are its advantages? What are its disadvantages?
8. List and describe the different types of databases.
9. What are the main components of a database system?
10. What is metadata?
11. Explain why database design is important.
12. What are the potential costs of implementing a database system?
13. Use examples to compare and contrast unstructured and structured data. Which type is more prevalent in a typical business environment?

PROBLEMS

ONLINE CONTENT

The file structures you see in this problem set are simulated in a Microsoft Access database named **Ch01_Problems**, available in the Student Online Companion for this book.

FIGURE P1.1   **The file structure for Problems 1–4**

PROJECT_CODE	PROJECT_MANAGER	MANAGER_PHONE	MANAGER_ADDRESS	PROJECT_BID_PRICE
21-5Z	Holly B. Parker	904-338-3416	3334 Lee Rd., Gainesville, FL 37123	16833460.00
25-2D	Jane D. Grant	615-898-9909	218 Clark Blvd., Nashville, TN 36362	12500000.00
25-5A	George F. Dorts	615-227-1245	124 River Dr., Franklin, TN 29185	32512420.00
25-9T	Holly B. Parker	904-338-3416	3334 Lee Rd., Gainesville, FL 37123	21563234.00
27-4Q	George F. Dorts	615-227-1245	124 River Dr., Franklin, TN 29185	10314545.00
29-2D	Holly B. Parker	904-338-3416	3334 Lee Rd., Gainesville, FL 37123	25559999.00
31-7P	William K. Moor	904-445-2719	216 Morton Rd., Stetson, FL 30155	56850000.00

Given the file structure shown in Figure P1.1, answer Problems 1–4.

1. How many records does the file contain? How many fields are there per record?
2. What problem would you encounter if you wanted to produce a listing by city? How would you solve this problem by altering the file structure?
3. If you wanted to produce a listing of the file contents by last name, area code, city, state, or zip code, how would you alter the file structure?
4. What data redundancies do you detect? How could those redundancies lead to anomalies?

FIGURE P1.5   **The file structure for Problems 5–8**

PROJ_NUM	PROJ_NAME	EMP_NUM	EMP_NAME	JOB_CODE	JOB_CHG_HOUR	PROJ_HOURS	EMP_PHONE
1	Hurricane	101	John D. Newson	CC	85.00	13.3	653-234-3245
1	Hurricane	105	David F. Schwann	CT	60.00	16.2	653-234-1123
1	Hurricane	110	Anne R. Ramoras	CT	60.00	14.3	615-233-5568
2	Coast	101	John D. Newson	EE	85.00	19.8	653-234-3254
2	Coast	108	June H. Sattlemeir	EE	85.00	17.5	905-554-7812
3	Satellite	110	Anne R. Ramoras	CT	62.00	11.6	615-233-5568
3	Satellite	105	David F. Schwann	CT	26.00	23.4	653-234-1123
3	Satelite	123	Mary D. Chen	EE	85.00	19.1	615-233-5432
3	Satelite	112	Allecia R. Smith	BE	85.00	20.7	615-678-6879

5. Identify and discuss the serious data redundancy problems exhibited by the file structure shown in Figure P1.5.
6. Looking at the EMP_NAME and EMP_PHONE contents in Figure P1.5, what change(s) would you recommend?
7. Identify the various data sources in the file you examined in Problem 5.
8. Given your answer to Problem 7, what new files should you create to help eliminate the data redundancies found in the file shown in Figure P1.5?

The file structure for Problems 9–10

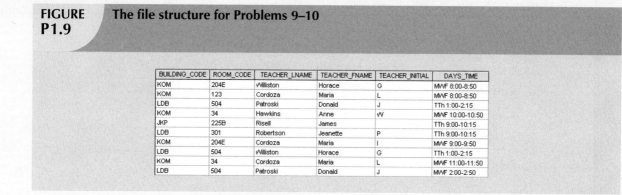

BUILDING_CODE	ROOM_CODE	TEACHER_LNAME	TEACHER_FNAME	TEACHER_INITIAL	DAYS_TIME
KOM	204E	Williston	Horace	G	MWF 8:00-8:50
KOM	123	Cordoza	Maria	L	MWF 8:00-8:50
LDB	504	Patroski	Donald	J	TTh 1:00-2:15
KOM	34	Hawkins	Anne	W	MWF 10:00-10:50
JKP	225B	Risell	James		TTh 9:00-10:15
LDB	301	Robertson	Jeanette	P	TTh 9:00-10:15
KOM	204E	Cordoza	Maria	I	MWF 9:00-9:50
LDB	504	Williston	Horace	G	TTh 1:00-2:15
KOM	34	Cordoza	Maria	L	MWF 11:00-11:50
LDB	504	Patroski	Donald	J	MWF 2:00-2:50

9. Identify and discuss the serious data redundancy problems exhibited by the file structure shown in Figure P1.9. (The file is meant to be used as a teacher class assignment schedule. One of the many problems with data redundancy is the likely occurrence of data inconsistencies—two different initials have been entered for the teacher named Maria Cordoza.)

10. Given the file structure shown in Figure P1.9, what problem(s) might you encounter if building KOM were deleted?

2

DATA MODELS

In this chapter, you will learn:

- About data modeling and why data models are important
- About the basic data-modeling building blocks
- What business rules are and how they influence database design
- How the major data models evolved
- How data models can be classified by level of abstraction

This chapter examines data modeling. Data modeling is the first step in the database design journey, serving as a bridge between real-world objects and the database that resides in the computer.

One of the most vexing problems of database design is that designers, programmers, and end users see data in different ways. Consequently, different views of the same data can lead to database designs that do not reflect an organization's actual operation, failing to meet end-user needs and data efficiency requirements. To avoid such failures, database designers must obtain a precise description of the nature of the data and of the many uses of that data within the organization. Communication among database designers, programmers, and end users should be frequent and clear. Data modeling clarifies such communication by reducing the complexities of database design to more easily understood abstractions that define entities and the relations among them.

First, you will learn what some of the basic data-modeling concepts are and how current data models developed from earlier models. Tracing the development of those database models will help you understand the database design and implementation issues that are addressed in the rest of this book. Second, you will be introduced to the Entity Relationship Diagram (ERD) as a data modeling tool. ER diagrams can be drawn using a variety of notations. Within this chapter you will be introduced to the traditional Chen notation, the more current Crows' Foot notation, and the newer class diagram notation, which is part of the Unified Modeling Language (UML). Finally, you will learn how various degrees of data abstraction help reconcile varying views of the same data.

Preview

2.1 DATA MODELING AND DATA MODELS

Database design focuses on how the database structure will be used to store and manage end-user data. Data modeling, the first step in designing a database, refers to the process of creating a specific data model for a determined problem domain. (A problem domain is a clearly defined area within the real world environment, with well defined scope and boundaries, that is to be systematically addressed.) A **data model** is a relatively simple representation, usually graphical, of more complex real-world data structures. In general terms, a *model* is an abstraction of a more complex real-world object or event. A model's main function is to help you understand the complexities of the real-world environment. Within the database environment, a data model represents data structures and their characteristics, relations, constraints, transformations, and other constructs with the purpose of supporting a specific problem domain.

> **NOTE**
>
> The terms *data model* and *database model* are often used interchangeably. In this book, the term *database model* is used to refer to the implementation of a *data model* in a specific database system.

Data modeling is an iterative, progressive process. You start with a simple understanding of the problem domain, and as your understanding of the problem domain increases, so does the level of detail of the data model. Done properly, the final data model is in effect a "blueprint" containing all the instructions to build a database that will meet all end-user requirements. This blueprint is narrative and graphical in nature, meaning that it contains both text descriptions in plain, unambiguous language and clear, useful diagrams depicting the main data elements.

> **NOTE**
>
> An implementation-ready data model should contain at least the following components:
>
> - A description of the data structure that will store the end-user data.
> - A set of enforceable rules to guarantee the integrity of the data.
> - A data manipulation methodology to support the real-world data transformations.

Traditionally, database designers relied on good judgment to help them develop a good data model. Unfortunately, good judgment is often in the eye of the beholder, and it often develops after much trial and error. For example, if each of the students in this class has to create a data model for a video store, it's very likely that each of them will come up with a different model. Which one would be the correct one? The simple answer is "the one that meets all the end-user requirements," and there may be more than one correct solution! Fortunately, database designers make use of existing data modeling constructs and powerful database design tools that substantially diminish the potential for errors in database modeling. In the following sections you will learn how existing data models are used to represent real world data and how the different degrees of data abstraction facilitate data modeling. But first, you must understand the importance of data models and their basic constructs.

2.2 THE IMPORTANCE OF DATA MODELS

Data models can facilitate interaction among the designer, the applications programmer, and the end user. A well-developed data model can even foster improved understanding of the organization for which the database design is developed. In short, data models are a communication tool. This important aspect of data modeling was summed up neatly by a client whose reaction was as follows: "I created this business, I worked with this business for years, and this is the first time I've really understood how all the pieces really fit together."

The importance of data modeling cannot be overstated. Data constitute the most basic information units employed by a system. Applications are created to manage data and to help transform data into information. But data are viewed in different ways by different people. For example, contrast the (data) view of a company manager with that of a company clerk. Although the manager and the clerk both work for the same company, the manager is more likely to have an enterprise-wide view of company data than the clerk.

Even different managers view data differently. For example, a company president is likely to take a universal view of the data because he or she must be able to tie the company's divisions to a common (database) vision. A purchasing manager in the same company is likely to have a more restricted view of the data, as is the company's inventory manager. In effect, each department manager works with a subset of the company's data. The inventory manager is more concerned about inventory levels, while the purchasing manager is more concerned about the cost of items and about personal/business relationships with the suppliers of those items.

Applications programmers have yet another view of data, being more concerned with data location, formatting, and specific reporting requirements. Basically, applications programmers translate company policies and procedures from a variety of sources into appropriate interfaces, reports, and query screens.

The different users and producers of data and information often reflect the "blind people and the elephant" analogy: the blind person who felt the elephant's trunk had quite a different view of the elephant from the one who felt the elephant's leg or tail. What is needed is a view of the whole elephant. Similarly, a house is not a random collection of rooms; if someone is going to build a house, he or she should first have the overall view that is provided by blueprints. Likewise, a sound data environment requires an overall database blueprint based on an appropriate data model.

When a good database blueprint is available, it does not matter that an applications programmer's view of the data is different from that of the manager and/or the end user. Conversely, when a good database blueprint is not available, problems are likely to ensue. For instance, an inventory management program or an order entry system may use conflicting product numbering schemes, thereby costing the company thousands (or even millions) of dollars.

Keep in mind that a house blueprint is an abstraction; you cannot live in the blueprint. Similarly, the data model is an abstraction; you cannot draw the required data out of the data model. Just as you are not likely to build a good house without a blueprint, you are equally unlikely to create a good database without first creating an appropriate data model.

2.3 DATA MODEL BASIC BUILDING BLOCKS

The basic building blocks of all data models are entities, attributes, relationships, and constraints. An **entity** is anything (a person, a place, a thing, or an event) about which data are to be collected and stored. An entity represents a particular type of object in the real world. Because an entity represents a particular type of object, entities are "distinguishable" that is, each entity occurrence is unique and distinct. For example, a CUSTOMER entity would have many distinguishable customer occurrences, such as John Smith, Pedro Dinamita, Tom Strickland, etc. Entities may be physical objects, such as customers or products, but entities may also be abstractions, such as flight routes or musical concerts.

An **attribute** is a characteristic of an entity. For example, a CUSTOMER entity would be described by attributes such as customer last name, customer first name, customer phone, customer address, and customer credit limit. Attributes are the equivalent of fields in file systems.

A **relationship** describes an association among entities. For example, a relationship exists between customers and agents that can be described as follows: an agent can serve many customers, and each customer may be served by one agent. Data models use three types of relationships: one-to-many, many-to-many, and one-to-one. Database designers usually use the shorthand notations 1:M or 1..*, M:N or *..*, and 1:1 or 1..1, respectively. (Although the M:N notation is a standard label for the many-to-many relationship, the label M:M may also be used.) The following examples illustrate the distinctions among the three.

- **One-to-many (1:M or 1..*) relationship**. A painter paints many different paintings, but each one of them is painted by only one painter. Thus, the painter (the "one") is related to the paintings (the "many"). Therefore, database designers label the relationship "PAINTER paints PAINTING" as 1:M. (Note that entity names are often capitalized as a convention so they are easily identified.) Similarly, a customer (the "one") may generate many invoices, but each invoice (the "many") is generated by only a single customer. The "CUSTOMER generates INVOICE" relationship would also be labeled 1:M.

- **Many-to-many (M:N or *..*) relationship**. An employee may learn many job skills, and each job skill may be learned by many employees. Database designers label the relationship "EMPLOYEE learns SKILL" as M:N. Similarly, a student can take many classes and each class can be taken by many students, thus yielding the M:N relationship label for the relationship expressed by "STUDENT takes CLASS."

- **One-to-one (1:1 or 1..1) relationship**. A retail company's management structure may require that each of its stores be managed by a single employee. In turn, each store manager, who is an employee, manages only a single store. Therefore, the relationship "EMPLOYEE manages STORE" is labeled 1:1.

The preceding discussion identified each relationship in both directions; that is, relationships are bidirectional:

- *One* CUSTOMER can generate *many* INVOICEs.
- Each of the *many* INVOICEs is generated by only *one* CUSTOMER.

A **constraint** is a restriction placed on the data. Constraints are important because they help to ensure data integrity. Constraints are normally expressed in the form of rules. For example:

- An employee's salary must have values that are between 6,000 and 350,000.
- A student's GPA must be between 0.00 and 4.00.
- Each class must have one and only one teacher.

How do you properly identify entities, attributes, relationships, and constraints? The first step is to clearly identify the business rules for the problem domain you are modeling.

2.4 BUSINESS RULES

When database designers go about selecting or determining the entities, attributes, and relationships that will be used to build a data model, they might start by gaining a thorough understanding of what types of data are in an organization, how the data are used, and in what time frames they are used. But such data and information do not, by themselves, yield the required understanding of the total business. From a database point of view, the collection of data becomes meaningful only when it reflects properly defined *business rules*. A **business rule** is a brief, precise, and unambiguous description of a policy, procedure, or principle within a specific organization. In a sense, business rules are misnamed: they apply to *any* organization, large or small—a business, a government unit, a religious group, or a research laboratory—that stores and uses data to generate information.

Business rules, derived from a detailed description of an organization's operations, help to create and enforce actions within that organization's environment. Business rules must be rendered in writing and updated to reflect any change in the organization's operational environment.

Properly written business rules are used to define entities, attributes, relationships, and constraints. Any time you see relationship statements such as "an agent can serve many customers, and each customer can be served by only one agent," you are seeing business rules at work. You will see the application of business rules throughout this book, especially in the chapters devoted to data modeling and database design.

To be effective, business rules must be easy to understand and widely disseminated to ensure that every person in the organization shares a common interpretation of the rules. Business rules describe, in simple language, the main and distinguishing characteristics of the data *as viewed by the company*. Examples of business rules are as follows:

- A customer may generate many invoices.
- An invoice is generated by only one customer.
- A training session cannot be scheduled for fewer than 10 employees or for more than 30 employees.

Note that those business rules establish entities, relationships, and constraints. For example, the first two business rules establish two entities (CUSTOMER and INVOICE) and a 1:M relationship between those two entities. The third business rule establishes a constraint (no fewer than 10 people and no more than 30 people), two entities (EMPLOYEE and TRAINING), and a relationship between EMPLOYEE and TRAINING.

2.4.1 DISCOVERING BUSINESS RULES

The main sources of business rules are company managers, policy makers, department managers, and written documentation such as a company's procedures, standards, or operations manuals. A faster and more direct source of business rules is direct interviews with end users. Unfortunately, because perceptions differ, end users sometimes are a less reliable source when it comes to specifying business rules. For example, a maintenance department mechanic might believe that any mechanic can initiate a maintenance procedure, when actually only mechanics with inspection authorization can perform such a task. Such a distinction might seem trivial, but it can have major legal consequences. Although end users are crucial contributors to the development of business rules, *it pays to verify end-user perceptions*. Too often, interviews with several people who perform the same job yield very different perceptions of what the job components are. While such a discovery may point to "management problems," that general diagnosis does not help the database designer. The database designer's job is to reconcile such differences and verify the results of the reconciliation to ensure that the business rules are appropriate and accurate.

The process of identifying and documenting business rules is essential to database design for several reasons:

- They help standardize the company's view of data.
- They can be a communications tool between users and designers.
- They allow the designer to understand the nature, role, and scope of the data.
- They allow the designer to understand business processes.
- They allow the designer to develop appropriate relationship participation rules and constraints and to create an accurate data model.

Of course, not all business rules can be modeled. For example, a business rule that specifies that "no pilot can fly more than 10 hours within any 24-hour period" cannot be modeled. However, such a business rule can be enforced by application software.

2.4.2 TRANSLATING BUSINESS RULES INTO DATA MODEL COMPONENTS

Business rules set the stage for the proper identification of entities, attributes, relationships, and constraints. In the real world, names are used to identify objects. If the business environment wants to keep track of the objects, there will be specific business rules for them. As a general rule, a noun in a business rule will translate into an entity in the model, and a verb (active or passive) associating nouns will translate into a relationship among the entities. For example, the business rule "a customer may generate many invoices" contains two nouns (*customer* and *invoices*) and a verb (*generate*) that associates the nouns. From this business rule, you could deduct that:

- Customer and invoice are objects of interest for the environment and should be represented by their respective entities.
- There is a "generate" relationship between customer and invoice.

To properly identify the type of relationship, you should consider that relationships are bidirectional; that is, they go both ways. For example, the business rule "a customer may generate many invoices" is complemented by the business rule "an invoice is generated by only one customer." In that case, the relationship is one-to-many (1:M). Customer is the "1" side, and invoice is the "many" side.

As a general rule, to properly identify the relationship type, you should ask two questions:

- How many instances of B are related to one instance of A?
- How many instances of A are related to one instance of B?

For example, you can assess the relationship between student and class by asking two questions:

- In how many classes can one student enroll? Answer: many classes.
- How many students can enroll in one class? Answer: many students.

Therefore, the relationship between student and class is many-to-many (M:N). You will have many opportunities to determine the relationships between entities as you proceed through this book, and soon the process will become second nature.

2.5 THE EVOLUTION OF DATA MODELS

The quest for better data management has led to several different models that attempt to resolve the file system's critical shortcomings. This section gives an overview of the major data models in roughly chronological order. You will discover that many of the "new" database concepts and structures bear a remarkable resemblance to some of the "old" data model concepts and structures. Table 2.1 traces the evolution of the major data models.

TABLE 2.1 Evolution of Major Data Models

GENERATION	TIME	MODEL	EXAMPLES	COMMENTS
First	1960s–1970s	File System	VMS/VSAM	Used mainly on IBM mainframe systems Managed records, not relationships
Second	1970s	Hierarchical and Network Data Model	IMS ADABAS IDS-II	Early database systems Navigational access
Third	Mid-1970s to present	Relational Data Model	DB2 Oracle MS SQL-Server MySQL	Conceptual simplicity Entity Relationship (ER) modeling and support for relational data modeling
Fourth	Mid-1980s to present	Object-Oriented Extended Relational	Versant FastObjects.Net Objectivity/DB DB/2 UDB Oracle 10g	Support complex data Extended relational products support objects and data warehousing Web databases become common
Next Generation	Present to future	XML	dbXML Tamino DB2 UDB Oracle 10g MS SQL Server	Organization and management of unstructured data Relational and object models add support for XML documents

ONLINE CONTENT

The hierarchical and network models are largely of historical interest, yet they do contain some elements and features that interest current database professionals. The technical details of those two models are discussed in detail in Appendixes K and L, respectively, in the Student Online Companion for this book. Appendix G is devoted to the object-oriented (OO) model. However, given the dominant market presence of the relational model, most of the book focuses on that model.

2.5.1 THE HIERARCHICAL MODEL

The **hierarchical model** was developed in the 1960s to manage large amounts of data for complex manufacturing projects such as the Apollo rocket that landed on the moon in 1969. Its basic logical structure is represented by an upside-down tree. The hierarchical structure contains levels, or segments. A **segment** is the equivalent of a file system's record type. Within the hierarchy, the top layer (the root) is perceived as the parent of the segment directly beneath it. For example, in Figure 2.1, the root segment is the parent of the Level 1 segments, which, in turn, are the parents of the Level 2 segments, etc. The segments below other segments are the children of the segment above. In short, the hierarchical model depicts a set of one-to-many (1:M) relationships between a parent and its children segments. (Each parent can have many children, but each child has only one parent.)

FIGURE 2.1   A hierarchical structure

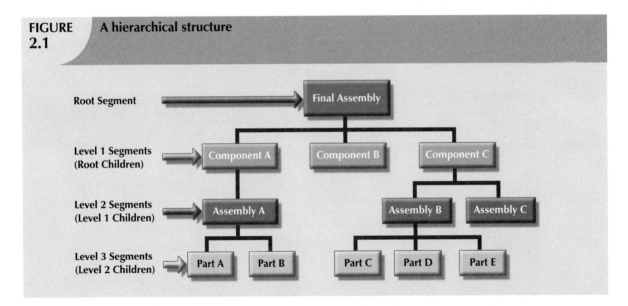

The hierarchical data model yielded many advantages over the file system model. In fact, many of the hierarchical data model's features formed the foundation for current data models. Many of its database application advantages are replicated, albeit in a different form, in current database environments. The hierarchical database quickly became dominant in the 1970s and generated a large installed base, which, in turn, created a pool of programmers who knew the systems and who developed numerous tried-and-true business applications. However, the hierarchical model had limitations: it was complex to implement, it was difficult to manage, and it lacked structural independence. Also, many common data relationships do not conform to the 1:M form, and there were no standards for how to implement the model.

In the 1970s, database professionals convened a set of meetings that culminated in the publication of a set of database standards that ultimately led to the development of alternative data models. The most prominent of those models is the network model.

2.5.2 THE NETWORK MODEL

The **network model** was created to represent complex data relationships more effectively than the hierarchical model, to improve database performance, and to impose a database standard. The lack of database standards was troublesome to programmers and application designers because it made database designs and applications less portable. Worse, the lack of even a standard set of database *concepts* impeded the search for better data models. Disorganization seldom fosters progress.

To help establish database standards, the **Conference on Data Systems Languages (CODASYL)** created the **Database Task Group (DBTG)** in the late 1960s. The DBTG was charged to define standard specifications for an environment that would facilitate database creation and data manipulation. The final DBTG report contained specifications for three crucial database components:

- The **schema**, which is the conceptual organization of the entire database as viewed by the database administrator. The schema includes a definition of the database name, the record type for each record, and the components that make up those records.

- The **subschema**, which defines the portion of the database "seen" by the application programs that actually produce the desired information from the data contained within the database. The existence of subschema definitions allows all application programs to simply invoke the subschema required to access the appropriate database file(s).

- A **data management language (DML)** that defines the environment in which data can be managed. To produce the desired standardization for each of the three components, the DBTG specified three distinct DML components:
  - A schema **data definition language (DDL)**, which enables the database administrator to define the schema components.
  - A subschema DDL, which allows the application programs to define the database components that will be used by the application.
  - A data manipulation language to work with the data in the database.

In the network model, the user perceives the network database as a collection of records in 1:M relationships. However, unlike the hierarchical model, the network model allows a record to have more than one parent. In network database terminology, a relationship is called a set. Each set is composed of at least two record types: an owner record and a member record. *A set represents a 1:M relationship between the owner and the member.* An example of such a relationship is depicted in Figure 2.2.

FIGURE 2.2   **A network data model**

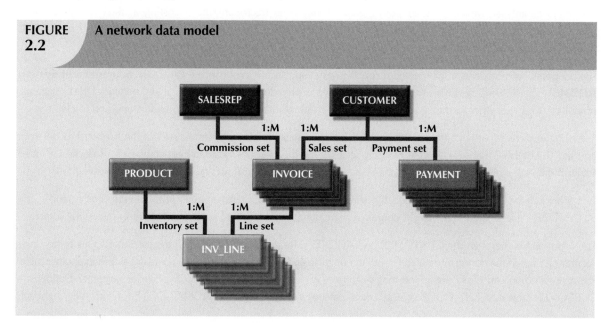

Figure 2.2 illustrates a network data model for a typical sales organization. In this model, CUSTOMER, SALESREP, INVOICE, INV_LINE, PRODUCT, and PAYMENT represent record types. Note that INVOICE is "owned" by both SALESREP and CUSTOMER. Similarly, INV_LINE has two owners, PRODUCT and INVOICE. Furthermore, the network model can also include one-owner relationships, such as CUSTOMER makes PAYMENT.

As information needs grew and as more sophisticated databases and applications were required, the network model became too cumbersome. The lack of ad hoc query capability put heavy pressure on programmers to generate the code required to produce even the simplest reports. And although the existing databases provided limited data independence, any structural change in the database still could produce havoc in all application programs that drew data from the database. Because of the disadvantages of the hierarchical and network models, they were largely replaced by the relational data model in the 1980s.

2.5.3 THE RELATIONAL MODEL

The **relational model** was introduced in 1970 by E. F. Codd (of IBM) in his landmark paper "A Relational Model of Data for Large Shared Databanks" (*Communications of the ACM*, June 1970, pp. 377–387). The relational model represented a major breakthrough for both users and designers. To use an analogy, the relational model produced an "automatic transmission" database to replace the "standard transmission" databases that preceded it. Its conceptual simplicity set the stage for a genuine database revolution.

> **NOTE**
>
> The relational database model presented in this chapter is an introduction and an overview. A more detailed discussion is in Chapter 3, The Relational Database Model. In fact, the relational model is so important that it will serve as the basis for discussions in most of the remaining chapters.

The relational model foundation is a mathematical concept known as a relation. To avoid the complexity of abstract mathematical theory, you can think of a **relation** (sometimes called a **table**) as a matrix composed of intersecting rows and columns. Each row in a relation is called a **tuple**. Each column represents an attribute. The relational model also describes a precise set of data manipulation constructs based on advanced mathematical concepts.

In 1970, Codd's work was considered ingenious but impractical. The relational model's conceptual simplicity was bought at the expense of computer overhead; computers at that time lacked the power to implement the relational model. Fortunately, computer power grew exponentially, as did operating system efficiency. Better yet, the cost of computers diminished rapidly as their power grew. Today even microcomputers, costing a fraction of what their mainframe ancestors did, can run sophisticated relational database software such as Oracle, DB2, Microsoft SQL Server, MySQL, and other mainframe relational software.

The relational data model is implemented through a very sophisticated **relational database management system (RDBMS)**. The RDBMS performs the same basic functions provided by the hierarchical and network DBMS systems, in addition to a host of other functions that make the relational data model easier to understand and implement.

Arguably the most important advantage of the RDBMS is its ability to hide the complexities of the relational model from the user. The RDBMS manages all of the physical details, while the user sees the relational database as a collection of tables in which data are stored. The user can manipulate and query the data in a way that seems intuitive and logical.

Tables are related to each other through the sharing of a common attribute (value in a column). For example, the CUSTOMER table in Figure 2.3 might contain a sales agent's number that is also contained in the AGENT table.

The common link between the CUSTOMER and AGENT tables enables you to match the customer to his or her sales agent even though the customer data are stored in one table and the sales representative data are stored in another table. For example, you can easily determine that customer Dunne's agent is Alex Alby because for customer Dunne, the CUSTOMER table's AGENT_CODE is 501, which matches the AGENT table's AGENT_CODE for Alex Alby. Although

FIGURE 2.3   Linking relational tables

Table name: AGENT (first six attributes)   **Database name: Ch02_InsureCo**

AGENT_CODE	AGENT_LNAME	AGENT_FNAME	AGENT_INITIAL	AGENT_AREACODE	AGENT_PHONE
501	Alby	Alex	B	713	228-1249
502	Hahn	Leah	F	615	882-1244
503	Okon	John	T	615	123-5589

Link through AGENT_CODE

Table name: CUSTOMER

CUS_CODE	CUS_LNAME	CUS_FNAME	CUS_INITIAL	CUS_AREACODE	CUS_PHONE	CUS_INSURE_TYPE	CUS_INSURE_AMT	CUS_RENEW_DATE	AGENT_CODE
10010	Ramas	Alfred	A	615	844-2573	T1	100.00	05-Apr-2008	502
10011	Dunne	Leona	K	713	894-1238	T1	250.00	16-Jun-2008	501
10012	Smith	Kathy	W	615	894-2285	S2	150.00	29-Jan-2009	502
10013	Olowski	Paul	F	615	894-2180	S1	300.00	14-Oct-2008	502
10014	Orlando	Myron		615	222-1672	T1	100.00	28-Dec-2008	501
10015	O'Brian	Amy	B	713	442-3381	T2	850.00	22-Sep-2008	503
10016	Brown	James	G	615	297-1228	S1	120.00	25-Mar-2009	502
10017	Williams	George		615	290-2556	S1	250.00	17-Jul-2008	503
10018	Farriss	Anne	G	713	382-7185	T2	100.00	03-Dec-2008	501
10019	Smith	Olette	K	615	297-3809	S2	500.00	14-Mar-2009	503

ONLINE CONTENT

This chapter's databases can be found in the Student Online Companion. For example, the contents of the AGENT and CUSTOMER tables shown in Figure 2.3 are found in the database named **Ch02_InsureCo**.

the tables are independent of one another, you can easily associate the data between tables. The relational model provides a minimum level of controlled redundancy to eliminate most of the redundancies commonly found in file systems.

The relationship type (1:1, 1:M, or M:N) is often shown in a relational schema, an example of which is shown in Figure 2.4. A **relational diagram** is a representation of the relational database's entities, the attributes within those entities, and the relationships between those entities.

FIGURE 2.4   A relational diagram

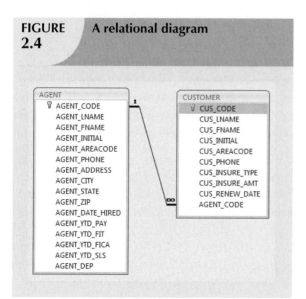

In Figure 2.4, the relational diagram shows the connecting fields (in this case, AGENT_CODE) and the relationship type, 1:M. Microsoft Access, the database software application used to generate Figure 2.4, employs the ∞ (infinity) symbol to indicate the "many" side. In this example, the CUSTOMER represents the "many" side because an AGENT can have many CUSTOMERs. The AGENT represents the "1" side because each CUSTOMER has only one AGENT.

A relational table stores a collection of related entities. In this respect, the relational database table resembles a file. But there is one crucial difference between a table and a file: a table yields complete data and structural independence because it is a purely logical structure. How the data are physically stored in the database is of no concern to the user or the designer; the perception is what counts.

And this property of the relational data model, explored in depth in the next chapter, became the source of a real database revolution.

Another reason for the relational data model's rise to dominance is its powerful and flexible query language. For most relational database software, the query language is Structured Query Language (SQL), which allows the user to specify what must be done without specifying how it must be done. The RDBMS uses SQL to translate user queries into instructions for retrieving the requested data. SQL makes it possible to retrieve data with far less effort than any other database or file environment.

From an end-user perspective, any SQL-based relational database application involves three parts: a user interface, a set of tables stored in the database, and the SQL "engine." Each of these parts is explained below.

- *The end-user interface.* Basically, the interface allows the end user to interact with the data (by auto-generating SQL code). Each interface is a product of the software vendor's idea of meaningful interaction with the data. You can also design your own customized interface with the help of application generators that are now standard fare in the database software arena.

- *A collection of tables stored in the database.* In a relational database, all data are perceived to be stored in tables. The tables simply "present" the data to the end user in a way that is easy to understand. Each table is independent from another. Rows in different tables are related based on common values in common attributes.

- *SQL engine.* Largely hidden from the end user, the SQL engine executes all queries, or data requests. Keep in mind that the SQL engine is part of the DBMS software. The end user uses SQL to create table structures and to perform data access and table maintenance. The SQL engine processes all user requests—largely behind the scenes and without the end user's knowledge. Hence, it's said that SQL is a declarative language that tells what must be done but not how it must be done. (You will learn more about the SQL engine in Chapter 11, Database Performance Tuning and Query Optimization.)

Because the RDBMS performs the behind-the-scenes tasks, it is not necessary to focus on the physical aspects of the database. Instead, the chapters that follow concentrate on the logical portion of the relational database and its design. Furthermore, SQL is covered in detail in Chapter 7, Introduction to Structured Query Language (SQL), and in Chapter 8, Advanced SQL.

2.5.4 THE ENTITY RELATIONSHIP MODEL

The conceptual simplicity of relational database technology triggered the demand for RDBMSs. In turn, the rapidly increasing requirements for transaction and information created the need for more complex database implementation structures, thus creating the need for more effective database design tools. (Building a skyscraper requires more detailed design activities than building a doghouse, for example.)

Complex design activities require conceptual simplicity to yield successful results. Although the relational model was a vast improvement over the hierarchical and network models, it still lacked the features that would make it an effective database *design* tool. Because it is easier to examine structures graphically than to describe them in text, database designers prefer to use a graphical tool in which entities and their relationships are pictured. Thus, the **entity relationship (ER) model**, or **ERM**, has become a widely accepted standard for data modeling.

Peter Chen first introduced the ER data model in 1976; it was the graphical representation of entities and their relationships in a database structure that quickly became popular because it *complemented* the relational data model concepts. The relational data model and ERM combined to provide the foundation for tightly structured database design. ER models are normally represented in an **entity relationship diagram (ERD)**, which uses graphical representations to model database components.

NOTE

Because this chapter's objective is to introduce data-modeling concepts, a simplified ERD is discussed in this section. You will learn how to use ERDs to design databases in Chapter 4, Entity Relationship (ER) Modeling.

The ER model is based on the following components:

- *Entity*. Earlier in this chapter, an entity was defined as anything about which data are to be collected and stored. An entity is represented in the ERD by a rectangle, also known as an entity box. The name of the entity, a noun, is written in the center of the rectangle. The entity name is generally written in capital letters and is written in the singular form: PAINTER rather than PAINTERS, and EMPLOYEE rather than EMPLOYEES. Usually, when applying the ERD to the relational model, an entity is mapped to a relational table. Each row in the relational table is known as an **entity instance** or **entity occurrence** in the ER model.

NOTE

A collection of like entities is known as an **entity set**. For example, you can think of the AGENT file in Figure 2.3 as a collection of three agents (*entities*) in the AGENT *entity set*. Technically speaking, the ERD depicts entity *sets*. Unfortunately, ERD designers use the word entity as a substitute for entity set, and this book will conform to that established practice when discussing any ERD and its components.

Each entity is described by a set of *attributes* that describes particular characteristics of the entity. For example, the entity EMPLOYEE will have attributes such as a Social Security number, a last name, and a first name. (Chapter 4 explains how attributes are included in the ERD.)

- *Relationships*. Relationships describe associations among data. Most relationships describe associations between two entities. When the basic data model components were introduced, three types of relationships among data were illustrated: one-to-many (1:M), many-to-many (M:N), and one-to-one (1:1). The ER model uses the term **connectivity** to label the relationship types. The name of the relationship usually is an active or passive verb. For example, a PAINTER *paints* many PAINTINGs; an EMPLOYEE *learns* many SKILLs; an EMPLOYEE *manages* a STORE.

Figure 2.5 shows the different types of relationships using two ER notations: the original **Chen notation** and the more current **Crow's Foot notation**.

The left side of the ER diagram shows the Chen notation, based on Peter Chen's landmark paper. In this notation, the connectivities are written next to each entity box. Relationships are represented by a diamond connected to the related entities through a relationship line. The relationship name is written inside the diamond.

The right side of Figure 2.5 illustrates the Crow's Foot notation. The name "Crow's Foot" is derived from the three-pronged symbol used to represent the "many" side of the relationship. As you examine the basic Crow's Foot ERD in Figure 2.5, note that the connectivities are represented by symbols. For example, the "1" is represented by a short line segment and the "M" is represented by the three-pronged "crow's foot." In this example the relationship name is written above the relationship line.

In Figure 2.5, entities and relationships are shown in a horizontal format, but they also may be oriented vertically. The entity location and the order in which the entities are presented are immaterial; just remember to read a 1:M relationship from the "1" side to the "M" side.

**FIGURE
2.5**   **The Chen and Crow's Foot notations**

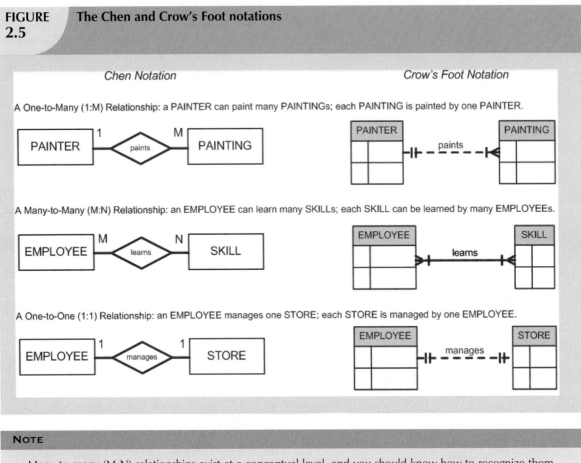

NOTE

Many-to-many (M:N) relationships exist at a conceptual level, and you should know how to recognize them. However, you will learn in Chapter 3 that M:N relationships are not appropriate in a relational model. For that reason, Microsoft Visio does not support the M:N relationship. Therefore, to illustrate the existence of a M:N relationship using Visio, two superimposed 1:M relationships have been used.

The Crow's Foot notation is used as the design standard in this book. However, the Chen notation is used to illustrate some of the ER modeling concepts whenever necessary. Most database modeling tools let you select the Crow's Foot notation. Microsoft Visio Professional software was used to generate the Crow's Foot designs you will see in subsequent chapters.

ONLINE CONTENT

Aside from the Chen and Crow's Foot notations, there are other ER model notations. For a summary of the symbols used by several additional ER model notations, see **Appendix D, Comparison of ER Model Notations,** in the Student Online Companion.

Its exceptional visual simplicity makes the ER model the dominant database modeling and design tool. Nevertheless, the search for better data-modeling tools continues as the data environment continues to evolve.

2.5.5 THE OBJECT-ORIENTED (OO) MODEL

Increasingly complex real-world problems demonstrated a need for a data model that more closely represented the real world. In the **object-oriented data model** (**OODM**), both data *and their relationships* are contained in a single structure known as an **object**. In turn, the OODM is the basis for the **object-oriented database management system** (**OODBMS**).

ONLINE CONTENT

This chapter introduces only basic OO concepts. You'll have a chance to examine object-orientation concepts and principles in detail in **Appendix G, Object-Oriented Databases,** found in the Student Online Companion for this book.

An OODM reflects a very different way to define and use entities. Like the relational model's entity, an object is described by its factual content. But quite *unlike* an entity, an object includes information about relationships between the facts within the object, as well as information about its relationships with other objects. Therefore, the facts within the object are given greater *meaning*. The OODM is said to be a **semantic data model** because *semantic* indicates meaning.

Subsequent OODM development has allowed an object also to contain all *operations* that can be performed on it, such as changing its data values, finding a specific data value, and printing data values. Because objects include data, various types of relationships, and operational procedures, the object becomes self-contained, thus making the object—at least potentially—a basic building block for autonomous structures.

The OO data model is based on the following components:

- An object is an abstraction of a real-world entity. In general terms, an object may be considered equivalent to an ER model's entity. More precisely, an object represents only one occurrence of an entity. (The object's semantic content is defined through several of the items in this list.)

- Attributes describe the properties of an object. For example, a PERSON object includes the attributes Name, Social Security Number, and Date of Birth.

- Objects that share similar characteristics are grouped in classes. A **class** is a collection of similar objects with shared structure (attributes) and behavior (methods). In a general sense, a class resembles the ER model's entity *set*. However, a class is different from an entity set in that it contains a set of procedures known as *methods*. A class's **method** represents a real-world action such as *finding* a selected PERSON's name, *changing* a PERSON's name, or *printing* a PERSON's address. In other words, methods are the equivalent of *procedures* in traditional programming languages. In OO terms, methods define an object's *behavior*.

- Classes are organized in a *class hierarchy*. The **class hierarchy** resembles an upside-down tree in which each class has only one parent. For example, the CUSTOMER class and the EMPLOYEE class share a parent PERSON class. (Note the similarity to the hierarchical data model in this respect.)

- **Inheritance** is the ability of an object within the class hierarchy to inherit the attributes and methods of the classes above it. For example, two classes, CUSTOMER and EMPLOYEE, can be created as subclasses from the class PERSON. In this case, CUSTOMER and EMPLOYEE will inherit all attributes and methods from PERSON.

Object-oriented data models are typically depicted using Unified Modeling Language (UML) class diagrams. **Unified Modeling Language (UML)** is a language based on OO concepts that describes a set of diagrams and symbols that can be used to graphically model a system. UML **class diagrams** are used to represent data and their relationships within the larger UML object-oriented systems modeling language. For a more complete description of UML *see* Appendix H, Unified Modeling Language (UML).

To illustrate the main concepts of the object-oriented data model, let's use a simple invoicing problem. In this case, invoices are generated by customers, each invoice references one or more lines, and each line represents an item purchased by a customer. Figure 2.6 illustrates the object representation for this simple invoicing problem, as well as the equivalent UML class diagram and ER model. The object representation is a simple way to visualize a single object occurrence.

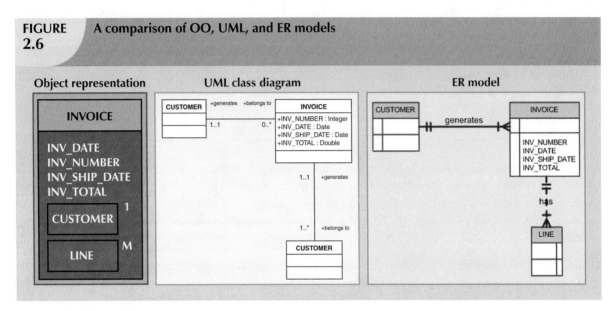

FIGURE 2.6   A comparison of OO, UML, and ER models

As you examine Figure 2.6, note that:

- The object representation of the INVOICE includes all related objects within the *same* object box. Note that the connectivities (1 and M) indicate the relationship of the related objects to the INVOICE. For example, the *1* next to the CUSTOMER object indicates that each INVOICE is related to only one CUSTOMER. The M next to the LINE object indicates that each INVOICE contains many LINEs.

- The UML class diagram uses three separate object classes (CUSTOMER, INVOICE, and LINE) and two relationships to represent this simple invoicing problem. Note that the relationship connectivities are represented by the 1..1, 0..* and 1..* symbols and that the relationships are named in both ends to represent the different "roles" that the objects play in the relationship.

- The ER model also uses three separate entities and two relationships to represent this simple invoice problem.

2.5.6 THE CONVERGENCE OF DATA MODELS

Another semantic data model was developed in response to the increasing complexity of applications—the **extended relational data model (ERDM)**. The ERDM, championed by many relational database researchers, constitutes the relational model's response to the OODM. This model includes many of the OO model's best features within an inherently simpler relational database structural environment. That's why a DBMS based on the ERDM is often described as an **object/relational database management system (O/RDBMS)**.

With the huge installed base of the relational database and the emergence of the ERDM, the OODM faces an uphill battle. Although the ERDM includes a strong semantic component, it is primarily based on the relational data model's concepts. In contrast, the OODM is wholly based on the OO and semantic data model concepts. The ERDM is primarily geared to business applications, while the OODM tends to focus on very specialized engineering and scientific applications. In the database arena, the most likely scenario appears to be an ever-increasing merging of OO and relational data model concepts and procedures, with an increasing emphasis on data models that facilitate Internet-age technologies.

2.5.7 DATABASE MODELS AND THE INTERNET

The use of the Internet as a prime business tool has drastically changed the role and scope of the database market. In fact, the Internet's impact on the database market has generated new database product strategies in which the OODM and ERDM-O/RDM have taken a backseat to Internet-age database development. Therefore, instead of an OODM vs. ERDM-O/RDM data-modeling duel occurring, vendors have been focusing their development efforts on creating database products that interface efficiently and easily with the Internet. The focus on effective Internet interfacing makes the underlying data model less important to the end user. If the database fits well into the Internet picture, its precise modeling heritage is of relatively little consequence. That's why the relational model has prospered by incorporating components from other data models. For example, Oracle Corporation's Oracle 10g database contains OO components *within a relational database structure*, as does IBM's current DB2 version. In any case, the Internet trumps all other aspects of data storage and access. Therefore, the Internet environment forces a focus on high levels of systems integration and development through new Internet-age technologies. Such technologies will be examined in detail in Chapter 14, Database Connectivity and Web Technologies.

With the dominance of the World Wide Web, there is a growing need to manage unstructured data, such as the data found in most of today's documents and Web pages. In response to this need, current databases now support Internet-age technologies such as Extensible Markup Language (XML). For example, extended relational databases such as Oracle 10g and IBM's DB2 support XML data types to store and manage unstructured data. Concurrently, native XML databases are now on the market to address similar needs. The importance of XML support cannot be underestimated, as XML is also the standard protocol for data exchange among different systems and Internet-based services (see Chapter 14).

2.5.8 DATA MODELS: A SUMMARY

The evolution of DBMSs has always been driven by the search for new ways of modeling increasingly complex real-world data. A summary of the most commonly recognized data models is shown in Figure 2.7.

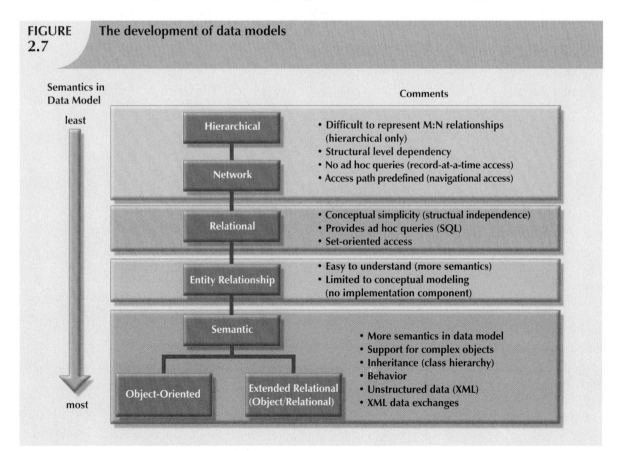

FIGURE 2.7   The development of data models

Semantics in Data Model — least ... most

Comments

Hierarchical / **Network**
- Difficult to represent M:N relationships (hierarchical only)
- Structural level dependency
- No ad hoc queries (record-at-a-time access)
- Access path predefined (navigational access)

Relational
- Conceptual simplicity (structual independence)
- Provides ad hoc queries (SQL)
- Set-oriented access

Entity Relationship
- Easy to understand (more semantics)
- Limited to conceptual modeling (no implementation component)

Semantic / **Object-Oriented** / **Extended Relational (Object/Relational)**
- More semantics in data model
- Support for complex objects
- Inheritance (class hierarchy)
- Behavior
- Unstructured data (XML)
- XML data exchanges

In the evolution of data models, there are some common characteristics that data models must have in order to be widely accepted:

- A data model must show some degree of conceptual simplicity without compromising the semantic completeness of the database. *It does not make sense to have a data model that is more difficult to conceptualize than the real world.*

- A data model must represent the real world as closely as possible. This goal is more easily realized by adding more semantics to the model's data representation. (Semantics concern the dynamic data behavior, while data representation constitutes the static aspect of the real-world scenario.)

- Representation of the real-world transformations (behavior) must be in compliance with the consistency and integrity characteristics of any data model.

Each new data model capitalizes on the shortcomings of previous models. The network model replaced the hierarchical model because the former made it much easier to represent complex (many-to-many) relationships. In turn, the relational model offers several advantages over the hierarchical and network models through its simpler data representation, superior data independence, and easy-to-use query language; the relational model also emerged as the dominant data model for business applications. Although the OO and ERDM have gained a substantial foothold, their attempts to dislodge the relational model have not been successful. And in the coming years, successful data models will have to facilitate the development of database products that incorporate unstructured data as well as provide support for easy data exchanges via XML.

It is important to note that not all data models are created equal; some data models are better suited than others for some tasks. For example, *conceptual* models are better suited for high-level data modeling, while *implementation* models are better for managing stored data for implementation purposes. The entity relationship model is an example of a conceptual model, while the hierarchical and network models are examples of implementation models. At the same time, some models, such as the relational model and the OODM, could be used as both conceptual and implementation models. Table 2.2 summarizes the advantages and disadvantages of the various database models.

TABLE 2.2 Advantages and Disadvantages of Various Database Models

DATA MODEL	DATA INDEPENDENCE	STRUCTURAL INDEPENDENCE	ADVANTAGES	DISADVANTAGES
Hierarchical	Yes	No	1. It promotes data sharing. 2. Parent/Child relationship promotes conceptual simplicity. 3. Database security is provided and enforced by DBMS. 4. Parent/Child relationship promotes data integrity. 5. It is efficient with 1:M relationships.	1. Complex implementation requires knowledge of physical data storage characteristics. 2. Navigational system yields complex application development, management, and use; requires knowledge of hierarchical path. 3. Changes in structure require changes in all application programs. 4. There are implementation limitations (no multiparent or M:N relationships). 5. There is no data definition or data manipulation language in the DBMS. 6. There is a lack of standards.
Network	Yes	No	1. Conceptual simplicity is at least equal to that of the hierarchical model. 2. It handles more relationship types, such as M:N and multiparent. 3. Data access is more flexible than in hierarchical and file system models. 4. Data Owner/Member relationship promotes data integrity. 5. There is conformance to standards. 6. It includes data definition language (DDL) and data manipulation language (DML) in DBMS.	1. System complexity limits efficiency—still a navigational system. 2. Navigational system yields complex implementation, application development, and management. 3. Structural changes require changes in all application programs.
Relational	Yes	Yes	1. Structural independence is promoted by the use of independent tables. Changes in a table's structure do not affect data access or application programs. 2. Tabular view substantially improves conceptual simplicity, thereby promoting easier database design, implementation, management, and use. 3. Ad hoc query capability is based on SQL. 4. Powerful RDBMS isolates the end user from physical-level details and improves implementation and management simplicity.	1. The RDBMS requires substantial hardware and system software overhead. 2. Conceptual simplicity gives relatively untrained people the tools to use a good system poorly, and if unchecked, it may produce the same data anomalies found in file systems. 3. It may promote "islands of information" problems as individuals and departments can easily develop their own applications.
Entity Relationship	Yes	Yes	1. Visual modeling yields exceptional conceptual simplicity. 2. Visual representation makes it an effective communication tool. 3. It is integrated with dominant relational model.	1. There is limited constraint representation. 2. There is limited relationship representation. 3. There is no data manipulation language. 4. Loss of information content occurs when attributes are removed from entities to avoid crowded displays. (This limitation has been addressed in subsequent graphical versions.)
Object-Oriented	Yes	Yes	1. Semantic content is added. 2. Visual representation includes semantic content. 3. Inheritance promotes data integrity.	1. Slow development of standards caused vendors to supply their own enhancements, thus eliminating a widely accepted standard. 2. It is a complex navigational system. 3. There is a steep learning curve. 4. High system overhead slows transactions.

Note: All databases assume the use of a common data pool within the database. Therefore, *all* database models promote data sharing, thus eliminating the potential problem of islands of information.

Thus far, you have been introduced to the basic constructs of the more prominent data models. Each model uses such constructs to capture the meaning of the real world data environment. Table 2.3 shows the basic terminology used by the various data models.

TABLE 2.3	Data Model Basic Terminology Comparison						
REAL WORLD	EXAMPLE	FILE PROCESSING	HIERARCHICAL MODEL	NETWORK MODEL	RELATIONAL MODEL	ER MODEL	OO MODEL
A group of vendors	Vendor file cabinet	File	Segment type	Record type	Table	Entity set	Class
A single vendor	Global Supplies	Record	Segment occurrence	Current record	Row (tuple)	Entity occurrence	Object instance
The contact name	Johnny Ventura	Field	Segment field	Record field	Table attribute	Entity attribute	Object attribute
The vendor identifier	G12987	Index	Sequence field	Record key	Key	Entity identifier	Object identifier

Note: For additional information about the terms used in this table please consult the corresponding chapters and online appendixes accompanying this book. For example, if you want to know more about the OO model, refer to Appendix G, Object-Oriented Databases.

2.6 DEGREES OF DATA ABSTRACTION

If you ask ten database designers what is a data model, you will end up with ten different answers—depending on the degree of data abstraction. To illustrate the meaning of data abstraction, consider the example of automotive design. A car designer begins by drawing the *concept* of the car that is to be produced. Next, engineers design the details that help transfer the basic concept into a structure that can be produced. Finally, the engineering drawings are translated into production specifications to be used on the factory floor. As you can see, the process of producing the car begins at a high level of abstraction and proceeds to an ever-increasing level of detail. The factory floor process cannot proceed unless the engineering details are properly specified, and the engineering details cannot exist without the basic conceptual framework created by the designer. Designing a usable database follows the same basic process. That is, a database designer starts with an abstract view of the overall data environment and adds details as the design comes closer to implementation. Using levels of abstraction can also be very helpful in integrating multiple (and sometimes conflicting) views of data as seen at different levels of an organization.

In the early 1970s, the **American National Standards Institute (ANSI)** Standards Planning and Requirements Committee (SPARC) defined a framework for data modeling based on degrees of data abstraction. The ANSI/SPARC architecture (as it is often referred to) defines three levels of data abstraction: external, conceptual, and internal. You can use this framework to better understand database models, as shown in Figure 2.8. In the figure, the ANSI/SPARC framework has been expanded with the addition of a *physical* model to explicitly address physical-level implementation details of the internal model.

2.6.1 THE EXTERNAL MODEL

The **external model** is the end users' view of the data environment. The term *end users* refers to people who use the application programs to manipulate the data and generate information. End users usually operate in an environment in which an application has a specific business unit focus. Companies are generally divided into several business units, such as sales, finance, and marketing. Each business unit is subject to specific constraints and requirements, and each one uses a data subset of the overall data in the organization. Therefore, end users working within those business units view their data subsets as separate from or external to other units within the organization.

FIGURE 2.8   Data abstraction levels

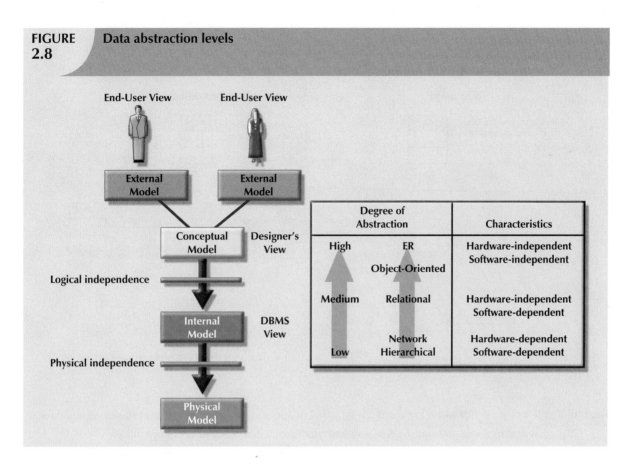

Because data is being modeled, ER diagrams will be used to represent the external views. A specific representation of an external view is known as an **external schema**. To illustrate the external model's view, examine the data environment of Tiny College. Figure 2.9 presents the external schemas for two Tiny College business units: student registration and class scheduling. Each external schema includes the appropriate entities, relationships, processes, and constraints imposed by the business unit. Also note that *although the application views are isolated from each other, each view shares a common entity with the other view.* For example, the registration and scheduling external schemas share the entities CLASS and COURSE.

Note the entity relationships represented in Figure 2.9. For example:

- A PROFESSOR may teach many CLASSes, and each CLASS is taught by only one PROFESSOR; that is, there is a 1:M relationship between PROFESSOR and CLASS.

- A CLASS may ENROLL many students, and each student may ENROLL in many CLASSes, thus creating an M:N relationship between STUDENT and CLASS. (You will learn about the precise nature of the ENROLL entity in Chapter 4.)

- Each COURSE may generate many CLASSes, but each CLASS references a single COURSE. For example, there may be several classes (sections) of a database course having a course code of CIS-420. One of those classes might be offered on MWF from 8:00 a.m. to 8:50 a.m., another might be offered on MWF from 1:00 p.m. to 1:50 p.m., while a third might be offered on Thursdays from 6:00 p.m. to 8:40 p.m. Yet all three classes have the course code CIS-420.

- Finally, a CLASS requires one ROOM, but a ROOM may be scheduled for many CLASSes. That is, each classroom may be used for several classes: one at 9:00 a.m., one at 11:00 a.m., and one at 1 p.m., for example. In other words, there is a 1:M relationship between ROOM and CLASS.

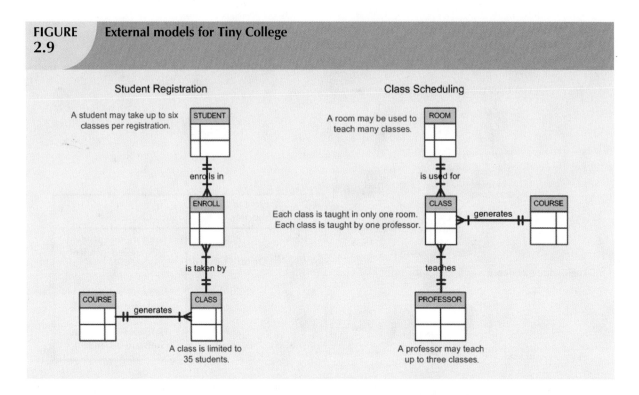

FIGURE 2.9   External models for Tiny College

The use of external views representing subsets of the database has some important advantages:

- It makes it easy to identify specific data required to support each business unit's operations.

- It makes the designer's job easy by providing feedback about the model's adequacy. Specifically, the model can be checked to ensure that it supports all processes as defined by their external models, as well as all operational requirements and constraints.

- It helps to ensure *security* constraints in the database design. Damaging an entire database is more difficult when each business unit works with only a subset of data.

- It makes application program development much simpler.

2.6.2 THE CONCEPTUAL MODEL

Having identified the external views, a conceptual model is used, graphically represented by an ERD (as in Figure 2.10), to integrate all external views into a single view. The **conceptual model** represents a global view of the entire database as viewed by the entire organization. That is, the conceptual model integrates all external views (entities, relationships, constraints, and processes) into a single global view of the entire data in the enterprise. Also known as a **conceptual schema**, it is the basis for the identification and high-level description of the main data objects (avoiding any database model-specific details).

The most widely used conceptual model is the ER model. Remember that the ER model is illustrated with the help of the ERD, which is, in effect, the basic database blueprint. The ERD is used to graphically *represent* the conceptual schema.

The conceptual model yields some very important advantages. First, it provides a relatively easily understood bird's-eye (macro level) view of the data environment. For example, you can get a summary of Tiny College's data environment by examining the conceptual model presented in Figure 2.10.

Second, the conceptual model is independent of both software and hardware. **Software independence** means that the model does not depend on the DBMS software used to implement the model. **Hardware independence** means that the model does not depend on the hardware used in the implementation of the model. Therefore, changes in

FIGURE 2.10	Conceptual model for Tiny College

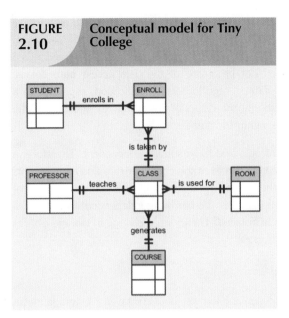

either the hardware or the DBMS software will have no effect on the database design at the conceptual level. Generally, the term **logical design** is used to refer to the task of creating a conceptual data model that could be implemented in any DBMS.

2.6.3 THE INTERNAL MODEL

Once a specific DBMS has been selected, the internal model maps the conceptual model to the DBMS. The **internal model** is the representation of the database as "seen" by the DBMS. In other words, the internal model requires the designer to match the conceptual model's characteristics and constraints to those of the selected implementation model. An **internal schema** depicts a specific representation of an internal model, using the database constructs supported by the chosen database.

Because this book focuses on the relational model, a relational database was chosen to implement the internal model. Therefore, the internal schema should map the conceptual model to the relational model constructs. In particular, the entities in the conceptual model are mapped to tables in the relational model. Likewise, because a relational database has been selected, the internal schema is expressed using SQL, the standard language for relational databases. In the case of the conceptual model for Tiny College depicted in Figure 2.11, the internal model was implemented by creating the tables PROFESSOR, COURSE, CLASS, STUDENT, ENROLL, and ROOM. A simplified version of the internal model for Tiny College is shown in Figure 2.11.

FIGURE 2.11	An internal model for Tiny College

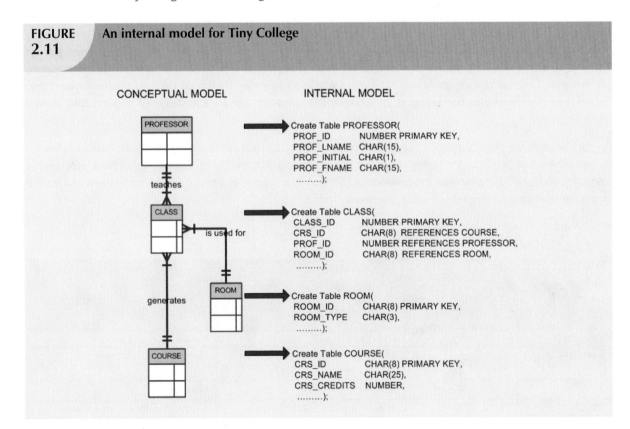

The development of a detailed internal model is especially important to database designers who work with hierarchical or network models because those models require very precise specification of data storage location and data access paths. In contrast, the relational model requires less detail in its internal model because most RDBMSs handle data access path definition *transparently*; that is, the designer need not be aware of the data access path details. Nevertheless, even relational database software usually requires data storage location specification, especially in a mainframe environment. For example, DB2 requires that you specify the data storage group, the location of the database within the storage group, and the location of the tables within the database.

Because the internal model depends on specific database software, it is said to be software-dependent. Therefore, a change in the DBMS software requires that the internal model be changed to fit the characteristics and requirements of the implementation database model. When you can change the internal model without affecting the conceptual model, you have **logical independence**. However, the internal model is still hardware-independent because it is unaffected by the choice of the computer on which the software is installed. Therefore, a change in storage devices or even a change in operating systems will not affect the internal model.

2.6.4 THE PHYSICAL MODEL

The **physical model** operates at the lowest level of abstraction, describing the way data are saved on storage media such as disks or tapes. The physical model requires the definition of both the physical storage devices and the (physical) access methods required to reach the data within those storage devices, making it both software- and hardware-dependent. The storage structures used are dependent on the software (the DBMS and the operating system) and on the type of storage devices that the computer can handle. The precision required in the physical model's definition demands that database designers who work at this level have a detailed knowledge of the hardware and software used to implement the database design.

Early data models forced the database designer to take the details of the physical model's data storage requirements into account. However, the now-dominant relational model is aimed largely at the logical rather than the physical level; therefore, it does not require the physical-level details common to its predecessors.

Although the relational model does not require the designer to be concerned about the data's physical storage characteristics, the *implementation* of a relational model may require physical-level fine-tuning for increased performance. Fine-tuning is especially important when very large databases are installed in a mainframe environment. Yet even such performance fine-tuning at the physical level does not require knowledge of physical data storage characteristics.

As noted earlier, the physical model is dependent on the DBMS, methods of accessing files, and types of hardware storage devices supported by the operating system. When you can change the physical model without affecting the internal model, you have **physical independence**. Therefore, a change in storage devices or methods and even a change in operating system will not affect the internal model.

A summary of the levels of data abstraction is given in Table 2.4.

TABLE 2.4	Levels of Data Abstraction		
MODEL	DEGREE OF ABSTRACTION	FOCUS	INDEPENDENT OF
External	High	End-user views	Hardware and software
Conceptual	↑ ↓	Global view of data (database model independent)	Hardware and software
Internal		Specific database model	Hardware
Physical	Low	Storage and access methods	Neither hardware nor software

SUMMARY

▶ A data model is an abstraction of a complex real-world data environment. Database designers use data models to communicate with applications programmers and end users. The basic data-modeling components are entities, attributes, relationships, and constraints. Business rules are used to identify and define the basic modeling components within a specific real-world environment.

▶ The hierarchical and network data models were early data models that are no longer used, but some of the concepts are found in current data models. The hierarchical model depicts a set of one-to-many (1:M) relationships between a parent and its children segments. The network model uses sets to represent 1:M relationships between record types.

▶ The relational model is the current database implementation standard. In the relational model, the end user perceives the data as being stored in tables. Tables are related to each other by means of common values in common attributes. The entity relationship (ER) model is a popular graphical tool for data modeling that complements the relational model. The ER model allows database designers to visually present different views of the data as seen by database designers, programmers, and end users and to integrate the data into a common framework.

▶ The object-oriented data model (OODM) uses objects as the basic modeling structure. An object resembles an entity in that it includes the facts that define it. But unlike an entity, the object also includes information about relationships between the facts as well as relationships with other objects, thus giving its data more meaning.

▶ The relational model has adopted many object-oriented (OO) extensions to become the extended relational data model (ERDM). At this point, the OODM is largely used in specialized engineering and scientific applications, while the ERDM is primarily geared to business applications. Although the most likely future scenario is an increasing merger of OODM and ERDM technologies, both are overshadowed by the need to develop Internet access strategies for databases. Usually OO data models are depicted using Unified Modeling Language (UML) class diagrams.

▶ Data-modeling requirements are a function of different data views (global vs. local) and the level of data abstraction. The American National Standards Institute Standards Planning and Requirements Committee (ANSI/SPARC) describes three levels of data abstraction: external, conceptual, and internal. There is also a fourth level of data abstraction (the physical level). This lowest level of data abstraction is concerned exclusively with physical storage methods.

KEY TERMS

American National Standards Institute (ANSI), 48

attribute, 32

business rule, 33

Chen notation, 41

class, 43

class diagram, 43

class hierarchy, 43

conceptual model, 50

conceptual schema, 50

Conference on Data Systems Languages (CODASYL), 37

connectivity, 41

constraint, 33

Crow's Foot notation, 41

data definition language (DDL), 37

data management language (DML), 37

data model, 31

Database Task Group (DBTG), 37

entity, 32

entity instance, 41

entity occurrence, 41

entity relationship diagram (ERD), 40

entity relationship (ER) model (ERM), 40

entity set, 41

extended relational data model (ERDM), 44

external model, 48

external schema, 49

hardware independence, 50

hierarchical model, 36

inheritance, 43

internal model, 51

internal schema, 51

logical design, 51

logical independence, 52

many-to-many (M:N or *..*) relationship, 33

method, 43

network model, 37

object, 43

object-oriented data model (OODM), 43

object-oriented database management system (OODBMS), 43

object/relational database management system (O/RDBMS), 38

one-to-many (1:M or 1..*) relationship, 33

one-to-one (1:1 or 1..1) relationship, 33

physical independence, 52

physical model, 52

relational database management system (RDBMS), 38

relational model, 38

relational diagram, 39

relation, 38

relationship, 32

segment, 36

schema, 37

semantic data model, 43

software independence, 50

subschema, 37

table, 38

tuple, 38

Unified Modeling Language (UML), 43

ONLINE CONTENT

Answers to selected Review Questions and Problems for this chapter are contained in the Student Online Companion for this book.

REVIEW QUESTIONS

1. Discuss the importance of data modeling.

2. What is a business rule, and what is its purpose in data modeling?

3. How do you translate business rules into data model components?

4. What does each of the following acronyms represent, and how is each one related to the birth of the network data model?

   a. CODASYL

   b. SPARC

   c. ANSI

   d. DBTG

5. What three languages were adopted by the DBTG to standardize the basic network data model, and why was such standardization important to users and designers?

6. Describe the basic features of the relational data model and discuss their importance to the end user and the designer.

7. Explain how the entity relationship (ER) model helped produce a more structured relational database design environment.

8. Use the scenario described by "A customer can make many payments, but each payment is made by only one customer" as the basis for an entity relationship diagram (ERD) representation.

9. Why is an object said to have greater semantic content than an entity?

10. What is the difference between an object and a class in the object-oriented data model (OODM)?

11. How would you model Question 8 with an OODM? (Use Figure 2.7 as your guide.)

12. What is an ERDM, and what role does it play in the modern (production) database environment?

13. In terms of data and structural independence, compare file system data management with the five data models discussed in this chapter.

14. What is a relationship, and what three types of relationships exist?

15. Give an example of each of the three types of relationships.

16. What is a table, and what role does it play in the relational model?

17. What is a relational diagram? Give an example.

18. What is logical independence?

19. What is physical independence?

20. What is connectivity? (Use a Crow's Foot ERD to illustrate connectivity.)

PROBLEMS

Use the contents of Figure 2.3 to work Problems 1–5.

1. Write the business rule(s) that govern the relationship between AGENT and CUSTOMER.

2. Given the business rule(s) you wrote in Problem 1, create the basic Crow's Foot ERD.

3. If the relationship between AGENT and CUSTOMER were implemented in a hierarchical model, what would the hierarchical structure look like? Label the structure fully, identifying the root segment and the Level 1 segment.

4. If the relationship between AGENT and CUSTOMER were implemented in a network model, what would the network model look like? (Identify the record types and the set.)

5. Using the ERD you drew in Problem 2, create the equivalent Object representation and UML class diagram. (Use Figure 2.7 as your guide.)

Using Figure P2.6 as your guide, work Problems 6–7. The DealCo relational diagram shows the initial entities and attributes for the DealCo stores, located in two regions of the country.

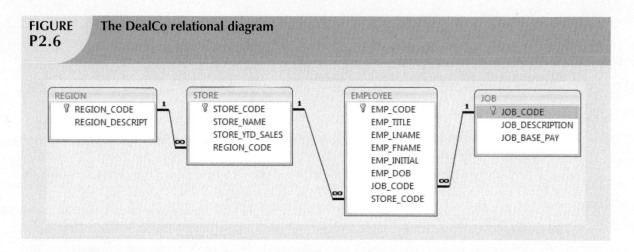

FIGURE P2.6   The DealCo relational diagram

6. Identify each relationship type and write all of the business rules.

7. Create the basic Crow's Foot ERD for DealCo.

Using Figure P2.8 as your guide, work Problems 8–11. The Tiny College relational diagram shows the initial entities and attributes for Tiny College.

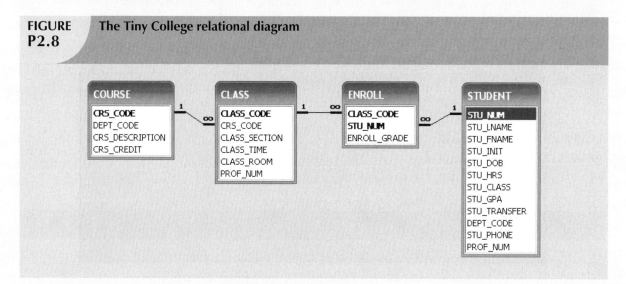

FIGURE P2.8   The Tiny College relational diagram

8. Identify each relationship type and write all of the business rules.

9. Create the basic Crow's Foot ERD for Tiny College.

10. Create the network model that reflects the entities and relationships you identified in the relational diagram.

11. Create the UML class diagram that reflects the entities and relationships you identified in the relational diagram.

12. Using the hierarchical representation shown in Figure P2.12, answer a, b, and c.

    a. Identify the segment types.

    b. Identify the components that are equivalent to the file system's fields.

    c. Describe the hierarchical path for the occurrence of the third PAINTING segment.

FIGURE P2.12   The hierarchical structure for the Artist database

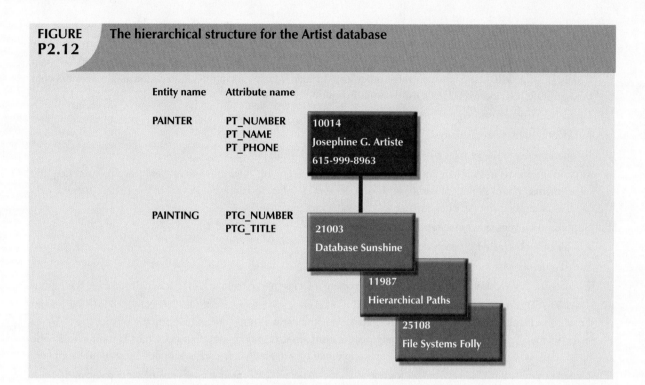

FIGURE P2.13   The hierarchical structure for the MedClinic database

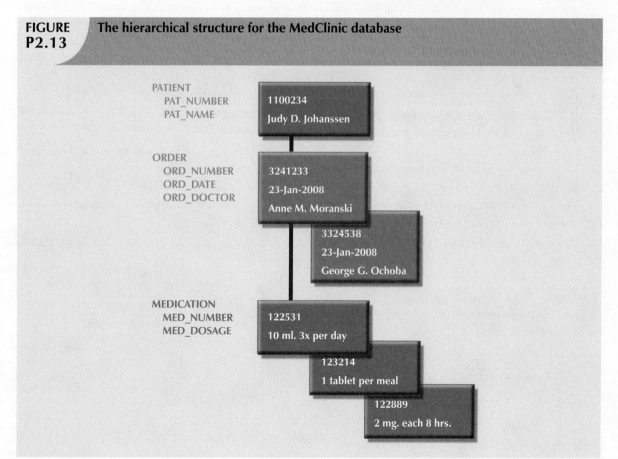

13. The hierarchical diagram shown in Figure P2.13 depicts a single record occurrence of a patient named Judy D. Johanssen. Typically, a patient staying in a hospital receives medications that have been ordered by a particular doctor. Because the patient often receives several medications per day, there is a 1:M relationship between PATIENT and ORDER. Similarly, each order can include several medications, creating a 1:M relationship between ORDER and MEDICATION.

   Given the structure shown in Figure P2.13:

   a. Identify the segment types.

   b. Identify the business rules for PATIENT, ORDER, and MEDICATION.

14. Expand the model in Problem 13 to include a DOCTOR segment; then draw its hierarchical structure. (Identify all segments.) (*Hint:* A patient can have several doctors assigned to his or her case, but the patient named Judy D. Johanssen occurs only once in each of those doctors' records.)

15. Suppose you want to write a report that shows:

   a. All patients treated by each doctor.

   b. All doctors who treated each patient.

   Evaluate the hierarchical structure you drew in Problem 14 in terms of its search efficiency in producing the report.

16. The PYRAID company wants to track each PART used in each specific piece of EQUIPMENT; each PART is bought from a specific SUPPLIER. Using that description, draw the network structure and identify the sets for the PYRAID company database. (*Hint:* A piece of equipment is composed of many parts, but each part is used in only one specific piece of equipment. A supplier can supply many parts, but each part is supplied by only one supplier.)

17. United Broke Artists (UBA) is a broker for not-so-famous painters. UBA maintains a small network database to track painters, paintings, and galleries. Using PAINTER, PAINTING, and GALLERY, write the network structure and identify appropriate sets within the UBA database. (*Hint 1:* A painting is painted by a particular artist, and that painting is exhibited in a particular gallery. *Hint 2:* A gallery can exhibit many paintings, but each painting can be exhibited in only one gallery. Similarly, a painting is painted by a single painter, but each painter can paint many paintings.)

18. If you decide to convert the network database in Problem 17 to a relational database:

   a. What tables would you create and what would the table components be?

   b. How might the (independent) tables be related to one another?

19. Using a Crow's Foot ERD, convert the network database model in Figure 2.2 into a design for a relational database model. Show all entities and relationships.

20. Using the ERD from Problem 19, create the relational schema. (Create an appropriate collection of attributes for each of the entities. Make sure you use the appropriate naming conventions to name the attributes.)

21. Convert the ERD from Problem 19 into the corresponding UML class diagram.

22. Describe the relationships (identify the business rules) depicted in the Crow's Foot ERD shown in Figure P2.22.

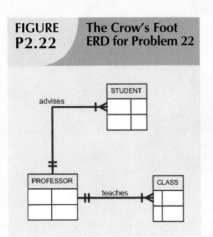

FIGURE P2.22   The Crow's Foot ERD for Problem 22

23. Create a Crow's Foot ERD to include the following business rules for the ProdCo company:

    a. Each sales representative writes many invoices.

    b. Each invoice is written by one sales representative.

    c. Each sales representative is assigned to one department.

    d. Each department has many sales representatives.

    e. Each customer can generate many invoices.

    f. Each invoice is generated by one customer.

24. Write the business rules that are reflected in the ERD shown in Figure P2.24. (Note that the ERD reflects some simplifying assumptions. For example, each book is written by only one author. Also, remember that the ERD is always read from the "1" to the "M" side, regardless of the orientation of the ERD components.)

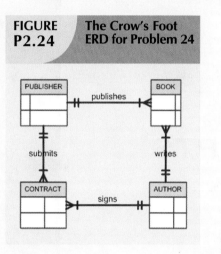

FIGURE P2.24 The Crow's Foot ERD for Problem 24

25. Create a Crow's Foot ERD for each of the following descriptions. (*Note:* The word *many* merely means "more than one" in the database modeling environment.)

    a. Each of the MegaCo Corporation's divisions is composed of many departments. Each department has many employees assigned to it, but each employee works for only one department. Each department is managed by one employee, and each of those managers can manage only one department at a time.

    b. During some period of time, a customer can rent many videotapes from the BigVid store. Each of the BigVid's videotapes can be rented to many customers during that period of time.

    c. An airliner can be assigned to fly many flights, but each flight is flown by only one airliner.

    d. The KwikTite Corporation operates many factories. Each factory is located in a region. Each region can be "home" to many of KwikTite's factories. Each factory employs many employees, but each of those employees is employed by only one factory.

    e. An employee may have earned many degrees, and each degree may have been earned by many employees.

NOTE

Many-to-many (M:N) relationships exist at a conceptual level, and you should know how to recognize them. However, you will learn in Chapter 3 that M:N relationships are not appropriate in a relational model. For that reason, Microsoft Visio does not support the M:N relationship. Therefore, to illustrate the existence of a M:N relationship using Visio, you must superimpose two 1:M relationships. (See Figure 2.5.)

PART

II

DESIGN CONCEPTS

DATABASE MODELING SUPPORTING COMMUNITIES

Business Vignette

Companies, governments, and organizations around the world turn to entity relationship diagrams and database modeling tools to help develop their databases. The advantages of using tools like Sybase PowerDesigner, Microsoft Visio Professional, ERwin Data Modeler, or Embarcadero ER/Studio significantly outweigh their expense. They improve database documentation. They facilitate staff communication, helping to ensure that the database will meet the needs of its users. They reduce development time. All these advantages translate into significant cost-savings. Yet sometimes this value goes well beyond anything that can be expressed in a dollar amount.

Rebuilding Together is a national nonprofit organization dedicated to preserving and revitalizing houses and communities for the elderly, disabled, and families with children. The national headquarters currently works with 255 affiliates serving over 1,897 communities. Based on the "barn-raising" tradition, local volunteers assemble on Rebuilding Day to help their neighbors. Over 267,000 volunteers have repaired or reconstructed approximately 9,000 houses and nonprofit facilities.

As the local affiliate in Des Moines, Iowa, founded in 1994, has grown rapidly, the organization has sought to document and improve their house selection and volunteer coordination processes. Several sources, including past participants, make referrals for potential housing projects. Each year, Rebuilding Together needs to evaluate the qualifications of each candidate, preview the site, select or reject the project, and finally implement the selected projects. Using modeling software ER/Studio, the staff built a database to keep track of these stages of the project and manage the volunteers that will work on each project.

By using the logical view of the data modeling software, the staff was able to understand the entities, their attributes, and the relationships that they were modeling prior to building the physical model. They also generated a short report and model diagram to educate all personnel involved in the project. The end result was that the company was able to develop an application process that is both more complex and user-friendly. As the organization continues to grow and the spirit of "barn-raising" spreads, the staff will be able to modify the design to accommodate its growing needs.

In this chapter, you will learn:

- That the relational database model offers a logical view of data
- About the relational model's basic component: relations
- That relations are logical constructs composed of rows (tuples) and columns (attributes)
- That relations are implemented as tables in a relational DBMS
- About relational database operators, the data dictionary, and the system catalog
- How data redundancy is handled in the relational database model
- Why indexing is important

In Chapter 2, Data Models, you learned that the relational data model's structural and data independence allow you to examine the model's logical structure without considering the physical aspects of data storage and retrieval. You also learned that entity relationship diagrams (ERDs) may be used to depict entities and their relationships graphically. In this chapter, you learn some important details about the relational model's logical structure and more about how the ERD can be used to design a relational database.

You learn how the relational database's basic data components fit into a logical construct known as a table. You discover that one important reason for the relational database model's simplicity is that its tables can be treated as logical rather than physical units. You also learn how the independent tables within the database can be related to one another.

After learning about tables, their components, and their relationships, you are introduced to the basic concepts that shape the design of tables. Because the table is such an integral part of relational database design, you also learn the characteristics of well-designed and poorly designed tables.

Finally, you are introduced to some basic concepts that will become your gateway to the next few chapters. For example, you examine different kinds of relationships and the way those relationships might be handled in the relational database environment.

Preview

NOTE

The relational model, introduced by E. F. Codd in 1970, is based on predicate logic and set theory. **Predicate logic**, used extensively in mathematics, provides a framework in which an assertion (statement of fact) can be verified as either true or false. For example, suppose that a student with a student ID of 12345678 is named Melissa Sanduski. This assertion can easily be demonstrated to be true or false. **Set theory** is a mathematical science that deals with sets, or groups of things, and is used as the basis for data manipulation in the relational model. For example, assume that set A contains three numbers: 16, 24, and 77. This set is represented as A(16, 24, 77). Furthermore, set B contains four numbers: 44, 77, 90, and 11, and so is represented as B(44, 77, 90, 11). Given this information, you can conclude that the intersection of A and B yields a result set with a single number, 77. This result can be expressed as $A \cap B = 77$. In other words, A and B share a common value, 77.

Based on these concepts, the relational model has three well-defined components:

1. A logical data structure represented by relations (Sections 3.1, 3.2, and 3.5).
2. A set of integrity rules to enforce that the data are and remain consistent over time (Sections 3.3, 3.6, 3.7, and 3.8).
3. A set of operations that define how data are manipulated (Section 3.4).

3.1 A LOGICAL VIEW OF DATA

In Chapter 1, Database Systems, you learned that a database stores and manages both data and metadata. You also learned that the DBMS manages and controls access to the data and the database structure. Such an arrangement—placing the DBMS between the application and the database—eliminates most of the file system's inherent limitations. The result of such flexibility, however, is a far more complex physical structure. In fact, the database structures required by both the hierarchical and network database models often become complicated enough to diminish efficient database design. The relational data model changed all of that by allowing the designer to focus on the logical representation of the data and its relationships, rather than on the physical storage details. To use an automotive analogy, the relational database uses an automatic transmission to relieve you of the need to manipulate clutch pedals and gearshifts. In short, the relational model enables you to view data *logically* rather than *physically*.

The practical significance of taking the logical view is that it serves as a reminder of the simple file concept of data storage. Although the use of a table, quite unlike that of a file, has the advantages of structural and data independence, a table does resemble a file from a conceptual point of view. Because you can think of related records as being stored in independent tables, the relational database model is much easier to understand than the hierarchical and network models. Logical simplicity tends to yield simple and effective database design methodologies.

Because the table plays such a prominent role in the relational model, it deserves a closer look. Therefore, our discussion begins with an exploration of the details of table structure and contents.

3.1.1 TABLES AND THEIR CHARACTERISTICS

The logical view of the relational database is facilitated by the creation of data relationships based on a logical construct known as a relation. Because a relation is a mathematical construct, end-users find it much easier to think of a relation as a table. A table is perceived as a two-dimensional structure composed of rows and columns. A table is also called a *relation* because the relational model's creator, E. F. Codd, used the term *relation* as a synonym for table. You can think of a table as a *persistent* representation of a logical relation, that is, a relation whose contents can be permanently saved for future use. As far as the table's user is concerned, *a table contains a group of related entity occurrences*, that is, an entity set. For example, a STUDENT table contains a collection of entity occurrences, each representing a student. For that reason, the terms *entity set* and *table* are often used interchangeably.

NOTE

The word *relation*, also known as a *dataset* in Microsoft Access, is based on the mathematical set theory from which Codd derived his model. Because the relational model uses attribute values to establish relationships among tables, many database users incorrectly assume that the term *relation* refers to such relationships. Many then incorrectly conclude that only the relational model permits the use of relationships.

You will discover that the table view of data makes it easy to spot and define entity relationships, thereby greatly simplifying the task of database design. The characteristics of a relational table are summarized in Table 3.1.

TABLE 3.1	Characteristics of a Relational Table
1	A table is perceived as a two-dimensional structure composed of rows and columns.
2	Each table row (**tuple**) represents a single entity occurrence within the entity set.
3	Each table column represents an attribute, and each column has a distinct name.
4	Each row/column intersection represents a single data value.
5	All values in a column must conform to the same data format.
6	Each column has a specific range of values known as the **attribute domain**.
7	The order of the rows and columns is immaterial to the DBMS.
8	Each table must have an attribute or a combination of attributes that uniquely identifies each row.

The tables shown in Figure 3.1 illustrate the characteristics listed in Table 3.1.

NOTE

Relational database terminology is very precise. Unfortunately, file system terminology sometimes creeps into the database environment. Thus, rows are sometimes referred to as *records* and columns are sometimes labeled as *fields*. Occasionally, tables are labeled *files*. Technically speaking, this substitution of terms is not always appropriate; the database table is a logical rather than a physical concept, and the terms *file, record,* and *field* describe physical concepts. Nevertheless, as long as you recognize that the table is actually a logical rather than a physical construct, you may (at the conceptual level) think of table rows as records and of table columns as fields. In fact, many database software vendors still use this familiar file system terminology.

ONLINE CONTENT

All of the databases used to illustrate the material in this chapter are found in the Student Online Companion for this book. The database names used in the folder match the database names used in the figures. For example, the source of the tables shown in Figure 3.1 is the **Ch03_TinyCollege** database.

Using the STUDENT table shown in Figure 3.1, you can draw the following conclusions corresponding to the points in Table 3.1:

1. The STUDENT table is perceived to be a two-dimensional structure composed of eight rows (tuples) and twelve columns (attributes).

2. Each row in the STUDENT table describes a single entity occurrence within the entity set. (The entity set is represented by the STUDENT table.) Note that the row (entity or record) defined by STU_NUM = 321452 defines the characteristics (attributes or fields) of a student named William C. Bowser. For example, row 4 in Figure 3.1 describes a student named Walter H. Oblonski. Similarly, row 3 describes a student named Juliette Brewer. Given the table contents, the STUDENT entity set includes eight distinct entities (rows), or students.

FIGURE
3.1

STUDENT table attribute values

Database name: Ch03_TinyCollege

Table name: STUDENT

STU_NUM	STU_LNAME	STU_FNAME	STU_INIT	STU_DOB	STU_HRS	STU_CLASS	STU_GPA	STU_TRANSFER	DEPT_CODE	STU_PHONE	PROF_NUM
321452	Bowser	William	C	12-Feb-1975	42	So	2.84	No	BIOL	2134	205
324257	Smithson	Anne	K	15-Nov-1981	81	Jr	3.27	Yes	CIS	2256	222
324258	Brewer	Juliette		23-Aug-1969	36	So	2.26	Yes	ACCT	2256	228
324269	Oblonski	Walter	H	16-Sep-1976	66	Jr	3.09	No	CIS	2114	222
324273	Smith	John	D	30-Dec-1958	102	Sr	2.11	Yes	ENGL	2231	199
324274	Katinga	Raphael	P	21-Oct-1979	114	Sr	3.15	No	ACCT	2267	228
324291	Robertson	Gerald	T	08-Apr-1973	120	Sr	3.87	No	EDU	2267	311
324299	Smith	John	B	30-Nov-1986	15	Fr	2.92	No	ACCT	2315	230

STU_CLASS	= Student classification
STU_DOB	= Student date of birth
STU_GPA	= Grade point average
STU_HRS	= Credit hours earned
STU_INIT	= Student middle initial
STU_PHONE	= 4-digit campus phone extension
STU_TRANSFER	= Student transferred from another institution
DEPT_CODE	= Department code
PROF_NUM	= Number of the professor who is the student's advisor

3. Each column represents an attribute, and each column has a distinct name.

4. All of the values in a column match the attribute's characteristics. For example, the grade point average (STU_GPA) column contains only STU_GPA entries for each of the table rows. Data must be classified according to their format and function. Although various DBMSs can support different data types, most support at least the following:

a. *Numeric.* Numeric data are data on which you can perform meaningful arithmetic procedures. For example, STU_HRS and STU_GPA in Figure 3.1 are numeric attributes. On the other hand, STU_PHONE is not a numeric attribute because adding or subtracting phone numbers does not yield an arithmetically meaningful result.

b. *Character.* Character data, also known as text data or string data, can contain any character or symbol not intended for mathematical manipulation. In Figure 3.1, for example, STU_LNAME, STU_FNAME, STU_INIT, STU_CLASS, and STU_PHONE are character attributes.

c. *Date.* Date attributes contain calendar dates stored in a special format known as the Julian date format. Although the physical storage of the Julian date is immaterial to the user and designer, the Julian date format allows you to perform a special kind of arithmetic known as Julian date arithmetic. Using Julian date arithmetic, you can determine the number of days that have elapsed between two dates, such as 12-May-1999 and 20-Mar-2008, by simply subtracting 12-May-1999 from 20-Mar-2008. In Figure 3.1, STU_DOB can properly be classified as a date attribute. Most relational database software packages support Julian date formats. While the database's internal date format is likely to be Julian, many different *presentation* formats are available. For example, in Figure 3.1, you could show Mr. Bowser's date of birth (STU_DOB) as 2/12/75. Most relational DBMSs allow you to define your own date presentation format. For instance, Access and Oracle users might specify the "dd-mmm-yyyy" date format to show the first STU_DOB value in Figure 3.1 as 12-Feb-1975. (As you can tell by examining the STU_DOB values in Figure 3.1, the "dd-mmm-yyyy" format was selected to present the output.)

    d. *Logical*. Logical data can have only a true or false (yes or no) condition. For example, is a student a junior college transfer? In Figure 3.1, the STU_TRANSFER attribute uses a logical data format. Most, but not all, relational database software packages support the logical data format. (Microsoft Access uses the label "Yes/No data type" to indicate a logical data type.)

5. The column's range of permissible values is known as its **domain**. Because the STU_GPA values are limited to the range 0–4, inclusive, the domain is [0,4].

6. The order of rows and columns is immaterial to the user.

7. Each table must have a primary key. In general terms, the **primary key (PK)** is an attribute (or a combination of attributes) that uniquely identifies any given row. In this case, STU_NUM (the student number) is the primary key. Using the data presented in Figure 3.1, observe that a student's last name (STU_LNAME) would not be a good primary key because it is possible to find several students whose last name is Smith. Even the combination of the last name and first name (STU_FNAME) would not be an appropriate primary key because, as Figure 3.1 shows, it is quite possible to find more than one student named John Smith.

3.2 KEYS

In the relational model, keys are important because they are used to ensure that each row in a table is uniquely identifiable. They are also used to establish relationships among tables and to ensure the integrity of the data. Therefore, a proper understanding of the concept and use of keys in the relational model is very important. A **key** consists of one or more attributes that determine other attributes. For example, an invoice number identifies all of the invoice attributes, such as the invoice date and the customer name.

One type of key, the primary key, has already been introduced. Given the structure of the STUDENT table shown in Figure 3.1, defining and describing the primary key seems simple enough. However, because the primary key plays such an important role in the relational environment, you will examine the primary key's properties more carefully. In this section, you also will become acquainted with superkeys, candidate keys, and secondary keys.

The key's role is based on a concept known as **determination**. In the context of a database table, the statement "A determines B" indicates that if you know the value of attribute A, you can look up (determine) the value of attribute B. For example, knowing the STU_NUM in the STUDENT table (see Figure 3.1) means that you are able to look up (determine) that student's last name, grade point average, phone number, and so on. The shorthand notation for "A determines B" is A → B. If A determines B, C, and D, you write A → B, C, D. Therefore, using the attributes of the STUDENT table in Figure 3.1, you can represent the statement "STU_NUM determines STU_LNAME" by writing:

STU_NUM → STU_LNAME

In fact, the STU_NUM value in the STUDENT table determines all of the student's attribute values. For example, you can write:

STU_NUM → STU_LNAME, STU_FNAME, STU_INIT

and

STU_NUM → STU_LNAME, STU_FNAME, STU_INIT, STU_DOB, STU_TRANSFER

In contrast, STU_NUM is not determined by STU_LNAME because it is quite possible for several students to have the last name Smith.

The principle of determination is very important because it is used in the definition of a central relational database concept known as functional dependence. The term **functional dependence** can be defined most easily this way: the attribute B is functionally dependent on A if A determines B. More precisely:

The attribute B is functionally dependent on the attribute A
if each value in column A determines *one and only one* value in column B.

Using the contents of the STUDENT table in Figure 3.1, it is appropriate to say that STU_PHONE is functionally dependent on STU_NUM. For example, the STU_NUM value 321452 determines the STU_PHONE value 2134. On the other hand, STU_NUM is not functionally dependent on STU_PHONE because the STU_PHONE value 2267 is associated with two STU_NUM values: 324274 and 324291. (This could happen in a dormitory situation, where students share a phone.) Similarly, the STU_NUM value 324273 determines the STU_LNAME value Smith. But the STU_NUM value is not functionally dependent on STU_LNAME because more than one student may have the last name Smith.

The functional dependence definition can be generalized to cover the case in which the determining attribute values occur more than once in a table. Functional dependence can then be defined this way:[1]

Attribute A determines attribute B (that is, B is functionally dependent on A) if all of the rows in the table that agree in value for attribute A also agree in value for attribute B.

Be careful when defining the dependency's direction. For example, Gigantic State University determines its student classification based on hours completed; these are shown in Table 3.2.

TABLE 3.2	Student Classification	
HOURS COMPLETED	**CLASSIFICATION**	
Less than 30	Fr	
30−59	So	
60−89	Jr	
90 or more	Sr	

Therefore, you can write:

STU_HRS → STU_CLASS

But the specific number of hours is not dependent on the classification. It is quite possible to find a junior with 62 completed hours or one with 84 completed hours. In other words, the classification (STU_CLASS) does not determine one and only one value for completed hours (STU_HRS).

Keep in mind that it might take more than a single attribute to define functional dependence; that is, a key may be composed of more than one attribute. Such a multi-attribute key is known as a **composite key**.

Any attribute that is part of a key is known as a **key attribute**. For instance, in the STUDENT table, the student's last name would not be sufficient to serve as a key. On the other hand, the combination of last name, first name, initial, and home phone is very likely to produce unique matches for the remaining attributes. For example, you can write:

STU_LNAME, STU_FNAME, STU_INIT, STU_PHONE → STU_HRS, STU_CLASS

or

STU_LNAME, STU_FNAME, STU_INIT, STU_PHONE → STU_HRS, STU_CLASS, STU_GPA

or

STU_LNAME, STU_FNAME, STU_INIT, STU_PHONE → STU_HRS, STU_CLASS, STU_GPA, STU_DOB

---

[1] *SQL:2003 ANSI standard specification.* ISO/IEC 9075-2:2003 - SQL/Foundation.

Given the possible existence of a composite key, the notion of functional dependence can be further refined by specifying **full functional dependence**:

If the attribute (B) is functionally dependent on a composite key (A) but not on any subset of that composite key, the attribute (B) is fully functionally dependent on (A).

Within the broad key classification, several specialized keys can be defined. For example, a **superkey** is any key that uniquely identifies each row. In short, the superkey functionally determines all of a row's attributes. In the STUDENT table, the superkey could be any of the following:

STU_NUM

STU_NUM, STU_LNAME

STU_NUM, STU_LNAME, STU_INIT

In fact, STU_NUM, with or without additional attributes, can be a superkey even when the additional attributes are redundant.

A **candidate key** can be described as a superkey without unnecessary attributes, that is, a minimal superkey. Using this distinction, note that the composite key

STU_NUM, STU_LNAME

is a superkey, but it is not a candidate key because STU_NUM by itself is a candidate key! The combination

STU_LNAME, STU_FNAME, STU_INIT, STU_PHONE

might also be a candidate key, as long as you discount the possibility that two students share the same last name, first name, initial, and phone number.

If the student's Social Security number had been included as one of the attributes in the STUDENT table in Figure 3.1—perhaps named STU_SSN—both it and STU_NUM would have been candidate keys because either one would uniquely identify each student. In that case, the selection of STU_NUM as the primary key would be driven by the designer's choice or by end-user requirements. In short, the primary key is the candidate key chosen to be the unique row identifier. Note, incidentally, that a primary key is a superkey as well as a candidate key.

Within a table, each primary key value must be unique to ensure that each row is uniquely identified by the primary key. In that case, the table is said to exhibit **entity integrity**. To maintain entity integrity, a **null** (that is, no data entry at all) is not permitted in the primary key.

> **NOTE**
>
> A null is no value at all. It does *not* mean a zero or a space. A null is created when you press the Enter key or the Tab key to move to the next entry without making a prior entry of any kind. Pressing the Spacebar creates a blank (or a space).

Nulls can *never* be part of a primary key, and they should be avoided—to the greatest extent possible—in other attributes, too. There are rare cases in which nulls cannot be reasonably avoided when you are working with nonkey attributes. For example, one of an EMPLOYEE table's attributes is likely to be the EMP_INITIAL. However, some employees do not have a middle initial. Therefore, some of the EMP_INITIAL values may be null. You will also discover later in this section that there may be situations in which a null exists because of the nature of the relationship between two entities. In any case, even if nulls cannot always be avoided, they must be used sparingly. In fact, the existence of nulls in a table is often an indication of poor database design.

Nulls, if used improperly, can create problems because they have many different meanings. For example, a null can represent:

- An unknown attribute value.
- A known, but missing, attribute value.
- A "not applicable" condition.

Depending on the sophistication of the application development software, nulls can create problems when functions such as COUNT, AVERAGE, and SUM are used. In addition, nulls can create logical problems when relational tables are linked.

Controlled redundancy makes the relational database work. Tables within the database share common attributes that enable the tables to be linked together. For example, note that the PRODUCT and VENDOR tables in Figure 3.2 share a common attribute named VEND_CODE. And note that the PRODUCT table's VEND_CODE value 232 occurs more than once, as does the VEND_CODE value 235. Because the PRODUCT table is related to the VENDOR table through these VEND_CODE values, the multiple occurrence of the values is *required* to make the 1:M relationship between VENDOR and PRODUCT work. Each VEND_CODE value in the VENDOR table is unique—the VENDOR is the "1" side in the VENDOR-PRODUCT relationship. But any given VEND_CODE value from the VENDOR table may occur more than once in the PRODUCT table, thus providing evidence that PRODUCT is the "M" side of the VENDOR-PRODUCT relationship. In database terms, the multiple occurrences of the VEND_CODE values in the PRODUCT table are not redundant because they are *required* to make the relationship work. You should recall from Chapter 2 that data redundancy exists only when there is *unnecessary* duplication of attribute values.

FIGURE 3.2   **An example of a simple relational database**

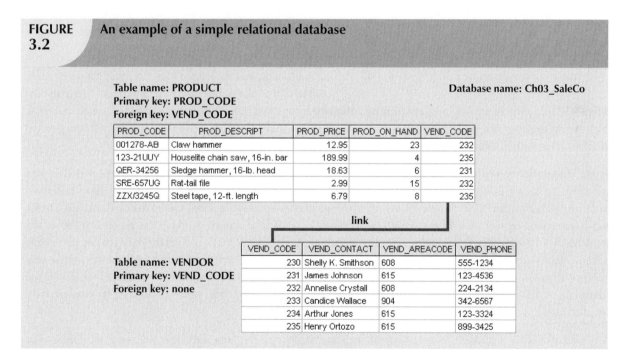

Table name: PRODUCT
Primary key: PROD_CODE
Foreign key: VEND_CODE

Database name: Ch03_SaleCo

PROD_CODE	PROD_DESCRIPT	PROD_PRICE	PROD_ON_HAND	VEND_CODE
001278-AB	Claw hammer	12.95	23	232
123-21UUY	Houselite chain saw, 16-in. bar	189.99	4	235
QER-34256	Sledge hammer, 16-lb. head	18.63	6	231
SRE-657UG	Rat-tail file	2.99	15	232
ZZX/3245Q	Steel tape, 12-ft. length	6.79	8	235

link

Table name: VENDOR
Primary key: VEND_CODE
Foreign key: none

VEND_CODE	VEND_CONTACT	VEND_AREACODE	VEND_PHONE
230	Shelly K. Smithson	608	555-1234
231	James Johnson	615	123-4536
232	Annelise Crystall	608	224-2134
233	Candice Wallace	904	342-6567
234	Arthur Jones	615	123-3324
235	Henry Ortozo	615	899-3425

As you examine Figure 3.2, note that the VEND_CODE value in one table can be used to point to the corresponding value in the other table. For example, the VEND_CODE value 235 in the PRODUCT table points to vendor Henry Ortozo in the VENDOR table. Consequently, you discover that the product "Houselite chain saw, 16-in. bar" is delivered by Henry Ortozo and that he can be contacted by calling 615-899-3425. The same connection can be made for the product "Steel tape, 12-ft. length" in the PRODUCT table.

Remember the naming convention—the prefix PROD was used in Figure 3.2 to indicate that the attributes "belong" to the PRODUCT table. Therefore, the prefix VEND in the PRODUCT table's VEND_CODE indicates that

VEND_CODE points to some other table in the database. In this case, the VEND prefix is used to point to the VENDOR table in the database.

A relational database can also be represented by a relational schema. A **relational schema** is a textual representation of the database tables where each table is listed by its name followed by the list of its attributes in parentheses. The primary key attribute(s) is (are) underlined. You will see such schemas in Chapter 5, Normalization of Database Tables. For example, the relational schema for Figure 3.2 would be shown as:

VENDOR (**VEND_CODE**, VEND_CONTACT, VEND_AREACODE, VEND_PHONE)

PRODUCT (**PROD_CODE**, PROD_DESCRIPT, PROD_PRICE, PROD_ON_HAND, VEND_CODE)

The link between the PRODUCT and VENDOR tables in Figure 3.2 can also be represented by the relational diagram shown in Figure 3.3. In this case, the link is indicated by the line that connects the VENDOR and PRODUCT tables.

Note that the link in Figure 3.3 is the equivalent of the relationship line in an ERD. This link is created when two tables share an attribute with common values. More specifically, the primary key of one table (VENDOR) appears as the *foreign key* in a related table (PRODUCT). A **foreign key (FK)** is an attribute whose values match the primary key values in the related table. For example, in Figure 3.2, the VEND_CODE is the primary key in the VENDOR table, and it occurs as a foreign key in the PRODUCT table. Because the VENDOR table is not linked to a third table, the VENDOR table shown in Figure 3.2 does not contain a foreign key.

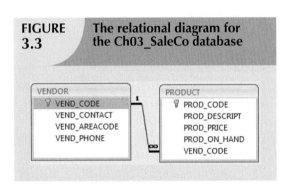

FIGURE 3.3   The relational diagram for the Ch03_SaleCo database

If the foreign key contains either matching values or nulls, the table that makes use of that foreign key is said to exhibit *referential integrity*. In other words, **referential integrity** means that if the foreign key contains a value, that value refers to an existing valid tuple (row) in another relation. Note that referential integrity is maintained between the PRODUCT and VENDOR tables shown in Figure 3.2.

Finally, a **secondary key** is defined as a key that is used strictly for data retrieval purposes. Suppose customer data are stored in a CUSTOMER table in which the customer number is the primary key. Do you suppose that most customers will remember their numbers? Data retrieval for a customer can be facilitated when the customer's last name and phone number are used. In that case, the primary key is the customer number; the secondary key is the combination of the customer's last name and phone number. Keep in mind that a secondary key does not necessarily yield a unique outcome. For example, a customer's last name and home telephone number could easily yield several matches where one family lives together and shares a phone line. A less efficient secondary key would be the combination of the last name and zip code; this could yield dozens of matches, which could then be combed for a specific match.

A secondary key's effectiveness in narrowing down a search depends on how restrictive that secondary key is. For instance, although the secondary key CUS_CITY is legitimate from a database point of view, the attribute values "New York" or "Sydney" are not likely to produce a usable return unless you want to examine millions of possible matches. (Of course, CUS_CITY is a better secondary key than CUS_COUNTRY.)

Table 3.3 summarizes the various relational database table keys.

TABLE 3.3	Relational Database Keys	
KEY TYPE	**DEFINITION**	
Superkey	An attribute (or combination of attributes) that uniquely identifies each row in a table.	
Candidate key	A minimal (irreducible) superkey. A superkey that does not contain a subset of attributes that is itself a superkey.	
Primary key	A candidate key selected to uniquely identify all other attribute values in any given row. Cannot contain null entries.	
Secondary key	An attribute (or combination of attributes) used strictly for data retrieval purposes.	
Foreign key	An attribute (or combination of attributes) in one table whose values must either match the primary key in another table or be null.	

3.3 INTEGRITY RULES

Relational database integrity rules are very important to good database design. Many (but by no means all) RDBMSs enforce integrity rules automatically. However, it is much safer to make sure that your application design conforms to the entity and referential integrity rules mentioned in this chapter. Those rules are summarized in Table 3.4.

TABLE 3.4	Integrity Rules	
ENTITY INTEGRITY	**DESCRIPTION**	
Requirement	All primary key entries are unique, and no part of a primary key may be null.	
Purpose	Each row will have a unique identity, and foreign key values can properly reference primary key values.	
Example	No invoice can have a duplicate number, nor can it be null. In short, all invoices are uniquely identified by their invoice number.	
REFERENTIAL INTEGRITY	**DESCRIPTION**	
Requirement	A foreign key may have either a null entry, as long as it is not a part of its table's primary key, or an entry that matches the primary key value in a table to which it is related. (Every non-null foreign key value *must* reference an *existing* primary key value.)	
Purpose	It is possible for an attribute NOT to have a corresponding value, but it will be impossible to have an invalid entry. The enforcement of the referential integrity rule makes it impossible to delete a row in one table whose primary key has mandatory matching foreign key values in another table.	
Example	A customer might not yet have an assigned sales representative (number), but it will be impossible to have an invalid sales representative (number).	

The integrity rules summarized in Table 3.4 are illustrated in Figure 3.4.

Note the following features of Figure 3.4.

1. *Entity integrity.* The CUSTOMER table's primary key is CUS_CODE. The CUSTOMER primary key column has no null entries, and all entries are unique. Similarly, the AGENT table's primary key is AGENT_CODE, and this primary key column also is free of null entries.

2. *Referential integrity.* The CUSTOMER table contains a foreign key, AGENT_CODE, which links entries in the CUSTOMER table to the AGENT table. The CUS_CODE row that is identified by the (primary key) number 10013 contains a null entry in its AGENT_CODE foreign key because Mr. Paul F. Olowski does not yet have a sales representative assigned to him. The remaining AGENT_CODE entries in the CUSTOMER table all match the AGENT_CODE entries in the AGENT table.

FIGURE 3.4   An illustration of integrity rules

Table name: CUSTOMER
Primary key: CUS_CODE
Foreign key: AGENT_CODE

Database name: Ch03_InsureCo

CUS_CODE	CUS_LNAME	CUS_FNAME	CUS_INITIAL	CUS_AREACODE	CUS_PHONE	CUS_INSURE_TYPE	CUS_INSURE_AMT	CUS_RENEW_DATE	AGENT_CODE
10010	Ramas	Alfred	A	615	844-2573	T1	100.00	05-Apr-2008	502
10011	Dunne	Leona	K	713	894-1238	T1	250.00	16-Jun-2008	501
10012	Smith	Kathy	W	615	894-2285	S2	150.00	29-Jan-2009	502
10013	Olowski	Paul	F	615	894-2180	S1	300.00	14-Oct-2008	502
10014	Orlando	Myron		615	222-1672	T1	100.00	28-Dec-2008	501
10015	O'Brian	Amy	B	713	442-3381	T2	850.00	22-Sep-2008	503
10016	Brown	James	G	615	297-1228	S1	120.00	25-Mar-2009	502
10017	Williams	George		615	290-2556	S1	250.00	17-Jul-2008	503
10018	Farriss	Anne	G	713	382-7185	T2	100.00	03-Dec-2008	501
10019	Smith	Olette	K	615	297-3809	S2	500.00	14-Mar-2009	503

Table name: AGENT
Primary key: AGENT_CODE
Foreign key: none

AGENT_CODE	AGENT_AREACODE	AGENT_PHONE	AGENT_LNAME	AGENT_YTD_SLS
501	713	228-1249	Alby	132735.75
502	615	882-1244	Hahn	138967.35
503	615	123-5589	Okon	127093.45

To avoid nulls, some designers use special codes, known as **flags**, to indicate the absence of some value. Using Figure 3.4 as an example, the code -99 could be used as the AGENT_CODE entry of the fourth row of the CUSTOMER table to indicate that customer Paul Olowski does not yet have an agent assigned to him. If such a flag is used, the AGENT table must contain a dummy row with an AGENT_CODE value of -99. Thus, the AGENT table's first record might contain the values shown in Table 3.5.

TABLE 3.5   A Dummy Variable Value Used as a Flag

AGENT_CODE	AGENT_AREACODE	AGENT_PHONE	AGENT_LNAME	AGENT_YTD_SALES
-99	000	000-0000	None	$0.00

Chapter 4, Entity Relationship (ER) Modeling, discusses several ways in which nulls may be handled.

Other integrity rules that can be enforced in the relational model are the *NOT NULL* and *UNIQUE* constraints. The NOT NULL constraint can be placed on a column to ensure that every row in the table has a value for that column. The UNIQUE constraint is a restriction placed on a column to ensure that no duplicate values exist for that column.

3.4 RELATIONAL SET OPERATORS

The data in relational tables are of limited value unless the data can be manipulated to generate useful information. This section describes the basic data manipulation capabilities of the relational model. **Relational algebra** defines the theoretical way of manipulating table contents using the eight relational operators: SELECT, PROJECT, JOIN, INTERSECT, UNION, DIFFERENCE, PRODUCT, and DIVIDE. In Chapter 7, Introduction to Structured Query Language (SQL), you will learn how SQL commands can be used to accomplish relational algebra operations.

NOTE

The degree of relational completeness can be defined by the extent to which relational algebra is supported. To be considered minimally relational, the DBMS must support the key relational operators SELECT, PROJECT, and JOIN. Very few DBMSs are capable of supporting all eight relational operators.

The relational operators have the property of **closure**; that is, the use of relational algebra operators on existing tables (relations) produces new relations. There is no need to examine the mathematical definitions, properties, and characteristics of those relational algebra operators. However, their *use* can easily be illustrated as follows:

1. UNION combines all rows from two tables, excluding duplicate rows. The tables must have the same attribute characteristics (the columns and domains must be identical) to be used in the UNION. When two or more tables share the same number of columns, when the columns have the same names, and when they share the same (or compatible) domains, they are said to be **union-compatible**. The effect of a UNION is shown in Figure 3.5.

FIGURE 3.5 UNION

P_CODE	P_DESCRIPT	PRICE
123456	Flashlight	5.26
123457	Lamp	25.15
123458	Box Fan	10.99
213345	9v battery	1.92
254467	100W bulb	1.47
311452	Powerdrill	34.99

UNION

P_CODE	P_DESCRIPT	PRICE
345678	Microwave	160.00
345679	Dishwasher	500.00

yields

P_CODE	P_DESCRIPT	PRICE
123456	Flashlight	5.26
123457	Lamp	25.15
123458	Box Fan	10.99
213345	9v battery	1.92
254467	100W bulb	1.47
311452	Powerdrill	34.99
345678	Microwave	160
345679	Dishwasher	500

2. INTERSECT yields only the rows that appear in both tables. As was true in the case of UNION, the tables must be union-compatible to yield valid results. For example, you cannot use INTERSECT if one of the attributes is numeric and one is character-based. The effect of an INTERSECT is shown in Figure 3.6.

FIGURE 3.6 INTERSECT

F_NAME
George
Jane
Elaine
Wilfred
Jorge

INTERSECT

F_NAME
Jane
William
Jorge
Dennis

yields

F_NAME
Jane
Jorge

3. DIFFERENCE yields all rows in one table that are not found in the other table; that is, it subtracts one table from the other. As was true in the case of UNION, the tables must be union-compatible to yield valid results.

The effect of a DIFFERENCE is shown in Figure 3.7. However, note that subtracting the first table from the second table is not the same as subtracting the second table from the first table.

FIGURE 3.7   **DIFFERENCE**

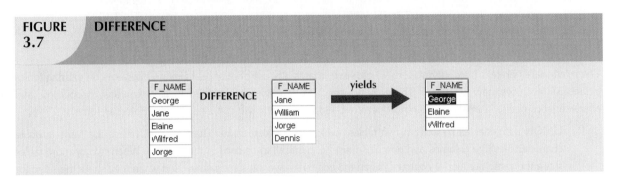

4. PRODUCT yields all possible pairs of rows from two tables—also known as the Cartesian product. Therefore, if one table has six rows and the other table has three rows, the PRODUCT yields a list composed of 6 × 3 = 18 rows. The effect of a PRODUCT is shown in Figure 3.8.

FIGURE 3.8   **PRODUCT**

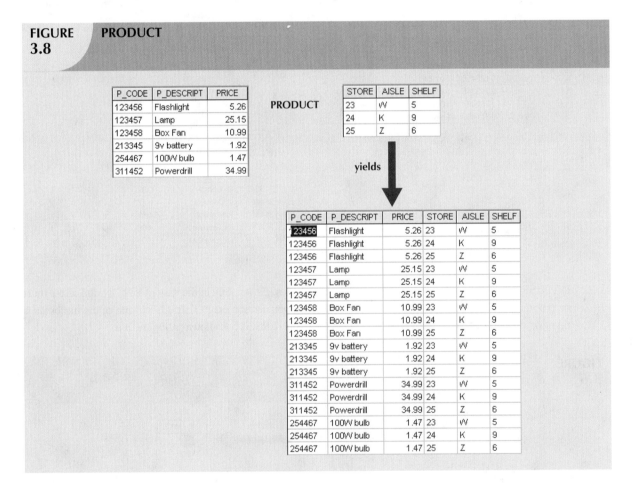

5. SELECT, also known as RESTRICT, yields values for all rows found in a table that satisfy a given condition. SELECT can be used to list all of the row values, or it can yield only those row values that match a specified criterion. In other words, SELECT yields a horizontal subset of a table. The effect of a SELECT is shown in Figure 3.9.

FIGURE 3.9   SELECT

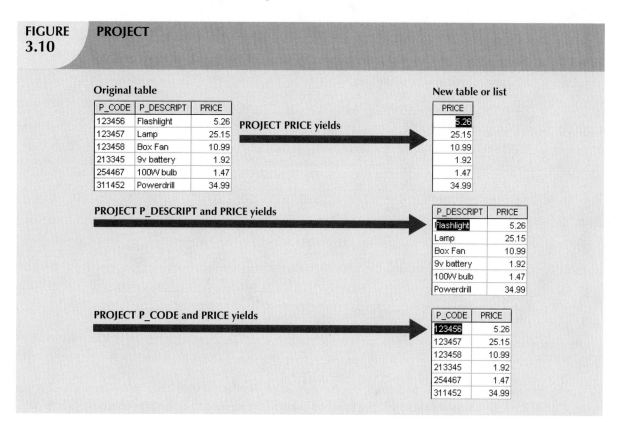

6. PROJECT yields all values for selected attributes. In other words, PROJECT yields a vertical subset of a table. The effect of a PROJECT is shown in Figure 3.10.

FIGURE 3.10   PROJECT

7. JOIN allows information to be combined from two or more tables. JOIN is the real power behind the relational database, allowing the use of independent tables linked by common attributes. The CUSTOMER and AGENT tables shown in Figure 3.11 will be used to illustrate several types of joins.

FIGURE
3.11
Two tables that will be used in join illustrations

Table name: CUSTOMER

CUS_CODE	CUS_LNAME	CUS_ZIP	AGENT_CODE
1132445	Walker	32145	231
1217782	Adares	32145	125
1312243	Rakowski	34129	167
1321242	Rodriguez	37134	125
1542311	Smithson	37134	421
1657399	Vanloo	32145	231

Table name: AGENT

AGENT_CODE	AGENT_PHONE
125	6152439887
167	6153426778
231	6152431124
333	9041234445

A **natural join** links tables by selecting only the rows with common values in their common attribute(s). A natural join is the result of a three-stage process:

a. First, a PRODUCT of the tables is created, yielding the results shown in Figure 3.12.

FIGURE
3.12
Natural join, Step 1: PRODUCT

CUS_CODE	CUS_LNAME	CUS_ZIP	CUSTOMER.AGENT_CODE	AGENT.AGENT_CODE	AGENT_PHONE
1132445	Walker	32145	231	125	6152439887
1132445	Walker	32145	231	167	6153426778
1132445	Walker	32145	231	231	6152431124
1132445	Walker	32145	231	333	9041234445
1217782	Adares	32145	125	125	6152439887
1217782	Adares	32145	125	167	6153426778
1217782	Adares	32145	125	231	6152431124
1217782	Adares	32145	125	333	9041234445
1312243	Rakowski	34129	167	125	6152439887
1312243	Rakowski	34129	167	167	6153426778
1312243	Rakowski	34129	167	231	6152431124
1312243	Rakowski	34129	167	333	9041234445
1321242	Rodriguez	37134	125	125	6152439887
1321242	Rodriguez	37134	125	167	6153426778
1321242	Rodriguez	37134	125	231	6152431124
1321242	Rodriguez	37134	125	333	9041234445
1542311	Smithson	37134	421	125	6152439887
1542311	Smithson	37134	421	167	6153426778
1542311	Smithson	37134	421	231	6152431124
1542311	Smithson	37134	421	333	9041234445
1657399	Vanloo	32145	231	125	6152439887
1657399	Vanloo	32145	231	167	6153426778
1657399	Vanloo	32145	231	231	6152431124
1657399	Vanloo	32145	231	333	9041234445

b. Second, a SELECT is performed on the output of Step a to yield only the rows for which the AGENT_CODE values are equal. The common columns are referred to as the **join columns**. Step b yields the results shown in Figure 3.13.

FIGURE 3.13 — Natural join, Step 2: SELECT

CUS_CODE	CUS_LNAME	CUS_ZIP	CUSTOMER.AGENT_CODE	AGENT.AGENT_CODE	AGENT_PHONE
1217782	Adares	32145	125	125	6152439887
1321242	Rodriguez	37134	125	125	6152439887
1312243	Rakowski	34129	167	167	6153426778
1132445	Walker	32145	231	231	6152431124
1657399	Vanloo	32145	231	231	6152431124

c. A PROJECT is performed on the results of Step b to yield a single copy of each attribute, thereby eliminating duplicate columns. Step c yields the output shown in Figure 3.14.

FIGURE 3.14 — Natural join, Step 3: PROJECT

CUS_CODE	CUS_LNAME	CUS_ZIP	AGENT_CODE	AGENT_PHONE
1217782	Adares	32145	125	6152439887
1321242	Rodriguez	37134	125	6152439887
1312243	Rakowski	34129	167	6153426778
1132445	Walker	32145	231	6152431124
1657399	Vanloo	32145	231	6152431124

The final outcome of a natural join yields a table that does not include unmatched pairs and provides only the copies of the matches.

Note a few crucial features of the natural join operation:

- If no match is made between the table rows, the new table does not include the unmatched row. In that case, neither AGENT_CODE 421 nor the customer whose last name is Smithson is included. Smithson's AGENT_CODE 421 does not match any entry in the AGENT table.

- The column on which the join was made—that is, AGENT_CODE—occurs only once in the new table.

- If the same AGENT_CODE were to occur several times in the AGENT table, a customer would be listed for each match. For example, if the AGENT_CODE 167 were to occur three times in the AGENT table, the customer named Rakowski, who is associated with AGENT_CODE 167, would occur three times in the resulting table. (A good AGENT table cannot, of course, yield such a result because it would contain unique primary key values.)

Another form of join, known as **equijoin**, links tables on the basis of an equality condition that compares specified columns of each table. The outcome of the equijoin does not eliminate duplicate columns, and the condition or criterion used to join the tables must be explicitly defined. The equijoin takes its name from the equality comparison operator (=) used in the condition. If any other comparison operator is used, the join is called a **theta join**.

In an **outer join**, the matched pairs would be retained and any unmatched values in the other table would be left null. More specifically, if an outer join is produced for tables CUSTOMER and AGENT, two scenarios are possible:

FIGURE 3.15 — Left outer join

CUS_CODE	CUS_LNAME	CUS_ZIP	AGENT_CODE	AGENT_PHONE
1217782	Adares	32145	125	6152439887
1321242	Rodriguez	37134	125	6152439887
1312243	Rakowski	34129	167	6153426778
1132445	Walker	32145	231	6152431124
1657399	Vanloo	32145	231	6152431124
1542311	Smithson	37134	421	

A **left outer join** yields all of the rows in the CUSTOMER table, including those that do not have a matching value in the AGENT table. An example of such a join is shown in Figure 3.15.

A **right outer join** yields all of the rows in the AGENT table, including those that do not have matching values in the CUSTOMER table. An example of such a join is shown in Figure 3.16.

FIGURE 3.16   Right outer join

CUS_CODE	CUS_LNAME	CUS_ZIP	AGENT_CODE	AGENT_PHONE
1217782	Adares	32145	125	6152439887
1321242	Rodriguez	37134	125	6152439887
1312243	Rakowski	34129	167	6153426778
1132445	Walker	32145	231	6152431124
1657399	Vanloo	32145	231	6152431124
			333	9041234445

Outer joins are especially useful when you are trying to determine what value(s) in related tables cause(s) referential integrity problems. Such problems are created when foreign key values do not match the primary key values in the related table(s). In fact, if you are asked to convert large spreadsheets or other nondatabase data into relational database tables, you will discover that the outer joins save you vast amounts of time and uncounted headaches when you encounter referential integrity errors after the conversions.

You may wonder why the outer joins are labeled *left* and *right*. The labels refer to the order in which the tables are listed in the SQL command. Chapter 7 explores such joins.

8. The DIVIDE operation uses one single-column table (i.e. column "a") as the divisor and one 2-column table (i.e. columns "a" and "b") as the dividend. The tables must have a common column (i.e. column "a".) The output of the DIVIDE operation is a single column with the values of column "a" from the dividend table rows where the value of the common column (i.e. column "a") in both tables match. Figure 3.17 shows a DIVIDE.

FIGURE 3.17   DIVIDE

CODE	LOC
A	5
A	9
A	4
B	5
B	3
C	6
D	7
D	8
E	8

DIVIDE

CODE
A
B

yields →

LOC
5

Using the example shown in Figure 3.17, note that:

a. Table 1 is "divided" by Table 2 to produce Table 3. Tables 1 and 2 both contain the column CODE but do not share LOC.

b. To be included in the resulting Table 3, a value in the unshared column (LOC) must be associated (in the dividing Table 2) with *every* value in Table 1.

c. The only value associated with both A and B is 5.

3.5 THE DATA DICTIONARY AND THE SYSTEM CATALOG

The **data dictionary** provides a detailed description of all tables found within the user/designer-created database. Thus, the data dictionary contains at least all of the attribute names and characteristics for each table in the system. In short, the data dictionary contains metadata—data about data. Using the small database presented in Figure 3.4, you might picture its data dictionary as shown in Table 3.6.

TABLE 3.6   A Sample Data Dictionary

TABLE NAME	ATTRIBUTE NAME	CONTENTS	TYPE	FORMAT	RANGE	REQUIRED	PK OR FK	FK REFERENCED TABLE
CUSTOMER	CUS_CODE	Customer account code	CHAR(5)	99999	10000–99999	Y	PK	
	CUS_LNAME	Customer last name	VARCHAR(20)	Xxxxxxxx		Y		
	CUS_FNAME	Customer first name	VARCHAR(20)	Xxxxxxxx		Y		
	CUS_INITIAL	Customer initial	CHAR(1)	X				
	CUS_RENEW_DATE	Customer insurance renewal date	DATE	dd-mmm-yyyy				
	AGENT_CODE	Agent code	CHAR(3)	999			FK	AGENT_CODE
AGENT	AGENT_CODE	Agent code	CHAR(3)	999		Y	PK	
	AGENT_AREACODE	Agent area code	CHAR(3)	999		Y		
	AGENT_PHONE	Agent telephone number	CHAR(8)	999-9999		Y		
	AGENT_LNAME	Agent last name	VARCHAR(20)	Xxxxxxxx		Y		
	AGENT_YTD_SLS	Agent year-to-date sales	NUMBER(9,2)	9,999,999.99		Y		

FK   =   Foreign key
PK   =   Primary key
CHAR   =   Fixed character length data (1–255 characters)
VARCHAR   =   Variable character length data (1–2,000 characters)
NUMBER   =   Numeric data (NUMBER(9,2)) is used to specify numbers with two decimal places and up to nine digits, including the decimal places.
Some RDBMSs permit the use of a MONEY or CURRENCY data type.

Note: Telephone area codes are always composed of digits 0–9. Because area codes are not used arithmetically; they are most efficiently stored as character data. Also, the area codes are always composed of three digits. Therefore, the area code data type is defined as CHAR(3). On the other hand, names do not conform to some standard length. Therefore, the customer first names are defined as VARCHAR(20), thus indicating that up to 20 characters may be used to store the names. Character data are shown as left-justified.

NOTE

The data dictionary in Table 3.6 is an example of the *human* view of the entities, attributes, and relationships. The purpose of this data dictionary is to ensure that all members of database design and implementation teams use the same table and attribute names and characteristics. The DBMS's internally stored data dictionary contains additional information about relationship types, entity and referential integrity checks and enforcement, and index types and components. This additional information is generated during the database implementation stage.

The data dictionary is sometimes described as "the database designer's database" because it records the design decisions about tables and their structures.

Like the data dictionary, the **system catalog** contains metadata. The system catalog can be described as a detailed system data dictionary that describes all objects within the database, including data about table names, the table's creator and creation date, the number of columns in each table, the data type corresponding to each column, index filenames, index creators, authorized users, and access privileges. Because the system catalog contains all required data dictionary information, the terms *system catalog* and *data dictionary* are often used interchangeably. In fact, current relational database software generally provides only a system catalog, from which the designer's data dictionary information may be derived. The system catalog is actually a system-created database whose tables store the user/designer-created database characteristics and contents. Therefore, the system catalog tables can be queried just like any user/designer-created table.

In effect, the system catalog automatically produces database documentation. As new tables are added to the database, that documentation also allows the RDBMS to check for and eliminate homonyms and synonyms. In general terms, **homonyms** are similar-sounding words with different meanings, such as *boar* and *bore*, or identically spelled words with different meanings, such as *fair* (meaning "just") and *fair* (meaning "festival"). In a database context, the word *homonym* indicates the use of the same attribute name to label different attributes. For example, you might use C_NAME to label a customer name attribute in a CUSTOMER table and also use C_NAME to label a consultant name attribute in a CONSULTANT table. To lessen confusion, you should avoid database homonyms; the data dictionary is very useful in this regard.

In a database context, a **synonym** is the opposite of a homonym and indicates the use of different names to describe the same attribute. For example, *car* and *auto* refer to the same object. Synonyms must be avoided. You will discover why using synonyms is a bad idea when you work through Problem 33 at the end of this chapter.

3.6 RELATIONSHIPS WITHIN THE RELATIONAL DATABASE

You already know that relationships are classified as one-to-one (1:1), one-to-many (1:M), and many-to-many (M:N or M:M). This section explores those relationships further to help you apply them properly when you start developing database designs, focusing on the following points:

- The 1:M relationship is the relational modeling ideal. Therefore, this relationship type should be the norm in any relational database design.
- The 1:1 relationship should be rare in any relational database design.
- M:N relationships cannot be implemented as such in the relational model. Later in this section, you will see how any M:N relationships can be changed into two 1:M relationships.

3.6.1 THE 1:M RELATIONSHIP

The 1:M relationship is the relational database norm. To see how such a relationship is modeled and implemented, consider the PAINTER paints PAINTING example that was used in Chapter 2. Compare the data model in Figure 3.18 with its implementation in Figure 3.19.

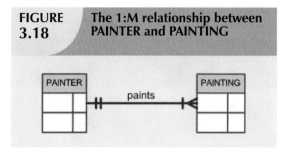

FIGURE 3.18 The 1:M relationship between PAINTER and PAINTING

As you examine the PAINTER and PAINTING table contents in Figure 3.19, note the following features:

- Each painting is painted by one and only one painter, but each painter could have painted many paintings. Note that painter 123 (Georgette P. Ross) has three paintings stored in the PAINTING table.

- There is only one row in the PAINTER table for any given row in the PAINTING table, but there may be many rows in the PAINTING table for any given row in the PAINTER table.

FIGURE 3.19 The implemented 1:M relationship between PAINTER and PAINTING

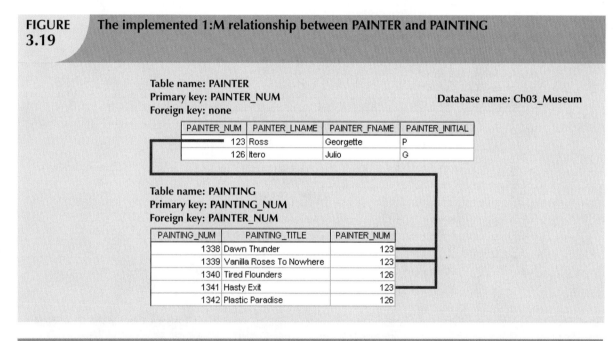

Table name: PAINTER
Primary key: PAINTER_NUM
Foreign key: none

Database name: Ch03_Museum

PAINTER_NUM	PAINTER_LNAME	PAINTER_FNAME	PAINTER_INITIAL
123	Ross	Georgette	P
126	Itero	Julio	G

Table name: PAINTING
Primary key: PAINTING_NUM
Foreign key: PAINTER_NUM

PAINTING_NUM	PAINTING_TITLE	PAINTER_NUM
1338	Dawn Thunder	123
1339	Vanilla Roses To Nowhere	123
1340	Tired Flounders	126
1341	Hasty Exit	123
1342	Plastic Paradise	126

NOTE

The one-to-many (1:M) relationship is easily implemented in the relational model by putting the *primary key of the "1" side in the table of the "many" side as a foreign key.*

The 1:M relationship is found in any database environment. Students in a typical college or university will discover that each COURSE can generate many CLASSes but that each CLASS refers to only one COURSE. For example, an Accounting II course might yield two classes: one offered on Monday, Wednesday, and Friday (MWF) from 10:00 a.m. to 10:50 a.m. and one offered on Thursday (Th) from 6:00 p.m. to 8:40 p.m. Therefore, the 1:M relationship between COURSE and CLASS might be described this way:

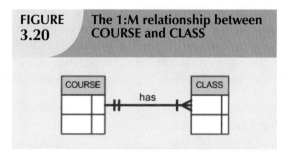

FIGURE 3.20 The 1:M relationship between COURSE and CLASS

- Each COURSE can have many CLASSes, but each CLASS references only one COURSE.

- There will be only one row in the COURSE table for any given row in the CLASS table, but there can be many rows in the CLASS table for any given row in the COURSE table.

Figure 3.20 maps the ERM for the 1:M relationship between COURSE and CLASS.

The 1:M relationship between COURSE and CLASS is further illustrated in Figure 3.21.

FIGURE 3.21   **The implemented 1:M relationship between COURSE and CLASS**

Table name: COURSE
Primary key: CRS_CODE
Foreign key: none   Database name: Ch03_TinyCollege

CRS_CODE	DEPT_CODE	CRS_DESCRIPTION	CRS_CREDIT
ACCT-211	ACCT	Accounting I	3
ACCT-212	ACCT	Accounting II	3
CIS-220	CIS	Intro. to Microcomputing	3
CIS-420	CIS	Database Design and Implementation	4
QM-261	CIS	Intro. to Statistics	3
QM-362	CIS	Statistical Applications	4

Table name: CLASS
Primary key: CLASS_CODE
Foreign key: CRS_CODE

CLASS_CODE	CRS_CODE	CLASS_SECTION	CLASS_TIME	CLASS_ROOM	PROF_NUM
10012	ACCT-211	1	MWF 8:00-8:50 a.m.	BUS311	105
10013	ACCT-211	2	MWF 9:00-9:50 a.m.	BUS200	105
10014	ACCT-211	3	TTh 2:30-3:45 p.m.	BUS252	342
10015	ACCT-212	1	MWF 10:00-10:50 a.m.	BUS311	301
10016	ACCT-212	2	Th 6:00-8:40 p.m.	BUS252	301
10017	CIS-220	1	MWF 9:00-9:50 a.m.	KLR209	228
10018	CIS-220	2	MWF 9:00-9:50 a.m.	KLR211	114
10019	CIS-220	3	MWF 10:00-10:50 a.m.	KLR209	228
10020	CIS-420	1	W 6:00-8:40 p.m.	KLR209	162
10021	QM-261	1	MWF 8:00-8:50 a.m.	KLR200	114
10022	QM-261	2	TTh 1:00-2:15 p.m.	KLR200	114
10023	QM-362	1	MWF 11:00-11:50 a.m.	KLR200	162
10024	QM-362	2	TTh 2:30-3:45 p.m.	KLR200	162

Using Figure 3.21, take a minute to review some important terminology. Note that CLASS_CODE in the CLASS table uniquely identifies each row. Therefore, CLASS_CODE has been chosen to be the primary key. However, the combination CRS_CODE and CLASS_SECTION will also uniquely identify each row in the class table. In other words, the *composite key* composed of CRS_CODE and CLASS_SECTION is a *candidate key*. Any candidate key must have the not null and unique constraints enforced. (You will see how this is done when you learn SQL in Chapter 7.)

For example, note in Figure 3.19 that the PAINTER table's primary key, PAINTER_NUM, is included in the PAINTING table as a foreign key. Similarly, in Figure 3.21, the COURSE table's primary key, CRS_CODE, is included in the CLASS table as a foreign key.

3.6.2 THE 1:1 RELATIONSHIP

As the 1:1 label implies, in this relationship, one entity can be related to only one other entity, and vice versa. For example, one department chair—a professor—can chair only one department and one department can have only one department chair. The entities PROFESSOR and DEPARTMENT thus exhibit a 1:1 relationship. (You might argue that not all professors chair a department and professors cannot be *required* to chair a department. That is, the relationship between the two entities is optional. However, at this stage of the discussion, you should focus your attention on the basic 1:1 relationship. Optional relationships will be addressed in Chapter 4.) The basic 1:1 relationship is modeled in Figure 3.22, and its implementation is shown in Figure 3.23.

As you examine the tables in Figure 3.23, note that there are several important features:

- Each professor is a Tiny College employee. Therefore, the professor identification is through the EMP_NUM. (However, note that not all employees are professors—there's another optional relationship.)

FIGURE 3.22	The 1:1 relationship between PROFESSOR and DEPARTMENT

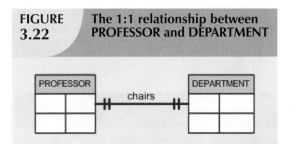

• The 1:1 PROFESSOR chairs DEPARTMENT relationship is implemented by having the EMP_NUM foreign key in the DEPARTMENT table. Note that the 1:1 relationship is treated as a special case of the 1:M relationship in which the "many" side is restricted to a single occurrence. In this case, DEPARTMENT contains the EMP_NUM as a foreign key to indicate that it is the *department* that has a chair.

FIGURE 3.23	The implemented 1:1 relationship between PROFESSOR and DEPARTMENT

Table name: PROFESSOR
Primary key: EMP_NUM
Foreign key: DEPT_CODE

Database name: Ch03_TinyCollege

EMP_NUM	DEPT_CODE	PROF_OFFICE	PROF_EXTENSION	PROF_HIGH_DEGREE
103	HIST	DRE 156	6783	Ph.D.
104	ENG	DRE 102	5561	MA
105	ACCT	KLR 229D	8665	Ph.D.
106	MKT/MGT	KLR 126	3899	Ph.D.
110	BIOL	AAK 160	3412	Ph.D.
114	ACCT	KLR 211	4436	Ph.D.
155	MATH	AAK 201	4440	Ph.D.
160	ENG	DRE 102	2248	Ph.D.
162	CIS	KLR 203E	2359	Ph.D.
191	MKT/MGT	KLR 409B	4016	DBA
195	PSYCH	AAK 297	3550	Ph.D.
209	CIS	KLR 333	3421	Ph.D.
228	CIS	KLR 300	3000	Ph.D.
297	MATH	AAK 194	1145	Ph.D.
299	ECON/FIN	KLR 284	2851	Ph.D.
301	ACCT	KLR 244	4683	Ph.D.
335	ENG	DRE 208	2000	Ph.D.
342	SOC	BBG 208	5514	Ph.D.
387	BIOL	AAK 230	8665	Ph.D.
401	HIST	DRE 156	6783	MA
425	ECON/FIN	KLR 284	2851	MBA
435	ART	BBG 185	2278	Ph.D.

↑ **The 1:M DEPARTMENT employs PROFESSOR relationship is implemented through the placement of the DEPT_CODE foreign key in the PROFESSOR table.**

The 1:1 PROFESSOR chairs DEPARTMENT relationship is implemented through the placement of the EMP_NUM foreign key in the DEPARTMENT table. ↓

Table name: DEPARTMENT
Primary key: DEPT_CODE
Foreign key: EMP_NUM

DEPT_CODE	DEPT_NAME	SCHOOL_CODE	EMP_NUM	DEPT_ADDRESS	DEPT_EXTENSION
ACCT	Accounting	BUS	114	KLR 211, Box 52	3119
ART	Fine Arts	A&SCI	435	BBG 185, Box 128	2278
BIOL	Biology	A&SCI	387	AAK 230, Box 415	4117
CIS	Computer Info. Systems	BUS	209	KLR 333, Box 56	3245
ECON/FIN	Economics/Finance	BUS	299	KLR 284, Box 63	3126
ENG	English	A&SCI	160	DRE 102, Box 223	1004
HIST	History	A&SCI	103	DRE 156, Box 284	1867
MATH	Mathematics	A&SCI	297	AAK 194, Box 422	4234
MKT/MGT	Marketing/Management	BUS	106	KLR 126, Box 55	3342
PSYCH	Psychology	A&SCI	195	AAK 297, Box 438	4110
SOC	Sociology	A&SCI	342	BBG 208, Box 132	2008

- Also note that the PROFESSOR table contains the DEPT_CODE foreign key to implement the 1:M DEPARTMENT employs PROFESSOR relationship. This is a good example of how two entities can participate in two (or even more) relationships simultaneously.

ONLINE CONTENT

If you open the **Ch03_TinyCollege** database in the Student Online Companion, you'll see that the STUDENT and CLASS entities still use PROF_NUM as their foreign key. PROF_NUM and EMP_NUM are labels for the same attribute, which is an example of the use of synonyms—different names for the same attribute. These synonyms will be eliminated in future chapters as the Tiny College database continues to be improved.

The preceding "PROFESSOR chairs DEPARTMENT" example illustrates a proper 1:1 relationship. *In fact, the use of a 1:1 relationship ensures that two entity sets are not placed in the same table when they should not be.* However, the existence of a 1:1 relationship sometimes means that the entity components were not defined properly. It could indicate that the two entities actually belong in the same table!

As rare as 1:1 relationships should be, certain conditions absolutely *require* their use. For example, suppose you manage the database for a company that employs pilots, accountants, mechanics, clerks, salespeople, service personnel, and more. Pilots have many attributes that the other employees don't have, such as licenses, medical certificates, flight experience records, dates of flight proficiency checks, and proof of required periodic medical checks. If you put all of the pilot-specific attributes in the EMPLOYEE table, you will have several nulls in that table for all employees who are not pilots. To avoid the proliferation of nulls, it is better to split the pilot attributes into a separate table (PILOT) that is linked to the EMPLOYEE table in a 1:1 relationship. Because pilots have many attributes that are shared by all employees—such as name, date of birth, and date of first employment—those attributes would be stored in the EMPLOYEE table.

ONLINE CONTENT

If you look at the **Ch03_AviaCo** database in the Student Online Companion, you will see the implementation of the 1:1 PILOT to EMPLOYEE relationship. This type of relationship will be examined in detail in Chapter 6, Advanced Data Modeling.

3.6.3 THE M:N RELATIONSHIP

A many-to-many (M:N) relationship is not supported directly in the relational environment. However, M:N relationships can be implemented by creating a new entity in 1:M relationships with the original entities.

FIGURE 3.24	The ERM's M:N relationship between STUDENT and CLASS

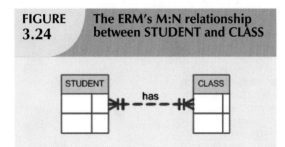

To explore the many-to-many (M:N) relationship, consider a rather typical college environment in which each STUDENT can take many CLASSes, and each CLASS can contain many STUDENTs. The ER model in Figure 3.24 shows this M:N relationship.

Note the features of the ERM in Figure 3.24.

- Each CLASS can have many STUDENTs, and each STUDENT can take many CLASSes.
- There can be many rows in the CLASS table for any given row in the STUDENT table, and there can be many rows in the STUDENT table for any given row in the CLASS table.

To examine the M:N relationship more closely, imagine a small college with two students, each of whom takes three classes. Table 3.7 shows the enrollment data for the two students.

TABLE 3.7

Sample Student Enrollment Data

STUDENT'S LAST NAME	SELECTED CLASSES
Bowser	Accounting 1, ACCT-211, code 10014 Intro to Microcomputing, CIS-220, code 10018 Intro to Statistics, QM-261, code 10021
Smithson	Accounting 1, ACCT-211, code 10014 Intro to Microcomputing, CIS-220, code 10018 Intro to Statistics, QM-261, code 10021

FIGURE 3.25

The M:N relationship between STUDENT and CLASS

Table name: STUDENT
Primary key: STU_NUM
Foreign key: none

Database name: Ch03_CollegeTry

STU_NUM	STU_LNAME	CLASS_CODE
321452	Bowser	10014
321452	Bowser	10018
321452	Bowser	10021
324257	Smithson	10014
324257	Smithson	10018
324257	Smithson	10021

Table name: CLASS
Primary key: CLASS_CODE
Foreign key: STU_NUM

CLASS_CODE	STU_NUM	CRS_CODE	CLASS_SECTION	CLASS_TIME	CLASS_ROOM	PROF_NUM
10014	321452	ACCT-211	3	TTh 2:30-3:45 p.m.	BUS252	342
10014	324257	ACCT-211	3	TTh 2:30-3:45 p.m.	BUS252	342
10018	321452	CIS-220	2	MWF 9:00-9:50 a.m.	KLR211	114
10018	324257	CIS-220	2	MWF 9:00-9:50 a.m.	KLR211	114
10021	321452	QM-261	1	MWF 8:00-8:50 a.m.	KLR200	114
10021	324257	QM-261	1	MWF 8:00-8:50 a.m.	KLR200	114

Although the M:N relationship is logically reflected in Figure 3.24, it should *not* be implemented as shown in Figure 3.25 for two good reasons:

- The tables create many redundancies. For example, note that the STU_NUM values occur many times in the STUDENT table. In a real-world situation, additional student attributes such as address, classification, major, and home phone would also be contained in the STUDENT table, and each of those attribute values would be repeated in each of the records shown here. Similarly, the CLASS table contains many duplications: each student taking the class generates a CLASS record. The problem would be even worse if the CLASS table included such attributes as credit hours and course description. Those redundancies lead to the anomalies discussed in Chapter 1.
- Given the structure and contents of the two tables, the relational operations become very complex and are likely to lead to system efficiency errors and output errors.

Fortunately, the problems inherent in the many-to-many (M:N) relationship can easily be avoided by creating a **composite entity** (also referred to as a **bridge entity** or an **associative entity)**. Because such a table is used to link the tables that originally were related in a M:N relationship, the composite entity structure includes—as foreign keys—*at least* the primary keys of the tables that are to be linked. The database designer has two main options when defining a composite table's primary key: use the combination of those foreign keys or create a new primary key.

Remember that each entity in the ERM is represented by a table. Therefore, you can create the composite ENROLL table shown in Figure 3.26 to link the tables CLASS and STUDENT. In this example, the ENROLL table's primary key is the combination of its foreign keys CLASS_CODE and STU_NUM. But the designer could have decided to create a single-attribute new primary key such as ENROLL_LINE, using a different line value to identify each ENROLL table row uniquely. (Microsoft Access users might use the *Autonumber* data type to generate such line values automatically.)

FIGURE 3.26	Converting the M:N relationship into two 1:M relationships

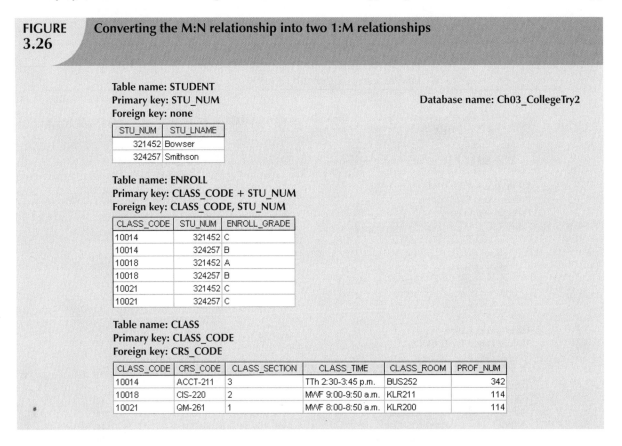

Table name: STUDENT
Primary key: STU_NUM
Foreign key: none

Database name: Ch03_CollegeTry2

STU_NUM	STU_LNAME
321452	Bowser
324257	Smithson

Table name: ENROLL
Primary key: CLASS_CODE + STU_NUM
Foreign key: CLASS_CODE, STU_NUM

CLASS_CODE	STU_NUM	ENROLL_GRADE
10014	321452	C
10014	324257	B
10018	321452	A
10018	324257	B
10021	321452	C
10021	324257	C

Table name: CLASS
Primary key: CLASS_CODE
Foreign key: CRS_CODE

CLASS_CODE	CRS_CODE	CLASS_SECTION	CLASS_TIME	CLASS_ROOM	PROF_NUM
10014	ACCT-211	3	TTh 2:30-3:45 p.m.	BUS252	342
10018	CIS-220	2	MWF 9:00-9:50 a.m.	KLR211	114
10021	QM-261	1	MWF 8:00-8:50 a.m.	KLR200	114

Because the ENROLL table in Figure 3.26 links two tables, STUDENT and CLASS, it is also called a **linking table**. In other words, a linking table is the implementation of a composite entity.

> **NOTE**
>
> In addition to the linking attributes, the composite ENROLL table can also contain such relevant attributes as the grade earned in the course. In fact, a composite table can contain any number of attributes that the designer wants to track. Keep in mind that the composite entity, *although it is implemented as an actual table*, is *conceptually* a logical entity that was created as a means to an end: to eliminate the potential for multiple redundancies in the original M:N relationship.

The linking table (ENROLL) shown in Figure 3.26 yields the required M:N to 1:M conversion. Observe that the composite entity represented by the ENROLL table must contain at least the primary keys of the CLASS and

STUDENT tables (CLASS_CODE and STU_NUM, respectively) for which it serves as a connector. Also note that the STUDENT and CLASS tables now contain only one row per entity. The linking ENROLL table contains multiple occurrences of the foreign key values, but those controlled redundancies are incapable of producing anomalies as long as referential integrity is enforced. Additional attributes may be assigned as needed. In this case, ENROLL_GRADE is selected to satisfy a reporting requirement. Also note that the ENROLL table's primary key consists of the two attributes CLASS_CODE and STU_NUM because both the class code and the student number are needed to define a particular student's grade. Naturally, the conversion is reflected in the ERM, too. The revised relationship is shown in Figure 3.27.

As you examine Figure 3.27, note that the composite entity named ENROLL represents the linking table between STUDENT and CLASS.

The 1:M relationship between COURSE and CLASS was first illustrated in Figure 3.20 and Figure 3.21. With the help of this relationship, you can increase the amount of available information even as you control the database's redundancies. Thus, Figure 3.27 can be expanded to include the 1:M relationship between COURSE and CLASS shown in Figure 3.28. Note that the model is able to handle multiple sections of a CLASS while controlling redundancies by making sure that all of the COURSE data common to each CLASS are kept in the COURSE table.

The relational diagram that corresponds to the ERD in Figure 3.28 is shown in Figure 3.29.

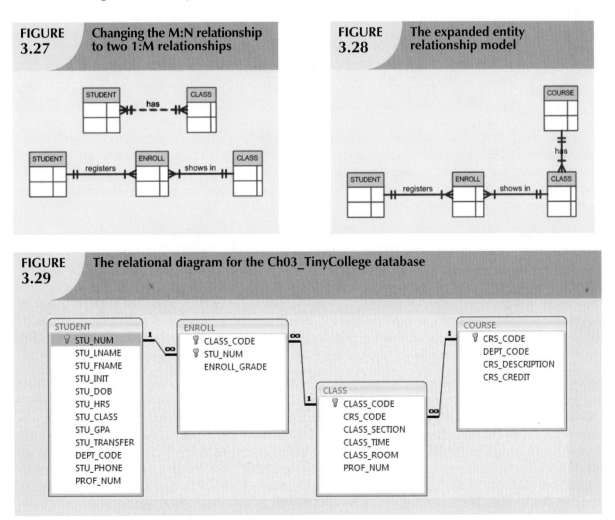

FIGURE 3.27 Changing the M:N relationship to two 1:M relationships

FIGURE 3.28 The expanded entity relationship model

FIGURE 3.29 The relational diagram for the Ch03_TinyCollege database

The ERD will be examined in greater detail in Chapter 4 to show you how it is used to design more complex databases. The ERD will also be used as the basis for the development and implementation of a realistic database design in Appendixes B and C (see the Student Online Companion Web site) for a university computer lab.

3.7 DATA REDUNDANCY REVISITED

In Chapter 1 you learned that data redundancy leads to data anomalies. Those anomalies can destroy the effectiveness of the database. You also learned that the relational database makes it possible to control data redundancies by using common attributes that are shared by tables, called foreign keys.

The proper use of foreign keys is crucial to controlling data redundancy. Although the use of foreign keys does not totally eliminate data redundancies because the foreign key values can be repeated many times, the proper use of foreign keys *minimizes* data redundancies, thus minimizing the chance that destructive data anomalies will develop.

> **NOTE**
>
> The real test of redundancy is *not* how many copies of a given attribute are stored, *but whether the elimination of an attribute will eliminate information*. Therefore, if you delete an attribute and the original information can still be generated through relational algebra, the inclusion of that attribute would be redundant. Given that view of redundancy, proper foreign keys are clearly not redundant in spite of their multiple occurrences in a table. However, even when you use this less restrictive view of redundancy, keep in mind that *controlled* redundancies are often designed as part of the system to ensure transaction speed and/or information requirements. Exclusive reliance on relational algebra to produce required information may lead to elegant designs that fail the test of practicality.

You will learn in Chapter 4 that database designers must reconcile three often contradictory requirements: design elegance, processing speed, and information requirements. And you will learn in Chapter 13, Business Intelligence and Data Warehouses, that proper data warehousing design requires carefully defined and controlled data redundancies to function properly. Regardless of how you describe data redundancies, the potential for damage is limited by proper implementation and careful control.

As important as data redundancy control is, there are times when the level of data redundancy must actually be increased to make the database serve crucial information purposes. You will learn about such redundancies in Chapter 13. There are also times when data redundancies *seem* to exist to preserve the historical accuracy of the data. For example, consider a small invoicing system. The system includes the CUSTOMER, who may buy one or more PRODUCTs, thus generating an INVOICE. Because a customer may buy more than one product at a time, an invoice may contain several invoice LINEs, each providing details about the purchased product. The PRODUCT table should contain the product price to provide a consistent pricing input for each product that appears on the invoice. The tables that are part of such a system are shown in Figure 3.30. The system's relational diagram is shown in Figure 3.31.

As you examine the tables in the invoicing system in Figure 3.30 and the relationships depicted in Figure 3.31, note that you can keep track of typical sales information. For example, by tracing the relationships among the four tables, you discover that customer 10014 (Myron Orlando) bought two items on March 8, 2006 that were written to invoice number 1001: one Houselite chain saw with a 16-inch bar and three rat-tail files. (*Note*: Trace the CUS_CODE number 10014 in the CUSTOMER table to the matching CUS_CODE value in the INVOICE table. Next, take the INV_NUMBER 1001 and trace it to the first two rows in the LINE table. Finally, match the two PROD_CODE values in LINE with the PROD_CODE values in PRODUCT.) Application software will be used to write the correct bill by multiplying each invoice line item's LINE_UNITS by its LINE_PRICE, adding the results, applying appropriate taxes, etc. Later, other application software might use the same technique to write sales reports that track and compare sales by week, month, or year.

FIGURE 3.30   A small invoicing system

Table name: **CUSTOMER**
Primary key: **CUS_CODE**
Foreign key: **none**

Database name: Ch03_SaleCo

CUS_CODE	CUS_LNAME	CUS_FNAME	CUS_INITIAL	CUS_AREACODE	CUS_PHONE
10010	Ramas	Alfred	A	615	844-2573
10011	Dunne	Leona	K	713	894-1238
10012	Smith	Kathy	W	615	894-2285
10013	Olowski	Paul	F	615	894-2180
10014	Orlando	Myron		615	222-1672
10015	O'Brian	Amy	B	713	442-3381
10016	Brown	James	G	615	297-1228
10017	Williams	George		615	290-2556
10018	Farriss	Anne	G	713	382-7185
10019	Smith	Olette	K	615	297-3809

Table name: **INVOICE**
Primary key: **INV_NUMBER**
Foreign key: **CUS_CODE**

INV_NUMBER	CUS_CODE	INV_DATE
1001	10014	08-Mar-08
1002	10011	08-Mar-08
1003	10012	08-Mar-08
1004	10011	09-Mar-08

Table name: **LINE**
Primary key: **INV_NUMBER + LINE_NUMBER**
Foreign keys: **INV_NUMBER, PROD_CODE**

INV_NUMBER	LINE_NUMBER	PROD_CODE	LINE_UNITS	LINE_PRICE
1001	1	123-21UUY	1	189.99
1001	2	SRE-657UG	3	2.99
1002	1	QER-34256	2	18.63
1003	1	ZZX/3245Q	1	6.79
1003	2	SRE-657UG	1	2.99
1003	3	001278-AB	1	12.95
1004	1	001278-AB	1	12.95
1004	2	SRE-657UG	2	2.99

Table name: **PRODUCT**
Primary key: **PROD_CODE**
Foreign key: **none**

PROD_CODE	PROD_DESCRIPT	PROD_PRICE	PROD_ON_HAND	VEND_CODE
001278-AB	Claw hammer	12.95	23	232
123-21UUY	Houselite chain saw, 16-in. bar	189.99	4	235
QER-34256	Sledge hammer, 16-lb. head	18.63	6	231
SRE-657UG	Rat-tail file	2.99	15	232
ZZX/3245Q	Steel tape, 12-ft. length	6.79	8	235

FIGURE 3.31   The relational diagram for the invoicing system

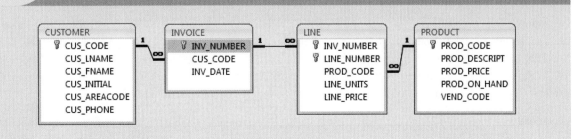

As you examine the sales transactions in Figure 3.30, you might reasonably suppose that the product price billed to the customer is derived from the PRODUCT table because that's where the product data are stored. *But why does that same product price occur again in the LINE table? Isn't that a data redundancy?* It certainly *appears* to be. But this time, the apparent redundancy is crucial to the system's success. Copying the product price from the PRODUCT table to the LINE table maintains the *historical accuracy of the transactions*. Suppose, for instance, that you fail to write the LINE_PRICE in the LINE table and that you use the PROD_PRICE from the PRODUCT table to calculate the sales revenue. Now suppose that the PRODUCT table's PROD_PRICE changes, as prices frequently do. This price change will be properly reflected in all subsequent sales revenue calculations. However, the calculations of past sales revenues will also reflect the new product price that was not in effect when the transaction took place! As a result, the revenue calculations for all past transactions will be incorrect, thus eliminating the possibility of making proper sales comparisons over time. On the other hand, if the price data are copied from the PRODUCT table and stored with the transaction in the LINE table, that price will always accurately reflect the transaction that took place *at that time*. You will discover that such planned "redundancies" are common in good database design.

Finally, you might wonder why the LINE_NUMBER attribute was used in the LINE table in Figure 3.30. Wouldn't the combination of INV_NUMBER and PROD_CODE be a sufficient composite primary key—and, therefore, isn't the LINE_NUMBER redundant? Yes, the LINE_NUMBER is redundant, but this redundancy is quite commonly created by invoicing software that generates such line numbers automatically. In this case, the redundancy is not necessary. But given its automatic generation, the redundancy is not a source of anomalies. The inclusion of LINE_NUMBER also adds another benefit: the order of the retrieved invoicing data will always match the order in which the data were entered. If product codes are used as part of the primary key, indexing will arrange those product codes as soon as the invoice is completed and the data are stored. You can imagine the potential confusion when a customer calls and says, "The second item on my invoice has an incorrect price" and you are looking at an invoice whose lines show a different order from those on the customer's copy!

3.8 INDEXES

Suppose you want to locate a particular book in a library. Does it make sense to look through *every* book in the library until you find the one you want? Of course not; you use the library's catalog, which is indexed by title, topic, and author. The index (in either a manual or a computer system) points you to the book's location, thereby making retrieval of the book a quick and simple matter. An **index** is an orderly arrangement used to logically access rows in a table.

Or suppose you want to find a topic, such as "ER model," in this book. Does it make sense to read through *every* page until you stumble across the topic? Of course not; it is much simpler to go to the book's index, look up the phrase *ER model*, and read the page references that point you to the appropriate page(s). In each case, an index is used to locate a needed item quickly.

Indexes in the relational database environment work like the indexes described in the preceding paragraphs. From a conceptual point of view, an index is composed of an index key and a set of pointers. The **index key** is, in effect, the index's reference point. More formally, an index is an ordered arrangement of keys and pointers. Each key points to the location of the data identified by the key.

For example, suppose you want to look up all of the paintings created by a given painter in the Ch03_Museum database in Figure 3.19. Without an index, you must read each row in the PAINTING table and see if the PAINTER_NUM matches the requested painter. However, if you index the PAINTER table and use the index key PAINTER_NUM, you merely need to look up the appropriate PAINTER_NUM in the index and find the matching pointers. Conceptually speaking, the index would resemble the presentation depicted in Figure 3.32.

As you examine Figure 3.32 and compare it to the Ch03_Museum database tables shown in Figure 3.19, note that the first PAINTER_NUM index key value (123) is found in records 1, 2, and 4 of the PAINTING table in Figure 3.19. The second PAINTER_NUM index key value (126) is found in records 3 and 5 of the PAINTING table in Figure 3.19.

DBMSs use indexes for many different purposes. You just learned that an index can be used to retrieve data more efficiently. But indexes can also be used by a DBMS to retrieve data ordered by a specific attribute or attributes. For example, creating an index on a customer's last name will allow you to retrieve the customer data alphabetically by the

FIGURE 3.32   Components of an index

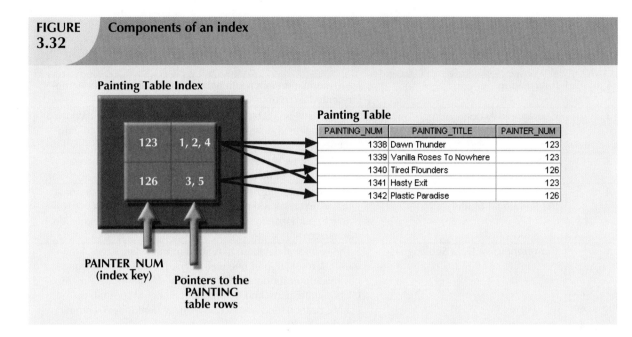

Painting Table Index

PAINTER_NUM (index key)

Pointers to the PAINTING table rows

Painting Table

PAINTING_NUM	PAINTING_TITLE	PAINTER_NUM
1338	Dawn Thunder	123
1339	Vanilla Roses To Nowhere	123
1340	Tired Flounders	126
1341	Hasty Exit	123
1342	Plastic Paradise	126

customer's last name. Also, an index key can be composed of one or more attributes. For example, in Figure 3.30, you can create an index on VEND_CODE and PROD_CODE to retrieve all rows in the PRODUCT table ordered by vendor, and within vendor, ordered by product.

Indexes play an important role in DBMSs for the implementation of primary keys. When you define a table's primary key, the DBMS automatically creates a unique index on the primary key column(s) you declared. For example, in Figure 3.30, when you declare CUS_CODE to be the primary key of the CUSTOMER table, the DBMS automatically creates a unique index on that attribute. A **unique index**, as its name implies, is an index in which the index key can have only one pointer value (row) associated with it. (The index in Figure 3.32 is not a unique index because the PAINTER_NUM has multiple pointer values associated with it. For example, painter number 123 points to three rows—1, 2, and 4—in the PAINTING table.)

A table can have many indexes, but each index is associated with only one table. The index key can have multiple attributes (composite index). Creating an index is easy. You learn in Chapter 7 that a simple SQL command produces any required index.

3.9 CODD'S RELATIONAL DATABASE RULES

In 1985, Dr. E. F. Codd published a list of 12 rules to define a relational database system.[2] The reason Dr. Codd published the list was his concern that many vendors were marketing products as "relational" even though those products did not meet minimum relational standards. Dr. Codd's list, shown in Table 3.8, serves as a frame of reference for what a truly relational database should be. Bear in mind that even the dominant database vendors do not fully support all 12 rules.

[2] Codd, E., "Is Your DBMS Really Relational?" and "Does Your DBMS Run by the Rules?" *Computerworld*, October 14 and October 21, 1985.

TABLE 3.8   Dr. Codd's 12 Relational Database Rules

RULE	RULE NAME	DESCRIPTION
1	Information	All information in a relational database must be logically represented as column values in rows within tables.
2	Guaranteed Access	Every value in a table is guaranteed to be accessible through a combination of table name, primary key value, and column name.
3	Systematic Treatment of Nulls	Nulls must be represented and treated in a systematic way, independent of data type.
4	Dynamic On-Line Catalog Based on the Relational Model	The metadata must be stored and managed as ordinary data, that is, in tables within the database. Such data must be available to authorized users using the standard database relational language.
5	Comprehensive Data Sublanguage	The relational database may support many languages. However, it must support one well defined, declarative language with support for data definition, view definition, data manipulation (interactive and by program), integrity constraints, authorization, and transaction management (begin, commit, and rollback).
6	View Updating	Any view that is theoretically updatable must be updatable through the system.
7	High-Level Insert, Update and Delete	The database must support set-level inserts, updates, and deletes.
8	Physical Data Independence	Application programs and ad hoc facilities are logically unaffected when physical access methods or storage structures are changed.
9	Logical Data Independence	Application programs and ad hoc facilities are logically unaffected when changes are made to the table structures that preserve the original table values (changing order of column or inserting columns).
10	Integrity Independence	All relational integrity constraints must be definable in the relational language and stored in the system catalog, not at the application level.
11	Distribution Independence	The end users and application programs are unaware and unaffected by the data location (distributed vs. local databases).
12	Nonsubversion	If the system supports low-level access to the data, there must not be a way to bypass the integrity rules of the database.
	Rule Zero	All preceding rules are based on the notion that in order for a database to be considered relational, it must use its relational facilities exclusively to manage the database.

SUMMARY

▶ Tables are the basic building blocks of a relational database. A grouping of related entities, known as an entity set, is stored in a table. Conceptually speaking, the relational table is composed of intersecting rows (tuples) and columns. Each row represents a single entity, and each column represents the characteristics (attributes) of the entities.

▶ Keys are central to the use of relational tables. Keys define functional dependencies; that is, other attributes are dependent on the key and can, therefore, be found if the key value is known. A key can be classified as a superkey, a candidate key, a primary key, a secondary key, or a foreign key.

▶ Each table row must have a primary key. The primary key is an attribute or a combination of attributes that uniquely identifies all remaining attributes found in any given row. Because a primary key must be unique, no null values are allowed if entity integrity is to be maintained.

▶ Although the tables are independent, they can be linked by common attributes. Thus, the primary key of one table can appear as the foreign key in another table to which it is linked. Referential integrity dictates that the foreign key must contain values that match the primary key in the related table or must contain nulls.

▶ The relational model supports relational algebra functions: SELECT, PROJECT, JOIN, INTERSECT, UNION, DIFFERENCE, PRODUCT, and DIVIDE. A relational database performs much of the data manipulation work behind the scenes. For example, when you create a database, the RDBMS automatically produces a structure to house a data dictionary for your database. Each time you create a new table within the database, the RDBMS updates the data dictionary, thereby providing the database documentation.

▶ Once you know the relational database basics, you can concentrate on design. Good design begins by identifying appropriate entities and their attributes and then the relationships among the entities. Those relationships (1:1, 1:M, and M:N) can be represented using ERDs. The use of ERDs allows you to create and evaluate simple logical design. The 1:M relationship is most easily incorporated in a good design; you just have to make sure that the primary key of the "1" is included in the table of the "many."

KEY TERMS

associative entity, 86	full functional dependence, 68	primary key (PK), 66
attribute domain, 32	functional dependence, 67	referential integrity, 70
bridge entity, 86	homonyms, 80	relational algebra, 72
candidate key, 68	index, 90	relational schema, 70
closure, 73	index key, 90	right outer join, 77
composite entity, 86	join column(s), 76	secondary key, 70
composite key, 67	key, 66	set theory, 63
data dictionary, 78	key attribute, 67	superkey, 68
determination, 66	left outer join, 77	synonym, 80
domain, 66	linking table, 86	system catalog, 80
entity integrity, 68	natural join, 76	theta join, 77
equijoin, 77	null, 68	tuple, 38
flags, 72	outer join, 77	union-compatible, 73
foreign key (FK), 70	predicate logic, 63	unique index, 91

ONLINE CONTENT

Answers to selected Review Questions and Problems for this chapter are contained in the Student Online Companion for this book.

REVIEW QUESTIONS

1. What is the difference between a database and a table?

2. What does it mean to say that a database displays both entity integrity and referential integrity?

3. Why are entity integrity and referential integrity important in a database?

4. A database user manually notes that "The file contains two hundred records, each record containing nine fields." Use appropriate relational database terminology to "translate" that statement.

5. Use the small database shown in Figure Q3.5 to illustrate the difference between a natural join, an equijoin, and an outer join.

FIGURE Q3.5	The Ch03_CollegeQue database tables

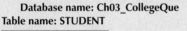

Database name: Ch03_CollegeQue
Table name: STUDENT

STU_CODE	PROF_CODE
100278	
128569	2
512272	4
531235	2
531268	
553427	1

Table name: PROFESSOR

		PROF_CODE	DEPT_CODE
▶	+	1	2
	+	2	6
	+	3	6
	+	4	4

6. Create the basic ERD for the database shown in Figure Q3.5.

7. Create the relational diagram for the database shown in Figure Q3.5.

8. Suppose you have the ERM shown in Figure Q3.8. How would you convert this model into an ERM that displays only 1:M relationships? (Make sure you create the revised ERM.)

9. What are homonyms and synonyms, and why should they be avoided in database design?

10. How would you implement a 1:M relationship in a database composed of two tables? Give an example.

11. Identify and describe the components of the table shown in Figure Q3.11, using correct terminology. Use your knowledge of naming conventions to identify the table's probable foreign key(s).

ONLINE CONTENT

All of the databases used in the questions and problems are found in the Student Online Companion for this book. The database names used in the folder match the database names used in the figures. For example, the source of the tables shown in Figure Q3.5 is the **Ch03_CollegeQue** database.

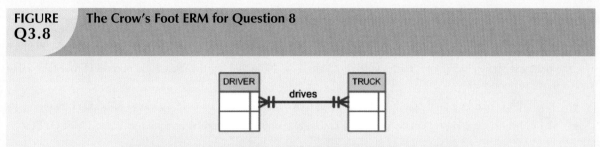

FIGURE Q3.8 The Crow's Foot ERM for Question 8

During some time interval, a DRIVER can drive many TRUCKs and any TRUCK can be driven by many DRIVERs.

FIGURE Q3.11 The Ch03_NoComp database EMPLOYEE table

Table name: EMPLOYEE   Database name: Ch03_NoComp

EMP_NUM	EMP_LNAME	EMP_INITIAL	EMP_FNAME	DEPT_CODE	JOB_CODE
11234	Friedman	K	Robert	MKTG	12
11238	Olanski	D	Delbert	MKTG	12
11241	Fontein		Juliette	INFS	5
11242	Cruazona	J	Maria	ENG	9
11245	Smithson	B	Bernard	INFS	6
11248	Washington	G	Oleta	ENGR	8
11256	McBride		Randall	ENGR	8
11257	Kachinn	D	Melanie	MKTG	14
11258	Smith	W	William	MKTG	14
11260	Ratula	A	Katrina	INFS	5

Use the database composed of the two tables shown in Figure Q3.12 to answer Questions 12-17.

FIGURE Q3.12 The Ch03_Theater database tables

Database name: Ch03_Theater

Table name: DIRECTOR

DIR_NUM	DIR_LNAME	DIR_DOB
100	Broadway	12-Jan-65
101	Hollywoody	18-Nov-53
102	Goofy	21-Jun-62

Table name: PLAY

PLAY_CODE	PLAY_NAME	DIR_NUM
1001	Cat On a Cold, Bare Roof	102
1002	Hold the Mayo, Pass the Bread	101
1003	I Never Promised You Coffee	102
1004	Silly Putty Goes To Washington	100
1005	See No Sound, Hear No Sight	101
1006	Starstruck in Biloxi	102
1007	Stranger In Parrot Ice	101

12. Identify the primary keys.

13. Identify the foreign keys.

14. Create the ERM.

15. Create the relational diagram to show the relationship between DIRECTOR and PLAY.

16. Suppose you wanted quick lookup capability to get a listing of all plays directed by a given director. Which table would be the basis for the INDEX table, and what would be the index key?

17. What would be the conceptual view of the INDEX table that is described in Question 16? Depict the contents of the conceptual INDEX table.

P R O B L E M S

Use the database shown in Figure P3.1 to work Problems 1–7. Note that the database is composed of four tables that reflect these relationships:

- An EMPLOYEE has only one JOB_CODE, but a JOB_CODE can be held by many EMPLOYEEs.
- An EMPLOYEE can participate in many PLANs, and any PLAN can be assigned to many EMPLOYEEs.

Note also that the M:N relationship has been broken down into two 1:M relationships for which the BENEFIT table serves as the composite or bridge entity.

FIGURE P3.1   **The Ch03_BeneCo database tables**

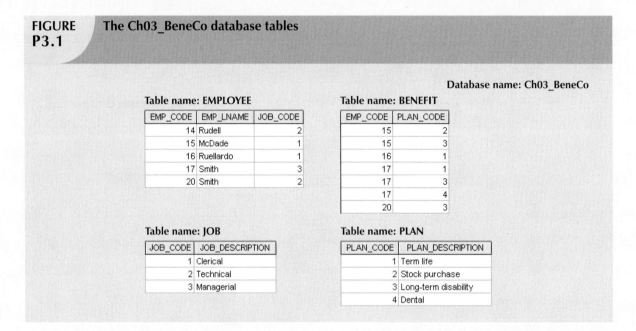

Database name: Ch03_BeneCo

Table name: EMPLOYEE

EMP_CODE	EMP_LNAME	JOB_CODE
14	Rudell	2
15	McDade	1
16	Ruellardo	1
17	Smith	3
20	Smith	2

Table name: BENEFIT

EMP_CODE	PLAN_CODE
15	2
15	3
16	1
17	1
17	3
17	4
20	3

Table name: JOB

JOB_CODE	JOB_DESCRIPTION
1	Clerical
2	Technical
3	Managerial

Table name: PLAN

PLAN_CODE	PLAN_DESCRIPTION
1	Term life
2	Stock purchase
3	Long-term disability
4	Dental

1. For each table in the database, identify the primary key and the foreign key(s). If a table does not have a foreign key, write *None* in the space provided.

TABLE	PRIMARY KEY	FOREIGN KEY(S)
EMPLOYEE		
BENEFIT		
JOB		
PLAN		

2. Create the ERD to show the relationship between EMPLOYEE and JOB.
3. Create the relational diagram to show the relationship between EMPLOYEE and JOB.
4. Do the tables exhibit entity integrity? Answer yes or no, and then explain your answer.

TABLE	ENTITY INTEGRITY	EXPLANATION
EMPLOYEE		
BENEFIT		
JOB		
PLAN		

5. Do the tables exhibit referential integrity? Answer yes or no, and then explain your answer. Write *NA* (Not Applicable) if the table does not have a foreign key.

TABLE	REFERENTIAL INTEGRITY	EXPLANATION
EMPLOYEE		
BENEFIT		
JOB		
PLAN		

6. Create the ERD to show the relationships among EMPLOYEE, BENEFIT, JOB, and PLAN.

7. Create the relational diagram to show the relationships among EMPLOYEE, BENEFIT, JOB, and PLAN.

Use the database shown in Figure P3.8 to answer Problems 8–16.

FIGURE P3.8   **The Ch03_StoreCo database tables**

Table name: EMPLOYEE   Database name: Ch03_StoreCo

EMP_CODE	EMP_TITLE	EMP_LNAME	EMP_FNAME	EMP_INITIAL	EMP_DOB	STORE_CODE
1	Mr.	Williamson	John	W	21-May-64	3
2	Ms.	Ratula	Nancy		09-Feb-69	2
3	Ms.	Greenboro	Lottie	R	02-Oct-61	4
4	Mrs.	Rumpersfro	Jennie	S	01-Jun-71	5
5	Mr.	Smith	Robert	L	23-Nov-59	3
6	Mr.	Renselaer	Cary	A	25-Dec-65	1
7	Mr.	Ogallo	Roberto	S	31-Jul-62	3
8	Ms.	Johnsson	Elizabeth	I	10-Sep-68	1
9	Mr.	Eindsmar	Jack	W	19-Apr-55	2
10	Mrs.	Jones	Rose	R	06-Mar-66	4
11	Mr.	Broderick	Tom		21-Oct-72	3
12	Mr.	Washington	Alan	Y	08-Sep-74	2
13	Mr.	Smith	Peter	N	25-Aug-64	3
14	Ms.	Smith	Sherry	H	25-May-66	4
15	Mr.	Olenko	Howard	U	24-May-64	5
16	Mr.	Archialo	Barry	V	03-Sep-60	5
17	Ms.	Grimaldo	Jeanine	K	12-Nov-70	4
18	Mr.	Rosenberg	Andrew	D	24-Jan-71	4
19	Mr.	Rosten	Peter	F	03-Oct-68	4
20	Mr.	Mckee	Robert	S	06-Mar-70	1
21	Ms.	Baumann	Jennifer	A	11-Dec-74	3

Table name: STORE

STORE_CODE	STORE_NAME	STORE_YTD_SALES	REGION_CODE	EMP_CODE
1	Access Junction	1003455.76	2	8
2	Database Corner	1421987.39	2	12
3	Tuple Charge	986783.22	1	7
4	Attribute Alley	944568.56	2	3
5	Primary Key Point	2930098.45	1	15

Table name: REGION

REGION_CODE	REGION_DESCRIPT
1	East
2	West

8. For each table, identify the primary key and the foreign key(s). If a table does not have a foreign key, write *None* in the space provided.

TABLE	PRIMARY KEY	FOREIGN KEY(S)
EMPLOYEE		
STORE		
REGION		

9. Do the tables exhibit entity integrity? Answer yes or no, and then explain your answer.

TABLE	ENTITY INTEGRITY	EXPLANATION
EMPLOYEE		
STORE		
REGION		

10. Do the tables exhibit referential integrity? Answer yes or no, and then explain your answer. Write *NA* (Not Applicable) if the table does not have a foreign key.

TABLE	REFERENTIAL INTEGRITY	EXPLANATION
EMPLOYEE		
STORE		
REGION		

11. Describe the type(s) of relationship(s) between STORE and REGION.

12. Create the ERD to show the relationship between STORE and REGION.

13. Create the relational diagram to show the relationship between STORE and REGION.

14. Describe the type(s) of relationship(s) between EMPLOYEE and STORE. (*Hint:* Each store employs many employees, one of whom manages the store.)

15. Create the ERD to show the relationships among EMPLOYEE, STORE, and REGION.

16. Create the relational diagram to show the relationships among EMPLOYEE, STORE, and REGION.

Use the database shown in Figure P3.17 to answer Problems 17–22.

FIGURE P3.17 The Ch03_CheapCo database tables

Table name: **PRODUCT**
Primary key: **PROD_CODE**
Foreign key: **VEND_CODE**

Database name: Ch03_CheapCo

PROD_CODE	PROD_DESCRIPTION	PROD_STOCK_DATE	PROD_ON_HAND	PROD_PRICE	VEND_CODE
12-WW/P2	7.25-in. power saw blade	07-Apr-08	12	1.19	123
1QQ23-55	2.5-in. wood screw, 100	19-Mar-08	123	4.49	123
231-78-W	PVC pipe, 3.5-in., 8 ft.	07-Dec-07	45	8.87	121
33564/U	Rat-tail file, 0.125-in., fine	08-Mar-08	18	1.19	123
AR/3/TYR	Cordless drill, 0.25-in.	29-Nov-07	8	45.99	121
DT-34-WW	Phillips screwdriver pack	20-Dec-07	11	23.29	123
EE3-67/WV	Sledge hammer, 12 lb.	25-Feb-08	9	17.99	121
ER-56/DF	Houselite chain saw, 16-in.	28-Dec-07	7	235.49	125
FRE-TRY9	Jigsaw, 12-in blade	12-Aug-07	67	1.45	125
SE-67-89	Jigsaw, 8-in. blade	11-Oct-07	34	1.35	125
ZW-QR/AV	Hardware cloth, 0.25-in.	23-Apr-08	14	12.99	123
ZX-WR/FR	Claw hammer	01-Mar-08	15	8.95	121

Table name: **VENDOR**
Primary key: **VEND_CODE**
Foreign key: **none**

VEND_CODE	VEND_NAME	VEND_CONTACT	VEND_AREACODE	VEND_PHONE
120	BargainSnapper, Inc.	Melanie T. Travis	615	899-1234
121	Cut'nGlow Co.	Henry J. Olero	615	342-9896
122	Rip & Rattle Supply Co.	Anne R. Morrins	901	225-1127
123	Tools 'R Us	Juliette G. McHenry	615	546-7894
124	Trowel & Dowel, Inc.	George F. Frederick	901	453-4567
125	Bow & Wow Tools	Bill S. Sedwick	904	324-9988

17. For each table, identify the primary key and the foreign key(s). If a table does not have a foreign key, write *None* in the space provided.

TABLE	PRIMARY KEY	FOREIGN KEY(S)
PRODUCT		
VENDOR		

18. Do the tables exhibit entity integrity? Answer yes or no, and then explain your answer.

TABLE	ENTITY INTEGRITY	EXPLANATION
PRODUCT		
VENDOR		

19. Do the tables exhibit referential integrity? Answer yes or no, and then explain your answer. Write *NA* (Not Applicable) if the table does not have a foreign key.

TABLE	REFERENTIAL INTEGRITY	EXPLANATION
PRODUCT		
VENDOR		

20. Create the ERD for this database.

21. Create the relational diagram for this database.

22. Create the data dictionary for this database.

Use the database shown in Figure P3.23 to answer Problems 23–29.

FIGURE P3.23   The Ch03_TransCo database tables

Table name: **TRUCK**
Primary key: **TRUCK_NUM**
Foreign key: **BASE_CODE, TYPE_CODE**

Database name: Ch03_TransCo

TRUCK_NUM	BASE_CODE	TYPE_CODE	TRUCK_MILES	TRUCK_BUY_DATE	TRUCK_SERIAL_NUM
1001	501	1	32123.5	23-Sep-07	AA-322-12212-W11
1002	502	1	76984.3	05-Feb-06	AC-342-22134-Q23
1003	501	2	12346.6	11-Nov-06	AC-445-78656-Z99
1004		1	2894.3	06-Jan-07	WQ-112-23144-T34
1005	503	2	45673.1	01-Mar-06	FR-998-32245-W12
1006	501	2	193245.7	15-Jul-03	AD-456-00845-R45
1007	502	3	32012.3	17-Oct-04	AA-341-96573-Z84
1008	502	3	44213.6	07-Aug-05	DR-559-22189-D33
1009	503	2	10932.9	12-Feb-08	DE-887-98456-E94

Table name: **BASE**
Primary key: **BASE_CODE**
Foreign key: **none**

BASE_CODE	BASE_CITY	BASE_STATE	BASE_AREA_CODE	BASE_PHONE	BASE_MANAGER
501	Murfreesboro	TN	615	123-4567	Andrea D. Gallager
502	Lexington	KY	568	234-5678	George H. Delarosa
503	Cape Girardeau	MO	456	345-6789	Maria J. Talindo
504	Dalton	GA	901	456-7890	Peter F. McAvee

Table name: **TYPE**
Primary key: **TYPE_CODE**
Foreign key: **none**

TYPE_CODE	TYPE_DESCRIPTION
1	Single box, double-axle
2	Single box, single-axle
3	Tandem trailer, single-axle

23. For each table, identify the primary key and the foreign key(s). If a table does not have a foreign key, write *None* in the space provided.

TABLE	PRIMARY KEY	FOREIGN KEY(S)
TRUCK		
BASE		
TYPE		

24. Do the tables exhibit entity integrity? Answer yes or no, and then explain your answer.

TABLE	ENTITY INTEGRITY	EXPLANATION
TRUCK		
BASE		
TYPE		

25. Do the tables exhibit referential integrity? Answer yes or no, and then explain your answer. Write *NA* (Not Applicable) if the table does not have a foreign key.

TABLE	REFERENTIAL INTEGRITY	EXPLANATION
TRUCK		
BASE		
TYPE		

26. Identify the TRUCK table's candidate key(s).

27. For each table, identify a superkey and a secondary key.

TABLE	SUPERKEY	SECONDARY KEY
TRUCK		
BASE		
TYPE		

28. Create the ERD for this database.

29. Create the relational diagram for this database.

Use the database shown in Figure P3.30 to answer Problems 30–34. ROBCOR is an aircraft charter company that supplies on-demand charter flight services using a fleet of four aircraft. Aircrafts are identified by a unique registration number. Therefore, the aircraft registration number is an appropriate primary key for the AIRCRAFT table.

FIGURE P3.30   **The Ch03_AviaCo database tables**

Table name: CHARTER   Database name: Ch03_AviaCo

CHAR_TRIP	CHAR_DATE	CHAR_PILOT	CHAR_COPILOT	CHAR_DESTINATION	CHAR_DISTANCE	CHAR_HOURS_FLOWN	CHAR_HOURS_WAIT	CUS_CODE
10001	05-Feb-08	104		ATL	936.0	5.1	2.2	10011
10002	05-Feb-08	101		BNA	320.0	1.6	0.0	10016
10003	05-Feb-08	105	109	GNV	1574.0	7.8	0.0	10014
10004	06-Feb-08	106		STL	472.0	2.9	4.9	10019
10005	06-Feb-08	101		ATL	1023.0	5.7	3.5	10011
10006	06-Feb-08	109		STL	472.0	2.6	5.2	10017
10007	06-Feb-08	104	105	GNV	1574.0	7.9	0.0	10012
10008	07-Feb-08	106		TYS	644.0	4.1	0.0	10014
10009	07-Feb-08	105		GNV	1574.0	6.6	23.4	10017
10010	07-Feb-08	109		ATL	998.0	6.2	3.2	10016
10011	07-Feb-08	101	104	BNA	352.0	1.9	5.3	10012
10012	08-Feb-08	101		MOB	884.0	4.8	4.2	10010
10013	08-Feb-08	105		TYS	644.0	3.9	4.5	10011
10014	09-Feb-08	106		ATL	936.0	6.1	2.1	10017
10015	09-Feb-08	104	101	GNV	1645.0	6.7	0.0	10016
10016	09-Feb-08	109	105	MQY	312.0	1.5	0.0	10011
10017	10-Feb-08	101		STL	508.0	3.1	0.0	10014
10018	10-Feb-08	105	104	TYS	644.0	3.8	4.5	10017

The destinations are indicated by standard three-letter airport codes. For example,
STL = St. Louis, MO   ATL = Atlanta, GA   BNA = Nashville, TN

Table name: AIRCRAFT

AC_NUMBER	MOD_CODE	AC_TTAF	AC_TTEL	AC_TTER
1484P	PA23-250	1833.1	1833.1	101.8
2289L	C-90A	4243.8	768.9	1123.4
2778V	PA31-350	7992.9	1513.1	789.5
4278Y	PA31-350	2147.3	622.1	243.2

AC-TTAF = Aircraft total time, airframe (hours)
AC-TTEL = Total time, left engine (hours)
AC_TTER = Total time, right engine (hours)

In a fully developed system, such attribute values would be updated by application software when the CHARTER table entries are posted.

Table name: MODEL

MOD_CODE	MOD_MANUFACTURER	MOD_NAME	MOD_SEATS	MOD_CHG_MILE
C-90A	Beechcraft	KingAir	8	2.67
PA23-250	Piper	Aztec	6	1.93
PA31-350	Piper	Navajo Chieftain	10	2.35

Customers are charged per round-trip mile, using the MOD_CHG_MILE rate. The MOD_SEAT gives the total number of seats in the airplane, including the pilot and copilot seats. Therefore, a PA31-350 trip that is flown by a pilot and a copilot has six passenger seats available.

FIGURE P3.30   **The Ch03_AviaCo database tables (continued)**

Table name: PILOT   Database name: Ch03_AviaCo

EMP_NUM	PIL_LICENSE	PIL_RATINGS	PIL_MED_TYPE	PIL_MED_DATE	PIL_PT135_DATE
101	ATP	ATP/SEL/MEL/Instr/CFII	1	20-Jan-08	11-Jan-08
104	ATP	ATP/SEL/MEL/Instr	1	18-Dec-07	17-Jan-08
105	COM	COMM/SEL/MEL/Instr/CFI	2	05-Jan-08	02-Jan-08
106	COM	COMM/SEL/MEL/Instr	2	10-Dec-07	02-Feb-08
109	COM	ATP/SEL/MEL/SES/Instr/CFII	1	22-Jan-08	15-Jan-08

The pilot licenses shown in the PILOT table include the ATP = Airline Transport Pilot and COM = Commercial Pilot. Businesses that operate on-demand air services are governed by Part 135 of the Federal Air Regulations (FARs) that are enforced by the Federal Aviation Administration (FAA). Such businesses are known as "Part 135 operators." Part 125 operations require that pilots successfully complete flight proficiency checks every six months. The "Part 135" flight proficiency check data is recorded in PIL_PT135_DATE. To fly commercially, pilots must have at least a commercial license and a second-class medical certificate (PIL_MED_TYPE = 2).

The PIL_RATINGS include

SEL	= Single Engine, Land	MEL	= Multiengine, Land
SES	= Single Engine, Sea	Instr.	= Instrument
CFI	= Certified Flight Instructor	CFII	= Certified Flight Instructor, Instrument

Table name: EMPLOYEE

EMP_NUM	EMP_TITLE	EMP_LNAME	EMP_FNAME	EMP_INITIAL	EMP_DOB	EMP_HIRE_DATE
100	Mr.	Kolmycz	George	D	15-Jun-42	15-Mar-88
101	Ms.	Lewis	Rhonda	G	19-Mar-65	25-Apr-86
102	Mr.	Vandam	Rhett		14-Nov-58	18-May-93
103	Ms.	Jones	Anne	M	11-May-74	26-Jul-99
104	Mr.	Lange	John	P	12-Jul-71	20-Aug-90
105	Mr.	Williams	Robert	D	14-Mar-75	19-Jun-03
106	Mrs.	Duzak	Jeanine	K	12-Feb-68	13-Mar-89
107	Mr.	Diante	Jorge	D	01-May-75	02-Jul-97
108	Mr.	Wiesenbach	Paul	R	14-Feb-66	03-Jun-93
109	Ms.	Travis	Elizabeth	K	18-Jun-61	14-Feb-06
110	Mrs.	Genkazi	Leighla	W	19-May-70	29-Jun-90

Table name: CUSTOMER

CUS_CODE	CUS_LNAME	CUS_FNAME	CUS_INITIAL	CUS_AREACODE	CUS_PHONE	CUS_BALANCE
10010	Ramas	Alfred	A	615	844-2573	0.00
10011	Dunne	Leona	K	713	894-1238	0.00
10012	Smith	Kathy	W	615	894-2285	896.54
10013	Olowski	Paul	F	615	894-2180	1285.19
10014	Orlando	Myron		615	222-1672	673.21
10015	O'Brian	Amy	B	713	442-3381	1014.56
10016	Brown	James	G	615	297-1228	0.00
10017	Williams	George		615	290-2556	0.00
10018	Farriss	Anne	G	713	382-7185	0.00
10019	Smith	Olette	K	615	297-3809	453.98

The nulls in the CHARTER table's CHAR_COPILOT column indicate that a copilot is not required for some charter trips or for some aircraft. Federal Aviation Administration (FAA) rules require a copilot on jet aircraft and on aircraft having a gross take-off weight over 12,500 pounds. None of the aircraft in the AIRCRAFT table are governed by this requirement; however, some customers may require the presence of a copilot for insurance reasons. All charter trips are recorded in the CHARTER table.

NOTE

Earlier in the chapter, it was stated that it is best to avoid homonyms and synonyms. In this problem, both the pilot and the copilot are pilots in the PILOT table, but EMP_NUM cannot be used for both in the CHARTER table. Therefore, the synonyms CHAR_PILOT and CHAR_COPILOT were used in the CHARTER table.

Although the solution works in this case, it is very restrictive and it generates nulls when a copilot is not required. Worse, such nulls proliferate as crew requirements change. For example, if the AviaCo charter company grows and starts using larger aircraft, crew requirements may increase to include flight engineers and load masters. The CHARTER table would then have to be modified to include the additional crew assignments; such attributes as CHAR_FLT_ENGINEER and CHAR_LOADMASTER would have to be added to the CHARTER table. Given this change, each time a smaller aircraft flew a charter trip without the number of crew members required in larger aircraft, the missing crew members would yield additional nulls in the CHARTER table.

You will have a chance to correct those design shortcomings in Problem 33. The problem illustrates two important points:

1. Don't use synonyms. If your design requires the use of synonyms, revise the design!

2. To the greatest possible extent, design the database to accommodate growth without requiring structural changes in the database tables. Plan ahead and try to anticipate the effects of change on the database.

30. For each table, where possible, identify:

    a. The primary key.

    b. A superkey.

    c. A candidate key.

    d. The foreign key(s).

    e. A secondary key.

31. Create the ERD. (*Hint*: Look at the table contents. You will discover that an AIRCRAFT can fly many CHARTER trips but that each CHARTER trip is flown by one AIRCRAFT, that a MODEL references many AIRCRAFT but that each AIRCRAFT references a single MODEL, etc.)

32. Create the relational diagram.

33. Modify the ERD you created in Problem 31 to eliminate the problems created by the use of synonyms. (*Hint*: Modify the CHARTER table structure by eliminating the CHAR_PILOT and CHAR_COPILOT attributes; then create a composite table named CREW to link the CHARTER and EMPLOYEE tables. Some crew members, such as flight attendants, may not be pilots. That's why the EMPLOYEE table enters into this relationship.)

34. Create the relational diagram for the design you revised in Problem 33. (After you have had a chance to revise the design, your instructor will show you the results of the design change, using a copy of the revised database named **Ch03_AviaCo_2**.)

ENTITY RELATIONSHIP (ER) MODELING

In this chapter, you will learn:

- The main characteristics of entity relationship components
- How relationships between entities are defined, refined, and incorporated into the database design process
- How ERD components affect database design and implementation
- That real-world database design often requires the reconciliation of conflicting goals

This chapter expands coverage of the data modeling aspect of database design. Data modeling is the first step in the database design journey, serving as a bridge between real-world objects and the database model that is implemented in the computer. Therefore, the importance of data modeling details, expressed graphically through entity relationship diagrams (ERDs), cannot be overstated.

Most of the basic concepts and definitions used in the entity relationship model (ERM) were introduced in Chapter 2, Data Models. For example, the basic components of entities and relationships and their representation should now be familiar to you. This chapter goes much deeper and broader, analyzing the graphic depiction of relationships among the entities and showing how those depictions help you summarize the wealth of data required to implement a successful design.

Finally, the chapter illustrates how conflicting goals can be a challenge in database design, possibly requiring you to make design compromises.

Preview

NOTE

Because this book generally focuses on the relational model, you might be tempted to conclude that the ERM is exclusively a relational tool. Actually, conceptual models such as the ERM can be used to understand and design the data requirements of an organization. Therefore, the ERM is independent of the database type. Conceptual models are used in the conceptual design of databases, while relational models are used in the logical design of databases. However, because you are now familiar with the relational model from the previous chapter, the relational model is used extensively in this chapter to explain ER constructs and the way they are used to develop database designs.

4.1 THE ENTITY RELATIONSHIP MODEL (ERM)

You should remember from Chapter 2 and Chapter 3, The Relational Database Model, that the ERM forms the basis of an ERD. The ERD represents the conceptual database as viewed by the end user. ERDs depict the database's main components: entities, attributes, and relationships. Because an entity represents a real-world object, the words *entity* and *object* are often used interchangeably. Thus, the entities (objects) of the Tiny College database design developed in this chapter include students, classes, teachers, and classrooms. The order in which the ERD components are covered in the chapter is dictated by the way the modeling tools are used to develop ERDs that can form the basis for successful database design and implementation.

In Chapter 2, you also learned about the various notations used with ERDs—the original Chen notation and the newer Crow's Foot and UML notations. The first two notations are used at the beginning of this chapter to introduce some basic ER modeling concepts. Some conceptual database modeling concepts can be expressed only using the Chen notation. However, because the emphasis is on *design and implementation* of databases, the Crow's Foot and UML class diagram notations were used for the final Tiny College ER diagram example. Because of its implementation emphasis, the Crow's Foot notation can represent only what could be implemented. In other words:

- The Chen notation favors conceptual modeling.
- The Crow's Foot notation favors a more implementation-oriented approach.
- The UML notation can be used for both conceptual and implementation modeling.

4.1.1 ENTITIES

Recall that an entity is an object of interest to the end user. In Chapter 2, you learned that at the ER modeling level, an entity actually refers to the *entity set* and not to a single entity occurrence. In other words, the word *entity* in the ERM corresponds to a table—not to a row—in the relational environment. The ERM refers to a table row as an *entity instance* or *entity occurrence*. In both the Chen and Crow's Foot notations, an entity is represented by a rectangle containing the entity's name. The entity name, a noun, is usually written in all capital letters.

4.1.2 ATTRIBUTES

Attributes are characteristics of entities. For example, the STUDENT entity includes, among many others, the attributes STU_LNAME, STU_FNAME, and STU_INITIAL. In the original Chen notation, attributes are represented by ovals and are connected to the entity rectangle with a line. Each oval contains the name of the attribute it represents. In the Crow's Foot notation, the attributes are written in the attribute box below the entity rectangle. See Figure 4.1. Because the Chen representation is rather space-consuming, software vendors have adopted the Crow's Foot style attribute display.

Required and Optional Attributes

A **required attribute** is an attribute that must have a value; in other words, it cannot be left empty. As shown in Figure 4.1, there are two boldfaced attributes in the Crow's Foot notation. This indicates that a data entry will be

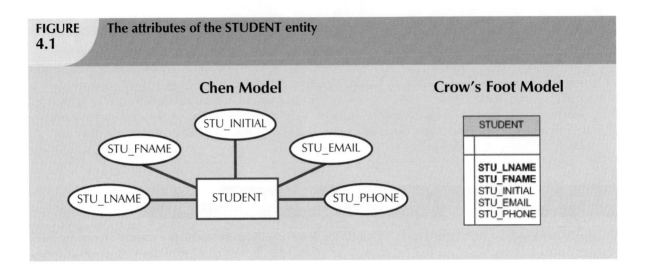

FIGURE 4.1   The attributes of the STUDENT entity

required. In this example, STU_LNAME and STU_FNAME require data entries because of the assumption that all students have a last name and a first name. But students might not have a middle name, and perhaps they do not (yet) have a phone number and an e-mail address. Therefore, those attributes are not presented in boldface in the entity box. An **optional attribute** is an attribute that does not require a value; therefore, it can be left empty.

ONLINE CONTENT

To learn how to create ER diagrams with the help of Microsoft Visio, see the Student Online Companion:

- **Appendix A, Designing Database with Visio Professional: A Tutorial** shows you how to create Crow's Foot ERDs.

- **Appendix H, Unified Modeling Language (UML)**, shows you how to create UML class diagrams.

Domains

Attributes have a domain. As you learned in Chapter 3, a *domain* is the set of possible values for a given attribute. For example, the domain for the grade point average (GPA) attribute is written (0,4) because the lowest possible GPA value is 0 and the highest possible value is 4. The domain for the gender attribute consists of only two possibilities: M or F (or some other equivalent code). The domain for a company's date of hire attribute consists of all dates that fit in a range (for example, company startup date to current date).

Attributes may share a domain. For instance, a student address and a professor address share the same domain of all possible addresses. In fact, the data dictionary may let a newly declared attribute inherit the characteristics of an existing attribute if the same attribute name is used. For example, the PROFESSOR and STUDENT entities may each have an attribute named ADDRESS and could therefore share a domain.

Identifiers (Primary Keys)

The ERM uses **identifiers**, that is, one or more attributes that uniquely identify each entity instance. In the relational model, such identifiers are mapped to primary keys (PKs) in tables. Identifiers are underlined in the ERD. Key attributes are also underlined in a frequently used table structure shorthand notation using the format:

TABLE NAME (**KEY_ATTRIBUTE 1**, ATTRIBUTE 2, ATTRIBUTE 3, . . . ATTRIBUTE K)

For example, a CAR entity may be represented by:

CAR (**CAR_VIN**, MOD_CODE, CAR_YEAR, CAR_COLOR)

(Each car is identified by a unique vehicle identification number, or CAR_VIN.)

Composite Identifiers

Ideally, an entity identifier is composed of only a single attribute. For example, the table in Figure 4.2 uses a single-attribute primary key named CLASS_CODE. However, it is possible to use a **composite identifier**, that is, a primary key composed of more than one attribute. For instance, the Tiny College database administrator may decide to identify each CLASS entity instance (occurrence) by using a composite primary key composed of the combination of CRS_CODE and CLASS_SECTION instead of using CLASS_CODE. Either approach uniquely identifies each entity instance. Given the current structure of the CLASS table shown in Figure 4.2, CLASS_CODE is the primary key and the combination of CRS_CODE and CLASS_SECTION is a proper candidate key. If the CLASS_CODE attribute is deleted from the CLASS entity, the candidate key (CRS_CODE and CLASS_SECTION) becomes an acceptable composite primary key.

FIGURE 4.2   **The CLASS table (entity) components and contents**

CLASS_CODE	CRS_CODE	CLASS_SECTION	CLASS_TIME	ROOM_CODE	PROF_NUM
10012	ACCT-211	1	MWF 8:00-8:50 a.m.	BUS311	105
10013	ACCT-211	2	MWF 9:00-9:50 a.m.	BUS200	105
10014	ACCT-211	3	TTh 2:30-3:45 p.m.	BUS252	342
10015	ACCT-212	1	MWF 10:00-10:50 a.m.	BUS311	301
10016	ACCT-212	2	Th 6:00-8:40 p.m.	BUS252	301
10017	CIS-220	1	MWF 9:00-9:50 a.m.	KLR209	228
10018	CIS-220	2	MWF 9:00-9:50 a.m.	KLR211	114
10019	CIS-220	3	MWF 10:00-10:50 a.m.	KLR209	228
10020	CIS-420	1	W 6:00-8:40 p.m.	KLR209	162
10021	QM-261	1	MWF 8:00-8:50 a.m.	KLR200	114
10022	QM-261	2	TTh 1:00-2:15 p.m.	KLR200	114
10023	QM-362	1	MWF 11:00-11:50 a.m.	KLR200	162
10024	QM-362	2	TTh 2:30-3:45 p.m.	KLR200	162
10025	MATH-243	1	Th 6:00-8:40 p.m.	DRE155	325

NOTE

Remember that Chapter 3 made a commonly accepted distinction between COURSE and CLASS. A CLASS constitutes a specific time and place of a COURSE offering. A class is defined by the course description and its time and place, or section. Consider a professor who teaches Database I, Section 2; Database I, Section 5; Database I, Section 8; and Spreadsheet II, Section 6. That instructor teaches two courses (Database I and Spreadsheet II), but four classes. Typically, the COURSE offerings are printed in a course catalog, while the CLASS offerings are printed in a class schedule for each semester, trimester, or quarter.

If the CLASS_CODE in Figure 4.2 is used as the primary key, the CLASS entity may be represented in shorthand form by:

CLASS (**CLASS_CODE**, CRS_CODE, CLASS_SECTION, CLASS_TIME, ROOM_CODE, PROF_NUM)

On the other hand, if CLASS_CODE is deleted, and the composite primary key is the combination of CRS_CODE and CLASS_SECTION, the CLASS entity may be represented by:

CLASS (**CRS_CODE**, **CLASS_SECTION**, CLASS_TIME, ROOM_CODE, PROF_NUM)

Note that *both* key attributes are underlined in the entity notation.

Composite and Simple Attributes

Attributes are classified as simple or composite. A **composite attribute**, not to be confused with a composite key, is an attribute that can be further subdivided to yield additional attributes. For example, the attribute ADDRESS can be subdivided into street, city, state, and zip code. Similarly, the attribute PHONE_NUMBER can be subdivided into area code and exchange number. A **simple attribute** is an attribute that cannot be subdivided. For example, age, sex, and marital status would be classified as simple attributes. To facilitate detailed queries, it is wise to change composite attributes into a series of simple attributes.

Single-Valued Attributes

A **single-valued attribute** is an attribute that can have only a single value. For example, a person can have only one Social Security number, and a manufactured part can have only one serial number. *Keep in mind that a single-valued attribute is not necessarily a simple attribute.* For instance, a part's serial number, such as SE-08-02-189935, is single-valued, but it is a composite attribute because it can be subdivided into the region in which the part was produced (SE), the plant within that region (08), the shift within the plant (02), and the part number (189935).

Multivalued Attributes

Multivalued attributes are attributes that can have many values. For instance, a person may have several college degrees, and a household may have several different phones, each with its own number. Similarly, a car's color may be subdivided into many colors (that is, colors for the roof, body, and trim). In the Chen ERM, the multivalued attributes are shown by a double line connecting the attribute to the entity. The Crow's Foot notation does not identify multivalued attributes. The ERD in Figure 4.3 contains all of the components introduced thus far. In Figure 4.3, note that CAR_VIN is the primary key, and CAR_COLOR is a multivalued attribute of the CAR entity.

FIGURE 4.3   A multivalued attribute in an entity

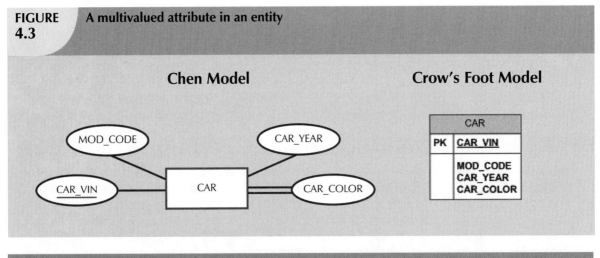

NOTE

In the ERD models in Figure 4.3, the CAR entity's foreign key (FK) has been typed as MOD_CODE. This attribute was manually added to the entity. Actually, proper use of a database modeling software will automatically produce the FK when the relationship is defined. In addition, the software will label the FK appropriately and write the FKs implementation details in a data dictionary. Therefore, when you use database modeling software like Visio Professional, *never type the FK attribute yourself*; let the software handle that task when the relationship between the entities is defined. (You can see how that's done in **Appendix A, Designing Databases with Visio Professional: A Tutorial**, in the Student Online Companion).

Implementing Multivalued Attributes

Although the conceptual model can handle M:N relationships and multivalued attributes, *you should not implement them in the RDBMS.* Remember from Chapter 3 that in the relational table, each column/row intersection represents a single data value. So if multivalued attributes exist, the designer must decide on one of two possible courses of action:

1. Within the original entity, create several new attributes, one for each of the original multivalued attribute's components. For example, the CAR entity's attribute CAR_COLOR can be split to create the new attributes CAR_TOPCOLOR, CAR_BODYCOLOR, and CAR_TRIMCOLOR, which are then assigned to the CAR entity. See Figure 4.4.

FIGURE 4.4	Splitting the multivalued attribute into new attributes

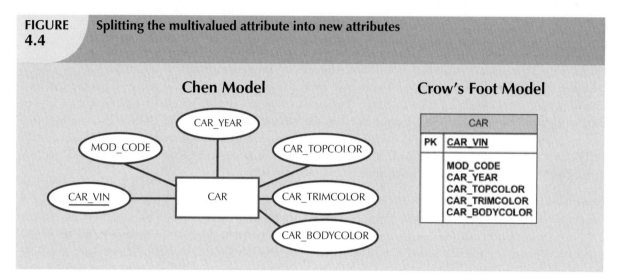

Although this solution seems to work, its adoption can lead to major structural problems in the table. For example, if additional color components—such as a logo color—are added for some cars, the table structure must be modified to accommodate the new color section. In that case, cars that do not have such color sections generate nulls for the nonexisting components, or their color entries for those sections are entered as N/A to indicate "not applicable." (Imagine how the solution in Figure 4.4—splitting a multivalued attribute into new attributes—would cause problems when it is applied to an employee entity containing employee degrees and certifications. If some employees have 10 degrees and certifications while most have fewer or none, the number of degree/certification attributes would number 10 and most of those attribute values would be null for most of the employees.) In short, although you have seen solution 1 applied, it is not an acceptable solution.

2. Create a new entity composed of the original multivalued attribute's components. (See Figure 4.5.) The new (independent) CAR_COLOR entity is then related to the original CAR entity in a 1:M relationship. Note that such a change allows the designer to define color for different sections of the car. (See Table 4.1.) Using the approach illustrated in Table 4.1, you even get a fringe benefit: you are now able to assign as many colors as necessary without having to change the table structure. Note that the ERM in Figure 4.5 reflects the components listed in Table 4.1. This is the preferred way to deal with multivalued attributes. Creating a new entity in a 1:M relationship with the original entity yields several benefits: it's a more flexible, expandable solution, and it is compatible with the relational model!

TABLE 4.1	Components of the Multivalued Attribute	
SECTION	**COLOR**	
Top	White	
Body	Blue	
Trim	Gold	
Interior	Blue	

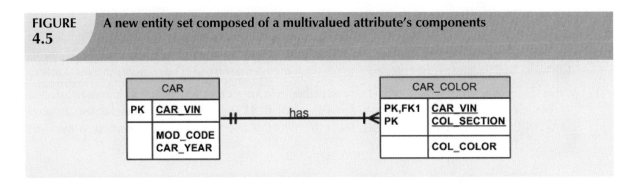

FIGURE 4.5   A new entity set composed of a multivalued attribute's components

Derived Attributes

Finally, an attribute may be classified as a derived attribute. A **derived attribute** is an attribute whose value is calculated (derived) from other attributes. The derived attribute need not be physically stored within the database; instead, it can be derived by using an algorithm. For example, an employee's age, EMP_AGE, may be found by computing the integer value of the difference between the current date and the EMP_DOB. If you use Microsoft Access, you would use the formula INT((DATE() – EMP_DOB)/365). In Microsoft SQL Server, you would use SELECT DATEDIFF("YEAR", EMP_DOB, GETDATE()); where DATEDIFF is a function that computes the difference between dates. The first parameter indicates the measurement, in this case, years.

If you use Oracle, you would use SYSDATE instead of DATE(). (You are assuming, of course, that the EMP_DOB was stored in the Julian date format.) Similarly, the total cost of an order can be derived by multiplying the quantity ordered by the unit price. Or the estimated average speed can be derived by dividing trip distance by the time spent en route. A derived attribute is indicated in the Chen notation by a dashed line connecting the attribute and the entity. See Figure 4.6. The Crow's Foot notation does not have a method for distinguishing the derived attribute from other attributes.

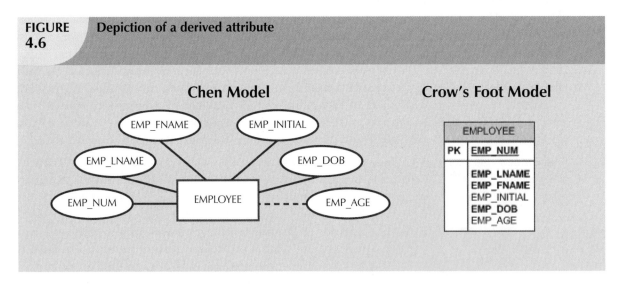

FIGURE 4.6   Depiction of a derived attribute

Derived attributes are sometimes referred to as *computed attributes*. A derived attribute computation can be as simple as adding two attribute values located on the same row, or it can be the result of aggregating the sum of values located on many table rows (from the same table or from a different table). The decision to store derived attributes in database tables depends on the processing requirements and the constraints placed on a particular application. The designer should be able to balance the design in accordance with such constraints. Table 4.2 shows the advantages and disadvantages of storing (or not storing) derived attributes in the database.

TABLE 4.2	Advantages and Disadvantages of Storing Derived Attributes	
	DERIVED ATTRIBUTE	
	STORED	NOT STORED
Advantage	Saves CPU processing cycles Saves data access time Data value is readily available Can be used to keep track of historical data	Saves storage space Computation always yields current value
Disadvantage	Requires constant maintenance to ensure derived value is current, especially if any values used in the calculation change	Uses CPU processing cycles Increases data access time Adds coding complexity to queries

4.1.3 RELATIONSHIPS

Recall from Chapter 2 that a relationship is an association between entities. The entities that participate in a relationship are also known as **participants**, and each relationship is identified by a name that describes the relationship. The relationship name is an active or passive verb; for example, a STUDENT *takes* a CLASS, a PROFESSOR *teaches* a CLASS, a DEPARTMENT *employs* a PROFESSOR, a DIVISION *is managed by* an EMPLOYEE, and an AIRCRAFT *is flown by* a CREW.

Relationships between entities always operate in both directions. That is, to define the relationship between the entities named CUSTOMER and INVOICE, you would specify that:

- A CUSTOMER may generate many INVOICEs.
- Each INVOICE is generated by one CUSTOMER.

Because you know both directions of the relationship between CUSTOMER and INVOICE, it is easy to see that this relationship can be classified as 1:M.

The relationship classification is difficult to establish if you know only one side of the relationship. For example, if you specify that:

A DIVISION is managed by one EMPLOYEE.

you don't know if the relationship is 1:1 or 1:M. Therefore, you should ask the question "Can an employee manage more than one division?" If the answer is yes, the relationship is 1:M, and the second part of the relationship is then written as:

An EMPLOYEE may manage many DIVISIONs.

If an employee cannot manage more than one division, the relationship is 1:1, and the second part of the relationship is then written as:

An EMPLOYEE may manage only one DIVISION.

4.1.4 CONNECTIVITY AND CARDINALITY

You learned in Chapter 2 that entity relationships may be classified as one-to-one, one-to-many, or many-to-many. You also learned how such relationships were depicted in the Chen and Crow's Foot notations. The term **connectivity** is used to describe the relationship classification.

Cardinality expresses the minimum and maximum number of entity occurrences associated with one occurrence of the related entity. In the ERD, cardinality is indicated by placing the appropriate numbers beside the entities, using the format (x,y). The first value represents the minimum number of associated entities, while the second value represents the maximum number of associated entities. Some Crow's Foot ER modeling tools do not print the numeric cardinality range in the diagram; instead, you could add it as text. In Crow's Foot notation, cardinality is implied by the use of symbols in Figure 4.7. The numeric cardinality range has been added using the Visio text drawing tool.

FIGURE 4.7	Connectivity and cardinality in an ERD

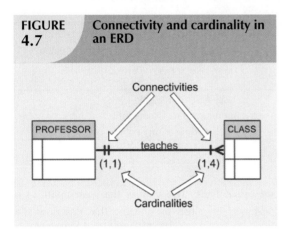

Knowing the minimum and maximum number of entity occurrences is very useful at the application software level. For example, Tiny College might want to ensure that a class is not taught unless it has at least 10 students enrolled. Similarly, if the classroom can hold only 30 students, the application software should use that cardinality to limit enrollment in the class. However, keep in mind that the DBMS cannot handle the implementation of the cardinalities at the table level—that capability is provided by the application software or by triggers. You will learn how to create and execute triggers in Chapter 8, Advanced SQL.

As you examine the Crow's Foot diagram in Figure 4.7, keep in mind that the cardinalities represent the number of occurrences in the *related* entity. For example, the cardinality (1,4) written next to the CLASS entity in the "PROFESSOR teaches CLASS" relationship indicates that the PROFESSOR table's primary key value occurs at least once and no more than four times as foreign key values in the CLASS table. If the cardinality had been written as (1,N), there would be no upper limit to the number of classes a professor might teach. Similarly, the cardinality (1,1) written next to the PROFESSOR entity indicates that each class is taught by one and only one professor. That is, each CLASS entity occurrence is associated with one and only one entity occurrence in PROFESSOR.

Connectivities and cardinalities are established by very concise statements known as business rules, which were introduced in Chapter 2. Such rules, derived from a precise and detailed description of an organization's data environment, also establish the ERM's entities, attributes, relationships, connectivities, cardinalities, and constraints. Because business rules define the ERM's components, making sure that all appropriate business rules are identified is a very important part of a database designer's job.

NOTE

The placement of the cardinalities in the ER diagram is a matter of convention. The Chen notation places the cardinalities on the side of the related entity. The Crow's Foot and UML diagrams place the cardinalities next to the entity to which the cardinalities apply.

ONLINE CONTENT

Because the careful definition of complete and accurate business rules is crucial to good database design, their derivation is examined in detail in **Appendix B, The University Lab: Conceptual Design.** The modeling skills you are learning in this chapter are applied in the development of a real database design in Appendix B. The initial design shown in Appendix B is then modified in **Appendix C, The University Lab: Conceptual Design Verification, Logical Design, and Implementation.** (Both appendixes are found in the Student Online Companion.)

4.1.5   EXISTENCE DEPENDENCE

An entity is said to be **existence-dependent** if it can exist in the database only when it is associated with another related entity occurrence. In implementation terms, an entity is existence-dependent if it has a mandatory foreign key—that is, a foreign key attribute that cannot be null. For example, if an employee wants to claim one or more dependents for tax-withholding purposes, the relationship "EMPLOYEE claims DEPENDENT" would be appropriate. In that case, the DEPENDENT entity is clearly existence-dependent on the EMPLOYEE entity because it is impossible for the dependent to exist apart from the EMPLOYEE in the database.

If an entity can exist apart from one or more related entities, it is said to be **existence-independent**. (Sometimes designers refer to such an entity as a *strong* or *regular* entity.) For example, suppose that the XYZ Corporation uses parts to produce its products. Further, suppose that some of those parts are produced in-house and other parts are bought from vendors. In that scenario, it is quite possible for a PART to exist independently from a VENDOR in the relationship "PART is supplied by VENDOR," because at least some of the parts are not supplied by a vendor. Therefore, PART is existence-independent from VENDOR.

NOTE

The relationship strength concept is not part of the original ERM. Instead, this concept applies directly to Crow's Foot diagrams. Because Crow's Foot diagrams are used extensively to design relational databases, it is important to understand relationship strength as it affects database inplementation. The Chen ERD notation is oriented towared conceptual modeling and therefore does not distinguish between weak and strong relationships.

4.1.6   RELATIONSHIP STRENGTH

The concept of relationship strength is based on how the primary key of a related entity is defined. To implement a relationship, the primary key of one entity appears as a foreign key in the related entity. For example, the 1:M relationship between VENDOR and PRODUCT in Chapter 3, Figure 3.3, is implemented by using the VEND_CODE primary key in VENDOR as a foreign key in PRODUCT. There are times when the foreign key also is a primary key component in the related entity. For example, in Figure 4.5, the CAR entity primary key (CAR_VIN) appears as both a primary key component and a foreign key in the CAR_COLOR entity. In this section, you learn how various relationship strength decisions affect primary key arrangement in database design.

Weak (Non-identifying) Relationships

A **weak relationship**, also known as a **non-identifying relationship**, exists if the PK of the related entity does not contain a PK component of the parent entity. By default, relationships are established by having the PK of the parent entity appear as an FK on the related entity. For example, suppose that the COURSE and CLASS entities are defined as:

COURSE(**CRS_CODE**, DEPT_CODE, CRS_DESCRIPTION, CRS_CREDIT)

CLASS(**CLASS_CODE**, CRS_CODE, CLASS_SECTION, CLASS_TIME, ROOM_CODE, PROF_NUM)

In this case, a weak relationship exists between COURSE and CLASS because the CLASS_CODE is the CLASS entity's PK, while the CRS_CODE in CLASS is only an FK. In this example, the CLASS PK did not inherit the PK component from the COURSE entity.

Figure 4.8 shows how the Crow's Foot notation depicts a weak relationship by placing a dashed relationship line between the entities. The tables shown below the ERD illustrate how such a relationship is implemented.

FIGURE 4.8   A weak (non-identifying) relationship between COURSE and CLASS

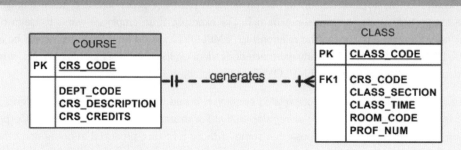

Table name: COURSE   **Database name: Ch04_TinyCollege**

CRS_CODE	DEPT_CODE	CRS_DESCRIPTION	CRS_CREDIT
ACCT-211	ACCT	Accounting I	3
ACCT-212	ACCT	Accounting II	3
CIS-220	CIS	Intro. to Microcomputing	3
CIS-420	CIS	Database Design and Implementation	4
MATH-243	MATH	Mathematics for Managers	3
QM-261	CIS	Intro. to Statistics	3
QM-362	CIS	Statistical Applications	4

Table name: CLASS

CLASS_CODE	CRS_CODE	CLASS_SECTION	CLASS_TIME	ROOM_CODE	PROF_NUM
10012	ACCT-211	1	MWF 8:00-8:50 a.m.	BUS311	105
10013	ACCT-211	2	MWF 9:00-9:50 a.m.	BUS200	105
10014	ACCT-211	3	TTh 2:30-3:45 p.m.	BUS252	342
10015	ACCT-212	1	MWF 10:00-10:50 a.m.	BUS311	301
10016	ACCT-212	2	Th 6:00-8:40 p.m.	BUS252	301
10017	CIS-220	1	MWF 9:00-9:50 a.m.	KLR209	228
10018	CIS-220	2	MWF 9:00-9:50 a.m.	KLR211	114
10019	CIS-220	3	MWF 10:00-10:50 a.m.	KLR209	228
10020	CIS-420	1	W 6:00-8:40 p.m.	KLR209	162
10021	QM-261	1	MWF 8:00-8:50 a.m.	KLR200	114
10022	QM-261	2	TTh 1:00-2:15 p.m.	KLR200	114
10023	QM-362	1	MWF 11:00-11:50 a.m.	KLR200	162
10024	QM-362	2	TTh 2:30-3:45 p.m.	KLR200	162
10025	MATH-243	1	Th 6:00-8:40 p.m.	DRE155	325

ONLINE CONTENT

All of the databases used to illustrate the material in this chapter are found in the Student Online Companion.

NOTE

If you are used to looking at relational diagrams such as the ones produced by Microsoft Access, you expect to see the relationship line *in the relational diagram* drawn from the PK to the FK. However, the relational diagram convention is not necessarily reflected in the ERD. In an ERD, the focus is on the entities and the relationships between them, rather than on the way those relationships are anchored graphically. You will discover that the placement of the relationship lines in a complex ERD that includes both horizontally and vertically placed entities is largely dictated by the designer's decision to improve the readability of the design. (Remember that the ERD is used for communication between the designer(s) and end users.)

Strong (Identifying) Relationships

A **strong relationship**, also known as an **identifying relationship**, exists when the PK of the related entity contains a PK component of the parent entity. For example, the definitions of the COURSE and CLASS entities

COURSE(**CRS_CODE**, DEPT_CODE, CRS_DESCRIPTION, CRS_CREDIT)

CLASS(**CRS_CODE**, **CLASS_SECTION**, CLASS_TIME, ROOM_CODE, PROF_NUM)

indicate that a strong relationship exists between COURSE and CLASS, because the CLASS entity's composite PK is composed of CRS_CODE + CLASS_SECTION. (Note that the CRS_CODE in CLASS is *also* the FK to the COURSE entity.)

The Crow's Foot notation depicts the strong (identifying) relationship with a solid line between the entities, shown in Figure 4.9. Whether the relationship between COURSE and CLASS is strong or weak depends on how the CLASS entity's primary key is defined.

FIGURE 4.9	A strong (identifying) relationship between COURSE and CLASS

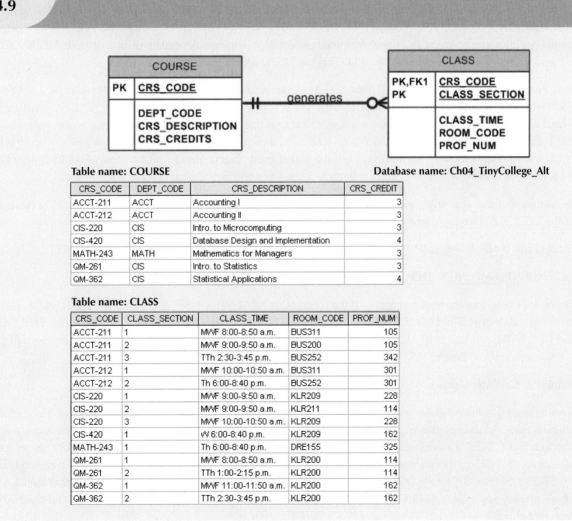

Table name: COURSE   Database name: Ch04_TinyCollege_Alt

CRS_CODE	DEPT_CODE	CRS_DESCRIPTION	CRS_CREDIT
ACCT-211	ACCT	Accounting I	3
ACCT-212	ACCT	Accounting II	3
CIS-220	CIS	Intro. to Microcomputing	3
CIS-420	CIS	Database Design and Implementation	4
MATH-243	MATH	Mathematics for Managers	3
QM-261	CIS	Intro. to Statistics	3
QM-362	CIS	Statistical Applications	4

Table name: CLASS

CRS_CODE	CLASS_SECTION	CLASS_TIME	ROOM_CODE	PROF_NUM
ACCT-211	1	MWF 8:00-8:50 a.m.	BUS311	105
ACCT-211	2	MWF 9:00-9:50 a.m.	BUS200	105
ACCT-211	3	TTh 2:30-3:45 p.m.	BUS252	342
ACCT-212	1	MWF 10:00-10:50 a.m.	BUS311	301
ACCT-212	2	Th 6:00-8:40 p.m.	BUS252	301
CIS-220	1	MWF 9:00-9:50 a.m.	KLR209	228
CIS-220	2	MWF 9:00-9:50 a.m.	KLR211	114
CIS-220	3	MWF 10:00-10:50 a.m.	KLR209	228
CIS-420	1	W 6:00-8:40 p.m.	KLR209	162
MATH-243	1	Th 6:00-8:40 p.m.	DRE155	325
QM-261	1	MWF 8:00-8:50 a.m.	KLR200	114
QM-261	2	TTh 1:00-2:15 p.m.	KLR200	114
QM-362	1	MWF 11:00-11:50 a.m.	KLR200	162
QM-362	2	TTh 2:30-3:45 p.m.	KLR200	162

Keep in mind that *the order in which the tables are created and loaded is very important.* For example, in the "COURSE generates CLASS" relationship, the COURSE table must be created before the CLASS table. After all, it would not be acceptable to have the CLASS table's foreign key reference a COURSE table that did not yet exist. In fact, *you must load the data of the "1" side first in a 1:M relationship to avoid the possibility of referential integrity errors*, regardless of whether the relationships are weak or strong.

As you examine Figure 4.9 you might wonder what the O symbol next to the CLASS entity signifies. You will discover the meaning of this cardinality in Section 4.1.8, Relationship Participation.

Remember that the nature of the relationship is often determined by the database designer, who must use professional judgment to determine which relationship type and strength best suits the database transaction, efficiency, and information requirements. That point will often be emphasized in detail!

4.1.7 WEAK ENTITIES

A **weak entity** is one that meets two conditions:

1. The entity is existence-dependent; that is, it cannot exist without the entity with which it has a relationship.

2. The entity has a primary key that is partially or totally derived from the parent entity in the relationship.

For example, a company insurance policy insures an employee and his/her dependents. For the purpose of describing an insurance policy, an EMPLOYEE might or might not have a DEPENDENT, but the DEPENDENT must be associated with an EMPLOYEE. Moreover, the DEPENDENT cannot exist without the EMPLOYEE; that is, a person cannot get insurance coverage as a dependent unless s(he) happens to be a dependent of an employee. DEPENDENT is the weak entity in the relationship "EMPLOYEE has DEPENDENT."

Note that the Chen notation in Figure 4.10 identifies the weak entity by using a double-walled entity rectangle. The Crow's Foot notation generated by Visio Professional uses the relationship line and the PK/FK designation to indicate whether the related entity is weak. A strong (identifying) relationship indicates that the related entity is weak. Such a relationship means that both conditions for the weak entity definition have been met—the related entity is existence-dependent, and the PK of the related entity contains a PK component of the parent entity. (Some versions of the Crow's Foot ERD depict the weak entity by drawing a short line segment in each of the four corners of the weak entity box.)

Remember that the weak entity inherits part of its primary key from its strong counterpart. For example, at least part of the DEPENDENT entity's key shown in Figure 4.10 was inherited from the EMPLOYEE entity:

EMPLOYEE (**EMP_NUM**, EMP_LNAME, EMP_FNAME, EMP_INITIAL, EMP_DOB, EMP_HIREDATE)

DEPENDENT (**EMP_NUM**, **DEP_NUM**, DEP_FNAME, DEP_DOB)

Figure 4.11 illustrates the implementation of the relationship between the weak entity (DEPENDENT) and its parent or strong counterpart (EMPLOYEE). Note that DEPENDENT's primary key is composed of two attributes, EMP_NUM and DEP_NUM, and that EMP_NUM was inherited from EMPLOYEE. Given this scenario, and with the help of this relationship, you can determine:

Jeanine J. Callifante claims two dependents, Annelise and Jorge.

Keep in mind that the database designer usually determines whether an entity can be described as weak based on the business rules. An examination of the relationship between COURSE and CLASS in Figure 4.8 might cause you to conclude that CLASS is a weak entity to COURSE. After all, in Figure 4.8, it seems clear that a CLASS cannot exist without a COURSE; so there is existence dependency. For example, a student cannot enroll in the Accounting I class ACCT-211, Section 3 (CLASS_CODE 10014) unless there is an ACCT_211 course. However, note that the CLASS table's primary key is CLASS_CODE, which is not derived from the COURSE parent entity. That is, CLASS may be represented by:

CLASS (**CLASS_CODE**, CRS_CODE, CLASS_SECTION, CLASS_TIME, ROOM_CODE, PROF_NUM)

FIGURE 4.10	A weak entity in an ERD

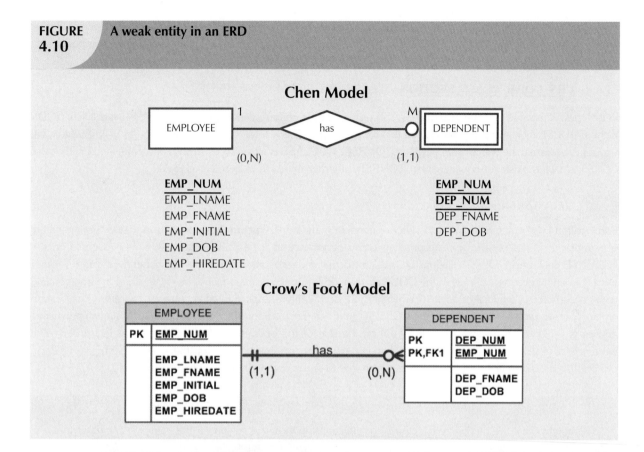

FIGURE 4.11	A weak entity in a strong relationship

Table name: EMPLOYEE   Database name: Ch04_ShortCo

EMP_NUM	EMP_LNAME	EMP_FNAME	EMP_INITIAL	EMP_DOB	EMP_HIREDATE
1001	Callifante	Jeanine	J	12-Mar-64	25-May-97
1002	Smithson	William	K	23-Nov-70	28-May-97
1003	Washington	Herman	H	15-Aug-68	28-May-97
1004	Chen	Lydia	B	23-Mar-74	15-Oct-98
1005	Johnson	Melanie		28-Sep-66	20-Dec-98
1006	Ortega	Jorge	G	12-Jul-79	05-Jan-02
1007	O'Donnell	Peter	D	10-Jun-71	23-Jun-02
1008	Brzenski	Barbara	A	12-Feb-70	01-Nov-03

Table name: DEPENDENT

EMP_NUM	DEP_NUM	DEP_FNAME	DEP_DOB
1001	1	Annelise	05-Dec-97
1001	2	Jorge	30-Sep-02
1003	1	Suzanne	25-Jan-04
1006	1	Carlos	25-May-01
1008	1	Michael	19-Feb-95
1008	2	George	27-Jun-98
1008	3	Katherine	18-Aug-03

The second weak entity requirement has not been met; therefore, by definition, the CLASS entity in Figure 4.8 may not be classified as weak. On the other hand, if the CLASS entity's primary key had been defined as a composite key, composed of the combination CRS_CODE and CLASS_SECTION, CLASS could be represented by:

CLASS (**CRS_CODE**, **CLASS_SECTION**, CLASS_TIME, ROOM_CODE, PROF_NUM)

In that case, illustrated in Figure 4.9, the CLASS primary key is partially derived from COURSE because CRS_CODE is the COURSE table's primary key. Given this decision, CLASS is a weak entity by definition. (In Visio Professional Crow's Foot terms, the relationship between COURSE and CLASS is classified as strong, or identifying.) In any case, CLASS is always existence-dependent on COURSE, *whether or not it is defined as weak.*

4.1.8 RELATIONSHIP PARTICIPATION

Participation in an entity relationship is either optional or mandatory. **Optional participation** means that one entity occurrence does not *require* a corresponding entity occurrence in a particular relationship. For example, in the "COURSE generates CLASS" relationship, you noted that at least some courses do not generate a class. In other words, an entity occurrence (row) in the COURSE table does not necessarily require the existence of a corresponding entity occurrence in the CLASS table. (Remember that each entity is implemented as a table.) Therefore, the CLASS entity is considered to be *optional* to the COURSE entity. In Crow's Foot notation, an optional relationship between entities is shown by drawing a small circle (O) on the side of the optional entity, as illustrated in Figure 4.9. The existence of an *optional entity* indicates that the minimum cardinality is 0 for the optional entity. (The term *optionality* is used to label any condition in which one or more optional relationships exist.)

> **NOTE**
>
> Remember that the burden of establishing the relationship is always placed on the entity that contains the foreign key. In most cases, that will be the entity on the many side of the relationship.

Mandatory participation means that one entity occurrence *requires* a corresponding entity occurrence in a particular relationship. If no optionality symbol is depicted with the entity, the entity exists in a mandatory relationship with the related entity. The existence of a mandatory relationship indicates that the minimum cardinality is 1 for the mandatory entity.

> **NOTE**
>
> You might be tempted to conclude that relationships are weak when they occur between entities in an optional relationship and that relationships are strong when they occur between entities in a mandatory relationship. However, this conclusion is not warranted. Keep in mind that relationship participation and relationship strength do not describe the same thing. You are likely to encounter a strong relationship when one entity is optional to another. For example, the relationship between EMPLOYEE and DEPENDENT is clearly a strong one, but DEPENDENT is clearly optional to EMPLOYEE. After all, you cannot *require* employees to have dependents. And it is just as possible for a weak relationship to be established when one entity is mandatory to another. *The relationship strength depends on how the PK of the related entity is formulated, while the relationship participation depends on how the business rule is written.* For example, the business rules "Each part must be supplied by a vendor" and "A part may or may not be supplied by a vendor" create different optionalities for the same entities! Failure to understand this distinction may lead to poor design decisions that cause major problems when table rows are inserted or deleted.

Because relationship participation turns out to be a very important component of the database design process, let's examine a few more scenarios. Suppose that Tiny College employs some professors who conduct research without teaching classes. If you examine the "PROFESSOR teaches CLASS" relationship, it is quite possible for a PROFESSOR not to teach a CLASS. Therefore, CLASS is *optional* to PROFESSOR. On the other hand, a CLASS must be taught by a PROFESSOR. Therefore, PROFESSOR is *mandatory* to CLASS. Note that the ERD model in Figure 4.12 shows the cardinality next to CLASS to be (0,3), thus indicating that a professor may teach no classes at all or as many as three classes. And each CLASS table row will reference one and only one PROFESSOR row—assuming each class is taught by one and only one professor, represented by the (1,1) cardinality next to the PROFESSOR table.

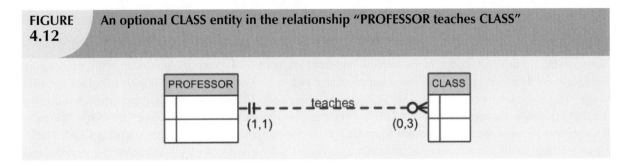

FIGURE 4.12   **An optional CLASS entity in the relationship "PROFESSOR teaches CLASS"**

Failure to understand the distinction between *mandatory* and *optional* participation in relationships might yield designs in which awkward (and unnecessary) temporary rows (entity instances) must be created just to accommodate the creation of required entities. Therefore, it is important that you clearly understand the concepts of mandatory and optional participation.

It is also important to understand that the semantics of a problem might determine the type of participation in a relationship. For example, suppose that Tiny College offers several courses; each course has several classes. Note again the distinction between *class* and *course* in this discussion: a CLASS constitutes a specific offering (or section) of a COURSE. (Typically, courses are listed in the university's course catalog, while classes are listed in the class schedules that students use to register for their classes.)

Analyzing the CLASS entity's contribution to the "COURSE generates CLASS" relationship, it is easy to see that a CLASS cannot exist without a COURSE. Therefore, you can conclude that the COURSE entity is *mandatory* in the relationship. But two scenarios for the CLASS entity may be written, shown in Figures 4.13 and 4.14. The different scenarios are a function of the semantics of the problem; that is, they depend on how the relationship is defined.

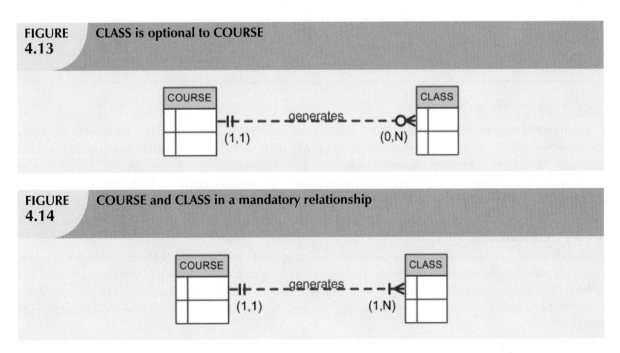

FIGURE 4.13   **CLASS is optional to COURSE**

FIGURE 4.14   **COURSE and CLASS in a mandatory relationship**

1. *CLASS is optional.* It is possible for the department to create the entity COURSE first and then create the CLASS entity after making the teaching assignments. In the real world, such a scenario is very likely; there may be courses for which sections (classes) have not yet been defined. In fact, some courses are taught only once a year and do not generate classes each semester.

2. *CLASS is mandatory.* This condition is created by the constraint that is imposed by the semantics of the statement "Each COURSE generates one or more CLASSes." In ER terms, each COURSE in the "generates" relationship must have at least one CLASS. Therefore, a CLASS must be created as the COURSE is created in order to comply with the semantics of the problem.

Keep in mind the practical aspects of the scenario presented in Figure 4.14. Given the semantics of this relationship, the system should not accept a course that is not associated with at least one class section. Is such a rigid environment desirable from an operational point of view? For example, when a new COURSE is created, the database first updates the COURSE table, thereby inserting a COURSE entity that does not yet have a CLASS associated with it. Naturally, the apparent problem seems to be solved when CLASS entities are inserted into the corresponding CLASS table. However, because of the mandatory relationship, the system will be in temporary violation of the business rule constraint. For practical purposes, it would be desirable to classify the CLASS as optional in order to produce a more flexible design.

Finally, as you examine the scenarios presented in Figures 4.13 and 4.14, keep in mind the role of the DBMS. To maintain data integrity, the DBMS must ensure that the "many" side (CLASS) is associated with a COURSE through the foreign key rules.

When you create a relationship in Visio, the default relationship will be mandatory on the "1" side and optional on the "many" side. Table 4.3 shows the various cardinalities that are supported by the Crow's Foot notation.

TABLE 4.3	Crow's Foot Symbols	
CROW'S FOOT SYMBOL	**CARDINALITY**	**COMMENT**
O€	(0,N)	Zero or many. Many side is optional.
I€	(1,N)	One or many. Many side is mandatory.
II	(1,1)	One and only one. 1 side is mandatory.
OH	(0,1)	Zero or one. 1 side is optional.

4.1.9 RELATIONSHIP DEGREE

A **relationship degree** indicates the number of entities or participants associated with a relationship. A **unary relationship** exists when an association is maintained within a single entity. A **binary relationship** exists when two entities are associated. A **ternary relationship** exists when three entities are associated. Although higher degrees exist, they are rare and are not specifically named. (For example, an association of four entities is described simply as a *four-degree relationship*.) Figure 4.15 shows these types of relationship degrees.

Unary Relationships

In the case of the unary relationship shown in Figure 4.15, an employee within the EMPLOYEE entity is the manager for one or more employees within that entity. In this case, the existence of the "manages" relationship means that EMPLOYEE requires another EMPLOYEE to be the manager—that is, EMPLOYEE has a relationship with itself. Such a relationship is known as a **recursive relationship**. The various cases of recursive relationships will be explored in Section 4.1.10.

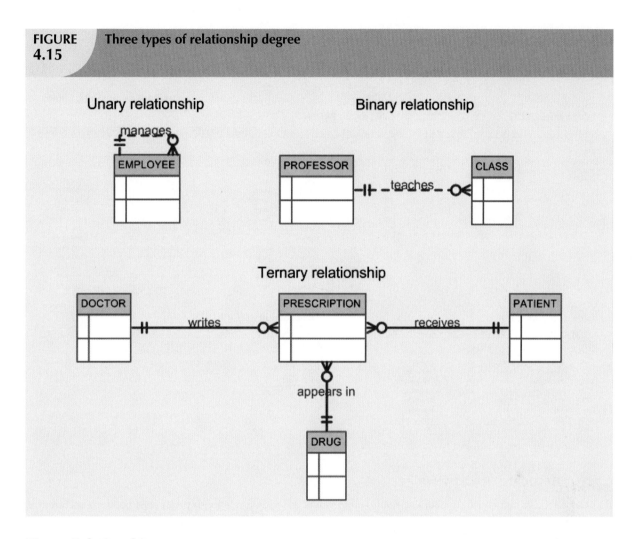

FIGURE 4.15   Three types of relationship degree

Binary Relationships

A binary relationship exists when two entities are associated in a relationship. Binary relationships are most common. In fact, to simplify the conceptual design, whenever possible, most higher-order (ternary and higher) relationships are decomposed into appropriate equivalent binary relationships. In Figure 4.15, the relationship "a PROFESSOR teaches one or more CLASSes" represents a binary relationship.

Ternary and Higher-Degree Relationships

Although most relationships are binary, the use of ternary and higher-order relationships does allow the designer some latitude regarding the semantics of a problem. A ternary relationship implies an association among three different entities. For example, note the relationships (and their consequences) in Figure 4.16, which are represented by the following business rules:

- A DOCTOR writes one or more PRESCRIPTIONs.
- A PATIENT may receive one or more PRESCRIPTIONs.
- A DRUG may appear in one or more PRESCRIPTIONs. (To simplify this example, assume that the business rule states that each prescription contains only one drug. In short, if a doctor prescribes more than one drug, a separate prescription must be written for each drug.)

As you examine the table contents in Figure 4.16, note that it is possible to track all transactions. For instance, you can tell that the first prescription was written by doctor 32445 for patient 102, using the drug DRZ.

FIGURE
4.16

The implementation of a ternary relationship

Database name: Ch04_Clinic

Table name: DRUG

DRUG_CODE	DRUG_NAME	DRUG_PRICE
AF15	Afgapan-15	25.00
AF25	Afgapan-25	35.00
DRO	Droalene Chloride	111.89
DRZ	Druzocholar Cryptolene	18.99
KO15	Koliabar Oxyhexalene	65.75
OLE	Oleander-Drizapan	123.95
TRYP	Tryptolac Heptadimetric	79.45

Table name: PATIENT

PAT_NUM	PAT_TITLE	PAT_LNAME	PAT_FNAME	PAT_INITIAL	PAT_DOB	PAT_AREACODE	PAT_PHONE
100	Mr.	Kolmycz	George	D	15-Jun-1942	615	324-5456
101	Ms.	Lewis	Rhonda	G	19-Mar-2005	615	324-4472
102	Mr.	Vandam	Rhett		14-Nov-1958	901	675-8993
103	Ms.	Jones	Anne	M	16-Oct-1974	615	898-3456
104	Mr.	Lange	John	P	08-Nov-1971	901	504-4430
105	Mr.	Williams	Robert	D	14-Mar-1975	615	890-3220
106	Mrs.	Smith	Jeanine	K	12-Feb-2003	615	324-7883
107	Mr.	Diante	Jorge	D	21-Aug-1974	615	890-4567
108	Mr.	Wiesenbach	Paul	R	14-Feb-1966	615	897-4358
109	Mr.	Smith	George	K	18-Jun-1961	901	504-3339
110	Mrs.	Genkazi	Leighla	W	19-May-1970	901	569-0093
111	Mr.	Washington	Rupert	E	03-Jan-1966	615	890-4925
112	Mr.	Johnson	Edward	E	14-May-1961	615	898-4387
113	Ms.	Smythe	Melanie	P	15-Sep-1970	615	324-9006
114	Ms.	Brandon	Marie	G	02-Nov-1932	901	882-0845
115	Mrs.	Saranda	Hermine	R	25-Jul-1972	615	324-5505
116	Mr.	Smith	George	A	08-Nov-1965	615	890-2984

Table name: DOCTOR

DOC_ID	DOC_LNAME	DOC_FNAME	DOC_INITIAL	DOC_SPECIALTY
29827	Sanchez	Julio	J	Dermatology
32445	Jorgensen	Annelise	G	Neurology
33456	Korenski	Anatoly	A	Urology
33989	LeGrande	George		Pediatrics
34409	Washington	Dennis	F	Orthopaedics
36221	McPherson	Katye	H	Dermatology
36712	Dreifag	Herman	G	Psychiatry
38995	Minh	Tran		Neurology
40004	Chin	Ming	D	Orthopaedics
40028	Feinstein	Denise	L	Gynecology

Table name: PRESCRIPTION

DOC_ID	PAT_NUM	DRUG_CODE	PRES_DOSAGE	PRES_DATE
32445	102	DRZ	2 tablets every four hours -- 50 tablets total	12-Nov-07
32445	113	OLE	1 teaspoon with each meal -- 250 ml total	14-Nov-07
34409	101	KO15	1 tablet every six hours -- 30 tablets total	14-Nov-07
36221	109	DRO	2 tablets with every meal -- 60 tablets total	14-Nov-07
38995	107	KO15	1 tablet every six hours -- 30 tablets total	14-Nov-07

4.1.10 RECURSIVE RELATIONSHIPS

As was previously mentioned, a *recursive relationship* is one in which a relationship can exist between occurrences of the same entity set. (Naturally, such a condition is found within a unary relationship.)

For example, a 1:M unary relationship can be expressed by "an EMPLOYEE may manage many EMPLOYEEs, and each EMPLOYEE is managed by one EMPLOYEE." And as long as polygamy is not legal, a 1:1 unary relationship may be expressed by "an EMPLOYEE may be married to one and only one other EMPLOYEE." Finally, the M:N unary relationship may be expressed by "a COURSE may be a prerequisite to many other COURSEs, and each COURSE may have many other COURSEs as prerequisites." Those relationships are shown in Figure 4.17.

The 1:1 relationship shown in Figure 4.17 can be implemented in the single table shown in Figure 4.18. Note that you can determine that James Ramirez is married to Louise Ramirez, who is married to James Ramirez. And Anne Jones is married to Anton Shapiro, who is married to Anne Jones.

Unary relationships are common in manufacturing industries. For example, Figure 4.19 illustrates that a rotor assembly (C-130) is composed of many parts, but each part is used to create only one rotor assembly. Figure 4.19 indicates that a rotor assembly is composed of four 2.5-cm washers, two cotter pins, one 2.5-cm steel shank, four 10.25-cm rotor blades, and two 2.5-cm hex nuts. The relationship implemented in Figure 4.19 thus enables you to track each part within each rotor assembly.

FIGURE 4.17   **An ER representation of recursive relationships**

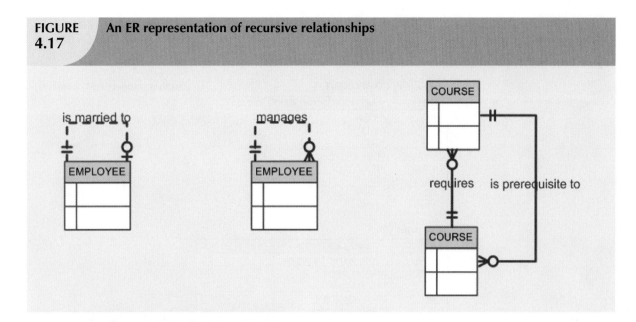

FIGURE 4.18   **The 1:1 recursive relationship "EMPLOYEE is married to EMPLOYEE"**

Database name: CH04_PartCo
Table name: EMPLOYEE_V1

EMP_NUM	EMP_LNAME	EMP_FNAME	EMP_SPOUSE
345	Ramirez	James	347
346	Jones	Anne	349
347	Ramirez	Louise	345
348	Delaney	Robert	
349	Shapiro	Anton	346

If a part can be used to assemble several different kinds of other parts and is itself composed of many parts, two tables are required to implement the "PART contains PART" relationship. Figure 4.20 illustrates such an environment. Parts tracking is increasingly important as managers become more aware of the legal ramifications of producing more complex output. In fact, in many industries, especially those involving aviation, full parts tracking is required by law.

FIGURE 4.19   **Another unary relationship: "PART contains PART"**

Table name: PART_V1   Database name; CH04_PartCo

PART_CODE	PART_DESCRIPTION	PART_IN_STOCK	PART_UNITS_NEEDED	PART_OF_PART
AA21-6	2.5 cm. washer, 1.0 mm. rim	432	4	C-130
AB-121	Cotter pin, copper	1034	2	C-130
C-130	Rotor assembly	36		
E129	2.5 cm. steel shank	128	1	C-130
X10	10.25 cm. rotor blade	345	4	C-130
X34AW	2.5 cm. hex nut	879	2	C-130

FIGURE 4.20   Implementation of the M:N recursive "PART contains PART" relationship

Table name: COMPONENT   Database name: Ch04_PartCo

COMP_CODE	PART_CODE	COMP_PARTS_NEEDED
C-130	AA21-6	4
C-130	AB-121	2
C-130	E129	1
C-131A2	E129	1
C-130	X10	4
C-131A2	X10	1
C-130	X34AW	2
C-131A2	X34AW	2

Table name: PART

PART_CODE	PART_DESCRIPTION	PART_IN_STOCK
AA21-6	2.5 cm. washer, 1.0 mm. rim	432
AB-121	Cotter pin, copper	1034
C-130	Rotor assembly	36
E129	2.5 cm. steel shank	128
X10	10.25 cm. rotor blade	345
X34AW	2.5 cm. hex nut	879

FIGURE 4.21   Implementation of the M:N "COURSE requires COURSE" recursive relationship

Table name: COURSE   Database name: Ch04_TinyCollege

CRS_CODE	DEPT_CODE	CRS_DESCRIPTION	CRS_CREDIT
ACCT-211	ACCT	Accounting I	3
ACCT-212	ACCT	Accounting II	3
CIS-220	CIS	Intro. to Microcomputing	3
CIS-420	CIS	Database Design and Implementation	4
MATH-243	MATH	Mathematics for Managers	3
QM-261	CIS	Intro. to Statistics	3
QM-362	CIS	Statistical Applications	4

Table name: PREREQ

CRS_CODE	PRE_TAKE
CIS-420	CIS-220
QM-261	MATH-243
QM-362	MATH-243
QM-362	QM-261

FIGURE 4.22   Implementation of the 1:M "EMPLOYEE manages EMPLOYEE" recursive relationship

Database name: Ch04_PartCo
Table name: EMPLOYEE_V2

EMP_CODE	EMP_LNAME	EMP_MANAGER
101	Waddell	102
102	Orincona	
103	Jones	102
104	Reballoh	102
105	Robertson	102
106	Deltona	102

The M:N recursive relationship might be more familiar in a school environment. For instance, note how the M:N "COURSE requires COURSE" relationship illustrated in Figure 4.17 is implemented in Figure 4.21. In this example, MATH-243 is a prerequisite to QM-261 and QM-362, while both MATH-243 and QM-261 are prerequisites to QM-362.

Finally, the 1:M recursive relationship "EMPLOYEE manages EMPLOYEE," shown in Figure 4.17, is implemented in Figure 4.22.

4.1.11 ASSOCIATIVE (COMPOSITE) ENTITIES

In the original ERM described by Chen, relationships do not contain attributes. You should recall from Chapter 3 that the relational model generally requires the use of 1:M relationships. (Also, recall that the 1:1 relationship has its place, but it should be used with caution and proper justification.) If M:N relationships are encountered, you must create a bridge between the entities that display such relationships. The associative entity is used to implement a M:M relationship between two or more entities. This associative entity (also known as a *composite or bridge entity*) is composed of the primary keys of each of the entities to be connected. An example of such a bridge is shown in Figure 4.23. The Crow's Foot notation does not identify the composite entity as such. Instead, the composite entity is identified by the solid relationship line between the parent and child entities, thereby indicating the presence of a strong (identifying) relationship.

FIGURE 4.23	Converting the M:N relationship into two 1:M relationships

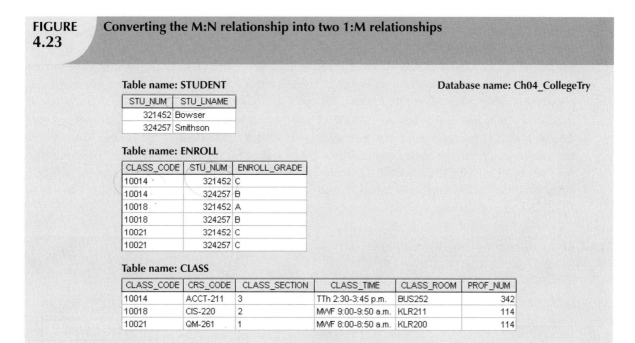

Table name: STUDENT   Database name: Ch04_CollegeTry

STU_NUM	STU_LNAME
321452	Bowser
324257	Smithson

Table name: ENROLL

CLASS_CODE	STU_NUM	ENROLL_GRADE
10014	321452	C
10014	324257	B
10018	321452	A
10018	324257	B
10021	321452	C
10021	324257	C

Table name: CLASS

CLASS_CODE	CRS_CODE	CLASS_SECTION	CLASS_TIME	CLASS_ROOM	PROF_NUM
10014	ACCT-211	3	TTh 2:30-3:45 p.m.	BUS252	342
10018	CIS-220	2	MWF 9:00-9:50 a.m.	KLR211	114
10021	QM-261	1	MWF 8:00-8:50 a.m.	KLR200	114

Note that the composite ENROLL entity in Figure 4.23 is existence-dependent on the other two entities; the composition of the ENROLL entity is based on the primary keys of the entities that are connected by the composite entity. The composite entity may also contain additional attributes that play no role in the connective process. For example, although the entity must be composed of at least the STUDENT and CLASS primary keys, it may also include such additional attributes as grades, absences, and other data uniquely identified by the student's performance in a specific class.

Finally, keep in mind that the ENROLL table's key (CLASS_CODE and STU_NUM) is composed entirely of the primary keys of the CLASS and STUDENT tables. Therefore, no null entries are possible in the ENROLL table's key attributes.

Implementing the small database shown in Figure 4.23 requires that you define the relationships clearly. Specifically, you must know the "1" and the "M" sides of each relationship, and you must know whether the relationships are mandatory or optional. For example, note the following points:

- A class may exist (at least at the start of registration) even though it contains no students. Therefore, if you examine Figure 4.24, an optional symbol should appear on the STUDENT side of the M:N relationship between STUDENT and CLASS.

  You might argue that to be classified as a STUDENT, a person must be enrolled in at least one CLASS. Therefore, CLASS is mandatory to STUDENT from a purely conceptual point of view. However, when a

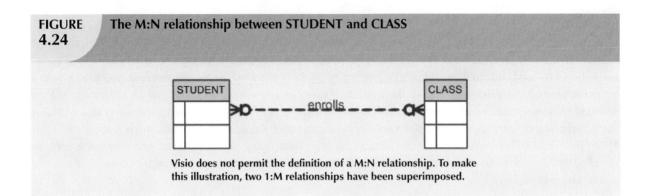

FIGURE 4.24   The M:N relationship between STUDENT and CLASS

Visio does not permit the definition of a M:N relationship. To make this illustration, two 1:M relationships have been superimposed.

student is admitted to college, that student has not (yet) signed up for any classes. Therefore, *at least initially,* CLASS is optional to STUDENT. Note that the practical considerations in the data environment help dictate the use of optionalities. If CLASS is *not* optional to STUDENT—from a database point of view—a class assignment must be made when the student is admitted. But that's *not* how the process actually works, and the database design must reflect this. In short, the optionality reflects practice.

Because the M:N relationship between STUDENT and CLASS is decomposed into two 1:M relationships through ENROLL, the optionalities must be transferred to ENROLL. See Figure 4.25. In other words, it now becomes possible for a class not to occur in ENROLL if no student has signed up for that class. Because a class need not occur in ENROLL, the ENROLL entity becomes optional to CLASS. And because the ENROLL entity is created before any students have signed up for a class, the ENROLL entity is also optional to STUDENT, at least initially.

FIGURE 4.25   A composite entity in an ERD

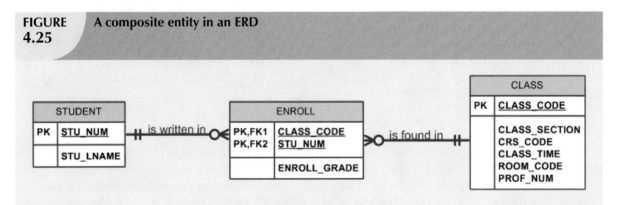

- As students begin to sign up for their classes, they will be entered into the ENROLL entity. Naturally, if a student takes more than one class, that student will occur more than once in ENROLL. For example, note that in the ENROLL table in Figure 4.23, STU_NUM = 321452 occurs three times . On the other hand, each student occurs only once in the STUDENT entity. (Note that the STUDENT table in Figure 4.23 has only one STU_NUM = 321452 entry.) Therefore, in Figure 4.25, the relationship between STUDENT and ENROLL is shown to be 1:M, with the M on the ENROLL side.

- As you can see in Figure 4.23, a class can occur more than once in the ENROLL table. For example, CLASS_CODE = 10014 occurs twice. However, CLASS_CODE = 10014 occurs only once in the CLASS table to reflect that the relationship between CLASS and ENROLL is 1:M. Note that in Figure 4.25, the M is located on the ENROLL side, while the 1 is located on the CLASS side.

4.2 DEVELOPING AN ER DIAGRAM

The process of database design is an iterative rather than a linear or sequential process. The verb *iterate* means "to do again or repeatedly." An **iterative process** is, thus, one based on repetition of processes and procedures. Building an ERD usually involves the following activities:

- Create a detailed narrative of the organization's description of operations.
- Identify the business rules based on the description of operations.
- Identify the main entities and relationships from the business rules.
- Develop the initial ERD.
- Identify the attributes and primary keys that adequately describe the entities.
- Revise and review the ERD.

During the review process, it is likely that additional objects, attributes, and relationships will be uncovered. Therefore, the basic ERM will be modified to incorporate the newly discovered ER components. Subsequently, another round of reviews might yield additional components or clarification of the existing diagram. The process is repeated until the end users and designers agree that the ERD is a fair representation of the organization's activities and functions.

During the design process, the database designer does not depend simply on interviews to help define entities, attributes, and relationships. A surprising amount of information can be gathered by examining the business forms and reports that an organization uses in its daily operations.

To illustrate the use of the iterative process that ultimately yields a workable ERD, let's start with an initial interview with the Tiny College administrators. The interview process yields the following business rules:

1. Tiny College (TC) is divided into several schools: a school of business, a school of arts and sciences, a school of education, and a school of applied sciences. Each school is administered by a dean who is a professor. Each dean can administer only one school. Therefore, a 1:1 relationship exists between PROFESSOR and SCHOOL. Note that the cardinality can be expressed by (1,1) for the entity PROFESSOR and by (1,1) for the entity SCHOOL. (The smallest number of deans per school is one, as is the largest number, and each dean is assigned to only one school.)

2. Each school is composed of several departments. For example, the school of business has an accounting department, a management/marketing department, an economics/finance department, and a computer information systems department. Note again the cardinality rules: the smallest number of departments operated by a school is one, and the largest number of departments is indeterminate (N). On the other hand, each department belongs to only a single school; thus, the cardinality is expressed by (1,1). That is, the minimum number of schools that a department belongs to is one, as is the maximum number. Figure 4.26 illustrates these first two business rules.

3. Each department may offer courses. For example, the management/marketing department offers courses such as Introduction to Management, Principles of Marketing, and Production Management. The ERD segment for this condition is shown in Figure 4.27. Note that this relationship is based on the way Tiny College operates. If, for example, Tiny College had some departments that were classified as "research only," those departments would not offer courses; therefore, the COURSE entity would be optional to the DEPARTMENT entity.

FIGURE
4.26
The first Tiny College ERD segment

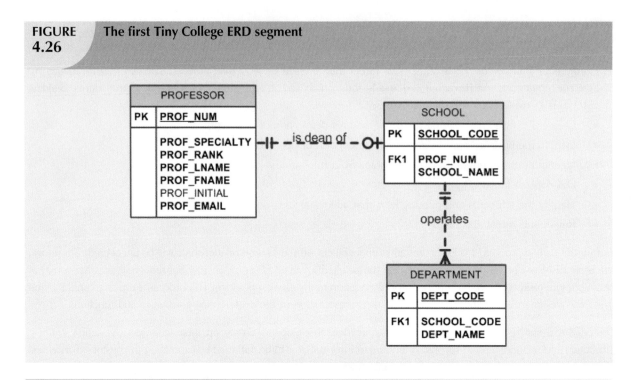

NOTE

It is again appropriate to evaluate the reason for maintaining the 1:1 relationship between PROFESSOR and SCHOOL in the "PROFESSOR is dean of SCHOOL" relationship. It is worth repeating that the existence of 1:1 relationships often indicates a misidentification of attributes as entities. In this case, the 1:1 relationship could easily be eliminated by storing the deans attributes in the SCHOOL entity. This solution also would make it easier to answer the queries, "Who is the school's dean" and What are that dean's credentials?" The downside of this solution is that it requires the duplication of data that are already stored in the PROFESSOR table, thus setting the stage for anomalies. However, because each school is run by a single dean, the problem of data duplication is rather minor. The selection of one approach over another often depends on information requirements, transaction speed, and the database designer's professional judgment. In short, do not use 1:1 relationships lightly, and make sure that each 1:1 relationship within the database design is defensible.

FIGURE
4.27
The second Tiny College ERD segment

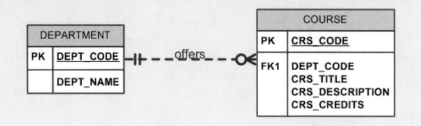

4. The relationship between COURSE and CLASS was illustrated in Figure 4.9. Nevertheless, it is worth repeating that a CLASS is a section of a COURSE. That is, a department may offer several sections (classes) of the same database course. Each of those classes is taught by a professor at a given time in a given place. In short, a 1:M relationship exists between COURSE and CLASS. However, because a course may exist in Tiny College's course catalog even when it is not offered as a class in a current class schedule, CLASS is optional to COURSE. Therefore, the relationship between COURSE and CLASS looks like Figure 4.28.

FIGURE 4.28   **The third Tiny College ERD segment**

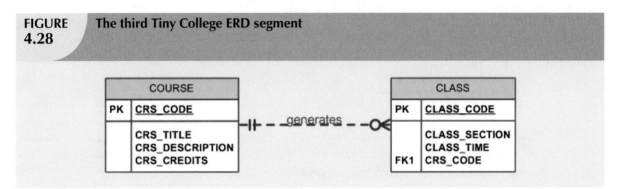

5. Each department may have many professors assigned to it. One and only one of those professors chairs the department, and no professor is required to accept the chair position. Therefore, DEPARTMENT is optional to PROFESSOR in the "chairs" relationship. Those relationships are summarized in the ER segment shown in Figure 4.29.

FIGURE 4.29   **The fourth Tiny College ERD segment**

6. Each professor may teach up to four classes; each class is a section of a course. A professor may also be on a research contract and teach no classes at all. The ERD segment in Figure 4.30 depicts those conditions.

7. A student may enroll in several classes but takes each class only once during any given enrollment period. For example, during the current enrollment period, a student may decide to take five classes—Statistics, Accounting, English, Database, and History—but that student would not be enrolled in the same Statistics class five times during the enrollment period! Each student may enroll in up to six classes, and each class may have up to 35 students, thus creating an M:N relationship between STUDENT and CLASS. Because a CLASS can initially exist (at the start of the enrollment period) even though no students have enrolled in it, STUDENT is optional to CLASS in the M:N relationship. This M:N relationship must be divided into two 1:M relationships through the use of the ENROLL entity, shown in the ERD segment in Figure 4.31. But note that the optional symbol is shown next to ENROLL. If a class exists but has no students enrolled in it, that class doesn't occur in the ENROLL table. Note also that the ENROLL entity is weak: it is existence-dependent, and its (composite) PK is composed of the PKs of the

FIGURE 4.30   **The fifth Tiny College ERD segment**

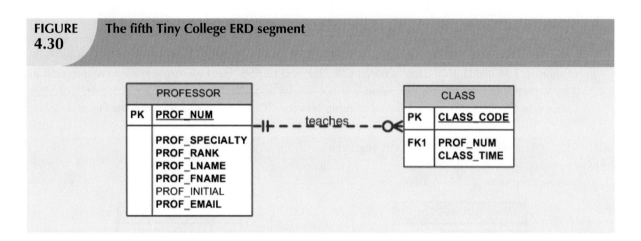

STUDENT and CLASS entities. You can add the cardinalities (0,6) and (0,35) next to the ENROLL entity to reflect the business rule constraints as shown in Figure 4.31. (Visio Professional does not automatically generate such cardinalities, but you can use a text box to accomplish that task.)

FIGURE 4.31   **The sixth Tiny College ERD segment**

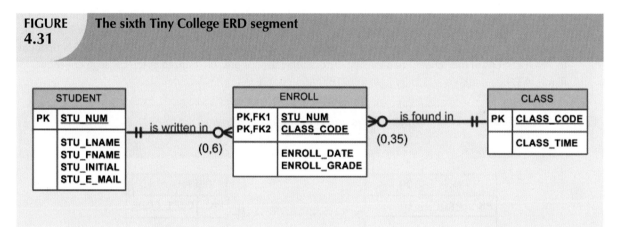

8.   Each department has several (or many) students whose major is offered by that department. However, each student has only a single major and is, therefore, associated with a single department. See Figure 4.32. However, in the Tiny College environment, it is possible—at least for a while—for a student not to declare a major field of study. Such a student would not be associated with a department; therefore, DEPARTMENT is optional to STUDENT. It is worth repeating that the relationships between entities and the entities themselves reflect the organization's operating environment. That is, the business rules define the ERD components.

FIGURE
4.32

FIGURE 4.32   **The seventh Tiny College ERD segment**

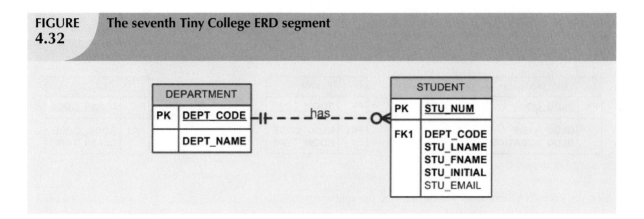

9. Each student has an advisor in his or her department; each advisor counsels several students. An advisor is also a professor, but not all professors advise students. Therefore, STUDENT is optional to PROFESSOR in the "PROFESSOR advises STUDENT" relationship. See Figure 4.33.

FIGURE 4.33   **The eighth Tiny College ERD segment**

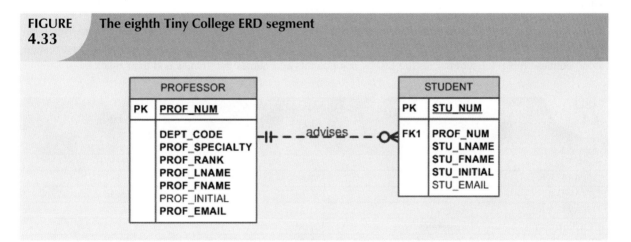

10. As you can see in Figure 4.34, the CLASS entity contains a ROOM_CODE attribute. Given the naming conventions, it is clear that ROOM_CODE is an FK to another entity. Clearly, because a class is taught in a room, it is reasonable to assume that the ROOM_CODE in CLASS is the FK to an entity named ROOM. In turn, each room is located in a building. So the last Tiny College ERD is created by observing that a BUILDING can contain many ROOMs, but each ROOM is found in a single BUILDING. In this ERD segment, it is clear that some buildings do not contain (class) rooms. For example, a storage building might not contain any named rooms at all.

**FIGURE
4.34**   **The ninth Tiny College ERD segment**

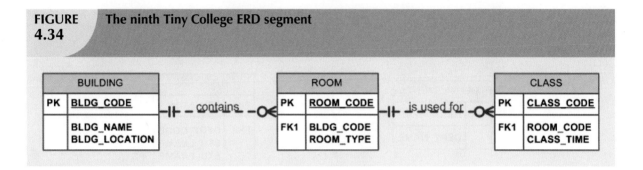

Using the preceding summary, you can identify the following entities:

SCHOOL	COURSE
DEPARTMENT	CLASS
ENROLL (the bridge entity between STUDENT and CLASS)	
PROFESSOR	STUDENT
BUILDING	ROOM

Once you have discovered the relevant entities, you can define the initial set of relationships among them. Next, you describe the entity attributes. Identifying the attributes of the entities helps you better understand the relationships among entities. Table 4.4 summarizes the ERM's components and names the entities and their relations.

**TABLE
4.4**   **Components of the ERM**

ENTITY	RELATIONSHIP	CONNECTIVITY	ENTITY
SCHOOL	operates	1:M	DEPARTMENT
DEPARTMENT	has	1:M	STUDENT
DEPARTMENT	employs	1:M	PROFESSOR
DEPARTMENT	offers	1:M	COURSE
COURSE	generates	1:M	CLASS
PROFESSOR	is dean of	1:1	SCHOOL
PROFESSOR	chairs	1:1	DEPARTMENT
PROFESSOR	teaches	1:M	CLASS
PROFESSOR	advises	1:M	STUDENT
STUDENT	enrolls in	M:N	CLASS
BUILDING	contains	1:M	ROOM
ROOM	is used for	1:M	CLASS
Note: ENROLL is the composite entity that implements the M:N relationship "STUDENT enrolls in CLASS."			

You must also define the connectivity and cardinality for the just-discovered relations based on the business rules. However, to avoid crowding the diagram, the cardinalities are not shown. Figure 4.35 shows the Crow's Foot ERD for Tiny College. Note that this is an implementation-ready model. Therefore it shows the ENROLL composite entity.

**FIGURE
4.35**   **The completed Tiny College ERD**

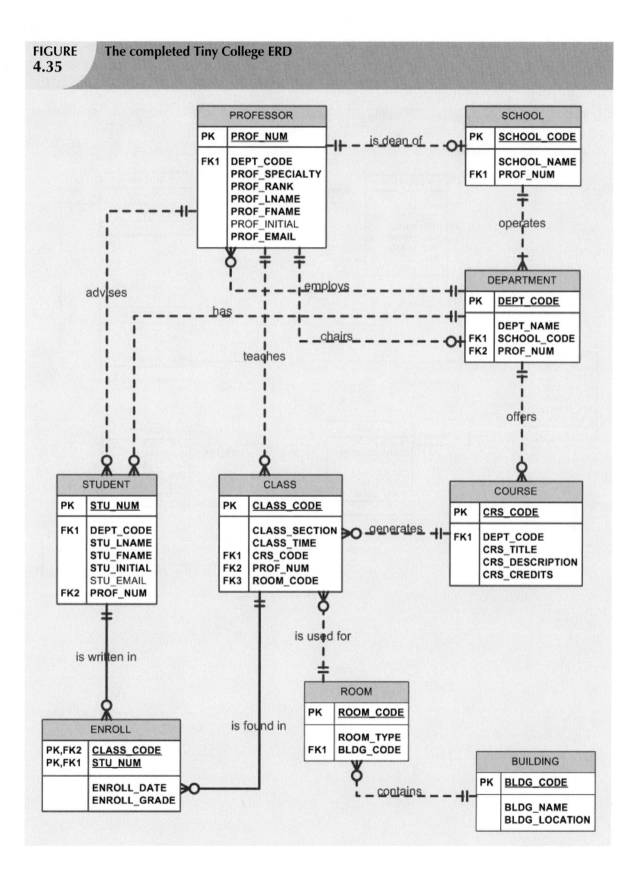

Figure 4.36 shows the conceptual UML class diagram for Tiny College. Note that this class diagram depicts the M:N relationship between STUDENT and CLASS. Figure 4.37 shows the implementation-ready UML class diagram for Tiny College (note that the ENROLL composite entity is shown in this class diagram.)

FIGURE 4.36   **The conceptual UML class diagram for Tiny College**

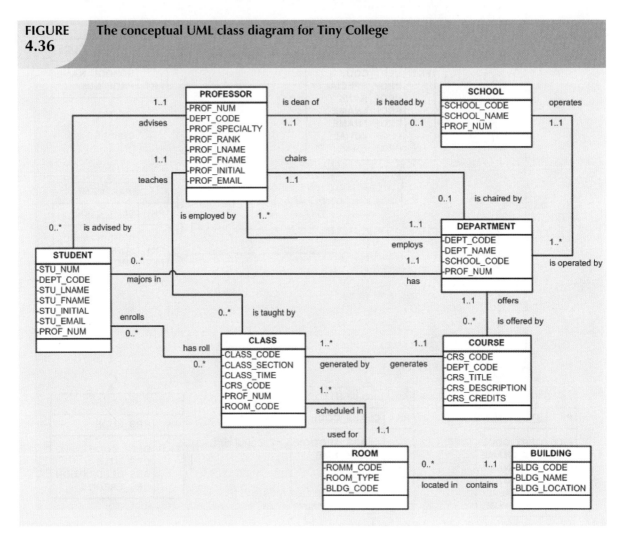

**FIGURE
4.37**   **The implementation-ready UML class diagram for Tiny College**

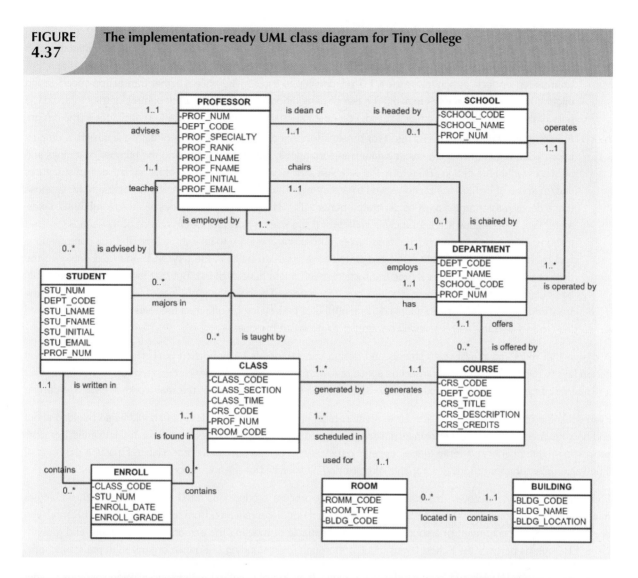

4.3 DATABASE DESIGN CHALLENGES: CONFLICTING GOALS

Database designers often must make design compromises that are triggered by conflicting goals, such as adherence to design standards (design elegance), processing speed, and information requirements.

- *Design standards.* The database design must conform to design standards. Such standards have guided you in developing logical structures that minimize data redundancies, thereby minimizing the likelihood that destructive data anomalies will occur. You have also learned how standards prescribed avoiding nulls to the greatest extent possible. In fact, you have learned that design standards govern the presentation of all components within the database design. In short, design standards allow you to work with well-defined components and to evaluate the interaction of those components with some precision. Without design standards, it is nearly impossible to formulate a proper design process, to evaluate an existing design, or to trace the likely logical impact of changes in design.

- *Processing speed.* In many organizations, particularly those generating large numbers of transactions, high processing speeds are often a top priority in database design. High processing speed means minimal access time, which may be achieved by minimizing the number and complexity of logically desirable relationships. For example, a "perfect" design might use a 1:1 relationship to avoid nulls, while a higher transaction-speed design might combine the two tables to avoid the use of an additional relationship, using dummy entries to avoid the nulls. If the focus is on data-retrieval speed, you might also be forced to include derived attributes in the design.

- *Information requirements.* The quest for timely information might be the focus of database design. Complex information requirements may dictate data transformations, and they may expand the number of entities and attributes within the design. Therefore, the database may have to sacrifice some of its "clean" design structures and/or some of its high transaction speed to ensure maximum information generation. For example, suppose that a detailed sales report must be generated periodically. The sales report includes all invoice subtotals, taxes, and totals; even the invoice lines include subtotals. If the sales report includes hundreds of thousands (or even millions) of invoices, computing the totals, taxes, and subtotals is likely to take some time. If those computations had been made and the results had been stored as derived attributes in the INVOICE and LINE tables at the time of the transaction, the real-time transaction speed might have declined. But that loss of speed would only be noticeable if there were many simultaneous transactions. The cost of a slight loss of transaction speed at the front end and the addition of multiple derived attributes is likely to pay off when the sales reports are generated (not to mention the fact that it will be simpler to generate the queries).

A design that meets all logical requirements and design conventions is an important goal. However, if this perfect design fails to meet the customer's transaction speed and/or information requirements, the designer will not have done a proper job from the end user's point of view. Compromises are a fact of life in the real world of database design.

Even while focusing on the entities, attributes, relationships, and constraints, the designer should begin thinking about end-user requirements such as performance, security, shared access, and data integrity. The designer must consider processing requirements and verify that all update, retrieval, and deletion options are available. Finally, a design is of little value unless the end product is capable of delivering all specified query and reporting requirements.

You are quite likely to discover that even the best design process produces an ERD that requires further changes mandated by operational requirements. Such changes should not discourage you from using the process. ER modeling is essential in the development of a sound design that is capable of meeting the demands of adjustment and growth. Using ERDs yields perhaps the richest bonus of all: a thorough understanding of how an organization really functions.

There are occasional design and implementation problems that do not yield "clean" implementation solutions. To get a sense of the design and implementation choices a database designer faces, let's revisit the 1:1 recursive relationship "EMPLOYEE is married to EMPLOYEE" first examined in Figure 4.18. Figure 4.38 shows three different ways of implementing such a relationship.

FIGURE 4.38	Various implementations of the 1:1 recursive relationship

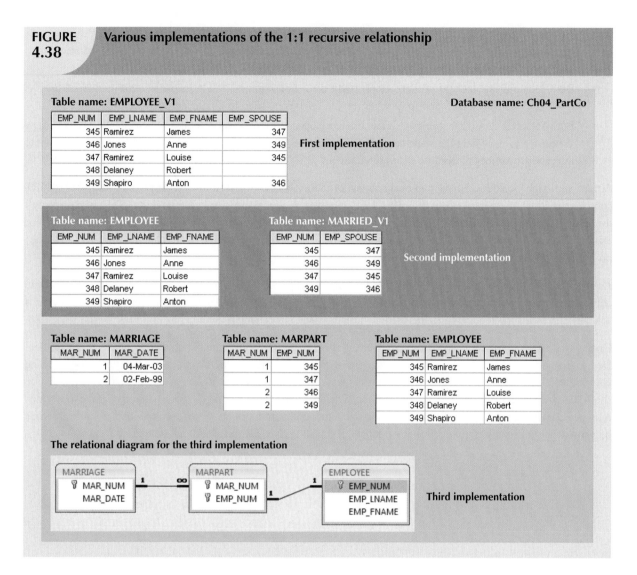

Table name: EMPLOYEE_V1   Database name: Ch04_PartCo

EMP_NUM	EMP_LNAME	EMP_FNAME	EMP_SPOUSE
345	Ramirez	James	347
346	Jones	Anne	349
347	Ramirez	Louise	345
348	Delaney	Robert	
349	Shapiro	Anton	346

First implementation

Table name: EMPLOYEE

EMP_NUM	EMP_LNAME	EMP_FNAME
345	Ramirez	James
346	Jones	Anne
347	Ramirez	Louise
348	Delaney	Robert
349	Shapiro	Anton

Table name: MARRIED_V1

EMP_NUM	EMP_SPOUSE
345	347
346	349
347	345
349	346

Second implementation

Table name: MARRIAGE

MAR_NUM	MAR_DATE
1	04-Mar-03
2	02-Feb-99

Table name: MARPART

MAR_NUM	EMP_NUM
1	345
1	347
2	346
2	349

Table name: EMPLOYEE

EMP_NUM	EMP_LNAME	EMP_FNAME
345	Ramirez	James
346	Jones	Anne
347	Ramirez	Louise
348	Delaney	Robert
349	Shapiro	Anton

The relational diagram for the third implementation

Third implementation

Note that the EMPLOYEE_V1 table in Figure 4.38 is likely to yield data anomalies. For example, if Anne Jones divorces Anton Shapiro, two records must be updated—by setting the respective EMP_SPOUSE values to null—to properly reflect that change. If only one record is updated, inconsistent data occur. The problem becomes even worse if several of the divorced employees then marry each other. In addition, that implementation also produces undesirable nulls for employees who are *not* married to other employees in the company.

Another approach would be to create a new entity shown as MARRIED_V1 in a 1:M relationship with EMPLOYEE. (See Figure 4.38.) This second implementation does eliminate the nulls for employees who are not married to somebody working for the same company. (Such employees would not be entered in the MARRIED_V1 table.) However, this approach still yields possible duplicate values. For example, the marriage between employees 345 and 347 may still appear twice, once as 345,347 and once as 347,345. (Since each of those permutations is unique the first time it appears, the creation of a unique index will not solve the problem.)

As you can see, the first two implementations yield several problems:

- Both solutions use synonyms. The EMPLOYEE_V1 table uses EMP_NUM and EMP_SPOUSE to refer to an employee. The MARRIED_V1 table uses the same synonyms.

- Both solutions are likely to produce inconsistent data. For example, it is possible to enter employee 345 as married to employee 347 and to enter employee 348 as married to employee 345.

- Both solutions allow data entries to show one employee married to several other employees. For example, it is possible to have data pairs such as 345,347 and 348,347 and 349,347, none of which will violate entity integrity requirements, because they are all unique.

A third approach would be to have two new entities, MARRIAGE and MARPART, in a 1:M relationship. MARPART contains the EMP_NUM foreign key to EMPLOYEE. (See the relational diagram in Figure 4.38.) This third approach would be the preferred solution in a relational environment. But even this approach requires some fine-tuning. For example, to ensure that an employee occurs only once in any given marriage, you would have to use a unique index on the EMP_NUM attribute in the MARPART table.

As you can see, a recursive 1:1 relationship yields many different solutions with varying degrees of effectiveness and adherence to basic design principles. Your job as a database designer is to use your professional judgment to yield a solution that meets the requirements imposed by business rules, processing requirements, and basic design principles.

Finally, document, document, and document! Put all design activities in writing. Then review what you've written. Documentation not only helps you stay on track during the design process, but also enables you (or those following you) to pick up the design thread when the time comes to modify the design. Although the need for documentation should be obvious, one of the most vexing problems in database and systems analysis work is that the "put it in writing" rule often is not observed in all of the design and implementation stages. The development of organizational documentation standards is a very important aspect of ensuring data compatibility and coherence.

SUMMARY

▶ The ERM uses ERDs to represent the conceptual database as viewed by the end user. The ERM's main components are entities, relationships, and attributes. The ERD also includes connectivity and cardinality notations. An ERD can also show relationship strength, relationship participation (optional or mandatory), and degree of relationship (unary, binary, ternary, etc.).

▶ Connectivity describes the relationship classification (1:1, 1:M, or M:N). Cardinality expresses the specific number of entity occurrences associated with an occurrence of a related entity. Connectivities and cardinalities are usually based on business rules.

▶ In the ERM, an M:N relationship is valid at the conceptual level. However, when implementing the ERM in a relational database, the M:N relationship must be mapped to a set of 1:M relationships through a composite entity.

▶ ERDs may be based on many different ERMs. However, regardless of which model is selected, the modeling logic remains the same. Because no ERM can accurately portray all real-world data and action constraints, application software must be used to augment the implementation of at least some of the business rules.

▶ Unified Modeling Language (UML) class diagrams are used to represent the static data structures in a data model. The symbols used in the UML class and ER diagrams are very similar. The UML class diagrams can be used to depict data models at the conceptual or implementation abstraction levels.

▶ Database designers, no matter how well they are able to produce designs that conform to all applicable modeling conventions, often are forced to make design compromises. Those compromises are required when end users have vital transaction speed and/or information requirements that prevent the use of "perfect" modeling logic and adherence to all modeling conventions. Therefore, database designers must use their professional judgment to determine how and to what extent the modeling conventions are subject to modification. To ensure that their professional judgments are sound, database designers must have detailed and in-depth knowledge of data-modeling conventions. It is also important to document the design process from beginning to end, which helps keep the design process on track and allows for easy modifications down the road.

KEY TERMS

binary relationship, 120

cardinality, 111

composite attribute, 108

composite identifier, 107

connectivity, 111

derived attribute, 110

existence-dependent, 113

existence-independent, 113

identifiers, 106

identifying relationship, 115

iterative process, 127

mandatory participation, 118

multivalued attribute, 108

non-identifying relationship, 113

optional attribute, 106

optional participation, 118

participants, 111

recursive relationship, 120

relationship degree, 120

required attribute, 105

simple attribute, 108

single-valued attribute, 108

strong relationship, 115

ternary relationship, 120

unary relationship, 120

weak entity, 113

weak relationship, 113

ONLINE CONTENT

Answers to selected Review Questions and Problems for this chapter are contained in the Student Online Companion for this book.

REVIEW QUESTIONS

1. What two conditions must be met before an entity can be classified as a weak entity? Give an example of a weak entity.

2. What is a strong (or identifying) relationship, and how is it depicted in a Crow's Foot ERD?

3. Given the business rule "an employee may have many degrees," discuss its effect on attributes, entities, and relationships. (*Hint:* Remember what a multivalued attribute is and how it might be implemented.)

4. What is a composite entity, and when is it used?

5. Suppose you are working within the framework of the conceptual model in Figure Q4.5.

FIGURE Q4.5   **The conceptual model for Question 5**

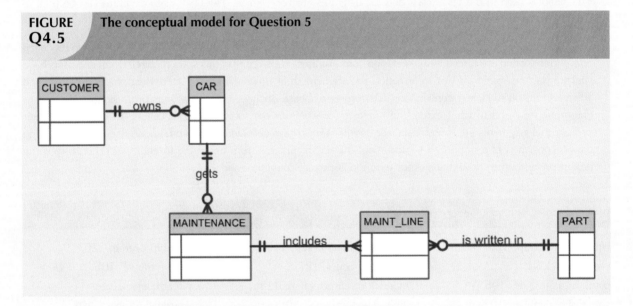

Given the conceptual model in Figure Q4.5:

a. Write the business rules that are reflected in it.

b. Identify all of the cardinalities.

6. What is a recursive relationship? Give an example.

7. How would you (graphically) identify each of the following ERM components in a Crow's Foot notation?

a. an entity

b. the cardinality (0,N)

c. a weak relationship

d. a strong relationship

8. Discuss the difference between a composite key and a composite attribute. How would each be indicated in an ERD?

9. What two courses of action are available to a designer encountering a multivalued attribute?

10. What is a derived attribute? Give an example.

11. How is a relationship between entities indicated in an ERD? Give an example, using the Crow's Foot notation.

12. Discuss two ways in which the 1:M relationship between COURSE and CLASS can be implemented. (*Hint:* Think about relationship strength.)

13. How is a composite entity represented in an ERD, and what is its function? Illustrate the Crow's Foot notation.

14. What three (often conflicting) database requirements must be addressed in database design?

15. Briefly, but precisely, explain the difference between single-valued attributes and simple attributes. Give an example of each.

16. What are multivalued attributes, and how can they be handled within the database design?

The final four questions are based on the ERD in Figure Q4.17.

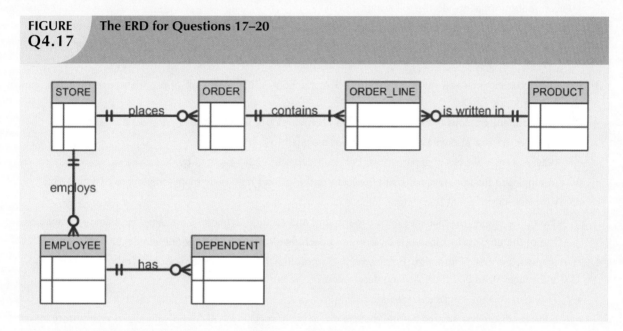

FIGURE Q4.17   The ERD for Questions 17–20

17. Write the ten cardinalities that are appropriate for this ERD.

18. Write the business rules reflected in this ERD.

19. What two attributes must be contained in the composite entity between STORE and PRODUCT? Use proper terminology in your answer.

20. Describe precisely the composition of the DEPENDENT weak entity's primary key. Use proper terminology in your answer.

PROBLEMS

1. Given the following business rules, create the appropriate Crow's Foot ERD.
   a. A company operates many departments.
   b. Each department employs one or more employees.
   c. Each of the employees might or might not have one or more dependents.
   d. Each employee might or might not have an employment history.

2. The Hudson Engineering Group (HEG) has contacted you to create a conceptual model whose application will meet the expected database requirements for the company's training program. The HEG administrator gives you the description (see below) of the training group's operating environment.

(*Hint:* Some of the following sentences identify the volume of data rather than cardinalities. Can you tell which ones?)

The HEG has 12 instructors and can handle up to 30 trainees per class. HEG offers five Advanced Technology courses, each of which may generate several classes. If a class has fewer than 10 trainees, it will be canceled. Therefore, it is possible for a course not to generate any classes. Each class is taught by one instructor. Each instructor may teach up to two classes or may be assigned to do research only. Each trainee may take up to two classes per year.

Given that information, do the following:

a. Define all of the entities and relationships. (Use Table 4.4 as your guide.)

b. Describe the relationship between instructor and class in terms of connectivity, cardinality, and existence-dependence.

3. Use the following business rules to create a Crow's Foot ERD. Write all appropriate connectivities and cardinalities in the ERD.

a. A department employs many employees, but each employee is employed by one department.

b. Some employees, known as "rovers," are not assigned to any department.

c. A division operates many departments, but each department is operated by one division.

d. An employee may be assigned many projects, and a project may have many employees assigned to it.

e. A project must have at least one employee assigned to it.

f. One of the employees manages each department, and each department is managed by only one employee.

g. One of the employees runs each division, and each division is run by only one employee.

4. During peak periods, Temporary Employment Corporation (TEC) places temporary workers in companies. TEC's manager gives you the following description of the business:

• TEC has a file of candidates who are willing to work.

• If the candidate has worked before, that candidate has a specific job history. (Naturally, no job history exists if the candidate has never worked.) Each time the candidate works, one additional job history record is created.

• Each candidate has earned several qualifications. Each qualification may be earned by more than one candidate. (For example, it is possible for more than one candidate to have earned a BBA degree or a Microsoft Network Certification. And clearly, a candidate may have earned both a BBA and a Microsoft Network Certification.)

• TEC also has a list of companies that request temporaries.

• Each time a company requests a temporary employee, TEC makes an entry in the Openings folder. That folder contains an opening number, a company name, required qualifications, a starting date, an anticipated ending date, and hourly pay.

• Each opening requires only one specific or main qualification.

• When a candidate matches the qualification, the job is assigned, and an entry is made in the Placement Record folder. That folder contains an opening number, a candidate number, the total hours worked, etc. In addition, an entry is made in the job history for the candidate.

• An opening can be filled by many candidates, and a candidate can fill many openings.

• TEC uses special codes to describe a candidate's qualifications for an opening. The list of codes is shown in table P4.4.

TABLE
P4.4

CODE	DESCRIPTION
SEC-45	Secretarial work, at least 45 words per minute
SEC-60	Secretarial work, at least 60 words per minute
CLERK	General clerking work
PRG-VB	Programmer, Visual Basic
PRG-C++	Programmer, C++
DBA-ORA	Database Administrator, Oracle
DBA-DB2	Database Administrator, IBM DB2
DBA-SQLSERV	Database Administrator, MS SQL Server
SYS-1	Systems Analyst, level 1
SYS-2	Systems Analyst, level 2
NW-NOV	Network Administrator, Novell experience
WD-CF	Web Developer, ColdFusion

TEC's management wants to keep track of the following entities:

- COMPANY
- OPENING
- QUALIFICATION
- CANDIDATE
- JOB_HISTORY
- PLACEMENT

Given that information, do the following:

    a. Draw the Crow's Foot ERDs for this enterprise.

    b. Identify all possible relationships.

    c. Identify the connectivity for each relationship.

    d. Identify the mandatory/optional dependencies for the relationships.

    e. Resolve all M:N relationships.

5. The Jonesburgh County Basketball Conference (JCBC) is an amateur basketball association. Each city in the county has one team as its representative. Each team has a maximum of 12 players and a minimum of 9 players. Each team also has up to three coaches (offensive, defensive, and physical training coaches). During the season, each team plays two games (home and visitor) against each of the other teams. Given those conditions, do the following:

    a. Identify the connectivity of each relationship.

    b. Identify the type of dependency that exists between CITY and TEAM.

    c. Identify the cardinality between teams and players and between teams and city.

    d. Identify the dependency between coach and team and between team and player.

    e. Draw the Chen and Crow's Foot ERDs to represent the JCBC database.

    f. Draw the UML class diagram to depict the JCBC database.

6. Automata Inc. produces specialty vehicles by contract. The company operates several departments, each of which builds a particular vehicle, such as a limousine, a truck, a van, or an RV.

Before a new vehicle is built, the department places an order with the purchasing department to request specific components. Automata's purchasing department is interested in creating a database to keep track of orders and to accelerate the process of delivering materials.

The order received by the purchasing department may contain several different items. An inventory is maintained so that the most frequently requested items are delivered almost immediately. When an order comes in, it is checked to determine whether the requested item is in inventory. If an item is not in inventory, it must be ordered from a supplier. Each item may have several suppliers.

Given that functional description of the processes encountered at Automata's purchasing department, do the following:

a. Identify all of the main entities.

b. Identify all of the relations and connectivities among entities.

c. Identify the type of existence dependency in all the relationships.

d. Give at least two examples of the types of reports that can be obtained from the database.

7. Create an ERD based on the Crow's Foot notation, using the following requirements:

- An INVOICE is written by a SALESREP. Each sales representative can write many invoices, but each invoice is written by a single sales representative.

- The INVOICE is written for a single CUSTOMER. However, each customer can have many invoices.

- An INVOICE can include many detail lines (LINE), each of which describes one product bought by the customer.

- The product information is stored in a PRODUCT entity.

- The product's vendor information is found in a VENDOR entity.

8. Given the following brief summary of business rules for the ROBCOR catering service and using the Crow's Foot ER notation, draw the fully labeled ERD. Make sure you include all appropriate entities, relationships, connectivities, and cardinalities.

NOTE

Limit your ERD to entities and relationships based on the business rules shown here. In other words, do *not* add realism to your design by expanding or refining the business rules. However, make sure you include the attributes that would permit the model to be successfully implemented.

Each dinner is based on a single entree, but each entree can be served at many dinners. A guest can attend many dinners, and each dinner can be attended by many guests. Each dinner invitation can be mailed to many guests, and each guest can receive many invitations.

9. Using the Crow's Foot notation, create an ERD that can be implemented for a medical clinic, using at least the following business rules:

a. A patient can make many appointments with one or more doctors in the clinic, and a doctor can accept appointments with many patients. However, each appointment is made with only one doctor and one patient.

b. Emergency cases do not require an appointment. However, for appointment management purposes, an emergency is entered in the appointment book as "unscheduled."

c. If kept, an appointment yields a visit with the doctor specified in the appointment. The visit yields a diagnosis and, when appropriate, treatment.

d. With each visit, the patient's records are updated to provide a medical history.

e. Each patient visit creates a bill. Each patient visit is billed by one doctor, and each doctor can bill many patients.

f. Each bill must be paid. However, a bill may be paid in many installments, and a payment may cover more than one bill.

g. A patient may pay the bill directly, or the bill may be the basis for a claim submitted to an insurance company.

h. If the bill is paid by an insurance company, the deductible is submitted to the patient for payment.

10. The administrators of Tiny College are so pleased with your design and implementation of their student registration/tracking system that they want you to expand the design to include the database for their motor vehicle pool. A brief description of operations follows:

- Faculty members may use the vehicles owned by Tiny College for officially sanctioned travel. For example, the vehicles may be used by faculty members to travel to off-campus learning centers, to travel to locations at which research papers are presented, to transport students to officially sanctioned locations, and to travel for public service purposes. The vehicles used for such purposes are managed by Tiny College's TFBS (Travel Far But Slowly) Center.

- Using reservation forms, each department can reserve vehicles for its faculty, who are responsible for filling out the appropriate trip completion form at the end of a trip. The reservation form includes the expected departure date, vehicle type required, destination, and name of the authorized faculty member. The faculty member arriving to pick up a vehicle must sign a checkout form to log out the vehicle and pick up a trip completion form. (The TFBS employee who releases the vehicle for use also signs the checkout form.) The faculty member's trip completion form includes the faculty member's identification code, the vehicle's identification, the odometer readings at the start and end of the trip, maintenance complaints (if any), gallons of fuel purchased (if any), and the Tiny College credit card number used to pay for the fuel. If fuel is purchased, the credit card receipt must be stapled to the trip completion form. Upon receipt of the faculty trip completion form, the faculty member's department is billed at a mileage rate based on the vehicle type (sedan, station wagon, panel truck, minivan, or minibus) used. (*Hint:* Do *not* use more entities than are necessary. Remember the difference between attributes and entities!)

- All vehicle maintenance is performed by TFBS. Each time a vehicle requires maintenance, a maintenance log entry is completed on a prenumbered maintenance log form. The maintenance log form includes the vehicle identification, a brief description of the type of maintenance required, the initial log entry date, the date on which the maintenance was completed, and the identification of the mechanic who released the vehicle back into service. (Only mechanics who have an inspection authorization may release the vehicle back into service.)

- As soon as the log form has been initiated, the log form's number is transferred to a maintenance detail form; the log form's number is also forwarded to the parts department manager, who fills out a parts usage form on which the maintenance log number is recorded. The maintenance detail form contains separate lines for each maintenance item performed, for the parts used, and for identification of the mechanic who performed the maintenance item. When all maintenance items have been completed, the maintenance detail form is stapled to the maintenance log form, the maintenance log form's completion date is filled out, and the mechanic who releases the vehicle back into service signs the form. The stapled forms are then filed, to be used later as the source for various maintenance reports.

- TFBS maintains a parts inventory, including oil, oil filters, air filters, and belts of various types. The parts inventory is checked daily to monitor parts usage and to reorder parts that reach the "minimum quantity on hand" level. To track parts usage, the parts manager requires each mechanic to sign out the parts that are used to perform each vehicle's maintenance; the parts manager records the maintenance log number under which the part is used.

- Each month TFBS issues a set of reports. The reports include the mileage driven by vehicle, by department, and by faculty members within a department. In addition, various revenue reports are generated by vehicle and department. A detailed parts usage report is also filed each month. Finally, a vehicle maintenance summary is created each month.

Given that brief summary of operations, draw the appropriate (and fully labeled) ERD. Use the Chen methodology to indicate entities, relationships, connectivities, and cardinalities.

11. Given the following information, produce an ERD—based on the Crow's Foot notation—that can be implemented. Make sure you include all appropriate entities, relationships, connectivities, and cardinalities.

- EverFail company is in the oil change and lube business. Although customers bring in their cars for what is described as "quick oil changes," EverFail also replaces windshield wipers, oil filters, and air filters, subject to customer approval. The invoice contains the charges for the oil and all parts used and a standard labor charge. When the invoice is presented to customers, they pay cash, use a credit card, or write a check. EverFail does not extend credit. EverFail's database is to be designed to keep track of all components in all transactions.

- Given the high parts usage of the business operations, EverFail must maintain careful control of its parts inventory (oil, wipers, oil filters, and air filters). Therefore, if parts reach their minimum on-hand quantity, the parts in low supply must be reordered from an appropriate vendor. EverFail maintains a vendor list, which contains vendors actually used and potential vendors.

- Periodically, based on the date of the car's service, EverFail mails updates to customers. EverFail also tracks each customer's car mileage.

NOTE

Problems 12 and 13 may be used as the basis for class projects. These problems illustrate the challenge of translating a description of operations to a set of business rules that will define the components for an ERD that can be successfully implemented. These problems can also be used as the basis for discussions about the components and contents of a proper description of operations. One of the things you must learn if you want to create databases that can be successfully implemented is to separate the generic background material from the details that directly affect database design. You must also keep in mind that many constraints cannot be incorporated into the database design; instead, such constraints are handled by the applications software. Although the description of operations in Problem 12 deals with a Web-based business, the focus should be on the *database* aspects of the design, rather than on its interface and the transaction management details. In fact, the argument can easily be made that the existence of Web-based businesses has made database design more important than ever. (You might be able to get away with a bad database design if you sell only a few items per day, but the problems of poorly designed databases are compounded as the number of transactions increases.)

12. Use the following descriptions of the operations of RC_Models Company to complete this exercise.

   RC_Models Company sells its products—plastic models (aircraft, ships, and cars) and "add-on" decals for those models—through its Internet Web site, *www.rc_models.com*. Models and decals are available in scales that vary from 1/144 to 1/32.

   Customers use the Web site to select the products and to pay by credit card. If a product is not currently available, it is placed on back order at the customer's discretion. (Back orders are not charged to a customer until the order is shipped.) When a customer completes a transaction, the invoice is printed and the products listed on the invoice are pulled from inventory for shipment. (The invoice includes a shipping charge.) The printed invoice is enclosed in the shipping container. The customer credit card charges are transmitted to the CC Bank, at which RC_Models Company maintains a commercial account. (*Note:* The CC Bank is *not* part of the RC_Models database.)

   RC_Models Company tracks customer purchases and periodically sends out promotional materials. Because the management at RC_Models Company requires detailed information to conduct its operations, numerous reports are available. Those reports include, but are not limited to, customer purchases by product category and amount, product turnover, and revenues by product and customer. If a product has not recorded a sale within four weeks of being stocked, it is removed from inventory and scrapped.

   Many of the customers on the RC_Models customer list have bought RC_Models products. However, RC_Models Company also has purchased a copy of the *FineScale Modeler* magazine subscription list to use in marketing its products to customers who have not yet bought from RC_Models Company. In addition, customer data are recorded when potential customers request product information.

   RC_Models Company orders its products directly from the manufacturers. For example, the plastic models are ordered from Tamiya, Academy, Revell/Monogram, and others. Decals are ordered from Aeromaster, Tauro, WaterMark, and others. (*Note:* Not all manufacturers in the RC_Models Company database have received orders.) All orders are placed via the manufacturers' Web sites, and the order amounts are automatically handled through RC_Models' commercial bank account with the CC Bank. Orders are automatically placed when product inventory reaches the specified minimum quantity on hand. (The number of product units ordered depends on the minimum order quantity specified for each product.)

   a. Given that brief and incomplete description of operations for RC_Models Company, write all applicable business rules to establish entities, relationships, optionalities, connectivities, and cardinalities. (*Hint:* Use the following three business rules as examples, writing the remaining business rules in the same format.)

- A customer may generate many invoices.
- Each invoice is generated by only one customer.
- Some customers have not (yet) generated an invoice.

   b. Draw the fully labeled and implementable Crow's Foot ERD based on the business rules you wrote in Part (a) of this problem. Include all entities, relationships, optionalities, connectivities, and cardinalities.

13. Use the following description of the operations of the RC_Charter2 Company to complete this exercise.

The RC_Charter2 Company operates a fleet of aircraft under the Federal Air Regulations (FAR) Part 135 (air taxi or charter) certificate, enforced by the FAA. The aircraft are available for air taxi (charter) operations within the United States and Canada.

Charter companies provide so-called "unscheduled" operations—that is, charter flights take place only after a customer reserves the use of an aircraft to fly at a customer-designated date and time to one or more customer-designated destinations, transporting passengers, cargo, or some combination of passengers and cargo. A customer can, of course, reserve many different charter flights (trips) during any time frame. However, for billing purposes, each charter trip is reserved by one and only one customer. Some of RC_Charter2's customers do not use the company's charter operations; instead, they purchase fuel, use maintenance services, or use other RC_Charter2 services. However, this database design will focus on the charter operations only.

Each charter trip yields revenue for the RC_Charter2 Company. This revenue is generated by the charges a customer pays upon the completion of a flight. The charter flight charges are a function of aircraft model used, distance flown, waiting time, special customer requirements, and crew expenses. The distance flown charges are computed by multiplying the round-trip miles by the model's charge per mile. Round-trip miles are based on the actual navigational path flown. The sample route traced in Figure P4.13 illustrates the procedure. Note that the number of round-trip miles is calculated to be 130 + 200 + 180 + 390 = 900.

Depending on whether a customer has RC_Charter2 credit authorization, the customer may:

- Pay the entire charter bill upon the completion of the charter flight.
- Pay a part of the charter bill and charge the remainder to the account. The charge amount may not exceed the available credit.
- Charge the entire charter bill to the account. The charge amount may not exceed the available credit.

Customers may pay all or part of the existing balance for previous charter trips. Such payments may be made at any time and are not necessarily tied to a specific charter trip. The charter mileage charge includes the expense of the pilot(s) and other crew required by FAR 135. However, if customers request *additional* crew *not* required by FAR 135, those customers are charged for the crew members on an hourly basis. The hourly crew-member charge is based on each crew member's qualifications.

The database must be able to handle crew assignments. Each charter trip requires the use of an aircraft, and a crew flies each aircraft. The smaller piston engine-powered charter aircraft require a crew consisting of only a single pilot. Larger aircraft (that is, aircraft having a gross takeoff weight of 12,500 pounds or more) and jet-powered aircraft require a pilot and a copilot, while some of the larger aircraft used to transport passengers may require flight attendants as part of the crew. Some of the older aircraft require the assignment of a flight engineer, and larger cargo-carrying aircraft require the assignment of a loadmaster. In short, a crew can consist of more than one person, and not all crew members are pilots.

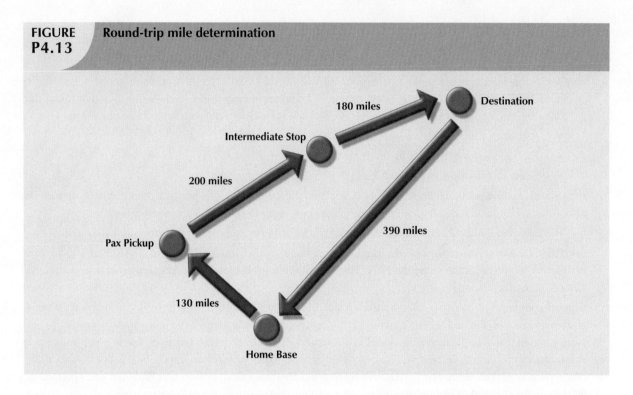

FIGURE P4.13   **Round-trip mile determination**

The charter flight's aircraft waiting charges are computed by multiplying the hours waited by the model's hourly waiting charge. Crew expenses are limited to meals, lodging, and ground transportation.

The RC_Charter2 database must be designed to generate a monthly summary of all charter trips, expenses, and revenues derived from the charter records. Such records are based on the data that each pilot in command is required to record for each charter trip: trip date(s) and time(s), destination(s), aircraft number, pilot (and other crew) data, distance flown, fuel usage, and other data pertinent to the charter flight. Such charter data are then used to generate monthly reports that detail revenue and operating cost information for customers, aircraft, and pilots. All pilots and other crew members are RC_Charter2 Company employees; that is, the company does not use contract pilots and crew.

FAR Part 135 operations are conducted under a strict set of requirements that govern the licensing and training of crew members. For example, pilots must have earned either a Commercial license or an Airline Transport Pilot (ATP) license. Both licenses require appropriate ratings. Ratings are specific competency requirements. For example:

- To operate a multiengine aircraft designed for takeoffs and landings on land only, the appropriate rating is MEL, or Multiengine Landplane. When a multiengine aircraft can take off and land on water, the appropriate rating is MES, or Multiengine Seaplane.

- The instrument rating is based on a demonstrated ability to conduct all flight operations with sole reference to cockpit instrumentation. The instrument rating is required to operate an aircraft under Instrument Meteorological Conditions (IMC), and all such operations are governed under FAR-specified Instrument Flight Rules (IFR). In contrast, operations conducted under "good weather" or *visual* flight conditions are based on the FAR Visual Flight Rules (VFR).

- The type rating is required for all aircraft with a takeoff weight of more than 12,500 pounds or for aircraft that are purely jet-powered. If an aircraft uses jet engines to drive propellers, that aircraft is said to be turboprop-powered. A turboprop—that is, a turbo propeller-powered aircraft—does not require a type rating unless it meets the 12,500-pound weight limitation.

Although pilot licenses and ratings are not time-limited, exercising the privilege of the license and ratings under Part 135 requires both *a current medical certificate and a current Part 135 checkride*. The following distinctions are important:

- The medical certificate may be Class I or Class II. The Class I medical is more stringent than the Class II, and it must be renewed *every* six months. The Class II medical must be renewed yearly. If the Class I medical is not renewed during the six-month period, it automatically reverts to a Class II certificate. If the Class II medical is not renewed within the specified period, it automatically reverts to a Class III medical, which is not valid for commercial flight operations.

- A Part 135 checkride is a practical flight examination that must be successfully completed every six months. The checkride includes all flight maneuvers and procedures specified in Part 135.

Nonpilot crew members must also have the proper certificates in order to meet specific job requirements. For example, loadmasters need an appropriate certificate, as do flight attendants. In addition, crew members such as loadmasters and flight attendants, who may be required in operations that involve large aircraft (more than a 12,500-pound takeoff weight and passenger configurations over 19) are also required periodically to pass a written and practical exam. The RC_Charter2 Company is required to keep a complete record of all test types, dates, and results for each crew member, as well as pilot medical certificate examination dates.

In addition, all flight crew members are required to submit to periodic drug testing; the results must be tracked, too. (Note that nonpilot crew members are not required to take pilot-specific tests such as Part 135 checkrides. Nor are pilots required to take crew tests such as loadmaster and flight attendant practical exams.) However, many crew members have licenses and/or certifications in several areas. For example, a pilot may have an ATP and a loadmaster certificate. If that pilot is assigned to be a loadmaster on a given charter flight, the loadmaster certificate is required. Similarly, a flight attendant may have earned a commercial pilot's license. Sample data formats are shown in Table P4.13.

TABLE
P4.13

PART A TESTS		
TEST CODE	TEST DESCRIPTION	TEST FREQUENCY
1	Part 135 Flight Check	6 months
2	Medical, Class 1	6 months
3	Medical, Class 2	12 months
4	Loadmaster Practical	12 months
5	Flight Attendant Practical	12 months
6	Drug test	Random
7	Operations, written exam	6 months

PART B RESULTS			
EMPLOYEE	TEST CODE	TEST DATE	TEST RESULT
101	1	12-Nov-07	Pass-1
103	6	23-Dec-07	Pass-1
112	4	23-Dec-07	Pass-2
103	7	11-Jan-08	Pass-1
112	7	16-Jan-08	Pass-1
101	7	16-Jan-08	Pass-1
101	6	11-Feb-08	Pass-2
125	2	15-Feb-08	Pass-1

PART C LICENSES AND CERTIFICATIONS	
LICENSE OR CERTIFICATE	LICENSE OR CERTIFICATE DESCRIPTION
ATP	Airline Transport Pilot
Comm	Commercial license
Med-1	Medical certificate, class 1
Med-2	Medical certificate, class 2
Instr	Instrument rating
MEL	Multiengine Land aircraft rating
LM	Load Master
FA	Flight Attendant

PART D LICENSES AND CERTIFICATES HELD BY EMPLOYEES		
EMPLOYEE	LICENSE OR CERTIFICATE	DATE EARNED
101	Comm	12-Nov-93
101	Instr	28-Jun-94
101	MEL	9-Aug-94
103	Comm	21-Dec-95
112	FA	23-Jun-02
103	Instr	18-Jan-96
112	LM	27-Nov-05

Pilots and other crew members must receive recurrency training appropriate to their work assignments. Recurrency training is based on an FAA-approved curriculum that is job-specific. For example, pilot recurrency training includes a review of all applicable Part 135 flight rules and regulations, weather data interpretation, company flight operations requirements, and specified flight procedures. The RC_Charter2 Company is required to keep a complete record of all recurrency training for each crew member subject to the training.

The RC_Charter2 Company is required to maintain a detailed record of all crew credentials and all training mandated by Part 135. The company must keep a complete record of each requirement and of all compliance data.

To conduct a charter flight, the company must have a properly maintained aircraft available. A pilot who meets all of the FAA's licensing and currency requirements must fly the aircraft as Pilot in Command (PIC). For those aircraft that are powered by piston engines or turboprops and have a gross takeoff weight under 12,500 pounds, single-pilot operations are permitted under Part 135 as long as a properly maintained autopilot is available. However, even if FAR Part 135 permits single-pilot operations, many customers require the presence of a copilot who is capable of conducting the flight operations under Part 135.

The RC_Charter2 operations manager anticipates the lease of turbojet-powered aircraft, and those aircraft are required to have a crew consisting of a pilot and copilot. Both pilot and copilot must meet the same Part 135 licensing, ratings, and training requirements.

The company also leases larger aircraft that exceed the 12,500-pound gross takeoff weight. Those aircraft can carry the number of passengers that requires the presence of one or more flight attendants. If those aircraft carry cargo weighing over 12,500 pounds, a loadmaster must be assigned as a crew member to supervise the loading and securing of the cargo. *The database must be designed to meet the anticipated additional charter crew assignment capability.*

a. Given this incomplete description of operations, write all applicable business rules to establish entities, relationships, optionalities, connectivities, and cardinalities. (*Hint:* Use the following five business rules as examples, writing the remaining business rules in the same format.)

- A customer may request many charter trips.
- Each charter trip is requested by only one customer.
- Some customers have not (yet) requested a charter trip.
- An employee may be assigned to serve as a crew member on many charter trips.
- Each charter trip may have many employees assigned to it to serve as crew members.

b. Draw the fully labeled and implementable Crow's Foot ERD based on the business rules you wrote in Part (a) of this problem. Include all entities, relationships, optionalities, connectivities, and cardinalities.

5

NORMALIZATION OF DATABASE TABLES

In this chapter, you will learn:

- What normalization is and what role it plays in the database design process
- About the normal forms 1NF, 2NF, 3NF, BCNF, and 4NF
- How normal forms can be transformed from lower normal forms to higher normal forms
- That normalization and ER modeling are used concurrently to produce a good database design
- That some situations require denormalization to generate information efficiently

Good database design must be matched to good table structures. In this chapter, you learn to evaluate and design good table structures to control data redundancies, thereby avoiding data anomalies. The process that yields such desirable results is known as normalization.

In order to recognize and appreciate the characteristics of a good table structure, it is useful to examine a poor one. Therefore, the chapter begins by examining the characteristics of a poor table structure and the problems it creates. You then learn how to correct a poor table structure. This methodology will yield important dividends: you will know how to design a good table structure and how to repair an existing poor one.

You will discover not only that data anomalies can be eliminated through normalization, but also that a properly normalized set of table structures is actually less complicated to use than an unnormalized set. In addition, you will learn that the normalized set of table structures more faithfully reflects an organization's real operations.

Preview

5.1 DATABASE TABLES AND NORMALIZATION

Having good relational database software is not enough to avoid the data redundancy discussed in Chapter 1, Database Systems. If the database tables are treated as though they are files in a file system, the RDBMS never has a chance to demonstrate its superior data-handling capabilities.

The table is the basic building block in the database design process. Consequently, the table's structure is of great interest. Ideally, the database design process explored in Chapter 4, Entity Relationship (ER) Modeling, yields good table structures. Yet it is possible to create poor table structures even in a good database design. So how do you recognize a poor table structure, and how do you produce a good table? The answer to both questions involves normalization. **Normalization** is a process for evaluating and correcting table structures to minimize data redundancies, thereby reducing the likelihood of data anomalies. The normalization process involves assigning attributes to tables based on the concept of determination you learned about in Chapter 3, The Relational Database Model.

Normalization works through a series of stages called normal forms. The first three stages are described as first normal form (1NF), second normal form (2NF), and third normal form (3NF). From a structural point of view, 2NF is better than 1NF, and 3NF is better than 2NF. For most purposes in business database design, 3NF is as high as you need to go in the normalization process. However, you will discover in Section 5.3 that properly designed 3NF structures also meet the requirements of fourth normal form (4NF).

Although normalization is a very important database design ingredient, you should not assume that the highest level of normalization is always the most desirable. Generally, the higher the normal form, the more relational join operations required to produce a specified output and the more resources required by the database system to respond to end-user queries. A successful design must also consider end-user demand for fast performance. Therefore, you will occasionally be expected to *denormalize* some portions of a database design in order to meet performance requirements. **Denormalization** produces a lower normal form; that is, a 3NF will be converted to a 2NF through denormalization. However, *the price you pay for increased performance through denormalization is greater data redundancy*.

> **NOTE**
>
> Although the word *table* is used throughout this chapter, formally, normalization is concerned with *relations*. In Chapter 3 you learned that the terms *table* and *relation* are frequently used interchangeably. In fact, you can say that a table is the "implementation view" of a logical relation that meets some specific conditions (see Table 3.1). However, being more rigorous, the mathematical relation does not allow duplicate tuples, whereas duplicate tuples could exist in tables (see Section 5.5).

5.2 THE NEED FOR NORMALIZATION

To get a better idea of the normalization process, consider the simplified database activities of a construction company that manages several building projects. Each project has its own project number, name, employees assigned to it, and so on. Each employee has an employee number, name, and job classification, such as engineer or computer technician.

The company charges its clients by billing the hours spent on each contract. The hourly billing rate is dependent on the employee's position. For example, one hour of computer technician time is billed at a different rate than one hour of engineer time. Periodically, a report is generated that contains the information displayed in Table 5.1.

The total charge in Table 5.1 is a derived attribute and, at this point, is not stored in the table.

TABLE 5.1   A Sample Report Layout

PROJECT NUMBER	PROJECT NAME	EMPLOYEE NUMBER	EMPLOYEE NAME	JOB CLASS	CHARGE/HOUR	HOURS BILLED	TOTAL CHARGE
15	Evergreen	103	June E. Arbough	Elec. Engineer	$ 85.50	23.8	$ 2,034.90
		101	John G. News	Database Designer	$105.00	19.4	$ 2,037.00
		105	Alice K. Johnson *	Database Designer	$105.00	35.7	$ 3,748.50
		106	William Smithfield	Programmer	$ 35.75	12.6	$ 450.45
		102	David H. Senior	Systems Analyst	$ 96.75	23.8	$ 2,302.65
				Subtotal			**$10,573.50**
18	Amber Wave	114	Annelise Jones	Applications Designer	$ 48.10	25.6	$ 1,183.26
		118	James J. Frommer	General Support	$ 18.36	45.3	$ 831.71
		104	Anne K. Ramoras *	Systems Analyst	$ 96.75	32.4	$ 3,134.70
		112	Darlene M. Smithson	DSS Analyst	$ 45.95	45.0	$ 2,067.75
				Subtotal			**$ 7,265.52**
22	Rolling Tide	105	Alice K. Johnson	Database Designer	$105.00	65.7	$ 6,998.50
		104	Anne K. Ramoras	Systems Analyst	$ 96.75	48.4	$ 4,682.70
		113	Delbert K. Joenbrood	Applications Designer	$ 48.10	23.6	$ 1,135.16
		111	Geoff B. Wabash	Clerical Support	$ 26.87	22.0	$ 591.14
		106	William Smithfield	Programmer	$ 35.75	12.8	$ 457.60
				Subtotal			**$13,765.10**
25	Starflight	107	Maria D. Alonzo	Programmer	$ 35.75	25.6	$ 915.20
		115	Travis B. Bawangi	Systems Analyst	$ 96.75	45.8	$ 4,431.15
		101	John G. News *	Database Designer	$105.00	56.3	$ 5,911.50
		114	Annelise Jones	Applications Designer	$ 48.10	33.1	$ 1,592.11
		108	Ralph B. Washington	Systems Analyst	$ 96.75	23.6	$ 2,283.30
		118	James J. Frommer	General Support	$ 18.36	30.5	$ 559.98
		112	Darlene M. Smithson	DSS Analyst	$ 45.95	41.4	$ 1,902.33
				Subtotal			**$17,595.57**
				Total			**$49,199.69**

Note: * indicates project leader

The easiest short-term way to generate the required report might seem to be a table whose contents correspond to the reporting requirements. See Figure 5.1.

ONLINE CONTENT

The databases used to illustrate the material in this chapter are found in the Student Online Companion.

FIGURE 5.1   **Tabular representation of the report format**

Table name: RPT_FORMAT   Database name: Ch05_ConstructCo

PROJ_NUM	PROJ_NAME	EMP_NUM	EMP_NAME	JOB_CLASS	CHG_HOUR	HOURS
15	Evergreen	103	June E. Arbough	Elect. Engineer	84.50	23.8
		101	John G. News	Database Designer	105.00	19.4
		105	Alice K. Johnson *	Database Designer	105.00	35.7
		106	William Smithfield	Programmer	35.75	12.6
		102	David H. Senior	Systems Analyst	96.75	23.8
18	Amber Wave	114	Annelise Jones	Applications Designer	48.10	24.6
		118	James J. Frommer	General Support	18.36	45.3
		104	Anne K. Ramoras *	Systems Analyst	96.75	32.4
		112	Darlene M. Smithson	DSS Analyst	45.95	44.0
22	Rolling Tide	105	Alice K. Johnson	Database Designer	105.00	64.7
		104	Anne K. Ramoras	Systems Analyst	96.75	48.4
		113	Delbert K. Joenbrood *	Applications Designer	48.10	23.6
		111	Geoff B. Wabash	Clerical Support	26.87	22.0
		106	William Smithfield	Programmer	35.75	12.8
25	Starflight	107	Maria D. Alonzo	Programmer	35.75	24.6
		115	Travis B. Bawangi	Systems Analyst	96.75	45.8
		101	John G. News *	Database Designer	105.00	56.3
		114	Annelise Jones	Applications Designer	48.10	33.1
		108	Ralph B. Washington	Systems Analyst	96.75	23.6
		118	James J. Frommer	General Support	18.36	30.5
		112	Darlene M. Smithson	DSS Analyst	45.95	41.4

Note that the data in Figure 5.1 reflects the assignment of employees to projects. Apparently, an employee can be assigned to more than one project. For example, Darlene Smithson (EMP_NUM = 112) has been assigned to two projects: Amber Wave and Starflight. Given the structure of the data set, each project includes only a single occurrence of any one employee. Therefore, knowing the PROJ_NUM and EMP_NUM value will let you find the job classification and its hourly charge. In addition, you will know the total number of hours each employee worked on each project. (The total charge—a derived attribute whose value can be computed by multiplying the hours billed and the charge per hour—has not been included in Figure 5.1. No structural harm is done if this derived attribute is included.)

Unfortunately, the structure of the data set in Figure 5.1 does not conform to the requirements discussed in Chapter 3, nor does it handle data very well. Consider the following deficiencies:

1. The project number (PROJ_NUM) is apparently intended to be a primary key or at least a part of a PK, but it contains nulls. (Given the preceding discussion, you know that PROJ_NUM + EMP_NUM will define each row.)

2. The table entries invite data inconsistencies. For example, the JOB_CLASS value "Elect. Engineer" might be entered as "Elect.Eng." in some cases, "El. Eng." in others, and "EE" in still others.

3. The table displays data redundancies. Those data redundancies yield the following anomalies:

    a. *Update anomalies*. Modifying the JOB_CLASS for employee number 105 requires (potentially) many alterations, one for each EMP_NUM = 105.

b. *Insertion anomalies*. Just to complete a row definition, a new employee must be assigned to a project. If the employee is not yet assigned, a phantom project must be created to complete the employee data entry.

c. *Deletion anomalies*. Suppose that only one employee is associated with a given project. If that employee leaves the company and the employee data are deleted, the project information will also be deleted. To prevent the loss of the project information, a fictitious employee must be created just to save the project information.

In spite of those structural deficiencies, the table structure *appears* to work; the report is generated with ease. Unfortunately, the report might yield varying results depending on what data anomaly has occurred. For example, if you want to print a report to show the total "hours worked" value by the job classification "Database Designer," that report will not include data for "DB Design" and "Database Design" data entries. Such reporting anomalies cause a multitude of problems for managers—and cannot be fixed through applications programming.

Even if very careful data entry auditing can eliminate most of the reporting problems (at a high cost), it is easy to demonstrate that even a simple data entry becomes inefficient. Given the existence of update anomalies, suppose Darlene M. Smithson is assigned to work on the Evergreen project. The data entry clerk must update the PROJECT file with the entry:

15   Evergreen   112   Darlene M. Smithson   DSS Analyst $45.95   0.0

to match the attributes PROJ_NUM, PROJ_NAME, EMP_NUM, EMP_NAME, JOB_CLASS, CHG_HOUR, and HOURS. (When Ms. Smithson has just been assigned to the project, she has not yet worked, so the total number of hours worked is 0.0.)

NOTE

Remember that the naming convention makes it easy to see what each attribute stands for and what its likely origin is. For example, PROJ_NAME uses the prefix PROJ to indicate that the attribute is associated with the PROJECT table, while the NAME component is self-documenting, too. However, keep in mind that name length is also an issue, especially in the prefix designation. For that reason, the prefix CHG was used rather than CHARGE. (Given the database's context, it is not likely that that prefix will be misunderstood.)

Each time another employee is assigned to a project, some data entries (such as PROJ_NAME, EMP_NAME, and CHG_HOUR) are unnecessarily repeated. Imagine the data entry chore when 200 or 300 table entries must be made! Note that the entry of the employee number should be sufficient to identify Darlene M. Smithson, her job description, and her hourly charge. Because there is only one person identified by the number 112, that person's characteristics (name, job classification, and so on) should not have to be typed in each time the main file is updated. Unfortunately, the structure displayed in Figure 5.1 does not make allowances for that possibility.

The data redundancy evident in Figure 5.1 leads to wasted disk space. What's more, data redundancy produces data anomalies. For example, suppose the data entry clerk had entered the data as:

15   Evergeen   112   Darla Smithson   DCS Analyst $45.95   0.0

At first glance, the data entry appears to be correct. But is Evergeen the same project as Evergreen? And is DCS Analyst supposed to be DSS Analyst? Is Darla Smithson the same person as Darlene M. Smithson? Such confusion is a data integrity problem that was caused because the data entry failed to conform to the rule that all copies of redundant data must be identical.

The possibility of introducing data integrity problems caused by data redundancy must be considered when a database is designed. The relational database environment is especially well suited to help the designer overcome those problems.

5.3 THE NORMALIZATION PROCESS

In this section, you learn how to use normalization to produce a set of normalized tables to store the data that will be used to generate the required information. The objective of normalization is to ensure that each table conforms to the concept of well-formed relations, that is, tables that have the following characteristics:

- Each table represents a single subject. For example, a course table will contain only data that directly pertains to courses. Similarly, a student table will contain only student data.

- No data item will be *unnecessarily* stored in more than one table (in short, tables have minimum controlled redundancy). The reason for this requirement is to ensure that the data are updated in only one place.

- All nonprime attributes in a table are dependent on the primary key—the entire primary key and nothing but the primary key. The reason for this requirement is to ensure that the data are uniquely identifiable by a primary key value.

- Each table is void of insertion, update, or deletion anomalies. This is to ensure the integrity and consistency of the data.

To accomplish the objective, the normalization process takes you through the steps that lead to successively higher normal forms. The most common normal forms and their basic characteristic are listed in Table 5.2. You will learn the details of these normal forms in the indicated sections.

TABLE 5.2 **Normal Forms**

NORMAL FORM	CHARACTERISTIC	SECTION
First normal form (1NF)	Table format, no repeating groups, and PK identified	5.3.1
Second normal form (2NF)	1NF and no partial dependencies	5.3.2
Third normal form (3NF)	2NF and no transitive dependencies	5.3.3
Boyce-Codd normal form (BCNF)	Every determinant is a candidate key (special case of 3NF)	5.6.1
Fourth normal form (4NF)	3NF and no independent multivalued dependencies	5.6.2

From the data modeler's point of view, the objective of normalization is to ensure that all tables are at least in third normal form (3NF). Even higher-level normal forms exist. However, normal forms such as the fifth normal form (5NF) and domain-key normal form (DKNF) are not likely to be encountered in a business environment and are mainly of theoretical interest. More often than not, such higher normal forms usually increase joins (slowing performance) without adding any value in the elimination of data redundancy. Some very specialized applications, such as statistical research, might require normalization beyond the 4NF, but those applications fall outside the scope of most business operations. Because this book focuses on practical applications of database techniques, the higher-level normal forms are not covered.

Functional Dependency

Before outlining the normalization process, it's a good idea to review the concepts of determination and functional dependency that were covered in detail in Chapter 3. Table 5.3 summarizes the main concepts.

TABLE 5.3	Functional Dependency Concepts
CONCEPT	**DEFINITION**
Functional dependency	The attribute B is fully functionally dependent on the attribute A if each value of A determines one and only one value of B. Example: PROJ_NUM → PROJ_NAME (read as "PROJ_NUM functionally determines PROJ_NAME) In this case, the attribute PROJ_NUM is known as the "determinant" attribute and the attribute PROJ_NAME is known as the "dependent" attribute.
Functional dependency (generalized definition)	Attribute A determines attribute B (that is, B is functionally dependent on A) if all of the rows in the table that agree in value for attribute A also agree in value for attribute B.
Fully functional dependency (composite key)	If attribute B is functionally dependent on a composite key A but not on any sub-set of that composite key, the attribute B is fully functionally dependent on A.

It is crucial to understand these concepts because they are used to derive the set of functional dependencies for a given relation. The normalization process works one relation at a time, identifying the dependencies on that relation and normalizing the relation. As you will see in the following sections, normalization starts by identifying the dependencies of a given relation and progressively breaking up the relation (table) into a set of new relations (tables) based on the identified dependencies.

5.3.1 CONVERSION TO FIRST NORMAL FORM

Because the relational model views data as part of a table or a collection of tables in which all key values must be identified, the data depicted in Figure 5.1 might not be stored as shown. Note that Figure 5.1 contains what is known as repeating groups. A **repeating group** derives its name from the fact that a group of multiple entries of the same type can exist for any *single* key attribute occurrence. In Figure 5.1, note that each single project number (PROJ_NUM) occurrence can reference a group of related data entries. For example, the Evergreen project (PROJ_NUM = 15) shows five entries at this point—and those entries are related because they each share the PROJ_NUM = 15 characteristic. Each time a new record is entered for the Evergreen project, the number of entries in the group grows by one.

A relational table must not contain repeating groups. The existence of repeating groups provides evidence that the RPT_FORMAT table in Figure 5.1 fails to meet even the lowest normal form requirements, thus reflecting data redundancies.

Normalizing the table structure will reduce the data redundancies. If repeating groups do exist, they must be eliminated by making sure that each row defines a single entity. In addition, the dependencies must be identified to diagnose the normal form. Identification of the normal form will let you know where you are in the normalization process. The normalization process starts with a simple three-step procedure.

Step 1: Eliminate the Repeating Groups

Start by presenting the data in a tabular format, where each cell has a single value and there are no repeating groups. To eliminate the repeating groups, eliminate the nulls by making sure that each repeating group attribute contains an appropriate data value. That change converts the table in Figure 5.1 to 1NF in Figure 5.2.

FIGURE 5.2	A table in first normal form

Table name: DATA_ORG_1NF   Database name: Ch05_ConstructCo

PROJ_NUM	PROJ_NAME	EMP_NUM	EMP_NAME	JOB_CLASS	CHG_HOUR	HOURS
15	Evergreen	103	June E. Arbough	Elect. Engineer	84.50	23.8
15	Evergreen	101	John G. News	Database Designer	105.00	19.4
15	Evergreen	105	Alice K. Johnson *	Database Designer	105.00	35.7
15	Evergreen	106	William Smithfield	Programmer	35.75	12.6
15	Evergreen	102	David H. Senior	Systems Analyst	96.75	23.8
18	Amber Wave	114	Annelise Jones	Applications Designer	48.10	24.6
18	Amber Wave	118	James J. Frommer	General Support	18.36	45.3
18	Amber Wave	104	Anne K. Ramoras *	Systems Analyst	96.75	32.4
18	Amber Wave	112	Darlene M. Smithson	DSS Analyst	45.95	44.0
22	Rolling Tide	105	Alice K. Johnson	Database Designer	105.00	64.7
22	Rolling Tide	104	Anne K. Ramoras	Systems Analyst	96.75	48.4
22	Rolling Tide	113	Delbert K. Joenbrood *	Applications Designer	48.10	23.6
22	Rolling Tide	111	Geoff B. Wabash	Clerical Support	26.87	22.0
22	Rolling Tide	106	William Smithfield	Programmer	35.75	12.8
25	Starflight	107	Maria D. Alonzo	Programmer	35.75	24.6
25	Starflight	115	Travis B. Bawangi	Systems Analyst	96.75	45.8
25	Starflight	101	John G. News *	Database Designer	105.00	56.3
25	Starflight	114	Annelise Jones	Applications Designer	48.10	33.1
25	Starflight	108	Ralph B. Washington	Systems Analyst	96.75	23.6
25	Starflight	118	James J. Frommer	General Support	18.36	30.5
25	Starflight	112	Darlene M. Smithson	DSS Analyst	45.95	41.4

Step 2: Identify the Primary Key

The layout in Figure 5.2 represents more than a mere cosmetic change. Even a casual observer will note that PROJ_NUM is not an adequate primary key because the project number does not uniquely identify all of the remaining entity (row) attributes. For example, the PROJ_NUM value 15 can identify any one of five employees. To maintain a proper primary key that will *uniquely* identify any attribute value, the new key must be composed of a *combination* of PROJ_NUM and EMP_NUM. For example, using the data shown in Figure 5.2, if you know that PROJ_NUM = 15 and EMP_NUM = 103, the entries for the attributes PROJ_NAME, EMP_NAME, JOB_CLASS, CHG_HOUR, and HOURS must be Evergreen, June E. Arbough, Elect. Engineer, $84.50, and 23.8, respectively.

Step 3: Identify All Dependencies

The identification of the PK in Step 2 means that you have already identified the following dependency:

PROJ_NUM, EMP_NUM → PROJ_NAME, EMP_NAME, JOB_CLASS, CHG_HOUR, HOURS

That is, the PROJ_NAME, EMP_NAME, JOB_CLASS, CHG_HOUR, and HOURS values are all dependent on—that is, they are determined by—the combination of PROJ_NUM and EMP_NUM. There are additional dependencies. For example, the project number identifies (determines) the project name. In other words, the project name is dependent on the project number. You can write that dependency as:

PROJ_NUM → PROJ_NAME

Also, if you know an employee number, you also know that employee's name, that employee's job classification, and that employee's charge per hour. Therefore, you can identify the dependency shown next:

EMP_NUM → EMP_NAME, JOB_CLASS, CHG_HOUR

However, given the previous dependency components, you can see that knowing the job classification means knowing the charge per hour for that job classification. In other words, you can identify one last dependency:

JOB_CLASS → CHG_HOUR

The dependencies you have just examined can also be depicted with the help of the diagram shown in Figure 5.3. Because such a diagram depicts all dependencies found within a given table structure, it is known as a **dependency diagram**. Dependency diagrams are very helpful in getting a bird's-eye view of all of the relationships among a table's attributes, and their use makes it less likely that you will overlook an important dependency.

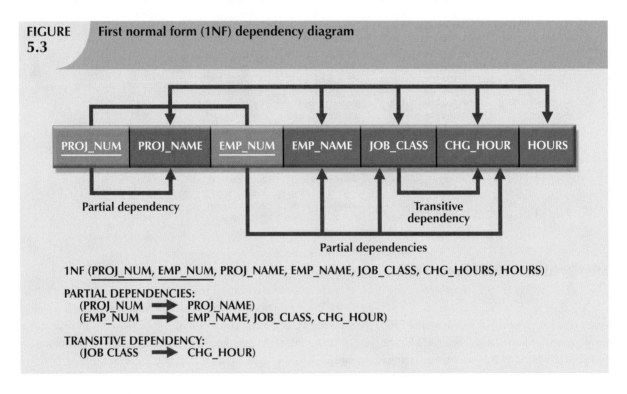

FIGURE 5.3   First normal form (1NF) dependency diagram

As you examine Figure 5.3, note the following dependency diagram features:

1. The primary key attributes are bold, underlined, and shaded in a different color.

2. The arrows above the attributes indicate all desirable dependencies, that is, dependencies that are based on the primary key. In this case, note that the entity's attributes are dependent on the *combination* of PROJ_NUM and EMP_NUM.

3. The arrows below the dependency diagram indicate less desirable dependencies. Two types of such dependencies exist:

    a. *Partial dependencies*. You need to know only the PROJ_NUM to determine the PROJ_NAME; that is, the PROJ_NAME is dependent on only part of the primary key. And you need to know only the EMP_NUM to find the EMP_NAME, the JOB_CLASS, and the CHG_HOUR. A dependency based on only a part of a composite primary key is called a **partial dependency**.

    b. *Transitive dependencies*. Note that CHG_HOUR is dependent on JOB_CLASS. Because neither CHG_HOUR nor JOB_CLASS is a prime attribute—that is, neither attribute is at least part of a key—the condition is known as a transitive dependency. In other words, a **transitive dependency** is a dependency of one nonprime attribute on another nonprime attribute. The problem with transitive dependencies is that they still yield data anomalies.

Note that Figure 5.3 includes the relational schema for the table in 1NF and a textual notation for each identified dependency.

NOTE

The term **first normal form** (**1NF**) describes the tabular format in which:

- All of the key attributes are defined.
- There are no repeating groups in the table. In other words, each row/column intersection contains one and only one value, not a set of values.
- All attributes are dependent on the primary key.

All relational tables satisfy the 1NF requirements. The problem with the 1NF table structure shown in Figure 5.3 is that it contains partial dependencies—that is, dependencies based on only a part of the primary key.

While partial dependencies are sometimes used for performance reasons, they should be used with caution. (If the information requirements seem to dictate the use of partial dependencies, it is time to evaluate the need for a data warehouse design, discussed in Chapter 13, Business Intelligence and Data Warehouses.) Such caution is warranted because a table that contains partial dependencies is still subject to data redundancies, and therefore, to various anomalies. The data redundancies occur because every row entry requires duplication of data. For example, if Alice K. Johnson submits her work log, then the user would have to make multiple entries during the course of a day. For each entry, the EMP_NAME, JOB_CLASS, and CHG_HOUR must be entered each time even though the attribute values are identical for each row entered. Such duplication of effort is very inefficient. What's more, the duplication of effort helps create data anomalies; nothing prevents the user from typing slightly different versions of the employee name, the position, or the hourly pay. For instance, the employee name for EMP_NUM = 102 might be entered as Dave Senior or D. Senior. The project name also might be entered correctly as Evergreen or misspelled as Evergeen. Such data anomalies violate the relational database's integrity and consistency rules.

5.3.2   CONVERSION TO SECOND NORMAL FORM

Converting to 2NF is done only when the 1NF has a composite primary key. If the 1NF has a single attribute primary key, then the table is automatically in 2NF. The 1NF-to-2NF conversion is simple. Starting with the 1NF format displayed in Figure 5.3, you do the following:

Step 1: Write Each Key Component on a Separate Line

Write each key component on a separate line; then write the original (composite) key on the last line. For example:

PROJ_NUM

EMP_NUM

PROJ_NUM   EMP_NUM

Each component will become the key in a new table. In other words, the original table is now divided into three tables (PROJECT, EMPLOYEE, and ASSIGNMENT).

Step 2: Assign Corresponding Dependent Attributes

Use Figure 5.3 to determine those attributes that are dependent on other attributes. The dependencies for the original key components are found by examining the arrows below the dependency diagram shown in Figure 5.3. In other words, the three new tables (PROJECT, EMPLOYEE, and ASSIGNMENT) are described by the following relational schemas:

PROJECT (**PROJ_NUM**, PROJ_NAME)

EMPLOYEE (**EMP_NUM**, EMP_NAME, JOB_CLASS, CHG_HOUR)

ASSIGNMENT (**PROJ_NUM**, **EMP_NUM**, ASSIGN_HOURS)

Because the number of hours spent on each project by each employee is dependent on both PROJ_NUM and EMP_NUM in the ASSIGNMENT table, you place those hours in the ASSIGNMENT table as ASSIGN_HOURS.

> **NOTE**
>
> The ASSIGNMENT table contains a composite primary key composed of the attributes PROJ_NUM and EMP_NUM. Any attribute that is at least part of a key is known as a **prime attribute** or a **key attribute**. Therefore, both PROJ_NUM and EMP_NUM are prime (or key) attributes. Conversely, a **nonprime attribute**, or a **nonkey attribute**, is not part of any key.

The results of Steps 1 and 2 are displayed in Figure 5.4. At this point, most of the anomalies discussed earlier have been eliminated. For example, if you now want to add, change, or delete a PROJECT record, you need to go only to the PROJECT table and make the change to only one row.

FIGURE 5.4   Second normal form (2NF) conversion results

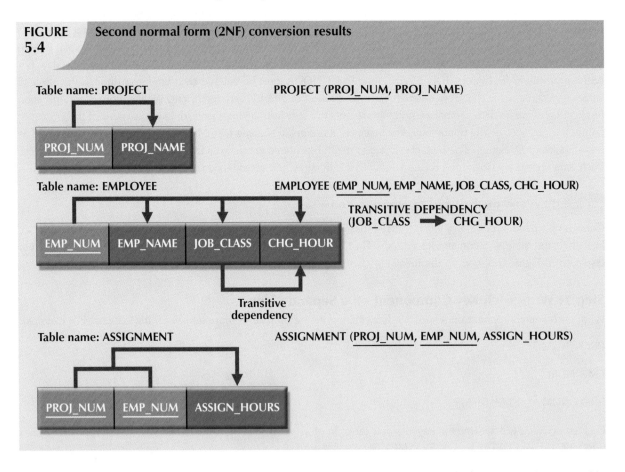

Because a partial dependency can exist only when a table's primary key is composed of several attributes, a table whose primary key consists of only a single attribute is automatically in 2NF once it is in 1NF.

Figure 5.4 still shows a transitive dependency, which can generate anomalies. For example, if the charge per hour changes for a job classification held by many employees, that change must be made for *each* of those employees. If you forget to update some of the employee records that are affected by the charge per hour change, different employees with the same job description will generate different hourly charges.

> **NOTE**
>
> A table is in **second normal form** (**2NF**) when:
>
> - It is in 1NF.
>
> *and*
>
> - It includes no partial dependencies; that is, no attribute is dependent on only a portion of the primary key. Note that it is still possible for a table in 2NF to exhibit transitive dependency; that is, one or more attributes may be functionally dependent on nonkey attributes.

5.3.3 CONVERSION TO THIRD NORMAL FORM

The data anomalies created by the database organization shown in Figure 5.4 are easily eliminated by completing the following three steps:

Step 1: Identify Each New Determinant

For every transitive dependency, write its determinant as a PK for a new table. A **determinant** is any attribute whose value determines other values within a row. If you have three different transitive dependencies, you will have three different determinants. Figure 5.4 shows only one table that contains a transitive dependency. Therefore, write the determinant for this transitive dependency as:

JOB_CLASS

Step 2: Identify the Dependent Attributes

Identify the attributes that are dependent on each determinant identified in Step 1 and identify the dependency. In this case, you write:

JOB_CLASS → CHG_HOUR

Name the table to reflect its contents and function. In this case, JOB seems appropriate.

Step 3: Remove the Dependent Attributes from Transitive Dependencies

Eliminate all dependent attributes in the transitive relationship(s) from each of the tables that have such a transitive relationship. In this example, eliminate CHG_HOUR from the EMPLOYEE table shown in Figure 5.4 to leave the EMPLOYEE table dependency definition as:

EMP_NUM → EMP_NAME, JOB_CLASS

Note that the JOB_CLASS remains in the EMPLOYEE table to serve as the FK.

Draw a new dependency diagram to show all of the tables you have defined in Steps 1–3. Check the new tables as well as the tables you modified in Step 3 to make sure that each table has a determinant and that no table contains inappropriate dependencies.

When you have completed Steps 1–3, you will see the results in Figure 5.5. (The usual procedure is to complete Steps 1–3 by simply drawing the revisions as you make them.)

FIGURE 5.5   Third normal form (3NF) conversion results

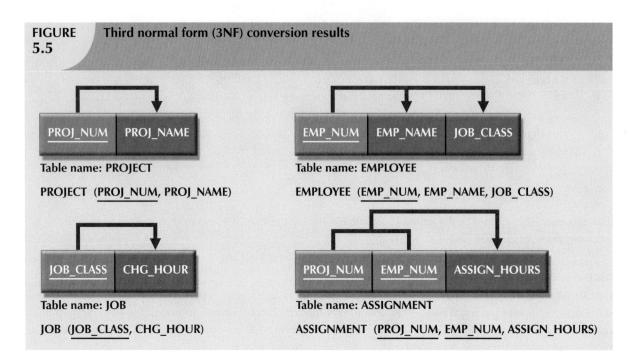

Table name: PROJECT

PROJECT (PROJ_NUM, PROJ_NAME)

Table name: EMPLOYEE

EMPLOYEE (EMP_NUM, EMP_NAME, JOB_CLASS)

Table name: JOB

JOB (JOB_CLASS, CHG_HOUR)

Table name: ASSIGNMENT

ASSIGNMENT (PROJ_NUM, EMP_NUM, ASSIGN_HOURS)

In other words, after the 3NF conversion has been completed, your database contains four tables:

PROJECT (**PROJ_NUM**, PROJ_NAME)

EMPLOYEE (**EMP_NUM**, EMP_NAME, JOB_CLASS)

JOB (**JOB_CLASS**, CHG_HOUR)

ASSIGNMENT (**PROJ_NUM**, **EMP_NUM**, ASSIGN_HOURS)

Note that this conversion has eliminated the original EMPLOYEE table's transitive dependency; the tables are now said to be in third normal form (3NF).

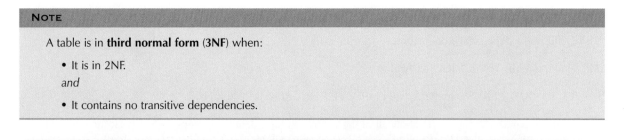

NOTE

A table is in **third normal form** (3NF) when:

- It is in 2NF.
  and
- It contains no transitive dependencies.

5.4 IMPROVING THE DESIGN

The table structures are cleaned up to eliminate the troublesome partial and transitive dependencies. You can now focus on improving the database's ability to provide information and on enhancing its operational characteristics. In the next few paragraphs, you will learn about the various types of issues you need to address to produce a good normalized set of tables. Please note that for space issues, each section presents just one example—the designer must apply the principle to all remaining tables in the design. Remember that normalization cannot, by itself, be relied on to make good designs. Instead, normalization is valuable because its use helps eliminate data redundancies.

Evaluate PK Assignments

Each time a new employee is entered into the EMPLOYEE table, a JOB_CLASS value must be entered. Unfortunately, it is too easy to make data-entry errors that lead to referential integrity violations. For example, entering DB Designer instead of Database Designer for the JOB_CLASS attribute in the EMPLOYEE table will trigger such a violation. Therefore, it would be better to add a JOB_CODE attribute to create a unique identifier. The addition of a JOB_CODE attribute produces the dependency:

JOB_CODE → JOB_CLASS, CHG_HOUR

If you assume that the JOB_CODE is a proper primary key, this new attribute does produce the transitive dependency:

JOB_CLASS → CHG_HOUR

A transitive dependency exists because a nonkey attribute—the JOB_CLASS—determines the value of another nonkey attribute—the CHG_HOUR. However, that transitive dependency is an easy price to pay; the presence of JOB_CODE greatly decreases the likelihood of referential integrity violations. Note that the new JOB table now has two candidate keys—JOB_CODE and JOB_CLASS. In this case, JOB_CODE is the chosen primary key as well as a surrogate key. A **surrogate key** is an artificial PK introduced by the designer with the purpose of simplifying the assignment of primary keys to tables. Surrogate keys are usually numeric, they are often automatically generated by the DBMS, they are free of semantic content (they have no special meaning), and they are usually hidden from the end users. You will learn more about PK characteristics and assignment in Chapter 6, Advanced Data Modeling.

Evaluate Naming Conventions

It is best to adhere to the naming conventions outlined in Chapter 2, Data Models. Therefore, CHG_HOUR will be changed to JOB_CHG_HOUR to indicate its association with the JOB table. In addition, the attribute name JOB_CLASS does not quite describe entries such as Systems Analyst, Database Designer, and so on; the label JOB_DESCRIPTION fits the entries better. Also, you might have noticed that HOURS was changed to ASSIGN_HOURS in the conversion from 1NF to 2NF. That change lets you associate the hours worked with the ASSIGNMENT table.

Refine Attribute Atomicity

It generally is good practice to pay attention to the *atomicity* requirement. An **atomic attribute** is one that cannot be further subdivided. Such an attribute is said to display **atomicity**. Clearly, the use of the EMP_NAME in the EMPLOYEE table is not atomic because EMP_NAME can be decomposed into a last name, a first name, and an initial. By improving the degree of atomicity, you also gain querying flexibility. For example, if you use EMP_LNAME, EMP_FNAME, and EMP_INITIAL, you can easily generate phone lists by sorting last names, first names, and initials. Such a task would be very difficult if the name components were within a single attribute. In general, designers prefer to use simple, single-valued attributes as indicated by the business rules and processing requirements.

Identify New Attributes

If the EMPLOYEE table were used in a real-world environment, several other attributes would have to be added. For example, year-to-date gross salary payments, Social Security payments, and Medicare payments would be desirable. Adding an employee hire date attribute (EMP_HIREDATE) could be used to track an employee's job longevity and serve as a basis for awarding bonuses to long-term employees and for other morale-enhancing measures. The same principle must be applied to all other tables in your design.

Identify New Relationships

The system's ability to supply detailed information about each project's manager is ensured by using the EMP_NUM as a foreign key in PROJECT. That action ensures that you can access the details of each PROJECT's manager data without producing unnecessary and undesirable data duplication. The designer must take care to place the right attributes in the right tables by using normalization principles.

Refine Primary Keys as Required for Data Granularity

Granularity refers to the level of detail represented by the values stored in a table's row. Data stored at their lowest level of granularity are said to be *atomic data, as explained earlier.* In Figure 5.5, the ASSIGNMENT table in 3NF uses the ASSIGN_HOURS attribute to represent the hours worked by a given employee on a given project. However, are those values recorded at their lowest level of granularity? In other words, do the ASSIGN_HOURS represent the *hourly* total, *daily* total, *weekly* total, *monthly* total, or *yearly* total? Clearly, ASSIGN_HOURS requires more careful definition. In this case, the relevant question would be as follows: For what time frame—hour, day, week, month, and so on—do you want to record the ASSIGN_HOURS data?

For example, assume that the combination of EMP_NUM and PROJ_NUM is an acceptable (composite) primary key in the ASSIGNMENT table. That primary key is useful in representing only the total number of hours an employee worked on a project since its start. Using a surrogate primary key such as ASSIGN_NUM provides lower granularity and yields greater flexibility. For example, assume that the EMP_NUM and PROJ_NUM combination is used as the primary key, and then an employee makes two "hours worked" entries in the ASSIGNMENT table. That action violates the entity integrity requirement. Even if you add the ASSIGN_DATE as part of a composite PK, an entity integrity violation is still generated if any employee makes two or more entries for the same project on the same day. (The employee might have worked on the project a few hours in the morning and then worked on it again later in the day.) The same data entry yields no problems when ASSIGN_NUM is used as the primary key.

> **NOTE**
>
> In an ideal (database design) world, the level of desired granularity is determined at the conceptual design or at the requirements gathering phase. However, as you have already seen in this chapter, many database designs involve the refinement of existing data requirements, thus triggering design modifications. In a real-world environment, changing granularity requirements might dictate changes in primary key selection, and those changes might ultimately require the use of surrogate keys.

Maintain Historical Accuracy

Writing the job charge per hour into the ASSIGNMENT table is crucial to maintaining the historical accuracy of the data in the ASSIGNMENT table. It would be appropriate to name this attribute ASSIGN_CHG_HOUR. Although this attribute would appear to have the same value as JOB_CHG_HOUR, that is true *only* if the JOB_CHG_HOUR value remains forever the same. However, it is reasonable to assume that the job charge per hour will change over time. But suppose that the charges to each project were figured (and billed) by multiplying the hours worked on the project, found in the ASSIGNMENT table, by the charge per hour, found in the JOB table. Those charges would always show the current charge per hour stored in the JOB table, rather than the charge per hour that was in effect at the time of the assignment.

Evaluate Using Derived Attributes

Finally, you can use a derived attribute in the ASSIGNMENT table to store the actual charge made to a project. That derived attribute, to be named ASSIGN_CHARGE, is the result of multiplying the ASSIGN_HOURS by the ASSIGN_CHG_HOUR. From a strictly database point of view, such derived attribute values can be calculated when they are needed to write reports or invoices. However, storing the derived attribute in the table makes it easy to write the application software to produce the desired results. Also, if many transactions must be reported and/or summarized, the availability of the derived attribute will save reporting time. (If the calculation is done at the time of data entry, it will be completed when the end user presses the Enter key, thus speeding up the process.)

The enhancements described in the preceding sections are illustrated in the tables and dependency diagrams shown in Figure 5.6.

FIGURE 5.6 The completed database

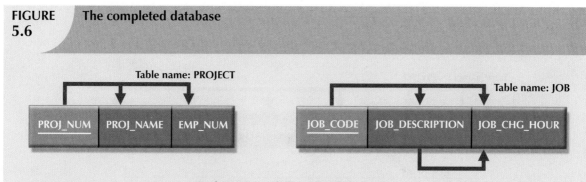

Table name: PROJECT

Table name: JOB

Database name: Ch05_ConstructCo

Table name: PROJECT

PROJ_NUM	PROJ_NAME	EMP_NUM
15	Evergreen	105
18	Amber Wave	104
22	Rolling Tide	113
25	Starflight	101

Table name: JOB

JOB_CODE	JOB_DESCRIPTION	JOB_CHG_HOUR
500	Programmer	35.75
501	Systems Analyst	96.75
502	Database Designer	105.00
503	Electrical Engineer	84.50
504	Mechanical Engineer	67.90
505	Civil Engineer	55.78
506	Clerical Support	26.87
507	DSS Analyst	45.95
508	Applications Designer	48.10
509	Bio Technician	34.55
510	General Support	18.36

Table name: ASSIGNMENT

ASSIGN_NUM	ASSIGN_DATE	PROJ_NUM	EMP_NUM	ASSIGN_HOURS	ASSIGN_CHG_HOUR	ASSIGN_CHARGE
1001	04-Mar-08	15	103	2.6	84.50	219.70
1002	04-Mar-08	18	118	1.4	18.36	25.70
1003	05-Mar-08	15	101	3.6	105.00	378.00
1004	05-Mar-08	22	113	2.5	48.10	120.25
1005	05-Mar-08	15	103	1.9	84.50	160.55
1006	05-Mar-08	25	115	4.2	96.75	406.35
1007	05-Mar-08	22	105	5.2	105.00	546.00
1008	05-Mar-08	25	101	1.7	105.00	178.50
1009	05-Mar-08	15	105	2.0	105.00	210.00
1010	06-Mar-08	15	102	3.8	96.75	367.65
1011	06-Mar-08	22	104	2.6	96.75	251.55
1012	06-Mar-08	15	101	2.3	105.00	241.50
1013	06-Mar-08	25	114	1.8	48.10	86.58
1014	06-Mar-08	22	111	4.0	26.87	107.48
1015	06-Mar-08	25	114	3.4	48.10	163.54
1016	06-Mar-08	18	112	1.2	45.95	55.14
1017	06-Mar-08	18	118	2.0	18.36	36.72
1018	06-Mar-08	18	104	2.6	96.75	251.55
1019	06-Mar-08	15	103	3.0	84.50	253.50
1020	07-Mar-08	22	105	2.7	105.00	283.50
1021	08-Mar-08	25	108	4.2	96.75	406.35
1022	07-Mar-08	25	114	5.8	48.10	278.98
1023	07-Mar-08	22	106	2.4	35.75	85.80

FIGURE 5.6   The completed database (continued)

Table name: EMPLOYEE

Table name: EMPLOYEE

EMP_NUM	EMP_LNAME	EMP_FNAME	EMP_INITIAL	EMP_HIREDATE	JOB_CODE
101	News	John	G	08-Nov-00	502
102	Senior	David	H	12-Jul-89	501
103	Arbough	June	E	01-Dec-97	503
104	Ramoras	Anne	K	15-Nov-88	501
105	Johnson	Alice	K	01-Feb-94	502
106	Smithfield	William		22-Jun-05	500
107	Alonzo	Maria	D	10-Oct-94	500
108	Washington	Ralph	B	22-Aug-89	501
109	Smith	Larry	W	18-Jul-99	501
110	Olenko	Gerald	A	11-Dec-96	505
111	Wabash	Geoff	B	04-Apr-89	506
112	Smithson	Darlene	M	23-Oct-95	507
113	Joenbrood	Delbert	K	15-Nov-94	508
114	Jones	Annelise		20-Aug-91	508
115	Bawangi	Travis	B	25-Jan-90	501
116	Pratt	Gerald	L	05-Mar-95	510
117	Williamson	Angie	H	19-Jun-94	509
118	Frommer	James	J	04-Jan-06	510

Figure 5.6 is a vast improvement over the original database design. If the application software is designed properly, the most active table (ASSIGNMENT) requires the entry of only the PROJ_NUM, EMP_NUM, and ASSIGN_HOURS values. The values for the attributes ASSIGN_NUM and ASSIGN_DATE can be generated by the application. For example, the ASSIGN_NUM can be created by using a counter, and the ASSIGN_DATE can be the system date read by the application and automatically entered into the ASSIGNMENT table. In addition, the application software can automatically insert the correct ASSIGN_CHG_HOUR value by writing the appropriate JOB table's JOB_CHG_HOUR value into the ASSIGNMENT table. (The JOB and ASSIGNMENT tables are related through the JOB_CODE attribute.) If the JOB table's JOB_CHG_HOUR value changes, the next insertion of that value into the ASSIGNMENT table will reflect the change automatically. The table structure thus minimizes the need for human intervention. In fact, if the system requires the employees to enter their own work hours, they can scan their EMP_NUM into the ASSIGNMENT table by using a magnetic card reader that enters their identity. Thus, the ASSIGNMENT table's structure can set the stage for maintaining some desired level of security.

5.5 SURROGATE KEY CONSIDERATIONS

Although this design meets the vital entity and referential integrity requirements, the designer still must address some concerns. For example, a composite primary key might become too cumbersome to use as the number of attributes grows. (It becomes difficult to create a suitable foreign key when the related table uses a composite primary key. In addition, a composite primary key makes it more difficult to write search routines.) Or a primary key attribute might simply have too much descriptive content to be usable—which is why the JOB_CODE attribute was added to the JOB table to serve as that table's primary key. When, for whatever reason, the primary key is considered to be unsuitable, designers use surrogate keys.

At the implementation level, a surrogate key is a system-defined attribute generally created and managed via the DBMS. Usually, a system-defined surrogate key is numeric, and its value is automatically incremented for each new row. For example, Microsoft Access uses an AutoNumber data type, Microsoft SQL Server uses an identity column, and Oracle uses a sequence object.

Recall from Section 5.4 that the JOB_CODE attribute was designated to be the JOB table's primary key. However, remember that the JOB_CODE does not prevent duplicate entries from being made, as shown in the JOB table in Table 5.4.

TABLE 5.4	Duplicate Entries in the Job Table	
JOB_CODE	JOB_DESCRIPTION	JOB_CHG_HOUR
511	Programmer	$35.75
512	Programmer	$35.75

Clearly, the data entries in Table 5.4 are inappropriate because they duplicate existing records—yet there has been no violation of either entity integrity or referential integrity. This "multiple duplicate records" problem was created when the JOB_CODE attribute was added as the PK. (When the JOB_DESCRIPTION was initially designated to be the PK, the DBMS would ensure unique values for all job description entries when it was asked to enforce entity integrity. But that option created the problems that caused use of the JOB_CODE attribute in the first place!) In any case, if JOB_CODE is to be the surrogate PK, you still must ensure the existence of unique values in the JOB_DESCRIPTION *through the use of a unique index.*

Note that all of the remaining tables (PROJECT, ASSIGNMENT, and EMPLOYEE) are subject to the same limitations. For example, if you use the EMP_NUM attribute in the EMPLOYEE table as the PK, you can make multiple entries for the same employee. To avoid that problem, you might create a unique index for EMP_LNAME, EMP_FNAME, and EMP_INITIAL. But how would you then deal with two employees named Joe B. Smith? In that case, you might use another (preferably externally defined) attribute to serve as the basis for a unique index.

It is worth repeating that database design often involves trade-offs and the exercise of professional judgment. In a real-world environment, you must strike a balance between design integrity and flexibility. For example, you might design the ASSIGNMENT table to use a unique index on PROJ_NUM, EMP_NUM, and ASSIGN_DATE if you want to limit an employee to only one ASSIGN_HOURS entry per date. That limitation would ensure that employees couldn't enter the same hours multiple times for any given date. Unfortunately, that limitation is likely to be undesirable from a managerial point of view. After all, if an employee works several different times on a project during any given day, it must be possible to make multiple entries for that same employee and the same project during that day. In that case, the best solution might be to add a new externally defined attribute—such as a stub, voucher, or ticket number—to ensure uniqueness. In any case, frequent data audits would be appropriate.

5.6 HIGHER-LEVEL NORMAL FORMS

Tables in 3NF will perform suitably in business transactional databases. However, there are occasions when higher normal forms are useful. In this section, you learn about a special case of 3NF, known as Boyce-Codd normal form (BCNF), and about fourth normal form (4NF).

5.6.1 THE BOYCE-CODD NORMAL FORM (BCNF)

A table is in **Boyce-Codd normal form** (**BCNF**) when every determinant in the table is a candidate key. (Recall from Chapter 3 that a candidate key has the same characteristics as a primary key, but for some reason, it was not chosen to be the primary key.) Clearly, when a table contains only one candidate key, the 3NF and the BCNF are equivalent. Putting that proposition another way, BCNF can be violated only when the table contains more than one candidate key.

NOTE

A table is in BCNF when every determinant in the table is a candidate key.

Most designers consider the BCNF to be a special case of the 3NF. In fact, if the techniques shown here are used, most tables conform to the BCNF requirements once the 3NF is reached. So how can a table be in 3NF and not be in BCNF? To answer that question, you must keep in mind that a transitive dependency exists when one nonprime attribute is dependent on another nonprime attribute.

In other words, a table is in 3NF when it is in 2NF and there are no transitive dependencies. But what about a case in which a nonkey attribute is the determinant of a key attribute? That condition does not violate 3NF, yet it fails to meet the BCNF requirements because BCNF requires that every determinant in the table be a candidate key.

The situation just described (a 3NF table that fails to meet BCNF requirements) is shown in Figure 5.7.

Note these functional dependencies in Figure 5.7:

$A + B \rightarrow C, D$

$C \rightarrow B$

FIGURE 5.7	A table that is in 3NF but not in BCNF

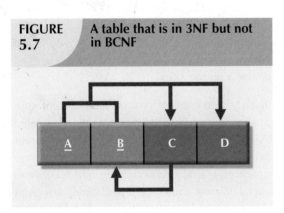

The table structure shown in Figure 5.7 has no partial dependencies, nor does it contain transitive dependencies. (The condition $C \rightarrow B$ indicates that *a nonkey attribute determines part of the primary key*—and *that* dependency is *not* transitive!) Thus, the table structure in Figure 5.7 meets the 3NF requirements. Yet the condition $C \rightarrow B$ causes the table to fail to meet the BCNF requirements.

To convert the table structure in Figure 5.7 into table structures that are in 3NF and in BCNF, first change the primary key to A + C. That is an appropriate action because the dependency $C \rightarrow B$ means that C is, in effect, a superset of B. At this point, the table is in 1NF because it contains a partial dependency $C \rightarrow B$. Next, follow the standard decomposition procedures to produce the results shown in Figure 5.8.

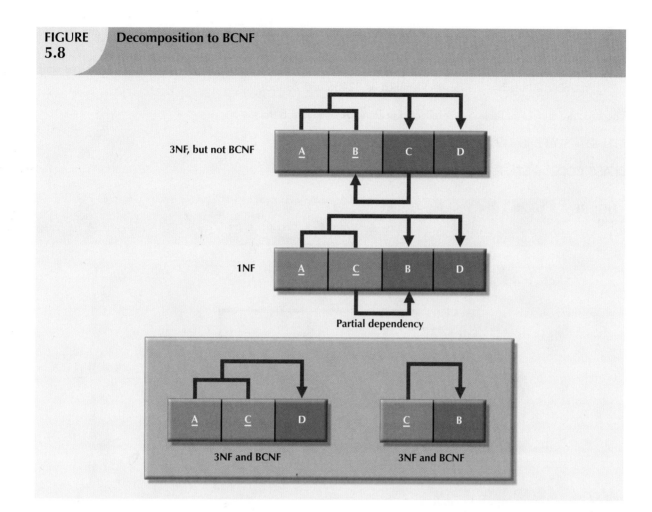

FIGURE 5.8   Decomposition to BCNF

To see how this procedure can be applied to an actual problem, examine the sample data in Table 5.5.

TABLE 5.5   Sample Data for a BCNF Conversion

STU_ID	STAFF_ID	CLASS_CODE	ENROLL_GRADE
125	25	21334	A
125	20	32456	C
135	20	28458	B
144	25	27563	C
144	20	32456	B

Table 5.5 reflects the following conditions:

- Each CLASS_CODE identifies a class uniquely. This condition illustrates the case in which a course might generate many classes. For example, a course labeled INFS 420 might be taught in two classes (sections), each identified by a unique code to facilitate registration. Thus, the CLASS_CODE 32456 might identify INFS 420, class section 1, while the CLASS_CODE 32457 might identify INFS 420, class section 2. Or the CLASS_CODE 28458 might identify QM 362, class section 5.

- A student can take many classes. Note, for example, that student 125 has taken both 21334 and 32456, earning the grades A and C, respectively.
- A staff member can teach many classes, but each class is taught by only one staff member. Note that staff member 20 teaches the classes identified as 32456 and 28458.

The structure shown in Table 5.5 is reflected in Panel A of Figure 5.9:

STU_ID + STAFF_ID → CLASS_CODE, ENROLL_GRADE

CLASS_CODE → STAFF_ID

FIGURE 5.9   Another BNCF decomposition

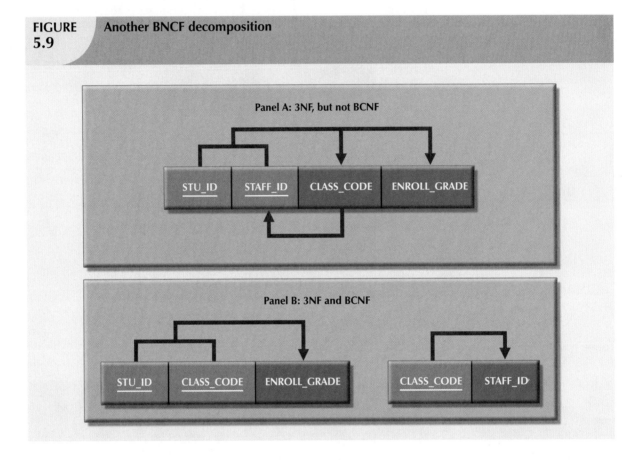

Panel A of Figure 5.9 shows a structure that is clearly in 3NF, but the table represented by this structure has a major problem, because it is trying to describe two things: staff assignments to classes and student enrollment information. Such a dual-purpose table structure will cause anomalies. For example, if a different staff member is assigned to teach class 32456, two rows will require updates, thus producing an update anomaly. And if student 135 drops class 28458, information about who taught that class is lost, thus producing a deletion anomaly. The solution to the problem is to decompose the table structure, following the procedure outlined earlier. Note that the decomposition of Panel B shown in Figure 5.9 yields two table structures that conform to both 3NF and BCNF requirements.

Remember that a table is in BCNF when every determinant in that table is a candidate key. Therefore, when a table contains only one candidate key, 3NF and BCNF are equivalent.

5.6.2 FOURTH NORMAL FORM (4NF)

You might encounter poorly designed databases, or you might be asked to convert spreadsheets into a database format in which multiple multivalued attributes exist. For example, consider the possibility that an employee can have multiple assignments and can also be involved in multiple service organizations. Suppose employee 10123 does volunteer work for the Red Cross and United Way. In addition, the same employee might be assigned to work on three projects: 1, 3, and 4. Figure 5.10 illustrates how that set of facts can be recorded in very different ways.

FIGURE 5.10   **Tables with multivalued dependencies**

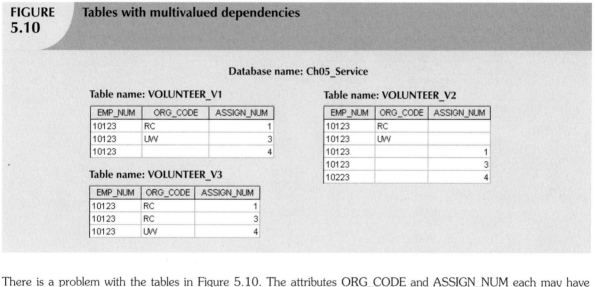

Database name: Ch05_Service

Table name: VOLUNTEER_V1

EMP_NUM	ORG_CODE	ASSIGN_NUM
10123	RC	1
10123	UW	3
10123		4

Table name: VOLUNTEER_V2

EMP_NUM	ORG_CODE	ASSIGN_NUM
10123	RC	
10123	UW	
10123		1
10123		3
10223		4

Table name: VOLUNTEER_V3

EMP_NUM	ORG_CODE	ASSIGN_NUM
10123	RC	1
10123	RC	3
10123	UW	4

There is a problem with the tables in Figure 5.10. The attributes ORG_CODE and ASSIGN_NUM each may have many different values. That is, the tables contain two sets of independent multivalued dependencies. (One employee can have many service entries and many assignment entries.) The presence of multiple sets of independent multivalued dependencies means that if versions 1 and 2 are implemented, the tables are likely to contain quite a few null values; in fact, the tables do not even have a viable candidate key. (The EMP_NUM values are not unique, so they cannot be PKs. No combination of the attributes in table versions 1 and 2 can be used to create a PK because some of them contain nulls.) Such a condition is not desirable, especially when there are thousands of employees, many of whom may have multiple job assignments and many service activities. Version 3 at least has a PK, but it is composed of all of the attributes in the table. In fact, version 3 meets 3NF requirements, yet it contains many redundancies that are clearly undesirable.

The solution is to eliminate the problems caused by independent multivalued dependencies. You do this by creating the ASSIGNMENT and SERVICE_V1 tables depicted in Figure 5.11. Note that in Figure 5.11, neither the ASSIGNMENT nor the SERVICE_V1 table contains independent multivalued dependencies. Those tables are said to be in 4NF.

If you follow the proper design procedures illustrated in this book, you shouldn't encounter the previously described problem. Specifically, the discussion of 4NF is largely academic if you make sure that your tables conform to the following two rules:

1. All attributes must be dependent on the primary key, but they must be independent of each other.

2. No row may contain two or more multivalued facts about an entity.

FIGURE 5.11   A set of tables in 4NF

Table name: EMPLOYEE   Database name: Ch05_Service

EMP_NUM	EMP_LNAME
10121	Rogers
10122	O'Leery
10123	Panera
10124	Johnson

Table name: PROJECT

PROJ_CODE	PROJ_NAME	PROJ_BUDGET
1	BeThere	1023245.00
2	BlueMoon	20198608.00
3	GreenThumb	3234456.00
4	GoFast	5674000.00
5	GoSlow	1002500.00

Table name: ORGANIZATION

ORG_CODE	ORG_NAME
RC	Red Cross
UW	United Way
WF	Wildlife Fund

Table name: ASSIGNMENT

ASSIGN_NUM	EMP_NUM	PROJ_CODE
1	10123	1
2	10121	2
3	10123	3
4	10123	4
5	10121	1
6	10124	2
7	10124	3
8	10124	5

Table name: SERVICE_V1

EMP_NUM	ORG_CODE
10123	RC
10123	UW
10123	WF

The relational diagram

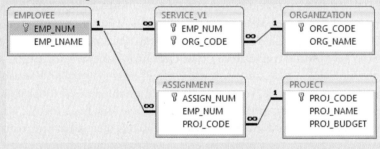

> **NOTE**
>
> A table is in **fourth normal form** (**4NF**) when it is in 3NF and has no multiple sets of multivalued dependencies.

5.7 NORMALIZATION AND DATABASE DESIGN

The tables shown in Figure 5.6 illustrate how normalization procedures can be used to produce good tables from poor ones. You will likely have ample opportunity to put this skill into practice when you begin to work with real-world databases. *Normalization should be part of the design process.* Therefore, make sure that proposed entities meet the required normal form *before* the table structures are created. Keep in mind that if you follow the design procedures discussed in Chapter 3 and Chapter 4 the likelihood of data anomalies will be small. But even the best database designers are known to make occasional mistakes that come to light during normalization checks. However, many of the real-world databases you encounter will have been improperly designed or burdened with anomalies if they were improperly modified during the course of time. And that means you might be asked to redesign and modify existing databases that are, in effect, anomaly traps. Therefore, you should be aware of good design principles and procedures as well as normalization procedures.

First, an ERD is created through an iterative process. You begin by identifying relevant entities, their attributes, and their relationships. Then you use the results to identify additional entities and attributes. The ERD provides the big picture, or macro view, of an organization's data requirements and operations.

Second, normalization focuses on the characteristics of specific entities; that is, normalization represents a micro view of the entities within the ERD. And as you learned in the previous sections of this chapter, the normalization process might yield additional entities and attributes to be incorporated into the ERD. Therefore, it is difficult to separate the normalization process from the ER modeling process; the two techniques are used in an iterative and incremental process.

To illustrate the proper role of normalization in the design process, let's reexamine the operations of the contracting company whose tables were normalized in the preceding sections. Those operations can be summarized by using the following business rules:

- The company manages many projects.
- Each project requires the services of many employees.
- An employee may be assigned to several different projects.
- Some employees are not assigned to a project and perform duties not specifically related to a project. Some employees are part of a labor pool, to be shared by all project teams. For example, the company's executive secretary would not be assigned to any one particular project.
- Each employee has a single primary job classification. That job classification determines the hourly billing rate.
- Many employees can have the same job classification. For example, the company employs more than one electrical engineer.

Given that simple description of the company's operations, two entities and their attributes are initially defined:

- PROJECT (**PROJ_NUM**, PROJ_NAME)
- EMPLOYEE (**EMP_NUM**, EMP_LNAME, EMP_FNAME, EMP_INITIAL, JOB_DESCRIPTION, JOB_CHG_HOUR)

Those two entities constitute the initial ERD shown in Figure 5.12.

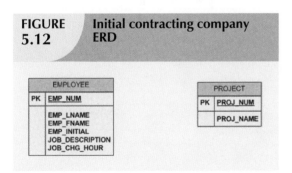

FIGURE 5.12   **Initial contracting company ERD**

After creating the initial ERD shown in Figure 5.12, the normal forms are defined:

- PROJECT is in 3NF and needs no modification at this point.
- EMPLOYEE requires additional scrutiny. The JOB_DESCRIPTION attribute defines job classifications such as Systems Analyst, Database Designer, and Programmer. In turn, those classifications determine the billing rate, JOB_CHG_HOUR. Therefore, EMPLOYEE contains a transitive dependency.

The removal of EMPLOYEE's transitive dependency yields three entities:

- PROJECT (**PROJ_NUM**, PROJ_NAME)
- EMPLOYEE (**EMP_NUM**, EMP_LNAME, EMP_FNAME, EMP_INITIAL, JOB_CODE)
- JOB (**JOB_CODE**, JOB_DESCRIPTION, JOB_CHG_HOUR)

Because the normalization process yields an additional entity (JOB), the initial ERD is modified as shown in Figure 5.13.

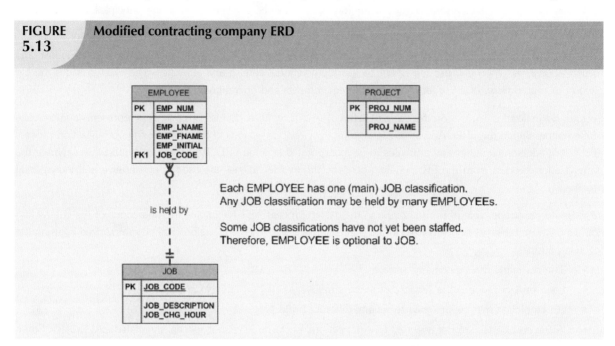

FIGURE 5.13   Modified contracting company ERD

To represent the M:N relationship between EMPLOYEE and PROJECT, you might think that two 1:M relationships could be used—an employee can be assigned to many projects, and each project can have many employees assigned to it. See Figure 5.14. Unfortunately, that representation yields a design that cannot be correctly implemented.

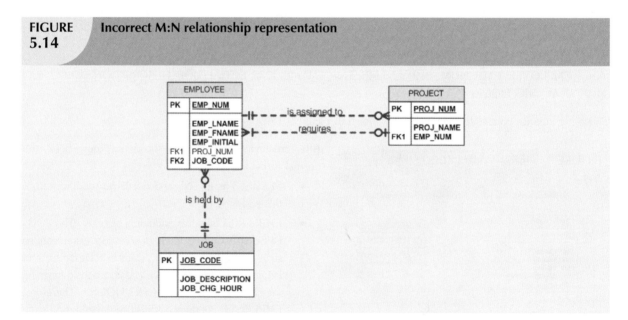

FIGURE 5.14   Incorrect M:N relationship representation

Because the M:N relationship between EMPLOYEE and PROJECT cannot be implemented, the ERD in Figure 5.14 must be modified to include the ASSIGNMENT entity to track the assignment of employees to projects, thus yielding the ERD shown in Figure 5.15. The ASSIGNMENT entity in Figure 5.15 uses the primary keys from the entities PROJECT and EMPLOYEE to serve as its foreign keys. However, note that in this implementation, the ASSIGNMENT entity's surrogate primary key is ASSIGN_NUM, to avoid the use of a composite primary key. Therefore, the *"enters"* relationship between EMPLOYEE and ASSIGNMENT and the *"requires"* relationship between PROJECT and ASSIGNMENT are shown as weak or nonidentifying.

FIGURE 5.15   **Final contracting company ERD**

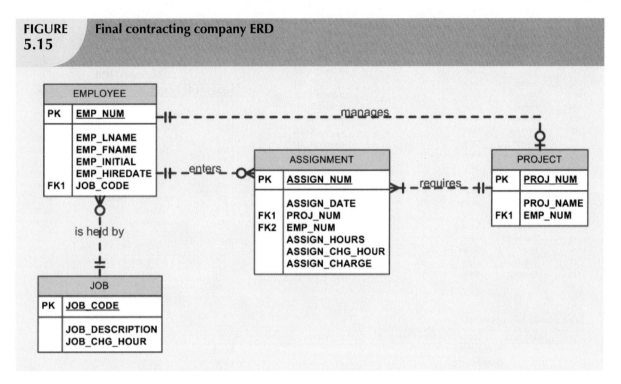

Note that in Figure 5.15, the ASSIGN_HOURS attribute is assigned to the composite entity named ASSIGNMENT. Because you will likely need detailed information about each project's manager, the creation of a "manages" relationship is useful. The "manages" relationship is implemented through the foreign key in PROJECT. Finally, some additional attributes may be created to improve the system's ability to generate additional information. For example, you may want to include the date on which the employee was hired (EMP_HIREDATE) to keep track of worker longevity. Based on this last modification, the model should include four entities and their attributes:

PROJECT (**PROJ_NUM**, PROJ_NAME, EMP_NUM)

EMPLOYEE (**EMP_NUM**, EMP_LNAME, EMP_FNAME, EMP_INITIAL, EMP_HIREDATE, JOB_CODE)

JOB (**JOB_CODE**, JOB_DESCRIPTION, JOB_CHG_HOUR)

ASSIGNMENT (**ASSIGN_NUM**, ASSIGN_DATE, PROJ_NUM, EMP_NUM, ASSIGN_HOURS, ASSIGN_CHG_HOUR, ASSIGN_CHARGE)

The design process is now on the right track. The ERD represents the operations accurately, and the entities now reflect their conformance to 3NF. The combination of normalization and ER modeling yields a useful ERD, whose entities may now be translated into appropriate table structures. In Figure 5.15, note that PROJECT is optional to EMPLOYEE in the "manages" relationship. This optionality exists because not all employees manage projects. The final database contents are shown in Figure 5.16.

FIGURE 5.16   The implemented database

Table name: EMPLOYEE

Database name: Ch05_ConstructCo

EMP_NUM	EMP_LNAME	EMP_FNAME	EMP_INITIAL	EMP_HIREDATE	JOB_CODE
101	News	John	G	08-Nov-00	502
102	Senior	David	H	12-Jul-89	501
103	Arbough	June	E	01-Dec-97	503
104	Ramoras	Anne	K	15-Nov-88	501
105	Johnson	Alice	K	01-Feb-94	502
106	Smithfield	William		22-Jun-05	500
107	Alonzo	Maria	D	10-Oct-94	500
108	Washington	Ralph	B	22-Aug-89	501
109	Smith	Larry	W	18-Jul-99	501
110	Olenko	Gerald	A	11-Dec-96	505
111	Wabash	Geoff	B	04-Apr-89	506
112	Smithson	Darlene	M	23-Oct-95	507
113	Joenbrood	Delbert	K	15-Nov-94	508
114	Jones	Annelise		20-Aug-91	508
115	Bawangi	Travis	B	25-Jan-90	501
116	Pratt	Gerald	L	05-Mar-95	510
117	Williamson	Angie	H	19-Jun-94	509
118	Frommer	James	J	04-Jan-06	510

Table name: JOB

JOB_CODE	JOB_DESCRIPTION	JOB_CHG_HOUR
500	Programmer	35.75
501	Systems Analyst	96.75
502	Database Designer	105.00
503	Electrical Engineer	84.50
504	Mechanical Engineer	67.90
505	Civil Engineer	55.78
506	Clerical Support	26.87
507	DSS Analyst	45.95
508	Applications Designer	48.10
509	Bio Technician	34.55
510	General Support	18.36

Table name: PROJECT

PROJ_NUM	PROJ_NAME	EMP_NUM
15	Evergreen	105
18	Amber Wave	104
22	Rolling Tide	113
25	Starflight	101

Table name: ASSIGNMENT

ASSIGN_NUM	ASSIGN_DATE	PROJ_NUM	EMP_NUM	ASSIGN_HOURS	ASSIGN_CHG_HOUR	ASSIGN_CHARGE
1001	04-Mar-08	15	103	2.6	84.50	219.70
1002	04-Mar-08	18	118	1.4	18.36	25.70
1003	05-Mar-08	15	101	3.6	105.00	378.00
1004	05-Mar-08	22	113	2.5	48.10	120.25
1005	05-Mar-08	15	103	1.9	84.50	160.55
1006	05-Mar-08	25	115	4.2	96.75	406.35
1007	05-Mar-08	22	105	5.2	105.00	546.00
1008	05-Mar-08	25	101	1.7	105.00	178.50
1009	05-Mar-08	15	105	2.0	105.00	210.00
1010	06-Mar-08	15	102	3.8	96.75	367.65
1011	06-Mar-08	22	104	2.6	96.75	251.55
1012	06-Mar-08	15	101	2.3	105.00	241.50
1013	06-Mar-08	25	114	1.8	48.10	86.58
1014	06-Mar-08	22	111	4.0	26.87	107.48
1015	06-Mar-08	25	114	3.4	48.10	163.54
1016	06-Mar-08	18	112	1.2	45.95	55.14
1017	06-Mar-08	18	118	2.0	18.36	36.72
1018	06-Mar-08	18	104	2.6	96.75	251.55
1019	06-Mar-08	15	103	3.0	84.50	253.50
1020	07-Mar-08	22	105	2.7	105.00	283.50
1021	08-Mar-08	25	108	4.2	96.75	406.35
1022	07-Mar-08	25	114	5.8	48.10	278.98
1023	07-Mar-08	22	106	2.4	35.75	85.80

5.8 DENORMALIZATION

It's important to remember that the optimal relational database implementation requires that all tables be at least in third normal form (3NF). A good relational DBMS excels at managing normalized relations; that is, relations void of any unnecessary redundancies that might cause data anomalies. Although the creation of normalized relations is an important database design goal, it is only one of many such goals. Good database design also considers processing (or reporting) requirements and processing speed. The problem with normalization is that as tables are decomposed to conform to normalization requirements, the number of database tables expands. Therefore, in order to generate information, data must be put together from various tables. Joining a large number of tables takes additional input/output (I/O) operations and processing logic, thereby reducing system speed. Most relational database systems are able to handle joins very efficiently. However, rare and occasional circumstances may allow some degree of denormalization so processing speed can be increased.

Keep in mind that the advantage of higher processing speed must be carefully weighed against the disadvantage of data anomalies. On the other hand, some anomalies are of only theoretical interest. For example, should people in a real-world database environment worry that a ZIP_CODE determines CITY in a CUSTOMER table whose primary key is the customer number? Is it really practical to produce a separate table for

ZIP (**ZIP_CODE**, CITY)

to eliminate a transitive dependency from the CUSTOMER table? (Perhaps your answer to that question changes if you are in the business of producing mailing lists.) As explained earlier, the problem with denormalized relations and redundant data is that the data integrity could be compromised due to the possibility of data anomalies (insert, update, and deletion anomalies.) The advice is simple: use common sense during the normalization process.

Furthermore, the database design process could, in some cases, introduce some small degree of redundant data in the model (as seen in the previous example). This, in effect, creates "denormalized" relations. Table 5.6 shows some common examples of data redundancy that are generally found in database implementations.

TABLE 5.6

Common Denormalization Examples

CASE	EXAMPLE	RATIONALE AND CONTROLS
Redundant data	Storing ZIP and CITY attributes in the CUSTOMER table when ZIP determines CITY. (See Table 1.3.)	• Avoid extra join operations • Program can validate city (drop-down box) based on the zip code.
Derived data	Storing STU_HRS and STU_CLASS (student classification) when STU_HRS determines STU_CLASS. (See Figure 3.29.)	• Avoid extra join operations • Program can validate classification (lookup) based on the student hours
Pre-aggregated data (also derived data)	Storing the student grade point average (STU_GPA) aggregate value in the STUDENT table when this can be calculated from the ENROLL and COURSE tables. (See Figure 3.29.)	• Avoid extra join operations • Program computes the GPA every time a grade is entered or updated. • STU_GPA can be updated only via administrative routine.
Information requirements	Using a temporary denormalized table to hold report data. This is required when creating a tabular report in which the columns represent data that is stored in the table as rows. (See Figure 5.17 and Figure 5.18.)	• Impossible to generate the data required by the report using plain SQL. • No need to maintain table. Temporary table is deleted once report is done. • Processing speed is not an issue.

A more comprehensive example of the need for denormalization due to reporting requirements is the case of a faculty evaluation report in which each row list the scores obtained during the last four semesters taught. See Figure 5.17.

FIGURE
5.17

The faculty evaluation report

Faculty Evaluation Report

Instructor	Department	I Semester	Mean	II Semester	Mean	III Semester	Mean	IV Semester	Mean	Last Two Sem. Avg.
Alton	INFS	2005S	2.91	2004F	2.84	2004S	2.55	2003F	2.51	2.875
Ames	INFS	2005S	3.24	2004F	3.26	2004S	3.31	2003F	3.19	3.250
Crandon	INFS	2005S	3.93	2004F	3.95	2004S	3.91	2003F	3.88	3.940
Dumas	MGMT	2004F	3.66	2004S	3.69	2003F	3.56	2003S	3.72	3.675
Landon	BMOM	2005S	3.57	2004F	3.64	2004S	3.39	2003F	3.57	3.605
Lohar	ECON	1999F	3.53	1998F	3.53					3.530
Rolman	INFS	1996S	3.50							3.500

Although this report seems simple enough, the problem arises from the fact that the data are stored in a normalized table in which each row represents a different score for a given faculty in a given semester. See Figure 5.18.

FIGURE
5.18

The EVALDATA and FACHIST tables

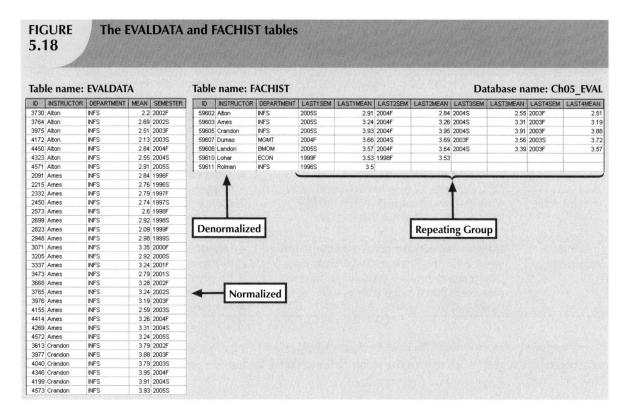

The difficulty of transposing multirow data to multicolumnar data is compounded by the fact that the last four semesters taught are not necessarily the same for all faculty members (some might have taken sabbaticals, some might have had research appointments, some might be new faculty with only two semesters on the job, etc.) To generate this report, the two tables you see in Figure 5.18 were used. The EVALDATA table is the master data table containing the evaluation scores for each faculty member for each semester taught; this table is normalized. The FACHIST table contains the last four data points—that is, evaluation score and semester—for each faculty member. The FACHIST table is a temporary denormalized table created from the EVALDATA table via a series of queries. (The FACHIST table is the basis for the report shown in Figure 5.17.)

As seen in the faculty evaluation report, the conflicts between design efficiency, information requirements, and performance are often resolved through compromises that may include denormalization. In this case and assuming there is enough storage space, the designer's choices could be narrowed down to:

- Store the data in a permanent denormalized table. This is not the recommended solution, because the denormalized table is subject to data anomalies (insert, update, and delete.) This solution is viable only if performance is an issue.

- Create a temporary denormalized table from the permanent normalized table(s). Because the denormalized table exists only as long as it takes to generate the report, it disappears after the report is produced. Therefore, there are no data anomaly problems. This solution is practical only if performance is not an issue and there are no other viable processing options.

As shown, normalization purity is often difficult to sustain in the modern database environment. You will learn in Chapter 13, Business Intelligence and Data Warehouses, that lower normalization forms occur (and are even required) in specialized databases known as data warehouses. Such specialized databases reflect the ever-growing demand for greater scope and depth in the data on which decision support systems increasingly rely. You will discover that the data warehouse routinely uses 2NF structures in its complex, multilevel, multisource data environment. In short, although normalization is very important, especially in the so-called production database environment, 2NF is no longer disregarded as it once was.

Although 2NF tables cannot always be avoided, the problem of working with tables that contain partial and/or transitive dependencies in a production database environment should not be minimized. Aside from the possibility of troublesome data anomalies being created, unnormalized tables in a production database tend to suffer from these defects:

- Data updates are less efficient because programs that read and update tables must deal with larger tables.

- Indexing is more cumbersome. It simply is not practical to build all of the indexes required for the many attributes that might be located in a single unnormalized table.

- Unnormalized tables yield no simple strategies for creating virtual tables known as *views*. You will learn how to create and use views in Chapter 7, Introduction to Structured Query Language (SQL).

Remember that good design cannot be created in the application programs that use a database. Also keep in mind that unnormalized database tables often lead to various data redundancy disasters in production databases such as the ones examined thus far. In other words, use denormalization cautiously and make sure that you can explain why the unnormalized tables are a better choice in certain situations than their normalized counterparts.

SUMMARY

▶ Normalization is a technique used to design tables in which data redundancies are minimized. The first three normal forms (1NF, 2NF, and 3NF) are most commonly encountered. From a structural point of view, higher normal forms are better than lower normal forms because higher normal forms yield relatively fewer data redundancies in the database. Almost all business designs use 3NF as the ideal normal form. A special, more restricted 3NF known as Boyce-Codd normal form, or BCNF, is also used.

▶ A table is in 1NF when all key attributes are defined and when all remaining attributes are dependent on the primary key. However, a table in 1NF can still contain both partial and transitive dependencies. (A partial dependency is one in which an attribute is functionally dependent on only a part of a multiattribute primary key. A transitive dependency is one in which one attribute is functionally dependent on another nonkey attribute.) A table with a single-attribute primary key cannot exhibit partial dependencies.

▶ A table is in 2NF when it is in 1NF and contains no partial dependencies. Therefore, a 1NF table is automatically in 2NF when its primary key is based on only a single attribute. A table in 2NF may still contain transitive dependencies.

▶ A table is in 3NF when it is in 2NF and contains no transitive dependencies. Given that definition of 3NF, the Boyce-Codd normal form (BCNF) is merely a special 3NF case in which all determinant keys are candidate keys. When a table has only a single candidate key, a 3NF table is automatically in BCNF.

▶ A table that is not in 3NF may be split into new tables until all of the tables meet the 3NF requirements. The process is illustrated in Figures 5.19 to 5.21.

FIGURE 5.19   The initial 1NF structure

The Initial 1NF Structure

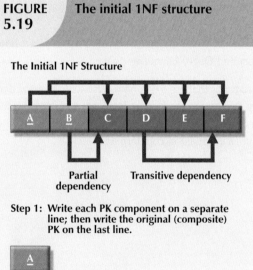

Partial dependency   Transitive dependency

Step 1: Write each PK component on a separate line; then write the original (composite) PK on the last line.

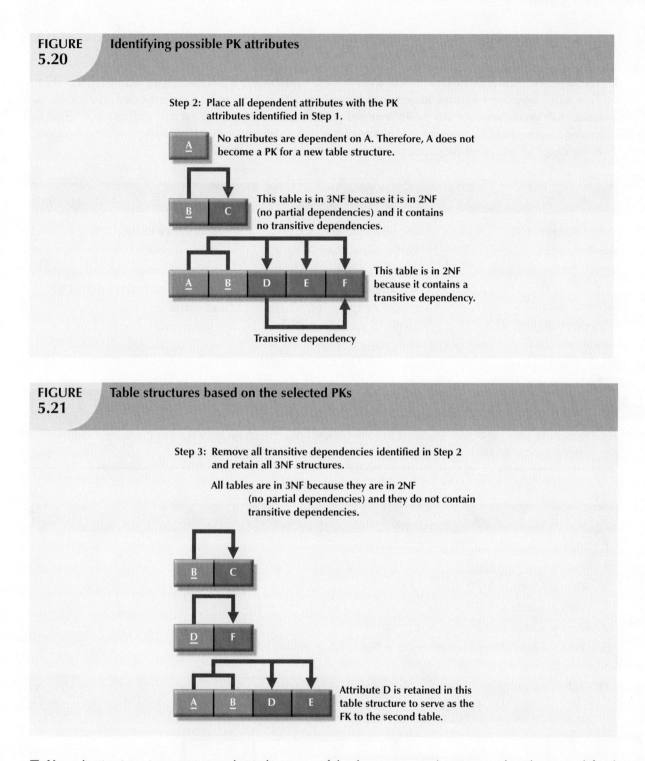

FIGURE 5.20   Identifying possible PK attributes

Step 2: Place all dependent attributes with the PK attributes identified in Step 1.

No attributes are dependent on A. Therefore, A does not become a PK for a new table structure.

This table is in 3NF because it is in 2NF (no partial dependencies) and it contains no transitive dependencies.

This table is in 2NF because it contains a transitive dependency.

Transitive dependency

FIGURE 5.21   Table structures based on the selected PKs

Step 3: Remove all transitive dependencies identified in Step 2 and retain all 3NF structures.

All tables are in 3NF because they are in 2NF (no partial dependencies) and they do not contain transitive dependencies.

Attribute D is retained in this table structure to serve as the FK to the second table.

- Normalization is an important part—but only a part—of the design process. As entities and attributes are defined during the ER modeling process, subject each entity (set) to normalization checks and form new entity (sets) as required. Incorporate the normalized entities into the ERD and continue the iterative ER process until all entities and their attributes are defined and all equivalent tables are in 3NF.

- A table in 3NF might contain multivalued dependencies that produce either numerous null values or redundant data. Therefore, it might be necessary to convert a 3NF table to the fourth normal form (4NF) by splitting the table to remove the multivalued dependencies. Thus, a table is in 4NF when it is in 3NF and contains no multivalued dependencies.

▸ The larger the number of tables, the more additional I/O operations and processing logic required to join them. Therefore, tables are sometimes denormalized to yield less I/O in order to increase processing speed. Unfortunately, with larger tables, you pay for the increased processing speed by making the data updates less efficient, by making indexing more cumbersome, and by introducing data redundancies that are likely to yield data anomalies. In the design of production databases, use denormalization sparingly and cautiously.

KEY TERMS

atomic attribute, 165
atomicity, 165
Boyce-Codd normal form
    (BCNF), 170
denormalization, 153
dependency diagram, 160
determinant, 163

first normal form (1NF), 161
fourth normal form (4NF), 174
granularity, 166
key attribute, 162
nonkey attribute, 162
nonprime attribute, 162
normalization, 153

partial dependency, 160
prime attribute, 162
repeating group, 158
second normal form (2NF), 163
surrogate key, 165
third normal form (3NF), 164
transitive dependency, 160

ONLINE CONTENT

Answers to selected Review Questions and Problems for this chapter are contained in the Student Online Companion for this book.

REVIEW QUESTIONS

1. What is normalization?
2. When is a table in 1NF?
3. When is a table in 2NF?
4. When is a table in 3NF?
5. When is a table in BCNF?
6. Given the dependency diagram shown in Figure Q5.6, answer Items 6a–6c.

FIGURE Q5.6   **Dependency diagram for Question 6**

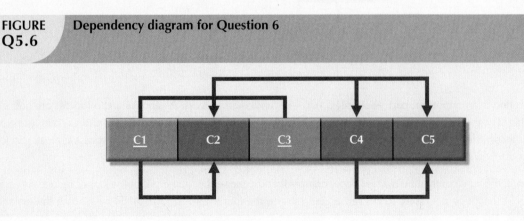

a. Identify and discuss each of the indicated dependencies.

b. Create a database whose tables are at least in 2NF, showing the dependency diagrams for each table.

c. Create a database whose tables are at least in 3NF, showing the dependency diagrams for each table.

7. What is a partial dependency? With what normal form is it associated?

8. What three data anomalies are likely to be the result of data redundancy? How can such anomalies be eliminated?

9. Define and discuss the concept of transitive dependency.

10. What is a surrogate key, and when should you use one?

11. Why is a table whose primary key consists of a single attribute automatically in 2NF when it is in 1NF?

12. How would you describe a condition in which one attribute is dependent on another attribute, when neither attribute is part of the primary key?

13. Suppose that someone tells you that an attribute that is part of a composite primary key is also a candidate key. How would you respond to that statement?

14. A table is in _____ normal form when it is in _____ and there are no transitive dependencies.

PROBLEMS

1. Using the INVOICE table structure shown below, write the relational schema, draw its dependency diagram, and identify all dependencies, including all partial and transitive dependencies. You can assume that the table does not contain repeating groups and that an invoice number references more than one product. (*Hint*: This table uses a composite primary key.)

TABLE
P5.1

ATTRIBUTE NAME	SAMPLE VALUE	SAMPLE VALUE	SAMPLE VALUE	SAMPLE VALUE	SAMPLE VALUE
INV_NUM	211347	211347	211347	211348	211349
PROD_NUM	AA-E3422QW	QD-300932X	RU-995748G	AA-E3422QW	GH-778345P
SALE_DATE	15-Jan-2008	15-Jan-2008	15-Jan-2008	15-Jan-2008	16-Jan-2008
PROD_LABEL	Rotary sander	0.25-in. drill bit	Band saw	Rotary sander	Power drill
VEND_CODE	211	211	309	211	157
VEND_NAME	NeverFail, Inc.	NeverFail, Inc.	BeGood, Inc.	NeverFail, Inc.	ToughGo, Inc.
QUANT_SOLD	1	8	1	2	1
PROD_PRICE	$49.95	$3.45	$39.99	$49.95	$87.75

2. Using the answer to Problem 1, remove all partial dependencies, write the relational schema, and draw the new dependency diagrams. Identify the normal forms for each table structure you created.

NOTE

You can assume that any given product is supplied by a single vendor, but a vendor can supply many products. Therefore, it is proper to conclude that the following dependency exists:

PROD_NUM → PROD_DESCRIPTION, PROD_PRICE, VEND_CODE, VEND_NAME

(*Hint*: Your actions should produce three dependency diagrams.)

3. Using the answer to Problem 2, remove all transitive dependencies, write the relational schema, and draw the new dependency diagrams. Also identify the normal forms for each table structure you created.

4. Using the results of Problem 3, draw the Crow's Foot ERD.

5. Using the STUDENT table structure shown in Table P5.5, write the relational schema and draw its dependency diagram. Identify all dependencies, including all transitive dependencies.

TABLE
P5.5

ATTRIBUTE NAME	SAMPLE VALUE	SAMPLE VALUE	SAMPLE VALUE	SAMPLE VALUE	SAMPLE VALUE
STU_NUM	211343	200128	199876	199876	223456
STU_LNAME	Stephanos	Smith	Jones	Ortiz	McKulski
STU_MAJOR	Accounting	Accounting	Marketing	Marketing	Statistics
DEPT_CODE	ACCT	ACCT	MKTG	MKTG	MATH
DEPT_NAME	Accounting	Accounting	Marketing	Marketing	Mathematics
DEPT_PHONE	4356	4356	4378	4378	3420
COLLEGE_NAME	Business Admin	Business Admin	Business Admin	Business Admin	Arts & Sciences
ADVISOR_LNAME	Grastrand	Grastrand	Gentry	Tillery	Chen
ADVISOR_OFFICE	T201	T201	T228	T356	J331
ADVISOR_BLDG	Torre Building	Torre Building	Torre Building	Torre Building	Jones Building
ADVISOR_PHONE	2115	2115	2123	2159	3209
STU_GPA	3.87	2.78	2.31	3.45	3.58
STU_HOURS	75	45	117	113	87
STU_CLASS	Junior	Sophomore	Senior	Senior	Junior

6. Using the answer to Problem 5, write the relational schema and draw the dependency diagram to meet the 3NF requirements to the greatest practical extent possible. If you believe that practical considerations dictate using a 2NF structure, explain why your decision to retain 2NF is appropriate. If necessary, add or modify attributes to create appropriate determinants and to adhere to the naming conventions.

NOTE

Although the completed student hours (STU_HOURS) do determine the student classification (STU_CLASS), this dependency is not as obvious as you might initially assume it to be. For example, a student is considered a junior if that student has completed between 61 and 90 credit hours. Therefore, a student who is classified as a junior may have completed 66, 72, or 87 hours or any other number of hours within the specified range of 61–90 hours. In short, any hour value within a specified range will define the classification.

7. Using the results of Problem 6, draw the Crow's Foot ERD.

NOTE

This ERD constitutes a small segment of a university's full-blown design. For example, this segment might be combined with the Tiny College presentation in Chapter 4.

8. To keep track of office furniture, computers, printers, and so on, the FOUNDIT company uses the table structure shown in Table P5.8.

TABLE
P5.8

ATTRIBUTE NAME	SAMPLE VALUE	SAMPLE VALUE	SAMPLE VALUE
ITEM_ID	231134-678	342245-225	254668-449
ITEM_LABEL	HP DeskJet 895Cse	HP Toner	DT Scanner
ROOM_NUMBER	325	325	123
BLDG_CODE	NTC	NTC	CSF
BLDG_NAME	Nottooclear	Nottoclear	Canseefar
BLDG_MANAGER	I. B. Rightonit	I. B. Rightonit	May B. Next

Given that information, write the relational schema and draw the dependency diagram. Make sure that you label the transitive and/or partial dependencies.

9. Using the answer to Problem 8, write the relational schema and create a set of dependency diagrams that meet 3NF requirements. Rename attributes to meet the naming conventions and create new entities and attributes as necessary.

10. Using the results of Problem 9, draw the Crow's Foot ERD.

NOTE

Problems 11–14 may be combined to serve as a case or a miniproject.

---

11. The table structure shown in Table P5.11 contains many unsatisfactory components and characteristics. For example, there are several multivalued attributes, naming conventions are violated, and some attributes are not atomic.

TABLE
P5.11

EMP_NUM	1003	1018	1019	1023
EMP_LNAME	Willaker	Smith	McGuire	McGuire
EMP_EDUCATION	BBA, MBA	BBA		BS, MS, Ph.D.
JOB_CLASS	SLS	SLS	JNT	DBA
EMP_DEPENDENTS	Gerald (spouse), Mary (daughter), John (son)		JoAnne (spouse)	George (spouse) Jill (daughter)
DEPT_CODE	MKTG	MKTG	SVC	INFS
DEPT_NAME	Marketing	Marketing	General Service	Info. Systems
DEPT_MANAGER	Jill H. Martin	Jill H. Martin	Hank B. Jones	Carlos G. Ortez
EMP_TITLE	Sales Agent	Sales Agent	Janitor	DB Admin
EMP_DOB	23-Dec-1968	28-Mar-1979	18-May-1982	20-Jul-1959
EMP_HIRE_DATE	14-Oct-1997	15-Jan-2006	21-Apr-2003	15-Jul-1999
EMP_TRAINING	L1, L2	L1	L1	L1, L3, L8, L15
EMP_BASE_SALARY	$38,255.00	$30,500.00	$19,750.00	$127,900.00
EMP_COMMISSION_RATE	0.015	0.010		

Given the structure shown in Table P5.11, write the relational schema and draw its dependency diagram. Label all transitive and/or partial dependencies.

12. Using the answer to Problem 11, draw the dependency diagrams that are in 3NF. (*Hint:* You might have to create a few new attributes. Also make sure that the new dependency diagrams contain attributes that meet proper

design criteria; that is, make sure that there are no multivalued attributes, that the naming conventions are met, and so on.)

13. Using the results of Problem 12, draw the relational diagram.

14. Using the results of Problem 13, draw the Crow's Foot ERD.

NOTE

Problems 15-17 may be combined to serve as a case or a miniproject.

15. Suppose you are given the following business rules to form the basis for a database design. The database must enable the manager of a company dinner club to mail invitations to the club's members, to plan the meals, to keep track of who attends the dinners, and so on.

- Each dinner serves many members, and each member may attend many dinners.
- A member receives many invitations, and each invitation is mailed to many members.
- A dinner is based on a single entree, but an entree may be used as the basis for many dinners. For example, a dinner may be composed of a fish entree, rice, and corn. Or the dinner may be composed of a fish entree, a baked potato, and string beans.
- A member may attend many dinners, and each dinner may be attended by many members.

Because the manager is not a database expert, the first attempt at creating the database uses the structure shown in Table P5.15.

TABLE P5.15

ATTRIBUTE NAME	SAMPLE VALUE	SAMPLE VALUE	SAMPLE VALUE
MEMBER_NUM	214	235	214
MEMBER_NAME	Alice B. VanderVoort	Gerald M. Gallega	Alice B. VanderVoort
MEMBER_ADDRESS	325 Meadow Park	123 Rose Court	325 Meadow Park
MEMBER_CITY	Murkywater	Highlight	Murkywater
MEMBER_ZIPCODE	12345	12349	12345
INVITE_NUM	8	9	10
INVITE_DATE	23-Feb-2008	12-Mar-2008	23-Feb-2008
ACCEPT_DATE	27-Feb-2008	15-Mar-2008	27-Feb-2008
DINNER_DATE	15-Mar-2008	17-Mar-2008	15-Mar-2008
DINNER_ATTENDED	Yes	Yes	No
DINNER_CODE	DI5	DI5	DI2
DINNER_DESCRIPTION	Glowing Sea Delight	Glowing Sea Delight	Ranch Superb
ENTREE_CODE	EN3	EN3	EN5
ENTREE_DESCRIPTION	Stuffed crab	Stuffed crab	Marinated steak
DESSERT_CODE	DE8	DE5	DE2
DESSERT_DESCRIPTION	Chocolate mousse with raspberry sauce	Cherries jubilee	Apple pie with honey crust

Given the table structure illustrated in Table P5.15, write the relational schema and draw its dependency diagram. Label all transitive and/or partial dependencies. (*Hint*: This structure uses a composite primary key.)

16. Break up the dependency diagram you drew in Problem 15 to produce dependency diagrams that are in 3NF and write the relational schema. (*Hint*: You might have to create a few new attributes. Also make sure that the new dependency diagrams contain attributes that meet proper design criteria; that is, make sure that there are no multivalued attributes, that the naming conventions are met, and so on.)

17. Using the results of Problem 16, draw the Crow's Foot ERD.

NOTE

Problems 18–20 may be combined to serve as a case or a miniproject.

18. The manager of a consulting firm has asked you to evaluate a database that contains the table structure shown in Table P5.18.

TABLE
P5.18

ATTRIBUTE NAME	SAMPLE VALUE	SAMPLE VALUE	SAMPLE VALUE
CLIENT_NUM	298	289	289
CLIENT_NAME	Marianne R. Brown	James D. Smith	James D. Smith
CLIENT_REGION	Midwest	Southeast	Southeast
CONTRACT_DATE	10-Feb-2008	15-Feb-2008	12-Mar-2008
CONTRACT_NUMBER	5841	5842	5843
CONTRACT_AMOUNT	$2,985,000.00	$670,300.00	$1,250,000.00
CONSULT_CLASS_1	Database Administration	Internet Services	Database Design
CONSULT_CLASS_2	Web Applications		Database Administration
CONSULT_CLASS_3			Network Installation
CONSULT_CLASS_4			
CONSULTANT_NUM_1	29	34	25
CONSULTANT_NAME_1	Rachel G. Carson	Gerald K. Ricardo	Angela M. Jamison
CONSULTANT_REGION_1	Midwest	Southeast	Southeast
CONSULTANT_NUM_2	56	38	34
CONSULTANT_NAME_2	Karl M. Spenser	Anne T. Dimarco	Gerald K. Ricardo
CONSULTANT_REGION_2	Midwest	Southeast	Southeast
CONSULTANT_NUM_3	22	45	
CONSULTANT_NAME_3	Julian H. Donatello	Geraldo J. Rivera	
CONSULTANT_REGION_3	Midwest	Southeast	
CONSULTANT_NUM_4		18	
CONSULTANT_NAME_4		Donald Chen	
CONSULTANT_REGION_4		West	

Table P5.18 was created to enable the manager to match clients with consultants. The objective is to match a client within a given region with a consultant in that region and to make sure that the client's need for specific consulting services is properly matched to the consultant's expertise. For example, if the client needs help with database design and is located in the Southeast, the objective is to make a match with a consultant who is located in the Southeast and whose expertise is in database design. (Although the consulting company manager tries to match consultant and client locations to minimize travel expense, it is not always possible to do so.) The following basic business rules are maintained:

- Each client is located in one region.
- A region can contain many clients.
- Each consultant can work on many contracts.
- Each contract might require the services of many consultants.
- A client can sign more than one contract, but each contract is signed by only one client.
- Each contract might cover multiple consulting classifications. (For example, a contract may list consulting services in database design and networking.)
- Each consultant is located in one region.

- A region can contain many consultants.

- Each consultant has one or more areas of expertise (class). For example, a consultant might be classified as an expert in both database design and networking.

- Each area of expertise (class) can have many consultants in it. For example, the consulting company might employ many consultants who are networking experts.

Given that brief description of the requirements and the business rules, write the relational schema and draw the dependency diagram for the preceding (and very poor) table structure. Label all transitive and/or partial dependencies.

19. Break up the dependency diagram you drew in Problem 18 to produce dependency diagrams that are in 3NF and write the relational schema. (*Hint*: You might have to create a few new attributes. Also make sure that the new dependency diagrams contain attributes that meet proper design criteria; that is, make sure that there are no multivalued attributes, that the naming conventions are met, and so on.)

20. Using the results of Problem 19, draw the Crow's Foot ERD.

21. Given the sample records in the CHARTER table shown in Table P5.21, write the relational schema and draw the dependency diagram for the table structure. Make sure that you label all dependencies. CHAR_PAX indicates the number of passengers carried. The CHAR_MILES entry is based on round-trip miles, including pickup points. (*Hint*: Look at the data values to determine the nature of the relationships. For example, note that employee Melton has flown two charter trips as pilot and one trip as copilot.)

TABLE
P5.21

ATTRIBUTE NAME	SAMPLE VALUE	SAMPLE VALUE	SAMPLE VALUE	SAMPLE VALUE
CHAR_TRIP	10232	10233	10234	10235
CHAR_DATE	15-Jan-2008	15-Jan-2008	16-Jan-2008	17-Jan-2008
CHAR_CITY	STL	MIA	TYS	ATL
CHAR_MILES	580	1,290	524	768
CUST_NUM	784	231	544	784
CUST_LNAME	Brown	Hanson	Bryana	Brown
CHAR_PAX	5	12	2	5
CHAR_CARGO	235 lbs.	18,940 lbs.	348 lbs.	155 lbs.
PILOT	Melton	Chen	Henderson	Melton
COPILOT		Henderson	Melton	
FLT_ENGINEER		O'Shaski		
LOAD_MASTER		Benkasi		
AC_NUMBER	1234Q	3456Y	1234Q	2256W
MODEL_CODE	PA31-350	CV-580	PA31-350	PA31-350
MODEL_SEATS	10	38	10	10
MODEL_CHG_MILE	$2.79	$23.36	$2.79	$2.79

22. Decompose the dependency diagram you drew to solve Problem 21 to create table structures that are in 3NF and write the relational schema. Make sure that you label all dependencies.

23. Draw the Crow's Foot ERD to reflect the properly decomposed dependency diagrams you created in Problem 22. Make sure that the ERD yields a database that can track all of the data shown in Problem 21. Show all entities, relationships, connectivities, optionalities, and cardinalities.

NOTE

Use the dependency diagram shown in Figure P5.24 to work Problems 24−26.

FIGURE P5.24   Initial dependency diagram for Problems 24—26

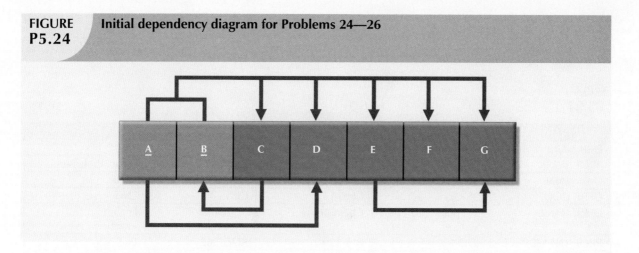

24. Break up the dependency diagram shown in Figure P5.24 to create two new dependency diagrams, one in 3NF and one in 2NF.

25. Modify the dependency diagrams you created in Problem 24 to produce a set of dependency diagrams that are in 3NF. To keep the entire collection of attributes together, copy the 3NF dependency diagram from Problem 24; then show the new dependency diagrams that are also in 3NF. (*Hint:* One of your dependency diagrams will be in 3NF but not in BCNF.)

26. Modify the dependency diagrams you created in Problem 25 to produce a collection of dependency diagrams that are in 3NF and BCNF. To ensure that all attributes are accounted for, copy the 3NF dependency diagrams from Problem 25; then show the new 3NF and BCNF dependency diagrams.

27. Suppose you have been given the table structure and data shown in Table P5.27, which was imported from an Excel spreadsheet. The data reflect that a professor can have multiple advisees, can serve on multiple committees, and can edit more than one journal.

TABLE
P5.27

ATTRIBUTE NAME	SAMPLE VALUE	SAMPLE VALUE	SAMPLE VALUE	SAMPLE VALUE
EMP_NUM	123	104	118	
PROF_RANK	Professor	Asst. Professor	Assoc. Professor	Assoc. Professor
EMP_NAME	Ghee	Rankin	Ortega	Smith
DEPT_CODE	CIS	CHEM	CIS	ENG
DEPT_NAME	Computer Info. Systems	Chemistry	Computer Info. Systems	English
PROF_OFFICE	KDD-567	BLF-119	KDD-562	PRT-345
ADVISEE	1215, 2312, 3233, 2218, 2098	3102, 2782, 3311, 2008, 2876, 2222, 3745, 1783, 2378	2134, 2789, 3456, 2002, 2046, 2018, 2764	2873, 2765, 2238, 2901, 2308
COMMITTEE_CODE	PROMO, TRAF, APPL, DEV	DEV	SPR, TRAF	PROMO, SPR, DEV
JOURNAL_CODE	JMIS, QED, JMGT		JCIS, JMGT	

Given the information in Table P5.27:

a. Draw the dependency diagram.

b. Identify the multivalued dependencies.

c. Create the dependency diagrams to yield a set of table structures in 3NF.

d. Eliminate the multivalued dependencies by converting the affected table structures to 4NF.

e. Draw the Crow's Foot ERD to reflect the dependency diagrams you drew in Part c. (*Note:* You might have to create additional attributes to define the proper PKs and FKs. Make sure that all of your attributes conform to the naming conventions.)

In this chapter, you will learn:

- About the extended entity relationship (EER) model
- How entity clusters are used to represent multiple entities and relationships
- The characteristics of good primary keys and how to select them
- How to use flexible solutions for special data modeling cases
- What issues to check for when developing data models based on EER diagrams

Preview

In the previous three chapters, you learned how to use entity relationship diagrams (ERDs) and normalization techniques to properly create a data model. In this chapter, you learn about the extended entity relationship (EER) model. The EER model builds on ER concepts and adds support for entity supertypes, subtypes, and entity clustering.

Most current database implementations are based on relational databases. Because the relational model uses keys to create associations among tables, it is essential to learn the characteristics of good primary keys and how to select them. Selecting a good primary key is too important to be left to chance, so in this chapter we cover the critical aspects of primary key identification and placement.

Focusing on practical database design, this chapter also illustrates some special design cases that highlight the importance of flexible designs, which can be adapted to meet the demands of changing data and information requirements. Data modeling is a vital step in the development of databases that in turn provide a good foundation for successful application development. Remember that good database applications cannot be based on bad database designs, and no amount of outstanding coding can overcome the limitations of poor database design.

To help you carry out data modeling tasks, the chapter concludes with a checklist that outlines basic database modeling principles.

6.1 THE EXTENDED ENTITY RELATIONSHIP MODEL

As the complexity of the data structures being modeled has increased and as application software requirements have become more stringent, there has been an increasing need to capture more information in the data model. The **extended entity relationship model (EERM)**, sometimes referred to as the enhanced entity relationship model, is the result of adding more semantic constructs to the original entity relationship (ER) model. As you might expect, a diagram using this model is called an **EER diagram (EERD)**. In the following sections, you will learn about the main EER model constructs—entity supertypes, entity subtypes, and entity clustering—and see how they are represented in ERDs.

6.1.1   ENTITY SUPERTYPES AND SUBTYPES

Because most employees possess a wide range of skills and special qualifications, data modelers must find a variety of ways to group employees based on employee characteristics. For instance, a retail company could group employees as salaried and hourly employees, while a university could group employees as faculty, staff, and administrators.

The grouping of employees to create various *types* of employees provides two important benefits:

- It avoids unnecessary nulls in the employee attributes when some employees have characteristics that are not shared by other employees.

- It enables a particular employee type to participate in relationships that are unique to that employee type.

To illustrate those benefits, let's explore the case of an aviation business. The aviation business employs pilots, mechanics, secretaries, accountants, database managers, and many other types of employees. Figure 6.1 illustrates how pilots share certain characteristics with other employees, such as a last name (EMP_LNAME) and hire date (EMP_HIRE_DATE). On the other hand, many pilot characteristics are not shared by other employees. For example, unlike other employees, pilots must meet special requirements such as flight hour restrictions, flight checks, and periodic training. Therefore, if all employee characteristics and special qualifications were stored in a single EMPLOYEE entity, you would have a lot of nulls or you would have to make a lot of needless dummy entries. In this case, special pilot characteristics such as EMP_LICENSE, EMP_RATINGS, and EMP_MED_TYPE will generate nulls for employees who are not pilots. In addition, pilots participate in some relationships that are unique to their qualifications. For example, not all employees can fly airplanes; only employees who are pilots can participate in the "employee flies airplane" relationship.

FIGURE 6.1   Nulls created by unique attributes

EMP_NUM	EMP_LNAME	EMP_FNAME	EMP_INITIAL	EMP_LICENSE	EMP_RATINGS	EMP_MED_TYPE	EMP_HIRE_DATE
100	Kolmycz	Xavier	T				15-Mar-88
101	Lewis	Marcos		ATP	SEL/MEL/Instr/CFII	1	25-Apr-89
102	Vandam	Jean					20-Dec-93
103	Jones	Victoria	R				28-Aug-03
104	Lange	Edith		ATP	SEL/MEL/Instr	1	20-Oct-97
105	Williams	Gabriel	U	COM	SEL/MEL/Instr/CFI	2	08-Nov-97
106	Duzak	Mario		COM	SEL/MEL/Instr	2	05-Jan-04
107	Diante	Venite	L				02-Jul-97
108	Wiesenbach	Joni					18-Nov-95
109	Travis	Brett	T	COM	SEL/MEL/SES/Instr/CFII	1	14-Apr-01
110	Genkazi	Stan					01-Dec-03

Based on the preceding discussion, you would correctly deduce that the PILOT entity stores only those attributes that are unique to pilots, and that the EMPLOYEE entity stores attributes that are common to all employees. Based on that hierarchy, you can conclude that PILOT is a *subtype* of EMPLOYEE, and that EMPLOYEE is the *supertype* of PILOT. In modeling terms, an **entity supertype** is a generic entity type that is related to one or more **entity subtypes**, where the entity supertype contains the common characteristics, and the entity subtypes contain the unique characteristics of each entity subtype. In the next section, you will learn how the entity supertypes and subtypes are related in a specialization hierarchy.

6.1.2 SPECIALIZATION HIERARCHY

Entity supertypes and subtypes are organized in a **specialization hierarchy**, which depicts the arrangement of higher-level entity supertypes (parent entities) and lower-level entity subtypes (child entities). Figure 6.2 shows the specialization hierarchy formed by an EMPLOYEE supertype and three entity subtypes—PILOT, MECHANIC, and ACCOUNTANT. The specialization hierarchy reflects the 1:1 relationship between EMPLOYEE and its subtypes. For example, a PILOT subtype occurrence is related to one instance of the EMPLOYEE supertype, and a MECHANIC subtype occurrence is related to one instance of the EMPLOYEE supertype. The terminology and symbols in Figure 6.2 are explained throughout this chapter.

FIGURE 6.2   A specialization hierarchy

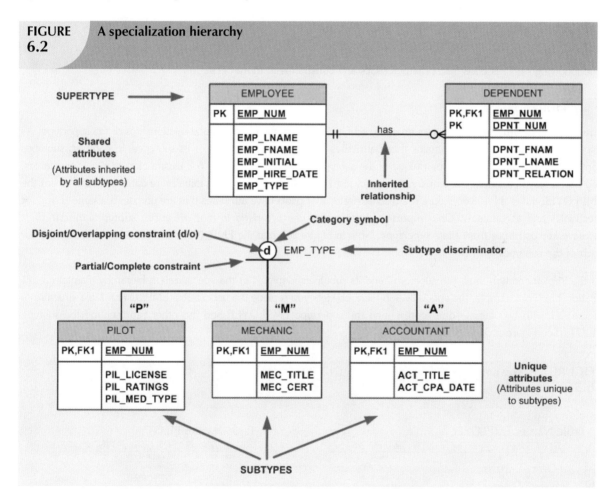

The relationships depicted within the specialization hierarchy are sometimes described in terms of "is-a" relationships. For example, a pilot *is an* employee, a mechanic *is an* employee, and an accountant *is an* employee. It is important to understand that within a specialization hierarchy, a subtype can exist only within the context of a supertype, and every subtype can have only one supertype to which it is directly related. However, a specialization hierarchy can have many levels of supertype/subtype relationships—that is, you can have a specialization hierarchy in which a supertype has many subtypes; in turn, one of the subtypes is the supertype to other lower-level subtypes.

ONLINE CONTENT

This chapter covers only specialization hierarchies. The EER model also supports specialization *lattices,* where a subtype can have multiple parents (supertypes). However, those concepts are better covered under the object-oriented model in **Appendix G, Object-Oriented Databases.** The appendix is available in the Student Online Companion for this book.

As you can see in Figure 6.2, the arrangement of entity supertypes and subtypes in a specialization hierarchy is more than a cosmetic convenience. Specialization hierarchies enable the data model to capture additional semantic content (meaning) into the ERD. A specialization hierarchy provides the means to:

- Support attribute inheritance.
- Define a special supertype attribute known as the subtype discriminator.
- Define disjoint/overlapping constraints and complete/partial constraints.

The following sections cover such characteristics and constraints in more detail.

6.1.3 INHERITANCE

The property of **inheritance** enables an entity subtype to inherit the attributes and relationships of the supertype. As discussed earlier, a supertype contains those attributes that are common to all of its subtypes. In contrast, subtypes contain only the attributes that are unique to the subtype. For example, Figure 6.2 illustrates that pilots, mechanics, and accountants all inherit the employee number, last name, first name, middle initial, hire date, and so on from the EMPLOYEE entity. However, Figure 6.2 also illustrates that pilots have attributes that are unique; the same is true for mechanics and accountants. *One important inheritance characteristic is that all entity subtypes inherit their primary key attribute from their supertype.* Note in Figure 6.2 that the EMP_NUM attribute is the primary key for each of the subtypes.

At the implementation level, the supertype and its subtype(s) depicted in the specialization hierarchy maintain a 1:1 relationship. For example, the specialization hierarchy lets you replace the undesirable EMPLOYEE table structure in Figure 6.1 with two tables—one representing the supertype EMPLOYEE and the other representing the subtype PILOT. (See Figure 6.3.)

**FIGURE
6.3**   The EMPLOYEE-PILOT supertype-subtype relationship

Table Name: EMPLOYEE

EMP_NUM	EMP_LNAME	EMP_FNAME	EMP_INITIAL	EMP_HIRE_DATE	EMP_TYPE
100	Kolmycz	Xavier	T	15-Mar-88	
101	Lewis	Marcos		25-Apr-89	P
102	Vandam	Jean		20-Dec-93	A
103	Jones	Victoria	R	28-Aug-03	
104	Lange	Edith		20-Oct-97	P
105	Williams	Gabriel	U	08-Nov-97	P
106	Duzak	Mario		05-Jan-04	P
107	Diante	Venite	L	02-Jul-97	M
108	Wiesenbach	Joni		18-Nov-95	M
109	Travis	Brett	T	14-Apr-01	P
110	Genkazi	Stan		01-Dec-03	A

Table Name: PILOT

EMP_NUM	PIL_LICENSE	PIL_RATINGS	PIL_MED_TYPE
101	ATP	SEL/MEL/Instr/CFII	1
104	ATP	SEL/MEL/Instr	1
105	COM	SEL/MEL/Instr/CFI	2
106	COM	SEL/MEL/Instr	2
109	COM	SEL/MEL/SES/Instr/CFII	1

Entity subtypes inherit all relationships in which the supertype entity participates. For example, Figure 6.2 shows the EMPLOYEE entity supertype participating in a 1:M relationship with a DEPENDENT entity. Through inheritance, all subtypes also participate in that relationship. In specialization hierarchies with multiple levels of supertype/subtypes, a lower-level subtype inherits all of the attributes and relationships from all of its upper-level supertypes.

6.1.4 SUBTYPE DISCRIMINATOR

A **subtype discriminator** is the attribute in the supertype entity that determines to which subtype the supertype occurrence is related. As seen in Figure 6.2, the subtype discriminator is the employee type (EMP_TYPE).

It is common practice to show the subtype discriminator and its value for each subtype in the ER diagram, as seen in Figure 6.2. However, not all ER modeling tools follow that practice. For example, MS Visio shows the subtype discriminator, but not its value. In Figure 6.2, the Visio text tool was used to manually add the discriminator value above the entity subtype, close to the connector line. Using Figure 6.2 as your guide, note that the supertype is related to a PILOT subtype if the EMP_TYPE has a value of "P." If the EMP_TYPE value is "M," the supertype is related to a MECHANIC subtype. And if the EMP_TYPE value is "A," the supertype is related to the ACCOUNTANT subtype.

It's important to note that the default comparison condition for the subtype discriminator attribute is the equality comparison. However, there may be situations in which the subtype discriminator is not necessarily based on an equality comparison. For example, based on business requirements, you might create two new pilot subtypes, PIC (pilot-in-command)-qualified and copilot-qualified only. A PIC-qualified pilot will be anyone with more than 1,500 PIC flight hours. In this case, the subtype discriminator would be "Flight_Hours," and the criteria would be > 1,500 or <= 1,500, respectively.

> **NOTE**
>
> In Visio, you select the subtype discriminator when creating a category using the Category shape from the available shapes. The Category shape is a small circle with a horizontal line under it that connects the supertype to its subtypes.

> **ONLINE CONTENT**
>
> For a tutorial on using MS Visio to create a specialization hierarchy, see **Appendix A, Designing Databases with Visio Professional: A Tutorial,** in the Student Online Companion for this book.

6.1.5 DISJOINT AND OVERLAPPING CONSTRAINTS

An entity supertype can have disjoint or overlapping entity subtypes. For example, in the aviation example, an employee can be a pilot or a mechanic or an accountant. Assume that one of the business rules dictates that an employee cannot belong to more than one subtype at a time; that is, an employee cannot be a pilot and a mechanic at the same time. **Disjoint subtypes**, also known as **non-overlapping subtypes**, are subtypes that contain a *unique* subset of the supertype entity set; in other words, each entity instance of the supertype can appear in only one of the subtypes. For example, in Figure 6.2, an employee (supertype) who is a pilot (subtype) can appear only in the PILOT subtype, not in any of the other subtypes. In Visio, such disjoint subtypes are indicated by the letter *d* inside the category shape.

On the other hand, if the business rule specifies that employees can have multiple classifications, the EMPLOYEE supertype may contain *overlapping* job classification subtypes. **Overlapping subtypes** are subtypes that contain nonunique subsets of the supertype entity set; that is, each entity instance of the supertype may appear in more than one subtype. For example, in a university environment, a person may be an employee or a student or both. In turn, an employee may be a professor as well as an administrator. Because an employee also may be a student, STUDENT and EMPLOYEE are overlapping subtypes of the supertype PERSON, just as PROFESSOR and ADMINISTRATOR are overlapping subtypes of the supertype EMPLOYEE. Figure 6.4 illustrates overlapping subtypes with the use of the letter *o* inside the category shape.

FIGURE 6.4   **Specialization hierarchy with overlapping subtypes**

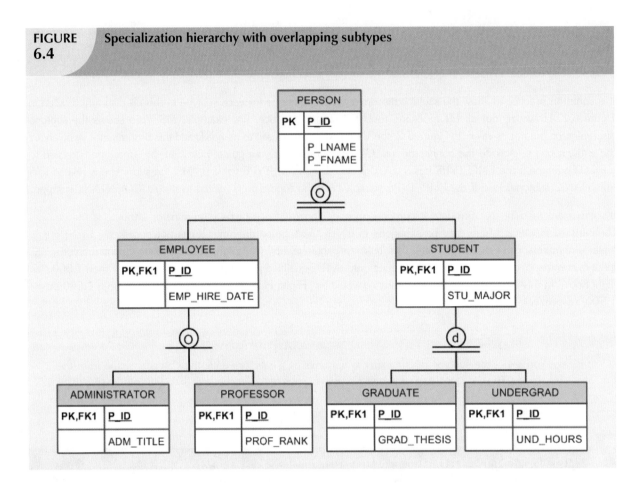

It is common practice to show the disjoint/overlapping symbols in the ERD. (See Figure 6.2 and Figure 6.4.) However, not all ER modeling tools follow that practice. For example, by default, Visio shows only the subtype discriminator (using the Category shape) but not the disjoint/overlapping symbol. Therefore, the Visio text tool was used to manually add the *d* and *o* symbols in Figures 6.2 and 6.4.

NOTE

Alternative notations exist for representing disjoint/overlapping subtypes. For example, Toby J. Teorey popularized the use of *G* and *Gs* to indicate disjoint and overlapping subtypes.

As you learned earlier in this section, the implementation of disjoint subtypes is based on the value of the subtype discriminator attribute in the supertype. However, *implementing* overlapping subtypes requires the use of one discriminator attribute for each subtype. For example, in the case of the Tiny College database design you saw in Chapter 4, Entity Relationship (ER) Modeling, a professor can also be an administrator. Therefore, the EMPLOYEE supertype would have the subtype discriminator attributes and values shown in Table 6.1.

TABLE 6.1   **Discriminator Attributes with Overlapping Subtypes**

DISCRIMINATOR ATTRIBUTES		COMMENT
PROFESSOR	ADMINISTRATOR	
"Y"	"N"	The Employee is a member of the Professor subtype.
"N"	"Y"	The Employee is a member of the Administrator subtype.
"Y"	"Y"	The Employee is both a Professor and an Administrator.

6.1.6 COMPLETENESS CONSTRAINT

The **completeness constraint** specifies whether each entity supertype occurrence must also be a member of at least one subtype. The completeness constraint can be partial or total. **Partial completeness** (symbolized by a circle over a single line) means that not every supertype occurrence is a member of a subtype; that is, there may be some supertype occurrences that are not members of any subtype. **Total completeness** (symbolized by a circle over a double line) means that every supertype occurrence must be a member of at least one subtype.

The ERDs in Figures 6.2 and 6.4 represent the completeness constraint based on the Visio Category shape. A single horizontal line under the circle represents a partial completeness constraint; a double horizontal line under the circle represents a total completeness constraint.

> **NOTE**
>
> Alternative notations exist to represent the completeness constraint. For example, some notations use a single line (partial) or double line (total) to connect the supertype to the Category shape.

Given the disjoint/overlapping subtypes and completeness constraints, it's possible to have the specialization hierarchy constraint scenarios shown in Table 6.2.

TABLE 6.2	Specialization Hierarchy Constraint Scenarios	
TYPE	DISJOINT CONSTRAINT	OVERLAPPING CONSTRAINT
Partial	Supertype has optional subtypes. Subtype discriminator can be null. Subtype sets are unique.	Supertype has optional subtypes. Subtype discriminators can be null. Subtype sets are not unique.
Total	Every supertype occurrence is a member of a (at least one) subtype. Subtype discriminator cannot be null. Subtype sets are unique.	Every supertype occurrence is a member of a (at least one) subtype. Subtype discriminators cannot be null. Subtype sets are not unique.

6.1.7 SPECIALIZATION AND GENERALIZATION

You can use various approaches to develop entity supertypes and subtypes. For example, you can first identify a regular entity, and then identify all entity subtypes based on their distinguishing characteristics. You also can start by identifying multiple entity types and then later extract the common characteristics of those entities to create a higher-level supertype entity.

Specialization is the top-down process of identifying lower-level, more specific entity subtypes from a higher-level entity supertype. Specialization is based on grouping unique characteristics and relationships of the subtypes. In the aviation example, you used specialization to identify multiple entity subtypes from the original employee supertype. **Generalization** is the bottom-up process of identifying a higher-level, more generic entity supertype from lower-level entity subtypes. Generalization is based on grouping common characteristics and relationships of the subtypes. For example, you might identify multiple types of musical instruments: piano, violin, and guitar. Using the generalization approach, you could identify a "string instrument" entity supertype to hold the common characteristics of the multiple subtypes.

6.2 ENTITY CLUSTERING

Developing an ER diagram entails the discovery of possibly hundreds of entity types and their respective relationships. Generally, the data modeler will develop an initial ERD containing a few entities. As the design approaches completion, the ERD will contain hundreds of entities and relationships that crowd the diagram to the point of making it unreadable and inefficient as a communication tool. In those cases, you can use entity clusters to minimize the number of entities shown in the ERD.

An **entity cluster** is a "virtual" entity type used to represent multiple entities and relationships in the ERD. An entity cluster is formed by combining multiple interrelated entities into a single abstract entity object. An entity cluster is considered "virtual" or "abstract" in the sense that it is not actually an entity in the final ERD. Instead, it is a temporary entity used to represent multiple entities and relationships, with the purpose of simplifying the ERD and thus enhancing its readability.

Figure 6.5 illustrates the use of entity clusters based on the Tiny College example in Chapter 4. Note that the ERD contains two entity clusters:

- OFFERING, which groups the COURSE and CLASS entities and relationships.
- LOCATION, which groups the ROOM and BUILDING entities and relationships.

Note also that the ERD in Figure 6.5 does not show attributes for the entities. When using entity clusters, the key attributes of the combined entities are no longer available. Without the key attributes, primary key inheritance rules change. In turn, the change in the inheritance rules can have undesirable consequences, such as changes in relationships—from identifying to nonidentifying or vice versa—and the loss of foreign key attributes from some entities. To eliminate those problems, the general rule is to *avoid the display of attributes when entity clusters are used*.

FIGURE 6.5 **Tiny College ERD using entity clusters**

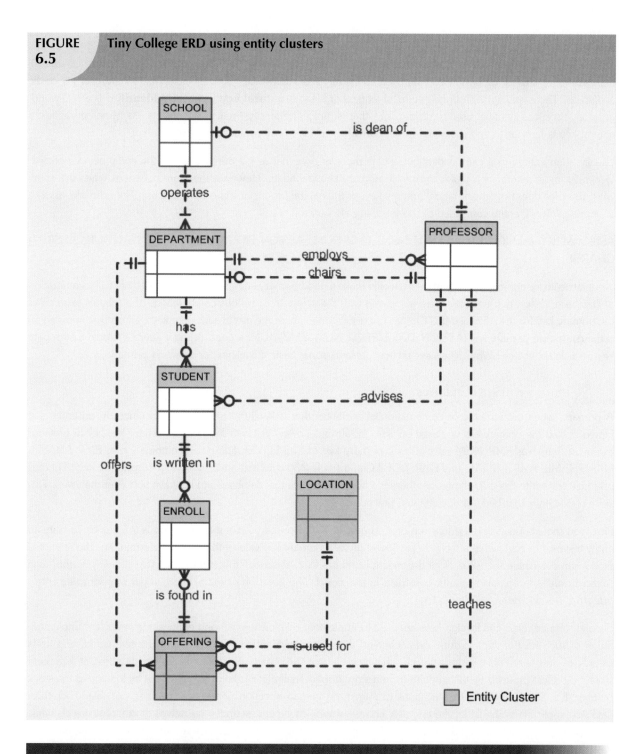

6.3 ENTITY INTEGRITY: SELECTING PRIMARY KEYS

Arguably, the most important characteristic of an entity is its primary key (a single attribute or some combination of attributes), which uniquely identifies each entity instance. The primary key's function is to guarantee entity integrity. Furthermore, primary keys and foreign keys work together to implement relationships in the relational model. Therefore, the importance of properly selecting the primary key has a direct bearing on the efficiency and effectiveness of database implementation.

6.3.1 NATURAL KEYS AND PRIMARY KEYS

The concept of a unique identifier is commonly encountered in the real world. For example, you use class (or section) numbers to register for classes, invoice numbers to identify specific invoices, account numbers to identify credit cards, and so on. Those examples illustrate natural identifiers or keys. A **natural key** or **natural identifier** is a real-world, generally accepted identifier used to distinguish—that is, uniquely identify—real-world objects. As its name implies, a natural key is familiar to end users and forms part of their day-to-day business vocabulary.

Usually, if an entity *has* a natural identifier, a data modeler uses that as the primary key of the entity being modeled. Generally, most natural keys make acceptable primary key identifiers. However, there are occasions when the entity being modeled does not have a natural primary key, or the natural key is not a *good* primary key. For example, assume an ASSIGNMENT entity composed of the following attributes:

ASSIGNMENT (ASSIGN_DATE, PROJ_NUM, EMP_NUM, ASSIGN_HOURS, ASSIGN_CHG_HOUR, ASSIGN_CHARGE)

What attribute (or combination of attributes) would make a good primary key? You learned in Chapter 5, Normalization of Database Tables, that tradeoffs were associated with the selection of various combinations of attributes to serve as the primary key for the ASSIGNMENT table. You also learned about the use of surrogate keys. Given that knowledge, is the composite primary key (ASSIGN_DATE, PROJ_NUM, EMP_NUM) a *good* primary key? Or would a surrogate key be a better choice? Why? The next section presents some basic guidelines for selecting primary keys.

6.3.2 PRIMARY KEY GUIDELINES

A primary key is the attribute or combination of attributes that uniquely identifies entity instances in an entity set. However, can the primary key be based on, say, 12 attributes? And just how long can a primary key be? In previous examples, why was EMP_NUM selected as a primary key of EMPLOYEE and not a combination of EMP_LNAME, EMP_FNAME, EMP_INITIAL, and EMP_DOB? Can a single 256-byte text attribute be a good primary key? There is no single answer to those questions, but there is a body of practice that database experts have built over the years. This section examines that body of documented practices.

First, you should understand the function of a primary key. The primary key's main function is to uniquely identify an entity instance or row within a table. In particular, given a primary key value—that is, the determinant—the relational model can determine the value of all dependent attributes that "describe" the entity. Note that "*identification*" and "*description*" are separate semantic constructs in the model. *The function of the primary key is to guarantee entity integrity, not to "describe" the entity.*

Second, primary keys and foreign keys are used to implement relationships among entities. However, the implementation of such relationships is done mostly behind the scenes, hidden from end users. In the real world, end users identify objects based on the characteristics they know about the objects. For example, when shopping at a grocery store, you select products by taking them from a store display shelf and reading the labels, not by looking at the stock number. It's wise for database applications to mimic the human selection process as much as possible. Therefore, database applications should let the end user choose among multiple descriptive narratives of different objects while using primary key values behind the scenes. Keeping those concepts in mind, look at Table 6.3, which summarizes desirable primary key characteristics.

TABLE 6.3	Desirable Primary Key Characteristics

PK CHARACTERISTIC	RATIONALE
Unique values	The PK must uniquely identify each entity instance. A primary key must be able to guarantee unique values. It cannot contain nulls.
Nonintelligent	The PK should not have embedded semantic meaning (factless). An attribute with embedded semantic meaning is probably better used as a descriptive characteristic of the entity rather than as an identifier. In other words, a student ID of 650973 would be preferred over Smith, Martha L. as a primary key identifier. In short, the PK should be factless.
No change over time	If an attribute has semantic meaning, it might be subject to updates. This is why names do not make good primary keys. If you have Vickie Smith as the primary key, what happens when she gets married? If a primary key is subject to change, the foreign key values must be updated, thus adding to the database work load. Furthermore, changing a primary key value means that you are basically changing the identity of an entity. In short, the PK should be permanent and unchangeable.
Preferably single-attribute	A primary key should have the minimum number of attributes possible (irreducible). Single-attribute primary keys are desirable but not required. Single-attribute primary keys simplify the implementation of foreign keys. Having multiple-attribute primary keys can cause primary keys of related entities to grow through the possible addition of many attributes, thus adding to the database work load and making (application) coding more cumbersome.
Preferably numeric	Unique values can be better managed when they are numeric, because the database can use internal routines to implement a counter-style attribute that automatically increments values with the addition of each new row. In fact, most database systems include the ability to use special constructs, such as Autonumber in Microsoft Access, to support self-incrementing primary key attributes.
Security compliant	The selected primary key must not be composed of any attribute(s) that might be considered a security risk or violation. For example, using a Social Security number as a PK in an EMPLOYEE table is not a good idea.

6.3.3 WHEN TO USE COMPOSITE PRIMARY KEYS

In the previous section, you learned about the desirable characteristics of primary keys. For example, you learned that the primary key should use the minimum number of attributes possible. However, that does *not* mean that composite primary keys are not permitted in a model. In fact, composite primary keys are particularly useful in two cases:

- As identifiers of composite entities, where each primary key combination is allowed only once in the M:N relationship.
- As identifiers of weak entities, where the weak entity has a strong identifying relationship with the parent entity.

To illustrate the first case, assume that you have a STUDENT entity set and a CLASS entity set. In addition, assume that those two sets are related in an M:N relationship via an ENROLL entity set in which each student/class combination may appear only once in the composite entity. Figure 6.6 shows the ERD to represent such a relationship.

As shown in Figure 6.6, the composite primary key automatically provides the benefit of ensuring that there cannot be duplicate values—that is, it ensures that the same student cannot enroll more than once in the same class.

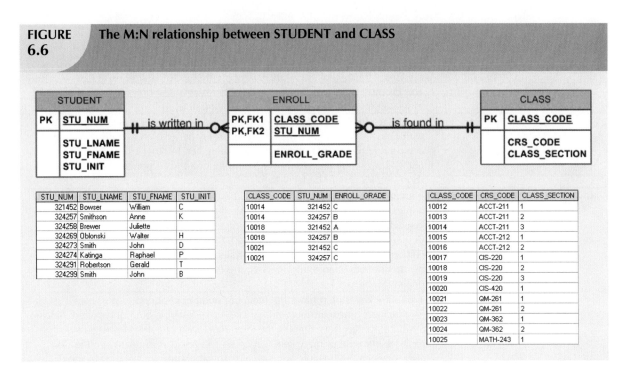

FIGURE 6.6   **The M:N relationship between STUDENT and CLASS**

In the second case, a weak entity in a strong identifying relationship with a parent entity is normally used to represent one of two situations:

1. *A real-world object that is existent-dependent on another real-world object.* Those types of objects are distinguishable in the real world. A dependent and an employee are two separate people who exist independent of each other. However, such objects can exist in the model only when they relate to each other in a strong identifying relationship. For example, the relationship between EMPLOYEE and DEPENDENT is one of existence dependency in which the primary key of the dependent entity is a composite key that contains the key of the parent entity.

2. *A real-world object that is represented in the data model as two separate entities in a strong identifying relationship.* For example, the real-world invoice object is represented by two entities in a data model: INVOICE and LINE. Clearly, the LINE entity does not exist in the real world as an independent object, but rather as part of an INVOICE.

In both situations, having a strong identifying relationship ensures that the dependent entity can exist only when it is related to the parent entity. In summary, the selection of a composite primary key for composite and weak entity types provides benefits that enhance the integrity and consistency of the model.

6.3.4 WHEN TO USE SURROGATE PRIMARY KEYS

There are some instances when a primary key doesn't exist in the real world or when the existing natural key might not be a suitable primary key. For example, consider the case of a park recreation facility that rents rooms for small parties. The manager of the facility keeps track of all events, using a folder with the format shown in Table 6.4.

TABLE 6.4	Data Used to Keep Track of Events					
DATE	TIME_START	TIME_END	ROOM	EVENT_NAME	PARTY_OF	
6/17/08	11:00AM	2:00PM	Allure	Burton Wedding	60	
6/17/08	11:00AM	2:00PM	Bonanza	Adams Office	12	
6/17/08	3:00PM	5:30PM	Allure	Smith Family	15	
6/17/08	3:30PM	5:30PM	Bonanza	Adams Office	12	
6/18/08	1:00PM	3:00PM	Bonanza	Boy Scouts	33	
6/18/08	11:00AM	2:00PM	Allure	March of Dimes	25	
6/18/08	11:00AM	12:30PM	Bonanza	Smith Family	12	

Given the data shown in Table 6.4, you would model the EVENT entity as:

EVENT (DATE, TIME_START, TIME_END, ROOM, EVENT_NAME, PARTY_OF)

What primary key would you suggest? In this case, there is no simple natural key that could be used as a primary key in the model. Based on the primary key concepts you learned about in previous chapters, you might suggest one of these options:

(**DATE**, **TIME_START**, **ROOM**) or (**DATE**, **TIME_END**, **ROOM**)

Assume you select the composite primary key (**DATE**, **TIME_START**, **ROOM**) for the EVENT entity. Next, you determine that one EVENT may use many RESOURCEs (such as tables, projectors, PCs, and stands), and that the same RESOURCE may be used for many EVENTs. The RESOURCE entity would be represented by the following attributes:

RESOURCE (**RSC_ID**, RSC_DESCRIPTION, RSC_TYPE, RSC_QTY, RSC_PRICE)

Given the business rules, the M:N relationship between RESOURCE and EVENT would be represented via the EVNTRSC composite entity with a composite primary key as follows:

EVNTRSC (**DATE, TIME_START, ROOM, RSC_ID,** QTY_USED)

You now have a lengthy four-attribute composite primary key. What would happen if the EVNTRSC entity's primary key were inherited by another existence-dependent entity? At this point, you can see that the composite primary key could make the implementation of the database and program coding unnecessarily complex.

As a data modeler, you probably noticed that the EVENT entity's selected primary key might not fare well, given the primary key guidelines in Table 6.3. In this case, the EVENT entity's selected primary key contains embedded semantic information and is formed by a combination of date, time, and text data columns. In addition, the selected primary key would cause lengthy primary keys for existence-dependent entities. The solution to the problem is to use a numeric single-attribute surrogate primary key.

Surrogate primary keys are accepted practice in today's complex data environments. They are especially helpful when there is no natural key, when the selected candidate key has embedded semantic contents, or when the selected candidate key is too long or cumbersome. However, there is a tradeoff: if you use a surrogate key, you must ensure that the candidate key of the entity in question performs properly through the use of "unique index" and "not null" constraints.

6.4 DESIGN CASES: LEARNING FLEXIBLE DATABASE DESIGN

Data modeling and design require skills that are acquired through experience. In turn, experience is acquired through practice—regular and frequent repetition, applying the concepts learned to specific and different design problems. This section presents four special design cases that highlight the importance of flexible designs, proper identification of primary keys, and placement of foreign keys.

> **NOTE**
>
> In describing the various modeling concepts throughout this book, the focus is on relational models. Also, given the focus on the practical nature of database design, all design issues are addressed with the implementation goal in mind. Therefore, there is no sharp line of demarcation between design and implementation.
>
> At the pure conceptual stage of the design, foreign keys are not part of an ER diagram. The ERD displays only entities and relationships. Entities are identified by identifiers that may become primary keys. During design, the modeler attempts to understand and define the entities and relationships. Foreign keys are the mechanism through which the relationship *designed* in an ERD is *implemented* in a relational model. If you use Visio Professional as your modeling tool, you will discover that this book's methodology is reflected in the Visio modeling practice.

6.4.1   DESIGN CASE #1: IMPLEMENTING 1:1 RELATIONSHIPS

Foreign keys work with primary keys to properly implement relationships in the relational model. The basic rule is very simple: put the primary key of the "one" side (the parent entity) on the "many" side (the dependent entity) as a foreign key. However, where do you place the foreign key when you are working with a 1:1 relationship? For example, take the case of a 1:1 relationship between EMPLOYEE and DEPARTMENT based on the business rule "one EMPLOYEE is the manager of one DEPARTMENT, and one DEPARTMENT is managed by one EMPLOYEE." In that case, there are two options for selecting and placing the foreign key:

1. *Place a foreign key in both entities.* This option is derived from the basic rule you learned in Chapter 4. Place EMP_NUM as a foreign key in DEPARTMENT, and place DEPT_ID as a foreign key in EMPLOYEE. However, that solution is not recommended, as it would create duplicated work, and it could conflict with other existing relationships. (Remember that DEPARTMENT and EMPLOYEE also participate in a 1:M relationship—one department employs many employees.)

2. *Place a foreign key in one of the entities.* In that case, the primary key of one of the two entities appears as a foreign key in the other entity. That is the preferred solution, but there is a remaining question: *which* primary key should be used as a foreign key? The answer to that question is found in Table 6.5. Table 6.5 shows the rationale for selecting the foreign key in a 1:1 relationship based on the relationship properties in the ERD.

TABLE 6.5   Selection of Foreign Key in a 1:1 Relationship

CASE	ER RELATIONSHIP CONSTRAINTS	ACTION
I	One side is mandatory and the other side is optional.	Place the PK of the entity on the mandatory side in the entity on the optional side as an FK, and make the FK mandatory.
II	Both sides are optional.	Select the FK that causes the fewest number of nulls, or place the FK in the entity in which the (relationship) role is played.
III	Both sides are mandatory.	See Case II, or consider revising your model to ensure that the two entities do not belong together in a single entity.

Figure 6.7 illustrates the "EMPLOYEE manages DEPARTMENT" relationship. Note that in this case, EMPLOYEE is mandatory to DEPARTMENT. Therefore, EMP_NUM is placed as the foreign key in DEPARTMENT. Alternatively, you might also argue that the "manager" role is played by the EMPLOYEE in the DEPARTMENT.

FIGURE 6.7	A 1:1 relationship between DEPARTMENT and EMPLOYEE

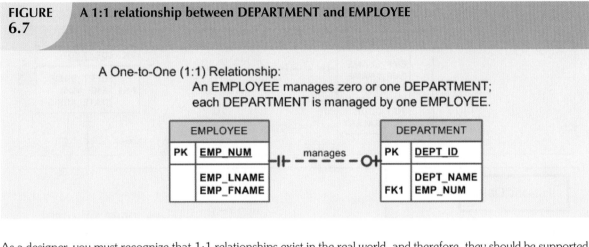

As a designer, you must recognize that 1:1 relationships exist in the real world, and therefore, they should be supported in the data model. In fact, a 1:1 relationship is used to ensure that two entity sets are not placed in the same table. In other words, EMPLOYEE and DEPARTMENT are clearly separate and unique entity types that do not belong together in a single entity. If you group them together in one entity, what would be the name of that entity?

6.4.2 DESIGN CASE #2: MAINTAINING HISTORY OF TIME-VARIANT DATA

Company managers generally realize that good decision making is based on the information that is generated through the data stored in databases. Such data reflect current as well as past events. In fact, company managers use the data stored in databases to answer questions such as, "How do the current company profits compare to those of previous years?" and, "What are XYZ product's sales trends?" In other words, the data stored on databases reflect not only current data, but also historic data.

Normally, data changes are managed by replacing the existing attribute value with the new value, without regard to the previous value. However, there are situations when the history of values for a given attribute must be preserved. From a data modeling point of view, **time-variant data** refer to data whose values change over time and for which you *must* keep a history of the data changes. You could argue that all data in a database are subject to change over time and are, therefore, time variant. However, some attribute values, such as your date of birth or your Social Security number, are not time variant. On the other hand, attributes such as your student GPA or your bank account balance are subject to change over time. Sometimes the data changes are externally originated and event driven, such as a product price change. On other occasions, changes are based on well-defined schedules, such as the daily stock quote "open" and "close" values.

In any case, keeping the history of time-variant data is equivalent to having a multivalued attribute in your entity. To model time-variant data, you must create a new entity in a 1:M relationship with the original entity. This new entity will contain the new value, the date of the change, and whatever other attribute is pertinent to the event being modeled. For example, if you want to keep track of the current manager as well as the history of all department managers over time, you could create the model shown in Figure 6.8.

Note that in Figure 6.8, the MGR_HIST entity has a 1:M relationship with EMPLOYEE and a 1:M relationship with DEPARTMENT to reflect the fact that, over time, an employee could be the manager of many different departments, and a department could have many different employee managers. Because you are recording time-variant data, you must store the DATE_ASSIGN attribute in the MGR_HIST entity to provide the date on which the employee (EMP_NUM) became the manager of the department. The primary key of MGR_HIST permits the same employee to

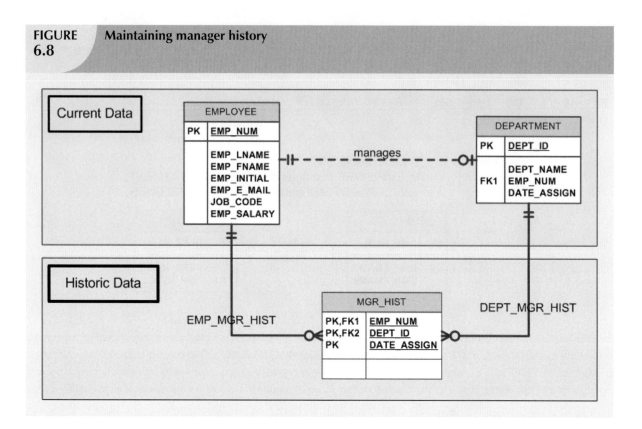

FIGURE 6.8   Maintaining manager history

be the manager of the same department, but on different dates. If that scenario is not the case in your environment—if, for example, an employee is the manager of a department only once—you could make DATE_ASSIGN a nonprime attribute in the MGR_HIST entity.

Note in Figure 6.8 that the "manages" relationship is optional in theory and redundant in practice. At any time, you could find out who the manager of a department is by retrieving the most recent DATE_ASSIGN date from MGR_HIST for a given department. On the other hand, the ERD in Figure 6.8 differentiates between current data and historic data. The *current* manager relationship is implemented by the "manages" relationship between EMPLOYEE and DEPARTMENT. Additionally, the historic data are managed through EMP_MGR_HIST and DEPT_MGR_HIST. The trade-off with that model is that each time a new manager is assigned to a department, there will be two data modifications: one update in the DEPARTMENT entity and one insert in the MGR_HIST entity.

The flexibility of the model proposed in Figure 6.8 becomes more apparent when you add the 1:M "one department employs many employees" relationship. In that case, the PK of the "1" side (DEPT_ID) appears in the "many" side (EMPLOYEE) as a foreign key. Now suppose you would like to keep track of the job history for each of the company's employees—you'd probably want to store the department, the job code, the date assigned, and the salary. To accomplish that task, you would modify the model in Figure 6.8 by adding a JOB_HIST entity. Figure 6.9 shows the use of the new JOB_HIST entity to maintain the employee's history.

Again, it's worth emphasizing that the "manages" and "employs" relationships are theoretically optional and redundant in practice. You can always find out where each employee works by looking at the job history and selecting only the most current data row for each employee. However, as you will discover in Chapter 7, An Introduction to Structured Query Language (SQL), and in Chapter 8, Advanced SQL, finding where each employee works is not a trivial task. Therefore, the model represented in Figure 6.9 includes the admittedly redundant but unquestionably useful "manages" and "employs" relationships to separate current data from historic data.

FIGURE 6.9	Maintaining job history

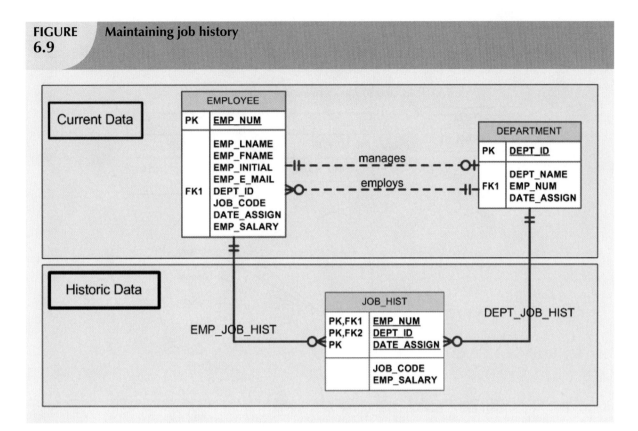

6.4.3 DESIGN CASE #3: FAN TRAPS

Creating a data model requires proper identification of the data relationships among entities. However, due to miscommunication or incomplete understanding of the business rules or processes, it is not uncommon to misidentify relationships among entities. Under those circumstances, the ERD may contain a design trap. A **design trap** occurs when a relationship is improperly or incompletely identified and is therefore represented in a way that is not consistent with the real world. The most common design trap is known as a *fan trap*.

A **fan trap** occurs when you have one entity in two 1:M relationships to other entities, thus producing an association among the other entities that is not expressed in the model. For example, assume the JCB basketball league has many divisions. Each division has many players, and each division has many teams. Given those "incomplete" business rules, you might create an ERD that looks like the one in Figure 6.10.

As you can see in Figure 6.10, DIVISION is in a 1:M relationship with TEAM and in a 1:M relationship with PLAYER. Although that representation is semantically correct, the relationships are not properly identified. For example, there is no way to identify what players belong to what team. Figure 6.10 also shows a sample instance relationship representation for the ERD. Note that the relationship lines for the DIVISION instances fan out to the TEAM and PLAYER entity instances—thus the "fan trap" label.

Figure 6.11 shows the correct ERD after the fan trap has been eliminated. Note that, in this case, DIVISION is in a 1:M relationship with TEAM. In turn, TEAM is in a 1:M relationship with PLAYER. Figure 6.11 also shows the instance relationship representation after eliminating the fan trap.

Given the design in Figure 6.11, note how easy it is to see which players play for which team. However, to find out which players play in which division, you first need to see what teams belong to each division; then you need to find out what players play on each team. In other words, there is a transitive relationship between DIVISION and PLAYER via the TEAM entity.

FIGURE 6.10   Incorrect ERD with fan trap problem

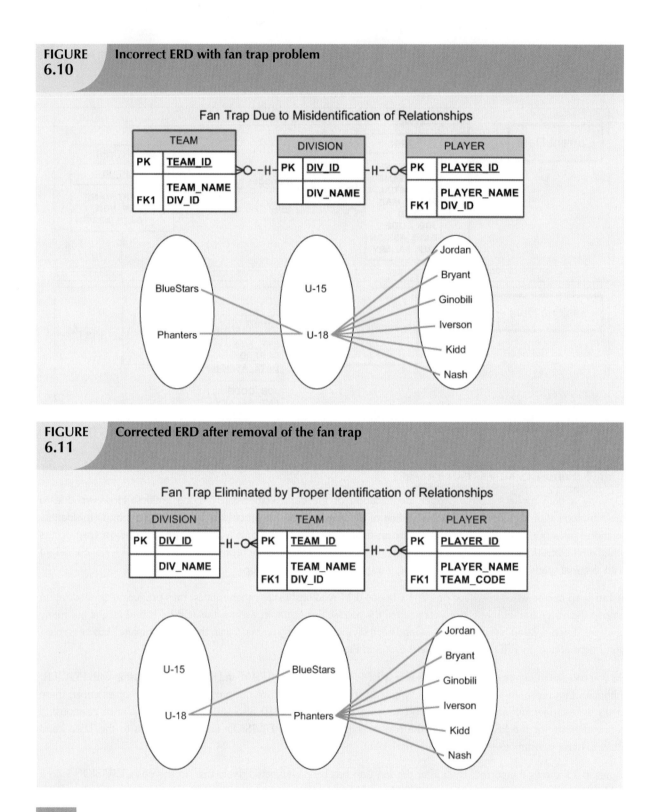

Fan Trap Due to Misidentification of Relationships

FIGURE 6.11   Corrected ERD after removal of the fan trap

Fan Trap Eliminated by Proper Identification of Relationships

6.5.4   DESIGN CASE #4: REDUNDANT RELATIONSHIPS

Although redundancy is often a good thing to have in computer environments (multiple backups in multiple places, for example), redundancy is seldom a good thing in the database environment. (As you learned in Chapter 3, The Relational Database Model, redundancies can cause data anomalies in a database.) Redundant relationships occur when there are multiple relationship paths between related entities. The main concern with redundant relationships is that they remain consistent across the model. However, it's important to note that some designs use redundant relationships as a way to simplify the design.

An example of redundant relationships was first introduced in Figure 6.8 during the discussion on maintaining a history of time-variant data. However, the use of the redundant "manages" and "employs" relationships was justified by the fact that such relationships were dealing with current data rather than historic data. Another more specific example of a redundant relationship is represented in Figure 6.12.

FIGURE 6.12   **A redundant relationship**

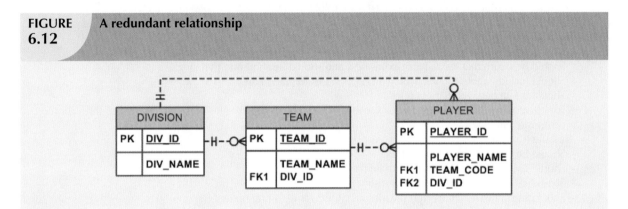

In Figure 6.12, note the transitive 1:M relationship between DIVISION and PLAYER through the TEAM entity set. Therefore, the relationship that connects DIVISION and PLAYER is, for all practical purposes, redundant. In that case, the relationship could be safely deleted without losing any information-generation capabilities in the model.

6.5 DATA MODELING CHECKLIST

Data modeling translates a specific real-world environment into a data model that represents the real-world data, users, processes, and interactions. You learned in this chapter how the EERM enables the designer to add more semantic content to the model. You also learned about the trade-offs and intricacies in the selection of primary keys, and you studied the modeling of time-variant data. The modeling techniques you have learned thus far give you the tools needed to produce successful database designs. However, just as any good pilot uses a checklist to ensure that all is in order for a successful flight, the data modeling checklist shown in Table 6.6 will help ensure that you perform data modeling tasks successfully. (The data modeling checklist in Table 6.6 is based on the concepts and tools you learned beginning in Chapter 3—the relational model, the entity relationship model, normalization, and the extended entity relationship model.) Therefore, it is assumed that you are familiar with the terms and labels used in the checklist, such as *synonyms, aliases,* and *3NF.*

TABLE 6.6   Data Modeling Checklist

BUSINESS RULES
- Properly document and verify all business rules with the end users.
- Ensure that all business rules are written precisely, clearly, and simply. The business rules must help identify entities, attributes, relationships, and constraints.
- Identify the source of all business rules, and ensure that each business rule is accompanied by the reason for its existence and by the date and person(s) responsible for verifying and approving the business rule.

DATA MODELING

Naming Conventions: All names should be limited in length (database-dependent size).
- Entity Names:
  - Should be nouns that are familiar to business and should be short and meaningful
  - Should include abbreviations, synonyms, and aliases for each entity
  - Should be unique within the model
  - For composite entities, may include a combination of abbreviated names of the entities linked through the composite entity
- Attribute Names:
  - Should be unique within the entity
  - Should use the entity abbreviation or prefix
  - Should be descriptive of the characteristic
  - Should use suffixes such as _ID, _NUM, or _CODE for the PK attribute
  - Should not be a reserved word
  - Should not contain spaces or special characters such as @, !, or &
- Relationship Names:
  - Should be active or passive verbs that clearly indicate the nature of the relationship

Entities:
- Each entity should represent a single subject.
- Each entity should represent a set of distinguishable entity instances.
- All entities should be in 3NF or higher.
- The granularity of the entity instance is clearly defined.
- The PK is clearly defined and supports the selected data granularity.

Attributes:
- Should be simple and single-valued (atomic data)
- Should include default values, constraints, synonyms, and aliases
- Derived attributes should be clearly identified and include source(s)
- Should not be redundant unless they are required for transaction accuracy or for maintaining a history or are used as a foreign key
- Non-key attributes must be fully dependent on the PK attribute

Relationships:
- Should clearly identify relationship participants
- Should clearly define participation and cardinality rules

ER Model:
- Should be validated against expected processes: inserts, updates, and deletes
- Should evaluate where, when, and how to maintain a history
- Should not contain redundant relationships except as required (see attributes)
- Should minimize data redundancy to ensure single-place updates
- Should conform to the minimal data rule: "All that is needed is there and all that is there is needed."

SUMMARY

- The extended entity relationship (EER) model adds semantics to the ER model via entity supertypes, subtypes, and clusters. An entity supertype is a generic entity type that is related to one or more entity subtypes.

- A specialization hierarchy depicts the arrangement and relationships between entity supertypes and entity subtypes. Inheritance means that an entity subtype inherits the attributes and relationships of the supertype. Subtypes can be disjoint or overlapping. A subtype discriminator is used to determine to which entity subtype the supertype occurrence is related. The subtypes can exhibit partial or total completeness. There are basically two approaches to developing a specialization hierarchy of entity supertypes and subtypes: specialization and generalization.

- An entity cluster is a "virtual" entity type used to represent multiple entities and relationships in the ERD. An entity cluster is formed by combining multiple interrelated entities and relationships into a single, abstract entity object.

- Natural keys are identifiers that exist in the real world. Natural keys sometimes make good primary keys, but this is not necessarily true. Primary keys should have these characteristics: They must have unique values, they should be nonintelligent, they must not change over time, and they are preferably numeric and composed of a single attribute.

- Composite keys are useful to represent M:N relationships and weak (strong-identifying) entities.

- Surrogate primary keys are useful when there is no natural key that makes a suitable primary key, when the primary key is a composite primary key with multiple different data types, or when the primary key is too long to be usable.

- In a 1:1 relationship, place the PK of the mandatory entity as a foreign key in the optional entity, as an FK in the entity that causes the least number of nulls, or as an FK where the role is played.

- Time-variant data refers to data whose values change over time and whose requirements mandate that you keep a history of data changes. To maintain the history of time-variant data, you must create an entity containing the new value, the date of change, and any other time-relevant data. This entity maintains a 1:M relationship with the entity for which the history is to be maintained.

- A fan trap occurs when you have one entity in two 1:M relationships to other entities and there is an association among the other entities that is not expressed in the model. Redundant relationships occur when there are multiple relationship paths between related entities. The main concern with redundant relationships is that they remain consistent across the model.

- The data modeling checklist provides a way for the designer to check that the ERD meets a set of minimum requirements.

KEY TERMS

completeness constraint, 199

design trap, 209

disjoint subtype (non-overlapping subtype), 197

EER diagram (EERD), 194

entity cluster, 200

entity subtype, 194

entity supertype, 194

extended entity relationship model (EERM), 194

fan trap, 209

inheritance, 196

natural key (natural identifier), 202

overlapping subtype, 197

partial completeness, 160

specialization hierarchy, 195

subtype discriminator, 197

time-variant data, 207

total completeness, 199

ONLINE CONTENT

Answers to selected Review Questions and Problems for this chapter are contained in the Student Online Companion for this book.

REVIEW QUESTIONS

1. What is an entity supertype, and why is it used?
2. What kinds of data would you store in an entity subtype?
3. What is a specialization hierarchy?
4. What is a subtype discriminator? Give an example of its use.
5. What is an overlapping subtype? Give an example.
6. What is the difference between partial completeness and total completeness?
7. What is an entity cluster, and what advantages are derived from its use?
8. What primary key characteristics are considered desirable? Explain *why* each characteristic is considered desirable.
9. Under what circumstances are composite primary keys appropriate?
10. What is a surrogate primary key, and when would you use one?
11. When implementing a 1:1 relationship, where should you place the foreign key if one side is mandatory and one side is optional? Should the foreign key be mandatory or optional?
12. What are time-variant data, and how would you deal with such data from a database design point of view?
13. What is the most common design trap, and how does it occur?
14. Using the design checklist shown in this chapter, what naming conventions should you use?
15. Using the design checklist shown in this chapter, what characteristics should entities have?

PROBLEMS

1. AVANTIVE Corporation is a company specializing in the commercialization of automotive parts. AVANTIVE has two types of customers: retail and wholesale. All customers have a customer ID, a name, an address, a phone number, a default shipping address, a date of last purchase, and a date of last payment. Retail customers have the customer attributes, plus the credit card type, credit card number, expiration date, and e-mail address. Wholesale customers have the customer attributes, plus a contact name, contact phone number, contact e-mail address, purchase order number and date, discount percentage, billing address, tax status (if exempt), and tax identification number. A retail customer cannot be a wholesale customer and vice versa. Given that information, create the ERD containing all primary keys, foreign keys, and main attributes.

2. AVANTIVE Corporation has five departments: administration, marketing, sales, shipping, and purchasing. Each department employs many employees. Each employee has an ID, a name, a home address, a home phone number, and a salary and tax ID (Social Security number). Some employees are classified as sales representatives, some as technical support, and some as administrators. Sales representatives receive a commission based on sales. Technical support employees are required to be certified in their areas of expertise. For example, some are certified as drivetrain specialists; others, as electrical systems specialists. All administrators have a title and a bonus. Given that information, create the ERD containing all primary keys, foreign keys, and main attributes.

3. AVANTIVE Corporation operates under the following business rules:

- AVANTIVE keeps a list of car models with information about the manufacturer, model, and year. AVANTIVE keeps several parts in stock. A part has a part ID, description, unit price, and quantity on hand. A part can be used for many car models, and a car model has many parts.

- A retail customer normally pays by credit card and is charged the list price for each purchased item. A wholesale customer normally pays via purchase order with terms of net 30 days and is charged a discounted price for each item purchased. (The discount varies from customer to customer.)

- A customer (retail or wholesale) can place many orders. Each order has an order number; a date; a shipping address; a billing address; and a list of part codes, quantities, unit prices, and extended line totals. Each order also has a sales representative ID (an employee) to identify the person who made the sale, an order subtotal, an order tax total, a shipping cost, a shipping date, an order total cost, an order total paid, and an order status (open, closed, or cancel).

Given that information, create the complete ERD containing all primary keys, foreign keys, and main attributes.

4. In Chapter 4, you saw the creation of the Tiny College database design. That design reflected such business rules as "a professor may advise many students" and "a professor may chair one department." Modify the design shown in Figure 4.36 to include these business rules:

- An employee could be staff or a professor or an administrator.

- A professor may also be an administrator.

- Staff employees have a work level classification, such a Level I and Level II.

- Only professors can chair a department. A department is chaired by only one professor.

- Only professors can serve as the dean of a college. Each of the university's colleges is served by one dean.

- A professor can teach many classes.

- Administrators have a position title.

Given that information, create the complete ERD containing all primary keys, foreign keys, and main attributes.

5. Tiny College wants to keep track of the history of all administrative appointments (date of appointment and date of termination). (*Hint*: Time variant data are at work.) The Tiny College chancellor may want to know how many deans worked in the College of Business between January 1, 1960 and January 1, 2008 or who the dean of the College of Education was in 1990. Given that information, create the complete ERD containing all primary keys, foreign keys, and main attributes.

6. Some Tiny College staff employees are information technology (IT) personnel. Some IT personnel provide technology support for academic programs. Some IT personnel provide technology infrastructure support. Some IT personnel provide technology support for academic programs and technology infrastructure support. IT personnel are not professors. IT personnel are required to take periodic training to retain their technical expertise. Tiny College tracks all IT personnel training by date, type, and results (completed vs. not completed). Given that information, create the complete ERD containing all primary keys, foreign keys, and main attributes.

7. The FlyRight Aircraft Maintenance (FRAM) division of the FlyRight Company (FRC) performs all maintenance for FRC's aircraft. Produce a data model segment that reflects the following business rules:

- All mechanics are FRC employees. Not all employees are mechanics.

- Some mechanics are specialized in engine (EN) maintenance. Some mechanics are specialized in airframe (AF) maintenance. Some mechanics are specialized in avionics (AV) maintenance. (Avionics are the electronic components of an aircraft that are used in communication and navigation.) All mechanics take periodic refresher courses to stay current in their areas of expertise. FRC tracks all courses taken by each mechanic—date, course type, certification (Y/N), and performance.

- FRC keeps a history of the employment of all mechanics. The history includes the date hired, date promoted, date terminated, and so on. (*Note:* The "and so on" component is, of course, not a real-world requirement. Instead, it has been used here to limit the number of attributes you will show in your design.)

Given those requirements, create the Crow's Foot ERD segment.

8. You have been asked to create a database design for the BoingX Aircraft Company (BAC), which has two HUD (heads-up display) products: TRX-5A and TRX-5B. The database must enable managers to track blueprints, parts, and software for each HUD, using the following business rules:

    *   For simplicity's sake, you may assume that the TRX-5A unit is based on two engineering blueprints and that the TRX-5B unit is based on three engineering blueprints. You are free to make up your own blueprint names.

    *   All parts used in the TRX-5A and TRX-5B are classified as hardware. For simplicity's sake, you may assume that the TRX-5A unit uses three parts and that the TRX-5B unit uses four parts. You are free to make up your own part names.

NOTE

Some parts are supplied by vendors, while others are supplied by the BoingX Aircraft Company. Parts suppliers must be able to meet the technical specification requirements (TCRs) set by the BoingX Aircraft Company. Any parts supplier that meets the BoingX Aircraft Company's TCRs may be contracted to supply parts. Therefore, any part may be supplied by multiple suppliers and a supplier can supply many different parts.

---

    *   BAC wants to keep track of all part price changes and the dates of those changes.

    *   BAC wants to keep track of all TRX-5A and TRX-5B software. For simplicity's sake, you may assume that the TRX-5A unit uses two named software components and that the TRX-5B unit also uses two named software components. You are free to make up your own software names.

    *   BAC wants to keep track of all changes made in blueprints and software. Those changes must reflect the date and time of the change, the description of the change, the person who authorized the change, the person who actually made the change, and the reason for the change.

    *   BAC wants to keep track of all HUD test data by test type, test date, and test outcome.

    Given those requirements, create the Crow's Foot ERD.

NOTE

Problem 9 is sufficiently complex to serve as a class project.

---

9. Global Computer Solutions (GCS) is an information technology consulting company with many offices located throughout the United States. The company's success is based on its ability to maximize its resources—that is, its ability to match highly skilled employees with projects according to region. To better manage its projects, GCS has contacted you to design a database so that GCS managers can keep track of their customers, employees, projects, project schedules, assignments, and invoices.

    The GCS database must support all of GCS's operations and information requirements. A basic description of the main entities follows:

    *   The *employees* working for GCS have an employee ID, an employee last name, a middle initial, a first name, a region, and a date of hire.

    *   Valid *regions* are as follows: Northwest (NW), Southwest (SW), Midwest North (MN), Midwest South (MS), Northeast (NE), and Southeast (SE).

    *   Each employee has many skills, and many employees have the same skill.

    *   Each *skill* has a skill ID, description, and rate of pay. Valid skills are as follows: data entry I, data entry II, systems analyst I, systems analyst II, database designer I, database designer II, Cobol I, Cobol II, C++ I, C++ II, VB I, VB II, ColdFusion I, ColdFusion II, ASP I, ASP II, Oracle DBA, MS SQL Server DBA, network engineer I, network engineer II, web administrator, technical writer, and project manager. Table P6.9a shows an example of the Skills Inventory.

TABLE
P6.9a

SKILL	EMPLOYEE
Data Entry I	Seaton Amy; Williams Josh; Underwood Trish
Data Entry II	Williams Josh; Seaton Amy
Systems Analyst I	Craig Brett; Sewell Beth; Robbins Erin; Bush Emily; Zebras Steve
Systems Analyst II	Chandler Joseph; Burklow Shane; Robbins Erin
DB Designer I	Yarbrough Peter; Smith Mary
DB Designer II	Yarbrough Peter; Pascoe Jonathan
Cobol I	Kattan Chris; Ephanor Victor; Summers Anna; Ellis Maria
Cobol II	Kattan Chris; Ephanor Victor, Batts Melissa
C++ I	Smith Jose; Rogers Adam; Cope Leslie
C++ II	Rogers Adam; Bible Hanah
VB I	Zebras Steve; Ellis Maria
VB II	Zebras Steve; Newton Christopher
ColdFusion I	Duarte Miriam; Bush Emily
ColdFusion II	Bush Emily; Newton Christopher
ASP I	Duarte Miriam; Bush Emily
ASP II	Duarte Miriam; Newton Christopher
Oracle DBA	Smith Jose; Pascoe Jonathan
SQL Server DBA	Yarbrough Peter; Smith Jose
Network Engineer I	Bush Emily; Smith Mary
Network Engineer II	Bush Emily; Smith Mary
Web Administrator	Bush Emily; Smith Mary; Newton Christopher
Technical Writer	Kilby Surgena; Bender Larry
Project Manager	Paine Brad; Mudd Roger; Kenyon Tiffany; Connor Sean

- GCS has many *customers*. Each customer has a customer ID, customer name, phone number, and region.

- GCS works by *projects*. A project is based on a contract between the customer and GCS to design, develop, and implement a computerized solution. Each project has specific characteristics such as the project ID, the customer to which the project belongs, a brief description, a project date (that is, the date on which the project's contract was signed), a project start date (an estimate), a project end date (also an estimate), a project budget (total estimated cost of project), an actual start date, an actual end date, an actual cost, and one employee assigned as manager of the project.

- The actual cost of the project is updated each Friday by adding that week's cost (computed by multiplying the hours each employee worked by the rate of pay for that skill) to the actual cost.

- The employee who is the manager of the project must complete a *project schedule*, which is, in effect, a design and development plan. In the project schedule (or plan), the manager must determine the tasks that will be performed to take the project from beginning to end. Each task has a task ID, a brief task description, the task's starting and ending date, the type of skill needed, and the number of employees (with the required skills) required to complete the task. General tasks are initial interview, database and system design, implementation, coding, testing, and final evaluation and sign-off. For example, GCS might have the project schedule shown in Table P6.9b.

TABLE
P6.9b

PROJECT ID: 1		DESCRIPTION: SALES MANAGEMENT SYSTEM		
COMPANY : SEE ROCKS		CONTRACT DATE: 2/12/2008	REGION: NW	
START DATE: 3/1/2008		END DATE: 7/1/2008	BUDGET: $15,500	
START DATE	END DATE	TASK DESCRIPTION	SKILL(S) REQUIRED	QUANTITY REQUIRED
3/1/08	3/6/08	Initial Interview	Project Manager Systems Analyst II DB Designer I	1 1 1
3/11/08	3/15/08	Database Design	DB Designer I	1
3/11/08	4/12/08	System Design	Systems Analyst II Systems Analyst I	1 2
3/18/08	3/22/08	Database Implementation	Oracle DBA	1
3/25/08	5/20/08	System Coding & Testing	Cobol I Cobol II Oracle DBA	2 1 1
3/25/08	6/7/08	System Documentation	Technical Writer	1
6/10/08	6/14/08	Final Evaluation	Project Manager Systems Analyst II DB Designer I Cobol II	1 1 1 1
6/17/08	6/21/08	On-Site System Online and Data Loading	Project Manager Systems Analyst II DB Designer I Cobol II	1 1 1 1
7/1/08	7/1/08	Sign-Off	Project Manager	1

- Assignments: GCS pools all of its employees by region, and from this pool, employees are assigned to a specific task scheduled by the project manager. For example, for the first project's schedule, you know that for the period 3/1/08 to 3/6/08, a Systems Analyst II, a Database Designer I, and a Project Manager are needed. (The project manager is assigned when the project is created and remains for the duration of the project). Using that information, GCS searches the employees who are located in the same region as the customer, matching the skills required and assigning them to the project task.

- Each project schedule task can have many employees assigned to it, and a given employee can work on multiple project tasks. However, an employee can work on only one project task at a time. For example, if an employee is already assigned to work on a project task from 2/20/08 to 3/3/08, (s)he cannot work on another task until the current assignment is closed (ends). The date on which an assignment is closed does not necessarily match the ending date of the project schedule task, because a task can be completed ahead of or behind schedule.

- Given all of the preceding information, you can see that the assignment associates an employee with a project task, using the project schedule. Therefore, to keep track of the *assignment*, you require at least the following information: assignment ID, employee, project schedule task, date assignment starts, and date assignment ends (which could be any date, as some projects run ahead of or behind schedule). Table P6.9c shows a sample assignment form.

TABLE
P6.9c

PROJECT ID: 1			DESCRIPTION: SALES MANAGEMENT SYSTEM			
COMPANY: SEEROCKS			CONTRACT DATE: 2/12/2008		AS OF: 03/29/08	
SCHEDULED				ACTUAL ASSIGNMENTS		
PROJECT TASK	START DATE	END DATE	SKILL	EMPLOYEE	START DATE	END DATE
Initial Interview	3/1/08	3/6/08	Project Mgr.	101—Connor S.	3/1/08	3/6/08
			Sys. Analyst II	102—Burklow S.	3/1/08	3/6/08
			DB Designer I	103—Smith M.	3/1/08	3/6/08
Database Design	3/11/08	3/15/08	DB Designer I	104—Smith M.	3/11/08	3/14/08
System Design	3/11/08	4/12/08	Sys. Analyst II	105—Burklow S.	3/11/08	
			Sys. Analyst I	106—Bush E.	3/11/08	
			Sys. Analyst I	107—Zebras S.	3/11/08	
Database Implementation	3/18/08	3/22/08	Oracle DBA	108—Smith J.	3/15/08	3/19/08
System Coding & Testing	3/25/08	5/20/08	Cobol I	109—Summers A.	3/21/08	
			Cobol I	110—Ellis M.	3/21/08	
			Cobol II	111—Ephanor V.	3/21/08	
			Oracle DBA	112—Smith J.	3/21/08	
System Documentation	3/25/08	6/7/08	Tech. Writer	113—Kilby S.	3/25/08	
Final Evaluation	6/10/08	6/14/08	Project Mgr.			
			Sys. Analyst II			
			DB Designer I			
			Cobol II			
On-Site System Online and Data Loading	6/17/08	6/21/08	Project Mgr.			
			Sys. Analyst II			
			DB Designer I			
			Cobol II			
Sign-Off	7/1/08	7/1/08	Project Mgr.			

(*Note:* The assignment number is shown as a prefix of the employee name; for example, 101, 102.) Assume that the assignments shown previously are the only ones existing as of the date of this design. The assignment number can be whatever number matches your database design.

- The hours an employee works are kept in a *work log* containing a record of the actual hours worked by an employee on a given assignment. The work log is a weekly form that the employee fills out at the end of each week (Friday) or at the end of each month. The form contains the date (of each Friday of the month or the last work day of the month if it doesn't fall on a Friday), the assignment ID, the total hours worked that week (or up to the end of the month), and the number of the bill to which the work log entry is charged. Obviously, each work log entry can be related to only one bill. A sample list of the current work log entries for the first sample project is shown in Table P6.9d.

TABLE
P6.9d

EMPLOYEE NAME	WEEK ENDING	ASSIGNMENT NUMBER	HOURS WORKED	BILL NUMBER
Burklow S.	3/1/08	1-102	4	xxx
Connor S.	3/1/08	1-101	4	xxx
Smith M.	3/1/08	1-103	4	xxx
Burklow S.	3/8/08	1-102	24	xxx
Connor S.	3/8/08	1-101	24	xxx
Smith M.	3/8/08	1-103	24	xxx
Burklow S.	3/15/08	1-105	40	xxx
Bush E.	3/15/08	1-106	40	xxx
Smith J.	3/15/08	1-108	6	xxx
Smith M.	3/15/08	1-104	32	xxx
Zebras S.	3/15/08	1-107	35	xxx
Burklow S.	3/22/08	1-105	40	
Bush E.	3/22/08	1-106	40	
Ellis M.	3/22/08	1-110	12	
Ephanor V.	3/22/08	1-111	12	
Smith J.	3/22/08	1-108	12	
Smith J.	3/22/08	1-112	12	
Summers A.	3/22/08	1-109	12	
Zebras S.	3/22/08	1-107	35	
Burklow S.	3/29/08	1-105	40	
Bush E.	3/29/08	1-106	40	
Ellis M.	3/29/08	1-110	35	
Ephanor V.	3/29/08	1-111	35	
Kilby S.	3/29/08	1-113	40	
Smith J.	3/29/08	1-112	35	
Summers A.	3/29/08	1-109	35	
Zebras S.	3/29/08	1-107	35	
Note: xxx represents the bill ID. Use the one that matches the bill number in your database.				

- Finally, every 15 days, a *bill* is written and sent to the customer, totaling the hours worked on the project that period. When GCS generates a bill, it uses the bill number to update the work-log entries that are part of that bill. In summary, a bill can refer to many work log entries, and each work log entry can be related to only one bill. GCS sent one bill on 3/15/08 for the first project (Xerox), totaling the hours worked between 3/1/08 and 3/15/08. Therefore, you can safely assume that there is only one bill in this table and that that bill covers the work-log entries shown in the above form.

Your assignment is to create a database that will fulfill the operations described in this problem. The minimum required entities are employee, skill, customer, region, project, project schedule, assignment, work log, and bill. (There are additional required entities that are not listed.)

- Create all of the required tables and all of the required relationships.
- Create the required indexes to maintain entity integrity when using surrogate primary keys.
- Populate the tables as needed (as indicated in the sample data and forms).

PART

III

ADVANCED DESIGN AND IMPLEMENTATION

USING QUERIES TO SCORE RUNS

Business Vignette

Today, we take for granted the ability to comb through vast amounts of data to find one item that meets a slew of requirements or to find several items that share common features. When we go to the library, retrieve bank account records, call Information to get a phone number, or search for a movie review or restaurant online, we are interacting with databases that did not exist 40 years ago. The impact of the information revolution is very apparent in our daily lives. What is less apparent is how this revolution is slowly changing society by determining who wins and who loses in school, in business, and even in sports.

In the old days, money was the major factor in establishing which teams went to the World Series. The rich teams could hunt for and buy the best players. As a result, the Yankees have dominated the event, playing in and winning many more championships than any other team in the Major League. Today, databases are being used to even the playing field.

In the late 1990s, the Yankees hired E Solutions, a Tampa-based IT company, to complete a customized software project to analyze scouting reports. E Solutions saw the potential and developed ScoutAdvisor, a program that runs queries on data on baseball players collected from many sources. Such sources include the Major League Baseball Scouting Bureau, which provides psychological profile data on players, while SportsTicker supplies game reports, and STATS provides game statistics such as where balls land in the field after being hit or types of pitches.

The ScoutAdvisor database stores information on prospective and current players, such as running speed, fielding ability, hitting ability, and plate discipline. Team managers can run queries to find a pitcher with high arm strength, arm accuracy, and pitch speed. They can check for injuries or discipline problems. They can run queries to determine if a player's performance justifies his cost. The database also stores automatic daily player updates. Managers can run queries to determine whether a pitcher's fastball speed is increasing or whether a hitter's tendency to swing at the first pitch is declining. ScoutAdvisor is customizable, so managers can also design their own queries.

The result is that more and more baseball teams are signing contracts with E Solutions as it becomes increasingly apparent that managing information is becoming as important to team success as managing money or players.

7

INTRODUCTION TO STRUCTURED QUERY LANGUAGE (SQL)

In this chapter, you will learn:

- The basic commands and functions of SQL
- How to use SQL for data administration (to create tables, indexes, and views)
- How to use SQL for data manipulation (to add, modify, delete, and retrieve data)
- How to use SQL to query a database for useful information

In this chapter, you learn the basics of Structured Query Language (SQL). SQL, pronounced S-Q-L by some and "sequel" by others, is composed of commands that enable users to create database and table structures, perform various types of data manipulation and data administration, and query the database to extract useful information. All relational DBMS software supports SQL, and many software vendors have developed extensions to the basic SQL command set.

Because SQL's vocabulary is simple, the language is relatively easy to learn. Its simplicity is enhanced by the fact that much of its work takes place behind the scenes. For example, a single command creates the complex table structures required to store and manipulate data successfully. Furthermore, SQL is a nonprocedural language; that is, the user specifies what must be done, but not how it is to be done. To issue SQL commands, end users and programmers do not need to know the physical data storage format or the complex activities that take place when a SQL command is executed.

Although quite useful and powerful, SQL is not meant to stand alone in the applications arena. Data entry with SQL is possible but awkward, as are data corrections and additions. SQL itself does not create menus, special report forms, overlays, pop-ups, or any of the other utilities and screen devices that end users usually expect. Instead, those features are available as vendor-supplied enhancements. SQL focuses on data definition (creating tables, indexes, and views) and data manipulation (adding, modifying, deleting, and retrieving data); we cover these basic functions in this chapter. In spite of its limitations, SQL is a powerful tool for extracting information and managing data.

Preview

7.1 INTRODUCTION TO SQL

Ideally, a database language allows you to create database and table structures, to perform basic data management chores (add, delete, and modify), and to perform complex queries designed to transform the raw data into useful information. Moreover, a database language must perform such basic functions with minimal user effort, and its command structure and syntax must be easy to learn. Finally, it must be portable; that is, it must conform to some basic standard so that an individual does not have to relearn the basics when moving from one RDBMS to another. SQL meets those ideal database language requirements well.

SQL functions fit into two broad categories:

- It is a *data definition language (DDL)*: SQL includes commands to create database objects such as tables, indexes, and views, as well as commands to define access rights to those database objects. The data definition commands you learn in this chapter are listed in Table 7.1.

- It is a *data manipulation language (DML)*: SQL includes commands to insert, update, delete, and retrieve data within the database tables. The data manipulation commands you learn in this chapter are listed in Table 7.2.

TABLE 7.1 **SQL Data Definition Commands**

COMMAND OR OPTION	DESCRIPTION
CREATE SCHEMA AUTHORIZATION	Creates a database schema
CREATE TABLE	Creates a new table in the user's database schema
NOT NULL	Ensures that a column will not have null values
UNIQUE	Ensures that a column will not have duplicate values
PRIMARY KEY	Defines a primary key for a table
FOREIGN KEY	Defines a foreign key for a table
DEFAULT	Defines a default value for a column (when no value is given)
CHECK	Validates data in an attribute
CREATE INDEX	Creates an index for a table
CREATE VIEW	Creates a dynamic subset of rows/columns from one or more tables
ALTER TABLE	Modifies a tables definition (adds, modifies, or deletes attributes or constraints)
CREATE TABLE AS	Creates a new table based on a query in the user's database schema
DROP TABLE	Permanently deletes a table (and its data)
DROP INDEX	Permanently deletes an index
DROP VIEW	Permanently deletes a view

TABLE 7.2 **SQL Data Manipulation Commands**

COMMAND OR OPTION	DESCRIPTION
INSERT	Inserts row(s) into a table
SELECT	Selects attributes from rows in one or more tables or views
WHERE	Restricts the selection of rows based on a conditional expression
GROUP BY	Groups the selected rows based on one or more attributes
HAVING	Restricts the selection of grouped rows based on a condition
ORDER BY	Orders the selected rows based on one or more attributes
UPDATE	Modifies an attribute's values in one or more table's rows
DELETE	Deletes one or more rows from a table
COMMIT	Permanently saves data changes
ROLLBACK	Restores data to their original values

TABLE 7.2	SQL Data Manipulation Commands (continued)
COMMAND OR OPTION	DESCRIPTION
COMPARISON OPERATORS	
=, <, >, <=, >=, <>	Used in conditional expressions
LOGICAL OPERATORS	
AND/OR/NOT	Used in conditional expressions
SPECIAL OPERATORS	Used in conditional expressions
BETWEEN	Checks whether an attribute value is within a range
IS NULL	Checks whether an attribute value is null
LIKE	Checks whether an attribute value matches a given string pattern
IN	Checks whether an attribute value matches any value within a value list
EXISTS	Checks whether a subquery returns any rows
DISTINCT	Limits values to unique values
AGGREGATE FUNCTIONS	Used with SELECT to return mathematical summaries on columns
COUNT	Returns the number of rows with non-null values for a given column
MIN	Returns the minimum attribute value found in a given column
MAX	Returns the maximum attribute value found in a given column
SUM	Returns the sum of all values for a given column
AVG	Returns the average of all values for a given column

You will be happy to know that SQL is relatively easy to learn. Its basic command set has a vocabulary of fewer than 100 words. Better yet, SQL is a nonprocedural language: you merely command *what* is to be done; you don't have to worry about *how* it is to be done. The American National Standards Institute (ANSI) prescribes a standard SQL—the current version is known as SQL-99 or SQL3. The ANSI SQL standards are also accepted by the International Organization for Standardization (ISO), a consortium composed of national standards bodies of more than 150 countries. Although adherence to the ANSI/ISO SQL standard is usually required in commercial and government contract database specifications, many RDBMS vendors add their own special enhancements. Consequently, it is seldom possible to move a SQL-based application from one RDBMS to another without making some changes.

However, even though there are several different SQL "dialects," the differences among them are minor. Whether you use Oracle, Microsoft SQL Server, MySQL, IBM's DB2, Microsoft Access, or any other well-established RDBMS, a software manual should be sufficient to get you up to speed if you know the material presented in this chapter.

At the heart of SQL is the query. In Chapter 1, Database Systems, you learned that a query is a spur-of-the-moment question. Actually, in the SQL environment, the word *query* covers both questions and actions. Most SQL queries are used to answer questions such as these: "What products currently held in inventory are priced over $100, and what is the quantity on hand for each of those products?" "How many employees have been hired since January 1, 2006 by each of the company's departments?" However, many SQL queries are used to perform actions such as adding or deleting table rows or changing attribute values within tables. Still other SQL queries create new tables or indexes. In short, for a DBMS, a query is simply a SQL statement that must be executed. But before you can use SQL to query a database, you must define the database environment for SQL with its data definition commands.

7.2 DATA DEFINITION COMMANDS

Before examining the SQL syntax for creating and defining tables and other elements, let's first examine the simple database model and the database tables that will form the basis for the many SQL examples you'll explore in this chapter.

THE DATABASE MODEL

A simple database composed of the following tables is used to illustrate the SQL commands in this chapter: CUSTOMER, INVOICE, LINE, PRODUCT, and VENDOR. This database model is shown in Figure 7.1.

FIGURE 7.1   The database model

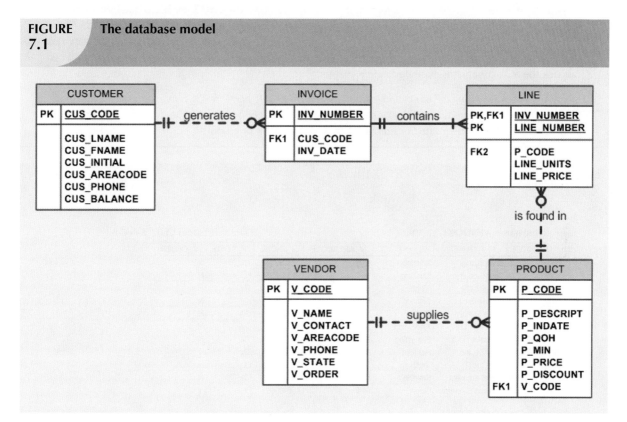

The database model in Figure 7.1 reflects the following business rules:

- A customer may generate many invoices. Each invoice is generated by one customer.
- An invoice contains one or more invoice lines. Each invoice line is associated with one invoice.
- Each invoice line references one product. A product may be found in many invoice lines. (You can sell more than one hammer to more than one customer.)
- A vendor *may* supply many products. Some vendors do not (yet?) supply products. (For example, a vendor list may include *potential* vendors.)
- If a product is vendor-supplied, that product is supplied by only a single vendor.
- Some products are not supplied by a vendor. (For example, some products may be produced in-house or bought on the open market.)

As you can see in Figure 7.1, the database model contains many tables. However, to illustrate the initial set of data definition commands, the focus of attention will be the PRODUCT and VENDOR tables. You will have the opportunity to use the remaining tables later in this chapter and in the problem section.

So that you have a point of reference for understanding the effect of the SQL queries, the contents of the PRODUCT and VENDOR tables are listed in Figure 7.2.

ONLINE CONTENT

The database model in Figure 7.1 is implemented in the Microsoft Access **Ch07_SaleCo** database located in the Student Online Companion. (This database contains a few additional tables that are not reflected in Figure 7.1. These tables are used for discussion purposes only.) If you use MS Access, you can use the database supplied online. However, it is strongly suggested that you create your own database structures so you can practice the SQL commands illustrated in this chapter.

SQL script files for creating the tables and loading the data in Oracle and MS SQL Server are also located in the Student Online Companion. How you connect to your database depends on how the software was installed on your computer. Follow the instructions provided by your instructor or school.

FIGURE 7.2   The VENDOR and PRODUCT tables

Table name: VENDOR   Database name: Ch07_SaleCo

V_CODE	V_NAME	V_CONTACT	V_AREACODE	V_PHONE	V_STATE	V_ORDER
21225	Bryson, Inc.	Smithson	615	223-3234	TN	Y
21226	SuperLoo, Inc.	Flushing	904	215-8995	FL	N
21231	D&E Supply	Singh	615	228-3245	TN	Y
21344	Gomez Bros.	Ortega	615	889-2546	KY	N
22567	Dome Supply	Smith	901	678-1419	GA	N
23119	Randsets Ltd.	Anderson	901	678-3998	GA	Y
24004	Brackman Bros.	Browning	615	228-1410	TN	N
24288	ORDVA, Inc.	Hakford	615	898-1234	TN	Y
25443	B&K, Inc.	Smith	904	227-0093	FL	N
25501	Damal Supplies	Smythe	615	890-3529	TN	N
25595	Rubicon Systems	Orton	904	456-0092	FL	Y

Table name: PRODUCT

P_CODE	P_DESCRIPT	P_INDATE	P_QOH	P_MIN	P_PRICE	P_DISCOUNT	V_CODE
11QER/31	Power painter, 15 psi., 3-nozzle	03-Nov-07	8	5	109.99	0.00	25595
13-Q2/P2	7.25-in. pwr. saw blade	13-Dec-07	32	15	14.99	0.05	21344
14-Q1/L3	9.00-in. pwr. saw blade	13-Nov-07	18	12	17.49	0.00	21344
1546-QQ2	Hrd. cloth, 1/4-in., 2x50	15-Jan-08	15	8	39.95	0.00	23119
1558-QW1	Hrd. cloth, 1/2-in., 3x50	15-Jan-08	23	5	43.99	0.00	23119
2232/QTY	B&D jigsaw, 12-in. blade	30-Dec-07	8	5	109.92	0.05	24288
2232/QWE	B&D jigsaw, 8-in. blade	24-Dec-07	6	5	99.87	0.05	24288
2238/QPD	B&D cordless drill, 1/2-in.	20-Jan-08	12	5	38.95	0.05	25595
23109-HB	Claw hammer	20-Jan-08	23	10	9.95	0.10	21225
23114-AA	Sledge hammer, 12 lb.	02-Jan-08	8	5	14.40	0.05	
54778-2T	Rat-tail file, 1/8-in. fine	15-Dec-07	43	20	4.99	0.00	21344
89-WRE-Q	Hicut chain saw, 16 in.	07-Feb-08	11	5	256.99	0.05	24288
PVC23DRT	PVC pipe, 3.5-in., 8-ft	20-Feb-08	188	75	5.87	0.00	
SM-18277	1.25-in. metal screw, 25	01-Mar-08	172	75	6.99	0.00	21225
SW-23116	2.5-in. wd. screw, 50	24-Feb-08	237	100	8.45	0.00	21231
WR3/TT3	Steel matting, 4'x8'x1/6", .5" mesh	17-Jan-08	18	5	119.95	0.10	25595

Note the following about these tables. (The features correspond to the business rules reflected in the ERD shown in Figure 7.1.)

- The VENDOR table contains vendors who are not referenced in the PRODUCT table. Database designers note that possibility by saying that PRODUCT is *optional* to VENDOR; a vendor may exist without a reference to a product. You examined such optional relationships in detail in Chapter 4, Entity Relationship (ER) Modeling.

- Existing V_CODE values in the PRODUCT table must (and do) have a match in the VENDOR table to ensure referential integrity.
- A few products are supplied factory-direct, a few are made in-house, and a few may have been bought in a warehouse sale. In other words, a product is not necessarily supplied by a vendor. Therefore, VENDOR is optional to PRODUCT.

A few of the conditions just described were made for the sake of illustrating specific SQL features. For example, null V_CODE values were used in the PRODUCT table to illustrate (later) how you can track such nulls using SQL.

7.2.2 CREATING THE DATABASE

Before you can use a new RDBMS, you must complete two tasks: first, create the database structure, and second, create the tables that will hold the end-user data. To complete the first task, the RDBMS creates the physical files that will hold the database. When you create a new database, the RDBMS automatically creates the data dictionary tables to store the metadata and creates a default database administrator. Creating the physical files that will hold the database means interacting with the operating system and the file systems supported by the operating system. Therefore, creating the database structure is the one feature that tends to differ substantially from one RDBMS to another. The good news is that it is relatively easy to create a database structure, regardless of which RDBMS you use.

If you use Microsoft Access, creating the database is simple: start Access, select **File/New/Blank Database**, specify the folder in which you want to store the database, and then name the database. However, if you work in a database environment typically used by larger organizations, you will probably use an enterprise RDBMS such as Oracle, SQL Server, MySQL or DB2. Given their security requirements and greater complexity, those database products require a more elaborate database creation process. (You will learn how to create and manage an Oracle database structure in Chapter 15, Database Administration and Security.)

You will be relieved to discover that, *with the exception of the database creation process*, most RDBMS vendors use SQL that deviates little from the ANSI standard SQL. For example, most RDBMSs require that each SQL command ends with a semicolon. However, some SQL implementations do not use a semicolon. Important syntax differences among implementations will be highlighted in Note boxes.

If you are using an enterprise RDBMS, before you can start creating tables you must be authenticated by the RDBMS. **Authentication** is the process through which the DBMS verifies that only registered users may access the database. To be authenticated, you must log on to the RDBMS using a user ID and a password created by the database administrator. In an enterprise RDBMS, every user ID is associated with a database schema.

7.2.3 THE DATABASE SCHEMA

In the SQL environment, a **schema** is a group of database objects—such as tables and indexes—that are related to each other. Usually, the schema belongs to a single user or application. A single database can hold multiple schemas belonging to different users or applications. Think of a schema as a logical grouping of database objects, such as tables, indexes, and views. Schemas are useful in that they group tables by owner (or function) and enforce a first level of security by allowing each user to see only the tables that belong to that user.

ANSI SQL standards define a command to create a database schema:

CREATE SCHEMA AUTHORIZATION {creator};

Therefore, if the creator is JONES, use the command:

CREATE SCHEMA AUTHORIZATION JONES;

Most enterprise RDBMSs support that command. However, the command is seldom used directly—that is, from the command line. (When a user is created, the DBMS automatically assigns a schema to that user.) When the DBMS *is* used, the CREATE SCHEMA AUTHORIZATION command must be issued by the user who owns the schema. That is, if you log on as JONES, you can use only CREATE SCHEMA AUTHORIZATION JONES.

For most RDBMSs, the CREATE SCHEMA AUTHORIZATION is optional. That is why this chapter focuses on the ANSI SQL commands required to create and manipulate tables.

7.2.4 DATA TYPES

After the database schema has been created, you are ready to define the PRODUCT and VENDOR table structures within the database. The table-creating SQL commands used in the example are based on the data dictionary shown in Table 7.3.

In the data dictionary in Table 7.3, note particularly the data types selected. Keep in mind that data type selection is usually dictated by the nature of the data and by the intended use. For example:

- P_PRICE clearly requires some kind of numeric data type; defining it as a character field is not acceptable.
- Just as clearly, a vendor name is an obvious candidate for a character data type. For example, VARCHAR2(35) fits well because vendor names are "variable-length" character strings, and in this case, such strings may be up to 35 characters long.
- U.S. state abbreviations are always two characters, so CHAR(2) is a logical choice.
- Selecting P_INDATE to be a (Julian) DATE field rather than a character field is desirable because the Julian dates allow you to make simple date comparisons and to perform date arithmetic. For instance, if you have used DATE fields, you can determine how many days are between them.

If you use DATE fields, you can also determine what the date will be in say, 60 days from a given P_INDATE by using P_INDATE + 60. Or you can use the RDBMS's system date—SYSDATE in Oracle, GETDATE() in MS SQL Server, and Date() in Access—to determine the answer to questions such as, "What will be the date 60 days from today?" For example, you might use SYSDATE + 60 (in Oracle); GETDATE() + 60 (in MS SQL Server) or Date() + 60 (in Access).

Date arithmetic capability is particularly useful in billing. Perhaps you want your system to start charging interest on a customer balance 60 days after the invoice is generated. Such simple date arithmetic would be impossible if you used a character data type.

Data type selection sometimes requires professional judgment. For example, you must make a decision about the V_CODE's data type as follows:

- If you want the computer to generate new vendor codes by adding 1 to the largest recorded vendor code, you must classify V_CODE as a numeric attribute. (You cannot perform mathematical procedures on character data.) The designation INTEGER will ensure that only the counting numbers (integers) can be used. Most SQL implementations also permit the use of SMALLINT for integer values up to six digits.
- If you do not want to perform mathematical procedures based on V_CODE, you should classify it as a character attribute, even though it is composed entirely of numbers. Character data are "quicker" to process in queries. Therefore, when there is no need to perform mathematical procedures on the attribute, store it as a character attribute.

The first option is used to demonstrate the SQL procedures in this chapter.

TABLE 7.3 Data Dictionary for the CH07_SALECO Database

TABLE NAME	ATTRIBUTE NAME	CONTENTS	TYPE	FORMAT	RANGE*	REQUIRED	PK OR FK	FK REFERENCED TABLE
PRODUCT	P_CODE	Product code	CHAR(10)	XXXXXXXXXX	NA	Y	PK	
	P_DESCRIPT	Product description	VARCHAR(35)	Xxxxxxxxxxx	NA	Y		
	P_INDATE	Stocking date	DATE	DD-MON-YYYY	NA	Y		
	P_QOH	Units available	SMALLINT	####	0-9999	Y		
	P_MIN	Minimum units	SMALLINT	####	0-9999	Y		
	P_PRICE	Product price	NUMBER(8,2)	####.##	0.00-9999.00	Y		
	P_DISCOUNT	Discount rate	NUMBER(5,2)	0.##	0.00-0.20	Y		
	V_CODE	Vendor code	INTEGER	###	100-999		FK	VENDOR
VENDOR	V_CODE	Vendor code	INTEGER	####	1000-9999	Y	PK	
	V_NAME	Vendor name	CHAR(35)	Xxxxxxxxxxx	NA	Y		
	V_CONTACT	Contact person	CHAR(25)	Xxxxxxxxxxx	NA	Y		
	V_AREACODE	Area code	CHAR(3)	999	NA	Y		
	V_PHONE	Phone number	CHAR(8)	999-9999	NA	Y		
	V_STATE	State	CHAR(2)	XX	NA	Y		
	V_ORDER	Previous order	CHAR(1)	X	Y or N	Y		

FK   = Foreign key
PK   = Primary key
CHAR   = Fixed character length data, 1 to 255 characters
VARCHAR   = Variable character length data, 1 to 2,000 characters. VARCHAR is automatically converted to VARCHAR2 in Oracle
NUMBER   = Numeric data. NUMBER(9,2) is used to specify numbers with two decimal places and up to nine digits long, including the decimal places. Some RDBMSs permit the use of a MONEY or a CURRENCY data type.
INT   = Integer values only
SMALLINT = Small integer values only
DATE formats vary. Commonly accepted formats are: 'DD-MON-YYYY', 'DD-MON-YY', 'MM/DD/YYYY' or 'MM/DD/YY'
* Not all the ranges shown here will be illustrated in this chapter. However, you can use these constraints to practice writing your own constraints.

When you define the attribute's data type, you must pay close attention to the expected use of the attributes for sorting and data retrieval purposes. For example, in a real estate application, an attribute that represents the numbers of bathrooms in a home (H_BATH_NUM) could be assigned the CHAR(3) data type because it is highly unlikely the application will do any addition, multiplication, or division with the number of bathrooms. Based on the CHAR(3) data type definition, valid H_BATH_NUM values would be '2','1','2.5','10'. However, this data type decision creates potential problems. For example, if an application sorts the homes by number of bathrooms, a query would "see" the value '10' as less than '2', which is clearly incorrect. So you must give some thought to the expected use of the data in order to properly define the attribute data type.

The data dictionary in Table 7.3 contains only a few of the data types supported by SQL. For teaching purposes, the selection of data types is limited to ensure that almost any RDBMS can be used to implement the examples. If your RDBMS is fully compliant with ANSI SQL, it will support many more data types than the ones shown in Table 7.4. And many RDBMSs support data types beyond the ones specified in ANSI SQL.

TABLE 7.4	Some Common SQL Data Types	
DATA TYPE	FORMAT	COMMENTS
Numeric	NUMBER(L,D)	The declaration NUMBER(7,2) indicates numbers that will be stored with two decimal places and may be up to seven digits long, including the sign and the decimal place. Examples: 12.32, −134.99.
	INTEGER	May be abbreviated as INT. Integers are (whole) counting numbers, so they cannot be used if you want to store numbers that require decimal places.
	SMALLINT	Like INTEGER, but limited to integer values up to six digits. If your integer values are relatively small, use SMALLINT instead of INT.
	DECIMAL(L,D)	Like the NUMBER specification, but the storage length is a *minimum* specification. That is, greater lengths are acceptable, but smaller ones are not. DECIMAL(9,2), DECIMAL(9), and DECIMAL are all acceptable.
Character	CHAR(L)	Fixed-length character data for up to 255 characters. If you store strings that are not as long as the CHAR parameter value, the remaining spaces are left unused. Therefore, if you specify CHAR(25), strings such as Smith and Katzenjammer are each stored as 25 characters. However, a U.S. area code is always three digits long, so CHAR(3) would be appropriate if you wanted to store such codes.
	VARCHAR(L) or VARCHAR2(L)	Variable-length character data. The designation VARCHAR2(25) will let you store characters up to 25 characters long. However, VARCHAR will not leave unused spaces. Oracle automatically converts VARCHAR to VARCHAR2.
Date	DATE	Stores dates in the Julian date format.

In addition to the data types shown in Table 7.4, SQL supports several other data types, including TIME, TIMESTAMP, REAL, DOUBLE, FLOAT, and intervals such as INTERVAL DAY TO HOUR. Many RDBMSs also have expanded the list to include other types of data, such as LOGICAL, CURRENCY, AutoNumber (Access), and sequence (Oracle). However, because this chapter is designed to introduce the SQL basics, the discussion is limited to the data types summarized in Table 7.4.

7.2.5 CREATING TABLE STRUCTURES

Now you are ready to implement the PRODUCT and VENDOR table structures with the help of SQL, using the **CREATE TABLE** syntax shown next.

```
CREATE TABLE tablename (
    column1          data type      [constraint] [,
    column2          data type      [constraint] ] [,
    PRIMARY KEY      (column1       [, column2]) ] [,
    FOREIGN KEY      (column1       [, column2]) REFERENCES tablename] [,
    CONSTRAINT       constraint ] );
```

ONLINE CONTENT

All the SQL commands you will see in this chapter are located in script files in the Student Online Companion for this book. You can copy and paste the SQL commands into your SQL program. Script files are provided for Oracle and SQL Server users.

To make the SQL code more readable, most SQL programmers use one line per column (attribute) definition. In addition, spaces are used to line up the attribute characteristics and constraints. Finally, both table and attribute names are fully capitalized. Those conventions are used in the following examples that create VENDOR and PRODUCT tables and throughout the book.

NOTE

SQL SYNTAX

Syntax notation for SQL commands used in this book:

CAPITALS	Required SQL command keywords
italics	An end-user-provided parameter (generally required)
{a | b | ..}	A mandatory parameter; use one option from the list separated by |
[.....]	An optional parameter—anything inside square brackets is optional
Tablename	The name of a table
Column	The name of an attribute in a table
data type	A valid data type definition
constraint	A valid constraint definition
condition	A valid conditional expression (evaluates to true or false)
columnlist	One or more column names or expressions separated by commas
tablelist	One or more table names separated by commas
conditionlist	One or more conditional expressions separated by logical operators
expression	A simple value (such as 76 or Married) or a formula (such as P_PRICE − 10)

```
CREATE TABLE VENDOR (
V_CODE              INTEGER          NOT NULL  UNIQUE,
V_NAME              VARCHAR(35)      NOT NULL,
V_CONTACT           VARCHAR(15)      NOT NULL,
V_AREACODE          CHAR(3)          NOT NULL,
V_PHONE             CHAR(8)          NOT NULL,
V_STATE             CHAR(2)          NOT NULL,
V_ORDER             CHAR(1)          NOT NULL,
PRIMARY KEY (V_CODE));
```

> **NOTE**
>
> - Because the PRODUCT table contains a foreign key that references the VENDOR table, create the VENDOR table first. (In fact, the M side of a relationship always references the 1 side. Therefore, in a 1:M relationship, you must *always* create the table for the 1 side first.)
> - If your RDBMS does not support the VARCHAR2 and FCHAR format, use CHAR.
> - Oracle accepts the VARCHAR data type and automatically converts it to VARCHAR2.
> - If your RDBMS does not support SINT or SMALLINT, use INTEGER or INT. If INTEGER is not supported, use NUMBER.
> - If you use Access, you can use the NUMBER data type, but you cannot use the number delimiters at the SQL level. For example, using NUMBER(8,2) to indicate numbers with up to eight characters and two decimal places is fine in Oracle, but you cannot use it in Access—you must use NUMBER without the delimiters.
> - If your RDBMS does not support primary and foreign key designations or the UNIQUE specification, delete them from the SQL code shown here.
> - If you use the PRIMARY KEY designation in Oracle, you do not need the NOT NULL and UNIQUE specifications.
> - The ON UPDATE CASCADE clause is part of the ANSI standard, but it may not be supported by your RDBMS. In that case, delete the ON UPDATE CASCADE clause.

```
CREATE TABLE PRODUCT (
P_CODE              VARCHAR(10)     NOT NULL          UNIQUE,
P_DESCRIPT          VARCHAR(35)     NOT NULL,
P_INDATE            DATE            NOT NULL,
P_QOH               SMALLINT        NOT NULL,
P_MIN               SMALLINT        NOT NULL,
P_PRICE             NUMBER(8,2)     NOT NULL,
P_DISCOUNT          NUMBER(5,2)     NOT NULL,
V_CODE              INTEGER,
PRIMARY KEY (P_CODE),
FOREIGN KEY (V_CODE) REFERENCES VENDOR ON UPDATE CASCADE);
```

As you examine the preceding SQL table-creating command sequences, note the following features:

- The NOT NULL specifications for the attributes ensure that a data entry will be made. When it is crucial to have the data available, the NOT NULL specification will not allow the end user to leave the attribute empty (with no data entry at all). Because this specification is made at the table level and stored in the data dictionary, application programs can use this information to create the data dictionary validation automatically.

- The UNIQUE specification creates a unique index in the respective attribute. Use it to avoid duplicated values in a column.

- The primary key attributes contain both a NOT NULL and a UNIQUE specification. Those specifications enforce the entity integrity requirements. If the NOT NULL and UNIQUE specifications are not supported, use PRIMARY KEY without the specifications. (For example, if you designate the PK in MS Access, the NOT NULL and UNIQUE specifications are automatically assumed and are not spelled out.)

- The entire table definition is enclosed in parentheses. A comma is used to separate each table element (attributes, primary key, and foreign key) definition.

> **NOTE**
>
> If you are working with a composite primary key, all of the primary keys attributes are contained within the parentheses and are separated with commas. For example, the LINE table in Figure 7.1 has a primary key that consists of the two attributes INV_NUMBER and LINE_NUMBER. Therefore, you would define the primary key by typing:
>
> PRIMARY KEY (INV_NUMBER, LINE_NUMBER),
>
> *The order of the primary key components is important* because the indexing starts with the first-mentioned attribute, then proceeds with the next attribute, and so on. In this example, the line numbers would be ordered within each of the invoice numbers:
>
> | INV_NUMBER | LINE_NUMBER |
> |------------|-------------|
> | 1001 | 1 |
> | 1001 | 2 |
> | 1002 | 1 |
> | 1003 | 1 |
> | 1003 | 2 |

- The ON UPDATE CASCADE specification ensures that if you make a change in any VENDOR's V_CODE, that change is automatically applied to all foreign key references throughout the system (cascade) to ensure that referential integrity is maintained. (Although the ON UPDATE CASCADE clause is part of the ANSI standard, some RDBMSs such as Oracle do not support ON UPDATE CASCADE. If your RDBMS does not support the clause, delete it from the code shown here.)

- An RDBMS will automatically enforce referential integrity for foreign keys. That is, you cannot have an invalid entry in the foreign key column; at the same time, you cannot delete a vendor row as long as a product row references that vendor.

- The command sequence ends with a semicolon. (Remember, your RDBMS may require that you omit the semicolon.)

> **NOTE**
>
> **NOTE ABOUT COLUMN NAMES**
>
> Do *not* use mathematical symbols such as +, −, and / in your column names; instead, use an underscore to separate words, if necessary. For example, PER-NUM might generate an error message, but PER_NUM is acceptable. Also, do *not* use reserved words. **Reserved words** are words used by SQL to perform specific functions. For example, in some RDBMSs, the column name INITIAL will generate the message invalid column name.

7.2.6 SQL CONSTRAINTS

In Chapter 3, The Relational Model, you learned that adherence to rules on entity integrity and referential integrity is crucial in a relational database environment. Fortunately, most SQL implementations support both integrity rules. Entity integrity is enforced automatically when the primary key is specified in the CREATE TABLE command sequence. For example, you can create the VENDOR table structure and set the stage for the enforcement of entity integrity rules by using:

PRIMARY KEY (V_CODE)

In the PRODUCT table's CREATE TABLE sequence, note that referential integrity has been enforced by specifying in the PRODUCT table:

FOREIGN KEY (V_CODE) REFERENCES VENDOR ON UPDATE CASCADE

NOTE

NOTE TO ORACLE USERS

When you press the Enter key after typing each line, a line number is automatically generated as long as you do not type a semicolon before pressing the Enter key. For example, Oracles execution of the CREATE TABLE command will look like this:

```
CREATE TABLE PRODUCT (
    2       P_CODE          VARCHAR2(10)
    3       CONSTRAINT      PRODUCT_P_CODE_PK PRIMARY KEY,
    4       P_DESCRIPT      VARCHAR2(35)    NOT NULL,
    5       P_INDATE        DATE            NOT NULL,
    6       P_QOH           NUMBER          NOT NULL,
    7       P_MIN           NUMBER          NOT NULL,
    8       P_PRICE         NUMBER(8,2)     NOT NULL,
    9       P_DISCOUNT      NUMBER(5,2)     NOT NULL,
   10       V_CODE          NUMBER,
   11       CONSTRAINT PRODUCT_V_CODE_FK
   12       FOREIGN KEYV_CODE REFERENCES VENDOR
   13
```

In the preceding SQL command sequence, note the following:

- The attribute definition for P_CODE starts in line 2 and ends with a comma at the end of line 3.

- The CONSTRAINT clause (line 3) allows you to define and name a constraint in Oracle. You can name the constraint to meet your own naming conventions. In this case, the constraint was named PRODUCT_P_CODE_PK.

- Examples of constraints are NOT NULL, UNIQUE, PRIMARY KEY, FOREIGN KEY, and CHECK. For additional details about constraints, see below.

- To define a PRIMARY KEY constraint, you could also use the following syntax: P_CODE VARCHAR2(10) PRIMARY KEY,.

- In this case, Oracle would automatically name the constraint.

- Lines 11 and 12 define a FOREIGN KEY constraint name PRODUCT_V_CODE_FK for the attribute V_CODE. The CONSTRAINT clause is generally used at the end of the CREATE TABLE command sequence.

- *If you do not name the constraints yourself, Oracle will automatically assign a name. Unfortunately, the Oracle-assigned name makes sense only to Oracle, so you will have a difficult time deciphering it later. You should assign a name that makes sense to human beings!*

That foreign key constraint definition ensures that:

- You cannot delete a vendor from the VENDOR table if at least one product row references that vendor. This is the default behavior for the treatment of foreign keys.

- On the other hand, if a change is made in an existing VENDOR table's V_CODE, that change must be reflected automatically in any PRODUCT table V_CODE reference (ON UPDATE CASCADE). That restriction makes it impossible for a V_CODE value to exist in the PRODUCT table pointing to a nonexistent VENDOR table V_CODE value. In other words, the ON UPDATE CASCADE specification ensures the preservation of referential integrity. (Oracle does not support ON UPDATE CASCADE.)

In general, ANSI SQL permits the use of ON DELETE and ON UPDATE clauses to cover CASCADE, SET NULL, or SET DEFAULT.

ONLINE CONTENT

For a more detailed discussion of the options for the ON DELETE and ON UPDATE clauses, see **Appendix D, Converting an ER Model into a Database Structure,** Section D.2, General Rules Governing Relationships Among Tables. Appendix D is in the Student Online Companion.

NOTE

NOTE ABOUT REFERENTIAL CONSTRAINT ACTIONS
The support for the referential constraints actions varies from product to product. For example:

- MS Access, SQL Server, and Oracle support ON DELETE CASCADE.
- MS Access and SQL Server support ON UPDATE CASCADE.
- Oracle does not support ON UPDATE CASCADE.
- Oracle supports SET NULL.
- MS Access and SQL Server do not support SET NULL.

Refer to your product manuals for additional information on referential constraints.

While MS Access does not support ON DELETE CASCADE or ON UPDATE CASCADE at the SQL command-line level, it does support them through the relationship window interface. In fact, whenever you try to establish a relationship between two tables in Access, the relationship window interface will automatically pop up.

Besides the PRIMARY KEY and FOREIGN KEY constraints, the ANSI SQL standard also defines the following constraints:

- The NOT NULL constraint ensures that a column does not accept nulls.
- The UNIQUE constraint ensures that all values in a column are unique.
- The DEFAULT constraint assigns a value to an attribute when a new row is added to a table. The end user may, of course, enter a value other than the default value.
- The CHECK constraint is used to validate data when an attribute value is entered. The CHECK constraint does precisely what its name suggests: it checks to see that a specified condition exists. Examples of such constraints include the following:
  - *The minimum order value must be at least 10.*
  - *The date must be after April 15, 2008.*

If the CHECK constraint is met for the specified attribute (that is, the condition is true), the data are accepted for that attribute. If the condition is found to be false, an error message is generated and the data are not accepted.

Note that the CREATE TABLE command lets you define constraints in two different places:

- When you create the column definition (known as a *column constraint*)
- When you use the CONSTRAINT keyword (known as a *table constraint*)

A column constraint applies to just one column; a table constraint may apply to many columns. Those constraints are supported at varying levels of compliance by enterprise RDBMSs.

In this chapter, Oracle is used to illustrate SQL constraints. For example, note that the following SQL command sequence uses the DEFAULT and CHECK constraints to define the table named CUSTOMER.

```
CREATE TABLE CUSTOMER (
CUS_CODE                NUMBER          PRIMARY KEY,
CUS_LNAME               VARCHAR(15)     NOT NULL,
CUS_FNAME               VARCHAR(15)     NOT NULL,
CUS_INITIAL             CHAR(1),
CUS_AREACODE            CHAR(3)         DEFAULT '615'       NOT NULL
                                        CHECK(CUS_AREACODE IN ('615','713','931')),
CUS_PHONE               CHAR(8)         NOT NULL,
CUS_BALANCE             NUMBER(9,2)     DEFAULT 0.00,
CONSTRAINT CUS_UI1 UNIQUE (CUS_LNAME, CUS_FNAME));
```

In this case, the CUS_AREACODE attribute is assigned a default value of '615'. Therefore, if a new CUSTOMER table row is added and the end user makes no entry for the area code, the '615' value will be recorded. Also note that the CHECK condition restricts the values for the customer's area code to 615, 713, and 931; any other values will be rejected.

It is important to note that the DEFAULT value applies only when new rows are added to a table and then only when no value is entered for the customer's area code. (The default value is not used when the table is modified.) In contrast, the CHECK condition is validated whether a customer row is added *or modified*. However, while the CHECK condition may include any valid expression, it applies only to the attributes in the table being checked. If you want to check for conditions that include attributes in other tables, you must use triggers. (See Chapter 8, Advanced SQL.) Finally, the last line of the CREATE TABLE command sequence creates a unique index constraint (named CUS_UI1) on the customer's last name and first name. The index will prevent the entry of two customers with the same last name and first name. (This index merely illustrates the process. Clearly, it should be possible to have more than one person named John Smith in the CUSTOMER table.)

> **NOTE**
>
> NOTE TO MS ACCESS USERS
> MS Access does not accept the DEFAULT or CHECK constraints. However, MS Access will accept the CONSTRAINT CUS_UI1 UNIQUE (CUS_LNAME, CUS_FNAME) line and create the unique index.

In the following SQL command to create the INVOICE table, the DEFAULT constraint assigns a default date to a new invoice, and the CHECK constraint validates that the invoice date is greater than January 1, 2008.

```
CREATE TABLE INVOICE (
INV_NUMBER          NUMBER      PRIMARY KEY,
CUS_CODE            NUMBER      NOT NULL REFERENCES CUSTOMER(CUS_CODE),
INV_DATE            DATE        DEFAULT SYSDATE NOT NULL,
CONSTRAINT INV_CK1 CHECK (INV_DATE > TO_DATE('01-JAN-2008','DD-MON-YYYY')));
```

In this case, notice the following:

- The CUS_CODE attribute definition contains REFERENCES CUSTOMER (CUS_CODE) to indicate that the CUS_CODE is a foreign key. This is another way to define a foreign key.
- The DEFAULT constraint uses the SYSDATE special function. This function always returns today's date.
- The invoice date (INV_DATE) attribute is automatically given today's date (returned by SYSDATE) when a new row is added and no value is given for the attribute.
- A CHECK constraint is used to validate that the invoice date is greater than 'January 1, 2008'. When comparing a date to a manually entered date in a CHECK clause, Oracle requires the use of the TO_DATE function. The TO_DATE function takes two parameters, the literal date and the date format used.

The final SQL command sequence creates the LINE table. The LINE table has a composite primary key (INV_NUMBER, LINE_NUMBER) and uses a UNIQUE constraint in INV_NUMBER and P_CODE to ensure that the same product is not ordered twice in the same invoice.

```
CREATE TABLE LINE (
INV_NUMBER          NUMBER          NOT NULL,
LINE_NUMBER         NUMBER(2,0)     NOT NULL,
P_CODE              VARCHAR(10)     NOT NULL,
LINE_UNITS          NUMBER(9,2)     DEFAULT 0.00        NOT NULL,
LINE_PRICE          NUMBER(9,2)     DEFAULT 0.00        NOT NULL,
PRIMARY KEY (INV_NUMBER, LINE_NUMBER),
FOREIGN KEY (INV_NUMBER) REFERENCES INVOICE ON DELETE CASCADE,
FOREIGN KEY (P_CODE) REFERENCES PRODUCT(P_CODE),
CONSTRAINT LINE_UI1 UNIQUE(INV_NUMBER, P_CODE));
```

In the creation of the LINE table, note that a UNIQUE constraint is added to prevent the duplication of an invoice line. A UNIQUE constraint is enforced through the creation of a unique index. Also note that the ON DELETE CASCADE foreign key action enforces referential integrity. The use of ON DELETE CASCADE is recommended for weak entities to ensure that the deletion of a row in the strong entity automatically triggers the deletion of the corresponding rows in the dependent weak entity. In that case, the deletion of an INVOICE row will automatically delete all of the LINE rows related to the invoice. In the following section, you will learn more about indexes and how to use SQL commands to create them.

7.2.7 SQL INDEXES

You learned in Chapter 3 that indexes can be used to improve the efficiency of searches and to avoid duplicate column values. In the previous section, you saw how to declare unique indexes on selected attributes when the table is created. In fact, when you declare a primary key, the DBMS automatically creates a unique index. Even with this feature, you often need additional indexes. The ability to create indexes quickly and efficiently is important. Using the **CREATE INDEX** command, SQL indexes can be created on the basis of any selected attribute. The syntax is:

CREATE [UNIQUE] INDEX *indexname* ON *tablename*(*column1* [, *column2*])

For example, based on the attribute P_INDATE stored in the PRODUCT table, the following command creates an index named P_INDATEX:

CREATE INDEX P_INDATEX ON PRODUCT(P_INDATE);

SQL does not let you write over an existing index without warning you first, thus preserving the index structure within the data dictionary. Using the UNIQUE index qualifier, you can even create an index that prevents you from using a value that has been used before. Such a feature is especially useful when the index attribute is a candidate key whose values must not be duplicated:

CREATE UNIQUE INDEX P_CODEX ON PRODUCT(P_CODE);

If you now try to enter a duplicate P_CODE value, SQL produces the error message "duplicate value in index." Many RDBMSs, including Access, automatically create a unique index on the PK attribute(s) when you declare the PK.

A common practice is to create an index on any field that is used as a search key, in comparison operations in a conditional expression, or when you want to list rows in a specific order. For example, if you want to create a report of all products by vendor, it would be useful to create an index on the V_CODE attribute in the PRODUCT table. Remember that a vendor can supply many products. Therefore, you should *not* create a UNIQUE index in this case. Better yet, to make the search as efficient as possible, a composite index is recommended.

Unique composite indexes are often used to prevent data duplication. For example, consider the case illustrated in Table 7.5, in which required employee test scores are stored. (An employee can take a test only once on a given date.) Given the structure of Table 7.5, the PK is EMP_NUM + TEST_NUM. The third test entry for employee 111 meets entity integrity requirements—the combination 111,3 is unique—yet the WEA test entry is clearly duplicated.

TABLE 7.5	A Duplicated Test Record			
EMP_NUM	**TEST_NUM**	**TEST_CODE**	**TEST_DATE**	**TEST_SCORE**
110	1	WEA	15-Jan-2008	93
110	2	WEA	12-Jan-2008	87
111	1	HAZ	14-Dec-2007	91
111	2	WEA	18-Feb-2008	95
111	3	WEA	18-Feb-2008	95
112	1	CHEM	17-Aug-2007	91

Such duplication could have been avoided through the use of a unique composite index, using the attributes EMP_NUM, TEST_CODE, and TEST_DATE:

CREATE UNIQUE INDEX EMP_TESTDEX ON TEST(EMP_NUM, TEST_CODE, TEST_DATE);

By default, all indexes produce results that are listed in ascending order, but you can create an index that yields output in descending order. For example, if you routinely print a report that lists all products ordered by price from highest to lowest, you could create an index named PROD_PRICEX by typing:

CREATE INDEX PROD_PRICEX ON PRODUCT(P_PRICE DESC);

To delete an index, use the **DROP INDEX** command:

DROP INDEX *indexname*

For example, if you want to eliminate the PROD_PRICEX index, type:

DROP INDEX PROD_PRICEX;

After creating the tables and some indexes, you are ready to start entering data. The following sections use two tables (VENDOR and PRODUCT) to demonstrate most of the data manipulation commands.

7.3 DATA MANIPULATION COMMANDS

In this section, you will learn how to use the basic SQL data manipulation commands INSERT, SELECT, COMMIT, UPDATE, ROLLBACK, and DELETE.

7.3.1 ADDING TABLE ROWS

SQL requires the use of the **INSERT** command to enter data into a table. The INSERT command's basic syntax looks like this:

INSERT INTO *tablename* VALUES (*value1, value2, … , valuen*)

Because the PRODUCT table uses its V_CODE to reference the VENDOR table's V_CODE, an integrity violation will occur if those VENDOR table V_CODE values don't yet exist. Therefore, you need to enter the VENDOR rows before the PRODUCT rows. Given the VENDOR table structure defined earlier and the sample VENDOR data shown in Figure 7.2, you would enter the first two data rows as follows:

INSERT INTO VENDOR
         VALUES (21225,'Bryson, Inc.','Smithson','615','223-3234','TN','Y');
INSERT INTO VENDOR
         VALUES (21226,'Superloo, Inc.','Flushing','904','215-8995','FL','N');

and so on, until all of the VENDOR table records have been entered.

(To see the contents of the VENDOR table, use the SELECT * FROM VENDOR; command.)

The PRODUCT table rows would be entered in the same fashion, using the PRODUCT data shown in Figure 7.2. For example, the first two data rows would be entered as follows, pressing the Enter key at the end of each line:

INSERT INTO PRODUCT
         VALUES ('11QER/31','Power painter, 15 psi., 3-nozzle','03-Nov-07',8,5,109.99,0.00,25595);
INSERT INTO PRODUCT
         VALUES ('13-Q2/P2','7.25-in. pwr. saw blade','13-Dec-07',32,15,14.99, 0.05, 21344);

(To see the contents of the PRODUCT table, use the SELECT * FROM PRODUCT; command.)

NOTE

Date entry is a function of the date format expected by the DBMS. For example, March 25, 2008 might be shown as 25-Mar-2008 in Access and Oracle, or it might be displayed in other presentation formats in another RDBMS. MS Access requires the use of # delimiters when performing any computations or comparisons based on date attributes, as in P_INDATE >= #25-Mar-08#.

In the preceding data entry lines, observe that:

- The row contents are entered between parentheses. Note that the first character after VALUES is a parenthesis and that the last character in the command sequence is also a parenthesis.
- Character (string) and date values must be entered between apostrophes (').
- Numerical entries are *not* enclosed in apostrophes.
- Attribute entries are separated by commas.
- A value is required for each column in the table.

This version of the INSERT commands adds one table row at a time.

Inserting Rows with Null Attributes

Thus far, you have entered rows in which all of the attribute values are specified. But what do you do if a product does not have a vendor or if you don't yet know the vendor code? In those cases, you would want to leave the vendor code null. To enter a null, use the following syntax:

INSERT INTO PRODUCT
         VALUES ('BRT-345','Titanium drill bit','18-Oct-07', 75, 10, 4.50, 0.06, NULL);

Incidentally, note that the NULL entry is accepted only because the V_CODE attribute is optional—the NOT NULL declaration was not used in the CREATE TABLE statement for this attribute.

Inserting Rows with Optional Attributes

There might be occasions when more than one attribute is optional. Rather than declaring each attribute as NULL in the INSERT command, you can indicate just the attributes that have required values. You do that by listing the attribute names inside parentheses after the table name. For the purpose of this example, assume that the only required attributes for the PRODUCT table are P_CODE and P_DESCRIPT:

INSERT INTO PRODUCT(P_CODE, P_DESCRIPT) VALUES ('BRT-345','Titanium drill bit');

7.3.2 SAVING TABLE CHANGES

Any changes made to the table contents are not saved on disk until you close the database, close the program you are using, or use the **COMMIT** command. If the database is open and a power outage or some other interruption occurs before you issue the COMMIT command, your changes will be lost and only the original table contents will be retained. The syntax for the COMMIT command is:

COMMIT [WORK]

The COMMIT command permanently saves *all* changes—such as rows added, attributes modified, and rows deleted—made to any table in the database. Therefore, if you intend to make your changes to the PRODUCT table permanent, it is a good idea to save those changes by using:

COMMIT;

> **NOTE**
>
> NOTE TO MS ACCESS USERS
> MS Access doesn't support the COMMIT command because it automatically saves changes after the execution of each SQL command.

However, the COMMIT command's purpose is not just to save changes. In fact, the ultimate purpose of the COMMIT and ROLLBACK commands (see Section 7.3.5) is to ensure database update integrity in transaction management. (You will see how such issues are addressed in Chapter 10, Transaction Management and Concurrency Control.)

7.3.3 LISTING TABLE ROWS

The **SELECT** command is used to list the contents of a table. The syntax of the SELECT command is as follows:

SELECT   *columnlist*   FROM   *tablename*

The *columnlist* represents one or more attributes, separated by commas. You could use the * (asterisk) as a wildcard character to list all attributes. A **wildcard character** is a symbol that can be used as a general substitute for other characters or commands. For example, to list all attributes and all rows of the PRODUCT table, use:

SELECT * FROM PRODUCT;

Figure 7.3 shows the output generated by that command. (Figure 7.3 shows all of the rows in the PRODUCT table that serve as the basis for subsequent discussions. If you entered only the PRODUCT table's first two records, as shown in the preceding section, the output of the preceding SELECT command would show only the rows you entered. Don't worry about the difference between your SELECT output and the output shown in Figure 7.3. When you complete the work in this section, you will have created and populated your VENDOR and PRODUCT tables with the correct rows for use in future sections.)

FIGURE 7.3   **The contents of the PRODUCT table**

P_CODE	P_DESCRIPT	P_INDATE	P_QOH	P_MIN	P_PRICE	P_DISCOUNT	V_CODE
11QER/31	Power painter, 15 psi., 3-nozzle	03-Nov-07	8	5	109.99	0.00	25595
13-Q2/P2	7.25-in. pwr. saw blade	13-Dec-07	32	15	14.99	0.05	21344
14-Q1/L3	9.00-in. pwr. saw blade	13-Nov-07	18	12	17.49	0.00	21344
1546-QQ2	Hrd. cloth, 1/4-in., 2x50	15-Jan-08	15	8	39.95	0.00	23119
1558-QW1	Hrd. cloth, 1/2-in., 3x50	15-Jan-08	23	5	43.99	0.00	23119
2232/QTY	B&D jigsaw, 12-in. blade	30-Dec-07	8	5	109.92	0.05	24288
2232/QWE	B&D jigsaw, 8-in. blade	24-Dec-07	6	5	99.87	0.05	24288
2238/QPD	B&D cordless drill, 1/2-in.	20-Jan-08	12	5	38.95	0.05	25595
23109-HB	Claw hammer	20-Jan-08	23	10	9.95	0.10	21225
23114-AA	Sledge hammer, 12 lb.	02-Jan-08	8	5	14.40	0.05	
54778-2T	Rat-tail file, 1/8-in. fine	15-Dec-07	43	20	4.99	0.00	21344
89-WRE-Q	Hicut chain saw, 16 in.	07-Feb-08	11	5	256.99	0.05	24288
PVC23DRT	PVC pipe, 3.5-in., 8-ft	20-Feb-08	188	75	5.87	0.00	
SM-18277	1.25-in. metal screw, 25	01-Mar-08	172	75	6.99	0.00	21225
SW-23116	2.5-in. wd. screw, 50	24-Feb-08	237	100	8.45	0.00	21231
WR3/TT3	Steel matting, 4'x8'x1/6", .5" mesh	17-Jan-08	18	5	119.95	0.10	25595

NOTE

Your listing may not be in the order shown in Figure 7.3. The listings shown in the figure are the result of system-controlled primary-key-based index operations. You will learn later how to control the output so that it conforms to the order you have specified.

NOTE

NOTE TO ORACLE USERS

Some SQL implementations (such as Oracle's) cut the attribute labels to fit the width of the column. However, Oracle lets you set the width of the display column to show the complete attribute name. You can also change the display format, regardless of how the data are stored in the table. For example, if you want to display dollar symbols and commas in the P_PRICE output, you can declare:

COLUMN P_PRICE FORMAT $99,999.99

to change the output 12347.67 to $12,347.67.

In the same manner, to display only the first 12 characters of the P_DESCRIPT attribute, use:

COLUMN P_DESCRIPT FORMAT A12 TRUNCATE

Although SQL commands can be grouped together on a single line, complex command sequences are best shown on separate lines, with space between the SQL command and the command's components. Using that formatting convention makes it much easier to see the components of the SQL statements, making it easy to trace the SQL logic, and if necessary, to make corrections. The number of spaces used in the indention is up to you. For example, note the following format for a more complex statement:

SELECT   P_CODE, P_DESCRIPT, P_INDATE, P_QOH, P_MIN, P_PRICE, P_DISCOUNT, V_CODE
FROM   PRODUCT;

When you run a SELECT command on a table, the RDBMS returns a set of one or more rows that have the same characteristics as a relational table. In addition, the SELECT command lists all rows from the table you specified in the FROM clause. This is a very important characteristic of SQL commands. By default, most SQL data manipulation commands operate over an entire table (or relation). That is why SQL commands are said to be *set-oriented* commands. A SQL set-oriented command works over a set of rows. The set may include one or more columns and zero or more rows from one or more tables.

7.3.4 UPDATING TABLE ROWS

Use the **UPDATE** command to modify data in a table. The syntax for this command is:

```
UPDATE      tablename
SET         columnname = expression [, columnname = expression]
[WHERE      conditionlist ];
```

For example, if you want to change P_INDATE from December 13, 2007, to January 18, 2008, in the second row of the PRODUCT table (see Figure 7.3), use the primary key (13-Q2/P2) to locate the correct (second) row. Therefore, type:

```
UPDATE      PRODUCT
SET         P_INDATE = '18-JAN-2008'
WHERE       P_CODE = '13-Q2/P2';
```

If more than one attribute is to be updated in the row, separate the corrections with commas:

```
UPDATE      PRODUCT
SET         P_INDATE = '18-JAN-2008', P_PRICE = 17.99, P_MIN = 10
WHERE       P_CODE = '13-Q2/P2';
```

What would have happened if the previous UPDATE command had not included the WHERE condition? The P_INDATE, P_PRICE, and P_MIN values would have been changed in *all* rows of the PRODUCT table. Remember, the UPDATE command is a set-oriented operator. Therefore, if you don't specify a WHERE condition, the UPDATE command will apply the changes to *all* rows in the specified table.

Confirm the correction(s) by using this SELECT command to check the PRODUCT table's listing:

SELECT * FROM PRODUCT;

7.3.5 RESTORING TABLE CONTENTS

If you have not yet used the COMMIT command to store the changes permanently in the database, you can restore the database to its previous condition with the **ROLLBACK** command. ROLLBACK undoes any changes since the last COMMIT command and brings the data back to the values that existed before the changes were made. To restore the data to their "pre-change" condition, type

ROLLBACK;

and then press the Enter key. Use the SELECT statement again to see that the ROLLBACK did, in fact, restore the data to their original values.

COMMIT and ROLLBACK work only with data manipulation commands that are used to add, modify, or delete table rows. For example, assume that you perform these actions:

1. CREATE a table called SALES.
2. INSERT 10 rows in the SALES table.
3. UPDATE two rows in the SALES table.
4. Execute the ROLLBACK command.

Will the SALES table be removed by the ROLLBACK command? No, the ROLLBACK command will undo *only* the results of the INSERT and UPDATE commands. All data definition commands (CREATE TABLE) are automatically committed to the data dictionary and cannot be rolled back. The COMMIT and ROLLBACK commands are examined in greater detail in Chapter 10.

NOTE

NOTE TO MS ACCESS USERS
MS Access doesn't support the ROLLBACK command.

Some RDBMSs, such as Oracle, automatically COMMIT data changes when issuing data definition commands. For example, if you had used the CREATE INDEX command after updating the two rows in the previous example, all previous changes would have been committed automatically; doing a ROLLBACK afterward wouldn't have undone anything. *Check your RDBMS manual to understand these subtle differences.*

7.3.6 DELETING TABLE ROWS

It is easy to delete a table row using the **DELETE** statement; the syntax is:

DELETE FROM   *tablename*
[WHERE   *conditionlist*];

For example, if you want to delete from the PRODUCT table the product that you added earlier whose code (P_CODE) is 'BRT-345', use:

DELETE FROM   PRODUCT
WHERE   P_CODE = 'BRT-345';

In that example, the primary key value lets SQL find the exact record to be deleted. However, deletions are not limited to a primary key match; any attribute may be used. For example, in your PRODUCT table, you will see that there are several products for which the P_MIN attribute is equal to 5. Use the following command to delete all rows from the PRODUCT table for which the P_MIN is equal to 5:

DELETE FROM   PRODUCT
WHERE   P_MIN = 5;

Check the PRODUCT table's contents again to verify that all products with P_MIN equal to 5 have been deleted.

Finally, remember that DELETE is a set-oriented command. And keep in mind that the WHERE condition is optional. Therefore, if you do not specify a WHERE condition, *all* rows from the specified table will be deleted!

7.3.7 INSERTING TABLE ROWS WITH A SELECT SUBQUERY

You learned in Section 7.3.1 how to use the INSERT statement to add rows to a table. In that section, you added rows one at a time. In this section, you learn how to add multiple rows to a table, using another table as the source of the data. The syntax for the INSERT statement is:

INSERT INTO *tablename*   SELECT *columnlist*   FROM *tablename*;

In that case, the INSERT statement uses a SELECT subquery. A **subquery**, also known as a **nested query** or an **inner query**, is a query that is embedded (or nested) inside another query. The inner query is always executed first by the RDBMS. Given the previous SQL statement, the INSERT portion represents the outer query, and the SELECT portion represents the subquery. You can nest queries (place queries inside queries) many levels deep; in every case, the output of the inner query is used as the input for the outer (higher-level) query. In Chapter 8 you will learn more about the various types of subqueries.

The values returned by the SELECT subquery should match the attributes and data types of the table in the INSERT statement. If the table into which you are inserting rows has one date attribute, one number attribute, and one character attribute, the SELECT subquery should return one or more rows in which the first column has date values, the second column has number values, and the third column has character values.

Populating the VENDOR and PRODUCT Tables

The following steps guide you through the process of populating the VENDOR and PRODUCT tables with the data to be used in the rest of the chapter. To accomplish that task, two tables named V and P are used as the data source. V and P have the same table structure (attributes) as the VENDOR and PRODUCT tables.

ONLINE CONTENT

Before you execute the following commands, you **MUST** do the following:

- If you are using Oracle, run the **create_P_V.sql** script file in the Online Student Companion to create the V and P tables used in the example below. To connect to the database, follow the instructions specific to your school's setup provided by your instructor.
- If you are using Access, copy the original **Ch07_SaleCo.mbd** file from the Online Student Companion.

Use the following steps to populate your VENDOR and PRODUCT tables. (If you haven't already created the PRODUCT and VENDOR tables to practice the SQL commands in the previous sections, do so before completing these steps.)

1. Delete all rows from the PRODUCT and VENDOR tables.
   - DELETE FROM PRODUCT;
   - DELETE FROM VENDOR;

2. Add the rows to VENDOR by copying all rows from V.
   - If you are using MS Access, type:
     INSERT INTO VENDOR SELECT * FROM V;
   - If you are using Oracle, type:
     INSERT INTO VENDOR SELECT * FROM TEACHER.V;

3. Add the rows to PRODUCT by copying all rows from P.
   - If you are using MS Access, type:
     INSERT INTO PRODUCT SELECT * FROM P;
   - If you are using Oracle, type:
     INSERT INTO PRODUCT SELECT * FROM TEACHER.P;
   - Oracle users must permanently save the changes by issuing the COMMIT; command.

If you followed those steps correctly, you now have the VENDOR and PRODUCT tables populated with the data that will be used in the remaining sections of the chapter.

ONLINE CONTENT

Before you execute the commands in the following sections, you **MUST** do the following:

- If you are using Oracle, run the **sqlintrodbinit.sql** script file in the Online Student Companion to create all tables and load the data in the database. To connect to the database, follow the instructions specific to your school's setup provided by your instructor.
- If you are using Access, copy the original **Ch07_SaleCo.mbd** file from the Online Student Companion.

7.4 SELECT QUERIES

In this section, you will learn how to fine-tune the SELECT command by adding restrictions to the search criteria. SELECT, coupled with appropriate search conditions, is an incredibly powerful tool that enables you to transform data into information. For example, in the following sections, you will learn how to create queries that can be used to answer questions such as these: "What products were supplied by a particular vendor?" "Which products are priced below $10?" "How many products supplied by a given vendor were sold between January 5, 2008 and March 20, 2008?"

7.4.1 SELECTING ROWS WITH CONDITIONAL RESTRICTIONS

You can select partial table contents by placing restrictions on the rows to be included in the output. This is done by using the WHERE clause to add conditional restrictions to the SELECT statement. The following syntax enables you to specify which rows to select:

SELECT	*columnlist*
FROM	*tablelist*
[WHERE	*conditionlist*];

The SELECT statement retrieves all rows that match the specified condition(s)—also known as the *conditional criteria*—you specified in the WHERE clause. The *conditionlist* in the WHERE clause of the SELECT statement is represented by one or more conditional expressions, separated by logical operators. The WHERE clause is optional. If no rows match the specified criteria in the WHERE clause, you see a blank screen or a message that tells you that no rows were retrieved. For example, the query:

SELECT	P_DESCRIPT, P_INDATE, P_PRICE, V_CODE
FROM	PRODUCT
WHERE	V_CODE = 21344;

returns the description, date, and price of products with a vendor code of 21344, as shown in Figure 7.4.

FIGURE 7.4	Selected PRODUCT table attributes for vendor code 21344

P_DESCRIPT	P_INDATE	P_PRICE	V_CODE
7.25-in. pwr. saw blade	13-Dec-07	14.99	21344
9.00-in. pwr. saw blade	13-Nov-07	17.49	21344
Rat-tail file, 1/8-in. fine	15-Dec-07	4.99	21344

MS Access users can use the Access QBE (query by example) query generator. Although the Access QBE generates its own "native" version of SQL, you can also elect to type standard SQL in the Access SQL window, as shown at the bottom of Figure 7.5. Figure 7.5 shows the Access QBE screen, the SQL window's QBE-generated SQL, and the listing of the modified SQL.

Numerous conditional restrictions can be placed on the selected table contents. For example, the comparison operators shown in Table 7.6 can be used to restrict output.

FIGURE 7.5   The Microsoft Access QBE and its SQL

Microsoft Access-generated SQL

```
SELECT PRODUCT.P_DESCRIPT, PRODUCT.P_INDATE, PRODUCT.P_PRICE, PRODUCT.V_CODE
FROM PRODUCT
WHERE (((PRODUCT.[V_CODE])=21344));
```

User-entered SQL

```
SELECT  P_DESCRIPT, P_INDATE, P_PRICE, V_CODE
FROM    PRODUCT
WHERE   V_CODE=21344;
```

NOTE

NOTE TO MS ACCESS USERS

The MS Access QBE interface automatically designates the data source by using the table name as a prefix. You will discover later that the table name prefix is used to avoid ambiguity when the same column name appears in multiple tables. For example, both the VENDOR and the PRODUCT tables contain the V_CODE attribute. Therefore, if both tables are used—as they would be in a join—the source of the V_CODE attribute must be specified.

TABLE 7.6   Comparison Operators

SYMBOL	MEANING
=	Equal to
<	Less than
<=	Less than or equal to
>	Greater than
>=	Greater than or equal to
<> or !=	Not equal to

The following example uses the "not equal to" operator:

```
SELECT  P_DESCRIPT, P_INDATE, P_PRICE, V_CODE
FROM    PRODUCT
WHERE   V_CODE <> 21344;
```

The output, shown in Figure 7.6, lists all of the rows for which the vendor code is *not* 21344.

Note that in Figure 7.6, rows with nulls in the V_CODE column (see Figure 7.3) are not included in the SELECT command's output.

FIGURE 7.6

Selected PRODUCT table attributes for vendor codes other than 21344

P_DESCRIPT	P_INDATE	P_PRICE	V_CODE
Power painter, 15 psi., 3-nozzle	03-Nov-07	109.99	25595
Hrd. cloth, 1/4-in., 2x50	15-Jan-08	39.95	23119
Hrd. cloth, 1/2-in., 3x50	15-Jan-08	43.99	23119
B&D jigsaw, 12-in. blade	30-Dec-07	109.92	24288
B&D jigsaw, 8-in. blade	24-Dec-07	99.87	24288
B&D cordless drill, 1/2-in.	20-Jan-08	38.95	25595
Claw hammer	20-Jan-08	9.95	21225
Hicut chain saw, 16 in.	07-Feb-08	256.99	24288
1.25-in. metal screw, 25	01-Mar-08	6.99	21225
2.5-in. wd. screw, 50	24-Feb-08	8.45	21231
Steel matting, 4'x8'x1/6", .5" mesh	17-Jan-08	119.95	25595

FIGURE 7.7

Selected PRODUCT table attributes with a P_PRICE restriction

P_DESCRIPT	P_QOH	P_MIN	P_PRICE
Claw hammer	23	10	9.95
Rat-tail file, 1/8-in. fine	43	20	4.99
PVC pipe, 3.5-in., 8-ft	188	75	5.87
1.25-in. metal screw, 25	172	75	6.99
2.5-in. wd. screw, 50	237	100	8.45

FIGURE 7.8

Selected PRODUCT table attributes: the ASCII code effect

P_CODE	P_DESCRIPT	P_QOH	P_MIN	P_PRICE
11QER/31	Power painter, 15 psi., 3-nozzle	8	5	109.99
13-Q2/P2	7.25-in. pwr. saw blade	32	15	14.99
14-Q1/L3	9.00-in. pwr. saw blade	18	12	17.49
1546-QQ2	Hrd. cloth, 1/4-in., 2x50	15	8	39.95

The command sequence:

```
SELECT     P_DESCRIPT, P_QOH, P_MIN, P_PRICE
FROM       PRODUCT
WHERE      P_PRICE <= 10;
```

yields the output shown in Figure 7.7.

Using Comparison Operators on Character Attributes

Because computers identify all characters by their (numeric) American Standard Code for Information Interchange (ASCII) codes, comparison operators may even be used to place restrictions on character-based attributes. Therefore, the command:

```
SELECT     P_CODE, P_DESCRIPT, P_QOH, P_MIN,
           P_PRICE
FROM       PRODUCT
WHERE      P_CODE < '1558-QW1';
```

would be correct and would yield a list of all rows in which the P_CODE is alphabetically less than 1558-QW1. (Because the ASCII code value for the letter *B* is greater than the value of the letter *A*, it follows that *A* is less than *B*.) Therefore, the output will be generated as shown in Figure 7.8.

String (character) comparisons are made from left to right. This left-to-right comparison is especially useful when attributes such as names are to be compared. For example, the string "Ardmore" would be judged *greater than* the string "Aarenson" but *less than* the string "Brown"; such results may be used to generate alphabetical listings like those found in a phone directory. If the characters 0−9 are stored as strings, the same left-to-right string comparisons can lead to apparent anomalies. For example, the ASCII code for the character "5" is, as expected, *greater than* the ASCII code for the character "4." Yet the same "5" will also be judged *greater than* the string "44" because the *first* character in the string "44" is less than the string "5." For that reason, you may get some unexpected results from comparisons when dates or other numbers are stored in character format. This also applies to date comparisons. For example, the left-to-right ASCII character comparison would force the conclusion that the date "01/01/2008" occurred *before* "12/31/2007." Because the leftmost character "0" in "01/01/2008" is *less than* the leftmost character "1" in "12/31/2007," "01/01/2008" is *less than* "12/31/2007." Naturally, if date strings are stored in a yyyy/mm/dd format, the comparisons will yield appropriate results, but this is a nonstandard date presentation. That's why all current RDBMSs support "date" data types; you should use them. In addition, using "date" data types gives you the benefit of date arithmetic.

Using Comparison Operators on Dates

Date procedures are often more software-specific than other SQL procedures. For example, the query to list all of the rows in which the inventory stock dates occur on or after January 20, 2008 will look like this:

```
SELECT     P_DESCRIPT, P_QOH, P_MIN, P_PRICE, P_INDATE
FROM       PRODUCT
WHERE      P_INDATE >= '20-Jan-2008';
```

(Remember that MS Access users must use the # delimiters for dates. For example, you would use #20-Jan-08# in the above WHERE clause.) The date-restricted output is shown in Figure 7.9.

FIGURE 7.9	Selected PRODUCT table attributes: date restriction

P_DESCRIPT	P_QOH	P_MIN	P_PRICE	P_INDATE
B&D cordless drill, 1/2-in.	12	5	38.95	20-Jan-08
Claw hammer	23	10	9.95	20-Jan-08
Hicut chain saw, 16 in.	11	5	256.99	07-Feb-08
PVC pipe, 3.5-in., 8-ft	188	75	5.87	20-Feb-08
1.25-in. metal screw, 25	172	75	6.99	01-Mar-08
2.5-in. wd. screw, 50	237	100	8.45	24-Feb-08

Using Computed Columns and Column Aliases

Suppose you want to determine the total value of each of the products currently held in inventory. Logically, that determination requires the multiplication of each product's quantity on hand by its current price. You can accomplish this task with the following command:

SELECT   P_DESCRIPT, P_QOH, P_PRICE, P_QOH * P_PRICE
FROM   PRODUCT;

Entering that SQL command in Access generates the output shown in Figure 7.10.

FIGURE 7.10	SELECT statement with a computed column

P_DESCRIPT	P_QOH	P_PRICE	Expr1
Power painter, 15 psi., 3-nozzle	8	109.99	879.92
7.25-in. pwr. saw blade	32	14.99	479.68
9.00-in. pwr. saw blade	18	17.49	314.82
Hrd. cloth, 1/4-in., 2x50	15	39.95	599.25
Hrd. cloth, 1/2-in., 3x50	23	43.99	1011.77
B&D jigsaw, 12-in. blade	8	109.92	879.36
B&D jigsaw, 8-in. blade	6	99.87	599.22
B&D cordless drill, 1/2-in.	12	38.95	467.40
Claw hammer	23	9.95	228.85
Sledge hammer, 12 lb.	8	14.40	115.20
Rat-tail file, 1/8-in. fine	43	4.99	214.57
Hicut chain saw, 16 in.	11	256.99	2826.89
PVC pipe, 3.5-in., 8-ft	188	5.87	1103.56
1.25-in. metal screw, 25	172	6.99	1202.28
2.5-in. wd. screw, 50	237	8.45	2002.65
Steel matting, 4'x8'x1/6", .5" mesh	18	119.95	2159.10

SQL accepts any valid expressions (or formulas) in the computed columns. Such formulas can contain any valid mathematical operators and functions that are applied to attributes in any of the tables specified in the FROM clause of the SELECT statement. Note also that Access automatically adds an Expr label to all computed columns. (The first computed column would be labeled Expr1; the second, Expr2; and so on.) Oracle uses the actual formula text as the label for the computed column.

To make the output more readable, the SQL standard permits the use of aliases for any column in a SELECT statement. An **alias** is an alternative name given to a column or table in any SQL statement.

For example, you can rewrite the previous SQL statement as:

SELECT   P_DESCRIPT, P_QOH, P_PRICE, P_QOH * P_PRICE AS TOTVALUE
FROM   PRODUCT;

The output of that command is shown in Figure 7.11.

FIGURE 7.11	SELECT statement with a computed column and an alias

P_DESCRIPT	P_QOH	P_PRICE	TOTVALUE
Power painter, 15 psi., 3-nozzle	8	109.99	879.92
7.25-in. pwr. saw blade	32	14.99	479.68
9.00-in. pwr. saw blade	18	17.49	314.82
Hrd. cloth, 1/4-in., 2x50	15	39.95	599.25
Hrd. cloth, 1/2-in., 3x50	23	43.99	1011.77
B&D jigsaw, 12-in. blade	8	109.92	879.36
B&D jigsaw, 8-in. blade	6	99.87	599.22
B&D cordless drill, 1/2-in.	12	38.95	467.40
Claw hammer	23	9.95	228.85
Sledge hammer, 12 lb.	8	14.40	115.20
Rat-tail file, 1/8-in. fine	43	4.99	214.57
Hicut chain saw, 16 in.	11	256.99	2826.89
PVC pipe, 3.5-in., 8-ft	188	5.87	1103.56
1.25-in. metal screw, 25	172	6.99	1202.28
2.5-in. wd. screw, 50	237	8.45	2002.65
Steel matting, 4'x8'x1/6", .5" mesh	18	119.95	2159.10

You could also use a computed column, an alias, and date arithmetic in a single query. For example, assume that you want to get a list of out-of-warranty products that have been stored more than 90 days. In that case, the P_INDATE is at least 90 days less than the current (system) date. The MS Access version of this query is shown as:

SELECT   P_CODE, P_INDATE, DATE() - 90 AS CUTDATE
FROM   PRODUCT
WHERE   P_INDATE <= DATE() - 90;

The Oracle version of the same query is shown below:

```
SELECT      P_CODE, P_INDATE, SYSDATE - 90 AS CUTDATE
FROM        PRODUCT
WHERE       P_INDATE <= SYSDATE - 90;
```

Note that DATE() and SYSDATE are special functions that return today's date in MS Access and Oracle, respectively. You could use the DATE() and SYSDATE functions anywhere a date literal is expected, such as in the value list of an INSERT statement, in an UPDATE statement when changing the value of a date attribute, or in a SELECT statement as shown here. Of course, the previous query output would change based on today's date.

Suppose a manager wants a list of all products, the dates they were received, and the warranty expiration date (90 days from when the product was received). To generate that list, type:

```
SELECT      P_CODE, P_INDATE, P_INDATE + 90 AS EXPDATE
FROM        PRODUCT;
```

Note that you can use all arithmetic operators with date attributes as well as with numeric attributes.

7.4.2 ARITHMETIC OPERATORS: THE RULE OF PRECEDENCE

As you saw in the previous example, you can use arithmetic operators with table attributes in a column list or in a conditional expression. In fact, SQL commands are often used in conjunction with the arithmetic operators shown in Table 7.7.

TABLE 7.7	The Arithmetic Operators
ARITHMETIC OPERATOR	**DESCRIPTION**
+	Add
-	Subtract
*	Multiply
/	Divide
^	Raise to the power of (some applications use ** instead of ^)

Do not confuse the multiplication symbol (*) with the wildcard symbol used by some SQL implementations such as MS Access; the latter is used only in string comparisons, while the former is used in conjunction with mathematical procedures.

As you perform mathematical operations on attributes, remember the rules of precedence. As the name suggests, the **rules of precedence** are the rules that establish the order in which computations are completed. For example, note the order of the following computational sequence:

1. Perform operations within parentheses.
2. Perform power operations.
3. Perform multiplications and divisions.
4. Perform additions and subtractions.

The application of the rules of precedence will tell you that $8 + 2 * 5 = 8 + 10 = 18$, but $(8 + 2) * 5 = 10 * 5 = 50$. Similarly, $4 + 5^2 * 3 = 4 + 25 * 3 = 79$, but $(4 + 5)^2 * 3 = 81 * 3 = 243$, while the operation expressed by $(4 + 5^2) * 3$ yields the answer $(4 + 25) * 3 = 29 * 3 = 87$.

7.4.3 LOGICAL OPERATORS: AND, OR, AND NOT

In the real world, a search of data normally involves multiple conditions. For example, when you are buying a new house, you look for a certain area, a certain number of bedrooms, bathrooms, stories, and so on. In the same way, SQL allows you to have multiple conditions in a query through the use of logical operators. The logical operators are

AND, OR, and NOT. For example, if you want a list of the table contents for either the V_CODE = 21344 **or** the V_CODE = 24288, you can use the **OR** operator, as in the following command sequence:

```
SELECT      P_DESCRIPT, P_INDATE, P_PRICE, V_CODE
FROM        PRODUCT
WHERE       V_CODE = 21344 OR V_CODE = 24288;
```

That command generates the six rows shown in Figure 7.12 that match the logical restriction.

The logical **AND** has the same SQL syntax requirement. The following command generates a list of all rows for which P_PRICE is less than $50 and for which P_INDATE is a date occurring after January 15, 2008:

```
SELECT      P_DESCRIPT, P_INDATE, P_PRICE, V_CODE
FROM        PRODUCT
WHERE       P_PRICE < 50
AND         P_INDATE > '15-Jan-2008';
```

This command will produce the output shown in Figure 7.13.

You can combine the logical OR with the logical AND to place further restrictions on the output. For example, suppose you want a table listing for the following conditions:

- The P_INDATE is after January 15, 2008, and the P_PRICE is less than $50.
- Or the V_CODE is 24288.

The required listing can be produced by using:

```
SELECT      P_DESCRIPT, P_INDATE, P_PRICE, V_CODE
FROM        PRODUCT
WHERE       (P_PRICE < 50 AND
             P_INDATE > '15-Jan-2008')
OR          V_CODE = 24288;
```

Note the use of parentheses to combine logical restrictions. Where you place the parentheses depends on how you want the logical restrictions to be executed. Conditions listed within parentheses are always executed first. The preceding query yields the output shown in Figure 7.14.

Note that the three rows with the V_CODE = 24288 are included regardless of the P_INDATE and P_PRICE entries for those rows.

FIGURE 7.12 Selected PRODUCT table attributes: the logical OR

P_DESCRIPT	P_INDATE	P_PRICE	V_CODE
7.25-in. pwr. saw blade	13-Dec-07	14.99	21344
9.00-in. pwr. saw blade	13-Nov-07	17.49	21344
B&D jigsaw, 12-in. blade	30-Dec-07	109.92	24288
B&D jigsaw, 8-in. blade	24-Dec-07	99.87	24288
Rat-tail file, 1/8-in. fine	15-Dec-07	4.99	21344
Hicut chain saw, 16 in.	07-Feb-08	256.99	24288

FIGURE 7.13 Selected PRODUCT table attributes: the logical AND

P_DESCRIPT	P_INDATE	P_PRICE	V_CODE
B&D cordless drill, 1/2-in.	20-Jan-08	38.95	25595
Claw hammer	20-Jan-08	9.95	21225
PVC pipe, 3.5-in., 8-ft	20-Feb-08	5.87	
1.25-in. metal screw, 25	01-Mar-08	6.99	21225
2.5-in. wd. screw, 50	24-Feb-08	8.45	21231

FIGURE 7.14 Selected PRODUCT table attributes: the logical AND and OR

P_DESCRIPT	P_INDATE	P_PRICE	V_CODE
B&D jigsaw, 12-in. blade	30-Dec-07	109.92	24288
B&D jigsaw, 8-in. blade	24-Dec-07	99.87	24288
B&D cordless drill, 1/2-in.	20-Jan-08	38.95	25595
Claw hammer	20-Jan-08	9.95	21225
Hicut chain saw, 16 in.	07-Feb-08	256.99	24288
PVC pipe, 3.5-in., 8-ft	20-Feb-08	5.87	
1.25-in. metal screw, 25	01-Mar-08	6.99	21225
2.5-in. wd. screw, 50	24-Feb-08	8.45	21231

The use of the logical operators OR and AND can become quite complex when numerous restrictions are placed on the query. In fact, a specialty field in mathematics known as **Boolean algebra** is dedicated to the use of logical operators.

The logical operator **NOT** is used to negate the result of a conditional expression. That is, in SQL, all conditional expressions evaluate to true or false. If an expression is true, the row is selected; if an expression is false, the row is not selected. The NOT logical operator is typically used to find the rows that *do not* match a certain condition. For example, if you want to see a listing of all rows for which the vendor code is not 21344, use the command sequence:

```
SELECT      *
FROM        PRODUCT
WHERE       NOT (V_CODE = 21344);
```

Note that the condition is enclosed in parentheses; that practice is optional, but it is highly recommended for clarity. The logical NOT can be combined with AND and OR.

NOTE

If your SQL version does not support the logical NOT, you can generate the required output by using the condition:

WHERE V_CODE <> 21344

If your version of SQL does not support <>, use:

WHERE V_CODE != 21344

7.4.4 SPECIAL OPERATORS

ANSI-standard SQL allows the use of special operators in conjunction with the WHERE clause. These special operators include:

BETWEEN—Used to check whether an attribute value is within a range.

IS NULL—Used to check whether an attribute value is null.

LIKE—Used to check whether an attribute value matches a given string pattern.

IN—Used to check whether an attribute value matches any value within a value list.

EXISTS—Used to check whether a subquery returns any rows.

The BETWEEN Special Operator

If you use software that implements a standard SQL, the operator BETWEEN may be used to check whether an attribute value is within a range of values. For example, if you want to see a listing for all products whose prices are between $50 and $100, use the following command sequence:

```
SELECT      *
FROM        PRODUCT
WHERE       P_PRICE BETWEEN 50.00 AND 100.00;
```

NOTE

NOTE TO ORACLE USERS
When using the BETWEEN special operator, always specify the lower range value first. If you list the higher range value first, Oracle will return an empty result set.

If your DBMS does not support BETWEEN, you can use:

```
SELECT      *
FROM        PRODUCT
WHERE       P_PRICE > 50.00 AND P_PRICE < 100.00;
```

The IS NULL Special Operator

Standard SQL allows the use of IS NULL to check for a null attribute value. For example, suppose you want to list all products that do not have a vendor assigned (V_CODE is null). Such a null entry could be found by using the command sequence:

```
SELECT      P_CODE, P_DESCRIPT, V_CODE
FROM        PRODUCT
WHERE       V_CODE IS NULL;
```

Similarly, if you want to check a null date entry, the command sequence is:

```
SELECT      P_CODE, P_DESCRIPT, P_INDATE
FROM        PRODUCT
WHERE       P_INDATE IS NULL;
```

Note that SQL uses a special operator to test for nulls. Why? Couldn't you just enter a condition such as "V_CODE = NULL"? No. Technically, NULL is not a "value" the way the number 0 (zero) or the blank space is, but instead a NULL is a special property of an attribute that represents precisely the absence of any value.

The LIKE Special Operator

The LIKE special operator is used in conjunction with wildcards to find patterns within string attributes. Standard SQL allows you to use the percent sign (%) and underscore (_) wildcard characters to make matches when the entire string is not known:

- % means any and all *following* or preceding characters are eligible. For example,
  'J%' includes Johnson, Jones, Jernigan, July, and J-231Q.
  'Jo%' includes Johnson and Jones.
  '%n' includes Johnson and Jernigan.
- _ means any *one* character may be substituted for the underscore. For example,
  '_23-456-6789' includes 123-456-6789, 223-456-6789, and 323-456-6789.
  '_23-_56-678_' includes 123-156-6781, 123-256-6782, and 823-956-6788.
  '_o_es' includes Jones, Cones, Cokes, totes, and roles.

> **NOTE**
>
> Some RDBMSs, such as Microsoft Access, use the wildcard characters * and ? instead of % and _.

For example, the following query would find all VENDOR rows for contacts whose last names begin with *Smith*.

```
SELECT      V_NAME, V_CONTACT, V_AREACODE, V_PHONE
FROM        VENDOR
WHERE       V_CONTACT LIKE 'Smith%';
```

If you check the original VENDOR data in Figure 7.2 again, you'll see that this SQL query yields three records: two Smiths and one Smithson.

Keep in mind that most SQL implementations yield case-sensitive searches. For example, Oracle will not yield a return that includes *Jones* if you use the wildcard search delimiter 'jo%' in a search for last names. The reason is because *Jones* begins with a capital *J* and your wildcard search starts with a lowercase *j*. On the other hand, MS Access searches are not case sensitive.

For example, suppose you typed the following query in Oracle:

```
SELECT     V_NAME, V_CONTACT, V_AREACODE, V_PHONE
FROM       VENDOR
WHERE      V_CONTACT LIKE 'SMITH%';
```

No rows will be returned because character-based queries may be case sensitive. That is, an uppercase character has a different ASCII code than a lowercase character, thus causing *SMITH, Smith,* and *smith* to be evaluated as different (unequal) entries. Because the table contains no vendor whose last name begins with (uppercase) *SMITH,* the (uppercase) 'SMITH%' used in the query cannot make a match. Matches can be made only when the query entry is written exactly like the table entry.

Some RDBMSs, such as Microsoft Access, automatically make the necessary conversions to eliminate case sensitivity. Others, such as Oracle, provide a special UPPER function to convert both table and query character entries to uppercase. (The conversion is done in the computer's memory only; the conversion has no effect on how the value is actually stored in the table.) So if you want to avoid a no-match result based on case sensitivity, and if your RDBMS allows the use of the UPPER function, you can generate the same results by using the query:

```
SELECT     V_NAME, V_CONTACT, V_AREACODE, V_PHONE
FROM       VENDOR
WHERE      UPPER(V_CONTACT) LIKE 'SMITH%';
```

The preceding query produces a list including all rows that contain a last name that begins with *Smith*, regardless of uppercase or lowercase letter combinations such as *Smith, smith,* and *SMITH*.

The logical operators may be used in conjunction with the special operators. For instance, the query:

```
SELECT     V_NAME, V_CONTACT, V_AREACODE, V_PHONE
FROM       VENDOR
WHERE      V_CONTACT NOT LIKE 'Smith%';
```

will yield an output of all vendors whose names do not start with *Smith*.

Suppose you do not know whether a person's name is spelled *Johnson* or *Johnsen*. The wildcard character _ lets you find a match for either spelling. The proper search would be instituted by the query:

```
SELECT     *
FROM       VENDOR
WHERE      V_CONTACT LIKE 'Johns_n';
```

Thus, the wildcards allow you to make matches when only approximate spellings are known. Wildcard characters may be used in combinations. For example, the wildcard search based on the string '_l%' can yield the strings Al, Alton, Elgin, Blakeston, blank, bloated, and eligible.

The IN Special Operator

Many queries that would require the use of the logical OR can be more easily handled with the help of the special operator IN. For example, the query:

```
SELECT     *
FROM       PRODUCT
WHERE      V_CODE = 21344
OR         V_CODE = 24288;
```

can be handled more efficiently with:

```
SELECT      *
FROM        PRODUCT
WHERE       V_CODE IN (21344, 24288);
```

Note that the IN operator uses a value list. All of the values in the list must be of the same data type. Each of the values in the value list is compared to the attribute—in this case, V_CODE. If the V_CODE value matches any of the values in the list, the row is selected. In this example, the rows selected will be only those in which the V_CODE is either 21344 or 24288.

If the attribute used is of a character data type, the list values must be enclosed in single quotation marks. For instance, if the V_CODE had been defined as CHAR(5) when the table was created, the preceding query would have read:

```
SELECT      *
FROM        PRODUCT
WHERE       V_CODE IN ('21344', '24288');
```

The IN operator is especially valuable when it is used in conjunction with subqueries. For example, suppose you want to list the V_CODE and V_NAME of only those vendors who provide products. In that case, you could use a subquery within the IN operator to automatically generate the value list. The query would be:

```
SELECT      V_CODE, V_NAME
FROM        VENDOR
WHERE       V_CODE IN (SELECT V_CODE FROM PRODUCT);
```

The preceding query will be executed in two steps:

1. The inner query or subquery will generate a list of V_CODE values from the PRODUCT tables. Those V_CODE values represent the vendors who supply products.

2. The IN operator will compare the values generated by the subquery to the V_CODE values in the VENDOR table and will select only the rows with matching values—that is, the vendors who provide products.

The IN special operator will receive additional attention in Chapter 8, where you will learn more about subqueries.

The EXISTS Special Operator

The EXISTS special operator can be used whenever there is a requirement to execute a command based on the result of another query. That is, if a subquery returns any rows, run the main query; otherwise, don't. For example, the following query will list all vendors, but only if there are products to order:

```
SELECT      *
FROM        VENDOR
WHERE       EXISTS (SELECT * FROM PRODUCT WHERE P_QOH <= P_MIN);
```

The EXISTS special operator is used in the following example to list all vendors, but only if there are products with the quantity on hand, less than double the minimum quantity:

```
SELECT      *
FROM        VENDOR
WHERE       EXISTS (SELECT * FROM PRODUCT WHERE P_QOH < P_MIN * 2);
```

The EXISTS special operator will receive additional attention in Chapter 8, where you will learn more about subqueries.

7.5 ADVANCED DATA DEFINITION COMMANDS

In this section, you learn how to change (alter) table structures by changing attribute characteristics and by adding columns. Then you will learn how to do advanced data updates to the new columns. Finally, you will learn how to copy tables or parts of tables and how to delete tables.

All changes in the table structure are made by using the **ALTER TABLE** command, followed by a keyword that produces the specific change you want to make. Three options are available: ADD, MODIFY, and DROP. You use ADD to add a column, MODIFY to change column characteristics, and DROP to delete a column from a table. Most RDBMSs do not allow you to delete a column (unless the column does not contain any values) because such an action might delete crucial data that are used by other tables. The basic syntax to add or modify columns is:

ALTER TABLE *tablename*
      {ADD | MODIFY} (*columnname datatype* [{ADD | MODIFY} *columnname datatype*]) ;

The ALTER TABLE command can also be used to add table constraints. In those cases, the syntax would be:

ALTER TABLE *tablename*
      ADD *constraint* [ADD *constraint*] ;

where *constraint* refers to a constraint definition similar to those you learned in Section 7.2.6.

You could also use the ALTER TABLE command to remove a column or table constraint. The syntax would be as follows:

ALTER TABLE *tablename*
      DROP{PRIMARY KEY | COLUMN *columnname* | CONSTRAINT *constraintname* };

Notice that when removing a constraint, you need to specify the name given to the constraint. That is one reason why you should always name your constraints in your CREATE TABLE or ALTER TABLE statement.

7.5.1 CHANGING A COLUMN'S DATA TYPE

Using the ALTER syntax, the (integer) V_CODE in the PRODUCT table can be changed to a character V_CODE by using:

ALTER TABLE PRODUCT
      MODIFY (V_CODE CHAR(5));

Some RDBMSs, such as Oracle, do not let you change data types unless the column to be changed is empty. For example, if you want to change the V_CODE field from the current number definition to a character definition, the above command will yield an error message, because the V_CODE column already contains data. The error message is easily explained. Remember that the V_CODE in PRODUCT references the V_CODE in VENDOR. If you change the V_CODE data type, the data types don't match, and there is a referential integrity violation, thus triggering the error message. If the V_CODE column does not contain data, the preceding command sequence will produce the expected table structure alteration (if the foreign key reference was not specified during the creation of the PRODUCT table).

7.5.2 CHANGING A COLUMN'S DATA CHARACTERISTICS

If the column to be changed already contains data, you can make changes in the column's characteristics if those changes do not alter the data *type*. For example, if you want to increase the width of the P_PRICE column to nine digits, use the command:

ALTER TABLE PRODUCT
      MODIFY (P_PRICE DECIMAL(9,2));

If you now list the table contents, you see that the column width of P_PRICE has increased by one digit.

> **NOTE**
>
> Some DBMSs impose limitations on when it's possible to change attribute characteristics. For example, Oracle lets you increase (but not decrease) the size of a column. The reason for this restriction is that an attribute modification will affect the integrity of the data in the database. In fact, some attribute changes can be done only when there are no data in any rows for the affected attribute.

7.5.3 ADDING A COLUMN

You can alter an existing table by adding one or more columns. In the following example, you add the column named P_SALECODE to the PRODUCT table. (This column will be used later to determine whether goods that have been in inventory for a certain length of time should be placed on special sale.)

Suppose you expect the P_SALECODE entries to be 1, 2, or 3. Because there will be no arithmetic performed with the P_SALECODE, the P_SALECODE will be classified as a single-character attribute. Note the inclusion of all required information in the following ALTER command:

ALTER TABLE PRODUCT
        ADD (P_SALECODE CHAR(1));

ONLINE CONTENT

If you are using the MS Access databases provided in the Student Online Companion, you can track each of the updates in the following sections. For example, look at the copies of the PRODUCT table in the **Ch07_SaleCo** database, one named Product_2 and one named PRODUCT_3. Each of the two copies includes the new P_SALECODE column. If you want to see the *cumulative* effect of all UPDATE commands, you can continue using the PRODUCT table with the P_SALECODE modification and all of the changes you will make in the following sections. (You might even want to use both options, first to examine the individual effects of the update queries and then to examine the cumulative effects.)

When adding a column, be careful not to include the NOT NULL clause for the new column. Doing so will cause an error message; if you add a new column to a table that already has rows, the existing rows will default to a value of null for the new column. Therefore, it is not possible to add the NOT NULL clause for this new column. (You can, of course, add the NOT NULL clause to the table structure after all of the data for the new column have been entered and the column no longer contains nulls.)

7.5.4 DROPPING A COLUMN

Occasionally, you might want to modify a table by deleting a column. Suppose you want to delete the V_ORDER attribute from the VENDOR table. To accomplish that, you would use the following command:

ALTER TABLE VENDOR
        DROP COLUMN V_ORDER;

Again, some RDBMSs impose restrictions on attribute deletion. For example, you may not drop attributes that are involved in foreign key relationships, nor may you delete an attribute of a table that contains only that one attribute.

7.5.5 ADVANCED DATA UPDATES

To make data entries in an existing row's columns, SQL allows the UPDATE command. The UPDATE command updates only data in existing rows. For example, to enter the P_SALECODE value '2' in the fourth row, use the UPDATE command together with the primary key P_CODE '1546-QQ2'. Enter the value by using the command sequence:

```
UPDATE      PRODUCT
SET         P_SALECODE = '2'
WHERE       P_CODE = '1546-QQ2';
```

Subsequent data can be entered the same way, defining each entry location by its primary key (P_CODE) and its column location (P_SALECODE). For example, if you want to enter the P_SALECODE value '1' for the P_CODE values '2232/QWE' and '2232/QTY', you use:

```
UPDATE      PRODUCT
SET         P_SALECODE = '1'
WHERE       P_CODE IN ('2232/QWE', '2232/QTY');
```

If your RDBMS does not support IN, use the following command:

```
UPDATE      PRODUCT
SET         P_SALECODE = '1'
WHERE       P_CODE = '2232/QWE' OR P_CODE = '2232/QTY';
```

The results of your efforts can be checked by using:

```
SELECT      P_CODE, P_DESCRIPT, P_INDATE, P_PRICE, P_SALECODE
FROM        PRODUCT;
```

Although the UPDATE sequences just shown allow you to enter values into specified table cells, the process is very cumbersome. Fortunately, if a relationship can be established between the entries and the existing columns, the relationship can be used to assign values to their appropriate slots. For example, suppose you want to place sales codes based on the P_INDATE into the table, using the following schedule:

P_INDATE	P_SALECODE
before December 25, 2007	2
between January 16, 2008, and February 10, 2008	1

Using the PRODUCT table, the following two command sequences make the appropriate assignments:

```
UPDATE      PRODUCT
SET         P_SALECODE = '2'
WHERE       P_INDATE < '25-Dec-2007';
```

```
UPDATE      PRODUCT
SET         P_SALECODE = '1'
WHERE       P_INDATE >= '16-Jan-2008'
            AND P_INDATE <='10-Feb-2008';
```

To check the results of those two command sequences, use:

```
SELECT      P_CODE, P_DESCRIPT, P_INDATE, P_PRICE, P_SALECODE
FROM        PRODUCT;
```

If you have made *all* of the updates shown in this section using Oracle, your PRODUCT table should look like Figure 7.15. *Make sure that you issue a COMMIT statement to save these changes.*

FIGURE
7.15

The cumulative effect of the multiple updates in the PRODUCT table (Oracle)

```
± Oracle SQL*Plus                                                    _|□|x|
File  Edit  Search  Options  Help
SQL> SELECT P_CODE, P_DESCRIPT, P_INDATE, P_PRICE, P_SALECODE FROM PRODUCT;  ▲

P_CODE     P_DESCRIPT                            P_INDATE   P_PRICE P_SALECODE
---------- ------------------------------------- ---------- ------- ----------
11QER/31   Power painter, 15 psi., 3-nozzle      03-NOV-07   109.99 2
13-Q2/P2   7.25-in. pwr. saw blade               13-DEC-07    14.99 2
14-Q1/L3   9.00-in. pwr. saw blade               13-NOV-07    17.49 2
1546-QQ2   Hrd. cloth, 1/4-in., 2x50             15-JAN-08    39.95 2
1558-QW1   Hrd. cloth, 1/2-in., 3x50             15-JAN-08    43.99
2232/QTY   B&D jigsaw, 12-in. blade              30-DEC-07   109.92 1
2232/QWE   B&D jigsaw, 8-in. blade               24-DEC-07    99.87 2
2238/QPD   B&D cordless drill, 1/2-in.           20-JAN-08    38.95 1
23109-HB   Claw hammer                           20-JAN-08     9.95 1
23114-AA   Sledge hammer, 12 lb.                 02-JAN-08    14.40
54778-2T   Rat-tail file, 1/8-in. fine           15-DEC-07     4.99 2
89-WRE-Q   Hicut chain saw, 16 in.               07-FEB-08   256.99 1
PVC23DRT   PVC pipe, 3.5-in., 8-ft               20-FEB-08     5.87
SM-18277   1.25-in. metal screw, 25              01-MAR-08     6.99
SW-23116   2.5-in. wd. screw, 50                 24-FEB-08     8.45
WR3/TT3    Steel matting, 4'x8'x1/6", .5" mesh   17-JAN-08   119.95 1

16 rows selected.

SQL>
```

The arithmetic operators are particularly useful in data updates. For example, if the quantity on hand in your PRODUCT table has dropped below the minimum desirable value, you'll order more of the product. Suppose, for example, you have ordered 20 units of product 2232/QWE. When the 20 units arrive, you'll want to add them to inventory, using:

```
UPDATE      PRODUCT
SET         P_QOH = P_QOH + 20
WHERE       P_CODE = '2232/QWE';
```

If you want to add 10 percent to the price for all products that have current prices below $50, you can use:

```
UPDATE      PRODUCT
SET         P_PRICE = P_PRICE * 1.10
WHERE       P_PRICE < 50.00;
```

If you are using Oracle, issue a ROLLBACK command to undo the changes made by the last two UPDATE statements.

NOTE

If you fail to roll back the changes of the preceding UPDATE queries, the output of the subsequent queries will not match the results shown in the figures. Therefore:

- If you are using Oracle, use the ROLLBACK command to restore the database to its previous state.
- If you are using Access, copy the original **Ch07_SaleCo.mdb** file from the Student Online Companion.

7.5.6   COPYING PARTS OF TABLES

As you will discover in later chapters on database design, sometimes it is necessary to break up a table structure into several component parts (or smaller tables). Fortunately, SQL allows you to copy the contents of selected table columns so that the data need not be reentered manually into the newly created table(s). For example, if you want to copy P_CODE, P_DESCRIPT, P_PRICE, and V_CODE from the PRODUCT table to a new table named PART, you create the PART table structure first, as follows:

```
CREATE TABLE PART(
PART_CODE            CHAR(8)            NOT NULL            UNIQUE,
PART_DESCRIPT        CHAR(35),
PART_PRICE           DECIMAL(8,2),
V_CODE               INTEGER,
PRIMARY KEY (PART_CODE));
```

Note that the PART column names need not be identical to those of the original table and that the new table need not have the same number of columns as the original table. In this case, the first column in the PART table is PART_CODE, rather than the original P_CODE found in the PRODUCT table. And the PART table contains only four columns rather than the seven columns found in the PRODUCT table. However, column characteristics must match; you cannot copy a character-based attribute into a numeric structure and vice versa.

Next, you need to add the rows to the new PART table, using the PRODUCT table rows. To do that, you use the INSERT command you learned in Section 7.3.7. The syntax is:

```
INSERT INTO        target_tablename[(target_columnlist)]
SELECT             source_columnlist
FROM               source_tablename;
```

Note that the target column list is required if the source column list doesn't match all of the attribute names and characteristics of the target table (including the order of the columns). Otherwise, you do not need to specify the target column list. In this example, you must specify the target column list in the INSERT command below because the column names of the target table are different:

```
INSERT INTO        PART (PART_CODE, PART_DESCRIPT, PART_PRICE, V_CODE)
SELECT             P_CODE, P_DESCRIPT, P_PRICE, V_CODE FROM PRODUCT;
```

The contents of the PART table can now be examined by using the query:

```
SELECT * FROM PART;
```

to generate the new PART table's contents, shown in Figure 7.16.

SQL also provides another way to rapidly create a new table based on selected columns and rows of an existing table. In this case, the new table will copy the attribute names, data characteristics, and rows of the original table. The Oracle version of the command is:

```
CREATE TABLE PART AS
SELECT             P_CODE AS PART_CODE, P_DESCRIPT AS PART_DESCRIPT,
                   P_PRICE AS PART_PRICE, V_CODE
FROM               PRODUCT;
```

If the PART table already exists, Oracle will not let you overwrite the existing table. To run this command, you must first delete the existing PART table. (See Section 7.5.8.)

FIGURE 7.16	PART table attributes copied from the PRODUCT table

PART_CODE	PART_DESCRIPT	PART_PRICE	V_CODE
11QER/31	Power painter, 15 psi., 3-nozzle	109.99	25595
13-Q2/P2	7.25-in. pwr. saw blade	14.99	21344
14-Q1/L3	9.00-in. pwr. saw blade	17.49	21344
1546-QQ2	Hrd. cloth, 1/4-in., 2x50	39.95	23119
1558-QW1	Hrd. cloth, 1/2-in., 3x50	43.99	23119
2232/QTY	B&D jigsaw, 12-in. blade	109.92	24288
2232/QWE	B&D jigsaw, 8-in. blade	99.87	24288
2238/QPD	B&D cordless drill, 1/2-in.	38.95	25595
23109-HB	Claw hammer	9.95	21225
23114-AA	Sledge hammer, 12 lb.	14.4	
54778-2T	Rat-tail file, 1/8-in. fine	4.99	21344
89-WRE-Q	Hicut chain saw, 16 in.	256.99	24288
PVC23DRT	PVC pipe, 3.5-in., 8-ft	5.87	
SM-18277	1.25-in. metal screw, 25	6.99	21225
SW-23116	2.5-in. wd. screw, 50	8.45	21231
WR3/TT3	Steel matting, 4'x8'x1/6", .5" mesh	119.95	25595

The MS Access version of this command is:

```
SELECT   P_CODE AS PART_CODE, P_DESCRIPT AS
         PART_DESCRIPT,
         P_PRICE AS PART_PRICE,
         V_CODE INTO PART
FROM     PRODUCT;
```

If the PART table already exists, MS Access will ask if you want to delete the existing table and continue with the creation of the new PART table.

The SQL command just shown creates a new PART table with PART_CODE, PART_DESCRIPT, PART_PRICE, and V_CODE columns. In addition, all of the data rows (for the selected columns) will be copied automatically. *But note that no entity integrity (primary key) or referential integrity (foreign key) rules are automatically applied to the new table.* In the next section, you will learn how to define the PK to enforce entity integrity and the FK to enforce referential integrity.

7.5.7 ADDING PRIMARY AND FOREIGN KEY DESIGNATIONS

When you create a new table based on another table, the new table does not include integrity rules from the old table. In particular, there is no primary key. To define the primary key for the new PART table, use the following command:

```
ALTER TABLE   PART
      ADD     PRIMARY KEY (PART_CODE);
```

Aside from the fact that the integrity rules are not automatically transferred to a new table that derives its data from one or more other tables, several other scenarios could leave you without entity and referential integrity. For example, you might have forgotten to define the primary and foreign keys when you created the original tables. Or if you imported tables from a different database, you might have discovered that the importing procedure did not transfer the integrity rules. In any case, you can reestablish the integrity rules by using the ALTER command. For example, if the PART table's foreign key has not yet been designated, it can be designated by:

```
ALTER TABLE   PART
      ADD     FOREIGN KEY (V_CODE) REFERENCES VENDOR;
```

Alternatively, if neither the PART table's primary key nor its foreign key has been designated, you can incorporate both changes at once, using:

```
ALTER TABLE   PART
      ADD     PRIMARY KEY (PART_CODE)
      ADD     FOREIGN KEY (V_CODE) REFERENCES VENDOR;
```

Even composite primary keys and multiple foreign keys can be designated in a single SQL command. For example, if you want to enforce the integrity rules for the LINE table shown in Figure 7.1, you can use:

```
ALTER TABLE   LINE
      ADD     PRIMARY KEY (INV_NUMBER, LINE_NUMBER)
      ADD     FOREIGN KEY (INV_NUMBER) REFERENCES INVOICE
      ADD     FOREIGN KEY (PROD_CODE) REFERENCES PRODUCT;
```

7.5.8 DELETING A TABLE FROM THE DATABASE

A table can be deleted from the database using the **DROP TABLE** command. For example, you can delete the PART table you just created with:

DROP TABLE PART;

You can drop a table only if that table is not the "one" side of any relationship. If you try to drop a table otherwise, the RDBMS will generate an error message indicating that a foreign key integrity violation has occurred.

7.6 ADVANCED SELECT QUERIES

One of the most important advantages of SQL is its ability to produce complex free-form queries. The logical operators that were introduced earlier to update table contents work just as well in the query environment. In addition, SQL provides useful functions that count, find minimum and maximum values, calculate averages, and so on. Better yet, SQL allows the user to limit queries to only those entries that have no duplicates or entries whose duplicates can be grouped.

7.6.1 ORDERING A LISTING

The **ORDER BY** clause is especially useful when the listing order is important to you. The syntax is:

SELECT   columnlist
FROM   tablelist
[WHERE   conditionlist]
[ORDER BY   columnlist [ASC | DESC]] ;

Although you have the option of declaring the order type—ascending or descending—the default order is ascending. For example, if you want the contents of the PRODUCT table listed by P_PRICE in ascending order, use:

SELECT   P_CODE, P_DESCRIPT, P_INDATE, P_PRICE
FROM   PRODUCT
ORDER BY   P_PRICE;

The output is shown in Figure 7.17. Note that ORDER BY yields an ascending price listing.

FIGURE 7.17 Selected PRODUCT table attributes: ordered by (ascending) P_PRICE

P_CODE	P_DESCRIPT	P_INDATE	P_PRICE
54778-2T	Rat-tail file, 1/8-in. fine	15-Dec-07	4.99
PVC23DRT	PVC pipe, 3.5-in., 8-ft	20-Feb-08	5.87
SM-18277	1.25-in. metal screw, 25	01-Mar-08	6.99
SW-23116	2.5-in. wd. screw, 50	24-Feb-08	8.45
23109-HB	Claw hammer	20-Jan-08	9.95
23114-AA	Sledge hammer, 12 lb.	02-Jan-08	14.40
13-Q2/P2	7.25-in. pwr. saw blade	13-Dec-07	14.99
14-Q1/L3	9.00-in. pwr. saw blade	13-Nov-07	17.49
2238/QPD	B&D cordless drill, 1/2-in.	20-Jan-08	38.95
1546-QQ2	Hrd. cloth, 1/4-in., 2x50	15-Jan-08	39.95
1558-QW1	Hrd. cloth, 1/2-in., 3x50	15-Jan-08	43.99
2232/QWE	B&D jigsaw, 8-in. blade	24-Dec-07	99.87
2232/QTY	B&D jigsaw, 12-in. blade	30-Dec-07	109.92
11QER/31	Power painter, 15 psi., 3-nozzle	03-Nov-07	109.99
WR3/TT3	Steel matting, 4'x8'x1/6", .5" mesh	17-Jan-08	119.95
89-WRE-Q	Hicut chain saw, 16 in.	07-Feb-08	256.99

Comparing the listing in Figure 7.17 to the actual table contents shown earlier in Figure 7.2, you will see that in Figure 7.17, the lowest-priced product is listed first, followed by the next lowest-priced product, and so on. However, although ORDER BY produces a sorted output, the actual table contents are unaffected by the ORDER command.

To produce the list in descending order, you would enter:

SELECT   P_CODE, P_DESCRIPT, P_INDATE, P_PRICE
FROM   PRODUCT
ORDER BY   P_PRICE DESC;

Ordered listings are used frequently. For example, suppose you want to create a phone directory. It would be helpful if you could produce an ordered sequence (last name, first name, initial) in three stages:

1. ORDER BY last name.

2. Within the last names, ORDER BY first name.

3. Within the first and last names, ORDER BY middle initial.

Such a multilevel ordered sequence is known as a **cascading order sequence**, and it can be created easily by listing several attributes, separated by commas, after the ORDER BY clause.

The cascading order sequence is the basis for any telephone directory. To illustrate a cascading order sequence, use the following SQL command on the EMPLOYEE table:

```
SELECT      EMP_LNAME, EMP_FNAME, EMP_INITIAL, EMP_AREACODE, EMP_PHONE
FROM        EMPLOYEE
ORDER BY    EMP_LNAME, EMP_FNAME, EMP_INITIAL;
```

That command yields the results shown in Figure 7.18.

FIGURE 7.18   **Telephone list query results**

EMP_LNAME	EMP_FNAME	EMP_INITIAL	EMP_AREACODE	EMP_PHONE
Brandon	Marie	G	901	882-0845
Diante	Jorge	D	615	890-4567
Genkazi	Leighla	W	901	569-0093
Johnson	Edward	E	615	898-4387
Jones	Anne	M	615	898-3456
Kolmycz	George	D	615	324-5456
Lange	John	P	901	504-4430
Lewis	Rhonda	G	615	324-4472
Saranda	Hermine	R	615	324-5505
Smith	George	A	615	890-2984
Smith	George	K	901	504-3339
Smith	Jeanine	K	615	324-7883
Smythe	Melanie	P	615	324-9006
Vandam	Rhett		901	675-8993
Washington	Rupert	E	615	890-4925
Wiesenbach	Paul	R	615	897-4358
Williams	Robert	D	615	890-3220

The ORDER BY clause is useful in many applications, especially because the DESC qualifier can be invoked. For example, listing the most recent items first is a standard procedure. Typically, invoice due dates are listed in descending order. Or if you want to examine budgets, it's probably useful to list the largest budget line items first.

You can use the ORDER BY clause in conjunction with other SQL commands, too. For example, note the use of restrictions on date and price in the following command sequence:

```
SELECT      P_DESCRIPT, V_CODE, P_INDATE, P_PRICE
FROM        PRODUCT
WHERE       P_INDATE < '21-Jan-2008' AND
            P_PRICE <= 50.00
ORDER BY    V_CODE, P_PRICE DESC;
```

The output is shown in Figure 7.19. Note that within each V_CODE, the P_PRICE values are in descending order.

FIGURE 7.19	A query based on multiple restrictions

P_DESCRIPT	V_CODE	P_INDATE	P_PRICE
B&D cordless drill, 1/2-in.	25595	20-Jan-08	38.95
Hrd. cloth, 1/2-in., 3x50	23119	15-Jan-08	43.99
Hrd. cloth, 1/4-in., 2x50	23119	15-Jan-08	39.95
9.00-in. pwr. saw blade	21344	13-Nov-07	17.49
7.25-in. pwr. saw blade	21344	13-Dec-07	14.99
Rat-tail file, 1/8-in. fine	21344	15-Dec-07	4.99
Claw hammer	21225	20-Jan-08	9.95
Sledge hammer, 12 lb.		02-Jan-08	14.40

NOTE

If the ordering column has nulls, they are listed either first or last, depending on the RDBMS.

The ORDER BY clause must always be listed last in the SELECT command sequence.

7.6.2   LISTING UNIQUE VALUES

FIGURE 7.20	A listing of distinct (different) V_CODE values in the PRODUCT table

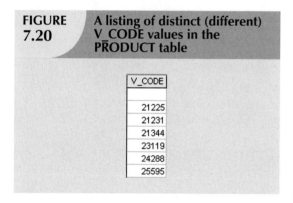

V_CODE
21225
21231
21344
23119
24288
25595

How many *different* vendors are currently represented in the PRODUCT table? A simple listing (SELECT) is not very useful if the table contains several thousand rows and you have to sift through the vendor codes manually. Fortunately, SQL's **DISTINCT** clause produces a list of only those values that are different from one another. For example, the command:

SELECT	DISTINCT V_CODE
FROM	PRODUCT;

yields only the different (distinct) vendor codes (V_CODE) that are encountered in the PRODUCT table, as shown in Figure 7.20. Note that the first output row shows the null. (By default, Access places the null V_CODE at the top of the list, while Oracle places it at the bottom. The placement of nulls does not affect the list contents. In Oracle, you could use ORDER BY V_CODE NULLS FIRST to place nulls at the top of the list.)

TABLE 7.8	Some Basic SQL Aggregate Functions

FUNCTION	OUTPUT
COUNT	The number of rows containing non-null values
MIN	The minimum attribute value encountered in a given column
MAX	The maximum attribute value encountered in a given column
SUM	The sum of all values for a given column
AVG	The arithmetic mean (average) for a specified column

7.6.3   AGGREGATE FUNCTIONS

SQL can perform various mathematical summaries for you, such as counting the number of rows that contain a specified condition, finding the minimum or maximum values for some specified attribute, summing the values in a specified column, and averaging the values in a specified column. Those aggregate functions are shown in Table 7.8.

To illustrate another standard SQL command format, most of the remaining input and output sequences are presented using the Oracle RDBMS.

COUNT

The **COUNT** function is used to tally the number of non-null values of an attribute. COUNT can be used in conjunction with the DISTINCT clause. For example, suppose you want to find out how many different vendors are in the PRODUCT table. The answer, generated by the first SQL code set shown in Figure 7.21, is 6. The answer indicates that six different VENDOR codes are found in the PRODUCT table. (Note that the nulls are not counted as V_CODE values.)

FIGURE 7.21	COUNT function output examples

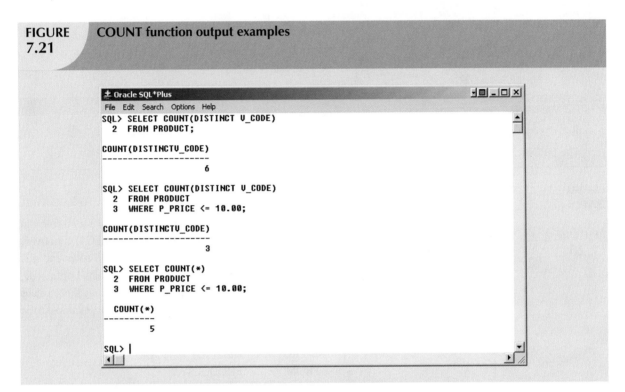

The aggregate functions can be combined with the SQL commands explored earlier. For example, the second SQL command set in Figure 7.21 supplies the answer to the question, "How many vendors referenced in the PRODUCT table have supplied products with prices that are less than or equal to $10?" The answer is three, indicating that three vendors referenced in the PRODUCT table have supplied products that meet the price specification.

The COUNT aggregate function uses one parameter within parentheses, generally a column name such as COUNT(V_CODE) or COUNT(P_CODE). The parameter may also be an expression such as COUNT(DISTINCT V_CODE) or COUNT(P_PRICE+10). Using that syntax, COUNT always returns the number of non-null values in the given column. (Whether the column values are computed or show stored table row values is immaterial). In contrast, the syntax COUNT(*) returns the number of total rows returned by the query, including the rows that contain nulls. In the example in Figure 7.21, SELECT COUNT(P_CODE) FROM PRODUCT and SELECT COUNT(*) FROM PRODUCT will yield the same answer because there are no null values in the P_CODE primary key column.

Note that the third SQL command set in Figure 7.21 uses the COUNT(*) command to answer the question, "How many rows in the PRODUCT table have a P_PRICE value less than or equal to $10?" The answer, five, indicates that five products have a listed price that meets the price specification. The COUNT(*) aggregate function is used to count rows in a query result set. In contrast, the COUNT(*column*) aggregate function counts the number of non-null values in a given column. For example, in Figure 7.20, the COUNT(*) function would return a value of 7 to indicate seven rows returned by the query. The COUNT(V_CODE) function would return a value of 6 to indicate the six non-null vendor code values.

> **NOTE**
>
> NOTE TO MS ACCESS USERS
>
> MS Access does not support the use of COUNT with the DISTINCT clause. If you want to use such queries in MS Access, you must create subqueries with DISTINCT and NOT NULL clauses. For example, the equivalent MS Access queries for the first two queries shown in Figure 7.21 are:
>
> | SELECT | COUNT(*) |
> |--------|----------|
> | FROM | (SELECT DISTINCT V_CODE FROM PRODUCT WHERE V_CODE IS NOT NULL) |
>
> and
>
> | SELECT | COUNT(*) |
> |--------|----------|
> | FROM | (SELECT DISTINCT(V_CODE) |
> | | FROM |
> | | (SELECT V_CODE, P_PRICE FROM PRODUCT |
> | | WHERE V_CODE IS NOT NULL AND P_PRICE < 10)) |
>
> Those two queries can be found in the Student Online Companion in the **Ch07_SaleCo** (Access) database. MS Access does add a trailer at the end of the query after you have executed it, but you can delete that trailer the next time you use the query.

MAX and MIN

The **MAX** and **MIN** functions help you find answers to problems such as the:

- Highest (maximum) price in the PRODUCT table.
- Lowest (minimum) price in the PRODUCT table.

The highest price, $256.99, is supplied by the first SQL command set in Figure 7.22. The second SQL command set shown in Figure 7.22 yields the minimum price of $4.99.

The third SQL command set in Figure 7.22 demonstrates that the numeric functions can be used in conjunction with more complex queries. However, you must remember that *the numeric functions yield only one value* based on all of the values found in the table: a single maximum value, a single minimum value, a single count, or a single average value. *It is easy to overlook this warning.* For example, examine the question, "Which product has the highest price?"

Although that query seems simple enough, the SQL command sequence:

```
SELECT     P_CODE, P_DESCRIPT, P_PRICE
FROM       PRODUCT
WHERE      P_PRICE = MAX(P_PRICE);
```

does not yield the expected results. This is because the use of MAX(P_PRICE) to the right side of a comparison operator is incorrect, thus producing an error message. The aggregate function MAX(*columnname*) can be used only in the column list of a SELECT statement. Also, in a comparison that uses an equality symbol, you can use only a single value to the right of the equals sign.

To answer the question, therefore, you must compute the maximum price first, then compare it to each price returned by the query. To do that, you need a nested query. In this case, the nested query is composed of two parts:

- The *inner query*, which is executed first.
- The *outer query*, which is executed last. (Remember that the outer query is always the first SQL command you encounter—in this case, SELECT.)

FIGURE
7.22
MAX and MIN function output examples

```
± Oracle SQL*Plus                                          _ □ X
File  Edit  Search  Options  Help
SQL> SELECT MAX(P_PRICE)
  2  FROM PRODUCT;

MAX(P_PRICE)
------------
      256.99

SQL> SELECT MIN(P_PRICE)
  2  FROM PRODUCT;

MIN(P_PRICE)
------------
        4.99

SQL> SELECT P_CODE, P_DESCRIPT, P_PRICE
  2  FROM PRODUCT
  3  WHERE P_PRICE = (SELECT MAX(P_PRICE) FROM PRODUCT);

P_CODE      P_DESCRIPT                               P_PRICE
----------  ---------------------------------------  ---------
89-WRE-Q    Hicut chain saw, 16 in.                   256.99

SQL>
```

Using the following command sequence as an example, note that the inner query first finds the maximum price value, which is stored in memory. Because the outer query now has a value to which to compare each P_PRICE value, the query executes properly.

SELECT	P_CODE, P_DESCRIPT, P_PRICE
FROM	PRODUCT
WHERE	P_PRICE = (SELECT MAX(P_PRICE) FROM PRODUCT);

The execution of that nested query yields the correct answer shown below the third (nested) SQL command set in Figure 7.22.

The MAX and MIN aggregate functions can also be used with date columns. For example, to find out the product that has the oldest date, you would use MIN(P_INDATE). In the same manner, to find out the most recent product, you would use MAX(P_INDATE).

NOTE

You can use expressions anywhere a column name is expected. Suppose you want to know what product has the highest inventory value. To find the answer, you can write the following query:

SELECT	*
FROM	PRODUCT
WHERE	P_QOH * P_PRICE = (SELECT MAX(P_QOH * P_PRICE) FROM PRODUCT);

SUM

The **SUM** function computes the total sum for any specified attribute, using whatever condition(s) you have imposed. For example, if you want to compute the total amount owed by your customers, you could use the following command:

SELECT   SUM(CUS_BALANCE) AS TOTBALANCE
FROM   CUSTOMER;

You could also compute the sum total of an expression. For example, if you want to find the total value of all items carried in inventory, you could use:

SELECT   SUM(P_QOH * P_PRICE) AS TOTVALUE
FROM   PRODUCT;

because the total value is the sum of the product of the quantity on hand and the price for all items. See Figure 7.23.

FIGURE 7.23	The total value of all items in the PRODUCT table

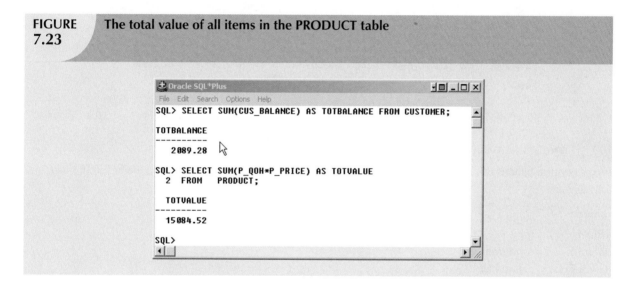

```
Oracle SQL*Plus                                        ▼□ _□×
File  Edit  Search  Options  Help
SQL> SELECT SUM(CUS_BALANCE) AS TOTBALANCE FROM CUSTOMER;

TOTBALANCE
----------
   2089.28

SQL> SELECT SUM(P_QOH*P_PRICE) AS TOTVALUE
  2  FROM    PRODUCT;

  TOTVALUE
----------
  15084.52

SQL>
```

AVG

The **AVG** function format is similar to that of MIN and MAX and is subject to the same operating restrictions. The first SQL command set shown in Figure 7.24 shows how a simple average P_PRICE value can be generated to yield the computed average price of 56.42125. The second SQL command set in Figure 7.24 produces five output lines that describe products whose prices exceed the average product price. Note that the second query uses nested SQL commands and the ORDER BY clause examined earlier.

FIGURE
7.24

AVG function output examples

```
Oracle SQL*Plus                                                    -|□| -|□| X|
File  Edit  Search  Options  Help
SQL> SELECT AVG(P_PRICE) FROM PRODUCT;

AVG(P_PRICE)
------------
    56.42125

SQL> SELECT P_CODE, P_DESCRIPT, P_QOH, P_PRICE, V_CODE
  2  FROM    PRODUCT
  3  WHERE   P_PRICE > (SELECT AVG(P_PRICE) FROM PRODUCT)
  4  ORDER   BY P_PRICE DESC;

P_CODE     P_DESCRIPT                                  P_QOH  P_PRICE   V_CODE
---------- ------------------------------------------- ------ --------- ----------
89-WRE-Q   Hicut chain saw, 16 in.                         11    256.99     24288
WR3/TT3    Steel matting, 4'x8'x1/6", .5" mesh             18    119.95     25595
11QER/31   Power painter, 15 psi., 3-nozzle                 8    109.99     25595
2232/QTY   B&D jigsaw, 12-in. blade                         8    109.92     24288
2232/QWE   B&D jigsaw, 8-in. blade                          6     99.87     24288

SQL> |
```

7.6.4 GROUPING DATA

Frequency distributions can be created quickly and easily using the **GROUP BY** clause within the SELECT statement. The syntax is:

SELECT	columnlist
FROM	tablelist
[WHERE	conditionlist]
[GROUP BY	columnlist]
[HAVING	conditionlist]
[ORDER BY	columnlist [ASC | DESC]] ;

The GROUP BY clause is generally used when you have attribute columns combined with aggregate functions in the SELECT statement. For example, to determine the minimum price for each sales code, use the first SQL command set shown in Figure 7.25.

The second SQL command set in Figure 7.25 generates the average price within each sales code. Note that the P_SALECODE nulls are included within the grouping.

The GROUP BY clause is valid only when used in conjunction with one of the SQL aggregate functions, such as COUNT, MIN, MAX, AVG, and SUM. For example, as shown in the first command set in Figure 7.26, if you try to group the output by using:

SELECT	V_CODE, P_CODE, P_DESCRIPT, P_PRICE
FROM	PRODUCT
GROUP BY	V_CODE;

you generate a "not a GROUP BY expression" error. However, if you write the preceding SQL command sequence in conjunction with some aggregate function, the GROUP BY clause works properly. The second SQL command sequence in Figure 7.26 properly answers the question, "How many products are supplied by each vendor?," because it uses a COUNT aggregate function.

FIGURE
7.25

GROUP BY clause output examples

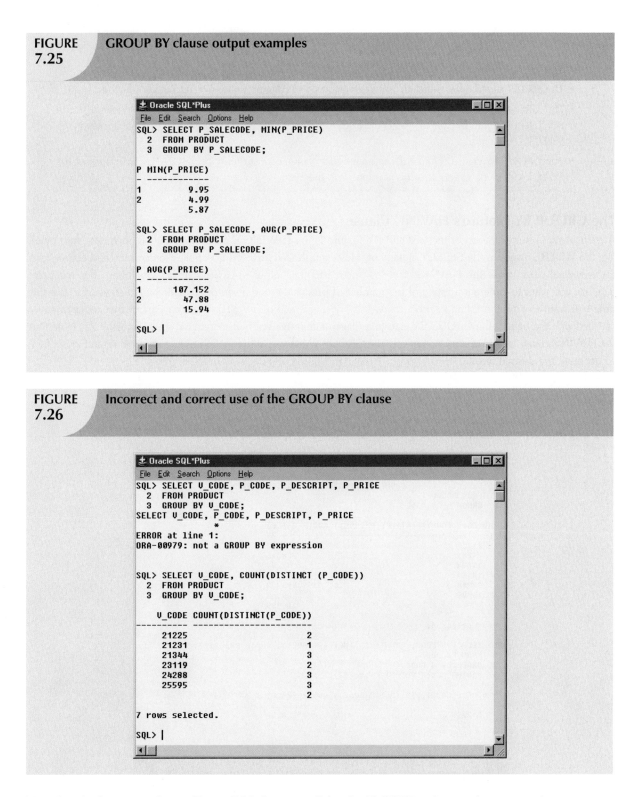

FIGURE
7.26

Incorrect and correct use of the GROUP BY clause

Note that the last output line in Figure 7.26 shows a null for the V_CODE, indicating that two products were not supplied by a vendor. Perhaps those products were produced in-house, or they might have been bought via a nonvendor channel, or the person making the data entry might have merely forgotten to enter a vendor code. (Remember that nulls can be the result of many things.)

> **NOTE**
>
> When using the GROUP BY clause with a SELECT statement:
>
> - The SELECT's *columnlist* must include a combination of column names and aggregate functions.
> - The GROUP BY clauses *columnlist* must include all nonaggregate function columns specified in the SELECTs *columnlist*. If required, you could also group by any aggregate function columns that appear in the SELECT's *columnlist*.
> - The GROUP BY clause *columnlist* can include any columns from the tables in the FROM clause of the SELECT statement, even if they do not appear in the SELECT's *columnlist*.

The GROUP BY Feature's HAVING Clause

A particularly useful extension of the GROUP BY feature is the **HAVING** clause. The HAVING operates very much like the WHERE clause in the SELECT statement. However, the WHERE clause applies to columns and expressions for individual rows, while the HAVING clause is applied to the output of a GROUP BY operation. For example, suppose you want to generate a listing of the number of products in the inventory supplied by each vendor. But this time you want to limit the listing to products whose prices average below $10. The first part of that requirement is satisfied with the help of the GROUP BY clause, as illustrated in the first SQL command set in Figure 7.27. Note that the HAVING clause is used in conjunction with the GROUP BY clause in the second SQL command set in Figure 7.27 to generate the desired result.

FIGURE 7.27   **An application of the HAVING clause**

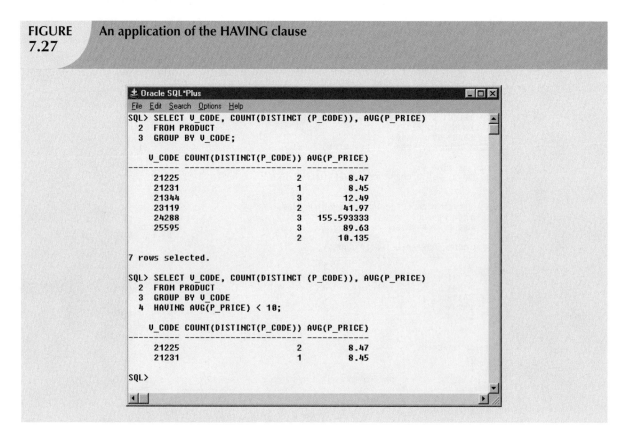

Using the WHERE clause instead of the HAVING clause— the second SQL command set in Figure 7.27 will produce an error message.

You can also combine multiple clauses and aggregate functions. For example, consider the following SQL statement:

```
SELECT      V_CODE, SUM(P_QOH * P_PRICE) AS TOTCOST
FROM        PRODUCT
GROUP BY    V_CODE
HAVING      (SUM(P_QOH * P_PRICE) > 500)
ORDER BY    SUM(P_QOH * P_PRICE) DESC;
```

This statement will do the following:

- Aggregate the total cost of products grouped by V_CODE.
- Select only the rows having totals that exceed $500.
- List the results in descending order by the total cost.

Note the syntax used in the HAVING and ORDER BY clauses; in both cases, you must specify the column expression (formula) used in the SELECT statement's column list, rather than the column alias (TOTCOST). Some RDBMSs allow you to substitute the column expression with the column alias, while others do not.

7.7 VIRTUAL TABLES: CREATING A VIEW

As you learned earlier, the output of a relational operator such as SELECT is another relation (or table). Suppose that at the end of every day, you would like to get a list of all products to reorder, that is, products with a quantity on hand that is less than or equal to the minimum quantity. Instead of typing the same query at the end of every day, wouldn't it be better to permanently save that query in the database? That's the function of a relational view. A **view** is a virtual table based on a SELECT query. The query can contain columns, computed columns, aliases, and aggregate functions from one or more tables. The tables on which the view is based are called **base tables**.

You can create a view by using the **CREATE VIEW** command:

CREATE VIEW *viewname* AS SELECT *query*

The CREATE VIEW statement is a data definition command that stores the subquery specification—the SELECT statement used to generate the virtual table—in the data dictionary.

The first SQL command set in Figure 7.28 shows the syntax used to create a view named PRICEGT50. This view contains only the designated three attributes (P_DESCRIPT, P_QOH, and P_PRICE) and only rows in which the price is over $50. The second SQL command sequence in Figure 7.28 shows the rows that make up the view.

NOTE

NOTE TO MS ACCESS USERS
The CREATE VIEW command is not directly supported in MS Access. To create a view in MS Access, you just need to create a SQL query and then save it.

> **FIGURE 7.28**   Creating a virtual table with the CREATE VIEW command

```
Oracle SQL*Plus
File  Edit  Search  Options  Help
SQL> CREATE VIEW PRICEGT50 AS
  2       SELECT P_DESCRIPT, P_QOH, P_PRICE
  3       FROM PRODUCT
  4       WHERE P_PRICE > 50.00;

View created.

SQL> SELECT * FROM PRICEGT50;

P_DESCRIPT                                  P_QOH  P_PRICE
------------------------------------------ ------ --------
Power painter, 15 psi., 3-nozzle               8   109.99
B&D jigsaw, 12-in. blade                       8   109.92
B&D jigsaw, 8-in. blade                        6    99.87
Hicut chain saw, 16 in.                       11   256.99
Steel matting, 4'x8'x1/6", .5" mesh           18   119.95

SQL>
```

A relational view has several special characteristics:

- You can use the name of a view anywhere a table name is expected in a SQL statement.

- Views are dynamically updated. That is, the view is re-created on demand each time it is invoked. Therefore, if new products are added (or deleted) to meet the criterion P_PRICE > 50.00, those new products will automatically appear (or disappear) in the PRICEGT50 view the next time the view is invoked.

- Views provide a level of security in the database because the view can restrict users to only specified columns and specified rows in a table. For example, if you have a company with hundreds of employees in several departments, you could give the secretary of each department a view of only certain attributes and for the employees that belong only to that secretary's department.

- Views may also be used as the basis for reports. For example, if you need a report that shows a summary of total product cost and quantity-on-hand statistics grouped by vendor, you could create a PROD_STATS view as:

```
CREATE VIEW PROD_STATS AS
SELECT      V_CODE, SUM(P_QOH*P_PRICE) AS TOTCOST,
            MAX(P_QOH) AS MAXQTY, MIN(P_QOH) AS MINQTY,
            AVG(P_QOH) AS AVGQTY
FROM        PRODUCT
GROUP BY    V_CODE;
```

In Chapter 8, you will learn more about views and, in particular, about updating data in base tables through views.

7.8 JOINING DATABASE TABLES

The ability to combine (join) tables on common attributes is perhaps the most important distinction between a relational database and other databases. A join is performed when data are retrieved from more than one table at a time. (If necessary, review the join definitions and examples in Chapter 3, The Relational Database Model.)

To join tables, you simply list the tables in the FROM clause of the SELECT statement. The DBMS will create the Cartesian product of every table in the FROM clause. (Review Chapter 3 to revisit these terms, if necessary.) However, to get the correct result—that is, a natural join—you must select only the rows in which the common attribute values match. To do this, use the WHERE clause to indicate the common attributes used to link the tables (this WHERE clause is sometimes referred to as the *join condition*).

The join condition is generally composed of an equality comparison between the foreign key and the primary key of related tables. For example, suppose you want to join the two tables VENDOR and PRODUCT. Because V_CODE is the foreign key in the PRODUCT table and the primary key in the VENDOR table, the link is established on V_CODE. (See Table 7.9.)

TABLE 7.9   Creating Links Through Foreign Keys

TABLE	ATTRIBUTES TO BE SHOWN	LINKING ATTRIBUTE
PRODUCT	P_DESCRIPT, P_PRICE	V_CODE
VENDOR	V_COMPANY, V_PHONE	V_CODE

When the same attribute name appears in more than one of the joined tables, the source table of the attributes listed in the SELECT command sequence must be defined. To join the PRODUCT and VENDOR tables, you would use the following, which produces the output shown in Figure 7.29:

```
SELECT      P_DESCRIPT, P_PRICE, V_NAME, V_CONTACT, V_AREACODE, V_PHONE
FROM        PRODUCT, VENDOR
WHERE       PRODUCT.V_CODE = VENDOR.V_CODE;
```

FIGURE 7.29   The results of a join

P_DESCRIPT	P_PRICE	V_NAME	V_CONTACT	V_AREACODE	V_PHONE
Claw hammer	9.95	Bryson, Inc.	Smithson	615	223-3234
1.25-in. metal screw, 25	6.99	Bryson, Inc.	Smithson	615	223-3234
2.5-in. wd. screw, 50	8.45	D&E Supply	Singh	615	228-3245
7.25-in. pwr. saw blade	14.99	Gomez Bros.	Ortega	615	889-2546
9.00-in. pwr. saw blade	17.49	Gomez Bros.	Ortega	615	889-2546
Rat-tail file, 1/8-in. fine	4.99	Gomez Bros.	Ortega	615	889-2546
Hrd. cloth, 1/4-in., 2x50	39.95	Randsets Ltd.	Anderson	901	678-3998
Hrd. cloth, 1/2-in., 3x50	43.99	Randsets Ltd.	Anderson	901	678-3998
B&D jigsaw, 12-in. blade	109.92	ORDVA, Inc.	Hakford	615	898-1234
B&D jigsaw, 8-in. blade	99.87	ORDVA, Inc.	Hakford	615	898-1234
Hicut chain saw, 16 in.	256.99	ORDVA, Inc.	Hakford	615	898-1234
Power painter, 15 psi., 3-nozzle	109.99	Rubicon Systems	Orton	904	456-0092
B&D cordless drill, 1/2-in.	38.95	Rubicon Systems	Orton	904	456-0092
Steel matting, 4'x8'x1/6", .5" mesh	119.95	Rubicon Systems	Orton	904	456-0092

Your output might be presented in a different order because the SQL command produces a listing in which the order of the columns is not relevant. In fact, you are likely to get a different order of the same listing the next time you execute the command. However, you can generate a more predictable list by using an ORDER BY clause:

```
SELECT      P_DESCRIPT, P_PRICE, V_NAME, V_CONTACT, V_AREACODE, V_PHONE
FROM        PRODUCT, VENDOR
WHERE       PRODUCT.V_CODE = VENDOR.V_CODE
ORDER BY    P_PRICE;
```

In that case, your listing will always be arranged from the lowest price to the highest price.

NOTE

Table names were used as prefixes in the preceding SQL command sequence. For example, PRODUCT.P_ PRICE was used rather than P_PRICE. Most current-generation RDBMSs do not require table names to be used as prefixes unless the same attribute name occurs in several of the tables being joined. In that case, V_CODE is used as a foreign key in PRODUCT and as a primary key in VENDOR; therefore, you must use the table names as prefixes in the WHERE clause. In other words, you can write the previous query as:

```
SELECT      P_DESCRIPT, P_PRICE, V_NAME, V_CONTACT, V_AREACODE, V_PHONE
FROM        PRODUCT, VENDOR WHERE PRODUCT.V_CODE = VENDOR.V_CODE;
```

Naturally, if an attribute name occurs in several places, its origin (table) must be specified. If you fail to provide such a specification, SQL will generate an error message to indicate that you have been ambiguous about the attributes origin.

The preceding SQL command sequence joins a row in the PRODUCT table with a row in the VENDOR table in which the V_CODE values of these rows are the same, as indicated in the WHERE clause's condition. Because any vendor can deliver any number of ordered products, the PRODUCT table might contain multiple V_CODE entries for each V_CODE entry in the VENDOR table. In other words, each V_CODE in VENDOR can be matched with many V_CODE rows in PRODUCT.

If you do not specify the WHERE clause, the result will be the Cartesian product of PRODUCT and VENDOR. Because the PRODUCT table contains 16 rows and the VENDOR table contains 11 rows, the Cartesian product would produce a listing of (16 × 11) = 176 rows. (Each row in PRODUCT would be joined to each row in the VENDOR table.)

All of the SQL commands can be used on the joined tables. For example, the following command sequence is quite acceptable in SQL and produces the output shown in Figure 7.30:

```
SELECT      P_DESCRIPT, P_PRICE, V_NAME, V_CONTACT, V_AREACODE, V_PHONE
FROM        PRODUCT, VENDOR
WHERE       PRODUCT.V_CODE = VENDOR.V_CODE
AND         P_INDATE > '15-Jan-2008';
```

FIGURE 7.30 An ordered and limited listing after a join

P_DESCRIPT	P_PRICE	V_NAME	V_CONTACT	V_AREACODE	V_PHONE
1.25-in. metal screw, 25	6.99	Bryson, Inc.	Smithson	615	223-3234
2.5-in. wd. screw, 50	8.45	D&E Supply	Singh	615	228-3245
Claw hammer	9.95	Bryson, Inc.	Smithson	615	223-3234
B&D cordless drill, 1/2-in.	38.95	Rubicon Systems	Orton	904	456-0092
Steel matting, 4'x8'x1/6", .5" mesh	119.95	Rubicon Systems	Orton	904	456-0092
Hicut chain saw, 16 in.	256.99	ORDVA, Inc.	Hakford	615	898-1234

When joining three or more tables, you need to specify a join condition for each pair of tables. The number of join conditions will always be N-1, where N represents the number of tables listed in the FROM clause. For example, if you have three tables, you must have two join conditions; if you have five tables, you must have four join conditions; and so on.

Remember, the join condition will match the foreign key of a table to the primary key of the related table. For example, using Figure 7.1, if you want to list the customer last name, invoice number, invoice date, and product descriptions for all invoices for customer 10014, you must type the following:

```
SELECT      CUS_LNAME, INV_NUMBER, INV_DATE, P_DESCRIPT
FROM        CUSTOMER, INVOICE, LINE, PRODUCT
WHERE       CUSTOMER.CUS_CODE = INVOICE.CUS_CODE
AND         INVOICE.INV_NUMBER = LINE.INV_NUMBER
AND         LINE.P_CODE = PRODUCT.P_CODE
AND         CUSTOMER.CUS_CODE = 10014
ORDER BY    INV_NUMBER;
```

Finally, be careful not to create circular join conditions. For example, if Table A is related to Table B, Table B is related to Table C, and Table C is also related to Table A, create only two join conditions: join A with B and B with C. Do not join C with A!

7.8.1 JOINING TABLES WITH AN ALIAS

An alias may be used to identify the source table from which the data are taken. The aliases P and V are used to label the PRODUCT and VENDOR tables in the next command sequence. Any legal table name may be used as an alias. (Also notice that there are no table name prefixes because the attribute listing contains no duplicate names in the SELECT statement.)

```
SELECT      P_DESCRIPT, P_PRICE, V_NAME, V_CONTACT, V_AREACODE, V_PHONE
FROM        PRODUCT P, VENDOR V
WHERE       P.V_CODE = V.V_CODE
ORDER BY    P_PRICE;
```

7.8.2 RECURSIVE JOINS

An alias is especially useful when a table must be joined to itself in a **recursive query**. For example, suppose you are working with the EMP table shown in Figure 7.31.

FIGURE 7.31   The contents of the EMP table

EMP_NUM	EMP_TITLE	EMP_LNAME	EMP_FNAME	EMP_INITIAL	EMP_DOB	EMP_HIRE_DATE	EMP_AREACODE	EMP_PHONE	EMP_MGR
100	Mr.	Kolmycz	George	D	15-Jun-42	15-Mar-85	615	324-5456	
101	Ms.	Lewis	Rhonda	G	19-Mar-65	25-Apr-86	615	324-4472	100
102	Mr.	Vandam	Rhett		14-Nov-58	20-Dec-90	901	675-8993	100
103	Ms.	Jones	Anne	M	16-Oct-74	28-Aug-94	615	898-3456	100
104	Mr.	Lange	John	P	08-Nov-71	20-Oct-94	901	504-4430	105
105	Mr.	Williams	Robert	D	14-Mar-75	08-Nov-98	615	890-3220	
106	Mrs.	Smith	Jeanine	K	12-Feb-68	05-Jan-89	615	324-7883	105
107	Mr.	Diante	Jorge	D	21-Aug-74	02-Jul-94	615	890-4567	105
108	Mr.	Wiesenbach	Paul	R	14-Feb-66	18-Nov-92	615	897-4358	
109	Mr.	Smith	George	K	18-Jun-61	14-Apr-89	901	504-3339	108
110	Mrs.	Genkazi	Leighla	W	19-May-70	01-Dec-90	901	569-0093	108
111	Mr.	Washington	Rupert	E	03-Jan-66	21-Jun-93	615	890-4925	105
112	Mr.	Johnson	Edward	E	14-May-61	01-Dec-83	615	898-4387	100
113	Ms.	Smythe	Melanie	P	15-Sep-70	11-May-99	615	324-9006	105
114	Ms.	Brandon	Marie	G	02-Nov-56	15-Nov-79	901	882-0845	108
115	Mrs.	Saranda	Hermine	R	25-Jul-72	23-Apr-93	615	324-5505	105
116	Mr.	Smith	George	A	08-Nov-65	10-Dec-88	615	890-2984	108

Using the data in the EMP table, you can generate a list of all employees with their managers' names by joining the EMP table to itself. In that case, you would also use aliases to differentiate the table from itself. The SQL command sequence would look like this:

```
SELECT      E.EMP_MGR, M.EMP_LNAME, E.EMP_NUM, E.EMP_LNAME
FROM        EMP E, EMP M
WHERE       E.EMP_MGR=M.EMP_NUM
ORDER BY    E.EMP_MGR;
```

The output of the above command sequence is shown in Figure 7.32.

FIGURE 7.32   Using an alias to join a table to itself

EMP_NUM	A.EMP_LNAME	EMP_MGR	B.EMP_LNAME
112	Johnson	100	Kolmycz
103	Jones	100	Kolmycz
102	Vandam	100	Kolmycz
101	Lewis	100	Kolmycz
115	Saranda	105	Williams
113	Smythe	105	Williams
111	Washington	105	Williams
107	Diante	105	Williams
106	Smith	105	Williams
104	Lange	105	Williams
116	Smith	108	Wiesenbach
114	Brandon	108	Wiesenbach
110	Genkazi	108	Wiesenbach
109	Smith	108	Wiesenbach

7.8.3 OUTER JOINS

Figure 7.29 showed the results of joining the PRODUCT and VENDOR tables. If you examine the output, note that 14 product rows are listed. Compare the output to the PRODUCT table in Figure 7.2, and note that two products are missing. Why? The reason is that there are two products with nulls in the V_CODE attribute. Because there is no matching null "value" in the VENDOR table's V_CODE attribute, the products do not show up in the final output based on the join. Also, note that in the VENDOR table in Figure 7.2, several vendors have no matching V_CODE in the PRODUCT table. To include those rows in the final join output, you must use an outer join.

NOTE

In MS Access, add AS to the previous SQL command sequence, making it read:

```
SELECT      E.EMP_MGR,M.EMP_LNAME,E.EMP_NUM,E.EMP_LNAME
FROM        EMP AS E, EMP AS M
WHERE       E.EMP_MGR = M.EMP_NUM
ORDER BY E.EMP_MGR;
```

There are two types of outer joins: left and right. (See Chapter 3.) Given the contents of the PRODUCT and VENDOR tables, the following left outer join will show all VENDOR rows and all matching PRODUCT rows:

SELECT   P_CODE, VENDOR.V_CODE, V_NAME
FROM   VENDOR LEFT JOIN PRODUCT
               ON VENDOR.V_CODE = PRODUCT.V_CODE;

Figure 7.33 shows the output generated by the left outer join command in MS Access. Oracle yields the same result, but shows the output in a different order.

The right outer join will join both tables and show all product rows with all matching vendor rows. The SQL command for the right outer join is:

SELECT   PRODUCT.P_CODE, VENDOR.V_CODE, V_NAME
FROM   VENDOR RIGHT JOIN PRODUCT
               ON VENDOR.V_CODE = PRODUCT.V_CODE;

Figure 7.34 shows the output generated by the right outer join command sequence in MS Access. Again, Oracle yields the same result, but shows the output in a different order.

In Chapter 8, you will learn more about joins and how to use the latest ANSI SQL standard syntax.

FIGURE 7.33   The left outer join results

P_CODE	V_CODE	V_NAME
23109-HB	21225	Bryson, Inc.
SM-18277	21225	Bryson, Inc.
	21226	SuperLoo, Inc.
SW-23116	21231	D&E Supply
13-Q2/P2	21344	Gomez Bros.
14-Q1/L3	21344	Gomez Bros.
54778-2T	21344	Gomez Bros.
	22567	Dome Supply
1546-QQ2	23119	Randsets Ltd.
1558-QW1	23119	Randsets Ltd.
	24004	Brackman Bros.
2232/QTY	24288	ORDVA, Inc.
2232/QWE	24288	ORDVA, Inc.
89-WRE-Q	24288	ORDVA, Inc.
	25443	B&K, Inc.
	25501	Damal Supplies
11QER/31	25595	Rubicon Systems
2238/QPD	25595	Rubicon Systems
WR3/TT3	25595	Rubicon Systems

FIGURE 7.34   The right outer join results

P_CODE	V_CODE	V_NAME
23114-AA		
PVC23DRT		
23109-HB	21225	Bryson, Inc.
SM-18277	21225	Bryson, Inc.
SW-23116	21231	D&E Supply
13-Q2/P2	21344	Gomez Bros.
14-Q1/L3	21344	Gomez Bros.
54778-2T	21344	Gomez Bros.
1546-QQ2	23119	Randsets Ltd.
1558-QW1	23119	Randsets Ltd.
2232/QTY	24288	ORDVA, Inc.
2232/QWE	24288	ORDVA, Inc.
89-WRE-Q	24288	ORDVA, Inc.
11QER/31	25595	Rubicon Systems
2238/QPD	25595	Rubicon Systems
WR3/TT3	25595	Rubicon Systems

ONLINE CONTENT

For a complete walk-through example of converting an ER model into a database structure and using SQL commands to create tables, see **Appendix D, Converting an ER Model into a Database Structure**, in the Student Online Companion.

▼ The SQL commands can be divided into two overall categories: data definition language (DDL) commands and data manipulation language (DML) commands.

▼ The ANSI standard data types are supported by all RDBMS vendors in different ways. The basic data types are NUMBER, INTEGER, CHAR, VARCHAR, and DATE.

▼ The basic data definition commands allow you to create tables, indexes, and views. Many SQL constraints can be used with columns. The commands are CREATE TABLE, CREATE INDEX, CREATE VIEW, ALTER TABLE, DROP TABLE, DROP VIEW, and DROP INDEX.

▼ DML commands allow you to add, modify, and delete rows from tables. The basic DML commands are SELECT, INSERT, UPDATE, DELETE, COMMIT, and ROLLBACK.

▼ The INSERT command is used to add new rows to tables. The UPDATE command is used to modify data values in existing rows of a table. The DELETE command is used to delete rows from tables. The COMMIT and ROLLBACK commands are used to permanently save or roll back changes made to the rows. Once you COMMIT the changes, you cannot undo them with a ROLLBACK command.

▼ The SELECT statement is the main data retrieval command in SQL. A SELECT statement has the following syntax:

SELECT	columnlist
FROM	tablelist
[WHERE	conditionlist]
[GROUP BY	columnlist]
[HAVING	conditionlist]
[ORDER BY	columnlist [ASC | DESC]] ;

▼ The column list represents one or more column names separated by commas. The column list may also include computed columns, aliases, and aggregate functions. A computed column is represented by an expression or formula (for example, P_PRICE * P_QOH). The FROM clause contains a list of table names or view names.

▼ The WHERE clause can be used with the SELECT, UPDATE, and DELETE statements to restrict the rows affected by the DDL command. The condition list represents one or more conditional expressions separated by logical operators (AND, OR, and NOT). The conditional expression can contain any comparison operators (=, >, <, >=, <=, and <>) as well as special operators (BETWEEN, IS NULL, LIKE, IN, and EXISTS).

▼ Aggregate functions (COUNT, MIN, MAX, and AVG) are special functions that perform arithmetic computations over a set of rows. The aggregate functions are usually used in conjunction with the GROUP BY clause to group the output of aggregate computations by one or more attributes. The HAVING clause is used to restrict the output of the GROUP BY clause by selecting only the aggregate rows that match a given condition.

▼ The ORDER BY clause is used to sort the output of a SELECT statement. The ORDER BY clause can sort by one or more columns and can use either ascending or descending order.

▼ You can join the output of multiple tables with the SELECT statement. The join operation is performed every time you specify two or more tables in the FROM clause and use a join condition in the WHERE clause to match the foreign key of one table to the primary key of the related table. If you do not specify a join condition, the DBMS will automatically perform a Cartesian product of the tables you specify in the FROM clause.

▼ The natural join uses the join condition to match only rows with equal values in the specified columns. You could also do a right outer join and left outer join to select the rows that have no matching values in the other related table.

ONLINE CONTENT

Answers to selected Review Questions and Problems for this chapter are contained in the Student Online Companion for this book.

ONLINE CONTENT

The Review Questions in this chapter are based on the **Ch07_Review** database located in the Student Online Companion. This database is stored in Microsoft Access format. If you use another DBMS such as Oracle, SQL Server, MySQL, or DB2, use its import utilities to move the Access database contents. The Student Online Companion provides Oracle and SQL script files.

REVIEW QUESTIONS

The **Ch07_Review** database stores data for a consulting company that tracks all charges to projects. The charges are based on the hours each employee works on each project. The structure and contents of the **Ch07_Review** database are shown in Figure Q7.1.

Note that the ASSIGNMENT table in Figure Q7.1 stores the JOB_CHG_HOUR values as an attribute (ASSIGN_CHG_HR) to maintain historical accuracy of the data. The JOB_CHG_HOUR values are likely to change over time. In fact, a JOB_CHG_HOUR change will be reflected in the ASSIGNMENT table. And, naturally, the employee primary job assignment might change, so the ASSIGN_JOB is also stored. Because those attributes are required to maintain the historical accuracy of the data, they are *not* redundant.

FIGURE Q7.1   Structure and contents of the Ch07_Review database

Database name: Ch07_Review

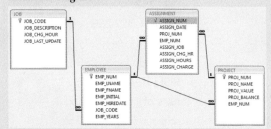

Relational diagram

Table name: EMPLOYEE

EMP_NUM	EMP_LNAME	EMP_FNAME	EMP_INITIAL	EMP_HIREDATE	JOB_CODE	EMP_YEARS
101	News	John	G	08-Nov-00	502	4
102	Senior	David	H	12-Jul-89	501	15
103	Arbough	June	E	01-Dec-96	503	8
104	Ramoras	Anne	K	15-Nov-87	501	17
105	Johnson	Alice	K	01-Feb-93	502	12
106	Smithfield	William		22-Jun-04	500	0
107	Alonzo	Maria	D	10-Oct-93	500	11
108	Washington	Ralph	B	22-Aug-91	501	13
109	Smith	Larry	W	18-Jul-97	501	7
110	Olenko	Gerald	A	11-Dec-95	505	9
111	Wabash	Geoff	B	04-Apr-91	506	14
112	Smithson	Darlene	M	23-Oct-94	507	10
113	Joenbrood	Delbert	K	15-Nov-96	508	8
114	Jones	Annelise		20-Aug-93	508	11
115	Bawangi	Travis	B	25-Jan-92	501	13
116	Pratt	Gerald	L	05-Mar-97	510	8
117	Williamson	Angie	H	19-Jun-96	509	8
118	Frommer	James	J	04-Jan-05	510	0

Table name: JOB

JOB_CODE	JOB_DESCRIPTION	JOB_CHG_HOUR	JOB_LAST_UPDATE
500	Programmer	35.75	20-Nov-07
501	Systems Analyst	96.75	20-Nov-07
502	Database Designer	125.00	24-Mar-08
503	Electrical Engineer	84.50	20-Nov-07
504	Mechanical Engineer	67.90	20-Nov-07
505	Civil Engineer	55.78	20-Nov-07
506	Clerical Support	26.87	20-Nov-07
507	DSS Analyst	45.95	20-Nov-07
508	Applications Designer	48.10	24-Mar-08
509	Bio Technician	34.55	20-Nov-07
510	General Support	18.36	20-Nov-07

Table name: ASSIGNMENT

ASSIGN_NUM	ASSIGN_DATE	PROJ_NUM	EMP_NUM	ASSIGN_JOB	ASSIGN_CHG_HR	ASSIGN_HOURS	ASSIGN_CHARGE
1001	22-Mar-08	18	103	503	84.50	3.5	295.75
1002	22-Mar-08	22	117	509	34.55	4.2	145.11
1003	22-Mar-08	18	117	509	34.55	2.0	69.10
1004	22-Mar-08	18	103	503	84.50	5.9	498.55
1005	22-Mar-08	25	108	501	96.75	2.2	212.85
1006	22-Mar-08	22	104	501	96.75	4.2	406.35
1007	22-Mar-08	25	113	508	50.75	3.8	192.85
1008	22-Mar-08	18	103	503	84.50	0.9	76.05
1009	23-Mar-08	15	115	501	96.75	5.6	541.80
1010	23-Mar-08	15	117	509	34.55	2.4	82.92
1011	23-Mar-08	25	105	502	105.00	4.3	451.50
1012	23-Mar-08	18	108	501	96.75	3.4	328.95
1013	23-Mar-08	25	115	501	96.75	2.0	193.50
1014	23-Mar-08	22	104	501	96.75	2.8	270.90
1015	23-Mar-08	15	103	503	84.50	6.1	515.45
1016	23-Mar-08	22	105	502	105.00	4.7	493.50
1017	23-Mar-08	18	117	509	34.55	3.8	131.29
1018	23-Mar-08	25	117	509	34.55	2.2	76.01
1019	24-Mar-08	25	104	501	110.50	4.9	541.45
1020	24-Mar-08	15	101	502	125.00	3.1	387.50
1021	24-Mar-08	22	108	501	110.50	2.7	298.35
1022	24-Mar-08	22	115	501	110.50	4.9	541.45
1023	24-Mar-08	22	105	502	125.00	3.5	437.50
1024	24-Mar-08	15	103	503	84.50	3.3	278.85
1025	24-Mar-08	18	117	509	34.55	4.2	145.11

Table name: PROJECT

PROJ_NUM	PROJ_NAME	PROJ_VALUE	PROJ_BALANCE	EMP_NUM
15	Evergreen	1453500.00	1002350.00	103
18	Amber Wave	3500500.00	2110346.00	108
22	Rolling Tide	805000.00	500345.20	102
25	Starflight	2650500.00	2309880.00	107

Given the structure and contents of the **Ch07_Review** database shown in Figure Q7.1, use SQL commands to answer Questions 1–25.

1. Write the SQL code that will create the table structure for a table named EMP_1. This table is a subset of the EMPLOYEE table. The basic EMP_1 table structure is summarized in the table below. (Note that the JOB_CODE is the FK to JOB.)

ATTRIBUTE (FIELD) NAME	DATA DECLARATION
EMP_NUM	CHAR(3)
EMP_LNAME	VARCHAR(15)
EMP_FNAME	VARCHAR(15)
EMP_INITIAL	CHAR(1)
EMP_HIREDATE	DATE
JOB_CODE	CHAR(3)

2. Having created the table structure in Question 1, write the SQL code to enter the first two rows for the table shown in Figure Q7.2.

3. Assuming the data shown in the EMP_1 table have been entered, write the SQL code that will list all attributes for a job code of 502.

4. Write the SQL code that will save the changes made to the EMP_1 table.

5. Write the SQL code to change the job code to 501 for the person whose employee number (EMP_NUM) is 107. After you have completed the task, examine the results, and then reset the job code to its original value.

FIGURE
Q7.2

The contents of the EMP_1 table

EMP_NUM	EMP_LNAME	EMP_FNAME	EMP_INITIAL	EMP_HIREDATE	JOB_CODE
101	News	John	G	08-Nov-00	502
102	Senior	David	H	12-Jul-89	501
103	Arbough	June	E	01-Dec-96	500
104	Ramoras	Anne	K	15-Nov-87	501
105	Johnson	Alice	K	01-Feb-93	502
106	Smithfield	William		22-Jun-04	500
107	Alonzo	Maria	D	10-Oct-93	500
108	Washington	Ralph	B	22-Aug-91	501
109	Smith	Larry	W	18-Jul-97	501

6. Write the SQL code to delete the row for the person named William Smithfield, who was hired on June 22, 2004, and whose job code classification is 500. (*Hint*: Use logical operators to include all of the information given in this problem.)

7. Write the SQL code that will restore the data to its original status; that is, the table should contain the data that existed before you made the changes in Questions 5 and 6.

8. Write the SQL code to create a copy of EMP_1, naming the copy EMP_2. Then write the SQL code that will add the attributes EMP_PCT and PROJ_NUM to its structure. The EMP_PCT is the bonus percentage to be paid to each employee. The new attribute characteristics are:

EMP_PCTNUMBER(4,2)

PROJ_NUMCHAR(3)

(*Note*: If your SQL implementation allows it, you may use DECIMAL(4,2) rather than NUMBER(4,2).)

9. Write the SQL code to change the EMP_PCT value to 3.85 for the person whose employee number (EMP_NUM) is 103. Next, write the SQL command sequences to change the EMP_PCT values as shown in Figure Q7.9.

FIGURE
Q7.9

The contents of the EMP_2 table

EMP_NUM	EMP_LNAME	EMP_FNAME	EMP_INITIAL	EMP_HIREDATE	JOB_CODE	EMP_PCT	PROJ_NUM
101	News	John	G	08-Nov-00	502	5.00	
102	Senior	David	H	12-Jul-89	501	8.00	
103	Arbough	June	E	01-Dec-96	500	3.85	
104	Ramoras	Anne	K	15-Nov-87	501	10.00	
105	Johnson	Alice	K	01-Feb-93	502	5.00	
106	Smithfield	William		22-Jun-04	500	6.20	
107	Alonzo	Maria	D	10-Oct-93	500	5.15	
108	Washington	Ralph	B	22-Aug-91	501	10.00	
109	Smith	Larry	W	18-Jul-97	501	2.00	

10. Using a single command sequence, write the SQL code that will change the project number (PROJ_NUM) to 18 for all employees whose job classification (JOB_CODE) is 500.

11. Using a single command sequence, write the SQL code that will change the project number (PROJ_NUM) to 25 for all employees whose job classification (JOB_CODE) is 502 or higher. When you finish Questions 10 and 11, the EMP_2 table will contain the data shown in Figure Q7.11.

(You may assume that the table has been saved again at this point.)

FIGURE
Q7.11

FIGURE
Q7.11

The EMP_2 table contents after the modifications

EMP_NUM	EMP_LNAME	EMP_FNAME	EMP_INITIAL	EMP_HIREDATE	JOB_CODE	EMP_PCT	PROJ_NUM
101	News	John	G	08-Nov-00	502	5.00	
102	Senior	David	H	12-Jul-89	501	8.00	
103	Arbough	June	E	01-Dec-96	500	3.85	
104	Ramoras	Anne	K	15-Nov-87	501	10.00	
105	Johnson	Alice	K	01-Feb-93	502	5.00	
106	Smithfield	William		22-Jun-04	500	6.20	
107	Alonzo	Maria	D	10-Oct-93	500	5.15	
108	Washington	Ralph	B	22-Aug-91	501	10.00	
109	Smith	Larry	W	18-Jul-97	501	2.00	

12. Write the SQL code that will change the PROJ_NUM to 14 for those employees who were hired before January 1, 1994 and whose job code is at least 501. (You may assume that the table will be restored to its condition preceding this question.)

13. Write the two SQL command sequences required to:

   a. Create a temporary table named TEMP_1 whose structure is composed of the EMP_2 attributes EMP_NUM and EMP_PCT.

   b. Copy the matching EMP_2 values into the TEMP_1 table.

14. Write the SQL command that will delete the newly created TEMP_1 table from the database.

15. Write the SQL code required to list all employees whose last names start with *Smith*. In other words, the rows for both Smith and Smithfield should be included in the listing. Assume case sensitivity.

16. Using the EMPLOYEE, JOB, and PROJECT tables in the **Ch07_Review** database (see Figure Q7.1), write the SQL code that will produce the results shown in Figure Q7.16.

FIGURE
Q7.16

The query results for Question 16

PROJ_NAME	PROJ_VALUE	PROJ_BALANCE	EMP_LNAME	EMP_FNAME	EMP_INITIAL	JOB_CODE	JOB_DESCRIPTION	JOB_CHG_HOUR
Rolling Tide	805000.00	500345.20	Senior	David	H	501	Systems Analyst	96.75
Evergreen	1453500.00	1002350.00	Arbough	June	E	500	Programmer	35.75
Starflight	2650500.00	2309880.00	Alonzo	Maria	D	500	Programmer	35.75
Amber Wave	3500500.00	2110346.00	Washington	Ralph	B	501	Systems Analyst	96.75

17. Write the SQL code that will produce a virtual table named REP_1. The virtual table should contain the same information that was shown in Question 16.

18. Write the SQL code to find the average bonus percentage in the EMP_2 table you created in Question 8.

19. Write the SQL code that will produce a listing for the data in the EMP_2 table in ascending order by the bonus percentage.

20. Write the SQL code that will list only the distinct project numbers found in the EMP_2 table.

21. Write the SQL code to calculate the ASSIGN_CHARGE values in the ASSIGNMENT table in the **Ch07_Review** database. (See Figure Q7.1.) Note that ASSIGN_CHARGE is a derived attribute that is calculated by multiplying ASSIGN_CHG_HR by ASSIGN_HOURS.

22. Using the data in the ASSIGNMENT table, write the SQL code that will yield the total number of hours worked for each employee and the total charges stemming from those hours worked. The results of running that query are shown in Figure Q7.22.

FIGURE Q7.22	Total hours and charges by employee

EMP_NUM	EMP_LNAME	SumOfASSIGN_HOURS	SumOfASSIGN_CHARGE
101	News	3.1	387.50
103	Arbough	19.7	1664.65
104	Ramoras	11.9	1218.70
105	Johnson	12.5	1382.50
108	Washington	8.3	840.15
113	Joenbrood	3.8	192.85
115	Bawangi	12.5	1276.75
117	Williamson	18.8	649.54

23. Write a query to produce the total number of hours and charges for each of the projects represented in the ASSIGNMENT table. The output is shown in Figure Q7.23.

FIGURE Q7.23	Total hour and charges by project

PROJ_NUM	SumOfASSIGN_HOURS	SumOfASSIGN_CHARGE
15	20.5	1806.52
18	23.7	1544.80
22	27.0	2593.16
25	19.4	1668.16

24. Write the SQL code to generate the total hours worked and the total charges made by all employees. The results are shown in Figure Q7.24. (*Hint:* This is a nested query. If you use Microsoft Access, you can generate the result by using the query output shown in Figure Q7.22 as the basis for the query that will produce the output shown in Figure Q7.24.)

25. Write the SQL code to generate the total hours worked and the total charges made to all projects. The results should be the same as those shown in Figure Q7.24. (*Hint:* This is a nested query. If you use Microsoft Access, you can generate the result by using the query output shown in Figure Q7.23 as the basis for this query.)

FIGURE Q7.24	Total hours and charges, all employees

SumOfSumOfASSIGN_HOURS	SumOfSumOfASSIGN_CHARGE
90.6	7612.64

PROBLEMS

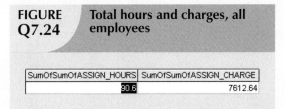

ONLINE CONTENT

Problems 1-15 are based on the **Ch07_AviaCo** database located in the Student Online Companion. This database is stored in Microsoft Access format. If you use another DBMS such as Oracle, SQL Server, MySQL, or DB2, use its import utilities to move the Access database contents. The Student Online Companion provides Oracle and SQL script files.

Before you attempt to write any SQL queries, familiarize yourself with the **Ch07_AviaCo** database structure and contents shown in Figure P7.1. Although the relational schema does not show optionalities, keep in mind that all pilots are employees, but not all employees are flight crew members. (Although in this database, the crew member assignments all involve pilots and copilots, the design is sufficiently flexible to accommodate crew member assignments—such as loadmasters and flight attendants—of people who are not pilots. That's why the relationship between CHARTER and EMPLOYEE is implemented through CREW.) Note also that this design implementation does not include multivalued attributes. For example, multiple ratings such as Instrument and Certified Flight Instructor ratings are stored in the (composite) EARNEDRATINGS table. Nor does the CHARTER table include multiple crew assignments, which are properly stored in the CREW table.

FIGURE P7.1   The Ch07_AviaCo database

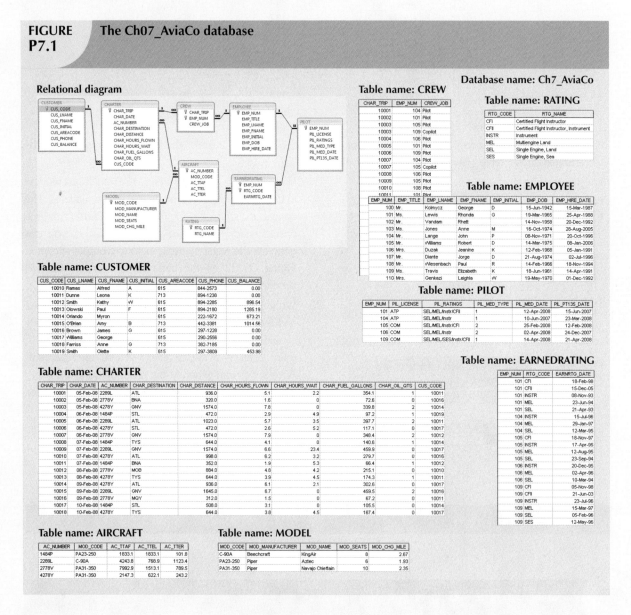

1. Write the SQL code that will list the values for the first four attributes in the CHARTER table.

2. Using the contents of the CHARTER table, write the SQL query that will produce the output shown in Figure P7.2. Note that the output is limited to selected attributes for aircraft number 2778V.

FIGURE P7.2   **Problem 2 query results**

CHAR_DATE	AC_NUMBER	CHAR_DESTINATION	CHAR_DISTANCE	CHAR_HOURS_FLOWN
05-Feb-08	2778V	BNA	320	1.6
06-Feb-08	2778V	GNV	1574	7.9
08-Feb-08	2778V	MOB	884	4.8
09-Feb-08	2778V	MQY	312	1.5

3. Create a virtual table (named AC2778V) containing the output presented in Problem 2.

4. Produce the output shown in Figure P7.4 for aircraft 2778V. Note that this output includes data from the CHARTER and CUSTOMER tables. (*Hint:* Use a JOIN in this query.)

FIGURE P7.4   **Problem 4 query results**

CHAR_DATE	AC_NUMBER	CHAR_DESTINATION	CUS_LNAME	CUS_AREACODE	CUS_PHONE
08-Feb-08	2778V	MOB	Ramas	615	844-2573
09-Feb-08	2778V	MQY	Dunne	713	894-1238
06-Feb-08	2778V	GNV	Smith	615	894-2285
05-Feb-08	2778V	BNA	Brown	615	297-1228

5. Produce the output shown in Figure P7.5. The output, derived from the CHARTER and MODEL tables, is limited to February 6, 2008. (*Hint:* The join passes through another table. Note that the "connection" between CHARTER and MODEL requires the existence of AIRCRAFT because the CHARTER table does not contain a foreign key to MODEL. However, CHARTER does contain AC_NUMBER, a foreign key to AIRCRAFT, which contains a foreign key to MODEL.)

FIGURE P7.5   **Problem 5 query results**

CHAR_DATE	CHAR_DESTINATION	AC_NUMBER	MOD_NAME	MOD_CHG_MILE
06-Feb-08	STL	1484P	Aztec	1.93
06-Feb-08	ATL	2289L	KingAir	2.67
06-Feb-08	STL	4278Y	Navajo Chieftain	2.35
06-Feb-08	GNV	2778V	Navajo Chieftain	2.35

6. Modify the query in Problem 5 to include data from the CUSTOMER table. This time the output is limited to charter records generated since February 9, 2008. (The query results are shown in Figure P7.6.)

FIGURE P7.6   Problem 6 query results

CHAR_DATE	CHAR_DESTINATION	AC_NUMBER	MOD_NAME	MOD_CHG_MILE	CUS_LNAME
09-Feb-08	ATL	4278Y	Navajo Chieftain	2.35	Williams
09-Feb-08	MQY	2778V	Navajo Chieftain	2.35	Dunne
09-Feb-08	GNV	2289L	KingAir	2.67	Brown
10-Feb-08	TYS	4278Y	Navajo Chieftain	2.35	Williams
10-Feb-08	STL	1484P	Aztec	1.93	Orlando

7. Modify the query in Problem 6 to produce the output shown in Figure P7.7. The date limitation in Problem 6 applies to this problem, too. Note that this query includes data from the CREW and EMPLOYEE tables. (*Note:* You might wonder why the date restriction seems to generate more records than it did in Problem 6. Actually, the number of (CHARTER) records is the same, but several records are listed twice to reflect a crew of two: a pilot and a copilot. For example, the record for the 09-Feb-2008 flight to GNV, using aircraft 2289L, required a crew consisting of a pilot (Lange) and a copilot (Lewis).)

FIGURE P7.7   Problem 7 query results

CHAR_DATE	CHAR_DESTINATION	AC_NUMBER	MOD_CHG_MILE	CHAR_DISTANCE	EMP_NUM	CREW_JOB	EMP_LNAME
09-Feb-08	GNV	2289L	2.67	1,645	104	Pilot	Lange
09-Feb-08	GNV	2289L	2.67	1,645	101	Copilot	Lewis
09-Feb-08	MQY	2778V	2.35	312	109	Pilot	Travis
09-Feb-08	MQY	2778V	2.35	312	105	Copilot	Williams
09-Feb-08	ATL	4278Y	2.35	936	106	Pilot	Duzak
10-Feb-08	STL	1484P	1.93	508	101	Pilot	Lewis
10-Feb-08	TYS	4278Y	2.35	644	105	Pilot	Williams
10-Feb-08	TYS	4278Y	2.35	644	104	Copilot	Lange

8. Modify the query in Problem 5 to include the computed (derived) attribute "fuel per hour." (*Hint:* It is possible to use SQL to produce computed "attributes" that are not stored in any table. For example, the SQL query:

SELECT   CHAR_DISTANCE, CHAR_FUEL_GALLONS/CHAR_DISTANCE
FROM   CHARTER;

is perfectly acceptable. The above query produces the "gallons per mile flown" value.) Use a similar technique on joined tables to produce the "gallons per hour" output shown in Figure P7.8. (Note that 67.2 gallons/1.5 hours produces 44.8 gallons per hour.)

FIGURE P7.8   Problem 8 query results

CHAR_DATE	AC_NUMBER	MOD_NAME	CHAR_HOURS_FLOWN	CHAR_FUEL_GALLONS	Expr1
09-Feb-08	2778V	Navajo Chieftain	1.5	67.2	44.8
09-Feb-08	2289L	KingAir	6.7	459.5	68.5820895522388
09-Feb-08	4278Y	Navajo Chieftain	6.1	302.6	49.6065573770492
10-Feb-08	4278Y	Navajo Chieftain	3.8	167.4	44.0526315789474
10-Feb-08	1484P	Aztec	3.1	105.5	34.0322580645161

Query output such as the "gallons per hour" result shown in Figure P7.8 provide managers with very important information. In this case, why is the fuel burn for the Navajo Chieftain 4278Y flown on 9-Feb-08 so much higher than the fuel burn for that aircraft on 10-Feb-08? Such a query result might lead to additional queries to find out who flew the aircraft or what special circumstances might have existed. Is the fuel burn difference due to poor fuel management by the pilot, does it reflect an engine fuel metering problem, or was there an error in the fuel recording? The ability to generate useful query output is an important management asset.

> **NOTE**
>
> The output format is determined by the RDBMS you use. In this example, the Access software defaulted to an output heading labeled Expr1 to indicate the expression resulting from the division:
>
> [CHARTER]![CHAR_FUEL_GALLONS]/[CHARTER]![CHAR_HOURS]
>
> created by its expression builder. Oracle defaults to the full division label. You should learn to control the output format with the help of your RDBMSs utility software.

9. Create a query to produce the output shown in Figure P7.9. Note that, in this case, the computed attribute requires data found in two different tables. (*Hint:* The MODEL table contains the charge per mile, and the CHARTER table contains the total miles flown.) Note also that the output is limited to charter records generated since February 9, 2008. In addition, the output is ordered by date and, within the date, by the customer's last name.

FIGURE P7.9	Problem 9 query results

CHAR_DATE	CUS_LNAME	CHAR_DISTANCE	MOD_CHG_MILE	Mileage Charge
09-Feb-08	Brown	1645	2.67	4392.15
09-Feb-08	Dunne	312	2.35	733.20
09-Feb-08	Williams	936	2.35	2199.60
10-Feb-08	Orlando	508	1.93	980.44
10-Feb-08	Williams	644	2.35	1513.40

10. Use the techniques that produced the output in Problem 9 to produce the charges shown in Figure P7.10. The total charge to the customer is computed by:

- Miles flown * charge per mile.
- Hours waited * $50 per hour.

The miles flown (CHAR_DISTANCE) value is found in the CHARTER table, the charge per mile (MOD_CHG_MILE) value is found in the MODEL table, and the hours waited (CHAR_HOURS_WAIT) value is found in the CHARTER table.

FIGURE P7.10	Problem 10 query results

CHAR_DATE	CUS_LNAME	Mileage Charge	Waiting Charge	Total Charge
09-Feb-08	Brown	4392.15	0.00	4392.15
09-Feb-08	Dunne	733.20	0.00	733.20
09-Feb-08	Williams	2199.60	105.00	2304.60
10-Feb-08	Orlando	980.44	0.00	980.44
10-Feb-08	Williams	1513.40	225.00	1738.40

11. Create the SQL query that will produce a list of customers who have an unpaid balance. The required output is shown in Figure P7.11. Note that the balances are listed in descending order.

FIGURE P7.11   A list of customers with unpaid balances

CUS_LNAME	CUS_FNAME	CUS_INITIAL	CUS_BALANCE
Olowski	Paul	F	1285.19
O'Brian	Amy	B	1014.56
Smith	Kathy	W	896.54
Orlando	Myron		673.21
Smith	Olette	K	453.98

12. Find the average customer balance, the minimum balance, the maximum balance, and the total of the unpaid balances. The resulting values are shown in Figure P7.12.

13. Using the CHARTER table as the source, group the aircraft data. Then use the SQL functions to produce the output shown in Figure P7.13. (Utility software was used to modify the headers, so your headers might look different.)

FIGURE P7.12   Customer balance summary

Average Balance	Minimum Balance	Maximum Balance	Total Unpaid Bills
432.35	0.00	1285.19	4323.48

FIGURE P7.13   The AIRCRAFT data summary statement

AC_NUMBER	Number of Trips	Total Distance	Average Distance	Total Hours	Average Hours
1484P	4	1976	494.0	12.0	3.0
2289L	4	5178	1294.5	24.1	6.0
2778V	4	3090	772.5	15.8	4.0
4278Y	6	5268	878.0	30.4	5.1

14. Write the SQL code to generate the output shown in Figure P7.14. Note that the listing includes all CHARTER flights that did not include a copilot crew assignment. (*Hint:* The crew assignments are listed in the CREW table. Also note that the pilot's last name requires access to the EMPLOYEE table, while the MOD_CODE requires access to the MODEL table.)

FIGURE P7.14   Charter flights that did not use a copilot

CHAR_TRIP	CHAR_DATE	AC_NUMBER	MOD_NAME	CHAR_HOURS_FLOWN	EMP_LNAME	CREW_JOB
10001	05-Feb-08	2289L	KingAir	5.1	Lange	Pilot
10002	05-Feb-08	2778V	Navajo Chieftain	1.6	Lewis	Pilot
10004	06-Feb-08	1484P	Aztec	2.9	Duzak	Pilot
10005	06-Feb-08	2289L	KingAir	5.7	Lewis	Pilot
10006	06-Feb-08	4278Y	Navajo Chieftain	2.6	Travis	Pilot
10008	07-Feb-08	1484P	Aztec	4.1	Duzak	Pilot
10009	07-Feb-08	2289L	KingAir	6.6	Williams	Pilot
10010	07-Feb-08	4278Y	Navajo Chieftain	6.2	Wiesenbach	Pilot
10012	08-Feb-08	2778V	Navajo Chieftain	4.8	Lewis	Pilot
10013	08-Feb-08	4278Y	Navajo Chieftain	3.9	Williams	Pilot
10014	09-Feb-08	4278Y	Navajo Chieftain	6.1	Duzak	Pilot
10017	10-Feb-08	1484P	Aztec	3.1	Lewis	Pilot

15. Write a query that will list the ages of the employees and the date the query was run. The required output is shown in Figure P7.15. (As you can tell, the query was run on May 16, 2007, so the ages of the employees are current as of that date.)

FIGURE P7.15   **Employee ages and date of query**

EMP_NUM	EMP_LNAME	EMP_FNAME	EMP_HIRE_DATE	EMP_DOB	Age	Query Date
100	Kolmycz	George	15-Mar-1987	15-Jun-1942	64	15-Apr-07
101	Lewis	Rhonda	25-Apr-1988	19-Mar-1965	42	15-Apr-07
102	Vandam	Rhett	20-Dec-1992	14-Nov-1958	48	15-Apr-07
103	Jones	Anne	28-Aug-2005	16-Oct-1974	32	15-Apr-07
104	Lange	John	20-Oct-1996	08-Nov-1971	35	15-Apr-07
105	Williams	Robert	08-Jan-2006	14-Mar-1975	32	15-Apr-07
106	Duzak	Jeanine	05-Jan-1991	12-Feb-1968	39	15-Apr-07
107	Diante	Jorge	02-Jul-1996	21-Aug-1974	32	15-Apr-07
108	Wiesenbach	Paul	18-Nov-1994	14-Feb-1966	41	15-Apr-07
109	Travis	Elizabeth	14-Apr-1991	18-Jun-1961	45	15-Apr-07
110	Genkazi	Leighla	01-Dec-1992	19-May-1970	36	15-Apr-07

ONLINE CONTENT

Problems 16–33 are based on the **Ch07_SaleCo** database located in the Student Online Companion. This database is stored in Microsoft Access format. If you use another DBMS such as Oracle, SQL Server, MySQL, or DB2, use its import utilities to move the Access database contents. The Student Online Companion provides Oracle and SQL script files.

The structure and contents of the **Ch07_SaleCo** database are shown in Figure P7.16. Use this database to answer the following problems. Save each query as QXX, where XX is the problem number.

FIGURE P7.16   The Ch07_SaleCo database

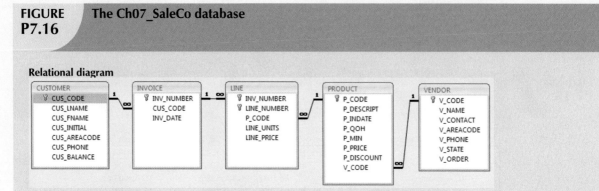

Relational diagram

Table name: CUSTOMER

CUS_CODE	CUS_LNAME	CUS_FNAME	CUS_INITIAL	CUS_AREACODE	CUS_PHONE	CUS_BALANCE
10010	Ramas	Alfred	A	615	844-2573	0.00
10011	Dunne	Leona	K	713	894-1238	0.00
10012	Smith	Kathy	W	615	894-2285	345.86
10013	Olowski	Paul	F	615	894-2180	536.75
10014	Orlando	Myron		615	222-1672	0.00
10015	O'Brian	Amy	B	713	442-3381	0.00
10016	Brown	James	G	615	297-1228	221.19
10017	Williams	George		615	290-2556	768.93
10018	Farriss	Anne	G	713	382-7185	216.55
10019	Smith	Olette	K	615	297-3809	0.00

Table name: VENDOR

V_CODE	V_NAME	V_CONTACT	V_AREACODE	V_PHONE	V_STATE	V_ORDER
21225	Bryson, Inc.	Smithson	615	223-3234	TN	Y
21226	SuperLoo, Inc.	Flushing	904	215-8995	FL	N
21231	D&E Supply	Singh	615	228-3245	TN	Y
21344	Gomez Bros.	Ortega	615	889-2546	KY	N
22567	Dome Supply	Smith	901	678-1419	GA	N
23119	Randsets Ltd.	Anderson	901	678-3998	GA	Y
24004	Brackman Bros.	Browning	615	228-1410	TN	N
24288	ORDVA, Inc.	Hakford	615	898-1234	TN	Y
25443	B&K, Inc.	Smith	904	227-0093	FL	N
25501	Damal Supplies	Smythe	615	890-3529	TN	N
25595	Rubicon Systems	Orton	904	456-0092	FL	Y

Table name: INVOICE

INV_NUMBER	CUS_CODE	INV_DATE
1001	10014	16-Mar-08
1002	10011	16-Mar-08
1003	10012	16-Mar-08
1004	10011	17-Mar-08
1005	10018	17-Mar-08
1006	10014	17-Mar-08
1007	10015	17-Mar-08
1008	10011	17-Mar-08

Table name: LINE

INV_NUMBER	LINE_NUMBER	P_CODE	LINE_UNITS	LINE_PRICE
1001	1	13-Q2/P2	1	14.99
1001	2	23109-HB	1	9.95
1002	1	54778-2T	2	4.99
1003	1	2238/QPD	1	38.95
1003	2	1546-QQ2	1	39.95
1003	3	13-Q2/P2	5	14.99
1004	1	54778-2T	3	4.99
1004	2	23109-HB	2	9.95
1005	1	PVC23DRT	12	5.87
1006	1	SM-18277	3	6.99
1006	2	2232/QTY	1	109.92
1006	3	23109-HB	1	9.95
1006	4	89-WRE-Q	1	256.99
1007	1	13-Q2/P2	2	14.99
1007	2	54778-2T	1	4.99
1008	1	PVC23DRT	5	5.87
1008	2	WR3/TT3	3	119.95
1008	3	23109-HB	1	9.95

Table name: PRODUCT

P_CODE	P_DESCRIPT	P_INDATE	P_QOH	P_MIN	P_PRICE	P_DISCOUNT	V_CODE
11QER/31	Power painter, 15 psi., 3-nozzle	03-Nov-07	8	5	109.99	0.00	25595
13-Q2/P2	7.25-in. pwr. saw blade	13-Dec-07	32	15	14.99	0.05	21344
14-Q1/L3	9.00-in. pwr. saw blade	13-Nov-07	18	12	17.49	0.00	21344
1546-QQ2	Hrd. cloth, 1/4-in., 2x50	15-Jan-08	15	8	39.95	0.00	23119
1558-QW1	Hrd. cloth, 1/2-in., 3x50	15-Jan-08	23	5	43.99	0.00	23119
2232/QTY	B&D jigsaw, 12-in. blade	30-Dec-07	8	5	109.92	0.05	24288
2232/QWE	B&D jigsaw, 8-in. blade	24-Dec-07	6	5	99.87	0.05	24288
2238/QPD	B&D cordless drill, 1/2-in.	20-Jan-08	12	5	38.95	0.05	25595
23109-HB	Claw hammer	20-Jan-08	23	10	9.95	0.10	21225
23114-AA	Sledge hammer, 12 lb.	02-Jan-08	8	5	14.40	0.05	
54778-2T	Rat-tail file, 1/8-in. fine	15-Dec-07	43	20	4.99	0.00	21344
89-WRE-Q	Hicut chain saw, 16 in.	07-Feb-08	11	5	256.99	0.05	24288
PVC23DRT	PVC pipe, 3.5-in., 8-ft	20-Feb-08	188	75	5.87	0.00	
SM-18277	1.25-in. metal screw, 25	01-Mar-08	172	75	6.99	0.00	21225
SW-23116	2.5-in. wd. screw, 50	24-Feb-08	237	100	8.45	0.00	21231
WR3/TT3	Steel matting, 4'x8'x1/6", .5" mesh	17-Jan-08	18	5	119.95	0.10	25595

16. Write a query to count the number of invoices.

17. Write a query to count the number of customers with a customer balance over $500.

18. Generate a listing of all purchases made by the customers, using the output shown in Figure P7.18 as your guide. (*Hint:* Use the ORDER BY clause to order the resulting rows shown in Figure P7.18.)

FIGURE P7.18 List of customer purchases

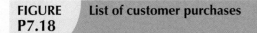

CUS_CODE	INV_NUMBER	INV_DATE	P_DESCRIPT	LINE_UNITS	LINE_PRICE
10011	1002	16-Mar-08	Rat-tail file, 1/8-in. fine	2	4.99
10011	1004	17-Mar-08	Claw hammer	2	9.95
10011	1004	17-Mar-08	Rat-tail file, 1/8-in. fine	3	4.99
10011	1008	17-Mar-08	Claw hammer	1	9.95
10011	1008	17-Mar-08	PVC pipe, 3.5-in., 8-ft	5	5.87
10011	1008	17-Mar-08	Steel matting, 4'x8'x1/6", .5" mesh	3	119.95
10012	1003	16-Mar-08	7.25-in. pwr. saw blade	5	14.99
10012	1003	16-Mar-08	B&D cordless drill, 1/2-in.	1	38.95
10012	1003	16-Mar-08	Hrd. cloth, 1/4-in., 2x50	1	39.95
10014	1001	16-Mar-08	7.25-in. pwr. saw blade	1	14.99
10014	1001	16-Mar-08	Claw hammer	1	9.95
10014	1006	17-Mar-08	1.25-in. metal screw, 25	3	6.99
10014	1006	17-Mar-08	B&D jigsaw, 12-in. blade	1	109.92
10014	1006	17-Mar-08	Claw hammer	1	9.95
10014	1006	17-Mar-08	Hicut chain saw, 16 in.	1	256.99
10015	1007	17-Mar-08	7.25-in. pwr. saw blade	2	14.99
10015	1007	17-Mar-08	Rat-tail file, 1/8-in. fine	1	4.99
10018	1005	17-Mar-08	PVC pipe, 3.5-in., 8-ft	12	5.87

19. Using the output shown in Figure P7.19 as your guide, generate a list of customer purchases, including the subtotals for each of the invoice line numbers. (*Hint:* Modify the query format used to produce the list of customer purchases in Problem 18, delete the INV_DATE column, and add the derived (computed) attribute LINE_UNITS * LINE_PRICE to calculate the subtotals.)

FIGURE P7.19 Summary of customer purchases with subtotals

CUS_CODE	INV_NUMBER	P_DESCRIPT	Units Bought	Unit Price	Subtotal
10011	1002	Rat-tail file, 1/8-in. fine	2	4.99	9.98
10011	1004	Claw hammer	2	9.95	19.90
10011	1004	Rat-tail file, 1/8-in. fine	3	4.99	14.97
10011	1008	Claw hammer	1	9.95	9.95
10011	1008	PVC pipe, 3.5-in., 8-ft	5	5.87	29.35
10011	1008	Steel matting, 4'x8'x1/6", .5" mesh	3	119.95	359.85
10012	1003	7.25-in. pwr. saw blade	5	14.99	74.95
10012	1003	B&D cordless drill, 1/2-in.	1	38.95	38.95
10012	1003	Hrd. cloth, 1/4-in., 2x50	1	39.95	39.95
10014	1001	7.25-in. pwr. saw blade	1	14.99	14.99
10014	1001	Claw hammer	1	9.95	9.95
10014	1006	1.25-in. metal screw, 25	3	6.99	20.97
10014	1006	B&D jigsaw, 12-in. blade	1	109.92	109.92
10014	1006	Claw hammer	1	9.95	9.95
10014	1006	Hicut chain saw, 16 in.	1	256.99	256.99
10015	1007	7.25-in. pwr. saw blade	2	14.99	29.98
10015	1007	Rat-tail file, 1/8-in. fine	1	4.99	4.99
10018	1005	PVC pipe, 3.5-in., 8-ft	12	5.87	70.44

20. Modify the query used in Problem 19 to produce the summary shown in Figure P7.20.

FIGURE P7.20   Customer purchase summary

CUS_CODE	CUS_BALANCE	Total Purchases
10011	0.00	444.00
10012	345.86	153.85
10014	0.00	422.77
10015	0.00	34.97
10018	216.55	70.44

21. Modify the query in Problem 20 to include the number of individual product purchases made by each customer. (In other words, if the customer's invoice is based on three products, one per LINE_NUMBER, you would count three product purchases. Note that in the original invoice data, customer 10011 generated three invoices, which contained a total of six lines, each representing a product purchase.) Your output values must match those shown in Figure P7.21.

22. Use a query to compute the average purchase amount per product made by each customer. (*Hint:* Use the results of Problem 21 as the basis for this query.) Your output values must match those shown in Figure P7.22. Note that the average purchase amount is equal to the total purchases divided by the number of purchases.

FIGURE P7.21   Customer total purchase amounts and number of purchases

CUS_CODE	CUS_BALANCE	Total Purchases	Number of Purchases
10011	0.00	444.00	6
10012	345.86	153.85	3
10014	0.00	422.77	6
10015	0.00	34.97	2
10018	216.55	70.44	1

23. Create a query to produce the total purchase per invoice, generating the results shown in Figure P7.23. The invoice total is the sum of the product purchases in the LINE that corresponds to the INVOICE.

FIGURE P7.22   Average purchase amount by customer

CUS_CODE	CUS_BALANCE	Total Purchases	Number of Purchases	Average Purchase Amount
10011	0.00	444.00	6	74.00
10012	345.86	153.85	3	51.28
10014	0.00	422.77	6	70.46
10015	0.00	34.97	2	17.48
10018	216.55	70.44	1	70.44

FIGURE P7.23   Invoice totals

INV_NUMBER	Invoice Total
1001	24.94
1002	9.98
1003	153.85
1004	34.87
1005	70.44
1006	397.83
1007	34.97
1008	399.15

FIGURE P7.24   Invoice totals by customer

CUS_CODE	INV_NUMBER	Invoice Total
10011	1002	9.98
10011	1004	34.87
10011	1008	399.15
10012	1003	153.85
10014	1001	24.94
10014	1006	397.83
10015	1007	34.97
10018	1005	70.44

24. Use a query to show the invoices and invoice totals as shown in Figure P7.24. (*Hint:* Group by the CUS_CODE.)

25. Write a query to produce the number of invoices and the total purchase amounts by customer, using the output shown in Figure P7.25 as your guide. (Compare this summary to the results shown in Problem 24.)

26. Using the query results in Problem 25 as your basis, write a query to generate the total number of invoices, the invoice total for all of the invoices, the smallest invoice amount, the largest invoice amount, and the average of all of the invoices. (*Hint:* Check the figure output in Problem 25.) Your output must match Figure P7.26.

FIGURE P7.25	Number of invoices and total purchase amounts by customer

CUS_CODE	Number of Invoices	Total Customer Purchases
10011	3	444.00
10012	1	153.85
10014	2	422.77
10015	1	34.97
10018	1	70.44

FIGURE P7.26	Number of invoices; invoice totals; minimum, maximum, and average sales

Total Invoices	Total Sales	Minimum Sale	Largest Sale	Average Sale
8	1126.03	34.97	444.00	225.21

27. List the balance characteristics of the customers who have made purchases during the current invoice cycle—that is, for the customers who appear in the INVOICE table. The results of this query are shown in Figure P7.27.

28. Using the results of the query created in Problem 27, provide a summary of the customer balance characteristics as shown in Figure P7.28.

FIGURE P7.27	Balances of customers who made purchases

CUS_CODE	CUS_BALANCE
10011	0.00
10012	345.86
10014	0.00
10015	0.00
10018	216.55

FIGURE P7.28	Balance summary for customers who made purchases

Minimum Balance	Maximum Balance	Average Balance
0	345.86	112.48

29. Create a query to find the customer balance characteristics for all customers, including the total of the outstanding balances. The results of this query are shown in Figure P7.29.

30. Find the listing of customers who did not make purchases during the invoicing period. Your output must match the output shown in Figure P7.30.

FIGURE P7.29	Balance summary for all customers

Total Balances	Minimum Balance	Maximum Balance	Average Balance
2089.28	0.00	768.93	208.93

FIGURE P7.30	Balances of customers who did not make purchases

CUS_CODE	CUS_BALANCE
10010	0.00
10013	536.75
10016	221.19
10017	768.93
10019	0.00

31. Find the customer balance summary for all customers who have not made purchases during the current invoicing period. The results are shown in Figure P7.31.

FIGURE P7.31   **Balance summary for customers who did not make purchases**

Total Balance	Minimum Balance	Maximum Balance	Average Balance
1526.87	0.00	768.93	305.37

32. Create a query to produce the summary of the value of products currently in inventory. Note that the value of each product is produced by the multiplication of the units currently in inventory and the unit price. Use the ORDER BY clause to match the order shown in Figure P7.32.

FIGURE P7.32   **Value of products currently in inventory**

P_DESCRIPT	P_QOH	P_PRICE	Subtotal
Power painter, 15 psi., 3-nozzle	8	109.99	879.92
7.25-in. pwr. saw blade	32	14.99	479.68
9.00-in. pwr. saw blade	18	17.49	314.82
Hrd. cloth, 1/4-in., 2x50	15	39.95	599.25
Hrd. cloth, 1/2-in., 3x50	23	43.99	1011.77
B&D jigsaw, 12-in. blade	8	109.92	879.36
B&D jigsaw, 8-in. blade	6	99.87	599.22
B&D cordless drill, 1/2-in.	12	38.95	467.40
Claw hammer	23	9.95	228.85
Sledge hammer, 12 lb.	8	14.40	115.20
Rat-tail file, 1/8-in. fine	43	4.99	214.57
Hicut chain saw, 16 in.	11	256.99	2826.89
PVC pipe, 3.5-in., 8-ft	188	5.87	1103.56
1.25-in. metal screw, 25	172	6.99	1202.28
2.5-in. wd. screw, 50	237	8.45	2002.65
Steel matting, 4'x8'x1/6", .5" mesh	18	119.95	2159.10

33. Using the results of the query created in Problem 32, find the total value of the product inventory. The results are shown in Figure P7.33.

FIGURE P7.33   **Total value of all products in inventory**

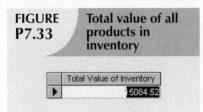

Total Value of Inventory
15084.52

8

In this chapter, you will learn:

- About the relational set operators UNION, UNION ALL, INTERSECT, and MINUS
- How to use the advanced SQL JOIN operator syntax
- About the different types of subqueries and correlated queries
- How to use SQL functions to manipulate dates, strings, and other data
- How to create and use updatable views
- How to create and use triggers and stored procedures
- How to create embedded SQL

Preview

In Chapter 7, Introduction to Structured Query Language (SQL), you learned the basic SQL data definition and data manipulation commands used to create and manipulate relational data. In this chapter, you build on what you learned in Chapter 7 and learn how to use more advanced SQL features.

In this chapter, you learn about the SQL relational set operators (UNION, INTERSECT, and MINUS) and how those operators are used to merge the results of multiple queries. Joins are at the heart of SQL, so you must learn how to use the SQL JOIN statement to extract information from multiple tables. In the previous chapter, you learned how cascading queries inside other queries can be useful in certain circumstances. In this chapter, you also learn about the different styles of subqueries that can be implemented in a SELECT statement. Finally, you learn more of SQL's many functions to extract information from data, including manipulation of dates and strings and computations based on stored or even derived data.

In the real world, business procedures require the execution of clearly defined actions when a specific event occurs, such as the addition of a new invoice or a student's enrollment in a class. Such procedures can be applied within the DBMS through the use of triggers and stored procedures. In addition, SQL facilitates the application of business procedures when it is embedded in a programming language such as Visual Basic .Net, C#, or COBOL.

ONLINE CONTENT

Although most of the examples used in this chapter are shown in Oracle, you could also use MS SQL Server. The Student Online companion provides you with the **ADVSQLDBINIT.SQL** script file (Oracle and MS SQL versions) to create the tables and load the data used in this chapter. There you will also find additional SQL script files to demonstrate each of the commands shown in this chapter.

8.1 RELATIONAL SET OPERATORS

In Chapter 3, The Relational Database Model, you learned about the eight general relational operators. In this section, you will learn how to use three SQL commands (UNION, INTERSECT, and MINUS) to implement the union, intersection, and difference relational operators.

In previous chapters, you learned that SQL data manipulation commands are set-oriented; that is, they operate over entire sets of rows and columns (tables) at once. Using sets, you can combine two or more sets to create new sets (or relations). That's precisely what the UNION, INTERSECT, and MINUS statements do. In relational database terms, you can use the words "sets," "relations," and "tables" interchangeably because they all provide a conceptual view of the data set as it is presented to the relational database user.

NOTE

The SQL standard defines the operations that all DBMSs must perform on data, but it leaves the implementation details to the DBMS vendors. Therefore, some advanced SQL features might not work on all DBMS implementations. Also, some DBMS vendors might implement additional features not found in the SQL standard.

UNION, INTERSECT, and MINUS are the names of the SQL statements implemented in Oracle. The SQL standard uses the keyword EXCEPT to refer to the difference (MINUS) relational operator. Other RDBMS vendors might use a different command name or might not implement a given command at all.

To learn more about the ANSI/ISO SQL standards, check the ANSI Web site (*www.ansi.org*) to find out how to obtain the latest standard documents in electronic form. As of this writing, the most recent published standard is SQL-2003. The SQL-2003 standard makes revisions and additions to the previous standard; most notable is support for XML data.

UNION, INTERSECT, and MINUS work properly only if relations are **union-compatible**, *which* means that the names of the relation attributes must be the same and their data types must be alike. In practice, some RDBMS vendors require the data types to be "compatible" but not necessarily "exactly the same." For example, compatible data types are VARCHAR (35) and CHAR (15). In that case, both attributes store character (string) values; the only difference is the string size. Another example of compatible data types is NUMBER and SMALLINT. Both data types are used to store numeric values.

NOTE

Some DBMS products might require union-compatible tables to have *identical* data types.

ONLINE CONTENT

The Student Online Companion provides you with SQL script files (Oracle and MS SQL Server) to demonstrate the UNION, INTERSECT, and MINUS commands. It also provides the **Ch08_SaleCo** MS Access database containing supported set operator alternative queries.

8.1.1 UNION

Suppose SaleCo has bought another company. SaleCo's management wants to make sure that the acquired company's customer list is properly merged with SaleCo's customer list. Because it is quite possible that some customers have purchased goods from both companies, the two lists might contain common customers. SaleCo's management wants to make sure that customer records are not duplicated when the two customer lists are merged. The UNION query is a perfect tool for generating a combined listing of customers—one that excludes duplicate records.

The UNION statement combines rows from two or more queries *without including duplicate rows*. The syntax of the UNION statement is:

query UNION *query*

In other words, the UNION statement combines the output of two SELECT queries. (Remember that the SELECT statements must be union-compatible. That is, they must return the same attribute names and similar data types.)

To demonstrate the use of the UNION statement in SQL, let's use the CUSTOMER and CUSTOMER_2 tables in the **Ch08_SaleCo** database. To show the combined CUSTOMER and CUSTOMER_2 records without the duplicates, the UNION query is written as follows:

```
SELECT     CUS_LNAME, CUS_FNAME, CUS_INITIAL, CUS_AREACODE, CUS_PHONE
FROM       CUSTOMER
UNION
SELECT     CUS_LNAME, CUS_FNAME, CUS_INITIAL, CUS_AREACODE, CUS_PHONE
FROM       CUSTOMER_2;
```

Figure 8.1 shows the contents of the CUSTOMER and CUSTOMER_2 tables and the result of the UNION query. Although MS Access is used to show the results here, similar results can be obtained with Oracle.

Note the following in Figure 8.1:

- The CUSTOMER table contains 10 rows, while the CUSTOMER_2 table contains 7 rows.
- Customers Dunne and Olowski are included in the CUSTOMER table as well as in the CUSTOMER_2 table.
- The UNION query yields 15 records because the duplicate records of customers Dunne and Olowski are not included. In short, the UNION query yields a unique set of records.

NOTE

The SQL standard calls for the elimination of duplicate rows when the UNION SQL statement is used. However, some DBMS vendors might not adhere to that standard. Check your DBMS manual to see if the UNION statement is supported and if so, *how* it is supported.

FIGURE 8.1   UNION query results

Database name: CH08_SaleCo

Table name: CUSTOMER

CUS_CODE	CUS_LNAME	CUS_FNAME	CUS_INITIAL	CUS_AREACODE	CUS_PHONE	CUS_BALANCE
10010	Ramas	Alfred	A	615	844-2573	0.00
10011	Dunne	Leona	K	713	894-1238	0.00
10012	Smith	Kathy	W	615	894-2285	345.86
10013	Olowski	Paul	F	615	894-2180	536.75
10014	Orlando	Myron		615	222-1672	0.00
10015	O'Brian	Amy	B	713	442-3381	0.00
10016	Brown	James	G	615	297-1228	221.19
10017	Williams	George		615	290-2556	768.93
10018	Farriss	Anne	G	713	382-7185	216.55
10019	Smith	Olette	K	615	297-3809	0.00

Query name: qryUNION-of-CUSTOMER-and-CUSTOMER_2

CUS_LNAME	CUS_FNAME	CUS_INITIAL	CUS_AREACODE	CUS_PHONE
Brown	James	G	615	297-1228
Dunne	Leona	K	713	894-1238
Farriss	Anne	G	713	382-7185
Hernandez	Carlos	J	723	123-7654
Lewis	Marie	J	734	332-1789
McDowell	George		723	123-7768
O'Brian	Amy	B	713	442-3381
Olowski	Paul	F	615	894-2180
Orlando	Myron		615	222-1672
Ramas	Alfred	A	615	844-2573
Smith	Kathy	W	615	894-2285
Smith	Olette	K	615	297-3809
Terrell	Justine	H	615	322-9870
Tirpin	Khaleed	G	723	123-9876
Williams	George		615	290-2556

Table name: CUSTOMER_2

CUS_CODE	CUS_LNAME	CUS_FNAME	CUS_INITIAL	CUS_AREACODE	CUS_PHONE
345	Terrell	Justine	H	615	322-9870
347	Olowski	Paul	F	615	894-2180
351	Hernandez	Carlos	J	723	123-7654
352	McDowell	George		723	123-7768
365	Tirpin	Khaleed	G	723	123-9876
368	Lewis	Marie	J	734	332-1789
369	Dunne	Leona	K	713	894-1238

The UNION statement can be used to unite more than just two queries. For example, assume that you have four union-compatible queries named T1, T2, T3, and T4. With the UNION statement, you can combine the output of all four queries into a single result set. The SQL statement will be similar to this:

SELECT column-list FROM T1
UNION
SELECT column-list FROM T2
UNION
SELECT column-list FROM T3
UNION
SELECT column-list FROM T4;

8.1.2 UNION ALL

If SaleCo's management wants to know how many customers are on *both* the CUSTOMER and CUSTOMER_2 lists, a UNION ALL query can be used to produce a relation that retains the duplicate rows. Therefore, the following query will keep all rows from both queries (including the duplicate rows) and return 17 rows.

SELECT   CUS_LNAME, CUS_FNAME, CUS_INITIAL, CUS_AREACODE, CUS_PHONE
FROM   CUSTOMER
UNION ALL
SELECT   CUS_LNAME, CUS_FNAME, CUS_INITIAL, CUS_AREACODE, CUS_PHONE
FROM   CUSTOMER_2;

Running the preceding UNION ALL query produces the result shown in Figure 8.2.

Like the UNION statement, the UNION ALL statement can be used to unite more than just two queries.

FIGURE 8.2   UNION ALL query results

Database name: CH08_SaleCo

Table name: CUSTOMER

CUS_CODE	CUS_LNAME	CUS_FNAME	CUS_INITIAL	CUS_AREACODE	CUS_PHONE	CUS_BALANCE
10010	Ramas	Alfred	A	615	844-2573	0.00
10011	Dunne	Leona	K	713	894-1238	0.00
10012	Smith	Kathy	W	615	894-2285	345.86
10013	Olowski	Paul	F	615	894-2180	536.75
10014	Orlando	Myron		615	222-1672	0.00
10015	O'Brian	Amy	B	713	442-3381	0.00
10016	Brown	James	G	615	297-1228	221.19
10017	Williams	George		615	290-2556	768.93
10018	Farriss	Anne	G	713	382-7185	216.55
10019	Smith	Olette	K	615	297-3809	0.00

Table name: CUSTOMER_2

CUS_CODE	CUS_LNAME	CUS_FNAME	CUS_INITIAL	CUS_AREACODE	CUS_PHONE
345	Terrell	Justine	H	615	322-9870
347	Olowski	Paul	F	615	894-2180
351	Hernandez	Carlos	J	723	123-7654
352	McDowell	George		723	123-7768
365	Tirpin	Khaleed	G	723	123-9876
368	Lewis	Marie	J	734	332-1789
369	Dunne	Leona	K	713	894-1238

Query name: qryUNION-ALL-of-CUSTOMER-and-CUSTOMER_2

CUS_LNAME	CUS_FNAME	CUS_INITIAL	CUS_AREACODE	CUS_PHONE
Ramas	Alfred	A	615	844-2573
Dunne	Leona	K	713	894-1238
Smith	Kathy	W	615	894-2285
Olowski	Paul	F	615	894-2180
Orlando	Myron		615	222-1672
O'Brian	Amy	B	713	442-3381
Brown	James	G	615	297-1228
Williams	George		615	290-2556
Farriss	Anne	G	713	382-7185
Smith	Olette	K	615	297-3809
Terrell	Justine	H	615	322-9870
Olowski	Paul	F	615	894-2180
Hernandez	Carlos	J	723	123-7654
McDowell	George		723	123-7768
Tirpin	Khaleed	G	723	123-9876
Lewis	Marie	J	734	332-1789
Dunne	Leona	K	713	894-1238

8.1.3 INTERSECT

If SaleCo's management wants to know which customer records are duplicated in the CUSTOMER and CUSTOMER_2 tables, the INTERSECT statement can be used to combine rows from two queries, returning only the rows that appear in both sets. The syntax for the INTERSECT statement is:

query INTERSECT *query*

To generate the list of duplicate customer records, you can use:

```
SELECT      CUS_LNAME, CUS_FNAME, CUS_INITIAL, CUS_AREACODE, CUS_PHONE
FROM        CUSTOMER
INTERSECT
SELECT      CUS_LNAME, CUS_FNAME, CUS_INITIAL, CUS_AREACODE, CUS_PHONE
FROM        CUSTOMER_2;
```

The INTERSECT statement can be used to generate additional useful customer information. For example, the following query returns the customer codes for all customers who are located in area code 615 and who have made purchases. (If a customer has made a purchase, there must be an invoice record for that customer.)

```
SELECT      CUS_CODE FROM CUSTOMER WHERE CUS_AREACODE = '615'
INTERSECT
SELECT      DISTINCT CUS_CODE FROM INVOICE;
```

Figure 8.3 shows both sets of SQL statements and their output.

8.1.4 MINUS

The MINUS statement in SQL combines rows from two queries and returns only the rows that appear in the first set but not in the second. The syntax for the MINUS statement is:

query MINUS *query*

FIGURE 8.3   INTERSECT query results

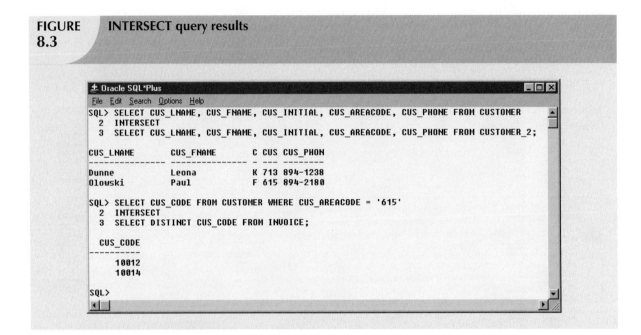

> **NOTE**
>
> MS Access does not support the INTERSECT query, nor does it support other complex queries you will explore in this chapter. At least in some cases, Access might be able to give you the desired results if you use an alternative query format or procedure. For example, although Access does not support SQL triggers and stored procedures, you can use Visual Basic code to perform similar actions. However, the objective here is to show you how some important standard SQL features may be used.

For example, if the SaleCo managers want to know what customers in the CUSTOMER table are not found in the CUSTOMER_2 table, they can use:

```
SELECT    CUS_LNAME, CUS_FNAME, CUS_INITIAL, CUS_AREACODE, CUS_PHONE
FROM      CUSTOMER
MINUS
SELECT    CUS_LNAME, CUS_FNAME, CUS_INITIAL, CUS_AREACODE, CUS_PHONE
FROM      CUSTOMER_2;
```

If the managers want to know what customers in the CUSTOMER_2 table are not found in the CUSTOMER table, they merely switch the table designations:

```
SELECT    CUS_LNAME, CUS_FNAME, CUS_INITIAL, CUS_AREACODE, CUS_PHONE
FROM      CUSTOMER_2
MINUS
SELECT    CUS_LNAME, CUS_FNAME, CUS_INITIAL, CUS_AREACODE, CUS_PHONE
FROM      CUSTOMER;
```

You can extract much useful information by combining MINUS with various clauses such as WHERE. For example, the following query returns the customer codes for all customers located in area code 615 minus the ones who have made purchases, leaving the customers in area code 615 who have not made purchases.

```
SELECT    CUS_CODE FROM CUSTOMER WHERE CUS_AREACODE = '615'
MINUS
SELECT    DISTINCT CUS_CODE FROM INVOICE;
```

Figure 8.4 shows the preceding three SQL statements and their output.

FIGURE 8.4   **MINUS query results**

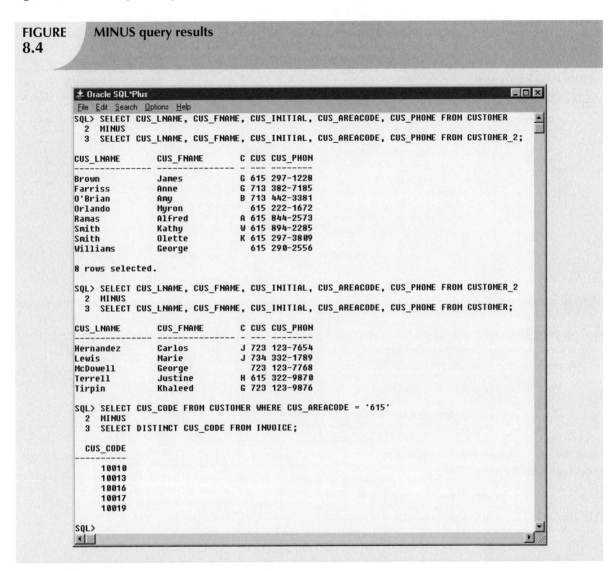

```
Oracle SQL*Plus                                                    _ □ ×
File  Edit  Search  Options  Help
SQL> SELECT CUS_LNAME, CUS_FNAME, CUS_INITIAL, CUS_AREACODE, CUS_PHONE FROM CUSTOMER
  2   MINUS
  3   SELECT CUS_LNAME, CUS_FNAME, CUS_INITIAL, CUS_AREACODE, CUS_PHONE FROM CUSTOMER_2;

CUS_LNAME       CUS_FNAME       C CUS CUS_PHON
--------------- --------------- - --- --------
Brown           James           G 615 297-1228
Farriss         Anne            G 713 382-7185
O'Brian         Amy             B 713 442-3381
Orlando         Myron             615 222-1672
Ramas           Alfred          A 615 844-2573
Smith           Kathy           W 615 894-2285
Smith           Olette          K 615 297-3809
Williams        George            615 290-2556

8 rows selected.

SQL> SELECT CUS_LNAME, CUS_FNAME, CUS_INITIAL, CUS_AREACODE, CUS_PHONE FROM CUSTOMER_2
  2   MINUS
  3   SELECT CUS_LNAME, CUS_FNAME, CUS_INITIAL, CUS_AREACODE, CUS_PHONE FROM CUSTOMER;

CUS_LNAME       CUS_FNAME       C CUS CUS_PHON
--------------- --------------- - --- --------
Hernandez       Carlos          J 723 123-7654
Lewis           Marie           J 734 332-1789
McDowell        George            723 123-7768
Terrell         Justine         H 615 322-9870
Tirpin          Khaleed         G 723 123-9876

SQL> SELECT CUS_CODE FROM CUSTOMER WHERE CUS_AREACODE = '615'
  2   MINUS
  3   SELECT DISTINCT CUS_CODE FROM INVOICE;

  CUS_CODE
----------
     10010
     10013
     10016
     10017
     10019

SQL>
```

NOTE

Some DBMS products do not support the INTERSECT or MINUS statements, while others might implement the difference relational operator in SQL as EXCEPT. Consult your DBMS manual to see if the statements illustrated here are supported by your DBMS.

8.1.5 SYNTAX ALTERNATIVES

If your DBMS doesn't support the INTERSECT or MINUS statements, you can use the IN and NOT IN subqueries to obtain similar results. For example, the following query will produce the same results as the INTERSECT query shown in Section 8.1.3.

```
SELECT      CUS_CODE FROM CUSTOMER
WHERE       CUS_AREACODE = '615' AND
            CUS_CODE IN (SELECT DISTINCT CUS_CODE FROM INVOICE);
```

Figure 8.5 shows the use of the INTERSECT alternative.

FIGURE 8.5 INTERSECT alternative

Database name: CH08_SaleCo

Table name: CUSTOMER

CUS_CODE	CUS_LNAME	CUS_FNAME	CUS_INITIAL	CUS_AREACODE	CUS_PHONE	CUS_BALANCE
10010	Ramas	Alfred	A	615	844-2573	0.00
10011	Dunne	Leona	K	713	894-1238	0.00
10012	Smith	Kathy	W	615	894-2285	345.86
10013	Olowski	Paul	F	615	894-2180	536.75
10014	Orlando	Myron		615	222-1672	0.00
10015	O'Brian	Amy	B	713	442-3381	0.00
10016	Brown	James	G	615	297-1228	221.19
10017	Williams	George		615	290-2556	768.93
10018	Farriss	Anne	G	713	382-7185	216.55
10019	Smith	Olette	K	615	297-3809	0.00

Table name: INVOICE

INV_NUMBER	CUS_CODE	INV_DATE
1001	10014	16-Jan-08
1002	10011	16-Jan-08
1003	10012	16-Jan-08
1004	10011	17-Jan-08
1005	10018	17-Jan-08
1006	10014	17-Jan-08
1007	10015	17-Jan-08
1008	10011	17-Jan-08

Query name: qry-INTERSECT-Alternative

CUS_CODE
10012
10014

NOTE

MS Access will generate an input request for the CUS_AREACODE if you use apostrophes around the area code. (If you supply the 615 area code, the query will execute properly.) You can eliminate that problem by using standard double quotation marks, writing the WHERE clause in the second line of the preceding SQL statement as:

WHERE CUS_AREACODE = "615" AND

MS Access will also accept single quotation marks.

Using the same alternative to the MINUS statement, you can generate the output for the third MINUS query shown in Section 8.1.4 by using:

SELECT   CUS_CODE FROM CUSTOMER
WHERE   CUS_AREACODE = '615' AND
            CUS_CODE NOT IN (SELECT DISTINCT CUS_CODE FROM INVOICE);

The results of that query are shown in Figure 8.6. Note that the query output includes only the customers in area code 615 who have not made any purchases and, therefore, have not generated invoices.

8.2 SQL JOIN OPERATORS

The relational join operation merges rows from two tables and returns the rows with one of the following conditions:
* Have common values in common columns (natural join).
* Meet a given join condition (equality or inequality).
* Have common values in common columns or have no matching values (outer join).

In Chapter 7, you learned how to use the SELECT statement in conjunction with the WHERE clause to join two or more tables. For example, you can join the PRODUCT and VENDOR tables through their common V_CODE by writing:

SELECT   P_CODE, P_DESCRIPT, P_PRICE, V_NAME
FROM   PRODUCT, VENDOR
WHERE   PRODUCT.V_CODE = VENDOR.V_CODE;

FIGURE 8.6   MINUS alternative

Database name: CH08_SaleCo

Table name: CUSTOMER

CUS_CODE	CUS_LNAME	CUS_FNAME	CUS_INITIAL	CUS_AREACODE	CUS_PHONE	CUS_BALANCE
10010	Ramas	Alfred	A	615	844-2573	0.00
10011	Dunne	Leona	K	713	894-1238	0.00
10012	Smith	Kathy	W	615	894-2285	345.86
10013	Olowski	Paul	F	615	894-2180	536.75
10014	Orlando	Myron		615	222-1672	0.00
10015	O'Brian	Amy	B	713	442-3381	0.00
10016	Brown	James	G	615	297-1228	221.19
10017	Williams	George		615	290-2556	768.93
10018	Farriss	Anne	G	713	382-7185	216.55
10019	Smith	Olette	K	615	297-3809	0.00

Table name: INVOICE

INV_NUMBER	CUS_CODE	INV_DATE
1001	10014	16-Jan-08
1002	10011	16-Jan-08
1003	10012	16-Jan-08
1004	10011	17-Jan-08
1005	10018	17-Jan-08
1006	10014	17-Jan-08
1007	10015	17-Jan-08
1008	10011	17-Jan-08

Query name: qry-MINUS-Alternative

CUS_CODE
10010
10013
10016
10017
10019

The preceding SQL join syntax is sometimes referred to as an "old-style" join. Note that the FROM clause contains the tables being joined and that the WHERE clause contains the condition(s) used to join the tables.

Note the following points about the preceding query:

- The FROM clause indicates which tables are to be joined. If three or more tables are included, the join operation takes place two tables at a time, starting from left to right. For example, if you are joining tables T1, T2, and T3, the first join is table T1 with T2; the results of that join are then joined to table T3.

- The join condition in the WHERE clause tells the SELECT statement which rows will be returned. In this case, the SELECT statement returns all rows for which the V_CODE values in the PRODUCT and VENDOR tables are equal.

- The number of join conditions is always equal to the number of tables being joined minus one. For example, if you join three tables (T1, T2, and T3), you will have two join conditions (j1 and j2). All join conditions are connected through an AND logical operator. The first join condition (j1) defines the join criteria for T1 and T2. The second join condition (j2) defines the join criteria for the output of the first join and T3.

- Generally, the join condition will be an equality comparison of the primary key in one table and the related foreign key in the second table.

Join operations can be classified as inner joins and outer joins. The **inner join** is the traditional join in which only rows that meet a given criteria are selected. The join criteria can be an equality condition (also called a natural join or an equijoin) or an inequality condition (also called a theta join). An **outer join** returns not only the matching rows, but also the rows with unmatched attribute values for one table or both tables to be joined. The SQL standard also introduces a special type of join that returns the same result as the Cartesian product of two sets or tables.

In this section, you will learn various ways to express join operations that meet the ANSI SQL standard. These are outlined in Table 8.1. It is useful to remember that not all DBMS vendors provide the same level of SQL support and that some do not support the join styles shown in this section. Oracle 10g is used to demonstrate the use of the following queries. Refer to your DBMS manual if you are using a different DBMS.

TABLE 8.1	SQL Join Expression Styles		
JOIN CLASSIFICATION	JOIN TYPE	SQL SYNTAX EXAMPLE	DESCRIPTION
CROSS	CROSS JOIN	SELECT * FROM T1, T2	Returns the Cartesian product of T1 and T2 (old style).
☛		SELECT * FROM T1 CROSS JOIN T2	Returns the Cartesian product of T1 and T2.
INNER	Old-Style JOIN	SELECT * FROM T1, T2 WHERE T1.C1=T2.C1	Returns only the rows that meet the join condition in the WHERE clause (old style). Only rows with matching values are selected.
	NATURAL JOIN	SELECT * FROM T1 NATURAL JOIN T2	Returns only the rows with matching values in the matching columns. The matching columns must have the same names and similar data types.
	JOIN USING	SELECT * FROM T1 JOIN T2 USING (C1)	Returns only the rows with matching values in the columns indicated in the USING clause.
	JOIN ON	SELECT * FROM T1 JOIN T2 ON T1.C1=T2.C1	Returns only the rows that meet the join condition indicated in the ON clause.
OUTER	LEFT JOIN	SELECT * FROM T1 LEFT OUTER JOIN T2 ON T1.C1=T2.C1	Returns rows with matching values and includes all rows from the left table (T1) with unmatched values.
	RIGHT JOIN	SELECT * FROM T1 RIGHT OUTER JOIN T2 ON T1.C1=T2.C1	Returns rows with matching values and includes all rows from the right table (T2) with unmatched values.
	FULL JOIN	SELECT * FROM T1 FULL OUTER JOIN T2 ON T1.C1=T2.C1	Returns rows with matching values and includes all rows from both tables (T1 and T2) with unmatched values.

8.2.1   CROSS JOIN

A **cross join** performs a relational product (also known as the Cartesian product) of two tables. The cross join syntax is:

SELECT column-list FROM table1 CROSS JOIN table2

For example,

SELECT * FROM INVOICE CROSS JOIN LINE;

performs a cross join of the INVOICE and LINE tables. That CROSS JOIN query generates 144 rows. (There were 8 invoice rows and 18 line rows, thus yielding 8 × 18 = 144 rows.)

You can also perform a cross join that yields only specified attributes. For example, you can specify:

SELECT   INVOICE.INV_NUMBER, CUS_CODE, INV_DATE, P_CODE
FROM   INVOICE CROSS JOIN LINE;

The results generated through that SQL statement can also be generated by using the following syntax:

SELECT   INVOICE.INV_NUMBER, CUS_CODE, INV_DATE, P_CODE
FROM   INVOICE, LINE;

8.2.2 NATURAL JOIN

Recall from Chapter 3 that a natural join returns all rows with matching values in the matching columns and eliminates duplicate columns. That style of query is used when the tables share one or more common attributes with common names. The natural join syntax is:

SELECT *column-list* FROM *table1* NATURAL JOIN *table2*

The natural join will perform the following tasks:

- Determine the common attribute(s) by looking for attributes with identical names and compatible data types.
- Select only the rows with common values in the common attribute(s).
- If there are no common attributes, return the relational product of the two tables.

The following example performs a natural join of the CUSTOMER and INVOICE tables and returns only selected attributes:

SELECT   CUS_CODE, CUS_LNAME, INV_NUMBER, INV_DATE
FROM   CUSTOMER NATURAL JOIN INVOICE;

The SQL code and its results are shown at the top of Figure 8.7.

FIGURE 8.7	NATURAL JOIN results

```
± Oracle SQL*Plus                                            _|□|×|
File  Edit  Search  Options  Help
SQL> SELECT CUS_CODE, CUS_LNAME, INV_NUMBER, INV_DATE
  2  FROM CUSTOMER NATURAL JOIN INVOICE;

  CUS_CODE CUS_LNAME       INV_NUMBER INV_DATE
---------- --------------- ---------- ---------
     10011 Dunne                 1008 17-JAN-08
     10011 Dunne                 1004 17-JAN-08
     10011 Dunne                 1002 16-JAN-08
     10012 Smith                 1003 16-JAN-08
     10014 Orlando               1006 17-JAN-08
     10014 Orlando               1001 16-JAN-08
     10015 O'Brian               1007 17-JAN-08
     10018 Farriss               1005 17-JAN-08

8 rows selected.

SQL> SELECT INV_NUMBER, P_CODE, P_DESCRIPT, LINE_UNITS, LINE_PRICE
  2  FROM INVOICE NATURAL JOIN LINE NATURAL JOIN PRODUCT;

INV_NUMBER P_CODE      P_DESCRIPT                      LINE_UNITS LINE_PRICE
---------- ----------- ------------------------------- ---------- ----------
      1001 13-Q2/P2    7.25-in. pwr. saw blade                  1      14.99
      1001 23109-HB    Claw hammer                              1       9.95
      1002 54778-2T    Rat-tail file, 1/8-in. fine              2       4.99
      1003 2238/QPD    B&D cordless drill, 1/2-in.              1      38.95
      1003 1546-QQ2    Hrd. cloth, 1/4-in., 2x50                1      39.95
      1003 13-Q2/P2    7.25-in. pwr. saw blade                  5      14.99
      1004 54778-2T    Rat-tail file, 1/8-in. fine              3       4.99
      1004 23109-HB    Claw hammer                              2       9.95
      1005 PVC23DRT    PVC pipe, 3.5-in., 8-ft                 12       5.87
      1006 SM-18277    1.25-in. metal screw, 25                 3       6.99
      1006 2232/QTY    B&D jigsaw, 12-in. blade                 1     109.92
      1006 23109-HB    Claw hammer                              1       9.95
      1006 89-WRE-Q    Hicut chain saw, 16 in.                  1     256.99
      1007 13-Q2/P2    7.25-in. pwr. saw blade                  2      14.99
      1007 54778-2T    Rat-tail file, 1/8-in. fine              1       4.99
      1008 PVC23DRT    PVC pipe, 3.5-in., 8-ft                  5       5.87
      1008 WR3/TT3     Steel matting, 4'x8'x1/6", .5" mesh      3     119.95
      1008 23109-HB    Claw hammer                              1       9.95

18 rows selected.
```

You are not limited to two tables when performing a natural join. For example, you can perform a natural join of the INVOICE, LINE, and PRODUCT tables and project only selected attributes by writing:

SELECT   INV_NUMBER, P_CODE, P_DESCRIPT, LINE_UNITS, LINE_PRICE
FROM   INVOICE NATURAL JOIN LINE NATURAL JOIN PRODUCT;

The SQL code and its results are shown at the bottom of Figure 8.7.

One important difference between the natural join and the "old-style" join syntax is that the natural join does not require the use of a table qualifier for the common attributes. In the first natural join example, you projected CUS_CODE. However, the projection did not require any table qualifier, even though the CUS_CODE attribute appeared in both CUSTOMER and INVOICE tables. The same can be said of the INV_NUMBER attribute in the second natural join example.

8.2.3 JOIN USING CLAUSE

A second way to express a join is through the USING keyword. That query returns only the rows with matching values in the column indicated in the USING clause—and that column must exist in both tables. The syntax is:

SELECT *column-list* FROM *table1* JOIN *table2* USING (*common-column*)

To see the JOIN USING query in action, let's perform a join of the INVOICE and LINE tables by writing:

SELECT   INV_NUMBER, P_CODE, P_DESCRIPT, LINE_UNITS, LINE_PRICE
FROM   INVOICE JOIN LINE USING (INV_NUMBER) JOIN PRODUCT USING (P_CODE);

The SQL statement produces the results shown in Figure 8.8.

FIGURE 8.8 **JOIN USING results**

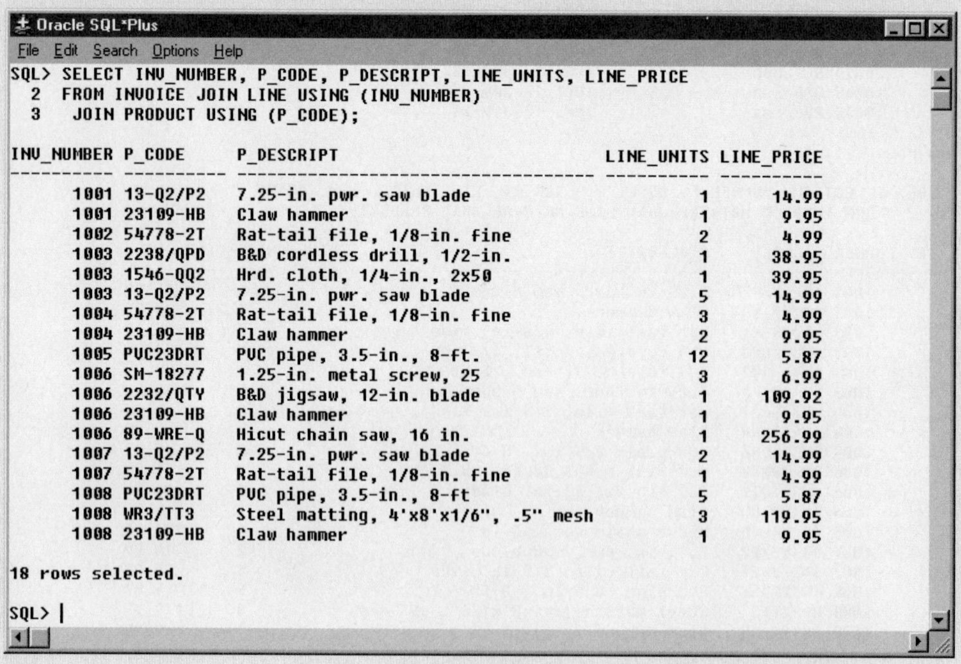

```
± Oracle SQL*Plus                                                    _ □ ×
File  Edit  Search  Options  Help
SQL> SELECT INV_NUMBER, P_CODE, P_DESCRIPT, LINE_UNITS, LINE_PRICE
  2  FROM INVOICE JOIN LINE USING (INV_NUMBER)
  3    JOIN PRODUCT USING (P_CODE);

INV_NUMBER P_CODE     P_DESCRIPT                            LINE_UNITS LINE_PRICE
---------- ---------- ------------------------------------- ---------- ----------
      1001 13-Q2/P2   7.25-in. pwr. saw blade                        1      14.99
      1001 23109-HB   Claw hammer                                    1       9.95
      1002 54778-2T   Rat-tail file, 1/8-in. fine                    2       4.99
      1003 2238/QPD   B&D cordless drill, 1/2-in.                    1      38.95
      1003 1546-QQ2   Hrd. cloth, 1/4-in., 2x50                      1      39.95
      1003 13-Q2/P2   7.25-in. pwr. saw blade                        5      14.99
      1004 54778-2T   Rat-tail file, 1/8-in. fine                    3       4.99
      1004 23109-HB   Claw hammer                                    2       9.95
      1005 PVC23DRT   PVC pipe, 3.5-in., 8-ft.                      12       5.87
      1006 SM-18277   1.25-in. metal screw, 25                       3       6.99
      1006 2232/QTY   B&D jigsaw, 12-in. blade                       1     109.92
      1006 23109-HB   Claw hammer                                    1       9.95
      1006 89-WRE-Q   Hicut chain saw, 16 in.                        1     256.99
      1007 13-Q2/P2   7.25-in. pwr. saw blade                        2      14.99
      1007 54778-2T   Rat-tail file, 1/8-in. fine                    1       4.99
      1008 PVC23DRT   PVC pipe, 3.5-in., 8-ft                        5       5.87
      1008 WR3/TT3    Steel matting, 4'x8'x1/6", .5" mesh            3     119.95
      1008 23109-HB   Claw hammer                                    1       9.95

18 rows selected.

SQL> |
```

As was the case with the NATURAL JOIN command, the JOIN USING operand does not require table qualifiers. As a matter of fact, Oracle will return an error if you specify the table name in the USING clause.

8.2.4 JOIN ON Clause

The previous two join styles used common attribute names in the joining tables. Another way to express a join when the tables have no common attribute names is to use the JOIN ON operand. That query will return only the rows that meet the indicated join condition. The join condition will typically include an equality comparison expression of two columns. (The columns may or may not share the same name but, obviously, must have comparable data types.) The syntax is:

SELECT column-list FROM table1 JOIN table2 ON join-condition

The following example performs a join of the INVOICE and LINE tables, using the ON clause. The result is shown in Figure 8.9.

```
SELECT      INVOICE.INV_NUMBER, P_CODE, P_DESCRIPT, LINE_UNITS, LINE_PRICE
FROM        INVOICE JOIN LINE ON INVOICE.INV_NUMBER = LINE.INV_NUMBER
            JOIN PRODUCT ON LINE.P_CODE = PRODUCT.P_CODE;
```

FIGURE 8.9   JOIN ON results

```
Oracle SQL*Plus
File  Edit  Search  Options  Help
SQL> SELECT INVOICE.INV_NUMBER, P_CODE, P_DESCRIPT, LINE_UNITS, LINE_PRICE
  2  FROM INVOICE JOIN LINE ON INVOICE.INV_NUMBER = LINE.INV_NUMBER
  3         JOIN PRODUCT ON LINE.P_CODE = PRODUCT.P_CODE;

INV_NUMBER P_CODE     P_DESCRIPT                          LINE_UNITS LINE_PRICE
---------- ---------- ----------------------------------- ---------- ----------
      1001 13-Q2/P2   7.25-in. pwr. saw blade                      1      14.99
      1001 23109-HB   Claw hammer                                  1       9.95
      1002 54778-2T   Rat-tail file, 1/8-in. fine                  2       4.99
      1003 2238/QPD   B&D cordless drill, 1/2-in.                  1      38.95
      1003 1546-QQ2   Hrd. cloth, 1/4-in., 2x50                    1      39.95
      1003 13-Q2/P2   7.25-in. pwr. saw blade                      5      14.99
      1004 54778-2T   Rat-tail file, 1/8-in. fine                  3       4.99
      1004 23109-HB   Claw hammer                                  2       9.95
      1005 PVC23DRT   PVC pipe, 3.5-in., 8-ft.                    12       5.87
      1006 SM-18277   1.25-in. metal screw, 25                     3       6.99
      1006 2232/QTY   B&D jigsaw, 12-in. blade                     1     109.92
      1006 23109-HB   Claw hammer                                  1       9.95
      1006 89-WRE-Q   Hicut chain saw, 16 in.                      1     256.99
      1007 13-Q2/P2   7.25-in. pwr. saw blade                      2      14.99
      1007 54778-2T   Rat-tail file, 1/8-in. fine                  1       4.99
      1008 PVC23DRT   PVC pipe, 3.5-in., 8-ft                      5       5.87
      1008 WR3/TT3    Steel matting, 4'x8'x1/6", .5" mesh          3     119.95
      1008 23109-HB   Claw hammer                                  1       9.95

18 rows selected.

SQL>
```

Note that unlike the NATURAL JOIN and the JOIN USING operands, the JOIN ON clause requires a table qualifier for the common attributes. If you do not specify the table qualifier, you will get a "column ambiguously defined" error message.

Keep in mind that the JOIN ON syntax lets you perform a join even when the tables do not share a common attribute name. For example, to generate a list of all employees with the managers' names, you can use the following (recursive) query:

```
SELECT      E.EMP_MGR, M.EMP_LNAME, E.EMP_NUM, E.EMP_LNAME
FROM        EMP E JOIN EMP M ON E.EMP_MGR = M.EMP_NUM
ORDER BY    E.EMP_MGR;
```

8.2.5   OUTER JOINS

An outer join returns not only the rows matching the join condition (that is, rows with matching values in the common columns), but also the rows with unmatched values. The ANSI standard defines three types of outer joins: left, right, and full. The left and right designations reflect the order in which the tables are processed by the DBMS. Remember that join operations take place two tables at a time. The first table named in the FROM clause will be the left side, and the second table named will be the right side. If three or more tables are being joined, the result of joining the first two tables becomes the left side, and the third table becomes the right side.

The left outer join returns not only the rows matching the join condition (that is, rows with matching values in the common column), but also the rows in the left side table with unmatched values in the right side table. The syntax is:

```
SELECT      column-list
FROM        table1 LEFT [OUTER] JOIN table2 ON join-condition
```

For example, the following query lists the product code, vendor code, and vendor name for all products and includes those vendors with no matching products:

```
SELECT      P_CODE, VENDOR.V_CODE, V_NAME
FROM        VENDOR LEFT JOIN PRODUCT ON VENDOR.V_CODE = PRODUCT.V_CODE;
```

The preceding SQL code and its results are shown in Figure 8.10.

The right outer join returns not only the rows matching the join condition (that is, rows with matching values in the common column), but also the rows in the right side table with unmatched values in the left side table. The syntax is:

```
SELECT      column-list
FROM        table1 RIGHT [OUTER] JOIN table2 ON join-condition
```

For example, the following query lists the product code, vendor code, and vendor name for all products and also includes those products that do not have a matching vendor code:

```
SELECT      P_CODE, VENDOR.V_CODE, V_NAME
FROM        VENDOR RIGHT JOIN PRODUCT ON VENDOR.V_CODE = PRODUCT.V_CODE;
```

FIGURE 8.10   **LEFT JOIN results**

```
± Oracle SQL*Plus                                              _□×
File  Edit  Search  Options  Help
SQL> SELECT P_CODE, VENDOR.V_CODE, V_NAME
  2  FROM VENDOR LEFT JOIN PRODUCT ON VENDOR.V_CODE = PRODUCT.V_CODE;

P_CODE        V_CODE V_NAME
----------    ------ --------------------------------
11QER/31       25595 Rubicon Systems
13-Q2/P2       21344 Gomez Bros.
14-Q1/L3       21344 Gomez Bros.
1546-QQ2       23119 Randsets Ltd.
1558-QW1       23119 Randsets Ltd.
2232/QTY       24288 ORDVA, Inc.
2232/QWE       24288 ORDVA, Inc.
2238/QPD       25595 Rubicon Systems
23109-HB       21225 Bryson, Inc.
54778-2T       21344 Gomez Bros.
89-WRE-Q       24288 ORDVA, Inc.
SM-18277       21225 Bryson, Inc.
SW-23116       21231 D&E Supply
WR3/TT3        25595 Rubicon Systems
               22567 Dome Supply
               21226 SuperLoo, Inc.
               24004 Brackman Bros.
               25501 Damal Supplies
               25443 B&K, Inc.

19 rows selected.

SQL>
```

The SQL code and its output are shown in Figure 8.11.

The full outer join returns not only the rows matching the join condition (that is, rows with matching values in the common column), but also all of the rows with unmatched values in either side table. The syntax is:

SELECT   column-list
FROM   *table1* FULL [OUTER] JOIN *table2* ON *join-condition*

For example, the following query lists the product code, vendor code, and vendor name for all products and includes all product rows (products without matching vendors) as well as all vendor rows (vendors without matching products).

SELECT   P_CODE, VENDOR.V_CODE, V_NAME
FROM   VENDOR FULL JOIN PRODUCT ON VENDOR.V_CODE = PRODUCT.V_CODE;

The SQL code and its results are shown in Figure 8.12.

**FIGURE
8.11**   **RIGHT JOIN results**

```
± Oracle SQL*Plus                                                    _ □ ×
File  Edit  Search  Options  Help
SQL> SELECT P_CODE, VENDOR.V_CODE, V_NAME
  2  FROM VENDOR RIGHT JOIN PRODUCT ON VENDOR.V_CODE = PRODUCT.V_CODE;

P_CODE       V_CODE V_NAME
----------   ------ ------------------------------
SM-18277      21225 Bryson, Inc.
23109-HB      21225 Bryson, Inc.
SW-23116      21231 D&E Supply
54778-2T      21344 Gomez Bros.
14-Q1/L3      21344 Gomez Bros.
13-Q2/P2      21344 Gomez Bros.
1558-QW1      23119 Randsets Ltd.
1546-QQ2      23119 Randsets Ltd.
89-WRE-Q      24288 ORDVA, Inc.
2232/QWE      24288 ORDVA, Inc.
2232/QTY      24288 ORDVA, Inc.
WR3/TT3       25595 Rubicon Systems
2238/QPD      25595 Rubicon Systems
11QER/31      25595 Rubicon Systems
PVC23DRT
23114-AA

16 rows selected.

SQL>
```

**FIGURE
8.12**   **FULL JOIN results**

```
± Oracle SQL*Plus                                                    _ □ ×
File  Edit  Search  Options  Help
SQL> SELECT P_CODE, VENDOR.V_CODE, V_NAME
  2  FROM VENDOR FULL JOIN PRODUCT ON VENDOR.V_CODE = PRODUCT.V_CODE;

P_CODE       V_CODE V_NAME
----------   ------ ------------------------------
11QER/31      25595 Rubicon Systems
13-Q2/P2      21344 Gomez Bros.
14-Q1/L3      21344 Gomez Bros.
1546-QQ2      23119 Randsets Ltd.
1558-QW1      23119 Randsets Ltd.
2232/QTY      24288 ORDVA, Inc.
2232/QWE      24288 ORDVA, Inc.
2238/QPD      25595 Rubicon Systems
23109-HB      21225 Bryson, Inc.
54778-2T      21344 Gomez Bros.
89-WRE-Q      24288 ORDVA, Inc.
SM-18277      21225 Bryson, Inc.
SW-23116      21231 D&E Supply
WR3/TT3       25595 Rubicon Systems
              22567 Dome Supply
              21226 SuperLoo, Inc.
              24004 Brackman Bros.
              25501 Damal Supplies
              25443 B&K, Inc.
23114-AA
PVC23DRT

21 rows selected.

SQL>
```

8.3 SUBQUERIES AND CORRELATED QUERIES

The use of joins in a relational database allows you to get information from two or more tables. For example, the following query allows you to get the customers' data with their respective invoices by joining the CUSTOMER and INVOICE tables.

```
SELECT      INV_NUMBER, INVOICE.CUS_CODE, CUS_LNAME, CUS_FNAME
FROM        CUSTOMER, INVOICE
WHERE       CUSTOMER.CUS_CODE = INVOICE.CUS_CODE;
```

In the previous query, the data from both tables (CUSTOMER and INVOICE) are processed at once, matching rows with shared CUS_CODE values.

However, it is often necessary to process data based on *other* processed data. Suppose, for example, you want to generate a list of vendors who provide products. (Recall that not all vendors in the VENDOR table have provided products—some of them are only *potential* vendors.) In Chapter 7, you learned that you could generate such a list by writing the following query:

```
SELECT      V_CODE, V_NAME FROM VENDOR
WHERE       V_CODE NOT IN (SELECT V_CODE FROM PRODUCT);
```

Similarly, to generate a list of all products with a price greater than or equal to the average product price, you can write the following query:

```
SELECT      P_CODE, P_PRICE FROM PRODUCT
WHERE       P_PRICE >= (SELECT AVG(P_PRICE) FROM PRODUCT);
```

In both of those cases, you needed to get information that was not previously known:

- What vendors provide products?
- What is the average price of all products?

In both cases, you used a subquery to generate the required information that could then be used as input for the originating query.

You learned how to use subqueries in Chapter 7; let's review their basic characteristics:

- A subquery is a query (SELECT statement) inside a query.
- A subquery is normally expressed inside parentheses.
- The first query in the SQL statement is known as the outer query.
- The query inside the SQL statement is known as the inner query.
- The inner query is executed first.
- The output of an inner query is used as the input for the outer query.
- The entire SQL statement is sometimes referred to as a nested query.

In this section, you learn more about the practical use of subqueries. You already know that a subquery is based on the use of the SELECT statement to return one or more values to another query. But subqueries have a wide range of uses. For example, you can use a subquery within an SQL data manipulation language (DML) statement (such as INSERT, UPDATE, or DELETE) where a value or a list of values (such as multiple vendor codes or a table) is expected. Table 8.2 uses simple examples to summarize the use of SELECT subqueries in DML statements.

TABLE 8.2	SELECT Subquery Examples

SELECT SUBQUERY EXAMPLES	EXPLANATION
INSERT INTO PRODUCT SELECT * FROM P;	Inserts all rows from Table P into the PRODUCT table. Both tables must have the same attributes. The subquery returns all rows from Table P.

UPDATE PRODUCT
SET   P_PRICE = (SELECT AVG(P_PRICE)
                 FROM PRODUCT)
WHERE V_CODE IN (SELECT V_CODE
                 FROM VENDOR
                 WHERE V_AREACODE = '615')
	Updates the product price to the average product price, but only for the products that are provided by vendors who have an area code equal to 615. The first subquery returns the average price; the second subquery returns the list of vendors with an area code equal to 615.

DELETE FROM PRODUCT
WHERE V_CODE IN (SELECT V_CODE
                 FROM VENDOR
                 WHERE V_AREACODE = '615')
	Deletes the PRODUCT table rows that are provided by vendors with area code equal to 615. The subquery returns the list of vendors codes with an area code equal to 615.

Using the examples shown in Table 8.2, note that the subquery is always at the right side of a comparison or assigning expression. Also, a subquery can return one value or multiple values. To be precise, the subquery can return:

- *One single value (one column and one row).* This subquery is used anywhere a single value is expected, as in the right side of a comparison expression (such as in the preceding UPDATE example when you assign the average price to the product's price). Obviously, when you assign a value to an attribute, that value is a single value, not a list of values. Therefore, the subquery must return only one value (one column, one row). If the query returns multiple values, the DBMS will generate an error.

- *A list of values (one column and multiple rows).* This type of subquery is used anywhere a list of values is expected, such as when using the IN clause (that is, when comparing the vendor code to a list of vendors). Again, in this case, there is only one column of data with multiple value instances. This type of subquery is used frequently in combination with the IN operator in a WHERE conditional expression.

- *A virtual table (multicolumn, multirow set of values).* This type of subquery can be used anywhere a table is expected, such as when using the FROM clause. You will see this type of query later in this chapter.

It's important to note that a subquery can return no values at all; it is a NULL. In such cases, the output of the outer query might result in an error or a null empty set, depending where the subquery is used (in a comparison, an expression, or a table set).

In the following sections, you will learn how to write subqueries within the SELECT statement to retrieve data from the database.

8.3.1   WHERE Subqueries

The most common type of subquery uses an inner SELECT subquery on the right side of a WHERE comparison expression. For example, to find all products with a price greater than or equal to the average product price, you write the following query:

```
SELECT      P_CODE, P_PRICE FROM PRODUCT
WHERE       P_PRICE >= (SELECT AVG(P_PRICE) FROM PRODUCT);
```

The output of the preceding query is shown in Figure 8.13. Note that this type of query, when used in a >, <, =, >=, or <= conditional expression, requires a subquery that returns only one single value (one column, one row). The value generated by the subquery must be of a "comparable" data type; if the attribute to the left of the comparison symbol is a character type, the subquery must return a character string. Also, if the query returns more than a single value, the DBMS will generate an error.

FIGURE 8.13   WHERE subquery example

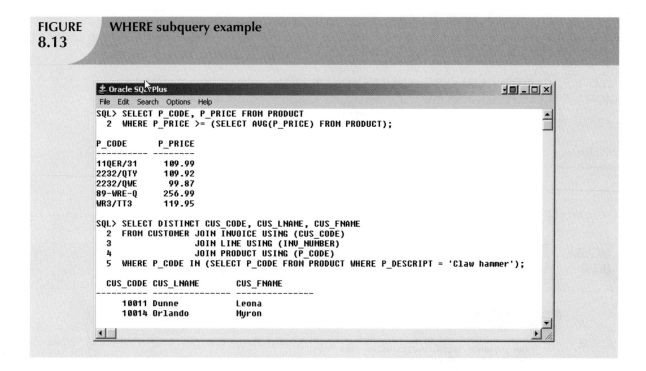

Subqueries can also be used in combination with joins. For example, the following query lists all of the customers who ordered the product "Claw hammer":

```
SELECT      DISTINCT CUS_CODE, CUS_LNAME, CUS_FNAME
FROM        CUSTOMER JOIN INVOICE USING (CUS_CODE)
                    JOIN LINE USING (INV_NUMBER)
                    JOIN PRODUCT USING (P_CODE)
WHERE       P_CODE = (SELECT P_CODE FROM PRODUCT WHERE P_DESCRIPT = 'Claw hammer');
```

The result of that query is also shown in Figure 8.13.

In the preceding example, the inner query finds the P_CODE for the product "Claw hammer." The P_CODE is then used to restrict the selected rows to only those where the P_CODE in the LINE table matches the P_CODE for "Claw hammer." Note that the previous query could have been written this way:

```
SELECT      DISTINCT CUS_CODE, CUS_LNAME, CUS_FNAME
FROM        CUSTOMER JOIN INVOICE USING (CUS_CODE)
                    JOIN LINE USING (INV_NUMBER)
                    JOIN PRODUCT USING (P_CODE)
WHERE       P_DESCRIPT = 'Claw hammer';
```

But what happens if the original query encounters the "Claw hammer" string in more than one product description? You get an error message. To compare one value to a list of values, you must use an IN operand, as shown in the next section.

8.3.2 IN SUBQUERIES

What would you do if you wanted to find all customers who purchased a "hammer" or any kind of saw or saw blade? Note that the product table has two different types of hammers: "Claw hammer" and "Sledge hammer." Also note that there are multiple occurrences of products that contain "saw" in their product descriptions. There are saw blades, jigsaws, and so on. In such cases, you need to compare the P_CODE not to one product code (single value), but to a list of product code values. When you want to compare a single attribute to a list of values, you use the IN operator. When the P_CODE values are not known beforehand but they can be derived using a query, you must use an IN subquery. The following example lists all customers who have purchased hammers, saws, or saw blades.

SELECT	DISTINCT CUS_CODE, CUS_LNAME, CUS_FNAME
FROM	CUSTOMER JOIN INVOICE USING (CUS_CODE)
	JOIN LINE USING (INV_NUMBER)
	JOIN PRODUCT USING (P_CODE)
WHERE	P_CODE IN (SELECT P_CODE FROM PRODUCT
	WHERE P_DESCRIPT LIKE '%hammer%'
	OR P_DESCRIPT LIKE '%saw%');

The result of that query is shown in Figure 8.14.

**FIGURE
8.14**   **IN subquery example**

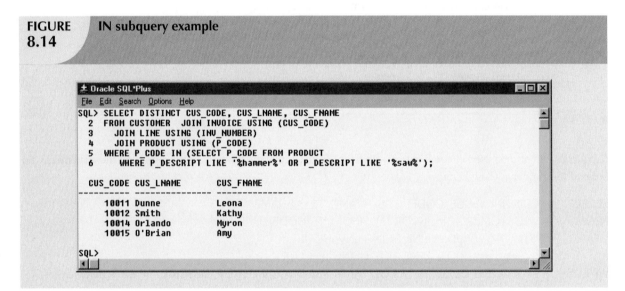

8.3.3 HAVING SUBQUERIES

Just as you can use subqueries with the WHERE clause, you can use a subquery with a HAVING clause. Remember that the HAVING clause is used to restrict the output of a GROUP BY query by applying a conditional criteria to the grouped rows. For example, to list all products with the total quantity sold greater than the average quantity sold, you would write the following query:

SELECT	P_CODE, SUM(LINE_UNITS)
FROM	LINE
GROUP BY	P_CODE
HAVING	SUM(LINE_UNITS) > (SELECT AVG(LINE_UNITS) FROM LINE);

The result of that query is shown in Figure 8.15.

FIGURE 8.15 HAVING subquery example

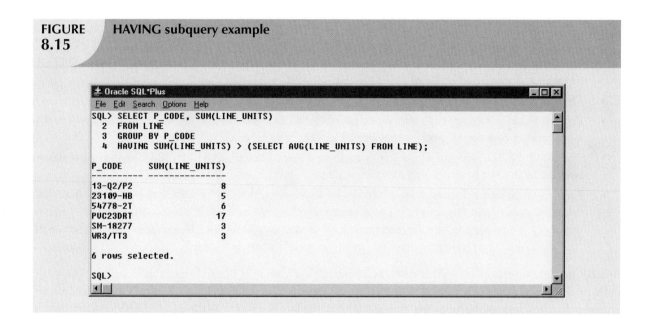

```
± Oracle SQL*Plus                                              _ □ X
 File  Edit  Search  Options  Help
SQL> SELECT P_CODE, SUM(LINE_UNITS)
  2   FROM LINE
  3   GROUP BY P_CODE
  4   HAVING SUM(LINE_UNITS) > (SELECT AVG(LINE_UNITS) FROM LINE);

P_CODE     SUM(LINE_UNITS)
---------- ---------------
13-Q2/P2                 8
23109-HB                 5
54778-2T                 6
PVC23DRT                17
SM-18277                 3
WR3/TT3                  3

6 rows selected.

SQL>
```

8.3.4 MULTIROW SUBQUERY OPERATORS: ANY AND ALL

So far, you have learned that you must use an IN subquery when you need to compare a value to a list of values. But the IN subquery uses an equality operator; that is, it selects only those rows that match (are equal to) at least one of the values in the list. What happens if you need to do an inequality comparison (> or <) of one value to a list of values?

For example, suppose you want to know what products have a product cost that is greater than all individual product costs for products provided by vendors from Florida.

```
SELECT      P_CODE, P_QOH * P_PRICE
FROM        PRODUCT
WHERE       P_QOH * P_PRICE > ALL (SELECT   P_QOH * P_PRICE
                                   FROM      PRODUCT
                                   WHERE     V_CODE IN (SELECT V_CODE
                                                        FROM    VENDOR
                                                        WHERE   V_STATE = 'FL'));
```

The result of that query is shown in Figure 8.16.

FIGURE 8.16 Multirow subquery operator example

```
± Oracle SQL*Plus                                              ◄□ _ □ X
 File  Edit  Search  Options  Help
SQL> SELECT P_CODE, P_QOH*P_PRICE
  2   FROM PRODUCT
  3   WHERE P_QOH*P_PRICE > ALL
  4   (SELECT P_QOH*P_PRICE FROM PRODUCT
  5    WHERE V_CODE IN (SELECT V_CODE FROM VENDOR WHERE V_STATE = 'FL'));

P_CODE     P_QOH*P_PRICE
---------- -------------
89-WRE-Q        2826.89

◄
```

It's important to note the following points about the query and its output in Figure 8.16:

- The query is a typical example of a nested query.

- The query has one outer SELECT statement with a SELECT subquery (call it sq^A) containing a second SELECT subquery (call it sq^B).

- The last SELECT subquery (sq^B) is executed first and returns a list of all vendors from Florida.

- The first SELECT subquery (sq^A) uses the output of the SELECT subquery (sq^B). The sq^A subquery returns the list of product costs for all products provided by vendors from Florida.

- The use of the ALL operator allows you to compare a single value (P_QOH * P_PRICE) with a list of values returned by the first subquery (sq^A) using a comparison operator other than equals.

- For a row to appear in the result set, it has to meet the criterion P_QOH * P_PRICE > ALL, of the individual values returned by the subquery sq^A. The values returned by sq^A are a list of product costs. In fact, "greater than ALL" is equivalent to "greater than the highest product cost of the list." In the same way, a condition of "less than ALL" is equivalent to "less than the lowest product cost of the list."

Another powerful operator is the ANY multirow operator (near cousin of the ALL multirow operator). The ANY operator allows you to compare a single value to a list of values, selecting only the rows for which the inventory cost is greater than any value of the list or less than any value of the list. You could use the equal to ANY operator, which would be the equivalent of the IN operator.

8.3.5 FROM SUBQUERIES

So far you have seen how the SELECT statement uses subqueries within WHERE, HAVING, and IN statements and how the ANY and ALL operators are used for multirow subqueries. In all of those cases, the subquery was part of a conditional expression and it always appeared at the right side of the expression. In this section, you will learn how to use subqueries in the FROM clause.

As you already know, the FROM clause specifies the table(s) from which the data will be drawn. Because the output of a SELECT statement is another table (or more precisely a "virtual" table), you could use a SELECT subquery in the FROM clause. For example, assume that you want to know all customers who have purchased products 13-Q2/P2 and 23109-HB. All product purchases are stored in the LINE table. It is easy to find out who purchased any given product by searching the P_CODE attribute in the LINE table. But in this case, you want to know all customers who purchased both products, not just one. You could write the following query:

```
SELECT      DISTINCT CUSTOMER.CUS_CODE, CUSTOMER.CUS_LNAME
FROM        CUSTOMER,
            (SELECT INVOICE.CUS_CODE FROM INVOICE NATURAL JOIN LINE
            WHERE P_CODE = '13-Q2/P2') CP1,
            (SELECT INVOICE.CUS_CODE FROM INVOICE NATURAL JOIN LINE
            WHERE P_CODE = '23109-HB') CP2
WHERE       CUSTOMER.CUS_CODE = CP1.CUS_CODE AND CP1.CUS_CODE = CP2.CUS_CODE;
```

The result of that query is shown in Figure 8.17.

Note in Figure 8.17 that the first subquery returns all customers who purchased product 13-Q2/P2, while the second subquery returns all customers who purchased product 23109-HB. So in this FROM subquery, you are joining the CUSTOMER table with two virtual tables. The join condition selects only the rows with matching CUS_CODE values in each table (base or virtual).

FIGURE 8.17 **FROM subquery example**

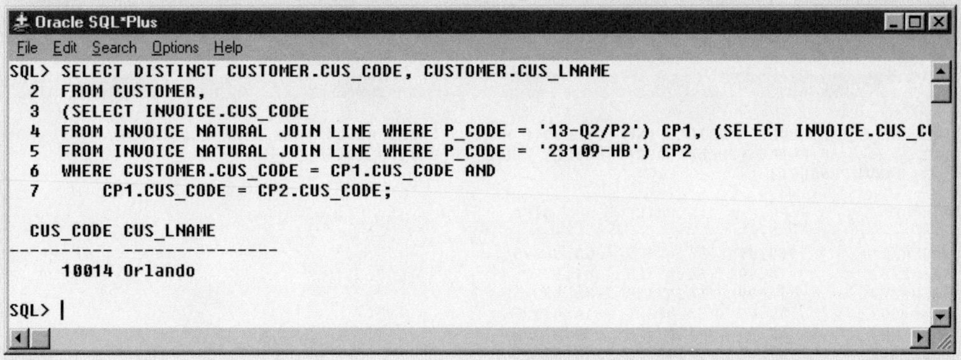

```
Oracle SQL*Plus                                                     _ □ ×
File  Edit  Search  Options  Help
SQL> SELECT DISTINCT CUSTOMER.CUS_CODE, CUSTOMER.CUS_LNAME
  2   FROM CUSTOMER,
  3   (SELECT INVOICE.CUS_CODE
  4   FROM INVOICE NATURAL JOIN LINE WHERE P_CODE = '13-Q2/P2') CP1, (SELECT INVOICE.CUS_C
  5   FROM INVOICE NATURAL JOIN LINE WHERE P_CODE = '23109-HB') CP2
  6   WHERE CUSTOMER.CUS_CODE = CP1.CUS_CODE AND
  7        CP1.CUS_CODE = CP2.CUS_CODE;

  CUS_CODE CUS_LNAME
---------- ----------------
     10014 Orlando

SQL> |
```

In the previous chapter, you learned that a view is also a virtual table; therefore, you can use a view name anywhere a table is expected. So in this example, you could create two views: one listing all customers who purchased product 13-Q2/P2 and another listing all customers who purchased product 23109-HB. Doing so, you would write the query as:

CREATE VIEW CP1 AS
> SELECT   INVOICE.CUS_CODE FROM INVOICE NATURAL JOIN LINE
> WHERE   P_CODE = '13-Q2/P2';

CREATE VIEW CP2 AS
> SELECT   INVOICE.CUS_CODE FROM INVOICE NATURAL JOIN LINE
> WHERE   P_CODE = '23109-HB';

SELECT   DISTINCT CUS_CODE, CUS_LNAME
FROM   CUSTOMER NATURAL JOIN CP1 NATURAL JOIN CP2;

You might speculate that the above query could also be written using the following syntax:

SELECT   CUS_CODE, CUS_LNAME
FROM   CUSTOMER NATURAL JOIN INVOICE NATURAL JOIN LINE
WHERE   P_CODE = '13-Q2/P2' AND P_CODE = '23109-HB';

But if you examine that query carefully, you will note that a P_CODE cannot be equal to two different values at the same time. Therefore, the query will not return any rows.

8.3.6 ATTRIBUTE LIST SUBQUERIES

The SELECT statement uses the attribute list to indicate what columns to project in the resulting set. Those columns can be attributes of base tables or computed attributes or the result of an aggregate function. The attribute list can also include a subquery expression, also known as an inline subquery. A subquery in the attribute list must return one single value; otherwise, an error code is raised. For example, a simple inline query can be used to list the difference between each product's price and the average product price:

SELECT   P_CODE, P_PRICE, (SELECT AVG(P_PRICE) FROM PRODUCT) AS AVGPRICE,
              P_PRICE – (SELECT AVG(P_PRICE) FROM PRODUCT) AS DIFF
FROM   PRODUCT;

Figure 8.18 shows the result of that query.

FIGURE 8.18   Inline subquery example

```
Oracle SQL*Plus                                                    _□X
File  Edit  Search  Options  Help
SQL> SELECT P_CODE, P_PRICE, (SELECT AVG(P_PRICE) FROM PRODUCT) AS AVGPRICE,
  2          P_PRICE-(SELECT AVG(P_PRICE) FROM PRODUCT) AS DIFF
  3  FROM PRODUCT;

P_CODE        P_PRICE   AVGPRICE       DIFF
----------  ----------  ----------  ----------
11QER/31      109.99    56.42125     53.56875
13-Q2/P2       14.99    56.42125    -41.43125
14-Q1/L3       17.49    56.42125    -38.93125
1546-QQ2       39.95    56.42125    -16.47125
1558-QW1       43.99    56.42125    -12.43125
2232/QTY      109.92    56.42125     53.49875
2232/QWE       99.87    56.42125     43.44875
2238/QPD       38.95    56.42125    -17.47125
23109-HB        9.95    56.42125    -46.47125
23114-AA       14.4     56.42125    -42.02125
54778-2T        4.99    56.42125    -51.43125
89-WRE-Q      256.99    56.42125    200.56875
PVC23DRT        5.87    56.42125    -50.55125
SM-18277        6.99    56.42125    -49.43125
SW-23116        8.45    56.42125    -47.97125
WR3/TT3       119.95    56.42125     63.52875

16 rows selected.

SQL>
```

In Figure 8.18, note that the inline query output returns one single value (the average product's price) and that the value is the same in every row. Note also that the query used the full expression instead of the column aliases when computing the difference. In fact, if you try to use the alias in the difference expression, you will get an error message. The column alias cannot be used in computations in the attribute list when the alias is defined in the same attribute list. That DBMS requirement is due to the way the DBMS parses and executes queries.

Another example will help you understand the use of attribute list subqueries and column aliases. For example, suppose you want to know the product code, the total sales by product, and the contribution by employee of each product's sales. To get the sales by product, you need to use only the LINE table. To compute the contribution by employee, you need to know the number of employees (from the EMPLOYEE table). As you study the tables' structures, you can see that the LINE and EMPLOYEE tables do not share a common attribute. In fact, you don't need a common attribute. You need to know only the total number of employees, not the total employees related to each product. So to answer the query, you would write the following code:

```
SELECT     P_CODE, SUM(LINE_UNITS * LINE_PRICE) AS SALES,
           (SELECT COUNT(*) FROM EMPLOYEE) AS ECOUNT,
           SUM(LINE_UNITS * LINE_PRICE)/(SELECT COUNT(*) FROM EMPLOYEE) AS CONTRIB
FROM       LINE
GROUP BY   P_CODE;
```

The result of that query is shown in Figure 8.19.

As you can see in Figure 8.19, the number of employees remains the same for each row in the result set. The use of that type of subquery is limited to certain instances where you need to include data from other tables that are not directly related to a main table or tables in the query. The value will remain the same for each row, like a constant in a programming language. (You will learn another use of inline subqueries in Section 8.3.7, Correlated Subqueries).

FIGURE 8.21   EXISTS correlated subquery examples

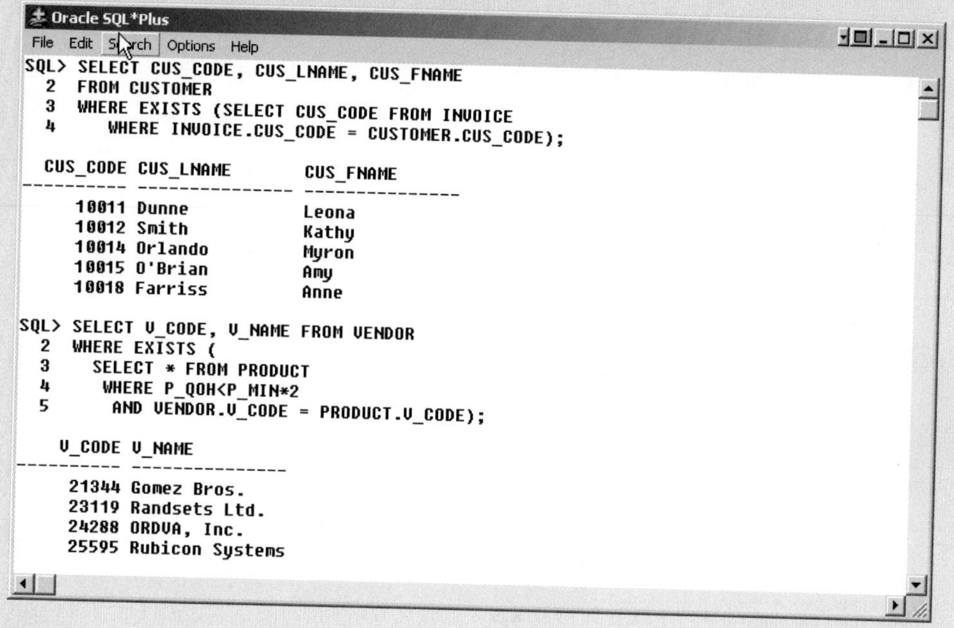

In the second query in Figure 8.21, note that:

1. The inner correlated subquery runs using the first vendor.

2. If any products match the condition (quantity on hand is less than double the minimum quantity), the vendor code and name are listed in the output.

3. The correlated subquery runs using the second vendor, and the process repeats itself until all vendors are used.

8.4 SQL FUNCTIONS

The data in databases are the basis of critical business information. Generating information from data often requires many data manipulations. Sometimes such data manipulation involves the decomposition of data elements. For example, an employee's date of birth can be subdivided into a day, a month, and a year. A product manufacturing code (for example, SE-05-2-09-1234-1-3/12/04-19:26:48) can be designed to record the manufacturing region, plant, shift, production line, employee number, date, and time. For years, conventional programming languages have had special functions that enabled programmers to perform data transformations like those data decompositions. If you know a modern programming language, it's very likely that the SQL functions in this section will look familiar.

SQL functions are very useful tools. You'll need to use functions when you want to list all employees ordered by year of birth or when your marketing department wants you to generate a list of all customers ordered by zip code and the first three digits of their telephone numbers. In both of those cases, you'll need to use data elements that are not present as such in the database; instead you'll need an SQL function that can be derived from an existing attribute. Functions always use a numerical, date, or string value. The value may be part of the command itself (a constant or literal) or it may be an attribute located in a table. Therefore, a function may appear anywhere in an SQL statement where a value or an attribute can be used.

FIGURE 8.19   Another example of an inline subquery

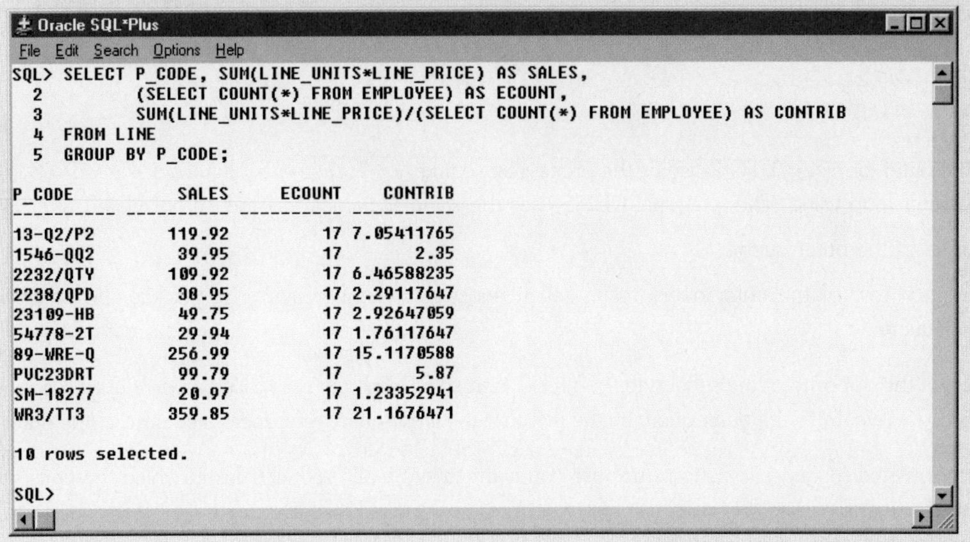

Note that you cannot use an alias in the attribute list to write the expression that computes the contribution per employee.

Another way to write the same query by using column aliases requires the use of a subquery in the FROM clause, as follows:

```
SELECT    P_CODE, SALES, ECOUNT, SALES/ECOUNT AS CONTRIB
FROM      (SELECT      P_CODE, SUM(LINE_UNITS * LINE_PRICE) AS SALES,
                       (SELECT COUNT(*) FROM EMPLOYEE) AS ECOUNT
          FROM         LINE
          GROUP BY     P_CODE);
```

In that case, you are actually using two subqueries. The subquery in the FROM clause executes first and returns a virtual table with three columns: P_CODE, SALES, and ECOUNT. The FROM subquery contains an inline subquery that returns the number of employees as ECOUNT. Because the outer query receives the output of the inner query, you can now refer to the columns in the outer subquery using the column aliases.

8.3.7 CORRELATED SUBQUERIES

Until now, all subqueries you have learned execute independently. That is, each subquery in a command sequence executes in a serial fashion, one after another. The inner subquery executes first; its output is used by the outer query, which then executes until the last outer query executes (the first SQL statement in the code).

In contrast, a **correlated subquery** is a subquery that executes once for each row in the outer query. That process is similar to the typical nested loop in a programming language. For example:

```
FOR X = 1 TO 2
      FOR Y = 1 TO 3
            PRINT "X = "X, "Y = "Y
      END
END
```

will yield the output

X = 1 Y = 1
X = 1 Y = 2
X = 1 Y = 3
X = 2 Y = 1
X = 2 Y = 2
X = 2 Y = 3

Note that the outer loop X = 1 TO 2 begins the process by setting X = 1; then the inner loop Y = 1 TO 3 is completed for each X outer loop value. The relational DBMS uses the same sequence to produce correlated subquery results:

1. It initiates the outer query.
2. For each row of the outer query result set, it executes the inner query by passing the outer row to the inner query.

That process is the opposite of the subqueries you have seen so far. The query is called a *correlated* subquery because the inner query is *related* to the outer query by the fact that the inner query references a column of the outer subquery.

To see the correlated subquery in action, suppose you want to know all product sales in which the units sold value is greater than the average units sold value *for that product* (as opposed to the average for *all* products). In that case, the following procedure must be completed:

1. Compute the average-units-sold value for a product.
2. Compare the average computed in Step 1 to the units sold in each sale row; then select only the rows in which the number of units sold is greater.

The following correlated query completes the preceding two-step process:

```
SELECT   INV_NUMBER, P_CODE, LINE_UNITS
FROM     LINE LS
WHERE    LS.LINE_UNITS > (SELECT   AVG(LINE_UNITS)
                          FROM     LINE LA
                          WHERE    LA.P_CODE = LS.P_CODE);
```

The first example in Figure 8.20 shows the result of that query.

In the top query and its result in Figure 8.20, note that the LINE table is used more than once; so you must use table aliases. In that case, the inner query computes the average units sold of the product that matches the P_CODE of the outer query P_CODE. That is, the inner query runs once using the first product code found in the (outer) LINE table and returns the average sale for that product. When the number of units sold in that (outer) LINE row is greater than the average computed, the row is added to the output. Then the inner query runs again, this time using the second product code found in the (outer) LINE table. The process repeats until the inner query has run for all rows in the (outer) LINE table. In that case, the inner query will be repeated as many times as there are rows in the outer query.

To verify the results and to provide an example of how you can combine subqueries, you can add a correlated inline subquery to the previous query. That correlated inline subquery will show the average units sold column for each product. (See the second query and its results in Figure 8.20.) As you can see, the new query contains a correlated inline subquery that computes the average units sold for each product. You not only get an answer, but you also can verify that the answer is correct.

FIGURE 8.20 Correlated subquery examples

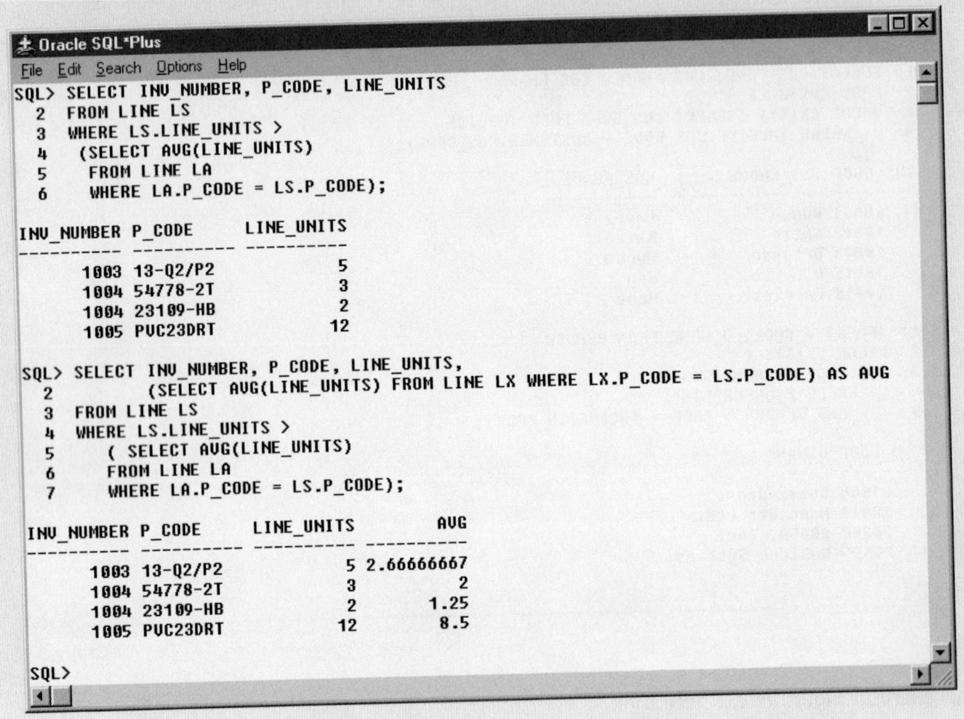

```
Oracle SQL*Plus
File Edit Search Options Help
SQL> SELECT INV_NUMBER, P_CODE, LINE_UNITS
  2  FROM LINE LS
  3  WHERE LS.LINE_UNITS >
  4   (SELECT AVG(LINE_UNITS)
  5    FROM LINE LA
  6    WHERE LA.P_CODE = LS.P_CODE);

INV_NUMBER P_CODE     LINE_UNITS
---------- ---------- ----------
      1003 13-Q2/P2            5
      1004 54778-2T            3
      1004 23109-HB            2
      1005 PVC23DRT           12

SQL> SELECT INV_NUMBER, P_CODE, LINE_UNITS,
  2         (SELECT AVG(LINE_UNITS) FROM LINE LX WHERE LX.P_CODE = LS.P_CODE) AS AVG
  3  FROM LINE LS
  4  WHERE LS.LINE_UNITS >
  5     ( SELECT AVG(LINE_UNITS)
  6     FROM LINE LA
  7     WHERE LA.P_CODE = LS.P_CODE);

INV_NUMBER P_CODE     LINE_UNITS        AVG
---------- ---------- ---------- ----------
      1003 13-Q2/P2            5 2.66666667
      1004 54778-2T            3          2
      1004 23109-HB            2       1.25
      1005 PVC23DRT           12        8.5

SQL>
```

Correlated subqueries can also be used with the EXISTS special operator. For example, suppose you want to know all customers who have placed an order lately. In that case, you could use a correlated subquery like the first one shown in Figure 8.21:

```
SELECT   CUS_CODE, CUS_LNAME, CUS_FNAME
FROM     CUSTOMER
WHERE EXISTS (SELECT   CUS_CODE FROM INVOICE
              WHERE    INVOICE.CUS_CODE = CUSTOMER.CUS_CODE);
```

The second example of an EXISTS correlated subquery in Figure 8.21 will help you understand how to use correlated queries. For example, suppose you want to know what vendors you must contact to start ordering products that are approaching the minimum quantity-on-hand value. In particular, you want to know the vendor code and name of vendors for products having a quantity on hand that is less than double the minimum quantity. The query that answers that question is as follows:

```
SELECT   V_CODE, V_NAME
FROM     VENDOR
WHERE EXISTS (SELECT *
              FROM    PRODUCT
              WHERE   P_QOH < P_MIN * 2
              AND     VENDOR.V_CODE = PRODUCT.V_CODE);
```

There are many types of SQL functions, such as arithmetic, trigonometric, string, date, and time functions. This section will not explain all of those types of functions in detail, but it will give you a brief overview of the most useful ones.

NOTE

Although the main DBMS vendors support the SQL functions covered here, the syntax or degree of support will probably differ. In fact, DBMS vendors invariably add their own functions to products to lure new customers. The functions covered in this section represent just a small portion of functions supported by your DBMS. Read your DBMS SQL reference manual for a complete list of available functions.

8.4.1 DATE AND TIME FUNCTIONS

All SQL-standard DBMSs support date and time functions. All date functions take one parameter (of a date or character data type) and return a value (character, numeric, or date type). Unfortunately, date/time data types are implemented differently by different DBMS vendors. The problem occurs because the ANSI SQL standard defines date data types, but it does not say how those data types are to be stored. Instead, it lets the vendor deal with that issue.

Because date/time functions differ from vendor to vendor, this section will cover basic date/time functions for MS Access/SQL Server and for Oracle. Table 8.3 shows a list of selected MS Access/SQL Server date/time functions.

TABLE 8.3   Selected MS Access/SQL Server Date/Time Functions

FUNCTION	EXAMPLE(S)
YEAR Returns a four-digit year Syntax: YEAR(date_value)	Lists all employees born in 1966: SELECT EMP_LNAME, EMP_FNAME, EMP_DOB, YEAR(EMP_DOB) AS YEAR FROM EMPLOYEE WHERE YEAR(EMP_DOB) = 1966;
MONTH Returns a two-digit month code Syntax: MONTH(date_value)	Lists all employees born in November: SELECT EMP_LNAME, EMP_FNAME, EMP_DOB, MONTH(EMP_DOB) AS MONTH FROM EMPLOYEE WHERE MONTH(EMP_DOB) = 11;
DAY Returns the number of the day Syntax: DAY(date_value)	Lists all employees born on the 14th day of the month: SELECT EMP_LNAME, EMP_FNAME, EMP_DOB, DAY(EMP_DOB) AS DAY FROM EMPLOYEE WHERE DAY(EMP_DOB) = 14;
DATE() – MS Access **GETDATE() – SQL Server** Returns today's date	Lists how many days are left until Christmas: SELECT #25-Dec-2008# – DATE(); Note two features: • There is no FROM clause, which is acceptable in MS Access. • The Christmas date is enclosed in # signs because you are doing date arithmetic. In MS SQL Server: Use GETDATE() to get the current system date. To compute the difference between dates, use the DATEDIFF function (see below).

TABLE 8.3   **Selected MS Access/SQL Server Date/Time Functions (continued)**

FUNCTION	EXAMPLE(S)
DATEADD – SQL Server Adds a number of selected time periods to a date Syntax: **DATEADD(datepart, number, date)**	Adds a *number* of *dateparts* to a given *date*. Dateparts can be minutes, hours, days, weeks, months, quarters, or years. For example: SELECT DATEADD(day,90, P_INDATE) AS DueDate FROM PRODUCT; The above example adds 90 days to P_INDATE. In MS Access use: SELECT P_INDATE+90 AS DueDate FROM PRODUCT;
DATEDIFF – SQL Server Subtracts two dates Syntax: **DATEDIFF(datepart, startdate, enddate)**	Returns the difference between two dates expressed in a selected *datepart*. For example: SELECT DATEDIFF(day, P_INDATE, GETDATE()) AS DaysAgo FROM PRODUCT; In MS Access use: SELECT DATE() - P_INDATE AS DaysAgo FROM PRODUCT;

Table 8.4 shows the equivalent date/time functions used in Oracle. Note that Oracle uses the same function (TO_CHAR) to extract the various parts of a date. Also, another function (TO_DATE) is used to convert character strings to a valid Oracle date format that can be used in date arithmetic.

TABLE 8.4   **Selected Oracle Date/Time Functions**

FUNCTION	EXAMPLE(S)
TO_CHAR Returns a character string or a formatted string from a date value Syntax: TO_CHAR(date_value, fmt) fmt = format used; can be: MONTH: name of month MON: three-letter month name MM: two-digit month name D: number for day of week DD: number day of month DAY: name of day of week YYYY: four-digit year value YY: two-digit year value	Lists all employees born in 1982: SELECT EMP_LNAME, EMP_FNAME, EMP_DOB, TO_CHAR(EMP_DOB, 'YYYY') AS YEAR FROM EMPLOYEE WHERE TO_CHAR(EMP_DOB, 'YYYY') = '1982'; Lists all employees born in November: SELECT EMP_LNAME, EMP_FNAME, EMP_DOB, TO_CHAR(EMP_DOB, 'MM') AS MONTH FROM EMPLOYEE WHERE TO_CHAR(EMP_DOB, 'MM') = '11'; Lists all employees born on the 14th day of the month: SELECT EMP_LNAME, EMP_FNAME, EMP_DOB, TO_CHAR(EMP_DOB, 'DD') AS DAY FROM EMPLOYEE WHERE TO_CHAR(EMP_DOB, 'DD') = '14';

TABLE 8.4	Selected Oracle Date/Time Functions (continued)

FUNCTION	EXAMPLE(S)
TO_DATE Returns a date value using a character string and a date format mask; also used to translate a date between formats Syntax: TO_DATE(char_value, fmt) fmt = format used; can be: MONTH: name of month MON: three-letter month name MM: two-digit month name D: number for day of week DD: number day of month DAY: name of day of week YYYY: four-digit year value YY: two-digit year value	Lists the approximate age of the employees on the company's tenth anniversary date (11/25/2008): SELECT EMP_LNAME, EMP_FNAME, EMP_DOB, '11/25/2008' AS ANIV_DATE, (TO_DATE('11/25/1998','MM/DD/YYYY') - EMP_DOB)/365 AS YEARS FROM EMPLOYEE ORDER BY YEARS; Note the following: • '11/25/2008' is a text string, not a date. • The TO_DATE function translates the text string to a valid Oracle date used in date arithmetic. How many days between Thanksgiving and Christmas 2008? SELECT TO_DATE('2008/12/25','YYYY/MM/DD') − TO_DATE('NOVEMBER 27, 2008','MONTH DD, YYYY') FROM DUAL; Note the following: • The TO_DATE function translates the text string to a valid Oracle date used in date arithmetic. • DUAL is Oracle's pseudo table used only for cases where a table is not really needed.
SYSDATE Returns today's date	Lists how many days are left until Christmas: SELECT TO_DATE('25-Dec-2008','DD-MON-YYYY') SYSDATE FROM DUAL; Notice two things: • DUAL is Oracle's pseudo table used only for cases where a table is not really needed. • The Christmas date is enclosed in a TO_DATE function to translate the date to a valid date format.
ADD_MONTHS Adds a number of months to a date; useful for adding months or years to a date Syntax: ADD_MONTHS(date_value, n) n = number of months	Lists all products with their expiration date (two years from the purchase date): SELECT P_CODE, P_INDATE, ADD_MONTHS(P_INDATE,24) FROM PRODUCT ORDER BY ADD_MONTHS(P_INDATE,24);
LAST_DAY Returns the date of the last day of the month given in a date Syntax: LAST_DAY(date_value)	Lists all employees who were hired within the last seven days of a month: SELECT EMP_LNAME, EMP_FNAME, EMP_HIRE_DATE FROM EMPLOYEE WHERE EMP_HIRE_DATE >=LAST_DAY(EMP_HIRE_DATE)-7;

8.4.2 NUMERIC FUNCTIONS

Numeric functions can be grouped in many different ways, such as algebraic, trigonometric, and logarithmic. In this section, you will learn two very useful functions. Do not confuse the SQL aggregate functions you saw in the previous chapter with the numeric functions in this section. The first group operates over a set of values (multiple rows—hence, the name *aggregate functions*), while the numeric functions covered here operate over a single row. Numeric functions take one numeric parameter and return one value. Table 8.5 shows a selected group of numeric functions available.

TABLE 8.5	**Selected Numeric Functions**

FUNCTION	EXAMPLE(S)
ABS Returns the absolute value of a number Syntax: ABS(numeric_value)	In Oracle use: SELECT 1.95, -1.93, ABS(1.95), ABS(-1.93) FROM DUAL; In MS Access/SQL Server use: SELECT 1.95, -1.93, ABS(1.95), ABS(-1.93);

ROUND
Rounds a value to a specified precision
(number of digits)
Syntax:
ROUND(numeric_value, p)
p = precision
	Lists the product prices rounded to one and zero decimal places:
	SELECT P_CODE, P_PRICE,
	                ROUND(P_PRICE,1) AS PRICE1,
	                ROUND(P_PRICE,0) AS PRICE0
	FROM PRODUCT;

CEIL/CEILING/FLOOR Returns the smallest integer greater than or equal to a number or returns the largest integer equal to or less than a number, respectively Syntax: CEIL(numeric_value) − Oracle CEILING(numeric_value) − SQL Server FLOOR(numeric_value)	Lists the product price, smallest integer greater than or equal to the product price, and the largest integer equal to or less than the product price. In Oracle use: SELECT P_PRICE, CEIL(P_PRICE), FLOOR(P_PRICE) FROM PRODUCT; In SQL Server use: SELECT P_PRICE, CEILING(P_PRICE), FLOOR(P_PRICE) FROM PRODUCT; MS Access does not support these functions.

8.4.3 STRING FUNCTIONS

String manipulations are among the most-used functions in programming. If you have ever created a report using any programming language, you know the importance of properly concatenating strings of characters, printing names in uppercase, or knowing the length of a given attribute. Table 8.6 shows a subset of useful string manipulation functions.

TABLE
8.6

Selected String Functions

FUNCTION	EXAMPLE(S)
Concatenation **|| − Oracle** **+ − MS Access/SQL Server** Concatenates data from two different character columns and returns a single column Syntax: strg_value || strg_value strg_value + strg_value	Lists all employee names (concatenated). In Oracle use: SELECT EMP_LNAME || ', ' || EMP_FNAME AS NAME FROM EMPLOYEE; In MS Access / SQL Server use: SELECT EMP_LNAME + ', ' + EMP_FNAME AS NAME FROM EMPLOYEE;
UPPER/LOWER Returns a string in all capital or all lowercase letters Syntax: UPPER(strg_value) LOWER(strg_value)	Lists all employee names in all capital letters (concatenated). In Oracle use: SELECT UPPER(EMP_LNAME) || ', ' || UPPER(EMP_FNAME) AS NAME FROM EMPLOYEE; In SQL Server use: SELECT UPPER(EMP_LNAME) + ', ' + UPPER(EMP_FNAME) AS NAME FROM EMPLOYEE; Lists all employee names in all lowercase letters (concatenated). In Oracle use: SELECT LOWER(EMP_LNAME) || ', ' || LOWER(EMP_FNAME) AS NAME FROM EMPLOYEE; In SQL Server use: SELECT LOWER(EMP_LNAME) + ', ' + LOWER(EMP_FNAME) AS NAME FROM EMPLOYEE; Not supported by MS Access.
SUBSTRING Returns a substring or part of a given string parameter Syntax: SUBSTR(strg_value, p, l) − Oracle SUBSTRING(strg_value,p,l) − SQL Server p = start position l = length of characters	Lists the first three characters of all employee phone numbers. In Oracle use: SELECT EMP_PHONE, SUBSTR(EMP_PHONE,1,3) AS PREFIX FROM EMPLOYEE; In SQL Server use: SELECT EMP_PHONE, SUBSTRING(EMP_PHONE,1,3) AS PREFIX FROM EMPLOYEE; Not supported by MS Access.
LENGTH Returns the number of characters in a string value Syntax: LENGTH(strg_value) − Oracle LEN(strg_value) − SQL Server	Lists all employee last names and the length of their names; ordered descended by last name length. In Oracle use: SELECT EMP_LNAME, LENGTH(EMP_LNAME) AS NAMESIZE FROM EMPLOYEE; In MS Access / SQL Server use: SELECT EMP_LNAME, LEN(EMP_LNAME) AS NAMESIZE FROM EMPLOYEE;

8.4.4 CONVERSION FUNCTIONS

Conversion functions allow you to take a value of a given data type and convert it to the equivalent value in another data type. In Section 8.4.1, you learned about two of the basic Oracle SQL conversion functions: TO_CHAR and TO_DATE. Note that the TO_CHAR function takes a date value and returns a character string representing a day, a month, or a year. In the same way, the TO_DATE function takes a character string representing a date and returns an actual date in Oracle format. SQL Server uses the CAST and CONVERT functions to convert one data type to another. A summary of the selected functions is shown in Table 8.7.

TABLE 8.7   Selected Conversion Functions

FUNCTION	EXAMPLE(S)

Numeric to Character:
TO_CHAR – Oracle
CAST – SQL Server
CONVERT – SQL Server
Returns a character string from a numeric value.
Syntax:
Oracle: TO_CHAR(numeric_value, fmt)
SQL Server:
CAST (numeric AS varchar(length))
CONVERT(varchar(length), numeric)
	Lists all product prices, quantity on hand, percent discount, and total inventory cost using formatted values.
	In Oracle use:
	SELECT P_CODE,
	                TO_CHAR(P_PRICE,'999.99') AS PRICE,
	                TO_CHAR(P_QOH,'9,999.99') AS QUANTITY,
	                TO_CHAR(P_DISCOUNT,'0.99') AS DISC,
	                TO_CHAR(P_PRICE*P_QOH,'99,999.99')
	                AS TOTAL_COST
	FROM PRODUCT;
	In SQL Server use:
	SELECT P_CODE, CAST(P_PRICE AS VARCHAR(8)) AS PRICE,
	                CONVERT(VARCHAR(4),P_QOH) AS QUANTITY,
	                CAST(P_DISCOUNT AS VARCHAR(4)) AS DISC,
	                CAST(P_PRICE*P_QOH AS VARCHAR(10)) AS TOTAL_COST
	FROM PRODUCT;
	Not supported in MS Access.

Date to Character:
TO_CHAR – Oracle
CAST – SQL Server
CONVERT – SQL Server
Returns a character string or a formatted character string from a date value
Syntax:
Oracle: TO_CHAR(date_value, fmt)
SQL Server:
CAST (date AS varchar(length))
CONVERT(varchar(length), date)
	Lists all employee dates of birth, using different date formats.
	In Oracle use:
	SELECT EMP_LNAME, EMP_DOB,
	                TO_CHAR(EMP_DOB, 'DAY, MONTH DD, YYYY')
	                AS 'DATEOFBIRTH'
	FROM EMPLOYEE;
	SELECT EMP_LNAME, EMP_DOB,
	                TO_CHAR(EMP_DOB, 'YYYY/MM/DD')
	                AS 'DATEOFBIRTH'
	FROM EMPLOYEE;
	In SQL Server use:
	SELECT EMP_LNAME, EMP_DOB,
	                CONVERT(varchar(11),EMP_DOB) AS "DATE OF BIRTH"
	FROM EMPLOYEE;
	SELECT EMP_LNAME, EMP_DOB,
	                CAST(EMP_DOB as varchar(11)) AS "DATE OF BIRTH"
	FROM EMPLOYEE;
	Not supported in MS Access.

TABLE 8.7 Selected Conversion Functions (continued)

FUNCTION	EXAMPLE(S)

String to Number:
TO_NUMBER
Returns a formatted number from a character string, using a given format
Syntax:
Oracle:
TO_NUMBER(char_value, fmt)
fmt = format used; can be:
9 = displays a digit
0 = displays a leading zero
, = displays the comma
. = displays the decimal point
$ = displays the dollar sign
B = leading blank
S = leading sign
MI = trailing minus sign
	Converts text strings to numeric values when importing data to a table from another source in text format; for example, the query shown below uses the TO_NUMBER function to convert text formatted to Oracle default numeric values using the format masks given.
	In Oracle use:
	SELECT TO_NUMBER('-123.99', 'S999.99'),
	                  TO_NUMBER('99.78-','B999.99MI')
	FROM DUAL;
	In SQL Server use:
	SELECT CAST('-123.99' AS NUMERIC(8,2)),
	                  CAST('-99.78' AS NUMERIC(8,2))
	The SQL Server CAST function does not support the trailing sign on the character string.
	Not supported in MS Access.

CASE — SQL Server
DECODE — Oracle
Compares an attribute or expression with a series of values and returns an associated value or a default value if no match is found
Syntax:
Oracle:
DECODE(e, x, y, d)
e = attribute or expression
x = value with which to compare e
y = value to return in e = x
d = default value to return if e is not equal to x
SQL Server:
CASE When condition
THEN value1 ELSE value2 END
	The following example returns the sales tax rate for specified states:
	• Compares V_STATE to 'CA'; if the values match, it returns .08.
	• Compares V_STATE to 'FL'; if the values match, it returns .05.
	• Compares V_STATE to 'TN'; if the values match, it returns .085.
	If there is no match, it returns 0.00 (the default value).
	SELECT V_CODE, V_STATE,
	                  DECODE(V_STATE,'CA',.08,'FL',.05, 'TN',.085, 0.00)
	                  AS TAX
	FROM VENDOR;
	In SQL Server use:
	SELECT V_CODE, V_STATE,
	                  CASE WHEN V_STATE = 'CA' THEN .08
	                           WHEN V_STATE = 'FL' THEN .05
	                           WHEN V_STATE = 'TN' THEN .085
	                  ELSE 0.00 END AS TAX
	FROM VENDOR
	Not supported in MS Access.

8.5 ORACLE SEQUENCES

If you use MS Access, you might be familiar with the AutoNumber data type, which you can use to define a column in your table that will be automatically populated with unique numeric values. In fact, if you create a table in MS Access and forget to define a primary key, MS Access will offer to create a primary key column; if you accept, you will notice that MS Access creates a column named *ID* with an AutoNumber data type. After you define a column as an AutoNumber type, every time you insert a row in the table, MS Access will automatically add a value to that column, starting with 1 and increasing the value by 1 in every new row you add. Also, you cannot include that column in your INSERT statements—Access will not let you edit that value at all. MS SQL Server uses the Identity column property to serve a similar purpose. In MS SQL Server a table can have at most one column defined as an Identity column. This column behaves similarly to an MS Access column with the AutoNumber data type.

Oracle does not support the AutoNumber data type or the Identity column property. Instead, you can use a "sequence" to assign values to a column on a table. But an Oracle sequence is very different from the Access AutoNumber data type and deserves close scrutiny:

- Oracle sequences are an independent object in the database. (Sequences are not a data type.)
- Oracle sequences have a name and can be used anywhere a value is expected.

- Oracle sequences are not tied to a table or a column.
- Oracle sequences generate a numeric value that can be assigned to any column in any table.
- The table attribute to which you assigned a value based on a sequence can be edited and modified.
- An Oracle sequence can be created and deleted anytime.

The basic syntax to create a sequence in Oracle is:

CREATE SEQUENCE name [START WITH n] [INCREMENT BY n] [CACHE | NOCACHE]

where:

- name is the name of the sequence.
- n is an integer value that can be positive or negative.
- START WITH specifies the initial sequence value. (The default value is 1.)
- INCREMENT BY determines the value by which the sequence is incremented. (The default increment value is 1. The sequence increment can be positive or negative to enable you to create ascending or descending sequences.)
- The CACHE or NOCACHE clause indicates whether Oracle will preallocate sequence numbers in memory. (Oracle preallocates 20 values by default.)

For example, you could create a sequence to automatically assign values to the customer code each time a new customer is added and create another sequence to automatically assign values to the invoice number each time a new invoice is added. The SQL code to accomplish those tasks is:

CREATE SEQUENCE CUS_CODE_SEQ START WITH 20010 NOCACHE;
CREATE SEQUENCE INV_NUMBER_SEQ START WITH 4010 NOCACHE;

You can check all of the sequences you have created by using the following SQL command, illustrated in Figure 8.22:

SELECT * FROM USER_SEQUENCES;

FIGURE 8.22   Oracle sequence

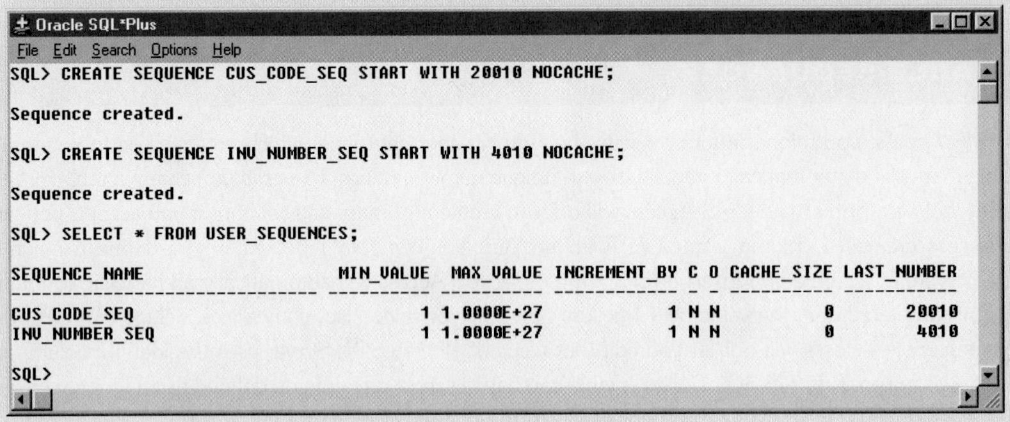

To use sequences during data entry, you must use two special pseudo columns: NEXTVAL and CURRVAL. NEXTVAL retrieves the next available value from a sequence, and CURRVAL retrieves the current value of a sequence. For example, you can use the following code to enter a new customer:

INSERT INTO CUSTOMER
VALUES (CUS_CODE_SEQ.NEXTVAL, 'Connery', 'Sean', NULL, '615', '898-2008', 0.00);

The preceding SQL statement adds a new customer to the CUSTOMER table and assigns the value 20010 to the CUS_CODE attribute. Let's examine some important sequence characteristics:

- CUS_CODE_SEQ.NEXTVAL retrieves the next available value from the sequence.
- Each time you use NEXTVAL, the sequence is incremented.
- Once a sequence value is used (through NEXTVAL), it cannot be used again. If, for some reason, your SQL statement rolls back, *the sequence value does not roll back*. If you issue another SQL statement (with another NEXTVAL), the next available sequence value will be returned to the user—it will look as though the sequence skips a number.
- You can issue an INSERT statement without using the sequence.

CURRVAL retrieves the current value of a sequence—that is, the last sequence number used, which was generated with a NEXTVAL. You cannot use CURRVAL unless a NEXTVAL was issued previously in the same session. The main use for CURRVAL is to enter rows in dependent tables. For example, the INVOICE and LINE tables are related in a one-to-many relationship through the INV_NUMBER attribute. You can use the INV_NUMBER_SEQ sequence to automatically generate invoice numbers. Then, using CURRVAL, you can get the latest INV_NUMBER used and assign it to the related INV_NUMBER foreign key attribute in the LINE table. For example:

INSERT INTO INVOICE   VALUES (INV_NUMBER_SEQ.NEXTVAL, 20010, SYSDATE);
INSERT INTO LINE   VALUES (INV_NUMBER_SEQ.CURRVAL, 1,'13-Q2/P2', 1, 14.99);
INSERT INTO LINE   VALUES (INV_NUMBER_SEQ.CURRVAL, 2,'23109-HB', 1, 9.95);
COMMIT;

The results are shown in Figure 8.23.

In the example shown in Figure 8.23, INV_NUMBER_SEQ.NEXTVAL retrieves the next available sequence number (4010) and assigns it to the INV_NUMBER column in the INVOICE table. Also note the use of the SYSDATE attribute to automatically insert the current date in the INV_DATE attribute. Next, the following two INSERT statements add the products being sold to the LINE table. In this case, INV_NUMBER_SEQ.CURRVAL refers to the last-used INV_NUMBER_SEQ sequence number (4010). In this way, the relationship between INVOICE and LINE is established automatically. The COMMIT statement at the end of the command sequence makes the changes permanent. Of course, you can also issue a ROLLBACK statement, in which case the rows you inserted in INVOICE and LINE tables would be rolled back (but remember that the sequence number would not). Once you use a sequence number (with NEXTVAL), there is no way to reuse it! This "no-reuse" characteristic is designed to guarantee that the sequence will always generate unique values.

Remember these points when you think about sequences:

- The use of sequences is optional. You can enter the values manually.
- A sequence is not associated with a table. As in the examples in Figure 8.23, two distinct sequences were created (one for customer code values and one for invoice number values), but you could have created just one sequence and used it to generate unique values for both tables.

Finally, you can drop a sequence from a database with a DROP SEQUENCE command. For example, to drop the sequences created earlier, you would type:

DROP SEQUENCE CUS_CODE_SEQ;
DROP SEQUENCE INV_NUMBER_SEQ;

FIGURE
8.23

Oracle sequence examples

```
Oracle SQL*Plus                                                    _ □ ×
File  Edit  Search  Options  Help
SQL> INSERT INTO CUSTOMER
  2  VALUES (CUS_CODE_SEQ.NEXTVAL, 'Connery', 'Sean', NULL, '615', '898-2007', 0.00);

1 row created.

SQL> SELECT * FROM CUSTOMER WHERE CUS_CODE = 20010;

  CUS_CODE CUS_LNAME        CUS_FNAME        C CUS CUS_PHON CUS_BALANCE
---------- ---------------  ---------------- - --- -------- -----------
     20010 Connery          Sean               615 898-2007           0

SQL> INSERT INTO INVOICE
  2  VALUES (INV_NUMBER_SEQ.NEXTVAL, 20010, SYSDATE);

1 row created.

SQL> SELECT * FROM INVOICE WHERE INV_NUMBER = 4010;

INV_NUMBER   CUS_CODE INV_DATE
---------- ---------- ---------
      4010      20010 27-MAY-08

SQL> INSERT INTO LINE
  2  VALUES (INV_NUMBER_SEQ.CURRVAL, 1,'13-Q2/P2', 1, 14.99);

1 row created.

SQL> INSERT INTO LINE
  2  VALUES (INV_NUMBER_SEQ.CURRVAL, 2,'23109-HB', 1, 9.95);

1 row created.

SQL> SELECT * FROM LINE WHERE INV_NUMBER = 4010;

INV_NUMBER LINE_NUMBER P_CODE      LINE_UNITS LINE_PRICE
---------- ----------- ----------  ---------- ----------
      4010           1 13-Q2/P2             1      14.99
      4010           2 23109-HB             1       9.95

SQL> COMMIT;

Commit complete.
```

NOTE

The latest SQL standard (SQL-2003) defines the use of Identity columns and sequence objects. However, some DBMS vendors might not adhere to the standard. Check your DBMS documentation.

Dropping a sequence does not delete the values you assigned to table attributes (CUS_CODE and INV_NUMBER); it deletes only the sequence object from the database. The *values* you assigned to the table columns (CUS_CODE and INV_NUMBER) remain in the database.

Because the CUSTOMER and INVOICE tables are used in the following examples, you'll want to keep the original data set. Therefore, you can delete the customer, invoice, and line rows you just added by using the following commands:

DELETE FROM INVOICE WHERE INV_NUMBER = 4010;
DELETE FROM CUSTOMER WHERE CUS_CODE = 20010;
COMMIT;

Those commands delete the recently added invoice and all of the invoice line rows associated with the invoice (the LINE table's INV_NUMBER foreign key was defined with the ON DELETE CASCADE option) and the recently added customer. The COMMIT statement saves all changes to permanent storage.

NOTE

At this point, you'll need to re-create the CUS_CODE_SEQ and INV_NUMBER_SEQ sequences, as they will be used again later in the chapter. Enter:

CREATE SEQUENCE CUS_CODE_SEQ START WITH 20010 NOCACHE;
CREATE SEQUENCE INV_NUMBER_SEQ START WITH 4010 NOCACHE;

8.6 UPDATABLE VIEWS

In Chapter 7, you learned how to create a view and why and how views are used. You will now take a look at how views can be made to serve common data management tasks executed by database administrators.

One of the most common operations in production database environments is using batch update routines to update a master table attribute (field) with transaction data. As the name implies, a **batch update routine** pools multiple transactions into a single batch to update a master table field *in a single operation*. For example, a batch update routine is commonly used to update a product's quantity on hand based on summary sales transactions. Such routines are typically run as overnight batch jobs to update the quantity on hand of products in inventory. The sales transactions performed, for example, by traveling salespeople were entered during periods when the system was offline.

To demonstrate a batch update routine, let's begin by defining the master product table (PRODMASTER) and the product monthly sales totals table (PRODSALES) shown in Figure 8.24. Note the 1:1 relationship between the two tables.

FIGURE 8.24	The PRODMASTER and PRODSALES tables

Database name: CH08_UV

Table name: PRODMASTER

PROD_ID	PROD_DESC	PROD_QOH
A123	SCREWS	60
BX34	NUTS	37
C583	BOLTS	50

Table name: PRODSALES

PROD_ID	PS_QTY
A123	7
BX34	3

ONLINE CONTENT

For MS Access users, the PRODMASTER and PRODSALES tables are located in the **Ch08_UV** database, which is located in the Student Online Companion.

ONLINE CONTENT

For Oracle users, all SQL commands you see in this section are located in the Student Online Companion. After you locate the script files (**uv-01.sql** through **uv-04.sql**), you can copy and paste the command sequences into your SQL*Plus program.

Using the tables in Figure 8.24, let's update the PRODMASTER table by subtracting the PRODSALES table's product monthly sales quantity (PS_QTY) from the PRODMASTER table's PROD_QOH. To produce the required update, the update query would be written like this:

```
UPDATE      PRODMASTER, PRODSALES
SET         PRODMASTER.PROD_QOH = PROD_QOH – PS_QTY
WHERE       PRODMASTER.PROD_ID = PRODSALES.PROD_ID;
```

Note that the update statement reflects the following sequence of events:

- Join the PRODMASTER and PRODSALES tables.
- Update the PROD_QOH attribute (using the PS_QTY value in the PRODSALES table) for each row of the PRODMASTER table with matching PROD_ID values in the PRODSALES table.

To be used in a batch update, the PRODSALES data must be stored in a base table rather than in a view. That query will work fine in Access, but Oracle will return the error message shown in Figure 8.25.

FIGURE 8.25   The Oracle UPDATE error message

Oracle produced the error message because Oracle expects to find a single table name in the UPDATE statement. In fact, you cannot join tables in the UPDATE statement in Oracle. To solve that problem, you have to create an *updatable* view. As its name suggests, an **updatable view** is a view that can be used to update attributes in the base table(s) that is (are) used in the view. You must realize that *not all views are updatable*. Actually, several restrictions govern updatable views, and some of them are vendor-specific.

> **NOTE**
>
> Keep in mind that the examples in this section are generated in Oracle. To see what restrictions are placed on updatable views by the DBMS you are using, check the appropriate DBMS documentation.

The most common updatable view restrictions are as follows:

- GROUP BY expressions or aggregate functions cannot be used.
- You cannot use set operators such as UNION, INTERSECT, and MINUS.
- Most restrictions are based on the use of JOINs or group operators in views.

To meet the Oracle limitations, an updatable view named PSVUPD has been created, as shown in Figure 8.26.

One easy way to determine whether a view can be used to update a base table is to examine the view's output. If the primary key columns of the base table you want to update still have unique values in the view, the base table is updatable. For example, if the PROD_ID column of the view returns the A123 or BX34 values more than once, the PRODMASTER table cannot be updated through the view.

FIGURE 8.26   Creating an updatable view in Oracle

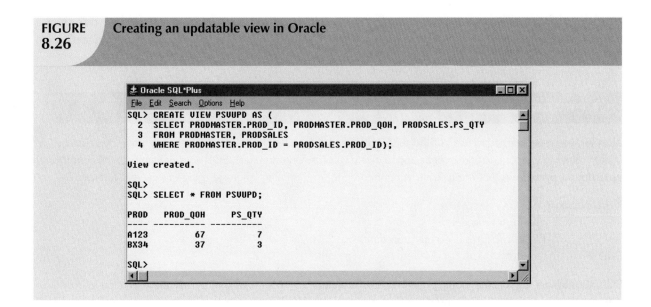

After creating the updatable view shown in Figure 8.26, you can use the UPDATE command to update the view, thereby updating the PRODMASTER table. Figure 8.27 shows how the UPDATE command is used and what the final contents of the PRODMASTER table are after the UPDATE has been executed.

FIGURE 8.27   PRODMASTER table update, using an updatable view

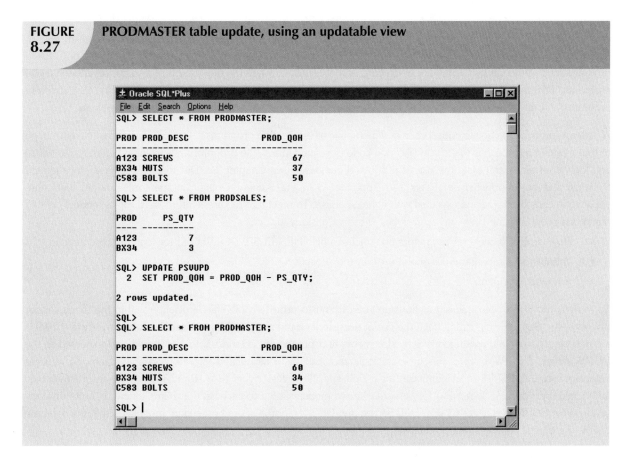

Although the batch update procedure just illustrated meets the goal of updating a master table with data from a transaction table, the preferred real-world solution to the update problem is to use procedural SQL, which you'll learn about next.

8.7 PROCEDURAL SQL

Thus far, you have learned to use SQL to read, write, and delete data in the database. For example, you learned to update values in a record, to add records, and to delete records. Unfortunately, SQL does not support the *conditional* execution of procedures that are typically supported by a programming language using the general format:

IF <condition>
      THEN <perform procedure>
          ELSE <perform alternate procedure>
END IF

SQL also fails to support the looping operations in programming languages that permit the execution of repetitive actions typically encountered in a programming environment. The typical format is:

DO WHILE
      <perform procedure>
END DO

Traditionally, if you wanted to perform a conditional (IF-THEN-ELSE) or looping (DO-WHILE) type of operation (that is, a procedural type of programming), you would use a programming language such as Visual Basic.Net, C#, or COBOL. That's why many older (so-called "legacy") business applications are based on enormous numbers of COBOL program lines. Although that approach is still common, it usually involves the duplication of application code in many programs. Therefore, when procedural changes are required, program modifications must be made in many different programs. An environment characterized by such redundancies often creates data management problems.

A better approach is to isolate critical code and then have all application programs call the shared code. The advantage of that modular approach is that the application code is isolated in a single program, thus yielding better maintenance and logic control. In any case, the rise of distributed databases (see Chapter 12, Distributed Database Management Systems) and object-oriented databases (see Appendix G in the Student Online Companion) required that more application code be stored and executed within the database. To meet that requirement, most RDBMS vendors created numerous programming language extensions. Those extensions include:

- Flow-control procedural programming structures (IF-THEN-ELSE, DO-WHILE) for logic representation.
- Variable declaration and designation within the procedures.
- Error management.

To remedy the lack of procedural functionality in SQL and to provide some standardization within the many vendor offerings, the SQL-99 standard defined the use of persistent stored modules. A **persistent stored module (PSM)** is a block of code containing standard SQL statements and procedural extensions that is stored and executed at the DBMS server. The PSM represents business logic that can be encapsulated, stored, and shared among multiple database users. A PSM lets an administrator assign specific access rights to a stored module to ensure that only authorized users can use it. Support for persistent stored modules is left to each vendor to implement. In fact, for many years, some RDBMSs (such as Oracle, SQL Server, and DB2) supported stored procedure modules within the database before the official standard was promulgated.

MS SQL Server implements persistent stored modules via Transact-SQL and other language extensions, the most notable of which are the .NET family of programming languages. Oracle implements PSMs through its procedural SQL language. **Procedural SQL (PL/SQL)** is a language that makes it possible to use and store procedural code and SQL statements within the database and to merge SQL and traditional programming constructs, such as variables, conditional processing (IF-THEN-ELSE), basic loops (FOR and WHILE loops,) and error trapping. The procedural code is executed as a unit by the DBMS when it is invoked (directly or indirectly) by the end user. End users can use PL/SQL to create:

- Anonymous PL/SQL blocks.
- Triggers (covered in Section 8.7.1).
- Stored procedures (covered in Section 8.7.2 and Section 8.7.3).
- PL/SQL functions (covered in Section 8.7.4).

Do not confuse PL/SQL functions with SQL's built-in aggregate functions such as MIN and MAX. SQL built-in functions can be used only within SQL statements, while PL/SQL functions are mainly invoked within PL/SQL programs such as triggers and stored procedures. Functions can also be called within SQL statements, provided they conform to very specific rules that are dependent on your DBMS environment.

> **NOTE**
>
> PL/SQL, triggers, and stored procedures are illustrated within the context of an Oracle DBMS. All examples in the following sections assume the use of Oracle RDBMS.

Using Oracle SQL*Plus, you can write a PL/SQL code block by enclosing the commands inside BEGIN and END clauses. For example, the following PL/SQL block inserts a new row in the VENDOR table, as shown in Figure 8.28.

```
BEGIN
        INSERT INTO VENDOR
        VALUES (25678,'Microsoft Corp. ', 'Bill Gates','765','546-8484','WA','N');
END;
/
```

The PL/SQL block shown in Figure 8.28 is known as an **anonymous PL/SQL block** because it has not been given a specific name. (Incidentally, note that the block's last line uses a forward slash ("/") to indicate the end of the command-line entry.) That type of PL/SQL block executes as soon as you press the Enter key after typing the forward slash. Following the PL/SQL block's execution, you will see the message "PL/SQL procedure successfully completed."

But suppose you want a more specific message displayed on the SQL*Plus screen after a procedure is completed, such as "New Vendor Added." To produce a more specific message, you must do two things:

1. At the SQL > prompt, type SET SERVEROUTPUT ON. This SQL*Plus command enables the client console (SQL*Plus) to receive messages from the server side (Oracle DBMS). Remember, just like standard SQL, the PL/SQL code (anonymous blocks, triggers, and procedures) are executed at the server side, not at the client side. (To stop receiving messages from the server, you would enter SET SERVEROUT OFF.)

2. To send messages from the PL/SQL block to the SQL*Plus console, use the DBMS_OUTPUT.PUT_LINE function.

The following anonymous PL/SQL block inserts a row in the VENDOR table and displays the message "New Vendor Added!" (See Figure 8.28).

```
BEGIN
        INSERT INTO VENDOR
        VALUES (25772,'Clue Store', 'Issac Hayes', '456','323-2009', 'VA', 'N');
        DBMS_OUTPUT.PUT_LINE('New Vendor Added!');
END;
/
```

FIGURE
8.28

Anonymous PL/SQL block examples

```
Oracle SQL*Plus                                                    _ □ ×
File  Edit  Search  Options  Help
SQL> BEGIN
  2   INSERT INTO VENDOR
  3   VALUES (25678,'Microsoft Corp.', 'Bill Gates','765','546-8484','WA','N');
  4   END;
  5   /

PL/SQL procedure successfully completed.

SQL> SET SERVEROUTPUT ON
SQL>
SQL> BEGIN
  2   INSERT INTO VENDOR
  3   VALUES (25772,'Clue Store','Issac Hayes','456','323-2009','VA','N');
  4   DBMS_OUTPUT.PUT_LINE('New Vendor Added!');
  5   END;
  6   /
New Vendor Added!

PL/SQL procedure successfully completed.

SQL> SELECT * FROM VENDOR;

    V_CODE V_NAME                        V_CONTACT         V_A V_PHONE  V_ V
  --------- ---------------------------- ---------------- --- -------- -- -
     21225 Bryson, Inc.                   Smithson          615 223-3234 TN Y
     21226 SuperLoo, Inc.                 Flushing          904 215-8995 FL N
     21231 D&E Supply                     Singh             615 228-3245 TN Y
     21344 Gomez Bros.                    Ortega            615 889-2546 KY N
     22567 Dome Supply                    Smith             901 678-1419 GA N
     23119 Randsets Ltd.                  Anderson          901 678-3998 GA Y
     24004 Brackman Bros.                 Browning          615 228-1410 TN N
     24288 ORDVA, Inc.                    Hakford           615 898-1234 TN Y
     25443 B&K, Inc.                      Smith             904 227-0093 FL N
     25501 Damal Supplies                 Smythe            615 890-3529 TN N
     25595 Rubicon Systems                Orton             904 456-0092 FL Y
     25678 Microsoft Corp.                Bill Gates        765 546-8484 WA N
     25772 Clue Store                     Issac Hayes       456 323-2009 VA N

13 rows selected.

SQL>
```

In Oracle, you can use the SQL*Plus command SHOW ERRORS to help you diagnose errors found in PL/SQL blocks. The SHOW ERRORS command yields additional debugging information whenever you generate an error after creating or executing a PL/SQL block.

The following example of an anonymous PL/SQL block demonstrates several of the constructs supported by the procedural language. Remember that the exact syntax of the language is vendor-dependent; in fact, many vendors enhance their products with proprietary features.

```
DECLARE
W_P1 NUMBER(3) := 0;
W_P2 NUMBER(3) := 10;
W_NUM NUMBER(2) := 0;
BEGIN
WHILE W_P2 < 300 LOOP
        SELECT COUNT(P_CODE) INTO W_NUM FROM PRODUCT
        WHERE P_PRICE BETWEEN W_P1 AND W_P2;
        DBMS_OUTPUT.PUT_LINE('There are ' || W_NUM || ' Products with price between ' || W_P1 ||
                        ' and ' || W_P2);
```

```
        W_P1 := W_P2 + 1;
        W_P2 := W_P2 + 50;
END LOOP;
END;
/
```

The block's code and execution are shown in Figure 8.29.

FIGURE
8.29
Anonymous PL/SQL block with variables and loops

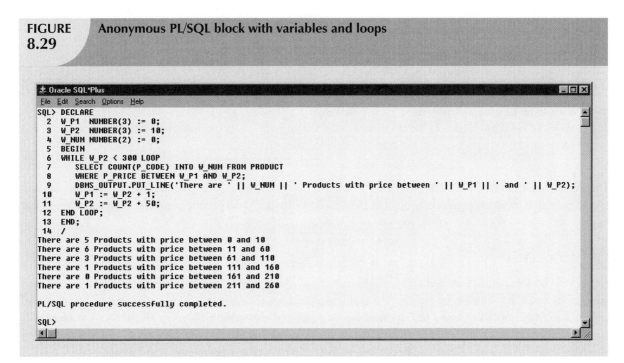

The PL/SQL block shown in Figure 8.29 has the following characteristics:

- The PL/SQL block starts with the DECLARE section in which you declare the variable names, the data types, and, if desired, an initial value. Supported data types are shown in Table 8.8.

**TABLE
8.8**
PL/SQL Basic Data Types

DATA TYPE	DESCRIPTION
CHAR	Character values of a fixed length; for example: W_ZIPCHAR(5)
VARCHAR2	Variable length character values; for example: W_FNAMEVARCHAR2(15)
NUMBER	Numeric values; for example: W_PRICENUMBER(6,2)
DATE	Date values; for example: W_EMP_DOBDATE
%TYPE	Inherits the data type from a variable that you declared previously or from an attribute of a database table; for example: W_PRICEPRODUCT.P_PRICE%TYPE Assigns W_PRICE the same data type as the P_PRICE column in the PRODUCT table

- A WHILE loop is used. Note the syntax:

    WHILE *condition* LOOP
        PL/SQL statements;
    END LOOP

- The SELECT statement uses the INTO keyword to assign the output of the query to a PL/SQL variable. You can use the INTO keyword only inside a PL/SQL block of code. If the SELECT statement returns more than one value, you will get an error.

- Note the use of the string concatenation symbol " | | " to display the output.

- Each statement inside the PL/SQL code must end with a semicolon ";".

> **NOTE**
>
> PL/SQL blocks can contain only standard SQL data manipulation language (DML) commands such as SELECT, INSERT, UPDATE, and DELETE. The use of data definition language (DDL) commands is not directly supported in a PL/SQL block.

The most useful feature of PL/SQL blocks is that they let you create code that can be named, stored, and executed—either implicitly or explicitly—by the DBMS. That capability is especially desirable when you need to use triggers and stored procedures, which you will explore next.

8.7.1 TRIGGERS

Automating business procedures and automatically maintaining data integrity and consistency are critical in a modern business environment. One of the most critical business procedures is proper inventory management. For example, you want to make sure that current product sales can be supported with sufficient product availability. Therefore, it is necessary to ensure that a product order be written to a vendor when that product's inventory drops below its minimum allowable quantity on hand. Better yet, how about ensuring that the task is completed automatically?

To accomplish automatic product ordering, you first must make sure the product's quantity on hand reflects an up-to-date and consistent value. After the appropriate product availability requirements have been set, two key issues must be addressed:

1. Business logic requires an update of the product quantity on hand each time there is a sale of that product.

2. If the product's quantity on hand falls below its minimum allowable inventory (quantity-on-hand) level, the product must be reordered.

To accomplish those two tasks, you could write multiple SQL statements: one to update the product quantity on hand and another to update the product reorder flag. Next, you would have to run each statement in the correct order each time there was a new sale. Such a multistage process would be inefficient because a series of SQL statements must be written and executed each time a product is sold. Even worse, that SQL environment requires that somebody must remember to perform the SQL tasks.

A **trigger** is procedural SQL code that is *automatically* invoked by the RDBMS upon the occurrence of a given data manipulation event. It is useful to remember that:

- A trigger is invoked before or after a data row is inserted, updated, or deleted.

- A trigger is associated with a database table.

- Each database table may have one or more triggers.

- A trigger is executed as part of the transaction that triggered it.

Triggers are critical to proper database operation and management. For example:

- Triggers can be used to enforce constraints that cannot be enforced at the DBMS design and implementation levels.
- Triggers add functionality by automating critical actions and providing appropriate warnings and suggestions for remedial action. In fact, one of the most common uses for triggers is to facilitate the enforcement of referential integrity.
- Triggers can be used to update table values, insert records in tables, and call other stored procedures.

Triggers play a critical role in making the database truly useful; they also add processing power to the RDBMS and to the database system as a whole. Oracle recommends triggers for:

- Auditing purposes (creating audit logs).
- Automatic generation of derived column values.
- Enforcement of business or security constraints.
- Creation of replica tables for backup purposes.

To see how a trigger is created and used, let's examine a simple inventory management problem. For example, if a product's quantity on hand is updated when the product is sold, the system should automatically check whether the quantity on hand falls below its minimum allowable quantity. To demonstrate that process, let's use the PRODUCT table in Figure 8.30. Note the use of the minimum order quantity (P_MIN_ORDER) and the product reorder flag (P_REORDER) columns. The P_MIN_ORDER indicates the minimum quantity for restocking an order. The P_REORDER column is a numeric field that indicates whether the product needs to be reordered (1 = Yes, 0 = No). The initial P_REORDER values will be set to 0 (No) to serve as the basis for the initial trigger development.

FIGURE 8.30   The PRODUCT table

```
Oracle SQL*Plus                                                                  _|□|×|
File  Edit  Search  Options  Help
SQL> SELECT * FROM PRODUCT;

P_CODE     P_DESCRIPT                    P_INDATE    P_QOH    P_MIN   P_PRICE  P_DISCOUNT    V_CODE  P_MIN_ORDER   P_REORDER
---------- ----------------------------- ---------- ------- -------- -------- ----------- --------- ------------ -----------
11QER/31   Power painter, 15 psi., 3-nozz 03-NOV-07       8        5   109.99         .00     25595           25           0
13-Q2/P2   7.25-in. pwr. saw blade        13-DEC-07      32       15    14.99         .05     21344           50           0
14-Q1/L3   9.00-in. pwr. saw blade        13-NOV-07      18       12    17.49         .00     21344           50           0
1546-QQ2   Hrd. cloth, 1/4-in., 2x50      15-JAN-08      15        8    39.95         .00     23119           35           0
1558-QW1   Hrd. cloth, 1/2-in., 3x50      15-JAN-08      23        5    43.99         .00     23119           25           0
2232/QTY   B&D jigsaw, 12-in. blade       30-DEC-07       8        5   109.92         .05     24288           15           0
2232/QWE   B&D jigsaw, 8-in. blade        24-DEC-07       6        5    99.87         .05     24288           15           0
2238/QPD   B&D cordless drill, 1/2-in.    20-JAN-08      12        5    38.95         .05     25595           12           0
23109-HB   Claw hammer                    20-JAN-08      23       10     9.95         .10     21225           25           0
23114-AA   Sledge hammer, 12 lb.          02-JAN-08       8        5     14.4         .05                     12           0
54778-2T   Rat-tail file, 1/8-in. fine    15-DEC-07      43       20     4.99         .00     21344           25           0
89-WRE-Q   Hicut chain saw, 16 in.        07-FEB-08      11        5   256.99         .05     24288           10           0
PVC23DRT   PVC pipe, 3.5-in., 8-ft        20-FEB-08     188       75     5.87         .00                     50           0
SM-18277   1.25-in. metal screw, 25       01-MAR-08     172       75     6.99         .00     21225           50           0
SW-23116   2.5-in. wd. screw, 50          24-FEB-08     237      100     8.45         .00     21231          100           0
WR3/TT3    Steel matting, 4'x8'x1/6", .5" 17-JAN-08      18        5   119.95         .10     25595           10           0

16 rows selected.
```

ONLINE CONTENT

Oracle users can run the **PRODLIST.SQL** script file to format the output of the PRODUCT table shown in Figure 8.30. The script file is located in the Student Online Companion.

Given the PRODUCT table listing shown in Figure 8.30, let's create a trigger to evaluate the product's quantity on hand, P_QOH. If the quantity on hand is below the minimum quantity shown in P_MIN, the trigger will set the P_REORDER column to 1. (Remember that the number 1 in the P_REORDER column represents "Yes.") The syntax to create a trigger in Oracle is:

CREATE OR REPLACE TRIGGER *trigger_name*
[BEFORE / AFTER] [DELETE / INSERT / UPDATE OF *column_name*] ON *table_name*
[FOR EACH ROW]
[DECLARE]
      [*variable_namedata type*[:=*initial_value*]]
BEGIN
      PL/SQL instructions;

      ....
END;

As you can see, a trigger definition contains the following parts:

- The triggering timing: BEFORE or AFTER. This timing indicates when the trigger's PL/SQL code executes; in this case, before or after the triggering statement is completed.
- The triggering event: the statement that causes the trigger to execute (INSERT, UPDATE, or DELETE).
- The triggering level: There are two types of triggers: statement-level triggers and row-level triggers.
  - A **statement-level trigger** is assumed if you omit the FOR EACH ROW keywords. This type of trigger is executed once, before or after the triggering statement is completed. This is the default case.
  - A **row-level trigger** requires use of the FOR EACH ROW keywords. This type of trigger is executed once for each row affected by the triggering statement. (In other words, if you update 10 rows, the trigger executes 10 times.)
- The triggering action: The PL/SQL code enclosed between the BEGIN and END keywords. Each statement inside the PL/SQL code must end with a semicolon ";".

In the PRODUCT table's case, you will create a statement-level trigger that is implicitly executed AFTER an UPDATE of the P_QOH attribute for an existing row or AFTER an INSERT of a new row in the PRODUCT table. The trigger action executes an UPDATE statement that compares the P_QOH with the P_MIN column. If the value of P_QOH is equal to or less than P_MIN, the trigger updates the P_REORDER to 1. To create the trigger, Oracle's SQL*Plus will be used. The trigger code is shown in Figure 8.31.

FIGURE 8.31   **Creating the TRG_PRODUCT_REORDER trigger**

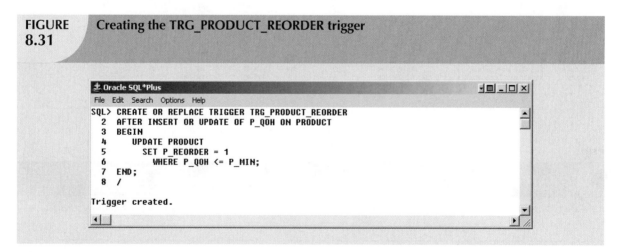

ONLINE CONTENT

The source code for all of the triggers shown in this section can be found in the Student Online Companion.

To test the TRG_PRODUCT_REORDER trigger, let's update the quantity on hand of product '11QER/31' to 4. After the UPDATE completes, the trigger is automatically fired and the UPDATE statement (inside the trigger code) sets the P_REORDER to 1 for all products that are below the minimum. See Figure 8.32.

FIGURE 8.32   Verifying the TRG_PRODUCT_REORDER trigger execution

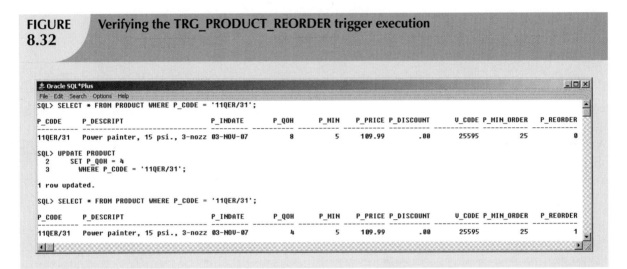

The trigger shown in Figure 8.32 seems to work fine, but what happens if you reduce the minimum quantity of product '2232/QWE'? Figure 8.33 shows that when you update the minimum quantity, the quantity on hand of the product '2232/QWE' falls below the new minimum, but the reorder flag is still 0. Why?

FIGURE 8.33   The P_REORDER value mismatch after update of the P_MIN attribute

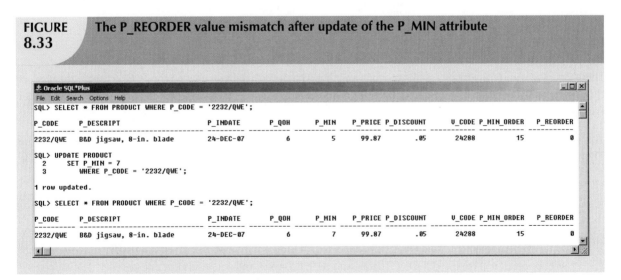

The answer is simple: you updated the P_MIN column, but the trigger is never executed. TRG_PRODUCT_REORDER executes only *after* an update of the P_QOH column! To avoid that inconsistency, you must modify the trigger event to execute after an update of the P_MIN field, too. The updated trigger code is shown in Figure 8.34.

FIGURE
8.34

Second version of the TRG_PRODUCT_REORDER trigger

```
Oracle SQL*Plus

File  Edit  Search  Options  Help

SQL> CREATE OR REPLACE TRIGGER TRG_PRODUCT_REORDER
  2    AFTER INSERT OR UPDATE OF P_QOH, P_MIN ON PRODUCT
  3    BEGIN
  4       UPDATE PRODUCT
  5          SET P_REORDER = 1
  6             WHERE P_QOH <= P_MIN;
  7    END;
  8    /

Trigger created.
```

To test this new trigger version, let's change the minimum quantity for product '23114-AA' to 8. After that update, the trigger makes sure that the reorder flag is properly set for all of the products in the PRODUCT table. See Figure 8.35.

FIGURE
8.35

Successful trigger execution after the P_MIN value is updated

```
Oracle SQL*Plus

File  Edit  Search  Options  Help

SQL> SELECT * FROM PRODUCT WHERE P_CODE = '23114-AA';

P_CODE    P_DESCRIPT           P_INDATE    P_QOH    P_MIN   P_PRICE P_DISCOUNT    V_CODE P_MIN_ORDER   P_REORDER
--------  -------------------  ---------  -------  -------  ------- ----------   ------- -----------  ----------
23114-AA  Sledge hammer, 12 lb. 02-JAN-08       8        5     14.4       .05                     12           0

SQL> UPDATE PRODUCT
  2    SET P_MIN = 10
  3       WHERE P_CODE = '23114-AA';

1 row updated.

SQL> SELECT * FROM PRODUCT WHERE P_CODE = '23114-AA';

P_CODE    P_DESCRIPT           P_INDATE    P_QOH    P_MIN   P_PRICE P_DISCOUNT    V_CODE P_MIN_ORDER   P_REORDER
--------  -------------------  ---------  -------  -------  ------- ----------   ------- -----------  ----------
23114-AA  Sledge hammer, 12 lb. 02-JAN-08       8       10     14.4       .05                     12           1
```

This second version of the trigger seems to work well, but what happens if you change the P_QOH value for product '11QER/31', as shown in Figure 8.36? Nothing! (Note that the reorder flag is *still* set to 1.) Why didn't the trigger change the reorder flag to 0?

The answer is that the trigger does not consider all possible cases. Let's examine the second version of the TRG_PRODUCT_REORDER trigger code (Figure 8.34) in more detail:

- The trigger fires after the triggering statement is completed. Therefore, the DBMS always executes two statements (INSERT plus UPDATE or UPDATE plus UPDATE). That is, after you do an update of P_MIN or P_QOH or you insert a new row in the PRODUCT table, the trigger executes another UPDATE statement automatically.

- The triggering action performs an UPDATE that updates *all* of the rows in the PRODUCT table, *even if the triggering statement updates just one row!* This can affect the performance of the database. Imagine what will happen if you have a PRODUCT table with 519,128 rows and you insert just one product. The trigger will update all 519,129 rows (519,128 original rows plus the one you inserted), including the rows that do not need an update!

- The trigger sets the P_REORDER value only to 1; it does not reset the value to 0, even if such an action is clearly required when the inventory level is back to a value greater than the minimum value.

FIGURE 8.36   The P_REORDER value mismatch after increasing the P_QOH value

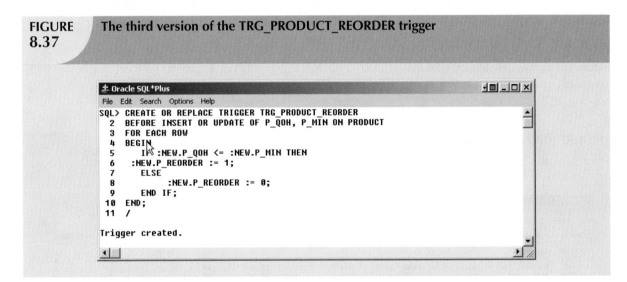

In short, the second version of the TRG_PRODUCT_REORDER trigger still does not complete all of the necessary steps. Now let's modify the trigger to handle all update scenarios, as shown in Figure 8.37.

FIGURE 8.37   The third version of the TRG_PRODUCT_REORDER trigger

```
SQL> CREATE OR REPLACE TRIGGER TRG_PRODUCT_REORDER
  2   BEFORE INSERT OR UPDATE OF P_QOH, P_MIN ON PRODUCT
  3   FOR EACH ROW
  4   BEGIN
  5      IF :NEW.P_QOH <= :NEW.P_MIN THEN
  6      :NEW.P_REORDER := 1;
  7      ELSE
  8           :NEW.P_REORDER := 0;
  9      END IF;
 10   END;
 11   /

Trigger created.
```

The trigger in Figure 8.37 sports several new features:

- The trigger is executed *before* the actual triggering statement is completed. In Figure 8.37, the triggering timing is defined in line 2, BEFORE INSERT OR UPDATE. This clearly indicates that the triggering statement is executed before the INSERT or UPDATE completes, unlike the previous trigger examples.

- The trigger is a row-level trigger instead of a statement-level trigger. The FOR EACH ROW keywords make the trigger a row-level trigger. Therefore, this trigger executes once for each row affected by the triggering statement.

- The trigger action uses the :NEW attribute reference to change the value of the P_REORDER attribute.

The use of the :NEW attribute references deserves a more detailed explanation. To understand its use, you must first consider a basic computing tenet: *all changes are done first in primary memory, then transferred to permanent memory*. In other words, the computer cannot change anything directly in permanent storage (disk). It must first read the data from permanent storage to primary memory; then it makes the change in primary memory; and finally, it writes the changed data back to permanent memory (disk).

The DBMS does the same thing, and one thing more. Because ensuring data integrity is critical, the DBMS makes two copies of every row being changed by a DML (INSERT, UPDATE, or DELETE) statement. (You will learn more about this in Chapter 10, Transaction Management and Concurrency Control.) The first copy contains the original ("old") values of the attributes before the changes. The second copy contains the changed ("new") values of the attributes that will be permanently saved to the database (after any changes made by an INSERT, UPDATE, or DELETE). You can use :OLD to refer to the original values; you can use :NEW to refer to the changed values (the values that will be stored in the table). You can use :NEW and :OLD attribute references only within the PL/SQL code of a database trigger action. For example:

- IF :NEW.P_QOH < = :NEW.P_MIN compares the quantity on hand with the minimum quantity of a product. Remember that this is a row-level trigger. Therefore, this comparison is done for each row that is updated by the triggering statement.
- Although the trigger is a BEFORE trigger, this does not mean that the triggering statement hasn't executed yet. To the contrary, the triggering statement has already taken place; otherwise, the trigger would not have fired and the :NEW values would not exist. Remember, BEFORE means *before* the changes are permanently saved to disk, but *after* the changes are made in memory.
- The trigger uses the :NEW reference to assign a value to the P_REORDER column before the UPDATE or INSERT results are permanently stored in the table. The assignment is always done to the :NEW value (never to the :OLD value), and the assignment always uses the " := " assignment operator. The :OLD values are *read-only* values; you cannot change them. Note that :NEW.P_REORDER := 1; assigns the value 1 to the P_REORDER column and :NEW.P_REORDER := 0; assigns the value 0 to the P_REORDER column.
- This new trigger version does not use any DML statement!

Before testing the new trigger, note that product '11QER/31' currently has a quantity on hand that is above the minimum quantity, yet the reorder flag is set to 1. Given that condition, the reorder flag must be 0. After creating the new trigger, you can execute an UPDATE statement to fire it, as shown in Figure 8.38.

FIGURE 8.38 Execution of the third trigger version

```
Oracle SQL*Plus
File Edit Search Options Help
SQL> SELECT * FROM PRODUCT;

P_CODE     P_DESCRIPT                          P_INDATE      P_QOH      P_MIN    P_PRICE P_DISCOUNT     V_CODE P_MIN_ORDER  P_REORDER
---------- ----------------------------------- --------- ---------- ---------- ---------- ---------- ---------- ----------- ----------
11QER/31   Power painter, 15 psi., 3-nozz      03-NOV-07         29          5     109.99        .00      25595          25          1
13-Q2/P2   7.25-in. pwr. saw blade             13-DEC-07         32         15      14.99        .05      21344          50          0
14-Q1/L3   9.00-in. pwr. saw blade             13-NOV-07         18         12      17.49        .00      21344          50          0
1546-QQ2   Hrd. cloth, 1/4-in., 2x50           15-JAN-08         15          8      39.95        .00      23119          35          0
1558-QW1   Hrd. cloth, 1/2-in., 3x50           15-JAN-08         23          5      43.99        .00      23119          25          0
2232/QTY   B&D jigsaw, 12-in. blade            30-DEC-07          8          5     109.92        .05      24288          15          0
2232/QWE   B&D jigsaw, 8-in. blade             24-DEC-07          6          7      99.87        .05      24288          15          1
2238/QPD   B&D cordless drill, 1/2-in.         20-JAN-08         12          5      38.95        .05      25595          12          0
23109-HB   Claw hammer                         20-JAN-08         23         10       9.95        .10      21225          25          0
23114-AA   Sledge hammer, 12 lb.               02-JAN-08          8         10       14.4        .05                     12          1
54778-2T   Rat-tail file, 1/8-in. fine         15-DEC-07         43         20       4.99        .00      21344          25          0
89-WRE-Q   Hicut chain saw, 16 in.             07-FEB-08         11          5     256.99        .05      24288          10          0
PVC23DRT   PVC pipe, 3.5-in., 8-ft             20-FEB-08        188         75       5.87        .00                     50          0
SM-18277   1.25-in. metal screw, 25            01-MAR-08        172         75       6.99        .00      21225          50          0
SW-23116   2.5-in. wd. screw, 50               24-FEB-08        237        100       8.45        .00      21231         100          0
WR3/TT3    Steel matting, 4'x8'x1/6", .5"      17-JAN-08         18          5     119.95        .10      25595          10          0

16 rows selected.

SQL> UPDATE PRODUCT SET P_QOH = P_QOH;

16 rows updated.

SQL> SELECT * FROM PRODUCT WHERE P_CODE = '11QER/31';

P_CODE     P_DESCRIPT                          P_INDATE      P_QOH      P_MIN    P_PRICE P_DISCOUNT     V_CODE P_MIN_ORDER  P_REORDER
---------- ----------------------------------- --------- ---------- ---------- ---------- ---------- ---------- ----------- ----------
11QER/31   Power painter, 15 psi., 3-nozz      03-NOV-07         29          5     109.99        .00      25595          25          0
```

Note the following important features of the code in Figure 8.38:

- The trigger is automatically invoked for each affected row—in this case, all rows of the PRODUCT table. If your triggering statement would have affected only three rows, not all PRODUCT rows would have the correct P_REORDER value set. That's the reason the triggering statement was set up as shown in Figure 8.38.
- The trigger will run only if you insert a new product row or update P_QOH or P_MIN. If you update any other attribute, the trigger won't run.

You can also use a trigger to update an attribute in a table other than the one being modified. For example, suppose you would like to create a trigger that automatically reduces the quantity on hand of a product with every sale. To accomplish that task, you must create a trigger for the LINE table that updates a row in the PRODUCT table. The sample code for that trigger is shown in Figure 8.39.

| FIGURE 8.39 | TRG_LINE_PROD trigger to update the PRODUCT quantity on hand |

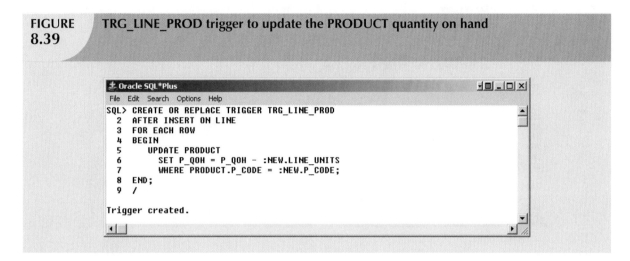

```
Oracle SQL*Plus
File  Edit  Search  Options  Help
SQL> CREATE OR REPLACE TRIGGER TRG_LINE_PROD
  2    AFTER INSERT ON LINE
  3    FOR EACH ROW
  4    BEGIN
  5       UPDATE PRODUCT
  6          SET P_QOH = P_QOH - :NEW.LINE_UNITS
  7          WHERE PRODUCT.P_CODE = :NEW.P_CODE;
  8    END;
  9  /

Trigger created.
```

Note that the TRG_LINE_PROD row-level trigger executes after inserting a new invoice's LINE and reduces the quantity on hand of the recently sold product by the number of units sold. This row-level trigger updates a row in a different table (PRODUCT), using the :NEW values of the recently added LINE row.

A third trigger example shows the use of variables within a trigger. In this case, you want to update the customer balance (CUS_BALANCE) in the CUSTOMER table after inserting every new LINE row. This trigger code is shown in Figure 8.40.

Let's carefully examine the trigger in Figure 8.40.

- The trigger is a row-level trigger that executes after each new LINE row is inserted.
- The DECLARE section in the trigger is used to declare any variables used inside the trigger code.
- You can declare a variable by assigning a name, a data type, and (optionally) an initial value, as in the case of the W_TOT variable.
- The first step in the trigger code is to get the customer code (CUS_CODE) from the related INVOICE table. Note that the SELECT statement returns only one attribute (CUS_CODE) from the INVOICE table. Also note that that attribute returns only one value as specified by the use of the WHERE clause *to restrict the query output to a single value.*
- Note the use of the INTO clause within the SELECT statement. You use the INTO clause to assign a value from a SELECT statement to a variable (W_CUS) used within a trigger.
- The second step in the trigger code computes the total of the line by multiplying the :NEW.LINE_UNITS times :NEW.LINE_PRICE and assigning the result to the W_TOT variable.

FIGURE 8.40   **TRG_LINE_CUS trigger to update the customer balance**

```
± Oracle SQL*Plus                                                    _ □ ×
File  Edit  Search  Options  Help
SQL> CREATE OR REPLACE TRIGGER TRG_LINE_CUS
  2  AFTER INSERT ON LINE
  3  FOR EACH ROW
  4  DECLARE
  5  W_CUS CHAR(5);
  6  W_TOT NUMBER:= 0;    -- to compute total cost
  7  BEGIN
  8     -- this trigger fires up after an INSERT of a LINE
  9     -- it will update the CUS_BALANCE in CUSTOMER
 10
 11     -- 1) get the CUS_CODE
 12     SELECT CUS_CODE INTO W_CUS
 13       FROM INVOICE
 14        WHERE INVOICE.INV_NUMBER = :NEW.INV_NUMBER;
 15
 16     -- 2) compute the total of the current line
 17     W_TOT := :NEW.LINE_PRICE * :NEW.LINE_UNITS;
 18
 19     -- 3) Update the CUS_BALANCE in CUSTOMER
 20     UPDATE CUSTOMER
 21       SET CUS_BALANCE = CUS_BALANCE + W_TOT
 22        WHERE CUS_CODE = W_CUS;
 23
 24     DBMS_OUTPUT.PUT_LINE(' * * * Balance updated for customer: ' || W_CUS);
 25
 26  END;
 27  /

Trigger created.

SQL>
```

- The final step updates the customer balance by using an UPDATE statement and the W_TOT and W_CUS trigger variables.
- Double dashes "--" are used to indicate comments within the PL/SQL block.

Let's summarize the triggers created in this section.

- The TRG_PROD_REORDER is a row-level trigger that updates P_REORDER in PRODUCT when a new product is added or when the P_QOH or P_MIN columns are updated.
- The TRG_LINE_PROD is a row-level trigger that automatically reduces the P_QOH in PRODUCT when a new row is added to the LINE table.
- TRG_LINE_CUS is a row-level trigger that automatically increases the CUS_BALANCE in CUSTOMER when a new row is added in the LINE table.

The use of triggers facilitates the automation of multiple data management tasks. Although triggers are independent objects, they are associated with database tables. When you delete a table, all its trigger objects are deleted with it. However, if you needed to delete a trigger without deleting the table, you could use the following command:

DROP TRIGGER trigger_name

Trigger Action Based on Conditional DML Predicates

You could also create triggers whose actions depend on the type of DML statement (INSERT, UPDATE, or DELETE) that fires the trigger. For example, you could create a trigger that executes after an insert, an update, or a delete on the PRODUCT table. But how do you know which one of the three statements caused the trigger to execute? In those cases, you could use the following syntax:

IF INSERTING THEN ... END IF;
IF UPDATING THEN ... END IF;
IF DELETING THEN ... END IF;

8.7.2 STORED PROCEDURES

A **stored procedure** is a named collection of procedural and SQL statements. Just like database triggers, stored procedures are stored in the database. One of the major advantages of stored procedures is that they can be used to encapsulate and represent business transactions. For example, you can create a stored procedure to represent a product sale, a credit update, or the addition of a new customer. By doing that, you can encapsulate SQL statements within a single stored procedure and execute them as a single transaction. There are two clear advantages to the use of stored procedures:

- Stored procedures substantially reduce network traffic and increase performance. Because the procedure is stored at the server, there is no transmission of individual SQL statements over the network. The use of stored procedures improves system performance because all transactions are executed locally on the RDBMS, so each SQL statement does not have to travel over the network.

- Stored procedures help reduce code duplication by means of code isolation and code sharing (creating unique PL/SQL modules that are called by application programs), thereby minimizing the chance of errors and the cost of application development and maintenance.

To create a stored procedure, you use the following syntax:

CREATE OR REPLACE PROCEDURE procedure_name [(argument [IN/OUT] data-type, ...)] [IS/AS]
                    [variable_name data type[:=initial_value]]
BEGIN
        PL/SQL or SQL statements;
        ...
END;

Note the following important points about stored procedures and their syntax:

- argument specifies the parameters that are passed to the stored procedure. A stored procedure could have zero or more arguments or parameters.
- IN/OUT indicates whether the parameter is for input, output, or both.
- data-type is one of the procedural SQL data types used in the RDBMS. The data types normally match those used in the RDBMS table-creation statement.
- Variables can be declared between the keywords IS and BEGIN. You must specify the variable name, its data type, and (optionally) an initial value.

To illustrate stored procedures, assume that you want to create a procedure (PRC_PROD_DISCOUNT) to assign an additional 5 percent discount for all products when the quantity on hand is more than or equal to twice the minimum quantity. Figure 8.41 shows how the stored procedure is created.

Note in Figure 8.41 that the PRC_PROD_DISCOUNT stored procedure uses the DBMS_OUTPUT.PUT_LINE function to display a message when the procedure executes. (This action assumes you previously ran SET SERVEROUTPUT ON.)

FIGURE
8.41

Creating the PRC_PROD_DISCOUNT stored procedure

```
Oracle SQL*Plus
File  Edit  Search  Options  Help
SQL> CREATE OR REPLACE PROCEDURE PRC_PROD_DISCOUNT
  2   AS BEGIN
  3      UPDATE PRODUCT
  4         SET P_DISCOUNT = P_DISCOUNT + .05
  5            WHERE P_QOH >= P_MIN*2;
  6      DBMS_OUTPUT.PUT_LINE ('* * Update finished * *');
  7   END;
  8   /

Procedure created.
```

ONLINE CONTENT

The source code for all of the stored procedures shown in this section can be found in the Student Online Companion.

To execute the stored procedure, you must use the following syntax:

EXEC procedure_name[(parameter_list)];

For example, to see the results of running the PRC_PROD_DISCOUNT stored procedure, you can use the EXEC PRC_PROD_DISCOUNT command shown in Figure 8.42.

Using Figure 8.42 as your guide, you can see how the product discount attribute for all products with a quantity on hand more than or equal to twice the minimum quantity was increased by 5 percent. (Compare the first PRODUCT table listing to the second PRODUCT table listing.)

FIGURE 8.42   **Results of the PRC_PROD_DISCOUNT stored procedure**

```
± Oracle SQL*Plus                                                                          _□×
File  Edit  Search  Options  Help
SQL> SELECT * FROM PRODUCT;

P_CODE     P_DESCRIPT                     P_INDATE     P_QOH    P_MIN   P_PRICE P_DISCOUNT     V_CODE P_MIN_ORDER   P_REORDER
--------   ---------------------------   ----------   ------   ------   -------  ----------   ------- -----------   ---------
11QER/31   Power painter, 15 psi., 3-nozz 03-NOV-07      29        5    109.99     .00         25595          25           0
13-Q2/P2   7.25-in. pwr. saw blade       13-DEC-07       32       15     14.99     .05         21344          50           0
14-Q1/L3   9.00-in. pwr. saw blade       13-NOV-07       18       12     17.49     .00         21344          50           0
1546-QQ2   Hrd. cloth, 1/4-in., 2x50     15-JAN-08       15        8     39.95     .00         23119          35           0
1558-QW1   Hrd. cloth, 1/2-in., 3x50     15-JAN-08       23        5     43.99     .00         23119          25           0
2232/QTY   B&D jigsaw, 12-in. blade      30-DEC-07        8        5    109.92     .05         24288          15           0
2232/QWE   B&D jigsaw, 8-in. blade       24-DEC-07        6        7     99.87     .05         24288          15           1
2238/QPD   B&D cordless drill, 1/2-in.   20-JAN-08       12        5     38.95     .05         25595          12           0
23109-HB   Claw hammer                   20-JAN-08       23       10      9.95     .10         21225          25           0
23114-AA   Sledge hammer, 12 lb.         02-JAN-08        8       10      14.4     .05                        12           1
54778-2T   Rat-tail file, 1/8-in. fine   15-DEC-07       43       20      4.99     .00         21344          25           0
89-WRE-Q   Hicut chain saw, 16 in.       07-FEB-08       11        5    256.99     .05         24288          10           0
PVC23DRT   PVC pipe, 3.5-in., 8-ft       20-FEB-08      188       75      5.87     .00                        50           0
SM-18277   1.25-in. metal screw, 25      01-MAR-08      172       75      6.99     .00         21225          50           0
SW-23116   2.5-in. wd. screw, 50         24-FEB-08      237      100      8.45     .00         21231         100           0
WR3/TT3    Steel matting, 4'x8'x1/6", .5" 17-JAN-08      18        5    119.95     .10         25595          10           0

16 rows selected.

SQL> EXEC PRC_PROD_DISCOUNT;

PL/SQL procedure successfully completed.

SQL> SELECT * FROM PRODUCT;

P_CODE     P_DESCRIPT                     P_INDATE     P_QOH    P_MIN   P_PRICE P_DISCOUNT     V_CODE P_MIN_ORDER   P_REORDER
--------   ---------------------------   ----------   ------   ------   -------  ----------   ------- -----------   ---------
11QER/31   Power painter, 15 psi., 3-nozz 03-NOV-07      29        5    109.99     .05         25595          25           0
13-Q2/P2   7.25-in. pwr. saw blade       13-DEC-07       32       15     14.99     .10         21344          50           0
14-Q1/L3   9.00-in. pwr. saw blade       13-NOV-07       18       12     17.49     .00         21344          50           0
1546-QQ2   Hrd. cloth, 1/4-in., 2x50     15-JAN-08       15        8     39.95     .00         23119          35           0
1558-QW1   Hrd. cloth, 1/2-in., 3x50     15-JAN-08       23        5     43.99     .05         23119          25           0
2232/QTY   B&D jigsaw, 12-in. blade      30-DEC-07        8        5    109.92     .05         24288          15           0
2232/QWE   B&D jigsaw, 8-in. blade       24-DEC-07        6        7     99.87     .05         24288          15           1
2238/QPD   B&D cordless drill, 1/2-in.   20-JAN-08       12        5     38.95     .10         25595          12           0
23109-HB   Claw hammer                   20-JAN-08       23       10      9.95     .15         21225          25           0
23114-AA   Sledge hammer, 12 lb.         02-JAN-08        8       10      14.4     .05                        12           1
54778-2T   Rat-tail file, 1/8-in. fine   15-DEC-07       43       20      4.99     .05         21344          25           0
89-WRE-Q   Hicut chain saw, 16 in.       07-FEB-08       11        5    256.99     .10         24288          10           0
PVC23DRT   PVC pipe, 3.5-in., 8-ft       20-FEB-08      188       75      5.87     .05                        50           0
SM-18277   1.25-in. metal screw, 25      01-MAR-08      172       75      6.99     .05         21225          50           0
SW-23116   2.5-in. wd. screw, 50         24-FEB-08      237      100      8.45     .05         21231         100           0
WR3/TT3    Steel matting, 4'x8'x1/6", .5" 17-JAN-08      18        5    119.95     .15         25595          10           0

16 rows selected.
```

One of the main advantages of procedures is that you can pass values to them. For example, the previous PRC_PRODUCT_DISCOUNT procedure worked fine, but what if you wanted to make the percentage increase an input variable? In that case, you can pass an argument to represent the rate of increase to the procedure. Figure 8.43 shows the code for that procedure.

FIGURE
8.43

Second version of the PRC_PROD_DISCOUNT stored procedure

```
Oracle SQL*Plus                                                    _|□|_|□|×|
File  Edit  Search  Options  Help
SQL> CREATE OR REPLACE PROCEDURE PRC_PROD_DISCOUNT(WPI IN NUMBER) AS
  2  BEGIN
  3     IF ((WPI <= 0) OR (WPI >= 1)) THEN -- validate WPI parameter
  4         DBMS_OUTPUT.PUT_LINE('Error: Value must be greater than 0 and less than 1');
  5     ELSE         -- if value is greater than 0 and less than 1
  6   UPDATE PRODUCT
  7    SET P_DISCOUNT = P_DISCOUNT + WPI
  8        WHERE P_QOH >= P_MIN*2;
  9   DBMS_OUTPUT.PUT_LINE ('* * Update finished * *');
 10      END IF;
 11  END;
 12  /

Procedure created.
```

Figure 8.44 shows the execution of the second version of the PRC_PROD_DISCOUNT stored procedure. Note that if the procedure requires arguments, those arguments must be enclosed in parentheses and they must be separated by commas.

FIGURE
8.44

Results of the second version of the PRC_PROD_DISCOUNT stored procedure

```
Oracle SQL*Plus                                            _|□|×|
File  Edit  Search  Options  Help
SQL> EXEC PRC_PROD_DISCOUNT(1.5);
Error: Value must be greater than 0 and less than 1

PL/SQL procedure successfully completed.

SQL> EXEC PRC_PROD_DISCOUNT(.05);
* * Update finished * *

PL/SQL procedure successfully completed.

SQL> |
```

Stored procedures are also useful to encapsulate shared code to represent business transactions. For example, you can create a simple stored procedure to add a new customer. By using a stored procedure, all programs can call the stored procedure by name each time a new customer is added. Naturally, if new customer attributes are added later, you would need to modify the stored procedure. However, the programs that use the stored procedure would not need to know the name of the newly added attribute and would need to add only a new parameter to the procedure call. (Notice the PRC_CUS_ADD stored procedure shown in Figure 8.45.)

As you examine Figure 8.45, note these features:

- The PRC_CUS_ADD procedure uses several parameters, one for each required attribute in the CUSTOMER table.
- The stored procedure uses the CUS_CODE_SEQ sequence to generate a new customer code.

FIGURE 8.45   The PRC_CUS_ADD stored procedure

```
Oracle SQL*Plus
File  Edit  Search  Options  Help
SQL> CREATE OR REPLACE PROCEDURE PRC_CUS_ADD
  2  (W_LN IN VARCHAR, W_FN IN VARCHAR, W_INIT IN VARCHAR, W_AC IN VARCHAR, W_PH IN VARCHAR)
  3  AS
  4  BEGIN
  5  -- note that the procedure uses the CUS_CODE_SEQ sequence created earlier
  6  -- attribute names are required when not giving values for all table attributes
  7      INSERT INTO CUSTOMER(CUS_CODE,CUS_LNAME, CUS_FNAME, CUS_INITIAL, CUS_AREACODE, CUS_PHONE)
  8          VALUES (CUS_CODE_SEQ.NEXTVAL, W_LN, W_FN, W_INIT, W_AC, W_PH);
  9      DBMS_OUTPUT.PUT_LINE ('Customer ' || W_LN || ', ' || W_FN || ' added.');
 10  END;
 11  /

Procedure created.

SQL> EXEC PRC_CUS_ADD('Walker','James',NULL,'615','84-HORSE');
Customer Walker, James added.

PL/SQL procedure successfully completed.

SQL> SELECT * FROM CUSTOMER WHERE CUS_LNAME = 'Walker';

  CUS_CODE CUS_LNAME        CUS_FNAME        C CUS CUS_PHON CUS_BALANCE
---------- ---------------  ---------------- - --- -------- -----------
     20010 Walker           James              615 84-HORSE           0

SQL> EXEC PRC_CUS_ADD('Lowery', 'Denisee', NULL, NULL, NULL);
BEGIN PRC_CUS_ADD('Lowery', 'Denisee', NULL, NULL, NULL); END;

*
ERROR at line 1:
ORA-01400: cannot insert NULL into ("STUDENT"."CUSTOMER"."CUS_AREACODE")
ORA-06512: at "STUDENT.PRC_CUS_ADD", line 7
ORA-06512: at line 1
```

- The required parameters—those specified in the table definition—must be included and can be null *only* when the table specifications permit nulls for that parameter. For example, note that the second customer addition was unsuccessful because the CUS_AREACODE is a required attribute and cannot be null.

- The procedure displays a message in the SQL*Plus console to let the user know that the customer was added.

The next two examples further illustrate the use of sequences within stored procedures. In this case, let's create two stored procedures:

1. The PRC_INV_ADD procedure adds a new invoice.

2. The PRC_LINE_ADD procedure adds a new product line row for a given invoice.

Both procedures are shown in Figure 8.46. Note the use of a variable in the PRC_LINE_ADD procedure to get the product price from the PRODUCT table.

To test the procedures shown in Figure 8.46:

1. Call the PRC_INV_ADD procedure with the new invoice data as arguments.

2. Call the PRC_LINE_ADD procedure and pass the product line arguments.

FIGURE
8.46

The PRC_INV_ADD and PRC_LINE_ADD stored procedures

```
± Oracle SQL*Plus                                                          _ □ X
File  Edit  Search  Options  Help
SQL> CREATE OR REPLACE PROCEDURE PRC_INV_ADD (W_CUS_CODE IN VARCHAR2, W_DATE IN DATE)
  2   AS BEGIN
  3      INSERT INTO INVOICE
  4           VALUES(INV_NUMBER_SEQ.NEXTVAL, W_CUS_CODE, W_DATE);
  5      DBMS_OUTPUT.PUT_LINE('Invoice added');
  6   END;
  7   /

Procedure created.

SQL> CREATE OR REPLACE PROCEDURE PRC_LINE_ADD (W_LN IN NUMBER, W_P_CODE IN VARCHAR2, W_LU NUMBER)
  2   AS
  3     W_LP NUMBER := 0.00;
  4   BEGIN
  5      -- GET THE PRODUCT PRICE
  6     SELECT P_PRICE INTO W_LP
  7          FROM PRODUCT
  8              WHERE P_CODE = W_P_CODE;
  9
 10     -- ADDS THE NEW LINE ROW
 11     INSERT INTO LINE
 12          VALUES(INV_NUMBER_SEQ.CURRVAL, W_LN, W_P_CODE, W_LU, W_LP);
 13
 14     DBMS_OUTPUT.PUT_LINE('Invoice line ' || W_LN || ' added');
 15   END;
 16   /

Procedure created.

SQL>
```

That process is illustrated in Figure 8.47.

FIGURE
8.47

Testing the PRC_INV_ADD and PRC_LINE_ADD procedures

```
± Oracle SQL*Plus                                                          _ □ X
File  Edit  Search  Options  Help
SQL> EXEC PRC_INV_ADD(20010,'09-APR-2008');
Invoice added

PL/SQL procedure successfully completed.

SQL> EXEC PRC_LINE_ADD(1,'13-Q2/P2',1);
* * * Balance updated for customer: 20010
Invoice line 1 added

PL/SQL procedure successfully completed.

SQL> EXEC PRC_LINE_ADD(2,'23109-HB',1);
* * * Balance updated for customer: 20010
Invoice line 2 added

PL/SQL procedure successfully completed.

SQL> SELECT * FROM INVOICE WHERE CUS_CODE = 20010;

INV_NUMBER   CUS_CODE INV_DATE
---------- ---------- ---------
      4010     20010 09-APR-08

SQL> SELECT * FROM LINE WHERE INV_NUMBER  = (SELECT INV_NUMBER FROM INVOICE WHERE CUS_CODE = 20010);

INV_NUMBER LINE_NUMBER P_CODE     LINE_UNITS LINE_PRICE
---------- ----------- ---------- ---------- ----------
      4010           1 13-Q2/P2            1      14.99
      4010           2 23109-HB            1       9.95

SQL> SELECT * FROM PRODUCT WHERE P_CODE IN ('13-Q2/P2', '23109-HB');

P_CODE   P_DESCRIPT          P_INDATE    P_QOH   P_MIN  P_PRICE P_DISCOUNT    V_CODE P_MIN_ORDER P_REORDER
-------- ------------------- --------- ------- ------- -------- ---------- --------- ----------- ---------
13-Q2/P2 7.25-in. pwr. saw blade 13-DEC-07    31      15    14.99        .15     21344          50         0
23109-HB Claw hammer         20-JAN-08      22      10     9.95        .20     21225          25         0

SQL> SELECT * FROM CUSTOMER WHERE CUS_CODE = 20010;

 CUS_CODE CUS_LNAME       CUS_FNAME       C CUS CUS_PHON CUS_BALANCE
--------- --------------- --------------- - --- -------- -----------
    20010 Walker          James             615 84-HORSE       24.94
```

8.7.3 PL/SQL Processing with Cursors

Until now, all of the SQL statements you have used inside a PL/SQL block (trigger or stored procedure) have returned a single value. If the SQL statement returns more than one value, you will generate an error. If you want to use an SQL statement that returns more than one value inside your PL/SQL code, you need to use a cursor. A **cursor** is a special construct used in procedural SQL to hold the data rows returned by an SQL query. You can think of a cursor as a reserved area of memory in which the output of the query is stored, like an array holding columns and rows. Cursors are held in a reserved memory area in the DBMS server, not in the client computer.

There are two types of cursors: implicit and explicit. An **implicit cursor** is automatically created in procedural SQL when the SQL statement returns only one value. Up to this point, all of the examples created an implicit cursor. An **explicit cursor** is created to hold the output of an SQL statement that may return two or more rows (but could return 0 or only one row). To create an explicit cursor, you use the following syntax inside a PL/SQL DECLARE section:

CURSOR cursor_name IS select-query;

Once you have declared a cursor, you can use specific PL/SQL cursor processing commands (OPEN, FETCH, and CLOSE) anywhere between the BEGIN and END keywords of the PL/SQL block. Table 8.9 summarizes the main use of each of those commands.

TABLE 8.9	Cursor Processing Commands	
CURSOR COMMAND	**EXPLANATION**	
OPEN	Opening the cursor executes the SQL command and populates the cursor with data, opening the cursor for processing. The cursor declaration command only reserves a named memory area for the cursor; it doesn't populate the cursor with the data. Before you can use a cursor, you need to open it. For example: OPEN cursor_name	
FETCH	Once the cursor is opened, you can use the FETCH command to retrieve data from the cursor and copy it to the PL/SQL variables for processing. The syntax is: FETCH cursor_name INTO variable1 [, variable2, ...] The PL/SQL variables used to hold the data must be declared in the DECLARE section and must have data types compatible with the columns retrieved by the SQL command. If the cursors SQL statement returns five columns, there must be five PL/SQL variables to receive the data from the cursor. This type of processing resembles the one-record-at-a-time processing used in previous database models. The first time you fetch a row from the cursor, the first row of data from the cursor is copied to the PL/SQL variables; the second time you fetch a row from the cursor, the second row of data is placed in the PL/SQL variables; and so on.	
CLOSE	The CLOSE command closes the cursor for processing.	

Cursor-style processing involves retrieving data from the cursor one row at a time. Once you open a cursor, it becomes an active data set. That data set contains a "current" row pointer. Therefore, after opening a cursor, the current row is the first row of the cursor.

When you fetch a row from the cursor, the data from the "current" row in the cursor is copied to the PL/SQL variables. After the fetch, the "current" row pointer moves to the next row in the set and continues until it reaches the end of the cursor.

How do you know what number of rows are in the cursor? Or how do you know when you have reached the end of the cursor data set? You know because cursors have special attributes that convey important information. Table 8.10 summarizes the cursor attributes.

TABLE 8.10	Cursor Attributes	
ATTRIBUTE	DESCRIPTION
%ROWCOUNT	Returns the number of rows fetched so far. If the cursor is not OPEN, it returns an error. If no FETCH has been done but the cursor is OPEN, it returns 0.
%FOUND	Returns TRUE if the last FETCH returned a row and FALSE if not. If the cursor is not OPEN, it returns an error. If no FETCH has been done, it contains NULL.
%NOTFOUND	Returns TRUE if the last FETCH did not return any row and FALSE if it did. If the cursor is not OPEN, it returns an error. If no FETCH has been done, it contains NULL.
%ISOPEN	Returns TRUE if the cursor is open (ready for processing) or FALSE if the cursor is closed. Remember, before you can use a cursor, you must open it.

To illustrate the use of cursors, let's use a simple stored procedure example that lists all products that have a quantity on hand greater than the average quantity on hand for all products. The code is shown in Figure 8.48.

FIGURE 8.48	A simple PRC_CURSOR_EXAMPLE

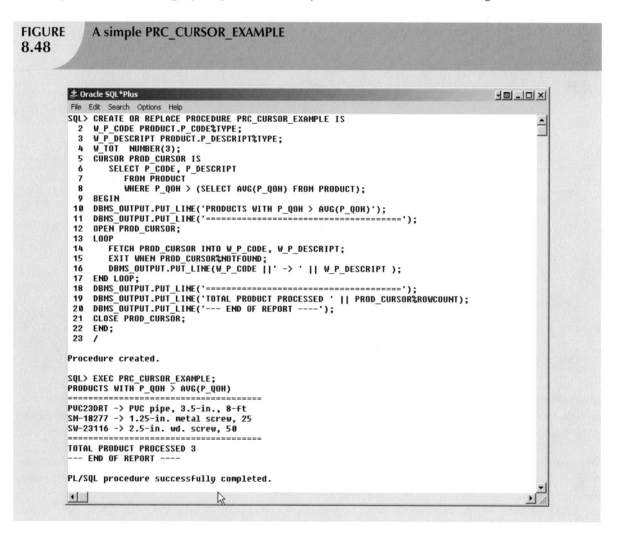

As you examine the stored procedure code shown in Figure 8.48, note the following important characteristics:

- Lines 2 and 3 use the %TYPE data type in the variable definition section. As indicated in Table 8.8, the %TYPE data type is used to indicate that the given variable inherits the data type from a variable previously declared or from an attribute of a database table. In this case, you are using the %TYPE to indicate that the W_P_CODE and W_P_DESCRIPT will have the same data type as the respective columns in the PRODUCT table. This way, you ensure that the PL/SQL variable will have a compatible data type.

- Line 5 declares the PROD_CURSOR cursor.

- Line 12 opens the PROD_CURSOR cursor and populates it.

- Line 13 uses the LOOP statement to loop through the data in the cursor, fetching one row at a time.

- Line 14 uses the FETCH command to retrieve a row from the cursor and place it in the respective PL/SQL variables.

- Line 15 uses the EXIT command to evaluate when there are no more rows in the cursor (using the %NOTFOUND cursor attribute) and to exit the loop.

- Line 19 uses the %ROWCOUNT cursor attribute to obtain the total number of rows processed.

- Line 21 issues the CLOSE PROD_CURSOR command to close the cursor.

The use of cursors, combined with standard SQL, makes relational databases very desirable because programmers can work in the best of both worlds: set-oriented processing and record-oriented processing. Any experienced programmer knows to use the tool that best fits the job. Sometimes you will be better off manipulating data in a set-oriented environment; at other times, it might be better to use a record-oriented environment. Procedural SQL lets you have your proverbial cake and eat it, too. Procedural SQL provides functionality that enhances the capabilities of the DBMS while maintaining a high degree of manageability.

8.7.4 PL/SQL STORED FUNCTIONS

Using programmable or procedural SQL, you can also create your own stored functions. Stored procedures and functions are very similar. A **stored function** is basically a named group of procedural and SQL statements that returns a value (indicated by a RETURN statement in its program code). To create a function, you use the following syntax:

```
CREATE FUNCTION function_name (argument IN data-type, … ) RETURN data-type [IS]
BEGIN
        PL/SQL statements;
        …
        RETURN (value or expression);
END;
```

Stored functions can be invoked only from within stored procedures or triggers and cannot be invoked from SQL statements (unless the function follows some very specific compliance rules). Remember not to confuse built-in SQL functions (such as MIN, MAX, and AVG) with stored functions.

8.8 EMBEDDED SQL

There is little doubt that SQL's popularity as a data manipulation language is in part due to its ease of use and its powerful data-retrieval capabilities. But in the real world, database systems are related to other systems and programs, and you still need a conventional programming language such as Visual Basic.Net, C#, or COBOL to integrate database systems with other programs and systems. If you are developing Web applications, you are most likely familiar with Visual Studio.Net, Java, ASP, or ColdFusion. Yet, almost regardless of the programming tools you use, if your

Web application or Windows-based GUI system requires access to a database such as MS Access, SQL Server, Oracle, or DB2, you will likely need to use SQL to manipulate the data in the database.

Embedded SQL is a term used to refer to SQL statements that are contained within an application programming language such as Visual Basic.Net, C#, COBOL, or Java. The program being developed might be a standard binary executable in Windows or Linux, or it might be a Web application designed to run over the Internet. No matter what language you use, if it contains embedded SQL statements, it is called the **host language**. Embedded SQL is still the most common approach to maintaining procedural capabilities in DBMS-based applications. However, mixing SQL with procedural languages requires that you understand some key differences between SQL and procedural languages.

- *Run-time mismatch*: Remember that SQL is a nonprocedural, interpreted language; that is, each instruction is parsed, its syntax is checked, and it is executed one instruction at a time.[1] All of the processing takes place at the server side. Meanwhile, the host language is generally a binary-executable program (also known as a compiled program). The host program typically runs at the client side in its own memory space (which is different from the DBMS environment).

- *Processing mismatch*: Conventional programming languages (COBOL, ADA, FORTRAN, PASCAL, C++, and PL/I) process one data element at a time. Although you can use arrays to hold data, you still process the array elements one row at a time. This is especially true for file manipulation, where the host language typically manipulates data one record at a time. However, newer programming environments (such as Visual Studio.Net) have adopted several object-oriented extensions that help the programmer manipulate data sets in a cohesive manner.

- *Data type mismatch*: SQL provides several data types, but some of those data types might not match data types used in different host languages (for example, the date and varchar2 data types).

To bridge the differences, the Embedded SQL standard[2] defines a framework to integrate SQL within several programming languages. The Embedded SQL framework defines the following:

- A standard syntax to identify embedded SQL code within the host language (EXEC SQL/END-EXEC).
- A standard syntax to identify host variables. Host variables are variables in the host language that receive data from the database (through the embedded SQL code) and process the data in the host language. All host variables are preceded by a colon (":").
- A communication area used to exchange status and error information between SQL and the host language. This communications area contains two variables—SQLCODE and SQLSTATE.

Another way to interface host languages and SQL is through the use of a call level interface (CLI)[3] , in which the programmer writes to an application programming interface (API). A common CLI in Windows is provided by the Open Database Connectivity (ODBC) interface.

ONLINE CONTENT

Additional coverage of CLIs and ODBC is found in **Appendix F, Client/Server Systems**, and **Appendix J, Web Database Development with ColdFusion** in the Student Online Companion.

---

[1] The authors are particularly grateful for the thoughtful comments provided by Emil T. Cipolla, who teaches at Mount Saint Mary College and whose IBM experience is the basis for his considerable and practical expertise.

[2] You can obtain more details about the Embedded SQL standard at *www.ansi.org*, SQL/Bindings is in the SQL Part II – SQL/Foundation section of the SQL 2003 standard.

[3] You can find additional information about the SQL Call Level Interface standard at *www.ansi.org*, in the SQL Part 3: Call Level Interface (SQL/CLI) section of the SQL 2003 standard.

Before continuing, let's explore the process required to create and run an executable program with embedded SQL statements. If you have ever programmed in COBOL or C++, you are familiar with the multiple steps required to generate the final executable program. Although the specific details vary among language and DBMS vendors, the following general steps are standard:

1. The programmer writes embedded SQL code within the host language instructions. The code follows the standard syntax required for the host language and embedded SQL.

2. A preprocessor is used to transform the embedded SQL into specialized procedure calls that are DBMS- and language-specific. The preprocessor is provided by the DBMS vendor and is specific to the host language.

3. The program is compiled using the host language compiler. The compiler creates an object code module for the program containing the DBMS procedure calls.

4. The object code is linked to the respective library modules and generates the executable program. This process binds the DBMS procedure calls to the DBMS run-time libraries. Additionally, the binding process typically creates an "access plan" module that contains instructions to run the embedded code at run time.

5. The executable is run, and the embedded SQL statement retrieves data from the database.

Note that you can embed individual SQL statements or even an entire PL/SQL block. Up to this point in the book, you have used a DBMS-provided application (SQL*Plus) to write SQL statements and PL/SQL blocks in an interpretive mode to address one-time or ad hoc data requests. However, it is extremely difficult and awkward to use ad hoc queries to process transactions inside a host language. Programmers typically embed SQL statements within a host language that it is compiled once and executed as often as needed. To embed SQL into a host language, follow this syntax:

```
EXEC SQL
        SQL statement;
END-EXEC.
```

The preceding syntax will work for SELECT, INSERT, UPDATE, and DELETE statements. For example, the following embedded SQL code will delete employee 109, George Smith, from the EMPLOYEE table:

```
EXEC SQL
        DELETE FROM EMPLOYEE WHERE EMP_NUM = 109;
END-EXEC.
```

Remember, the preceding embedded SQL statement is compiled to generate an executable statement. Therefore, the statement is fixed permanently and cannot change (unless, of course, the programmer changes it). Each time the program runs, it deletes the same row. In short, the preceding code is good only for the first run; all subsequent runs will likely generate an error. Clearly, this code would be more useful if you could specify a variable to indicate the employee number to be deleted.

In embedded SQL, all host variables are preceded by a colon (":"). The host variables may be used to send data from the host language to the embedded SQL, or they may be used to receive the data from the embedded SQL. To use a host variable, you must first declare it in the host language. Common practice is to use similar host variable names as the SQL source attributes. For example, if you are using COBOL, you would define the host variables in the Working Storage section. Then you would refer to them in the embedded SQL section by preceding them with a colon (":"). For example, to delete an employee whose employee number is represented by the host variable W_EMP_NUM, you would write the following code:

```
EXEC SQL
        DELETE FROM EMPLOYEE WHERE EMP_NUM = :W_EMP_NUM;
END-EXEC.
```

At run time, the host variable value will be used to execute the embedded SQL statement. What happens if the employee you are trying to delete doesn't exist in the database? How do you know that the statement has been completed without errors? As mentioned previously, the embedded SQL standard defines a SQL communication area to hold status and error information. In COBOL, such an area is known as the SQLCA area and is defined in the Data Division as follows:

```
EXEC SQL
        INCLUDE SQLCA
END-EXEC.
```

The SQLCA area contains two variables for status and error reporting. Table 8.11 shows some of the main values returned by the variables and their meaning.

TABLE 8.11	SQL Status and Error Reporting Variables	
VARIABLE NAME	VALUE	EXPLANATION
SQLCODE		Old-style error reporting supported for backward compatibility only; returns an integer value (positive or negative).
	0	Successful completion of command.
	100	No data; the SQL statement did not return any rows or did not select, update, or delete any rows.
	-999	Any negative value indicates that an error occurred.
SQLSTATE		Added by SQL-92 standard to provide predefined error codes; defined as a character string (5 characters long).
	00000	Successful completion of command.
		Multiple values in the format XXYYY where: XX-> represents the class code. YYY-> represents the subclass code.

The following embedded SQL code illustrates the use of the SQLCODE within a COBOL program.

```
EXEC SQL
EXEC SQL
        SELECT      EMP_LNAME, EMP_LNAME INTO :W_EMP_FNAME, :W_EMP_LNAME
                    WHERE EMP_NUM = :W_EMP_NUM;
END-EXEC.
IF SQLCODE = 0 THEN
        PERFORM DATA_ROUTINE
ELSE
        PERFORM ERROR_ROUTINE
END-IF.
```

In this example, the SQLCODE host variable is checked to determine whether the query completed successfully. If that is the case, the DATA_ROUTINE is performed; otherwise, the ERROR_ROUTINE is performed.

Just as with PL/SQL, embedded SQL requires the use of cursors to hold data from a query that returns more than one value. If COBOL is used, the cursor can be declared either in the Working Storage Section or in the Procedure Division. The cursor must be declared and processed as you learned earlier in Section 8.7.3. To declare a cursor, you use the syntax shown in the following example:

```
EXEC SQL
      DECLARE PROD_CURSOR FOR
              SELECT    P_CODE, P_DESCRIPT FROM PRODUCT
              WHERE     P_QOH > (SELECT AVG(P_QOH) FROM PRODUCT);
END-EXEC.
```

Next, you must open the cursor to make it ready for processing:

```
EXEC SQL
        OPEN PROD_CURSOR;
END-EXEC.
```

To process the data rows in the cursor, you use the FETCH command to retrieve one row of data at a time and place the values in the host variables. The SQLCODE must be checked to ensure that the FETCH command completed successfully. This section of code typically constitutes part of a routine in the COBOL program. Such a routine is executed with the PERFORM command. For example:

```
EXEC SQL
        FETCH PROD_CURSOR INTO :W_P_CODE, :W_P_DESCRIPT;
END-EXEC.
IF SQLCODE = 0 THEN
        PERFORM DATA_ROUTINE
ELSE
        PERFORM ERROR_ROUTINE
END-IF.
```

When all rows have been processed, you close the cursor as follows:

```
EXEC SQL
        CLOSE PROD_CURSOR;
END-EXEC.
```

Thus far, you have seen examples of embedded SQL in which the programmer used predefined SQL statements and parameters. Therefore, the end users of the programs are limited to the actions that were specified in the application programs. That style of embedded SQL is known as **static SQL**, meaning that the SQL statements will not change while the application is running. For example, the SQL statement might read like this:

```
SELECT    P_CODE, P_DESCRIPT, P_QOH, P_PRICE
FROM      PRODUCT
WHERE     P_PRICE > 100;
```

Note that the attributes, tables, and conditions are known in the preceding SQL statement. Unfortunately, end users seldom work in a static environment. They are more likely to require the flexibility of defining their data access requirements on the fly. Therefore, the end user requires that SQL be as dynamic as the data access requirements.

Dynamic SQL is a term used to describe an environment in which the SQL statement is not known in advance; instead, the SQL statement is generated at run time. At run time in a dynamic SQL environment, a program can generate the SQL statements that are required to respond to ad hoc queries. In such an environment, neither the programmer nor the end user is likely to know precisely what kind of queries are to be generated or how those queries are to be structured. For example, a dynamic SQL equivalent of the preceding example could be:

```
SELECT      :W_ATTRIBUTE_LIST
FROM        :W_TABLE
WHERE       :W_CONDITION;
```

Note that the attribute list and the condition are not known until the end user specifies them. W_TABLE, W_ATRIBUTE_LIST, and W_CONDITION are text variables that contain the end-user input values used in the query generation. Because the program uses the end-user input to build the text variables, the end user can run the same program multiple times to generate varying outputs. For example, in one instance, the end user might want to know what products have a price less than $100; in another case, the end user might want to know how many units of a given product are available for sale at any given moment.

Although dynamic SQL is clearly flexible, such flexibility carries a price. Dynamic SQL tends to be much slower than static SQL. Dynamic SQL also requires more computer resources (overhead). Finally, you are more likely to find inconsistent levels of support and incompatibilities among DBMS vendors.

SUMMARY

- SQL provides relational set operators to combine the output of two queries to generate a new relation. The UNION and UNION ALL set operators combine the output of two (or more) queries and produce a new relation with all unique (UNION) or duplicate (UNION ALL) rows from both queries. The INTERSECT relational set operator selects only the common rows. The MINUS set operator selects only the rows that are different. UNION, INTERSECT, and MINUS require union-compatible relations.

- Operations that join tables can be classified as inner joins and outer joins. An inner join is the traditional join in which only rows that meet a given criteria are selected. An outer join returns the matching rows as well as the rows with unmatched attribute values for one table or both tables to be joined.

- A natural join returns all rows with matching values in the matching columns and eliminates duplicate columns. This style of query is used when the tables share a common attribute with a common name. One important difference between the syntax for a natural join and for the "old-style" join is that the natural join does not require the use of a table qualifier for the common attributes.

- Joins may use keywords such as USING and ON. If the USING clause is used, the query will return only the rows with matching values in the column indicated in the USING clause; that column must exist in both tables. If the ON clause is used, the query will return only the rows that meet the specified join condition.

- Subqueries and correlated queries are used when it is necessary to process data based on *other* processed data. That is, the query uses results that were previously unknown and that are generated by another query. Subqueries may be used with the FROM, WHERE, IN, and HAVING clauses in a SELECT statement. A subquery may return a single row or multiple rows.

- Most subqueries are executed in a serial fashion. That is, the outer query initiates the data request, and then the inner subquery is executed. In contrast, a correlated subquery is a subquery that is executed once for each row in the outer query. That process is similar to the typical nested loop in a programming language. A correlated subquery is so named because the inner query is related to the outer query—the inner query references a column of the outer subquery.

- SQL functions are used to extract or transform data. The most frequently used functions are date and time functions. The results of the function output can be used to store values in a database table, to serve as the basis for the computation of derived variables, or to serve as a basis for data comparisons. Function formats can be vendor-specific. Aside from time and date functions, there are numeric and string functions as well as conversion functions that convert one data format to another.

- Oracle sequences may be used to generate values to be assigned to a record. For example, a sequence may be used to number invoices automatically. MS Access uses an AutoNumber data type to generate numeric sequences. MS SQL Server uses the Identity column property to designate the column that will have sequential numeric values automatically assigned to it. There can only be one Identity column per SQL Server table.

- Procedural SQL (PL/SQL) can be used to create triggers, stored procedures, and PL/SQL functions. A trigger is procedural SQL code that is automatically invoked by the DBMS upon the occurrence of a specified data manipulation event (UPDATE, INSERT, or DELETE). Triggers are critical to proper database operation and management. They help automate various transaction and data management processes, and they can be used to enforce constraints that are not enforced at the DBMS design and implementation levels.

- A stored procedure is a named collection of SQL statements. Just like database triggers, stored procedures are stored in the database. One of the major advantages of stored procedures is that they can be used to encapsulate and represent complete business transactions. Use of stored procedures substantially reduces network traffic and increases system performance. Stored procedures help reduce code duplication by creating unique PL/SQL modules that are called by the application programs, thereby minimizing the chance of errors and the cost of application development and maintenance.

▶ When SQL statements are designed to return more than one value inside the PL/SQL code, a cursor is needed. You can think of a cursor as a reserved area of memory in which the output of the query is stored, like an array holding columns and rows. Cursors are held in a reserved memory area in the DBMS server, rather than in the client computer. There are two types of cursors: implicit and explicit.

▶ Embedded SQL refers to the use of SQL statements within an application programming language such as Visual Basic.Net, C#, COBOL, or Java. The language in which the SQL statements are embedded is called the host language. Embedded SQL is still the most common approach to maintaining procedural capabilities in DBMS-based applications.

K E Y   T E R M S

anonymous PL/SQL block, 339

batch update routine, 335

correlated subquery, 321

cross join, 306

cursor, 357

dynamic SQL, 364

embedded SQL, 360

explicit cursor, 357

host language, 360

implicit cursor, 357

inner join, 305

outer join, 305

persistent stored module (PSM), 338

procedural SQL (PL/SQL), 338

row-level trigger, 344

statement-level trigger, 344

static SQL, 363

stored function, 359

stored procedure, 359

trigger, 342

union-compatible, 298

updatable view, 336

O N L I N E   C O N T E N T

Answers to selected Review Questions and Problems for this chapter are contained in the Student Online Companion for this book.

R E V I E W   Q U E S T I O N S

1. The relational set operators UNION, INTERSECT, and MINUS work properly only when the relations are union-compatible. What does *union-compatible* mean, and how would you check for this condition?

2. What is the difference between UNION and UNION ALL? Write the syntax for each.

3. Suppose you have two tables: EMPLOYEE and EMPLOYEE_1. The EMPLOYEE table contains the records for three employees: Alice Cordoza, John Cretchakov, and Anne McDonald. The EMPLOYEE_1 table contains the records for employees John Cretchakov and Mary Chen. Given that information, list the query output for the UNION query.

4. Given the employee information in Question 3, list the query output for the UNION ALL query.

5. Given the employee information in Question 3, list the query output for the INTERSECT query.

6. Given the employee information in Question 3, list the query output for the MINUS query.

7. What is a CROSS JOIN? Give an example of its syntax.

8. What three join types are included in the OUTER JOIN classification?

9. Using tables named T1 and T2, write a query example for each of the three join types you described in Question 8. Assume that T1 and T2 share a common column named C1.

10. What is a subquery, and what are its basic characteristics?

11. What is a correlated subquery? Give an example.

12. What MS Access/SQL Server function should you use to calculate the number of days between the current date and January 25, 1999?

13. What Oracle function should you use to calculate the number of days between the current date and January 25, 1999?

14. Suppose a PRODUCT table contains two attributes, PROD_CODE and VEND_CODE. Those two attributes have values of ABC, 125, DEF, 124, GHI, 124, and JKL, 123, respectively. The VENDOR table contains a single attribute, VEND_CODE, with values 123, 124, 125, and 126, respectively. (The VEND_CODE attribute in the PRODUCT table is a foreign key to the VEND_CODE in the VENDOR table.) Given that information, what would be the query output for:

    a. A UNION query based on the two tables?

    b. A UNION ALL query based on the two tables?

    c. An INTERSECT query based on the two tables?

    d. A MINUS query based on the two tables?

15. What string function should you use to list the first three characters of a company's EMP_LNAME values? Give an example using a table named EMPLOYEE. Provide examples for Oracle and SQL Server.

16. What is an Oracle sequence? Write its syntax.

17. What is a trigger, and what is its purpose? Give an example.

18. What is a stored procedure, and why is it particularly useful? Give an example.

19. What is embedded SQL, and how is it used?

20. What is dynamic SQL, and how does it differ from static SQL?

P R O B L E M S

Use the database tables in Figure P8.1 as the basis for Problems 1–18.

O N L I N E   C O N T E N T

The **Ch08_SimpleCo** database is located in the Student Online Companion, as are the script files to duplicate this data set in Oracle.

1. Create the tables. (Use the MS Access example shown in Figure P8.1 to see what table names and attributes to use.)

2. Insert the data into the tables you created in Problem 1.

3. Write the query that will generate a combined list of customers (from the tables CUSTOMER and CUSTOMER_2) that do not include the duplicate customer records. (Note that only the customer named Juan Ortega shows up in both customer tables.)

4. Write the query that will generate a combined list of customers to include the duplicate customer records.

5. Write the query that will show only the duplicate customer records.

6. Write the query that will generate only the records that are unique to the CUSTOMER_2 table.

7. Write the query to show the invoice number, the customer number, the customer name, the invoice date, and the invoice amount for all customers with a customer balance of $1,000 or more.

FIGURE P8.1   Ch08_SimpleCo database tables

Database name: CH08_SimpleCo

Table name: CUSTOMER

CUST_NUM	CUST_LNAME	CUST_FNAME	CUST_BALANCE
1000	Smith	Jeanne	1050.11
1001	Ortega	Juan	840.92

Table name: CUSTOMER_2

CUST_NUM	CUST_LNAME	CUST_FNAME
2000	McPherson	Anne
2001	Ortega	Juan
2002	Kowalski	Jan
2003	Chen	George

Table name: INVOICE

INV_NUM	CUST_NUM	INV_DATE	INV_AMOUNT
8000	1000	23-Mar-08	235.89
8001	1001	23-Mar-08	312.82
8002	1001	30-Mar-08	528.10
8003	1000	12-Apr-08	194.78
8004	1000	23-Apr-08	619.44

8. Write the query that will show (for all the invoices) the invoice number, the invoice amount, the average invoice amount, and the difference between the average invoice amount and the actual invoice amount.

9. Write the query that will write Oracle sequences to produce automatic customer number and invoice number values. Start the customer numbers at 1000 and the invoice numbers at 5000.

10. Modify the CUSTOMER table to included two new attributes: CUST_DOB and CUST_AGE. Customer 1000 was born on March 15, 1979, and customer 1001 was born on December 22, 1988.

11. Assuming you completed Problem 10, write the query that will list the names and ages of your customers.

12. Assuming the CUSTOMER table contains a CUST_AGE attribute, write the query to update the values in that attribute. (*Hint:* Use the results of the previous query.)

13. Write the query that lists the average age of your customers. (Assume that the CUSTOMER table has been modified to include the CUST_DOB and the derived CUST_AGE attribute.)

14. Write the trigger to update the CUST_BALANCE in the CUSTOMER table when a new invoice record is entered. (Assume that the sale is a credit sale.) Test the trigger, using the following new INVOICE record:

    8005, 1001, '27-APR-08', 225.40

    Name the trigger **trg_updatecustbalance**.

15. Write a procedure to add a new customer to the CUSTOMER table. Use the following values in the new record:

    1002, 'Rauthor', 'Peter', 0.00

    Name the procedure **prc_cust_add**. Run a query to see if the record has been added.

16. Write a procedure to add a new invoice record to the INVOICE table. Use the following values in the new record:

    8006, 1000, '30-APR-08', 301.72

    Name the procedure **prc_invoice_add**. Run a query to see if the record has been added.

17. Write a trigger to update the customer balance when an invoice is deleted. Name the trigger **trg_updatecustbalance2**.

18. Write a procedure to delete an invoice, giving the invoice number as a parameter. Name the procedure **prc_inv_delete**. Test the procedure by deleting invoices 8005 and 8006.

NOTE

The following problem sets can serve as the basis for a class project or case.

Use the **Ch08_SaleCo2** database to work Problems 19–22, shown in Figure P8.19.

FIGURE P8.19   Ch08_SaleCo2 database tables

Database name: CH08_SaleCo2

Table name: CUSTOMER

CUS_CODE	CUS_LNAME	CUS_FNAME	CUS_INITIAL	CUS_AREACODE	CUS_PHONE	CUS_BALANCE
10010	Ramas	Alfred	A	615	844-2573	0.00
10011	Dunne	Leona	K	713	894-1238	0.00
10012	Smith	Kathy	W	615	894-2285	345.86
10013	Olowski	Paul	F	615	894-2180	536.75
10014	Orlando	Myron		615	222-1672	0.00
10015	O'Brian	Amy	B	713	442-3381	0.00
10016	Brown	James	G	615	297-1228	221.19
10017	Williams	George		615	290-2556	768.93
10018	Farriss	Anne	G	713	382-7185	216.55
10019	Smith	Olette	K	615	297-3809	0.00

Table name: PRODUCT

P_CODE	P_DESCRIPT	P_INDATE	P_QOH	P_MIN	P_PRICE	P_DISCOUNT	V_CODE
11QER/31	Power painter, 15 psi., 3-nozzle	03-Nov-07	8	5	109.99	0.00	25595
13-Q2/P2	7.25-in. pwr. saw blade	13-Dec-07	32	15	14.99	0.05	21344
14-Q1/L3	9.00-in. pwr. saw blade	13-Nov-07	18	12	17.49	0.00	21344
1546-QQ2	Hrd. cloth, 1/4-in., 2x50	15-Jan-08	15	8	39.95	0.00	23119
1558-QW1	Hrd. cloth, 1/2-in., 3x50	15-Jan-08	23	5	43.99	0.00	23119
2232/QTY	B&D jigsaw, 12-in. blade	30-Dec-07	8	5	109.92	0.05	24288
2232/QWE	B&D jigsaw, 8-in. blade	24-Dec-07	6	5	99.87	0.05	24288
2238/QPD	B&D cordless drill, 1/2-in.	20-Jan-08	12	5	38.95	0.05	25595
23109-HB	Claw hammer	20-Jan-08	23	10	9.95	0.10	21225
23114-AA	Sledge hammer, 12 lb.	02-Jan-08	8	5	14.40	0.05	
54778-2T	Rat-tail file, 1/8-in. fine	15-Dec-07	43	20	4.99	0.00	21344
89-WRE-Q	Hicut chain saw, 16 in.	07-Feb-08	11	5	256.99	0.05	24288
PVC23DRT	PVC pipe, 3.5-in., 8-ft	20-Feb-08	188	75	5.87	0.00	
SM-18277	1.25-in. metal screw, 25	01-Mar-08	172	75	6.99	0.00	21225
SW-23116	2.5-in. wd. screw, 50	24-Feb-08	237	100	8.45	0.00	21231
WR3/TT3	Steel matting, 4'x8'x1/6", .5" mesh	17-Jan-08	18	5	119.95	0.10	25595

Table name: VENDOR

V_CODE	V_NAME	V_CONTACT	V_AREACODE	V_PHONE	V_STATE	V_ORDER
21225	Bryson, Inc.	Smithson	615	223-3234	TN	Y
21226	SuperLoo, Inc.	Flushing	904	215-8995	FL	N
21231	D&E Supply	Singh	615	228-3245	TN	Y
21344	Gomez Bros.	Ortega	615	889-2546	KY	N
22567	Dome Supply	Smith	901	678-1419	GA	N
23119	Randsets Ltd.	Anderson	901	678-3998	GA	Y
24004	Brackman Bros.	Browning	615	228-1410	TN	N
24288	ORDVA, Inc.	Hakford	615	898-1234	TN	Y
25443	B&K, Inc.	Smith	904	227-0093	FL	N
25501	Damal Supplies	Smythe	615	890-3529	TN	N
25595	Rubicon Systems	Orton	904	456-0092	FL	Y

Table name: INVOICE

INV_NUMBER	CUS_CODE	INV_DATE	INV_SUBTOTAL	INV_TAX	INV_TOTAL
1001	10014	16-Jan-08	24.90	1.99	26.89
1002	10011	16-Jan-08	9.98	0.80	10.78
1003	10012	16-Jan-08	153.85	12.31	166.16
1004	10011	17-Jan-08	34.97	2.80	37.77
1005	10018	17-Jan-08	70.44	5.64	76.08
1006	10014	17-Jan-08	397.83	31.83	429.66
1007	10015	17-Jan-08	34.97	2.80	37.77
1008	10011	17-Jan-08	399.15	31.93	431.08

Table name: LINE

INV_NUMBER	LINE_NUMBER	P_CODE	LINE_UNITS	LINE_PRICE	LINE_TOTAL
1001	1	13-Q2/P2	1	14.99	14.99
1001	2	23109-HB	1	9.95	9.95
1002	1	54778-2T	2	4.99	9.98
1003	1	2238/QPD	1	38.95	38.95
1003	2	1546-QQ2	1	39.95	39.95
1003	3	13-Q2/P2	5	14.99	74.95
1004	1	54778-2T	3	4.99	14.97
1004	2	23109-HB	2	9.95	19.90
1005	1	PVC23DRT	12	5.87	70.44
1006	1	SM-18277	3	6.99	20.97
1006	2	2232/QTY	1	109.92	109.92
1006	3	23109-HB	1	9.95	9.95
1006	4	89-WRE-Q	1	256.99	256.99
1007	1	13-Q2/P2	2	14.99	29.98
1007	2	54778-2T	1	4.99	4.99
1008	1	PVC23DRT	5	5.87	29.35
1008	2	WR3/TT3	3	119.95	359.85
1008	3	23109-HB	1	9.95	9.95

ONLINE CONTENT

The **Ch08_SaleCo2** database used in Problems 19–22 is located in the Student Online Companion for this book, as are the script files to duplicate this data set in Oracle.

19. Create a trigger named **trg_line_total** to write the LINE_TOTAL value in the LINE table every time you add a new LINE row. (The LINE_TOTAL value is the product of the LINE_UNITS and the LINE_PRICE values.)

20. Create a trigger named **trg_line_prod** that will automatically update the quantity on hand for each product sold after a new LINE row is added.

21. Create a stored procedure named **prc_inv_amounts** to update the INV_SUBTOTAL, INV_TAX, and INV_TOTAL. The procedure takes the invoice number as a parameter. The INV_SUBTOTAL is the sum of the LINE_TOTAL amounts for the invoice, the INV_TAX is the product of the INV_SUBTOTAL and the tax rate (8%), and the INV_TOTAL is the sum of the INV_SUBTOTAL and the INV_TAX.

22. Create a procedure named **prc_cus_balance_update** that will take the invoice number as a parameter and update the customer balance. (*Hint*: You can use the DECLARE section to define a TOTINV numeric variable that holds the computed invoice total.)

Use the **Ch08_AviaCo** database to work Problems 23–34, shown in Figure P8.23.

FIGURE P8.23   Ch08_AviaCo database tables

Table name: CHARTER

Database name: CH08_AviaCo

CHAR_TRIP	CHAR_DATE	AC_NUMBER	CHAR_DESTINATION	CHAR_DISTANCE	CHAR_HOURS_FLOWN	CHAR_HOURS_WAIT	CHAR_FUEL_GALLONS	CHAR_OIL_QTS	CUS_CODE
10001	05-Feb-08	2289L	ATL	936	5.1	2.2	354.1	1	10011
10002	05-Feb-08	2778V	BNA	320	1.6	0	72.6	0	10016
10003	05-Feb-08	4278Y	GNV	1574	7.8	0	339.8	2	10014
10004	06-Feb-08	1484P	STL	472	2.9	4.9	97.2	1	10019
10005	06-Feb-08	2289L	ATL	1023	5.7	3.5	397.7	2	10011
10006	06-Feb-08	4278Y	STL	472	2.6	5.2	117.1	0	10017
10007	06-Feb-08	2778V	GNV	1574	7.9	0	348.4	2	10012
10008	07-Feb-08	1484P	TYS	644	4.1	0	140.6	1	10014
10009	07-Feb-08	2289L	GNV	1574	6.6	23.4	459.9	0	10017
10010	07-Feb-08	4278Y	ATL	998	6.2	3.2	279.7	0	10016
10011	07-Feb-08	1484P	BNA	352	1.9	5.3	66.4	1	10012
10012	08-Feb-08	2778V	MOB	884	4.8	4.2	215.1	0	10010
10013	08-Feb-08	4278Y	TYS	644	3.9	4.5	174.3	1	10011
10014	09-Feb-08	4278Y	ATL	936	6.1	2.1	302.6	0	10017
10015	09-Feb-08	2289L	GNV	1645	6.7	0	459.5	2	10016
10016	09-Feb-08	2778V	MQY	312	1.5	0	67.2	0	10011
10017	10-Feb-08	1484P	STL	508	3.1	0	105.5	0	10014
10018	10-Feb-08	4278Y	TYS	644	3.8	4.5	167.4	0	10017

Table name: EARNEDRATING

EMP_NUM	RTG_CODE	EARNRTG_DATE
101	CFI	18-Feb-98
101	CFII	15-Dec-05
101	INSTR	08-Nov-93
101	MEL	23-Jun-94
101	SEL	21-Apr-93
104	INSTR	15-Jul-96
104	MEL	29-Jan-97
104	SEL	12-Mar-95
105	CFI	18-Nov-97
105	INSTR	17-Apr-95
105	MEL	12-Aug-95
105	SEL	23-Sep-94
106	INSTR	20-Dec-95
106	MEL	02-Apr-96
106	SEL	10-Mar-94
109	CFI	05-Nov-98
109	CFII	21-Jun-03
109	INSTR	23-Jul-96
109	MEL	15-Mar-97
109	SEL	05-Feb-96
109	SES	12-May-96

Table name: CREW

CHAR_TRIP	EMP_NUM	CREW_JOB
10001	104	Pilot
10002	101	Pilot
10003	105	Pilot
10003	109	Copilot
10004	106	Pilot
10005	101	Pilot
10006	109	Pilot
10007	104	Pilot
10007	105	Copilot
10008	106	Pilot
10009	105	Pilot
10010	108	Pilot
10011	101	Pilot
10011	104	Copilot
10012	101	Pilot
10013	105	Pilot
10014	106	Pilot
10015	101	Copilot
10015	104	Pilot
10016	105	Copilot
10016	109	Pilot
10017	101	Pilot
10018	104	Copilot
10018	105	Pilot

Table name: CREW

CUS_CODE	CUS_LNAME	CUS_FNAME	CUS_INITIAL	CUS_AREACODE	CUS_PHONE	CUS_BALANCE
10010	Ramas	Alfred	A	615	844-2573	0.00
10011	Dunne	Leona	K	713	894-1238	0.00
10012	Smith	Kathy	W	615	894-2285	896.54
10013	Olowski	Paul	F	615	894-2180	1285.19
10014	Orlando	Myron		615	222-1672	673.21
10015	O'Brian	Amy	B	713	442-3381	1014.56
10016	Brown	James	G	615	297-1228	0.00
10017	Williams	George		615	290-2556	0.00
10018	Farriss	Anne	G	713	382-7185	0.00
10019	Smith	Olette	K	615	297-3809	453.98

Table name: RATING

RTG_CODE	RTG_NAME
CFI	Certified Flight Instructor
CFII	Certified Flight Instructor, Instrument
INSTR	Instrument
MEL	Multiengine Land
SEL	Single Engine, Land
SES	Single Engine, Sea

Table name: CREW

EMP_NUM	EMP_TITLE	EMP_LNAME	EMP_FNAME	EMP_INITIAL	EMP_DOB	EMP_HIRE_DATE
100	Mr.	Kolmycz	George	D	15-Jun-1942	15-Mar-1987
101	Ms.	Lewis	Rhonda	G	19-Mar-1965	25-Apr-1988
102	Mr.	Vandam	Rhett		14-Nov-1958	20-Dec-1992
103	Ms.	Jones	Anne	M	16-Oct-1974	28-Aug-2005
104	Mr.	Lange	John	P	08-Nov-1971	20-Oct-1996
105	Mr.	Williams	Robert	D	14-Mar-1975	08-Jan-2006
106	Mrs.	Duzak	Jeanine	K	12-Feb-1968	05-Jan-1991
107	Mr.	Diante	Jorge	D	21-Aug-1974	02-Jul-1996
108	Mr.	Wiesenbach	Paul	R	14-Feb-1966	18-Nov-1994
109	Ms.	Travis	Elizabeth	K	18-Jun-1961	14-Apr-1991
110	Mrs.	Genkazi	Leighla	W	19-May-1970	01-Dec-1992

Table name: MODEL

MOD_CODE	MOD_MANUFACTURER	MOD_NAME	MOD_SEATS	MOD_CHG_MILE
C-90A	Beechcraft	KingAir	8	2.67
PA23-250	Piper	Aztec	6	1.93
PA31-350	Piper	Navajo Chieftain	10	2.35

Table name: AIRCRAFT

AC_NUMBER	MOD_CODE	AC_TTAF	AC_TTEL	AC_TTER
1484P	PA23-250	1833.1	1833.1	101.8
2289L	C-90A	4243.8	768.9	1123.4
2778V	PA31-350	7992.9	1513.1	789.5
4278Y	PA31-350	2147.3	622.1	243.2

Table name: PILOT

EMP_NUM	PIL_LICENSE	PIL_RATINGS	PIL_MED_TYPE	PIL_MED_DATE	PIL_PT135_DATE
101	ATP	ATP/SEL/MEL/Instr/CFII	1	20-Jan-08	11-Jan-08
104	ATP	ATP/SEL/MEL/Instr	1	18-Dec-07	17-Jan-08
105	COM	COMM/SEL/MEL/Instr/CFI	2	05-Jan-08	02-Jan-08
106	COM	COMM/SEL/MEL/Instr	2	10-Dec-07	02-Feb-08
109	COM	ATP/SEL/MEL/SES/Instr/CFII	1	22-Jan-08	15-Jan-08

ONLINE CONTENT

The **Ch08_AviaCo** database used for Problems 23–34 is located in the Student Online Companion for this book, as are the script files to duplicate this data set in Oracle.

23. Modify the MODEL table to add the attribute and insert the values shown in the following table.

ATTRIBUTE NAME	ATTRIBUTE DESCRIPTION	ATTRIBUTE TYPE	ATTRIBUTE VALUES
MOD_WAIT_CHG	Waiting charge per hour for each model	Numeric	$100 for C-90A $50 for PA23-250 $75 for PA31-350

24. Write the queries to update the MOD_WAIT_CHG attribute values based on Problem 23.

25. Modify the CHARTER table to add the attributes shown in the following table.

ATTRIBUTE NAME	ATTRIBUTE DESCRIPTION	ATTRIBUTE TYPE
CHAR_WAIT_CHG	Waiting charge for each model (copied from the MODEL table)	Numeric
CHAR_FLT_CHG_HR	Flight charge per mile for each model (copied from the MODEL table using the MOD_CHG_MILE attribute)	Numeric
CHAR_FLT_CHG	Flight charge (calculated by CHAR_HOURS_FLOWN x CHAR_FLT_CHG_HR)	Numeric
CHAR_TAX_CHG	CHAR_FLT_CHG x tax rate (8%)	Numeric
CHAR_TOT_CHG	CHAR_FLT_CHG + CHAR_TAX_CHG	Numeric
CHAR_PYMT	Amount paid by customer	Numeric
CHAR_BALANCE	Balance remaining after payment	Numeric

26. Write the sequence of commands required to update the CHAR_WAIT_CHG attribute values in the CHARTER table. (*Hint*: Use either an updatable view or a stored procedure.)

27. Write the sequence of commands required to update the CHAR_FLT_CHG_HR attribute values in the CHARTER table. (*Hint*: Use either an updatable view or a stored procedure.)

28. Write the command required to update the CHAR_FLT_CHG attribute values in the CHARTER table.

29. Write the command required to update the CHAR_TAX_CHG attribute values in the CHARTER table.

30. Write the command required to update the CHAR_TOT_CHG attribute values in the CHARTER table.

31. Modify the PILOT table to add the attribute shown in the following table.

ATTRIBUTE NAME	ATTRIBUTE DESCRIPTION	ATTRIBUTE TYPE
PIL_PIC_HRS	Pilot in command (PIC) hours; updated by adding the CHARTER table's CHAR_HOURS_FLOWN to the PIL_PIC_HRS when the CREW table shows the CREW_JOB to be pilot	Numeric

32. Create a trigger named **trg_char_hours** that will automatically update the AIRCRAFT table when a new CHARTER row is added. Use the CHARTER table's CHAR_HOURS_FLOWN to update the AIRCRAFT table's AC_TTAF, AC_TTEL, and AC_TTER values.

33. Create a trigger named **trg_pic_hours** that will automatically update the PILOT table when a new CREW row is added and the CREW table uses a 'pilot' CREW_JOB entry. Use the CHARTER table's CHAR_HOURS_FLOWN to update the PILOT table's PIL_PIC_HRS only when the CREW table uses a 'pilot' CREW_JOB entry.

34. Create a trigger named **trg_cust_balance** that will automatically update the CUSTOMER table's CUST_BALANCE when a new CHARTER row is added. Use the CHARTER table's CHAR_TOT_CHG as the update source. (Assume that all charter charges are charged to the customer balance.)

9 DATABASE DESIGN

N I N E

In this chapter, you will learn:

- That successful database design must reflect the information system of which the database is a part
- That successful information systems are developed within a framework known as the Systems Development Life Cycle (SDLC)
- That within the information system, the most successful databases are subject to frequent evaluation and revision within a framework known as the Database Life Cycle (DBLC)
- How to conduct evaluation and revision within the SDLC and DBLC frameworks
- About database design strategies: top-down vs. bottom-up design and centralized vs. decentralized design

Preview

Databases are a part of a larger picture called an information system. Database designs that fail to recognize that the database is part of this larger whole are not likely to be successful. That is, database designers must recognize that the database is a critical means to an end rather than an end in itself. (Managers want the database to serve their management needs, but too many databases seem to require that managers alter their routines to fit the database requirements.)

Information systems don't just happen; they are the product of a carefully staged development process. Systems analysis is used to determine the need for an information system and to establish its limits. Within systems analysis, the actual information system is created through a process known as systems development.

The creation and evolution of information systems follows an iterative pattern called the Systems Development Life Cycle, a continuous process of creation, maintenance, enhancement, and replacement of the information system. A similar cycle applies to databases. The database is created, maintained, and enhanced. When even enhancement can no longer stretch the database's usefulness and the database can no longer perform its functions adequately, it might have to be replaced. The Database Life Cycle is carefully traced in this chapter and is shown in the context of the larger Systems Development Life Cycle.

At the end of the chapter, you are introduced to some classical approaches to database design: top-down vs. bottom-up and centralized vs. decentralized.

9.1 THE INFORMATION SYSTEM

Basically, a database is a carefully designed and constructed repository of facts. The database is a part of a larger whole known as an **information system**, which provides for data collection, storage, and retrieval. The information system also facilitates the transformation of data into information, and it allows for the management of both data and information. Thus, a complete information system is composed of people, hardware, software, the database(s), application programs, and procedures. **Systems analysis** is the process that establishes the need for and the extent of an information system. The process of creating an information system is known as **systems development**.

One key characteristic of current information systems is the strategic value of information in the age of global business. Therefore, information systems should always be aligned with the strategic business goals; the view of isolated and independent information systems is no longer valid. Current information systems should always be integrated with the company's enterprise-wide information systems architecture.

> **NOTE**
>
> This chapter is not meant to cover all aspects of systems analysis and development—those usually are covered in a separate course or book. However, this chapter should help you develop a better understanding of the issues associated with database design, implementation, and management that are affected by the information system in which the database is a critical component.

Within the framework of systems development, applications transform data into the information that forms the basis for decision making. Applications usually produce formal reports, tabulations, and graphic displays designed to produce insight into the information. Figure 9.1 illustrates that every application is composed of two parts: the data and the code (program instructions) by which the data are transformed into information. Data and code work together to represent real-world business functions and activities. At any given moment, physically stored data represent a snapshot of the business. But the picture is not complete without an understanding of the business activities that are represented by the code.

FIGURE 9.1   **Generating information for decision making**

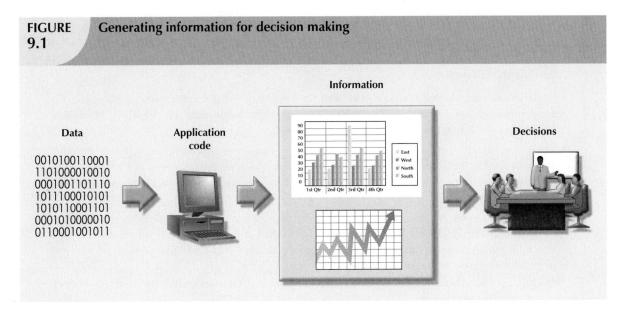

The performance of an information system depends on a triad of factors:

- Database design and implementation.
- Application design and implementation.
- Administrative procedures.

This book emphasizes the database design and implementation segment of the triad—arguably the most important of the three. However, failure to address the other two segments will likely yield a poorly functioning information system. Creating a sound information system is hard work: systems analysis and development require much planning to ensure that all of the activities will interface with each other, that they will complement each other, and that they will be completed on time.

In a broad sense, the term **database development** describes the process of database design and implementation. The primary objective in database design is to create complete, normalized, nonredundant (to the extent possible), and fully integrated conceptual, logical, and physical database models. The implementation phase includes creating the database storage structure, loading data into the database, and providing for data management.

To make the procedures discussed in this chapter broadly applicable, the chapter focuses on the elements that are common to all information systems. Most of the processes and procedures described in this chapter do not depend on the size, type, or complexity of the database being implemented. However, the procedures that would be used to design a small database, such as one for a neighborhood shoe store, do not precisely scale up to the procedures that would be needed to design a database for a large corporation or even a segment of such a corporation. To use an analogy, building a small house requires a blueprint, just as building the Golden Gate Bridge does, but the bridge requires more complex and further-ranging planning, analysis, and design than the house.

The next sections will trace the overall Systems Development Life Cycle and the related Database Life Cycle. Once you are familiar with those processes and procedures, you will learn about various approaches to database design, such as top-down vs. bottom-up and centralized vs. decentralized design.

> **NOTE**
>
> The Systems Development Life Cycle (SDLC) is a general framework through which you can track and come to understand the activities required to develop and maintain information systems. Within that framework, there are several ways to complete various tasks specified in the SDLC. For example, this texts focus is on ER modeling and on relational database implementation issues, and that focus is maintained in this chapter. However, there are alternative methodologies, such as:
>
> - Unified Modeling Language (UML) provides object-oriented tools to support the tasks associated with the development of information systems. UML is covered in **Appendix H, Unified Modeling Language (UML),** in the Student Online Companion.
> - Rapid Application Development (RAD)[1] is an iterative software development methodology that uses prototypes, CASE tools, and flexible management to develop application systems. RAD started as an alternative to traditional structured development which had long deliverable times and unfulfilled requirements.
> - Agile Software Development[2] is a framework for developing software applications that divides the work to be done in smaller subprojects to obtain valuable deliverables in shorter times and with better cohesion. This method emphasizes close communication among all users and continuous evaluation with the purpose of increasing customer satisfaction.
>
> Although the development *methodologies* may change, the basic framework within which those methodologies are used does not change.

[1]See *Rapid Application Development*, James Martin, Prentice-Hall, Macmillan College Division, 1991.
[2]Further information about Agile Software Development can be found at *www.agilealliance.org*.

9.2 THE SYSTEMS DEVELOPMENT LIFE CYCLE (SDLC)

The **Systems Development Life Cycle** (**SDLC**) traces the history (life cycle) of an information system. Perhaps more important to the system designer, the SDLC provides the big picture within which the database design and application development can be mapped out and evaluated.

As illustrated in Figure 9.2, the traditional SDLC is divided into five phases: planning, analysis, detailed systems design, implementation, and maintenance. The SDLC is an iterative rather than a sequential process. For example, the details of the feasibility study might help refine the initial assessment, and the details discovered during the user requirements portion of the SDLC might help refine the feasibility study.

FIGURE 9.2 The Systems Development Life Cycle (SDLC)

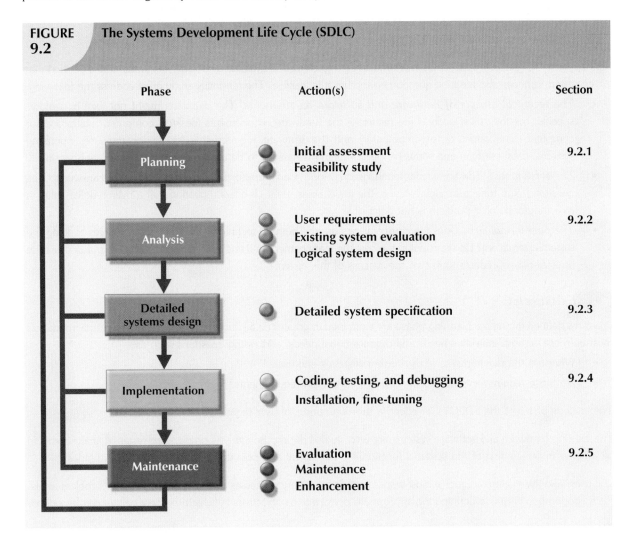

Phase	Action(s)	Section
Planning	Initial assessment Feasibility study	9.2.1
Analysis	User requirements Existing system evaluation Logical system design	9.2.2
Detailed systems design	Detailed system specification	9.2.3
Implementation	Coding, testing, and debugging Installation, fine-tuning	9.2.4
Maintenance	Evaluation Maintenance Enhancement	9.2.5

Because the Database Life Cycle (DBLC) fits into and resembles the Systems Development Life Cycle (SDLC), a brief description of the SDLC is in order.

9.2.1 PLANNING

The SDLC planning phase yields a general overview of the company and its objectives. An initial assessment of the information flow-and-extent requirements must be made during this discovery portion of the SDLC. Such an assessment should answer some important questions:

- *Should the existing system be continued?* If the information generator does its job well, there is no point in modifying or replacing it. To quote an old saying, "If it ain't broke, don't fix it."

- *Should the existing system be modified?* If the initial assessment indicates deficiencies in the extent and flow of the information, minor (or even major) modifications might be in order. When considering modifications, the participants in the initial assessment must keep in mind the distinction between wants and needs.

- *Should the existing system be replaced?* The initial assessment might indicate that the current system's flaws are beyond fixing. Given the effort required to create a new system, a careful distinction between wants and needs is perhaps even more important in this case than it is in modifying the system.

Participants in the SDLC's initial assessment must begin to study and evaluate alternative solutions. If it is decided that a new system is necessary, the next question is whether it is feasible. The feasibility study must address the following:

- *The technical aspects of hardware and software requirements.* The decisions might not (yet) be vendor-specific, but they must address the nature of the hardware requirements (desktop computer, multiprocessor computer, mainframe, or supercomputer) and the software requirements (single- or multiuser operating systems, database type and software, programming languages to be used by the applications, and so on).

- *The system cost.* The admittedly mundane question, "Can we afford it?" is crucial. (And the answer to that question might force a careful review of the initial assessment.) It bears repeating that a million-dollar solution to a thousand-dollar problem is not defensible.

- *The operational cost.* Does the company possess the human, technical, and financial resources to keep the system operational? Do we count in the cost the management and end-user support needed to put in place the operational procedures to ensure the success of this system?

9.2.2 ANALYSIS

Problems defined during the planning phase are examined in greater detail during the analysis phase. A macroanalysis must be made of both individual needs and organizational needs, addressing questions such as:

- What are the requirements of the current system's end users?
- Do those requirements fit into the overall information requirements?

The analysis phase of the SDLC is, in effect, a thorough *audit* of user requirements.

The existing hardware and software systems are also studied during the analysis phase. The result of analysis should be a better understanding of the system's functional areas, actual and potential problems, and opportunities.

End users and the system designer(s) must work together to identify processes and to uncover potential problem areas. Such cooperation is vital to defining the appropriate performance objectives by which the new system can be judged.

Along with a study of user requirements and the existing systems, the analysis phase also includes the creation of a logical systems design. The logical design must specify the appropriate conceptual data model, inputs, processes, and expected output requirements.

When creating a logical design, the designer might use tools such as data flow diagrams (DFDs), hierarchical input process output (HIPO) diagrams, and entity relationship (ER) diagrams. The database design's data-modeling activities take place at this point to discover and describe all entities and their attributes and the relationships among the entities within the database.

Defining the logical system also yields functional descriptions of the system's components (modules) for each process within the database environment. All data transformations (processes) are described and documented using such systems analysis tools as DFDs. The conceptual data model is validated against those processes.

9.2.3 DETAILED SYSTEMS DESIGN

In the detailed systems design phase, the designer completes the design of the system's processes. The design includes all the necessary technical specifications for the screens, menus, reports, and other devices that might be used to help make the system a more efficient information generator. The steps are laid out for conversion from the old to the new system. Training principles and methodologies are also planned and must be submitted for management's approval.

> **NOTE**
>
> Because attention has been focused on the details of the systems design process, the book has not until this point explicitly recognized the fact that management approval is needed at all stages of the process. Such approval is needed because a GO decision requires funding. There are many GO/NO GO decision points along the way to a completed systems design!

9.2.4 IMPLEMENTATION

During the implementation phase, the hardware, DBMS software, and application programs are installed, and the database design is implemented. During the initial stages of the implementation phase, the system enters into a cycle of coding, testing, and debugging until it is ready to be delivered. The actual database is created, and the system is customized by the creation of tables and views, user authorizations, and so on.

The database contents might be loaded interactively or in batch mode, using a variety of methods and devices:

- Customized user programs.
- Database interface programs.
- Conversion programs that import the data from a different file structure, using batch programs, a database utility, or both.

The system is subjected to exhaustive testing until it is ready for use. Traditionally, the implementation and testing of a new system took 50 to 60 percent of the total development time. However, the advent of sophisticated application generators and debugging tools has substantially decreased coding and testing time. After testing is concluded, the final documentation is reviewed and printed and end users are trained. The system is in full operation at the end of this phase but will be continuously evaluated and fine-tuned.

9.2.5 MAINTENANCE

Almost as soon as the system is operational, end users begin to request changes in it. Those changes generate system maintenance activities, which can be grouped into three types:

- *Corrective maintenance* in response to systems errors.
- *Adaptive maintenance* due to changes in the business environment.
- *Perfective maintenance* to enhance the system.

Because every request for structural change requires retracing the SDLC steps, the system is, in a sense, always at some stage of the SDLC.

Each system has a predetermined operational life span. The actual operational life span of a system depends on its perceived utility. There are several reasons for reducing the operational life of certain systems. Rapid technological change is one reason, especially for systems based on processing speed and expandability. Another common reason is the cost of maintaining a system.

If the system's maintenance cost is high, its value becomes suspect. **Computer-aided systems engineering (CASE)** technology, such as System Architect or Visio Professional, helps make it possible to produce better systems within a reasonable amount of time and at a reasonable cost. In addition, CASE-produced applications are more structured, documented, and especially *standardized*, which tends to prolong the operational life of systems by making them easier and cheaper to update and maintain. For example, if you have used Microsoft's Visio Professional to develop your database design, you already know that Visio Professional tests the internal consistency of your ERDs when you create the relationships. Visio Professional implements the FKs according to the design's entity types (weak, strong) and the nature of the relationship (identifying, non-identifying) between those entities. When you see the result of the implementation, you immediately see whether the results are what you intended them to be. In addition, if there are circular arguments in the design, Visio Professional will make that clear. Therefore, you will be able to spot design problems before they become permanent.

9.3 THE DATABASE LIFE CYCLE (DBLC)

Within the larger information system, the database, too, is subject to a life cycle. The **Database Life Cycle (DBLC)** contains six phases, as shown in Figure 9.3: database initial study, database design, implementation and loading, testing and evaluation, operation, and maintenance and evolution.

FIGURE 9.3   The Database Life Cycle (DBLC)

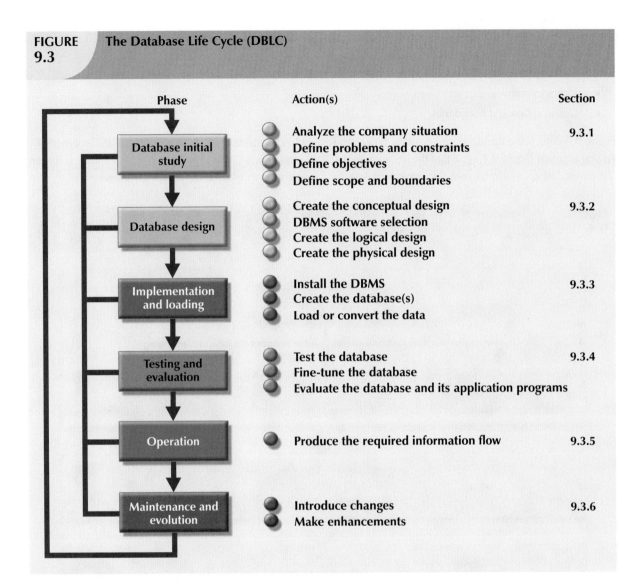

Phase	Action(s)	Section
Database initial study	Analyze the company situation Define problems and constraints Define objectives Define scope and boundaries	9.3.1
Database design	Create the conceptual design DBMS software selection Create the logical design Create the physical design	9.3.2
Implementation and loading	Install the DBMS Create the database(s) Load or convert the data	9.3.3
Testing and evaluation	Test the database Fine-tune the database Evaluate the database and its application programs	9.3.4
Operation	Produce the required information flow	9.3.5
Maintenance and evolution	Introduce changes Make enhancements	9.3.6

9.3.1 THE DATABASE INITIAL STUDY

If a designer has been called in, chances are the current system has failed to perform functions deemed vital by the company. (You don't call the plumber unless the pipes leak.) So in addition to examining the current system's operation within the company, the designer must determine how and why the current system fails. That means spending a lot of time talking with (but mostly listening to) end users. Although database design is a technical business, it is also people-oriented. Database designers must be excellent communicators, and they must have finely tuned interpersonal skills.

Depending on the complexity and scope of the database environment, the database designer might be a lone operator or part of a systems development team composed of a project leader, one or more senior systems analysts, and one or more junior systems analysts. The word *designer* is used generically here to cover a wide range of design team compositions.

The overall purpose of the database initial study is to:

- Analyze the company situation.
- Define problems and constraints.
- Define objectives.
- Define scope and boundaries.

Figure 9.4 depicts the interactive and iterative processes required to complete the first phase of the DBLC successfully. As you examine Figure 9.4, note that the database initial study phase leads to the development of the database system objectives. Using Figure 9.4 as a discussion template, let's examine each of its components in greater detail.

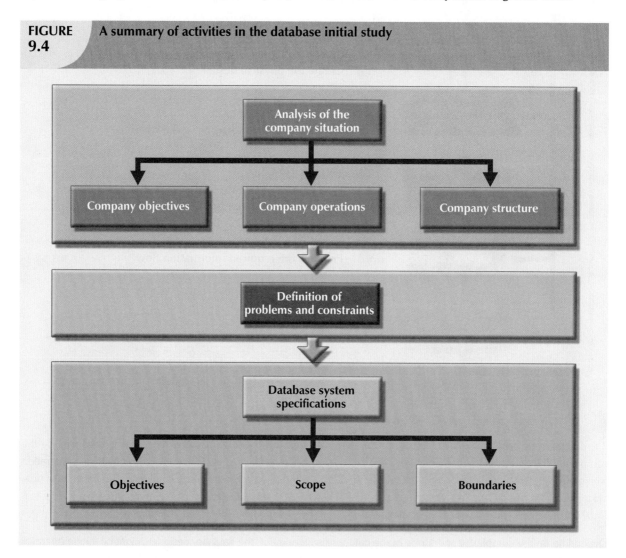

FIGURE 9.4   A summary of activities in the database initial study

Analyze the Company Situation

The *company situation* describes the general conditions in which a company operates, its organizational structure, and its mission. To analyze the company situation, the database designer must discover what the company's operational components are, how they function, and how they interact.

These issues must be resolved:

- *What is the organization's general operating environment, and what is its mission within that environment?* The design must satisfy the operational demands created by the organization's mission. For example, a mail-order business is likely to have operational requirements involving its database that are quite different from those of a manufacturing business.
- *What is the organization's structure?* Knowing who controls what and who reports to whom is quite useful when you are trying to define required information flows, specific report and query formats, and so on.

Define Problems and Constraints

The designer has both formal and informal sources of information. If the company has existed for any length of time, it already has some kind of system in place (either manual or computer-based). How does the existing system function? What input does the system require? What documents does the system generate? By whom and how is the system output used? Studying the paper trail can be very informative. In addition to the official version of the system's operation, there is also the more informal, real version; the designer must be shrewd enough to see how these differ.

The process of defining problems might initially appear to be unstructured. Company end users are often unable to describe precisely the larger scope of company operations or to identify the real problems encountered during company operations. Often the managerial view of a company's operation and its problems is different from that of the end users, who perform the actual routine work.

During the initial problem definition process, the designer is likely to collect very broad problem descriptions. For example, note these concerns expressed by the president of a fast-growing transnational manufacturing company:

Although the rapid growth is gratifying, members of the management team are concerned that such growth is beginning to undermine the ability to maintain a high customer service standard and, perhaps worse, to diminish manufacturing standards control.

The problem definition process quickly leads to a host of general problem descriptions. For example, the marketing manager comments:

I'm working with an antiquated filing system. We manufacture more than 1,700 specialty machine parts. When a regular customer calls in, we can't get a very quick inventory scan. If a new customer calls in, we can't do a current parts search by using a simple description, so we often do a machine setup for a part that we have in inventory. That's wasteful. And of course, some new customers get irritated when we can't give a quick response.

The production manager comments:

At best, it takes hours to generate the reports I need for scheduling purposes. I don't have hours for quick turnarounds. It's difficult to manage what I don't have information about.

I don't get quick product request routing. Take machine setup. Right now I've got operators either waiting for the right stock or getting it themselves when a new part is scheduled for production. I can't afford to have an operator doing chores that a much lower-paid worker ought to be doing. There's just too much waiting around with the current scheduling. I'm losing too much time, and my schedules back up. Our overtime bill is ridiculous.

I sometimes produce parts that are already in inventory because we don't seem to be able to match what we've got in inventory with what we have scheduled. Shipping yells at me because I can't turn out the parts, and often they've got them in inventory one bay down. That's costing us big bucks sometimes.

New reports can take days or even weeks to get to my office. And I need a ton of reports to schedule personnel, downtime, training, etc. I can't get new reports that I need NOW. What I need is the ability to get quick updates on percent defectives, percent rework, the effectiveness of training, you name it. I need such reports by shift, by date, by any characteristic I can think of to help me manage scheduling, training, you name it.

A machine operator comments:

It takes a long time to set my stuff up. If I get my schedule banged up because John doesn't get the paperwork on time, I wind up looking for setup specs, startup material, bin assignments, and other stuff. Sometimes I spend two or three hours just setting up. Now you know why I can't meet schedules. I try to be productive, but I'm spending too much time getting ready to do my job.

After the initial declarations, the database designer must continue to probe carefully in order to generate additional information that will help define the problems within the larger framework of company operations. How does the problem of the marketing manager's customer fit within the broader set of marketing department activities? How does the solution to the customer's problem help meet the objectives of the marketing department and the rest of the company? How do the marketing department's activities relate to those of the other departments? That last question is especially important. Note that there are common threads in the problems described by the marketing and production department managers. If the inventory query process can be improved, both departments are likely to find simple solutions to at least some of the problems.

Finding precise answers is important, especially concerning the operational relationships among business units. If a proposed system will solve the marketing department's problems but exacerbate those of the production department, not much progress will have been made. Using an analogy, suppose your home water bill is too high. You have determined the problem: the faucets leak. The solution? You step outside and cut off the water supply to the house. Is that an adequate solution? Or would the replacement of faucet washers do a better job of solving the problem? You might find the leaky faucet scenario simplistic, yet almost any experienced database designer can find similar instances of so-called database problem solving (admittedly more complicated and less obvious).

Even the most complete and accurate problem definition does not always lead to the perfect solution. The real world usually intrudes to limit the design of even the most elegant database by imposing constraints. Such constraints include time, budget, personnel, and more. If you must have a solution within a month and within a $12,000 budget, a solution that takes two years to develop at a cost of $100,000 is not a solution. *The designer must learn to distinguish between what's perfect and what's possible.*

Define Objectives

A proposed database system must be designed to help solve at least the major problems identified during the problem discovery process. As the list of problems unfolds, several common sources are likely to be discovered. In the previous example, both the marketing manager and the production manager seem to be plagued by inventory inefficiencies. If the designer can create a database that sets the stage for more efficient parts management, both departments gain. The initial objective, therefore, might be to create an efficient inventory query and management system.

> **NOTE**
>
> When trying to develop solutions, the database designer must look for the *source* of the problems. There are many cases of database systems that failed to satisfy the end users because they were designed to treat the *symptoms* of the problems rather than their source.

Note that the initial study phase also yields proposed problem solutions. The designer's job is to make sure that the database system objectives, as seen by the designer, correspond to those envisioned by the end user(s). In any case, the database designer must begin to address the following questions:

- What is the proposed system's initial objective?
- Will the system interface with other existing or future systems in the company?
- Will the system share the data with other systems or users?

Define Scope and Boundaries

The designer must recognize the existence of two sets of limits: scope and boundaries. The system's **scope** defines the extent of the design according to operational requirements. Will the database design encompass the entire organization, one or more departments within the organization, or one or more functions of a single department? The designer must know the "size of the ballpark." Knowing the scope helps in defining the required data structures, the type and number of entities, the physical size of the database, and so on.

The proposed system is also subject to limits known as **boundaries**, which are external to the system. Has any designer ever been told, "We have all the time in the world" or "Use an unlimited budget and use as many people as needed to make the design come together"? Boundaries are also imposed by existing hardware and software. Ideally, the designer can choose the hardware and software that will best accomplish the system goals. In fact, software selection is an important aspect of the Systems Development Life Cycle. Unfortunately, in the real world, a system often must be designed around existing hardware. Thus, the scope and boundaries become the factors that force the design into a specific mold, and the designer's job is to design the best system possible within those constraints. (Note that problem definitions and the objectives sometimes must be reshaped to meet the system scope and boundaries.)

9.3.2 DATABASE DESIGN

The second phase focuses on the design of the database model that will support company operations and objectives. This is arguably the most critical DBLC phase: making sure that the final product meets user and system requirements. In the process of database design, you must concentrate on the data characteristics required to build the database model. At this point, there are two views of the data within the system: the business view of data as a source of information and the designer's view of the data structure, its access, and the activities required to transform the data into information. Figure 9.5 contrasts those views. Note that you can summarize the different views by looking at the terms *what* and *how*. Defining data is an integral part of the DBLC's second phase.

As you examine the procedures required to complete the design phase in the DBLC, remember these points:

- The process of database design is loosely related to the analysis and design of a larger system. The data component is only one element of a larger information system.
- The systems analysts or systems programmers are in charge of designing the other system components. Their activities create the procedures that will help transform the data within the database into useful information.
- The database design does not constitute a sequential process. Rather, it is an iterative process that provides continuous feedback designed to trace previous steps.

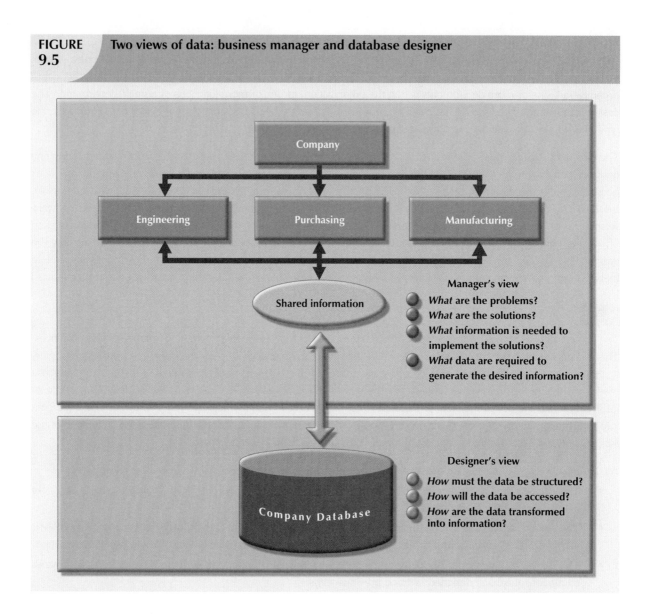

FIGURE 9.5   Two views of data: business manager and database designer

The database design process is depicted in Figure 9.6. Look at the procedure flow in the figure.

Now let's explore in detail each of the components in Figure 9.6. Knowing those details will help you successfully design and implement databases in a real-world setting.

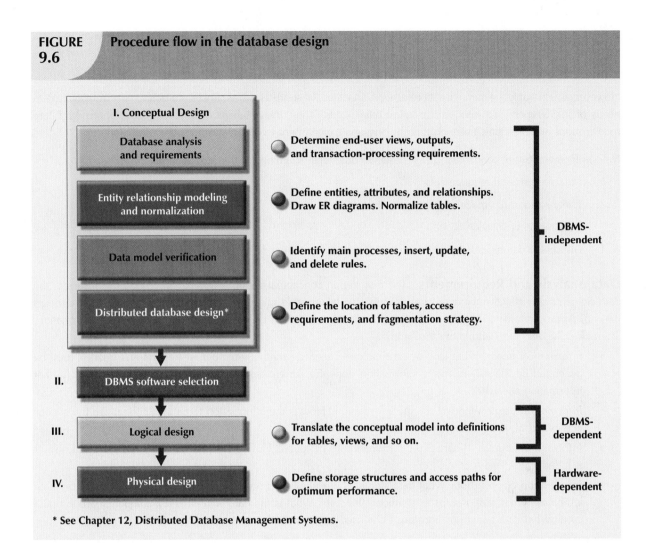

FIGURE 9.6   Procedure flow in the database design

I. Conceptual Design

- Database analysis and requirements
  - Determine end-user views, outputs, and transaction-processing requirements.
- Entity relationship modeling and normalization
  - Define entities, attributes, and relationships. Draw ER diagrams. Normalize tables.
- Data model verification
  - Identify main processes, insert, update, and delete rules.
- Distributed database design*
  - Define the location of tables, access requirements, and fragmentation strategy.

DBMS-independent

II.   DBMS software selection

III.   Logical design
- Translate the conceptual model into definitions for tables, views, and so on.

DBMS-dependent

IV.   Physical design
- Define storage structures and access paths for optimum performance.

Hardware-dependent

* See Chapter 12, Distributed Database Management Systems.

ONLINE CONTENT

In **Appendixes B** and **C** in the Student Online Companion, **The University Lab: Conceptual Design** and **The University Lab: Conceptual Design Verification, Logical Design, and Implementation,** respectively, you learn what happens during each of these stages in developing real databases.

I. Conceptual Design

In the **conceptual design** stage, data modeling is used to create an abstract database structure that represents real-world objects in the most realistic way possible. The conceptual model must embody a clear understanding of the business and its functional areas. At this level of abstraction, the type of hardware and/or database model to be used might not yet have been identified. Therefore, the design must be software- and hardware-independent so the system can be set up within any hardware and software platform chosen later.

Keep in mind the following **minimal data rule**:

All that is needed is there, and all that is there is needed.

In other words, make sure that all data needed are in the model and that all data in the model are needed. All data elements required by the database transactions must be defined in the model, and all data elements defined in the model must be used by at least one database transaction.

However, as you apply the minimal data rule, avoid an excessive short-term bias. Focus not only on the immediate data needs of the business, but also on the future data needs. Thus, the database design must leave room for future modifications and additions, ensuring that the business's investment in information resources will endure.

Note in Figure 9.6 that conceptual design requires four steps, examined in the next sections:

1. Data analysis and requirements
2. Entity relationship modeling and normalization
3. Data model verification
4. Distributed database design

Data Analysis and Requirements   The first step in conceptual design is to discover the characteristics of the data elements. An effective database is an information factory that produces key ingredients for successful decision making. Appropriate data element characteristics are those that can be transformed into appropriate information. Therefore, the designer's efforts are focused on:

- *Information needs*. What kind of information is needed—that is, what output (reports and queries) must be generated by the system, what information does the current system generate, and to what extent is that information adequate?

- *Information users*. Who will use the information? How is the information to be used? What are the various end-user data views?

- *Information sources*. Where is the information to be found? How is the information to be extracted once it is found?

- *Information constitution*. What data elements are needed to produce the information? What are the data attributes? What relationships exist among the data? What is the data volume? How frequently are the data used? What data transformations are to be used to generate the required information?

The designer obtains the answers to those questions from a variety of sources in order to compile the necessary information. Note these sources:

- *Developing and gathering end-user data views*. The database designer and the end user(s) interact to jointly develop a precise description of end-user data views. In turn, the end-user data views will be used to help identify the database's main data elements.

- *Directly observing the current system: existing and desired output*. The end user usually has an existing system in place, whether it's manual or computer-based. The designer reviews the existing system to identify the data and their characteristics. The designer examines the input forms and files (tables) to discover the data type and volume. If the end user already has an automated system in place, the designer carefully examines the current and desired reports to describe the data required to support the reports.

- *Interfacing with the systems design group*. As noted earlier in this chapter, the database design process is part of the Systems Development Life Cycle (SDLC). In some cases, the systems analyst in charge of designing the new system will also develop the conceptual database model. (This is usually true in a decentralized environment.) In other cases, the database design is considered part of the database administrator's job. The presence of a database administrator (DBA) usually implies the existence of a formal data-processing department. The DBA designs the database according to the specifications created by the systems analyst.

To develop an accurate data model, the designer must have a thorough understanding of the company's data types and their extent and uses. But data do not by themselves yield the required understanding of the total business. From a database point of view, the collection of data becomes meaningful only when business rules are defined. Remember from Chapter 2, Data Models, that a *business rule* is a brief and precise description of a policy, procedure, or principle within a specific organization's environment. Business rules, derived from a detailed description of an organization's operations, help to create and enforce actions within that organization's environment. When business rules are written properly, they define entities, attributes, relationships, connectivities, cardinalities, and constraints.

To be effective, business rules must be easy to understand and they must be widely disseminated to ensure that every person in the organization shares a common interpretation of the rules. Using simple language, business rules describe the main and distinguishing characteristics of the data *as viewed by the company*. Examples of business rules are as follows:

- A customer may make many payments on account.
- Each payment on account is credited to only one customer.
- A customer may generate many invoices.
- Each invoice is generated by only one customer.

Given their critical role in database design, business rules must not be established casually. Poorly defined or inaccurate business rules lead to database designs and implementations that fail to meet the needs of the organization's end users.

Ideally, business rules are derived from a formal **description of operations**, which is a document that provides a precise, up-to-date, and thoroughly reviewed description of the activities that define an organization's operating environment. (To the database designer, the operating environment is both the data sources and the data users.) Naturally, an organization's operating environment is dependent on the organization's mission. For example, the operating environment of a university would be quite different from that of a steel manufacturer, an airline, or a nursing home. Yet no matter how different the organizations may be, the *data analysis and requirements* component of the database design process is enhanced when the data environment and data use are described accurately and precisely within a description of operations.

In a business environment, the main sources of information for the description of operations—and, therefore, of business rules—are company managers, policy makers, department managers, and written documentation such as company procedures, standards, and operations manuals. A faster and more direct source of business rules is direct interviews with end users. Unfortunately, because perceptions differ, the end user can be a less reliable source when it comes to specifying business rules. For example, a maintenance department mechanic might believe that any mechanic can initiate a maintenance procedure, when actually only mechanics with inspection authorization should perform such a task. Such a distinction might seem trivial, but it has major legal consequences. Although end users are crucial contributors to the development of business rules, it pays to verify end-user perceptions. Often interviews with several people who perform the same job yield very different perceptions of their job components. While such a discovery might point to "management problems," that general diagnosis does not help the database designer. Given the discovery of such problems, the database designer's job is to reconcile the differences and verify the results of the reconciliation to ensure that the business rules are appropriate and accurate.

Knowing the business rules enables the designer to understand fully how the business works and what role the data plays within company operations. Consequently, the designer must identify the company's business rules and analyze their impact on the nature, role, and scope of data.

Business rules yield several important benefits in the design of new systems:
- They help standardize the company's view of data.
- They constitute a communications tool between users and designers.
- They allow the designer to understand the nature, role, and scope of the data.
- They allow the designer to understand business processes.
- They allow the designer to develop appropriate relationship participation rules and foreign key constraints. (See Chapter 4, Entity Relationship (ER) Modeling.)

The last point is especially noteworthy: whether a given relationship is mandatory or optional is usually a function of the applicable business rule.

Entity Relationship Modeling and Normalization   Before creating the ER model, the designer must communicate and enforce appropriate standards to be used in the documentation of the design. The standards include the use of diagrams and symbols, documentation writing style, layout, and any other conventions to be followed during documentation. Designers often overlook this very important requirement, especially when they are working as members of a design team. Failure to standardize documentation often means a failure to communicate later, and communications failures often lead to poor design work. In contrast, well-defined and enforced standards make design work easier and promise (but do not guarantee) a smooth integration of all system components.

Because the business rules usually define the nature of the relationship(s) among the entities, the designer must incorporate them into the conceptual model. The process of defining business rules and developing the conceptual model using ER diagrams can be described using the steps shown in Table 9.1.[3]

TABLE 9.1	Developing the Conceptual Model Using ER Diagrams	
STEP	ACTIVITY	
1	Identify, analyze, and refine the business rules.	
2	Identify the main entities, using the results of Step 1.	
3	Define the relationships among the entities, using the results of Steps 1 and 2.	
4	Define the attributes, primary keys, and foreign keys for each of the entities.	
5	Normalize the entities. (Remember that entities are implemented as tables in an RDBMS.)	
6	Complete the initial ER diagram.	
7	Have the main end users verify the model in Step 6 against the data, information, and processing requirements.	
8	Modify the ER diagram, using the results of Step 7.	

Some of the steps listed in Table 9.1 take place concurrently. And some, such as the normalization process, can generate a demand for additional entities and/or attributes, thereby causing the designer to revise the ER model. For example, while identifying two main entities, the designer might also identify the composite bridge entity that represents the many-to-many relationship between those two main entities.

To review, suppose you are creating a conceptual model for the JollyGood Movie Rental Corporation, whose end users want to track customers' movie rentals. The simple ER diagram presented in Figure 9.7 shows a composite entity that helps track customers and their video rentals. Business rules define the optional nature of the relationships between the entities VIDEO and CUSTOMER depicted in Figure 9.7. (For example, customers are not required to check out a video. A video need not be checked out in order to exist on the shelf. A customer may rent many videos, and a video may be rented by many customers.) In particular, note the composite RENTAL entity that connects the two main entities.

As you will likely discover, the initial ER model may be subjected to several revisions before it meets the system's requirements. Such a revision process is quite natural. Remember that the ER model is a communications tool as well as a design blueprint. Therefore, when you meet with the proposed system users, the initial ER model should give rise to questions such as, "Is this really what you meant?" For example, the ERD shown in Figure 9.7 is far from complete. Clearly, many more attributes must be defined and the dependencies must be checked before the design can be implemented. In addition, the design cannot yet support the typical video rental transactions environment. For example, each video is likely to have many copies available for rental purposes. However, if the VIDEO entity shown in Figure 9.7 is used to store the titles as well as the copies, the design triggers the data redundancies shown in Table 9.2.

---

[3] See "Linking Rules to Models," Alice Sandifer and Barbara von Halle, *Database Programming and Design*, 4(3), March 1991, pp. 13–16. Although the source seems dated, it remains the current standard. The technology has changed substantially, but the process has not.

FIGURE 9.7   A composite entity

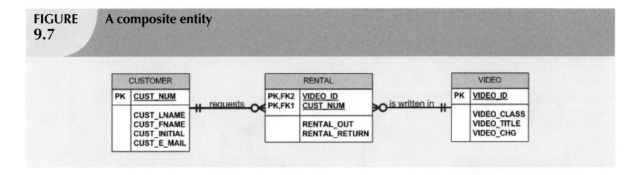

TABLE 9.2   Data Redundancies in the VIDEO Table

VIDEO_ID	VIDEO_TITLE	VIDEO_COPY	VIDEO_CHG	VIDEO_DAYS
SF-12345FT-1	Adventures on Planet III	1	$4.50	1
SF-12345FT-2	Adventures on Planet III	2	$4.50	1
SF-12345FT-3	Adventures on Planet III	3	$4.50	1
WE-5432GR-1	TipToe Canu and Tyler 2: A Journey	1	$2.99	2
WE-5432GR-2	TipToe Canu and Tyler 2: A Journey	2	$2.99	2

The initial ERD shown in Figure 9.7 must be modified to reflect the answer to the question, "Is more than one copy available for each title?" Also, payment transactions must be supported. (You will have an opportunity to modify this initial design in Problem 5 at the end of the chapter.)

From the preceding discussion, you might get the impression that ER modeling activities (entity/attribute definition, normalization, and verification) take place in a precise sequence. In fact, once you have completed the initial ER model, chances are you will move back and forth among the activities until you are satisfied that the ER model accurately represents a database design that is capable of meeting the required system demands. The activities often take place in parallel, and the process is iterative. Figure 9.8 summarizes the ER modeling process interactions. Figure 9.9 summarizes the array of design tools and information sources that the designer can use to produce the conceptual model.

All objects (entities, attributes, relations, views, and so on) are defined in a data dictionary, which is used in tandem with the normalization process to help eliminate data anomalies and redundancy problems. During this ER modeling process, the designer must:

- Define entities, attributes, primary keys, and foreign keys. (The foreign keys serve as the basis for the relationships among the entities.)
- Make decisions about adding new primary key attributes to satisfy end-user and/or processing requirements.
- Make decisions about the treatment of multivalued attributes.
- Make decisions about adding derived attributes to satisfy processing requirements.
- Make decisions about the placement of foreign keys in 1:1 relationships. (If necessary, review the supertype/ subtype relationships in Chapter 6, Advanced Data Modeling.)
- Avoid unnecessary ternary relationships.
- Draw the corresponding ER diagram.
- Normalize the entities.
- Include all data element definitions in the data dictionary.
- Make decisions about standard naming conventions.

FIGURE 9.8   ER modeling is an iterative process based on many activities

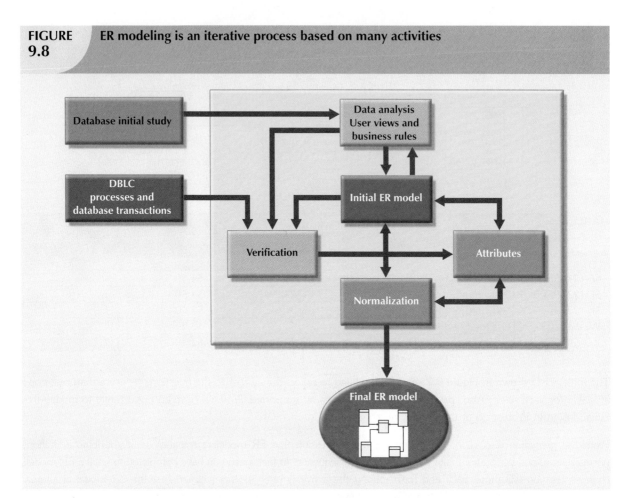

FIGURE 9.9   Conceptual design tools and information sources

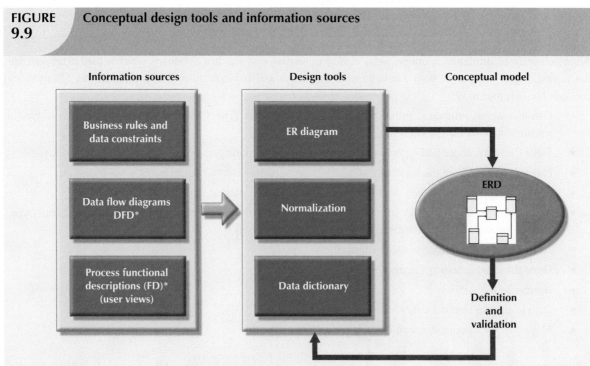

* Output generated by the systems analysis and design activities

The naming conventions requirement is important, yet it is frequently ignored at the designer's risk. Real database design is generally accomplished by teams. Therefore, it is important to ensure that the team members work in an environment in which naming standards are defined and enforced. Proper documentation is crucial to the successful completion of the design. Therefore, it is very useful to establish procedures that are, in effect, self-documenting.

Although some useful entity and attribute naming conventions were established in Chapter 4, they will be revisited in greater detail here. However, keep in mind that such conventions are sometimes subject to constraints imposed by the DBMS software. In an enterprise-wide database environment, the lowest common denominator rules. For example, Microsoft Access finds the attribute name LINE_ITEM_NUMBER to be perfectly acceptable. Many older DBMSs, however, are likely to truncate such long names when they are exported from one DBMS to another, thus making documentation more difficult. Therefore, table-exporting requirements might dictate the use of shorter names. (The same is true for data types. For example, many older DBMSs cannot handle OLE or memo formats.)

This book uses naming conventions that are likely to be acceptable across a reasonably broad range of DBMSs and will meet self-documentation requirements to the greatest extent possible. As the older DBMSs fade from the scene, the naming conventions will be more broadly applicable. You should try to adhere to the following conventions:

- Use descriptive entity and attribute names wherever possible. For example, in the University Computer Lab database, the USER entity contains data about the lab's users and the LOCATION entity is related to the location of the ITEMs that the lab director wants to track.

- Composite entities usually are assigned a name that describes the relationship they represent. For example, in the University Computer Lab database, an ITEM may be stored in many LOCATIONs and a LOCATION may have many ITEMs stored in it. Therefore, the composite (bridge) entity that links ITEM and LOCATION will be named STORAGE. Occasionally, the designer finds it necessary to show what entities are being linked by the composite entity. In such cases, the composite entity name may borrow segments of those entity names. For example, STU_CLASS may be the composite entity that links STUDENT and CLASS. However, that naming convention might make the next one more cumbersome, so it should be used sparingly. (A better choice would be the composite entity name ENROLL, to indicate that the STUDENT enrolls in a CLASS.)

- An attribute name should be descriptive, and it should contain a prefix that helps identify the table in which it is found. For the purposes here, the maximum prefix length will be five characters. For example, the VENDOR table might contain attributes such as VEND_ID and VEND_PHONE. Similarly, the ITEM table might contain attribute names such as ITEM_ID and ITEM_DESCRIPTION. The advantage of that naming convention is that it immediately identifies a table's foreign key(s). For example, if the EMPLOYEE table contains attributes such as EMP_ID, EMP_LNAME, and DEPT_CODE, it is immediately obvious that DEPT_CODE is the foreign key that probably links EMPLOYEE to DEPARTMENT. Naturally, the existence of relationships and table names that start with the same characters might dictate that you bend this naming convention occasionally, as you can see in the next bulleted item.

- If one table is named ORDER and its weak counterpart is named ORDER_ITEM, the prefix ORD will be used to indicate an attribute originating in the ORDER table. The ITEM prefix will identify an attribute originating in the ITEM table. Clearly, you cannot use ORD as a prefix to the attributes originating in the ORDER_ITEM table, so you should use a combination of characters, such as OI, as the prefix to the ORDER_ITEM attribute names. In spite of that limitation, it is generally possible to assign prefixes that identify an attribute's origin. (Keep in mind that some RDBMSs use a "reserved word" list. For example, ORDER might be interpreted as a reserved word in a SELECT statement. In that case, you should use a table name other than ORDER.)

As you can tell, it is not always possible to strictly adhere to the naming conventions. Sometimes the requirement to limit name lengths makes the attribute or entity names less descriptive. Also, with a large number of entities and attributes in a complex design, you might have to be somewhat inventive about using proper attribute name prefixes. But then those prefixes are less helpful in identifying the precise source of the attribute. Nevertheless, the consistent use of prefixes will reduce sourcing doubts significantly. For example, while the prefix CO does not obviously relate to the CHECK_OUT table, just as obvious is the fact that it does not originate in WITHDRAW, ITEM, or USER.

Adherence to the naming conventions just described serves database designers well. In fact, a common refrain from users seems to be this: "I didn't know why you made such a fuss over naming conventions, but now that I'm doing this stuff for real, I've become a true believer."

Data Model Verification   The ER model must be verified against the proposed system processes in order to corroborate that the intended processes can be supported by the database model. Verification requires that the model be run through a series of tests against:

- End-user data views and their required transactions: SELECT, INSERT, UPDATE, and DELETE operations and queries and reports.
- Access paths and security.
- Business-imposed data requirements and constraints.

Revision of the original database design starts with a careful reevaluation of the entities, followed by a detailed examination of the attributes that describe those entities. This process serves several important purposes:

- The emergence of the attribute details might lead to a revision of the entities themselves. Perhaps some of the components first believed to be entities will, instead, turn out to be attributes within other entities. Or what was originally considered to be an attribute might turn out to contain a sufficient number of subcomponents to warrant the introduction of one or more new entities.
- The focus on attribute details can provide clues about the nature of relationships as they are defined by the primary and foreign keys. Improperly defined relationships lead to implementation problems first and to application development problems later.
- To satisfy processing and/or end-user requirements, it might be useful to create a new primary key to replace an existing primary key. For example, in the invoicing example illustrated in Figure 3.30 in Chapter 3, The Relational Database Model, a primary key composed of INV_NUMBER and LINE_NUMBER replaced the original primary key composed of INV_NUMBER and PROD_CODE. That change ensured that the items in the invoice would always appear in the same order as they were entered. To simplify queries and to increase processing speed, you may create a single-attribute primary key to replace an existing multiple-attribute primary key.
- Unless the entity details (the attributes and their characteristics) are precisely defined, it is difficult to evaluate the extent of the design's normalization. Knowledge of the normalization levels helps guard against undesirable redundancies.
- A careful review of the rough database design blueprint is likely to lead to revisions. Those revisions will help ensure that the design is capable of meeting end-user requirements.

Because real-world database design is generally done by teams, you should strive to organize the design's major components into modules. A **module** is an information system component that handles a specific function, such as inventory, orders, payroll, and so on. At the design level, a module is an ER segment that is an integrated part of the overall ER model. Creating and using modules accomplishes several important ends:

- The modules (and even the segments within them) can be delegated to design groups within teams, greatly speeding up the development work.
- The modules simplify the design work. The large number of entities within a complex design can be daunting. Each module contains a more manageable number of entities.
- The modules can be prototyped quickly. Implementation and applications programming trouble spots can be identified more readily. (Quick prototyping is also a great confidence builder.)
- Even if the entire system can't be brought online quickly, the implementation of one or more modules will demonstrate that progress is being made and that at least part of the system is ready to begin serving the end users.

As useful as modules are, they represent ER model fragments. Fragmentation creates a potential problem: the fragments might not include all of the ER model's components and might not, therefore, be able to support all of the required processes. To avoid that problem, the modules must be verified against the complete ER model. That verification process is detailed in Table 9.3.

TABLE 9.3	The ER Model Verification Process	
STEP	**ACTIVITY**
1	Identify the ER model's central entity.
2	Identify each module and its components.
3	Identify each module's transaction requirements: Internal: Updates/Inserts/Deletes/Queries/Reports External: Module interfaces
4	Verify all processes against the ER model.
5	Make all necessary changes suggested in Step 4.
6	Repeat Steps 2−5 for all modules.

Keep in mind that the verification process requires the continuous verification of business transactions as well as system and user requirements. The verification sequence must be repeated for each of the system's modules. Figure 9.10 illustrates the iterative nature of the process.

The verification process starts with selecting the central (most important) entity. The central entity is defined in terms of its participation in most of the model's relationships, and it is the focus for most of the system's operations. In other words, to identify the central entity, the designer selects the entity involved in the greatest number of relationships. In the ER diagram, it is the entity that has more lines connected to it than any other.

FIGURE 9.10	Iterative ER model verification process

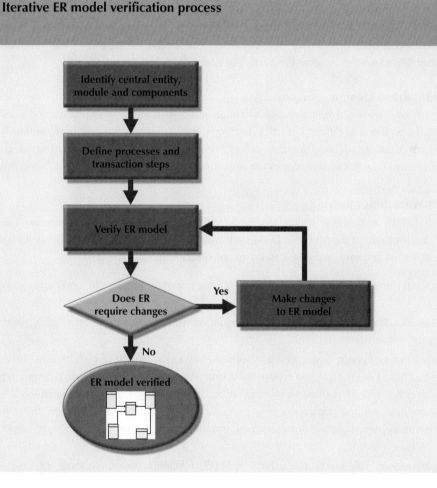

The next step is to identify the module or subsystem to which the central entity belongs and to define that module's boundaries and scope. The entity belongs to the module that uses it most frequently. Once each module is identified, the central entity is placed within the module's framework to let you focus your attention on the module's details.

Within the central entity/module framework, you must:

- *Ensure the module's cohesivity.* The term **cohesivity** describes the strength of the relationships found among the module's entities. A module must display *high cohesivity*—that is, the entities must be strongly related, and the module must be complete and self-sufficient.

- *Analyze each module's relationships with other modules to address module coupling.* **Module coupling** describes the extent to which modules are independent of one another. Modules must display *low coupling*, indicating that they are independent of other modules. Low coupling decreases unnecessary intermodule dependencies, thereby allowing the creation of a truly modular system and eliminating unnecessary relationships among entities.

Processes may be classified according to their:

- Frequency (daily, weekly, monthly, yearly, or exceptions).
- Operational type (INSERT or ADD, UPDATE or CHANGE, DELETE, queries and reports, batches, maintenance, and backups).

All identified processes must be verified against the ER model. If necessary, appropriate changes are implemented. The process verification is repeated for all of the model's modules. You can expect that additional entities and attributes will be incorporated into the conceptual model during its validation.

At this point, a conceptual model has been defined as hardware- and software-independent. Such independence ensures the system's portability across platforms. Portability can extend the database's life by making it possible to migrate to another DBMS and/or another hardware platform.

Distributed Database Design   Portions of a database may reside in several physical locations. Processes that access the database may also vary from one location to another. For example, a retail process and a warehouse storage process are likely to be found in different physical locations. If the database process is to be distributed across the system, the designer must also develop the data distribution and allocation strategies for the database. The design complications introduced by distributed processes are examined in detail in Chapter 12, Distributed Database Systems.

II. DBMS Software Selection

The selection of DBMS software is critical to the information system's smooth operation. Consequently, the advantages and disadvantages of the proposed DBMS software should be carefully studied. To avoid false expectations, the end user must be made aware of the limitations of both the DBMS and the database.

Although the factors affecting the purchasing decision vary from company to company, some of the most common are:

- *Cost.* This includes the original purchase price, along with maintenance, operational, license, installation, training, and conversion costs.

- *DBMS features and tools.* Some database software includes a variety of tools that facilitate the application development task. For example, the availability of query by example (QBE), screen painters, report generators, application generators, data dictionaries, and so on, helps to create a more pleasant work environment for both the end user and the application programmer. Database administrator facilities, query facilities, ease of use, performance, security, concurrency control, transaction processing, and third-party support also influence DBMS software selection.

- *Underlying model.* This can be hierarchical, network, relational, object/relational, or object-oriented.

- *Portability.* A DBMS can be portable across platforms, systems, and languages.

- *DBMS hardware requirements.* Items to consider include processor(s), RAM, disk space, and so on.

III. Logical Design

Logical design translates the conceptual design into the internal model for a selected database management system (DBMS) such as DB2, SQL Server, MySQL, Oracle, and Access. Therefore, the logical design is software-dependent.

Logical design requires that all objects in the model be mapped to the specific constructs used by the selected database software. For example, the logical design for a relational DBMS includes the specifications for the tables, indexes, views, transactions, access authorizations, and so on. In the following discussion, a small portion of the simple conceptual model shown in Figure 9.11 is converted into a logical design based on the relational model.

FIGURE 9.11	A simple conceptual model

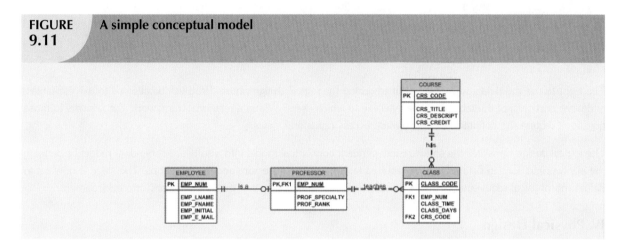

The translation of the conceptual model in Figure 9.11 requires the definition of the attribute domains, design of the required tables, and appropriate access restriction formats. For example, the domain definitions for the CLASS_CODE, CLASS_DAYS, and CLASS_TIME attributes displayed in the CLASS entity in Figure 9.11 are written this way:

CLASS_CODE   is a valid class code.
                Type: numeric
                Range: low value = 1000 high value = 9999
                Display format: 9999
                Length: 4
CLASS_DAYS   is a valid day code.
                Type: character
                Display format: XXX
                Valid entries: MWF, TTh, M, T, W, Th, F, S
                Length: 3
CLASS_TIME   is a valid time.
                Type: character
                Display format: 99:99 (24-hour clock)
                Display range: 06:00 to 22:00
                Length: 5

The logical design's tables must correspond to the entities (EMPLOYEE, PROFESSOR, COURSE, and CLASS) shown in the conceptual design of Figure 9.11, and the table columns must correspond to the attributes specified in the conceptual design. For example, the initial table layout for the COURSE table might look like Table 9.4.

TABLE 9.4	Sample Layout for the COURSE Table		
CRS_CODE	CRS_TITLE	CRS_DESCRIPT	CRS_CREDIT
CIS-4567	Database Systems Design	Design and implementation of database systems; includes conceptual design, logical design, implementation, and management; prerequisites: CIS 2040, CIS 2345, CIS 3680, and upper-division standing	4
QM-3456	Statistics II	Statistical applications; course requires use of statistical software (MINITAB and SAS) to interpret data; prerequisites: MATH 2345 and QM 2233	3

The right to use the database is also specified during the logical design phase. Who will be allowed to use the tables, and what portion(s) of the table(s) will be available to which users? Within a relational framework, the answers to those questions require the definition of appropriate access rights and views.

The logical design translates the software-independent conceptual model into a software-dependent model by defining the appropriate domain definitions, the required tables, and the necessary access restrictions. The stage is now set to define the physical requirements that allow the system to function within the selected hardware environment.

IV. Physical Design

Physical design is the process of selecting the data storage and data access characteristics of the database. The storage characteristics are a function of the types of devices supported by the hardware, the type of data access methods supported by the system, and the DBMS. Physical design affects not only the location of the data in the storage device(s), but also the performance of the system.

Physical design is a very technical job, more typical of the client/server and mainframe world than of the PC world. Yet even in the more complex midrange and mainframe environments, modern database software has assumed much of the burden of the physical portion of the design and its implementation.

ONLINE CONTENT

Physical design is particularly important in the older hierarchical and network models described in **Appendixes K** and **L, The Hierarchical Database Model** and **The Network Database Model,** respectively, in the Student Online Companion. Relational databases are more insulated from physical details than the older hierarchical and network models.

In spite of the fact that relational models tend to hide the complexities of the computer's physical characteristics, the performance of relational databases is affected by physical characteristics. For example, performance can be affected by the characteristics of the storage media, such as seek time, sector and block (page) size, buffer pool size, and the number of disk platters and read/write heads. In addition, factors such as the creation of an index can have a considerable effect on the relational database's performance, that is, data access speed and efficiency.

Even the type of data request must be analyzed carefully to determine the optimum access method for meeting the application requirements, establishing the data volume to be stored, and estimating the performance. Some DBMSs automatically reserve the space required to store the database definition and the user's data in permanent storage devices. This ensures that the data are stored in sequentially adjacent locations, thereby reducing data access time and increasing system performance. (Database performance tuning is covered in detail in Chapter 11, Database Performance Tuning and Query Optimization.)

Physical design becomes more complex when data are distributed at different locations because the performance is affected by the communication media's throughput. Given such complexities, it is not surprising that designers favor database software that hides as many of the physical-level activities as possible.

The preceding sections have separated the discussions of logical and physical design activities. In fact, logical and physical design can be carried out in parallel, on a table-by-table (or file-by-file) basis. Logical and physical design can also be carried out in parallel when the designer is working with hierarchical and network models. Such parallel activities require the designer to have a thorough understanding of the software and hardware in order to take full advantage of both software and hardware characteristics.

9.3.3 IMPLEMENTATION AND LOADING

In most modern relational DBMSs, such as IBM DB2, Microsoft SQL Server, and Oracle, a new database implementation requires the creation of special storage-related constructs to house the end-user tables. The constructs usually include the storage group, the table space, and the tables. See Figure 9.12. Note that a table space may contain more than one table.

FIGURE 9.12	Physical organization of a DB2 database environment

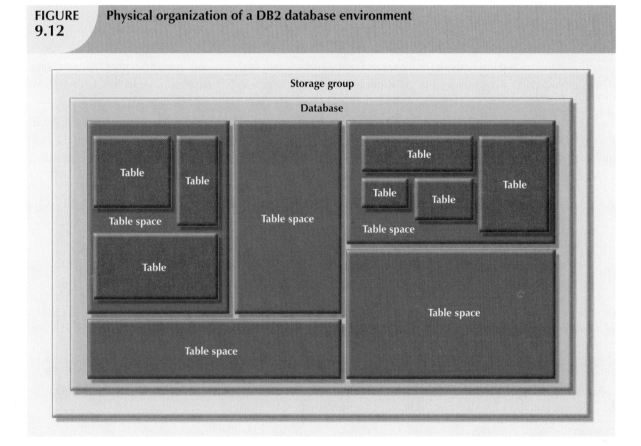

For example, the implementation of the logical design in IBM's DB2 would require that you:

1. *Create the database storage group.* This step (done by the system administrator or SYSADM) is mandatory for such mainframe databases as DB2. Other DBMS software may create equivalent storage groups automatically when a database is created. (See Step 2.) Consult your DBMS documentation to see if you must create a storage group and, if so, what the command syntax must be.

2. Create the database within the storage group (also done by the SYSADM).

3. Assign the rights to use the database to a database administrator (DBADM).

4. Create the table space(s) within the database (usually done by a DBADM).

5. Create the table(s) within the table space(s) (also usually done by a DBADM). A generic SQL table creation might look like this:

CREATE TABLE COURSE (
CRS_CODE   CHAR(10) NOT NULL,
CRS_TITLE   CHAR(C15) NOT NULL,
CRS_DESCRIPT   CHARC(8) NOT NULL
CRS_CREDIT NUMBER,
PRIMARY KEY (CRS_CODE));
CREATE TABLE CLASS (
CLASS_CODE   CHAR(4) NOT NULL,
CLASS_DAYS   CHAR(3) NOT NULL,
CLASS_TIME   CHAR(14) NOT NULL,
CLASS_DAY   CHAR(3) NOT NULL,
CRS_CODE   CHAR(10) NOT NULL,
PRIMARY KEY (CLASS_CODE),
FOREIGN KEY (CRS_CODE) REFERENCES COURSE;

(Note that the COURSE table is created first because it is referenced by the CLASS table.)

6. Assign access rights to the table spaces and to the tables within specified table spaces (another DBADM duty). Access rights may be limited to views rather than to whole tables. The creation of views is not required for database access in the relational environment, but views are desirable from a security standpoint.

Access rights to a table named PROFESSOR may be granted to a person whose identification code is PROB by typing:

GRANT USE OF TABLE PROFESSOR
TO PROB;

A view named PROF may be substituted for the PROFESSOR table:

CREATE VIEW PROF
SELEC   TEMP_LNAME
FROM   EMPLOYEE
WHERE   PROFESSOR.EMP_NUM = EMPLOYEE.EMP_NUM;

After the database has been created, the data must be loaded into the database tables. If the data are currently stored in a format different from that required by the new DBMS, the data must be converted prior to being loaded.

NOTE

The following summary of database implementation activities assumes the use of a sophisticated DBMS. All current generation DBMSs offer the features discussed next.

During the implementation and loading phase, you also must address performance, security, backup and recovery, integrity, and company standards. They will be discussed next.

Performance

Database performance is one of the most important factors in certain database implementations. Chapter 11 covers the subject in greater detail. However, not all DBMSs have performance-monitoring and fine-tuning tools embedded in their software, thus making it difficult to evaluate performance.

Performance evaluation is also rendered more difficult because there is no standard measurement for database performance. Performance varies according to the hardware and software environment used. Naturally, the database's size also affects database performance: a search of 10 tuples will be faster than a search of 100,000 tuples.

Important factors in database performance also include system and database configuration parameters, such as data placement, access path definition, the use of indexes, and buffer size.

Security

Data stored in the company database must be protected from access by unauthorized users. (It does not take much imagination to predict the likely results when students have access to a student database or when employees have access to payroll data!) Consequently, you must provide for (at least) the following:

- *Physical security* allows only authorized personnel physical access to specific areas. Depending on the type of database implementation, however, establishing physical security might not always be practical. For example, a university student research database is not a likely candidate for physical security. The existence of large multiserver PC networks often makes physical security impractical.
- *Password security* allows the assignment of access rights to specific authorized users. Password security is usually enforced at logon time at the operating system level.
- *Access rights* can be established through the use of database software. The assignment of access rights may restrict operations (CREATE, UPDATE, DELETE, and so on) on predetermined objects such as databases, tables, views, queries, and reports.
- *Audit trails* are usually provided by the DBMS to check for access violations. Although the audit trail is an after-the-fact device, its mere existence can discourage unauthorized use.
- *Data encryption* can be used to render data useless to unauthorized users who might have violated some of the database security layers.
- *Diskless workstations* allow end users to access the database without being able to download the information from their workstations.

For a more detailed discussion of security issues, please refer to Chapter 15, Database Administration and Security.

Backup and Recovery

Timely data availability is crucial for almost every database. Unfortunately, the database can be subject to data loss through unintended data deletion, power outages, and so on. Data backup and recovery (restore) procedures create a safety valve, allowing the database administrator to ensure the availability of consistent data. Typically, database vendors encourage the use of fault-tolerant components such as uninterruptible power supply (UPS) units, RAID storage devices, clustered servers, and data replication technologies to ensure the continuous operation of the database in case of a hardware failure. Even with these components, backup and restore functions constitute a very important component of daily database operations. Some DBMSs provide functions that allow the database administrator to schedule automatic database backups to permanent storage devices such as disks, DVDs, and tapes. Database backups can be performed at different levels:

- A **full backup** of the database, or *dump* of the entire database. In this case, all database objects are backed up in their entirety.

- A **differential backup** of the database, in which only the last modifications to the database (when compared with a previous full backup copy) are copied. In this case, only the objects that have been updated since the last full backup are backed up.

- A **transaction log backup**, which backs up only the transaction log operations that are not reflected in a previous backup copy of the database. In this case, only the transaction log is backed up; no other database objects are backed up. (For a complete explanation of the use of the transaction log see Chapter 10, Transaction Management and Concurrency Control.)

The database backup is stored in a secure place, usually in a different building from the database itself, and is protected against dangers such as fire, theft, flood, and other potential calamities. The main purpose of the backup is to guarantee database restoration following system (hardware/software) failures.

Failures that plague databases and systems are generally induced by software, hardware, programming exemptions, transactions, or external factors. Table 9.5 briefly summarizes the most common sources of database failure.

TABLE 9.5
Common Sources of Database Failure

SOURCE	DESCRIPTION	EXAMPLE
Software	Software-induced failures may be traceable to the operating system, the DBMS software, application programs, or viruses.	The SQL.Slammer worm affected many unpatched MS SQL Server systems in 2003 causing damages valued in millions of dollars.
Hardware	Hardware-induced failures may include memory chip errors, disk crashes, bad disk sectors, and "disk full" errors.	A bad memory module or a multiple hard disk failure in a database system can bring a database system to an abrupt stop.
Programming exemptions	Application programs or end users may roll back transactions when certain conditions are defined. Programming exemptions can also be caused by malicious or improperly tested code that can be exploited by hackers.	Hackers constantly searching for exploits in unprotected Web database systems.
Transactions	The system detects deadlocks and aborts one of the transactions. (See Chapter 10.)	Deadlock occurs when executing multiple simultaneous transactions.
External factors	Backups are especially important when a system suffers complete destruction due to fire, earthquake, flood, or other natural disaster.	The 2005 Katrina hurricane in New Orleans caused data losses in the millions of dollars.

Depending on the type and extent of the failure, the recovery process ranges from a minor short-term inconvenience to a major long-term rebuild. Regardless of the extent of the required recovery process, recovery is not possible without a usable backup.

The database recovery process generally follows a predictable scenario. First, the type and extent of the required recovery are determined. If the entire database needs to be recovered to a consistent state, the recovery uses the most recent backup copy of the database in a known consistent state. The backup copy is then rolled forward to restore all subsequent transactions by using the transaction log information. If the database needs to be recovered but the committed portion of the database is still usable, the recovery process uses the transaction log to "undo" all of the transactions that were not committed.

Integrity

Data integrity is enforced by the DBMS through the proper use of primary and foreign key rules. In addition, data integrity is also the result of properly implemented data management policies. Such policies are part of a comprehensive data administration framework. For a more detailed study of this topic, see The DBA's Managerial Role section in Chapter 15.

Company Standards

Database standards may be partially defined by specific company requirements. The database administrator must implement and enforce such standards.

9.3.4 TESTING AND EVALUATION

Once the data have been loaded into the database, the DBA tests and fine-tunes the database for performance, integrity, concurrent access, and security constraints. The testing and evaluation phase occurs in parallel with applications programming.

Programmers use database tools to *prototype* the applications during coding of the programs. Tools such as report generators, screen painters, and menu generators are especially useful to the applications programmers during the prototyping phase.

If the database implementation fails to meet some of the system's evaluation criteria, several options may be considered to enhance the system:

- For performance-related issues, the designer must consider fine-tuning specific system and DBMS configuration parameters. The best sources of information are the hardware and software technical reference manuals.
- Modify the physical design. (For example, the proper use of indexes tends to be particularly effective in facilitating pointer movements, thus enhancing performance.)
- Modify the logical design.
- Upgrade or change the DBMS software and/or the hardware platform.

9.3.5 OPERATION

Once the database has passed the evaluation stage, it is considered to be operational. At that point, the database, its management, its users, and its application programs constitute a complete information system.

The beginning of the operational phase invariably starts the process of system evolution. As soon as all of the targeted end users have entered the operations phase, problems that could not have been foreseen during the testing phase begin to surface. Some of the problems are serious enough to warrant emergency "patchwork," while others are merely minor annoyances. For example, if the database design is implemented to interface with the Web, the sheer volume of transactions might cause even a well-designed system to bog down. In that case, the designers have to identify the source(s) of the bottleneck(s) and produce alternative solutions. Those solutions may include using load-balancing software to distribute the transactions among multiple computers, increasing the available cache for the DBMS, and so on. In any case, the demand for change is the designer's constant concern, which leads to phase 6, maintenance and evolution.

9.3.6 MAINTENANCE AND EVOLUTION

The database administrator must be prepared to perform routine maintenance activities within the database. Some of the required periodic maintenance activities include:

- Preventive maintenance (backup).
- Corrective maintenance (recovery).
- Adaptive maintenance (enhancing performance, adding entities and attributes, and so on).
- Assignment of access permissions and their maintenance for new and old users.
- Generation of database access statistics to improve the efficiency and usefulness of system audits and to monitor system performance.
- Periodic security audits based on the system-generated statistics.
- Periodic (monthly, quarterly, or yearly) system-usage summaries for internal billing or budgeting purposes.

The likelihood of new information requirements and the demand for additional reports and new query formats require application changes and possible minor changes in the database components and contents. Those changes can be easily implemented only when the database design is flexible and when all documentation is updated and online. Eventually, even the best-designed database environment will no longer be capable of incorporating such evolutionary changes; then the whole DBLC process begins anew.

You should not be surprised to discover that many of the activities described in the Database Life Cycle (DBLC) remind you of those in the Systems Development Life Cycle (SDLC). After all, the SDLC represents the framework within which the DBLC activities take place. A summary of the parallel activities that take place within the SDLC and the DBLC is shown in Figure 9.13.

FIGURE 9.13   Parallel activities in the DBLC and the SDLC

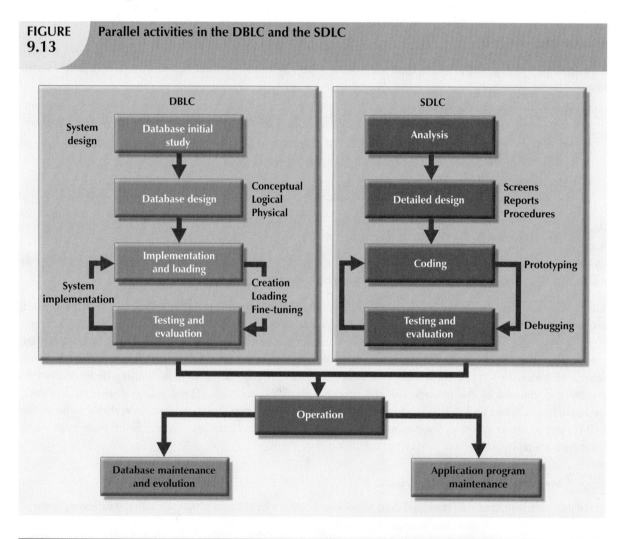

9.4 DATABASE DESIGN STRATEGIES

There are two classical approaches to database design:

- **Top-down design** starts by identifying the data sets, and then defines the data elements for each of those sets. This process involves the identification of different entity types and the definition of each entity's attributes.

- **Bottom-up design** first identifies the data elements (items), and then groups them together in data sets. In other words, it first defines attributes, and then groups them to form entities.

The two approaches are illustrated in Figure 9.14. The selection of a primary emphasis on top-down or bottom-up procedures often depends on the scope of the problem or on personal preferences. Although the two methodologies are complementary rather than mutually exclusive, a primary emphasis on a bottom-up approach may be more productive for small databases with few entities, attributes, relations, and transactions. For situations in which the number, variety, and complexity of entities, relations, and transactions is overwhelming, a primarily top-down approach may be more easily managed. Most companies have standards for systems development and database design already in place.

FIGURE 9.14   **Top-down vs. bottom-up design sequencing**

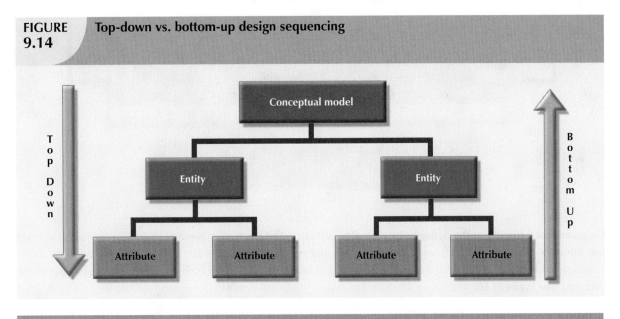

NOTE

Even when a primarily top-down approach is selected, the normalization process that revises existing table structures is (inevitably) a bottom-up technique. ER models constitute a top-down process even when the selection of attributes and entities can be described as bottom-up. Because both the ER model and normalization techniques form the basis for most designs, the top-down vs. bottom-up debate may be based on a theoretical distinction rather than an actual difference.

9.5 CENTRALIZED VS. DECENTRALIZED DESIGN

The two general approaches (bottom-up and top-down) to database design can be influenced by factors such as the scope and size of the system, the company's management style, and the company's structure (centralized or decentralized). Depending on such factors, the database design may be based on two very different design philosophies: centralized and decentralized.

Centralized design is productive when the data component is composed of a relatively small number of objects and procedures. The design can be carried out and represented in a fairly simple database. Centralized design is typical of relatively simple and/or small databases and can be successfully done by a single person (database administrator) or by a small, informal design team. The company operations and the scope of the problem are sufficiently limited to allow even a single designer to define the problem(s), create the conceptual design, verify the conceptual design with the user views, define system processes and data constraints to ensure the efficacy of the design, and ensure that the design will comply with all the requirements. (Although centralized design is typical for small companies, do not make the mistake of assuming that centralized design is limited to small companies. Even large companies can operate within a relatively simple database environment.) Figure 9.15 summarizes the centralized design option. Note that a single conceptual design is completed and then validated in the centralized design approach.

FIGURE 9.15   Centralized design

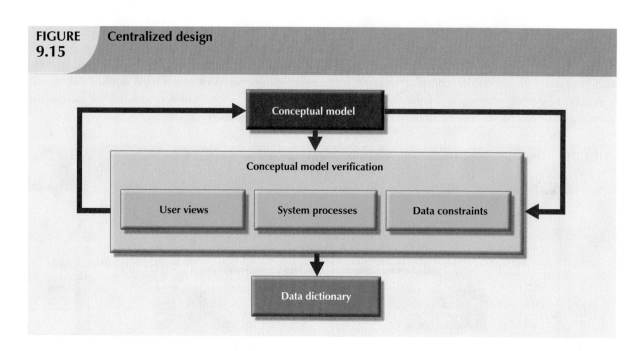

Decentralized design might be used when the data component of the system has a considerable number of entities and complex relations on which very complex operations are performed. Decentralized design is also likely to be employed when the problem itself is spread across several operational sites and each element is a subset of the entire data set. See Figure 9.16.

In large and complex projects, the database design typically cannot be done by only one person. Instead, a carefully selected team of database designers is employed to tackle a complex database project. Within the decentralized design framework, the database design task is divided into several modules. Once the design criteria have been established, the lead designer assigns design subsets or modules to design groups within the team.

FIGURE 9.16   **Decentralized design**

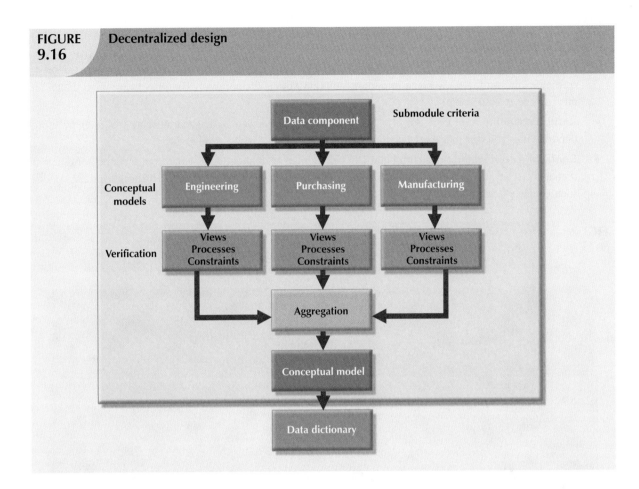

Because each design group focuses on modeling a subset of the system, the definition of boundaries and the interrelation among data subsets must be very precise. Each design group creates a conceptual data model corresponding to the subset being modeled. Each conceptual model is then verified individually against the user views, processes, and constraints for each of the modules. After the verification process has been completed, all modules are integrated into one conceptual model. Because the data dictionary describes the characteristics of all objects within the conceptual data model, it plays a vital role in the integration process. Naturally, after the subsets have been aggregated into a larger conceptual model, the lead designer must verify that the combined conceptual model is still able to support all of the required transactions.

Keep in mind that the aggregation process requires the designer to create a single model in which various aggregation problems must be addressed. See Figure 9.17.

- *Synonyms and homonyms.* Various departments might know the same object by different names (synonyms), or they might use the same name to address different objects (homonyms). The object can be an entity, an attribute, or a relationship.

- *Entity and entity subtypes.* An entity subtype might be viewed as a separate entity by one or more departments. The designer must integrate such subtypes into a higher-level entity.

- *Conflicting object definitions.* Attributes can be recorded as different types (character, numeric), or different domains can be defined for the same attribute. Constraint definitions, too, can vary. The designer must remove such conflicts from the model.

FIGURE 9.17   Summary of aggregation problems

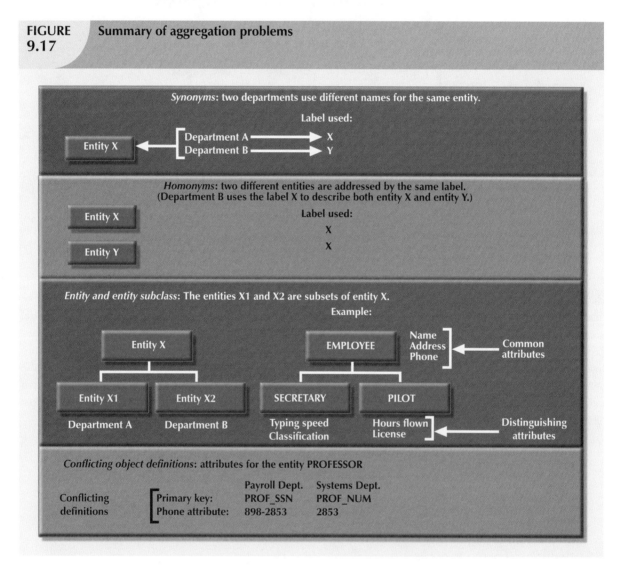

SUMMARY

- An information system is designed to facilitate the transformation of data into information and to manage both data and information. Thus, the database is a very important part of the information system. Systems analysis is the process that establishes the need for and the extent of an information system. Systems development is the process of creating an information system.

- The Systems Development Life Cycle (SDLC) traces the history (life cycle) of an application within the information system. The SDLC can be divided into five phases: planning, analysis, detailed systems design, implementation, and maintenance. The SDLC is an iterative rather than a sequential process.

- The Database Life Cycle (DBLC) describes the history of the database within the information system. The DBLC is composed of six phases: database initial study, database design, implementation and loading, testing and evaluation, operation, and maintenance and evolution. Like the SDLC, the DBLC is iterative rather than sequential.

- The database design and implementation process moves through a series of well-defined stages: database initial study, database design, implementation and loading, testing and evaluation, operation, and maintenance and evolution.

- The conceptual portion of the design may be subject to several variations based on two basic design philosophies: bottom-up vs. top-down and centralized vs. decentralized.

KEY TERMS

bottom-up design, 402

boundaries, 383

centralized design, 403

cohesivity, 394

computer-aided systems engineering (CASE), 378

conceptual design, 385

database development, 374

Database Life Cycle (DBLC), 378

decentralized design, 404

description of operations, 387

differential backup, 400

full backup, 399

information system, 373

logical design, 395

minimal data rule, 385

module, 392

module coupling, 392

physical design, 396

scope, 383

systems analysis, 373

systems development, 373

Systems Development Life Cycle (SDLC), 373

top-down design, 402

transaction log backup, 400

ONLINE CONTENT

Answers to selected Review Questions and Problems for this chapter are contained in the Student Online Companion for this book.

REVIEW QUESTIONS

1. What is an information system? What is its purpose?
2. How do systems analysis and systems development fit into a discussion about information systems?
3. What does the acronym SDLC mean, and what does an SDLC portray?
4. What does the acronym DBLC mean, and what does a DBLC portray?
5. Discuss the distinction between centralized and decentralized conceptual database design.

6. What is the minimal data rule in conceptual design? Why is it important?

7. Discuss the distinction between top-down and bottom-up approaches in database design.

8. What are business rules? Why are they important to a database designer?

9. What is the data dictionary's function in database design?

10. What steps are required in the development of an ER diagram? (*Hint*: See Table 9.1.)

11. List and briefly explain the activities involved in the verification of an ER model.

12. What factors are important in a DBMS software selection?

13. What three levels of backup may be used in database recovery management? Briefly describe what each of those three backup levels does.

P R O B L E M S

1. The ABC Car Service & Repair Centers are owned by the SILENT car dealer; ABC services and repairs only SILENT cars. Three ABC Car Service & Repair Centers provide service and repair for the entire state.

   Each of the three centers is independently managed and operated by a shop manager, a receptionist, and at least eight mechanics. Each center maintains a fully stocked parts inventory.

   Each center also maintains a manual file system in which each car's maintenance history is kept: repairs made, parts used, costs, service dates, owner, and so on. Files are also kept to track inventory, purchasing, billing, employees' hours, and payroll.

   You have been contacted by the manager of one of the centers to design and implement a computerized database system. Given the preceding information, do the following:

   a. Indicate the most appropriate sequence of activities by labeling each of the following steps in the correct order. (For example, if you think that "Load the database." is the appropriate first step, label it "1.")

      _____ Normalize the conceptual model.

      _____ Obtain a general description of company operations.

      _____ Load the database.

      _____ Create a description of each system process.

      _____ Test the system.

      _____ Draw a data flow diagram and system flowcharts.

      _____ Create a conceptual model using ER diagrams.

      _____ Create the application programs.

      _____ Interview the mechanics.

      _____ Create the file (table) structures.

      _____ Interview the shop manager.

   b. Describe the various modules that you believe the system should include.

   c. How will a data dictionary help you develop the system? Give examples.

   d. What general (system) recommendations might you make to the shop manager? (For example, if the system will be integrated, what modules will be integrated? What benefits would be derived from such an integrated system? Include several general recommendations.)

   e. What is the best approach to conceptual database design? Why?

   f. Name and describe at least four reports the system should have. Explain their use. Who will use those reports?

2. Suppose you have been asked to create an information system for a manufacturing plant that produces nuts and bolts of many shapes, sizes, and functions. What questions would you ask, and how would the answers to those questions affect the database design?

   a. What do you envision the SDLC to be?

   b. What do you envision the DBLC to be?

3. Suppose you perform the same functions noted in Problem 2 for a larger warehousing operation. How are the two sets of procedures similar? How and why are they different?

4. Using the same procedures and concepts employed in Problem 1, how would you create an information system for the Tiny College example in Chapter 4?

5. Write the proper sequence of activities in the design of a video rental database. (The initial ERD was shown in Figure 9.7.) The design must support all rental activities, customer payment tracking, and employee work schedules, as well as track which employees checked out the videos to the customers. After you finish writing the design activity sequence, complete the ERD to ensure that the database design can be successfully implemented. (Make sure that the design is normalized properly and that it can support the required transactions.)

PART
IV

ADVANCED DATABASE CONCEPTS

JetBlue's Database Crisis

During the Valentine's Day snowstorm of 2007, JetBlue, hailed as the discount airline model of success, nearly destroyed its reputation in a single day when, rather than cancel flights at JFK, the airline sent planes out to the tarmac hoping for a break in the weather. The weather worsened, and the airplanes were grounded. Passengers spent all morning, then the afternoon, waiting inside the planes. Finally, the airline sent busses out to the stranded planes to retrieve the passengers. Then the real disaster hit. As angry passengers arrived in the terminal, they had only one method of rebooking their flights: via telephone. Unlike JetBlue's larger competitors that offered airport kiosks and Internet booking, JetBlue relied exclusively on the Navitaire Open Skies reservation system that was set up to accommodate only 650 reservation agents who, working from home, logged into the system via the Internet. While JetBlue managers recruited off-duty reservation agents to pitch in during the crisis, Navitaire worked to boost the number of concurrent users the system could handle. Navitaire discovered that it could only increase the number of concurrent users to 950 before the system began to fail. Many passengers waited over an hour to reach a reservation agent. Others could not reach agents at all.

Navitaire had built Open Skies, a reservation system serving 50 airlines, on HP3000 mini mainframe computers with a proprietary HP operating system and database products. Prior to the crisis, the company knew the system was approaching the limits of its processing capabilities for its larger clients. In 2006, the company decided to boost processing capabilities by re-platforming Open Skies with Microsoft SQL Server 2005 on Intel-based 8-CPU 64-bit database servers. Navitaire created a new system with Microsoft Visual Studio and the Microsoft .NET Framework. The system is expected to enjoy faster development and easier database management capabilities.

Yet JetBlue's Open Skies problem was only one of several database crises the company faced. The database storing reservation and check-in information, for example, tracked bag tag identification numbers but not the location where bags were picked up. Lost bags had always been recovered manually. In the past, this approach worked because of JetBlue's policy of avoiding flight cancellations. Now, an IT team arrived at the airport, had the bags hauled offsite, and spent three days building a database application using Microsoft SQL Server and handheld scanning devices that agents accessed to locate lost luggage.

After six days and over 1000 canceled flights, the crisis abated. However, JetBlue's reputation had been deeply marred. To improve its public standing, JetBlue issued a Customer Bill of Rights. At the same time, internally JetBlue focused attention on revising its database systems to respond in a timely fashion to situations demanding flight cancellations and to efficiently track luggage, so that the airline would not be crippled in times of crisis.

10

TEN

TRANSACTION MANAGEMENT AND CONCURRENCY CONTROL

In this chapter, you will learn:

- About database transactions and their properties
- What concurrency control is and what role it plays in maintaining the database's integrity
- What locking methods are and how they work
- How stamping methods are used for concurrency control
- How optimistic methods are used for concurrency control
- How database recovery management is used to maintain database integrity

Preview

Database transactions reflect real-world transactions that are triggered by events such as buying a product, registering for a course, or making a deposit in a checking account. Transactions are likely to contain many parts. For example, a sales transaction might require updating the customer's account, adjusting the product inventory, and updating the seller's accounts receivable. All parts of a transaction must be successfully completed to prevent data integrity problems. Therefore, executing and managing transactions are important database system activities.

The main database transaction properties are atomicity, consistency, isolation, and durability. In addition, serializability is a characteristic of the schedule of operations for the execution of concurrent transactions. After defining those transaction properties, the chapter shows how SQL can be used to represent transactions and how transaction logs can ensure the DBMS's ability to recover transactions.

When many transactions take place at the same time, they are called concurrent transactions. Managing the execution of such transactions is called concurrency control. As you can imagine, concurrency control is especially important in a multiuser database environment. (Just imagine the number of transactions routinely handled by companies that conduct sales and provide services via the Web!) This chapter discusses some of the problems that can occur with concurrent transactions—lost updates, uncommitted data, and inconsistent summaries. And you discover that such problems can be solved when a DBMS scheduler enforces concurrency control.

In this chapter you learn about the most common algorithms for concurrency control: locks, time stamping, and optimistic methods. Because locks are the most widely used method, you examine various levels and types of locks. Locks can also create deadlocks, so you learn about strategies for managing deadlocks.

Database contents can be damaged or destroyed by critical operational errors, including transaction management failures. Therefore, in this chapter you also learn how database recovery management maintains a database's contents.

10.1 WHAT IS A TRANSACTION?

To illustrate what transactions are and how they work, let's use the **Ch10_SaleCo** database. The relational diagram for that database is shown in Figure 10.1.

FIGURE 10.1   The Ch10_SaleCo database relational diagram

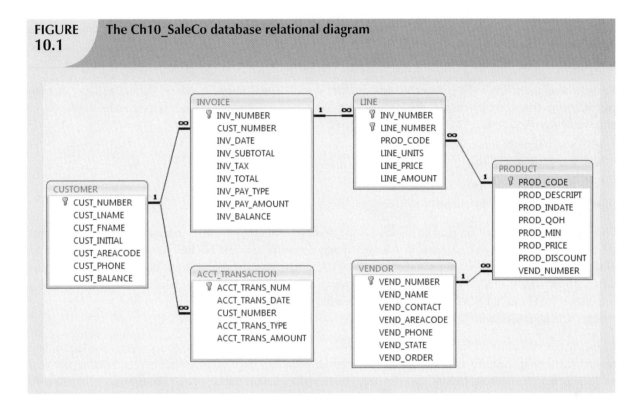

ONLINE CONTENT

The **Ch10_SaleCo** database used to illustrate the material in this chapter is found in the Student Online Companion for this book.

NOTE

Although SQL commands illustrate several transaction and concurrency control issues, you should be able to follow the discussions even if you have not studied Chapter 7, Introduction to Structured Query Language (SQL), and Chapter 8, Advanced SQL. If you don't know SQL, ignore the SQL commands and focus on the discussions. If you have a working knowledge of SQL, you can use the **Ch10_SaleCo** database to generate your own SELECT and UPDATE examples and to augment the material presented in Chapters 7 and 8 by writing your own triggers and stored procedures.

As you examine the relational diagram in Figure 10.1, note the following features:

- The design stores the customer balance (CUST_BALANCE) value in the CUSTOMER table to indicate the total amount owed by the customer. The CUST_BALANCE attribute is increased when the customer makes a purchase on credit, and it is decreased when the customer makes a payment. Including the current customer account balance in the CUSTOMER table makes it very easy to write a query to determine the current balance for any customer and to generate important summaries such as total, average, minimum, and maximum balances.

- The ACCT_TRANSACTION table records all customer purchases and payments to track the details of customer account activity.

Naturally, you can change the database design of the **Ch10_SaleCo** database to reflect accounting practice more precisely, but the implementation provided here will enable you to track the transactions well enough to serve the purpose of the chapter's discussions.

To understand the concept of a transaction, suppose that you sell a product to a customer. Further, suppose that the customer may charge the purchase to the customer's account. Given that scenario, your sales transaction consists of at least the following parts:

- You must write a new customer invoice.
- You must reduce the quantity on hand in the product's inventory.
- You must update the account transactions.
- You must update the customer balance.

The preceding sales transaction must be reflected in the database. In database terms, a **transaction** is any action that reads from and/or writes to a database. A transaction may consist of a simple SELECT statement to generate a list of table contents; it may consist of a series of related UPDATE statements to change the values of attributes in various tables; it may consist of a series of INSERT statements to add rows to one or more tables; or it may consist of a combination of SELECT, UPDATE, and INSERT statements. The sales transaction example includes a combination of INSERT and UPDATE statements.

Given the preceding discussion, you can now augment the definition of a transaction. A transaction is a *logical* unit of work that must be entirely completed or entirely aborted; no intermediate states are acceptable. In other words, a multicomponent transaction, such as the previously mentioned sale, must not be partially completed. Updating only the inventory or only the accounts receivable is not acceptable. All of the SQL statements in the transaction must be completed successfully. If any of the SQL statements fail, the entire transaction is rolled back to the original database state that existed before the transaction started. A successful transaction changes the database from one consistent state to another. A **consistent database state** is one in which all data integrity constraints are satisfied.

To ensure consistency of the database, every transaction must begin with the database in a known consistent state. If the database is not in a consistent state, the transaction will yield an inconsistent database that violates its integrity and business rules. For that reason, subject to limitations discussed later, all transactions are controlled and executed by the DBMS to guarantee database integrity.

Most real-world database transactions are formed by two or more database requests. A **database request** is the equivalent of a single SQL statement in an application program or transaction. For example, if a transaction is composed of two UPDATE statements and one INSERT statement, the transaction uses three database requests. In turn, each database request generates several input/output (I/O) operations that read from or write to physical storage media.

10.1.1 EVALUATING TRANSACTION RESULTS

Not all transactions update the database. Suppose you want to examine the CUSTOMER table to determine the current balance for customer number 10016. Such a transaction can be completed by using the SQL code:

```
SELECT      CUST_NUMBER, CUST_BALANCE
FROM        CUSTOMER
WHERE       CUST_NUMBER = 10016;
```

Although that query does not make any changes in the CUSTOMER table, the SQL code represents a transaction because it *accesses* the database. If the database existed in a consistent state before the access, the database remains in a consistent state after the access because the transaction did not alter the database.

Remember that a transaction may consist of a single SQL statement or a collection of related SQL statements. Let's revisit the previous sales example to illustrate a more complex transaction, using the **Ch10_SaleCo** database. Suppose that on January 18, 2008 you register the credit sale of one unit of product 89-WRE-Q to customer 10016 in the amount of $277.55. The required transaction affects the INVOICE, LINE, PRODUCT, CUSTOMER, and ACCT_TRANSACTION tables. The SQL statements that represent this transaction are as follows:

```
INSERT INTO INVOICE
        VALUES (1009, 10016,'18-Jan-2008', 256.99, 20.56, 277.55, 'cred', 0.00, 277.55);
INSERT INTO LINE
        VALUES (1009, 1, '89-WRE-Q', 1, 256.99, 256.99);
UPDATE PRODUCT
        SET PROD_QOH = PROD_QOH – 1
                WHERE        PROD_CODE = '89-WRE-Q';
UPDATE CUSTOMER
        SET CUST_BALANCE = CUST_BALANCE + 277.55
                WHERE        CUST_NUMBER = 10016;
INSERT INTO ACCT_TRANSACTION
        VALUES (10007, '18-Jan-08', 10016, 'charge', 277.55);
COMMIT;
```

The results of the successfully completed transaction are shown in Figure 10.2. (Note that all records involved in the transaction have been highlighted.)

To further your understanding of the transaction results, note the following:

- A new row 1009 was added to the INVOICE table. In this row, derived attribute values were stored for the invoice subtotal, the tax, the invoice total, and the invoice balance.

- The LINE row for invoice 1009 was added to reflect the purchase of one unit of product 89-WRE-Q with a price of $256.99. In this row, the derived attribute values for the line amount were stored.

- The product 89-WRE-Q's quantity on hand (PROD_QOH) in the PRODUCT table was reduced by one (the initial value was 12), thus leaving a quantity on hand of 11.

- The customer balance (CUST_BALANCE) for customer 10016 was updated by adding $277.55 to the existing balance (the initial value was $0.00).

- A new row was added to the ACCT_TRANSACTION table to reflect the new account transaction number 10007.

- The COMMIT statement is used to end a successful transaction. (See Section 10.1.3.)

Now suppose that the DBMS completes the first three SQL statements. Further, suppose that during the execution of the fourth statement (the UPDATE of the CUSTOMER table's CUST_BALANCE value for customer 10016), the computer system experiences a loss of electrical power. If the computer does not have a backup power supply, the transaction cannot be completed. Therefore, the INVOICE and LINE rows were added, the PRODUCT table was updated to represent the sale of product 89-WRE-Q, but customer 10016 was not charged, nor was the required record in the ACCT_TRANSACTION table written. The database is now in an inconsistent state, and it is not usable for subsequent transactions. Assuming that the DBMS supports transaction management, *the DBMS will roll back the database to a previous consistent state.*

FIGURE 10.2   Tracing the transaction in the Ch10_SaleCo database

Table name: INVOICE

INV_NUMBER	CUST_NUMBER	INV_DATE	INV_SUBTOTAL	INV_TAX	INV_TOTAL	INV_PAY_TYPE	INV_PAY_AMOUNT	INV_BALANCE
1001	10014	16-Jan-08	54.92	4.39	59.31	cc	59.31	0.00
1002	10011	16-Jan-08	9.98	0.80	10.78	cash	10.78	0.00
1003	10012	16-Jan-08	270.70	21.66	292.36	cc	292.36	0.00
1004	10011	17-Jan-08	34.87	2.79	37.66	cc	37.66	0.00
1005	10018	17-Jan-08	70.44	5.64	76.08	cc	76.08	0.00
1006	10014	17-Jan-08	397.83	31.83	429.66	cred	100.00	329.66
1007	10015	17-Jan-08	34.97	2.80	37.77	chk	37.77	0.00
1008	10011	17-Jan-08	1033.08	82.65	1115.73	cred	500.00	615.73
1009	10016	18-Jan-08	256.99	20.56	277.55	cred	0.00	277.55

Table name: LINE

INV_NUMBER	LINE_NUMBER	PROD_CODE	LINE_UNITS	LINE_PRICE	LINE_AMOUNT
1001	1	13-Q2/P2	3	14.99	44.97
1001	2	23109-HB	1	9.95	9.95
1002	1	54778-2T	2	4.99	9.98
1003	1	2238/QPD	4	38.95	155.80
1003	2	1546-QQ2	1	39.95	39.95
1003	3	13-Q2/P2	5	14.99	74.95
1004	1	54778-2T	3	4.99	14.97
1004	2	23109-HB	2	9.95	19.90
1005	1	PVC23DRT	12	5.87	70.44
1006	1	SM-18277	3	6.99	20.97
1006	2	2232/QTY	1	109.92	109.92
1006	3	23109-HB	1	9.95	9.95
1006	4	89-WRE-Q	1	256.99	256.99
1007	1	13-Q2/P2	2	14.99	29.98
1007	2	54778-2T	1	4.99	4.99
1008	1	PVC23DRT	5	5.87	29.35
1008	2	WR3/TT3	4	119.95	479.80
1008	3	23109-HB	1	9.95	9.95
1008	4	89-WRE-Q	2	256.99	513.98
1009	1	89-WRE-Q	1	256.99	256.99

Table name: PRODUCT

PROD_CODE	PROD_DESCRIPT	PROD_INDATE	PROD_QOH	PROD_MIN	PROD_PRICE	PROD_DISCOUNT	VEND_NUMBER
11QER/31	Power painter, 15 psi., 3-nozzle	03-Nov-07	8	5	109.99	0.00	25595
13-Q2/P2	7.25-in. pwr. saw blade	13-Dec-07	32	15	14.99	0.05	21344
14-Q1/L3	9.00-in. pwr. saw blade	13-Nov-07	18	12	17.49	0.00	21344
1546-QQ2	Hrd. cloth, 1/4-in., 2x50	15-Jan-08	15	8	39.95	0.00	23119
1558-QW1	Hrd. cloth, 1/2-in., 3x50	15-Jan-08	23	5	43.99	0.00	23119
2232/QTY	B&D jigsaw, 12-in. blade	30-Dec-07	8	5	109.92	0.05	24288
2232/QWE	B&D jigsaw, 8-in. blade	24-Dec-07	6	5	99.87	0.05	24288
2238/QPD	B&D cordless drill, 1/2-in.	20-Jan-08	12	5	38.95	0.05	25595
23109-HB	Claw hammer	20-Jan-08	23	10	9.95	0.10	21225
23114-AA	Sledge hammer, 12 lb.	02-Jan-08	8	5	14.40	0.05	
54778-2T	Rat-tail file, 1/8-in. fine	15-Dec-07	43	20	4.99	0.00	21344
89-WRE-Q	Hicut chain saw, 16 in.	07-Jan-08	11	5	256.99	0.05	24288
PVC23DRT	PVC pipe, 3.5-in., 8-ft	06-Jan-08	188	75	5.87	0.00	
SM-18277	1.25-in. metal screw, 25	01-Mar-08	172	75	6.99	0.00	21225
SW-23116	2.5-in. wd. screw, 50	24-Feb-08	237	100	8.45	0.00	21231
WR3/TT3	Steel matting, 4'x8'x1/6", .5" mesh	17-Jan-08	18	5	119.95	0.10	25595

Table name: CUSTOMER

CUST_NUMBER	CUST_LNAME	CUST_FNAME	CUST_INITIAL	CUST_AREACODE	CUST_PHONE	CUST_BALANCE
10010	Ramas	Alfred	A	615	844-2573	0.00
10011	Dunne	Leona	K	713	894-1238	615.73
10012	Smith	Kathy	W	615	894-2285	0.00
10013	Olowski	Paul	F	615	894-2180	0.00
10014	Orlando	Myron		615	222-1672	0.00
10015	O'Brian	Amy	B	713	442-3381	0.00
10016	Brown	James	G	615	297-1228	277.55
10017	Williams	George		615	290-2556	0.00
10018	Farriss	Anne		713	382-7185	0.00
10019	Smith	Olette	K	615	297-3809	0.00

Table name: ACCT TRANSACTION

ACCT_TRANS_NUM	ACCT_TRANS_DATE	CUST_NUMBER	ACCT_TRANS_TYPE	ACCT_TRANS_AMOUNT
10003	17-Jan-08	10014	charge	329.66
10004	17-Jan-08	10011	charge	615.73
10006	29-Jan-08	10014	payment	329.66
10007	18-Jan-08	10016	charge	277.55

NOTE

By default, MS Access does not support transaction management as discussed here. More sophisticated DBMSs, such as Oracle, SQL Server, and DB2, do support the transaction management components discussed in this chapter.

Although the DBMS is designed to recover a database to a previous consistent state when an interruption prevents the completion of a transaction, the transaction itself is defined by the end user or programmer and must be semantically correct. *The DBMS cannot guarantee that the semantic meaning of the transaction truly represents the real-world event.* For example, suppose that following the sale of 10 units of product 89-WRE-Q, the inventory UPDATE commands were written this way:

```
UPDATE      PRODUCT
SET         PROD_QOH = PROD_QOH + 10
WHERE       PROD_CODE = '89-WRE-Q';
```

The sale should have *decreased* the PROD_QOH value for product 89-WRE-Q by 10. Instead, the UPDATE *added* 10 to product 89-WRE-Q's PROD_QOH value.

Although the UPDATE command's syntax is correct, its use yields incorrect results. Yet the DBMS will execute the transaction anyway. The DBMS cannot evaluate whether the transaction represents the real-world event correctly; that is the end user's responsibility. End users and programmers are capable of introducing many errors in this fashion. Imagine the consequences of reducing the quantity on hand for product 1546-QQ2 instead of product 89-WRE-Q or of crediting the CUST_BALANCE value for customer 10012 rather than customer 10016.

Clearly, improper or incomplete transactions can have a devastating effect on database integrity. Some DBMSs—*especially* the relational variety—provide means by which the user can define enforceable constraints based on business rules. Other integrity rules, such as those governing referential and entity integrity, are enforced automatically by the DBMS when the table structures are properly defined, thereby letting the DBMS validate some transactions. For example, if a transaction inserts a new customer number into a customer table and the customer number being inserted already exists, the DBMS will end the transaction with an error code to indicate a violation of the primary key integrity rule.

10.1.2 TRANSACTION PROPERTIES

Each individual transaction must display *atomicity, consistency, isolation, and durability*. These properties are sometimes referred to as the ACID test. In addition, when executing multiple transactions, the DBMS must schedule the concurrent execution of the transaction's operations. The schedule of such transaction's operations must exhibit the property of *serializability*. Let's look briefly at each of the properties.

- **Atomicity** requires that *all* operations (SQL requests) of a transaction be completed; if not, the transaction is aborted. If a transaction T1 has four SQL requests, all four requests must be successfully completed; otherwise, the entire transaction is aborted. In other words, a transaction is treated as a single, indivisible, logical unit of work.

- **Consistency** indicates the permanence of the database's consistent state. A transaction takes a database from one consistent state to another consistent state. When a transaction is completed, the database must be in a consistent state; if any of the transaction parts violates an integrity constraint, the entire transaction is aborted.

- **Isolation** means that the data used during the execution of a transaction cannot be used by a second transaction until the first one is completed. In other words, if a transaction T1 is being executed and is using the data item X, that data item cannot be accessed by any other transaction (T2 ... T*n*) until T1 ends. This property is particularly useful in multiuser database environments because several users can access and update the database at the same time.

- **Durability** ensures that once transaction changes are done (committed), they cannot be undone or lost, even in the event of a system failure.

- **Serializability** ensures that the schedule for the concurrent execution of the transactions yields consistent results. This property is important in multiuser and distributed databases, where multiple transactions are likely to be executed concurrently. Naturally, if only a single transaction is executed, serializability is not an issue.

By its very nature, a single-user database system automatically ensures serializability and isolation of the database because only one transaction is executed at a time. The atomicity, consistency, and durability of transactions must be guaranteed by the single-user DBMSs. (Even a single-user DBMS must manage recovery from errors created by operating-system-induced interruptions, power interruptions, and improper application execution.)

Multiuser databases are typically subject to multiple concurrent transactions. Therefore, the multiuser DBMS must implement controls to ensure serializability and isolation of transactions—in addition to atomicity and durability—to guard the database's consistency and integrity. For example, if several concurrent transactions are executed over the same data set and the second transaction updates the database before the first transaction is finished, the isolation property is violated and the database is no longer consistent. The DBMS must manage the transactions by using concurrency control techniques to avoid such undesirable situations.

10.1.3 TRANSACTION MANAGEMENT WITH SQL

The American National Standards Institute (ANSI) has defined standards that govern SQL database transactions. Transaction support is provided by two SQL statements: COMMIT and ROLLBACK. The ANSI standards require that when a transaction sequence is initiated by a user or an application program, the sequence must continue through all succeeding SQL statements until one of the following four events occurs:

- A COMMIT statement is reached, in which case all changes are permanently recorded within the database. The COMMIT statement automatically ends the SQL transaction.
- A ROLLBACK statement is reached, in which case all changes are aborted and the database is rolled back to its previous consistent state.
- The end of a program is successfully reached, in which case all changes are permanently recorded within the database. This action is equivalent to COMMIT.
- The program is abnormally terminated, in which case the changes made in the database are aborted and the database is rolled back to its previous consistent state. This action is equivalent to ROLLBACK.

The use of COMMIT is illustrated in the following simplified sales example, which updates a product's quantity on hand (PROD_QOH) and the customer's balance when the customer buys two units of product 1558-QW1 priced at $43.99 per unit (for a total of $87.98) and charges the purchase to the customer's account:

```
UPDATE      PRODUCT
SET         PROD_QOH = PROD_QOH – 2
WHERE       PROD_CODE = '1558-QW1';
UPDATE      CUSTOMER
SET         CUST_BALANCE = CUST_BALANCE + 87.98
WHERE       CUST_NUMBER = '10011';
COMMIT;
```

(Note that the example is simplified to make it easy to trace the transaction. In the **Ch10_SaleCo** database, the transaction would involve several additional table updates.)

Actually, the COMMIT statement used in that example is not necessary if the UPDATE statement is the application's last action and the application terminates normally. However, good programming practice dictates that you include the COMMIT statement at the end of a transaction declaration.

A transaction begins implicitly when the first SQL statement is encountered. Not all SQL implementations follow the ANSI standard; some (such as SQL Server) use transaction management statements such as:

BEGIN TRANSACTION;

to indicate the beginning of a new transaction. Other SQL implementations allow you to assign characteristics for the transactions as parameters to the BEGIN statement. For example, the Oracle RDBMS uses the SET TRANSACTION statement to declare a new transaction start and its properties.

10.1.4 THE TRANSACTION LOG

A DBMS uses a **transaction log** to keep track of all transactions that update the database. The information stored in this log is used by the DBMS for a recovery requirement triggered by a ROLLBACK statement, a program's abnormal termination, or a system failure such as a network discrepancy or a disk crash. Some RDBMSs use the transaction log to recover a database *forward* to a currently consistent state. After a server failure, for example, Oracle automatically rolls back uncommitted transactions and rolls forward transactions that were committed but not yet written to the physical database. This behavior is required for transactional correctness and is typical of any transactional DBMS.

While the DBMS executes transactions that modify the database, it also automatically updates the transaction log. The transaction log stores:

- A record for the beginning of the transaction.
- For each transaction component (SQL statement):
    - The type of operation being performed (update, delete, insert).
    - The names of the objects affected by the transaction (the name of the table).
    - The "before" and "after" values for the fields being updated.
    - Pointers to the previous and next transaction log entries for the same transaction.
- The ending (COMMIT) of the transaction.

Although using a transaction log increases the processing overhead of a DBMS, the ability to restore a corrupted database is worth the price.

Table 10.1 illustrates a simplified transaction log that reflects a basic transaction composed of two SQL UPDATE statements. If a system failure occurs, the DBMS will examine the transaction log for all uncommitted or incomplete transactions and restore (ROLLBACK) the database to its previous state on the basis of that information. When the recovery process is completed, the DBMS will write in the log all committed transactions that were not physically written to the database before the failure occurred.

TABLE 10.1 A Transaction Log

TRL_ID	TRX_NUM	PREV PTR	NEXT PTR	OPERATION	TABLE	ROW ID	ATTRIBUTE	BEFORE VALUE	AFTER VALUE
341	101	Null	352	START	****Start Transaction				
352	101	341	363	UPDATE	PRODUCT	1558-QW1	PROD_QOH	25	23
363	101	352	365	UPDATE	CUSTOMER	10011	CUST_ BALANCE	525.75	615.73
365	101	363	Null	COMMIT	**** End of Transaction				

TRL_ID = Transaction log record ID   PTR = Pointer to a transaction log record ID
TRX_NUM = Transaction number
(*Note*: The transaction number is auto-matically assigned by the DBMS.)

If a ROLLBACK is issued before the termination of a transaction, the DBMS will restore the database only for that particular transaction, rather than for all transactions, to maintain the *durability* of the previous transactions. In other words, committed transactions are not rolled back.

The transaction log is a critical part of the database, and it is usually implemented as one or more files that are managed separately from the actual database files. The transaction log is subject to common dangers such as disk-full conditions and disk crashes. Because the transaction log contains some of the most critical data in a DBMS, some implementations support logs on several different disks to reduce the consequences of a system failure.

10.2 CONCURRENCY CONTROL

The coordination of the simultaneous execution of transactions in a multiuser database system is known as **concurrency control**. The objective of concurrency control is to ensure the serializability of transactions in a multiuser database environment. Concurrency control is important because the simultaneous execution of transactions over a shared database can create several data integrity and consistency problems. The three main problems are lost updates, uncommitted data, and inconsistent retrievals.

10.2.1 LOST UPDATES

The **lost update** problem occurs when two concurrent transactions, T1 and T2, are updating the same data element and one of the updates is lost (overwritten by the other transaction). To see an illustration of lost updates, let's examine a simple PRODUCT table. One of the PRODUCT table's attributes is a product's quantity on hand (PROD_QOH). Assume that you have a product whose current PROD_QOH value is 35. Also assume that two concurrent transactions, T1 and T2, occur that update the PROD_QOH value for some item in the PRODUCT table. The transactions are shown in Table 10.2:

TABLE 10.2 Two Concurrent Transactions to Update QOH

TRANSACTION	COMPUTATION
T1: Purchase 100 units	PROD_QOH = PROD_QOH + 100
T2: Sell 30 units	PROD_QOH = PROD_QOH - 30

Table 10.3 shows the serial execution of those transactions under normal circumstances, yielding the correct answer PROD_QOH = 105.

TABLE 10.3 Serial Execution of Two Transactions

TIME	TRANSACTION	STEP	STORED VALUE
1	T1	Read PROD_QOH	35
2	T1	PROD_QOH = 35 + 100	
3	T1	Write PROD_QOH	135
4	T2	Read PROD_QOH	135
5	T2	PROD_QOH = 135 − 30	
6	T2	Write PROD_QOH	105

But suppose that a transaction is able to read a product's PROD_QOH value from the table *before* a previous transaction (using the *same* product) has been committed. The sequence depicted in Table 10.4 shows how the lost update problem can arise. Note that the first transaction (T1) has not yet been committed when the second transaction (T2) is executed. Therefore, T2 still operates on the value 35, and its subtraction yields 5 in memory. In the meantime, T1 writes the value 135 to disk, which is promptly overwritten by T2. In short, the addition of 100 units is "lost" during the process.

TABLE 10.4	Lost Updates		
TIME	TRANSACTION	STEP	STORED VALUE
1	T1	Read PROD_QOH	35
2	T2	Read PROD_QOH	35
3	T1	PROD_QOH = 35 + 100	
4	T2	PROD_QOH = 35 − 30	
5	T1	Write PROD_QOH (**Lost update**)	135
6	T2	Write PROD_QOH	5

10.2.2 UNCOMMITTED DATA

The phenomenon of **uncommitted data** occurs when two transactions, T1 and T2, are executed concurrently and the first transaction (T1) is rolled back after the second transaction (T2) has already accessed the uncommitted data—thus violating the isolation property of transactions. To illustrate that possibility, let's use the same transactions described during the lost updates discussion. T1 has two atomic parts to it, one of which is the update of the inventory, the other possibly being the update of the invoice total (not shown). T1 is forced to roll back due to an error during the update of the invoice total; hence, it rolls back all the way, undoing the inventory update as well. This time the T1 transaction is rolled back to eliminate the addition of the 100 units (Table 10.5). Because T2 subtracts 30 from the original 35 units, the correct answer should be 5.

TABLE 10.5	Transactions Creating Uncommitted Data Problem	
TRANSACTION	COMPUTATION	
T1: Purchase 100 units	PROD_QOH = PROD_QOH + 100 (**Rolled back**)	
T2: Sell 30 units	PROD_QOH = PROD_QOH 30	

Table 10.6 shows how, under normal circumstances, the serial execution of those transactions yields the correct answer.

TABLE 10.6	Correct Execution of Two Transactions		
TIME	TRANSACTION	STEP	STORED VALUE
1	T1	Read PROD_QOH	35
2	T1	PROD_QOH = 35 + 100	
3	T1	Write PROD_QOH	135
4	T1	*****ROLLBACK *****	35
5	T2	Read PROD_QOH	35
6	T2	PROD_QOH = 35 − 30	
7	T2	Write PROD_QOH	5

Table 10.7 shows how the uncommitted data problem can arise when the ROLLBACK is completed after T2 has begun its execution.

TABLE 10.7	An Uncommitted Data Problem		

TIME	TRANSACTION	STEP	STORED VALUE
1	T1	Read PROD_QOH	35
2	T1	PROD_QOH = 35 + 100	
3	T1	Write PROD_QOH	135
4	T2	Read PROD_QOH **(Read uncommitted data)**	135
5	T2	PROD_QOH = 135 − 30	
6	T1	***** **ROLLBACK** *****	35
7	T2	Write PROD_QOH	105

10.2.3 INCONSISTENT RETRIEVALS

Inconsistent retrievals occur when a transaction accesses data before and after another transaction(s) finish working with such data. For example, an inconsistent retrieval would occur if transaction T1 calculates some summary (aggregate) function over a set of data while another transaction (T2) is updating the same data. The problem is that the transaction might read some data before they are changed and other data *after* they are changed, thereby yielding inconsistent results.

To illustrate that problem, assume the following conditions:

1.   T1 calculates the total quantity on hand of the products stored in the PRODUCT table.

2.   At the same time, T2 updates the quantity on hand (PROD_QOH) for two of the PRODUCT table's products.

The two transactions are shown in Table 10.8.

TABLE 10.8	Retrieval During Update

TRANSACTION T1	TRANSACTION T2
SELECT SUM(PROD_QOH) FROM PRODUCT	UPDATE PRODUCT SET PROD_QOH = PROD_QOH + 10 WHERE PROD_CODE = '1546-QQ2'
	UPDATE PRODUCT SET PROD_QOH = PROD_QOH − 10 WHERE PROD_CODE = '1558-QW1'
	COMMIT;

While T1 calculates the total quantity on hand (PROD_QOH) for all items, T2 represents the correction of a typing error: the user added 10 units to product 1558-QW1's PROD_QOH, but *meant* to add the 10 units to product 1546-QQ2's PROD_QOH. To correct the problem, the user adds 10 to product 1546-QQ2's PROD_QOH and subtracts 10 from product 1558-QW1's PROD_QOH. (See the two UPDATE statements in Table 10.7.) The initial and final PROD_QOH values are reflected in Table 10.9. (Only a few of the PROD_CODE values for the PRODUCT table are shown. To illustrate the point, the sum for the PROD_QOH values is given for those few products.)

TABLE 10.9	Transaction Results: Data Entry Correction	
	BEFORE	AFTER
PROD_CODE	PROD_QOH	PROD_QOH
11QER/31	8	8
13-Q2/P2	32	32
1546-QQ2	15	(15 + 10) ➝ 25
1558-QW1	23	(23 − 10) ➝ 13
2232-QTY	8	8
2232-QWE	6	6
Total	92	92

Although the final results shown in Table 10.8 are correct after the adjustment, Table 10.10 demonstrates that inconsistent retrievals are possible during the transaction execution, making the result of T1's execution incorrect. The "After" summation shown in Table 10.9 reflects the fact that the value of 25 for product 1546-QQ2 was read *after* the WRITE statement was completed. Therefore, the "After" total is 40 + 25 = 65. The "Before" total reflects the fact that the value of 23 for product 1558-QW1 was read *before* the next WRITE statement was completed to reflect the corrected update of 13. Therefore, the "Before" total is 65 + 23 = 88.

TABLE 10.10	Inconsistent Retrievals			
TIME	TRANSACTION	ACTION	VALUE	TOTAL
1	T1	Read PROD_QOH for PROD_CODE = '11QER/31'	8	8
2	T1	Read PROD_QOH for PROD_CODE = '13-Q2/P2'	32	40
3	T2	Read PROD_QOH for PROD_CODE = '1546-QQ2'	15	
4	T2	PROD_QOH = 15 + 10		
5	T2	Write PROD_QOH for PROD_CODE = '1546-QQ2'	25	
6	T1	Read PROD_QOH for PROD_CODE = '1546-QQ2'	25	(After) 65
7	T1	Read PROD_QOH for PROD_CODE = '1558-QW1'	23	(Before) 88
8	T2	Read PROD_QOH for PROD_CODE = '1558-QW1'	23	
9	T2	PROD_QOH = 23 − 10		
10	T2	Write PROD_QOH for PROD_CODE = '1558-QW1'	13	
11	T2	***** **COMMIT** *****		
12	T1	Read PROD_QOH for PROD_CODE = '2232-QTY'	8	96
13	T1	Read PROD_QOH for PROD_CODE = '2232-QWE'	6	102

The computed answer of 102 is obviously wrong because you know from Table 10.9 that the correct answer is 92. Unless the DBMS exercises concurrency control, a multiuser database environment can create havoc within the information system.

10.2.4 THE SCHEDULER

You now know that severe problems can arise when two or more concurrent transactions are executed. You also know that a database transaction involves a series of database I/O operations that take the database from one consistent state to another. Finally, you know that database consistency can be ensured only before and after the execution of transactions. A database always moves through an unavoidable temporary state of inconsistency during a transaction's execution if such transaction updates multiple tables/rows. (If the transaction contains only one update, then there is no temporary inconsistency.) That temporary inconsistency exists because a computer executes the operations serially, one after another. During this serial process, the isolation property of transactions prevents them from accessing the data not yet released by other transactions. The job of the scheduler is even more important today, with the use of multicore processors which have the capability of executing several instructions at the same time. What would happen if two transactions execute concurrently and they are accessing the same data?

In previous examples, the operations within a transaction were executed in an arbitrary order. As long as two transactions, T1 and T2, access *unrelated* data, there is no conflict and the order of execution is irrelevant to the final outcome. But if the transactions operate on related (or the same) data, conflict is possible among the transaction components and the selection of one execution order over another might have some undesirable consequences. So how is the correct order determined, and who determines that order? Fortunately, the DBMS handles that tricky assignment by using a built-in scheduler.

The **scheduler** is a special DBMS process that establishes the order in which the operations within concurrent transactions are executed. The scheduler *interleaves* the execution of database operations to ensure serializability and isolation of transactions. To determine the appropriate order, the scheduler bases its actions on concurrency control algorithms, such as locking or time stamping methods, which are explained in the next sections. However, it is important to understand that not all transactions are serializable. The DBMS determines what transactions are serializable and proceeds to interleave the execution of the transaction's operations. Generally, transactions that are not serializable are executed on a first-come, first-served basis by the DBMS. The scheduler's main job is to create a serializable schedule of a transaction's operations. A **serializable schedule** is a schedule of a transaction's operations in which the interleaved execution of the transactions (T1, T2, T3, etc.) yields the same results as if the transactions were executed in serial order (one after another).

The scheduler also makes sure that the computer's central processing unit (CPU) and storage systems are used efficiently. If there were no way to schedule the execution of transactions, all transactions would be executed on a first-come, first-served basis. The problem with that approach is that processing time is wasted when the CPU waits for a READ or WRITE operation to finish, thereby losing several CPU cycles. In short, first-come, first-served scheduling tends to yield unacceptable response times within the multiuser DBMS environment. Therefore, some other scheduling method is needed to improve the efficiency of the overall system.

Additionally, the scheduler facilitates data isolation to ensure that two transactions do not update the same data element at the same time. Database operations might require READ and/or WRITE actions that produce conflicts. For example, Table 10.11 shows the possible conflict scenarios when two transactions, T1 and T2, are executed concurrently over the same data. Note that in Table 10.11, two operations are in conflict when they access the same data and at least one of them is a WRITE operation.

TABLE 10.11	READ/WRITE Conflict Scenarios: Conflicting Database Operations Matrix		
	TRANSACTIONS		
	T1	T2	RESULT
Operations	Read	Read	No conflict
	Read	Write	Conflict
	Write	Read	Conflict
	Write	Write	Conflict

Several methods have been proposed to schedule the execution of conflicting operations in concurrent transactions. Those methods have been classified as locking, time stamping, and optimistic. Locking methods, discussed next, are used most frequently.

10.3 CONCURRENCY CONTROL WITH LOCKING METHODS

A **lock** guarantees exclusive use of a data item to a current transaction. In other words, transaction T2 does not have access to a data item that is currently being used by transaction T1. A transaction acquires a lock prior to data access; the lock is released (unlocked) when the transaction is completed so that another transaction can lock the data item for its exclusive use.

Recall from the earlier discussions (Evaluating Transaction Results and Transaction Properties) that data consistency cannot be guaranteed *during* a transaction; the database might be in a temporary inconsistent state when several updates are executed. Therefore, locks are required to prevent another transaction from reading inconsistent data.

Most multiuser DBMSs automatically initiate and enforce locking procedures. All lock information is managed by a **lock manager**, which is responsible for assigning and policing the locks used by the transactions.

10.3.1 LOCK GRANULARITY

Lock granularity indicates the level of lock use. Locking can take place at the following levels: database, table, page, row, or even field (attribute).

Database Level

In a **database-level lock**, the entire database is locked, thus preventing the use of any tables in the database by transaction T2 while transaction Tl is being executed. This level of locking is good for batch processes, but it is unsuitable for multiuser DBMSs. You can imagine how s-l-o-w the data access would be if thousands of transactions had to wait for the previous transaction to be completed before the next one could reserve the entire database. Figure 10.3 illustrates the database-level lock. Note that because of the database-level lock, transactions T1 and T2 cannot access the same database concurrently *even when they use different tables*.

FIGURE 10.3   Database-level locking sequence

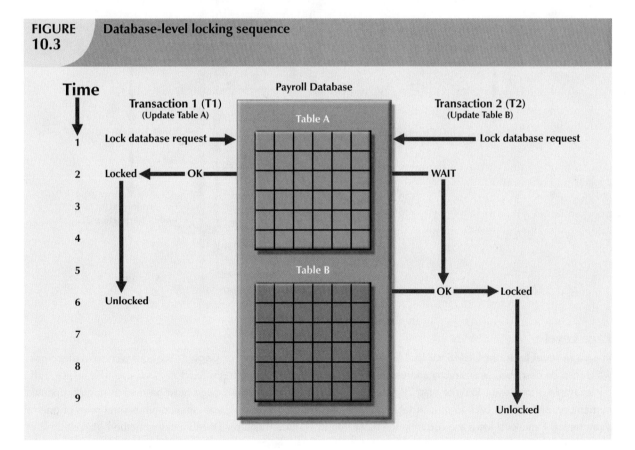

Table Level

In a **table-level lock**, the entire table is locked, preventing access to any row by transaction T2 while transaction T1 is using the table. If a transaction requires access to several tables, each table may be locked. However, two transactions can access the same database as long as they access different tables.

Table-level locks, while less restrictive than database-level locks, cause traffic jams when many transactions are waiting to access the same table. Such a condition is especially irksome if the lock forces a delay when different transactions require access to different parts of the same table, that is, when the transactions would not interfere with each other. Consequently, table-level locks are not suitable for multiuser DBMSs. Figure 10.4 illustrates the effect of a table-level lock. Note that in Figure 10.4, transactions T1 and T2 cannot access the same table even when they try to use different rows; T2 must wait until T1 unlocks the table.

FIGURE 10.4	An example of a table-level lock

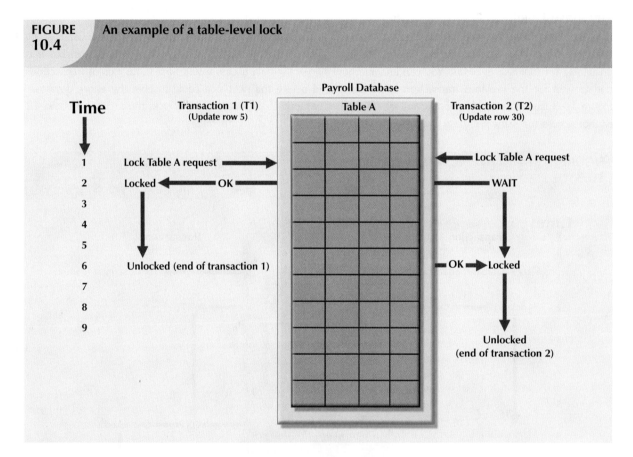

Page Level

In a **page-level lock**, the DBMS will lock an entire diskpage. A **diskpage**, or **page**, is the equivalent of a *diskblock*, which can be described as a directly addressable section of a disk. A page has a fixed size, such as 4K, 8K, or 16K. For example, if you want to write only 73 bytes to a 4K page, the entire 4K page must be read from disk, updated in memory, and written back to disk. A table can span several pages, and a page can contain several rows of one or more tables. Page-level locks are currently the most frequently used multiuser DBMS locking method. An example of a page-level lock is shown in Figure 10.5. Note that T1 and T2 access the same table while locking different diskpages. If T2 requires the use of a row located on a page that is locked by T1, T2 must wait until the page is unlocked by T1.

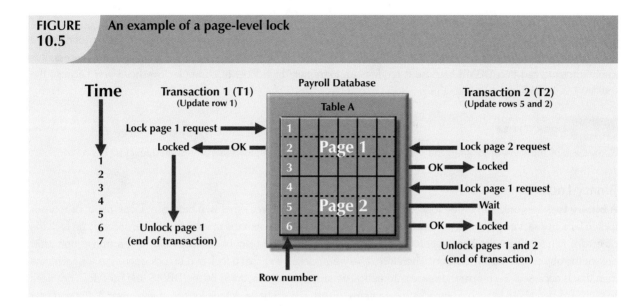

FIGURE 10.5   An example of a page-level lock

Row Level

A **row-level lock** is much less restrictive than the locks discussed earlier. The DBMS allows concurrent transactions to access different rows of the same table even when the rows are located on the same page. Although the row-level locking approach improves the availability of data, its management requires high overhead because a lock exists for each row in a table of the database involved in a conflicting transaction. Modern DBMS automatically escalate a lock from row level to page level lock when the application session requests multiple locks on the same page. Figure 10.6 illustrates the use of a row-level lock.

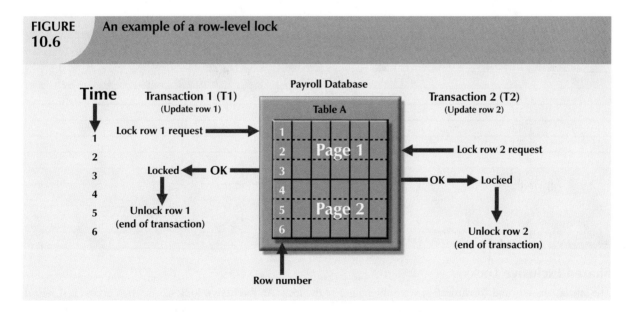

FIGURE 10.6   An example of a row-level lock

Note in Figure 10.6 that both transactions can execute concurrently, even when the requested rows are on the same page. T2 must wait only if it requests the same row as T1.

Field Level

The **field-level lock** allows concurrent transactions to access the same row as long as they require the use of different fields (attributes) within that row. Although field-level locking clearly yields the most flexible multiuser data access, it is rarely implemented in a DBMS because it requires an extremely high level of computer overhead and because the row-level lock is much more useful in practice.

10.3.2 LOCK TYPES

Regardless of the level of locking, the DBMS may use different lock types: binary or shared/exclusive.

Binary Locks

A **binary lock** has only two states: locked (1) or unlocked (0). If an object—that is, a database, table, page, or row—is locked by a transaction, no other transaction can use that object. If an object is unlocked, any transaction can lock the object for its use. Every database operation requires that the affected object be locked. As a rule, a transaction must unlock the object after its termination. Therefore, every transaction requires a lock and unlock operation for each data item that is accessed. Such operations are automatically managed and scheduled by the DBMS; the user does not need to be concerned about locking or unlocking data items. (Every DBMS has a default locking mechanism. If the end user wants to override the default, the LOCK TABLE and other SQL commands are available for that purpose.)

The binary locking technique is illustrated in Table 10.12, using the lost updates problem you encountered in Table 10.4. Note that the lock and unlock features eliminate the lost update problem because the lock is not released until the WRITE statement is completed. Therefore, a PROD_QOH value cannot be used until it has been properly updated. However, binary locks are now considered too restrictive to yield optimal concurrency conditions. For example, the DBMS will not allow two transactions to read the same database object even though neither transaction updates the database, and therefore, no concurrency problems can occur. Remember from Table 10.11 that concurrency conflicts occur only when two transactions execute concurrently and one of them updates the database.

TABLE 10.12	An Example of a Binary Lock		
TIME	**TRANSACTION**	**STEP**	**STORED VALUE**
1	T1	Lock PRODUCT	
2	T1	Read PROD_QOH	15
3	T1	PROD_QOH = 15 + 10	
4	T1	Write PROD_QOH	25
5	T1	Unlock PRODUCT	
6	T2	Lock PRODUCT	
7	T2	Read PROD_QOH	23
8	T2	PROD_QOH = 23 − 10	
9	T2	Write PROD_QOH	13
10	T2	Unlock PRODUCT	

Shared/Exclusive Locks

The labels "shared" and "exclusive" indicate the nature of the lock. An **exclusive lock** exists when access is reserved specifically for the transaction that locked the object. The exclusive lock must be used when the potential for conflict exists. (See Table 10.9, READ vs. WRITE.) A **shared lock** exists when concurrent transactions are granted read access on the basis of a common lock. A shared lock produces no conflict as long as all the concurrent transactions are read only.

A shared lock is issued when a transaction wants to read data from the database and no exclusive lock is held on that data item. An exclusive lock is issued when a transaction wants to update (write) a data item and no locks are currently held on that data item by any other transaction. Using the shared/exclusive locking concept, a lock can have three states: unlocked, shared (read), and exclusive (write).

As shown in Table 10.11, two transactions conflict only when at least one of them is a WRITE transaction. Because the two READ transactions can be safely executed at once, shared locks allow several READ transactions to read the same data item concurrently. For example, if transaction T1 has a shared lock on data item X and transaction T2 wants to read data item X, T2 may also obtain a shared lock on data item X.

If transaction T2 updates data item X, an exclusive lock is required by T2 over data item X. *The exclusive lock is granted if and only if no other locks are held on the data item.* Therefore, if a shared or exclusive lock is already held on data item X by transaction T1, an exclusive lock cannot be granted to transaction T2 and T2 must wait to begin until T1 commits. This condition is known as the **mutual exclusive rule**: only one transaction at a time can own an exclusive lock on the same object.

Although the use of shared locks renders data access more efficient, a shared/exclusive lock schema increases the lock manager's overhead, for several reasons:

- The type of lock held must be known before a lock can be granted.
- Three lock operations exist: READ_LOCK (to check the type of lock), WRITE_LOCK (to issue the lock), and UNLOCK (to release the lock).
- The schema has been enhanced to allow a lock upgrade (from shared to exclusive) and a lock downgrade (from exclusive to shared).

Although locks prevent serious data inconsistencies, they can lead to two major problems:

- The resulting transaction schedule might not be serializable.
- The schedule might create deadlocks. A database **deadlock**, which is equivalent to traffic gridlock in a big city, is caused when two or more transactions wait for each other to unlock data.

Fortunately, both problems can be managed: serializability is guaranteed through a locking protocol known as two-phase locking, and deadlocks can be managed by using deadlock detection and prevention techniques. Those techniques are examined in the next two sections.

10.3.3 TWO-PHASE LOCKING TO ENSURE SERIALIZABILITY

Two-phase locking defines how transactions acquire and relinquish locks. Two-phase locking guarantees serializability, but it does not prevent deadlocks. The two phases are:

1. A growing phase, in which a transaction acquires all required locks without unlocking any data. Once all locks have been acquired, the transaction is in its locked point.
2. A shrinking phase, in which a transaction releases all locks and cannot obtain any new lock.

The two-phase locking protocol is governed by the following rules:

- Two transactions cannot have conflicting locks.
- No unlock operation can precede a lock operation in the same transaction.
- No data are affected until all locks are obtained—that is, until the transaction is in its locked point.

Figure 10.7 depicts the two-phase locking protocol.

FIGURE 10.7   Two-phase locking protocol

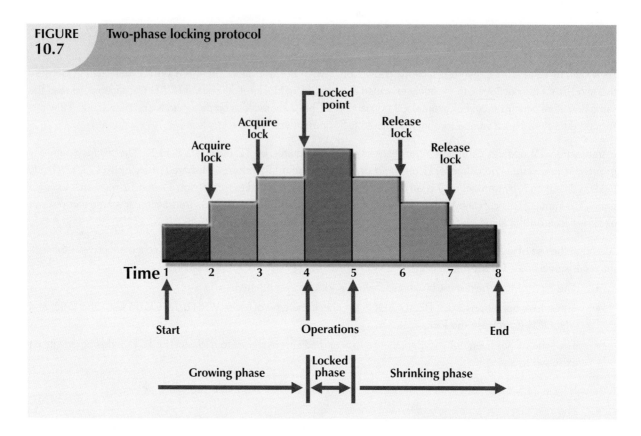

In this example, the transaction acquires all of the locks it needs until it reaches its locked point. (In this example, the transaction requires two locks.) When the locked point is reached, the data are modified to conform to the transaction requirements. Finally, the transaction is completed as it releases all of the locks it acquired in the first phase.

Two-phase locking increases the transaction processing cost and might cause additional undesirable effects. One undesirable effect is the possibility of creating deadlocks.

10.3.4 DEADLOCKS

A **deadlock** occurs when two transactions wait indefinitely for each other to unlock data. For example, a deadlock occurs when two transactions, T1 and T2, exist in the following mode:

$$T1 = \text{access data items X and Y}$$
$$T2 = \text{access data items Y and X}$$

If T1 has not unlocked data item Y, T2 cannot begin; if T2 has not unlocked data item X, T1 cannot continue. Consequently, T1 and T2 each wait for the other to unlock the required data item. Such a deadlock is also known as a **deadly embrace**. Table 10.13 demonstrates how a deadlock condition is created.

TABLE 10.13	How a Deadlock Condition Is Created				
TIME	TRANSACTION	REPLY	LOCK STATUS		
0			Data X		Data Y
1	T1:LOCK(X)	OK	Unlocked		Unlocked
2	T2: LOCK(Y)	OK	Locked		Unlocked
3	T1:LOCK(Y)	WAIT	Locked		Locked
4	T2:LOCK(X)	WAIT	Locked		Locked
5	T1:LOCK(Y)	WAIT	Locked	Deadlock	Locked
6	T2:LOCK(X)	WAIT	Locked		Locked
7	T1:LOCK(Y)	WAIT	Locked		Locked
8	T2:LOCK(X)	WAIT	Locked		Locked
9	T1:LOCK(Y)	WAIT	Locked		Locked
...	....	....	....		....
...	....	....	....		....
...	....	....	....		....
...	....	....	....		....

The preceding example used only two concurrent transactions to demonstrate a deadlock condition. In a real-world DBMS, many more transactions can be executed simultaneously, thereby increasing the probability of generating deadlocks. Note that deadlocks are possible only when one of the transactions wants to obtain an exclusive lock on a data item; no deadlock condition can exist among *shared* locks.

The three basic techniques to control deadlocks are:

- *Deadlock prevention.* A transaction requesting a new lock is aborted when there is the possibility that a deadlock can occur. If the transaction is aborted, all changes made by this transaction are rolled back and all locks obtained by the transaction are released. The transaction is then rescheduled for execution. Deadlock prevention works because it avoids the conditions that lead to deadlocking.

- *Deadlock detection.* The DBMS periodically tests the database for deadlocks. If a deadlock is found, one of the transactions (the "victim") is aborted (rolled back and restarted) and the other transaction continues.

- *Deadlock avoidance.* The transaction must obtain all of the locks it needs before it can be executed. This technique avoids the rollback of conflicting transactions by requiring that locks be obtained in succession. However, the serial lock assignment required in deadlock avoidance increases action response times.

The choice of the best deadlock control method to use depends on the database environment. For example, if the probability of deadlocks is low, deadlock detection is recommended. However, if the probability of deadlocks is high, deadlock prevention is recommended. If response time is not high on the system's priority list, deadlock avoidance might be employed. All current DBMSs support deadlock detention in transactional databases, while some DBMSs use a blend of prevention and avoidance techniques for other types of data, such as data warehouses or XML data.

10.4 CONCURRENCY CONTROL WITH TIME STAMPING METHODS

The **time stamping** approach to scheduling concurrent transactions assigns a global, unique time stamp to each transaction. The time stamp value produces an explicit order in which transactions are submitted to the DBMS. Time stamps must have two properties: uniqueness and monotonicity. **Uniqueness** ensures that no equal time stamp values can exist, and **monotonicity**[1] ensures that time stamp values always increase.

---

[1] The term *monotonicity* is part of the standard concurrency control vocabulary. The authors' first introduction to this term and its proper use was in an article written by W. H. Kohler, "A Survey of Techniques for Synchronization and Recovery in Decentralized Computer Systems," *Computer Surveys* 3(2), June 1981, pp. 149–283.

All database operations (READ and WRITE) within the same transaction must have the same time stamp. The DBMS executes conflicting operations in time stamp order, thereby ensuring serializability of the transactions. If two transactions conflict, one is stopped, rolled back, rescheduled, and assigned a new time stamp value.

The disadvantage of the time stamping approach is that each value stored in the database requires two additional time stamp fields: one for the last time the field was read and one for the last update. Time stamping thus increases memory needs and the database's processing overhead. Time stamping demands a lot of system resources because many transactions might have to be stopped, rescheduled, and restamped.

10.4.1   WAIT/DIE AND WOUND/WAIT SCHEMES

You have learned that time stamping methods are used to manage concurrent transaction execution. In this section, you will learn about two schemes used to decide which transaction is rolled back and which continues executing: the wait/die scheme and the wound/wait scheme.[2] An example illustrates the difference. Assume that you have two conflicting transactions: T1 and T2, each with a unique time stamp. Suppose T1 has a time stamp of 11548789 and T2 has a time stamp of 19562545. You can deduce from the time stamps that T1 is the older transaction (the lower time stamp value) and T2 is the newer transaction. Given that scenario, the four possible outcomes are shown in Table 10.14.

TABLE 10.14   Wait/Die and Wound/Wait Concurrency Control Schemes

TRANSACTION REQUESTING LOCK	TRANSACTION OWNING LOCK	WAIT/DIE SCHEME	WOUND/WAIT SCHEME
T1 (11548789)	T2 (19562545)	• T1 waits until T2 is completed and T2 releases its locks.	• T1 preempts (rolls back) T2. • T2 is rescheduled using the same time stamp.
T2 (19562545)	T1 (11548789)	• T2 dies (rolls back) • T2 is rescheduled using the same time stamp	• T2 waits until T1 is completed and T1 releases its locks.

Using the wait/die scheme:

- If the transaction requesting the lock is the older of the two transactions, it will *wait* until the other transaction is completed and the locks are released.
- If the transaction requesting the lock is the younger of the two transactions, it will *die* (roll back) and is rescheduled using the same time stamp.

In short, in the **wait/die** scheme, the older transaction waits for the younger to complete and release its locks.

In the wound/wait scheme:

- If the transaction requesting the lock is the older of the two transactions, it will preempt (*wound*) the younger transaction (by rolling it back). T1 preempts T2 when T1 rolls back T2. The younger, preempted transaction is rescheduled using the same time stamp.
- If the transaction requesting the lock is the younger of the two transactions, it will wait until the other transaction is completed and the locks are released.

In short, in the **wound/wait** scheme, the older transaction rolls back the younger transaction and reschedules it.

---

[2] The procedure was first described by R. E. Stearnes and P. M. Lewis II in "System-level Concurrency Control for Distributed Database Systems," *ACM Transactions on Database Systems*, No. 2, June 1978, pp. 178–198.

In both schemes, one of the transactions waits for the other transaction to finish and release the locks. However, in many cases, a transaction requests multiple locks. How long does a transaction have to wait for each lock request? Obviously, that scenario can cause some transactions to wait indefinitely, causing a deadlock. To prevent that type of deadlock, each lock request has an associated time-out value. If the lock is not granted before the time-out expires, the transaction is rolled back.

10.5 CONCURRENCY CONTROL WITH OPTIMISTIC METHODS

The **optimistic approach** is based on the assumption that the majority of the database operations do not conflict. The optimistic approach requires neither locking nor time stamping techniques. Instead, a transaction is executed without restrictions until it is committed. Using an optimistic approach, each transaction moves through two or three phases, referred to as *read*, *validation*, and *write*.[3]

- During the *read phase*, the transaction reads the database, executes the needed computations, and makes the updates to a private copy of the database values. All update operations of the transaction are recorded in a temporary update file, which is not accessed by the remaining transactions.

- During the *validation phase*, the transaction is validated to ensure that the changes made will not affect the integrity and consistency of the database. If the validation test is positive, the transaction goes to the write phase. If the validation test is negative, the transaction is restarted and the changes are discarded.

- During the *write phase*, the changes are permanently applied to the database.

The optimistic approach is acceptable for most read or query database systems that require few update transactions.

In a heavily used DBMS environment, the management of deadlocks—their prevention and detection—constitutes an important DBMS function. The DBMS will use one or more of the techniques discussed here, as well as variations on those techniques. However, the deadlock is sometimes worse than the disease that locks are supposed to cure. Therefore, it may be necessary to employ database recovery techniques to restore the database to a consistent state.

10.6 DATABASE RECOVERY MANAGEMENT

Database recovery restores a database from a given state (usually inconsistent) to a previously consistent state. Recovery techniques are based on the **atomic transaction property**: all portions of the transaction must be treated as a single, logical unit of work in which all operations are applied and completed to produce a consistent database. If, for some reason, any transaction operation cannot be completed, the transaction must be aborted and any changes to the database must be rolled back (undone). In short, transaction recovery reverses all of the changes that the transaction made to the database before the transaction was aborted.

Although this chapter has emphasized the recovery of *transactions*, recovery techniques also apply to the *database* and to the *system* after some type of critical error has occurred. Critical events can cause a database to become non-operational and compromise the integrity of the data. Examples of critical events are:

- *Hardware/software failures.* Failure of this type could be a hard disk media failure, a bad capacitor on a motherboard, or a failing memory bank. Other causes of errors under this category include application program or operating system errors that cause data to be overwritten, deleted, or lost. Some database administrators argue that this is one of the most common sources of database problems.

---

[3] The optimistic approach to concurrency control is described in an article by H. T. King and J. T. Robinson, "Optimistic Methods for Concurrency Control," *ACM Transactions on Database Systems* 6(2), June 1981, pp. 213–226. Even the most current software is built on conceptual standards that were developed more than two decades ago.

- *Human-caused incidents.* This type of event can be categorized as unintentional or intentional.
  - An unintentional failure is caused by carelessness by end-users. Such errors include deleting the wrong rows from a table, pressing the wrong key on the keyboard, or shutting down the main database server by accident.
  - Intentional events are of a more severe nature and normally indicate that the company data are at serious risk. Under this category are security threats caused by hackers trying to gain unauthorized access to data resources and virus attacks caused by disgruntled employees trying to compromise the database operation and damage the company.
- *Natural disasters.* This category includes fires, earthquakes, floods, and power failures.

Whatever the cause, a critical error can render the database in an inconsistent state. The following section introduces the various techniques used to recover the database from an inconsistent state to a consistent state.

10.6.1 TRANSACTION RECOVERY

In Section 10.1.4, you learned about the transaction log and how it contains data for database recovery purposes. Database transaction recovery uses data in the transaction log to recover a database from an inconsistent state to a consistent state.

Before continuing, let's examine four important concepts that affect the recovery process:

- The **write-ahead-log protocol** ensures that transaction logs are always written *before* any database data are actually updated. This protocol ensures that, in case of a failure, the database can later be recovered to a consistent state, using the data in the transaction log.
- **Redundant transaction logs** (several copies of the transaction log) ensure that a physical disk failure will not impair the DBMS's ability to recover data.
- Database **buffers** are temporary storage areas in primary memory used to speed up disk operations. To improve processing time, the DBMS software reads the data from the physical disk and stores a copy of it on a "buffer" in primary memory. When a transaction updates data, it actually updates the copy of the data in the buffer because that process is much faster than accessing the physical disk every time. Later on, all buffers that contain updated data are written to a physical disk during a single operation, thereby saving significant processing time.
- Database **checkpoints** are operations in which the DBMS writes all of its updated buffers to disk. While this is happening, the DBMS does not execute any other requests. A checkpoint operation is also registered in the transaction log. As a result of this operation, the physical database and the transaction log will be in sync. This synchronization is required because update operations update the copy of the data in the buffers and not in the physical database. Checkpoints are automatically scheduled by the DBMS several times per hour. As you will see next, checkpoints also play an important role in transaction recovery.

The database recovery process involves bringing the database to a consistent state after a failure. Transaction recovery procedures generally make use of deferred-write and write-through techniques.

When the recovery procedure uses a **deferred–write technique** (also called a **deferred update)**, the transaction operations do not immediately update the physical database. Instead, only the transaction log is updated. The database is physically updated only after the transaction reaches its commit point, using information from the transaction log. If the transaction aborts before it reaches its commit point, no changes (no ROLLBACK or undo) need to be made to the database because the database was never updated. The recovery process for all started and committed transactions (before the failure) follows these steps:

1. Identify the last checkpoint in the transaction log. This is the last time transaction data was physically saved to disk.

2. For a transaction that started and was committed before the last checkpoint, nothing needs to be done because the data are already saved.

3. For a transaction that performed a commit operation after the last checkpoint, the DBMS uses the transaction log records to redo the transaction and to update the database, using the "after" values in the transaction log. The changes are made in ascending order, from oldest to newest.

4. For any transaction that had a ROLLBACK operation after the last checkpoint or that was left active (with neither a COMMIT nor a ROLLBACK) before the failure occurred, nothing needs to be done because the database was never updated.

When the recovery procedure uses a **write-through technique** (also called an **immediate update**), the database is immediately updated by transaction operations during the transaction's execution, even before the transaction reaches its commit point. If the transaction aborts before it reaches its commit point, a ROLLBACK or undo operation needs to be done to restore the database to a consistent state. In that case, the ROLLBACK operation will use the transaction log "before" values. The recovery process follows these steps:

1. Identify the last checkpoint in the transaction log. This is the last time transaction data were physically saved to disk.

2. For a transaction that started and was committed before the last checkpoint, nothing needs to be done because the data are already saved.

3. For a transaction that was committed after the last checkpoint, the DBMS redoes the transaction, using the "after" values of the transaction log. Changes are applied in ascending order, from oldest to newest.

4. For any transaction that had a ROLLBACK operation after the last checkpoint or that was left active (with neither a COMMIT nor a ROLLBACK) before the failure occurred, the DBMS uses the transaction log records to ROLLBACK or undo the operations, using the "before" values in the transaction log. Changes are applied in reverse order, from newest to oldest.

Use the transaction log in Table 10.15 to trace a simple database recovery process. To make sure you understand the recovery process, a simple transaction log is used that includes three transactions and one checkpoint. This transaction log includes the transaction components used earlier in the chapter, so you should already be familiar with the basic process. Given the transaction, the transaction log has the following characteristics:

*   Transaction 101 consists of two UPDATE statements that reduce the quantity on hand for product 54778-2T and increase the customer balance for customer 10011 for a credit sale of two units of product 54778-2T.

*   Transaction 106 is the same credit sales event you saw in Section 10.1.1. This transaction represents the credit sale of one unit of product 89-WRE-Q to customer 10016 in the amount of $277.55. This transaction consists of five SQL DML statements: three INSERT statements and two UPDATE statements.

*   Transaction 155 represents a simple inventory update. This transaction consists of one UPDATE statement that increases the quantity on hand of product 2232/QWE from 6 units to 26 units.

*   A database checkpoint wrote all updated database buffers to disk. The checkpoint event writes only the changes for all previously committed transactions. In this case, the checkpoint applies all changes done by transaction 101 to the database data files.

TABLE 10.15 A Transaction Log for Transaction Recovery Examples

TRL ID	TRX NUM	PREV PTR	NEXT PTR	OPERATION	TABLE	ROW ID	ATTRIBUTE	BEFORE VALUE	AFTER VALUE
341	101	Null	352	START	****Start Transaction				
352	101	341	363	UPDATE	PRODUCT	54778-2T	PROD_QOH	45	43
363	101	352	365	UPDATE	CUSTOMER	10011	CUST_BALANCE	615.73	675.62
365	101	363	Null	COMMIT	**** End of Transaction				
397	106	Null	405	START	****Start Transaction				
405	106	397	415	INSERT	INVOICE	1009			1009,10016, ...
415	106	405	419	INSERT	LINE	1009,1			1009,1, 89-WRE-Q,1, ...
419	106	415	427	UPDATE	PRODUCT	89-WRE-Q	PROD_QOH	12	11
423				CHECKPOINT					
427	106	419	431	UPDATE	CUSTOMER	10016	CUST_BALANCE	0.00	277.55
431	106	427	457	INSERT	ACCT_TRANSACTION	10007			1007,18-JAN-2008, ...
457	106	431	Null	COMMIT	**** End of Transaction				
521	155	Null	525	START	****Start Transaction				
525	155	521	528	UPDATE	PRODUCT	2232/QWE	PROD_QOH	6	26
528	155	525	Null	COMMIT	**** End of Transaction				

*****C*R*A*S*H*****

Using Table 10.15, you can now trace the database recovery process for a DBMS, using the deferred update method as follows:

1. Identify the last checkpoint. In this case, the last checkpoint was TRL ID 423. This was the last time database buffers were physically written to disk.

2. Note that transaction 101 started and finished before the last checkpoint. Therefore, all changes were already written to disk, and no additional action needs to be taken.

3. For each transaction that committed after the last checkpoint (TRL ID 423), the DBMS will use the transaction log data to write the changes to disk, using the "after" values. For example, for transaction 106:

    a. Find COMMIT (TRL ID 457).

    b. Use the previous pointer values to locate the start of the transaction (TRL ID 397).

    c. Use the next pointer values to locate each DML statement and apply the changes to disk, using the "after" values. (Start with TRL ID 405, then 415, 419, 427 and 431.) Remember that TRL ID 457 was the COMMIT statement for this transaction.

    d. Repeat the process for transaction 155.

4. Any other transactions will be ignored. Therefore, for transactions that ended with ROLLBACK or that were left active (those that do not end with a COMMIT or ROLLBACK), nothing is done because no changes were written to disk.

SUMMARY

▶ A transaction is a sequence of database operations that access the database. A transaction represents a real-world event. A transaction must be a logical unit of work; that is, no portion of the transaction can exist by itself. Either all parts are executed or the transaction is aborted. A transaction takes a database from one consistent state to another. A consistent database state is one in which all data integrity constraints are satisfied.

▶ Transactions have four main properties: *atomicity* (all parts of the transaction are executed; otherwise, the transaction is aborted), *consistency* (the database's consistent state is maintained), *isolation* (data used by one transaction cannot be accessed by another transaction until the first transaction is completed), and *durability* (the changes made by a transaction cannot be rolled back once the transaction is committed). In addition, transaction schedules have the property of *serializability* (the result of the concurrent execution of transactions is the same as that of the transactions being executed in serial order).

▶ SQL provides support for transactions through the use of two statements: COMMIT (saves changes to disk) and ROLLBACK (restores the previous database state).

▶ SQL transactions are formed by several SQL statements or database requests. Each database request originates several I/O database operations.

▶ The transaction log keeps track of all transactions that modify the database. The information stored in the transaction log is used for recovery (ROLLBACK) purposes.

▶ Concurrency control coordinates the simultaneous execution of transactions. The concurrent execution of transactions can result in three main problems: lost updates, uncommitted data, and inconsistent retrievals.

▶ The scheduler is responsible for establishing the order in which the concurrent transaction operations are executed. The transaction execution order is critical and ensures database integrity in multiuser database systems. Locking, time stamping, and optimistic methods are used by the scheduler to ensure the serializability of transactions.

▶ A lock guarantees unique access to a data item by a transaction. The lock prevents one transaction from using the data item while another transaction is using it. There are several levels of locks: database, table, page, row, and field.

▶ Two types of locks can be used in database systems: binary locks and shared/exclusive locks. A binary lock can have only two states: locked (1) or unlocked (0). A shared lock is used when a transaction wants to read data from a database and no other transaction is updating the same data. Several shared or "read" locks can exist for a particular item. An exclusive lock is issued when a transaction wants to update (write to) the database and no other locks (shared or exclusive) are held on the data.

▶ Serializability of schedules is guaranteed through the use of two-phase locking. The two-phase locking schema has a growing phase, in which the transaction acquires all of the locks that it needs without unlocking any data, and a shrinking phase, in which the transaction releases all of the locks without acquiring new locks.

▶ When two or more transactions wait indefinitely for each other to release a lock, they are in a deadlock, also called a deadly embrace. There are three deadlock control techniques: prevention, detection, and avoidance.

▶ Concurrency control with time stamping methods assigns a unique time stamp to each transaction and schedules the execution of conflicting transactions in time-stamp order. Two schemes are used to decide which transaction is rolled back and which continues executing: the wait/die scheme and the wound/wait scheme.

▶ Concurrency control with optimistic methods assumes that the majority of database transactions do not conflict and that transactions are executed concurrently, using private, temporary copies of the data. At commit time, the private copies are updated to the database.

▶ Database recovery restores the database from a given state to a previous consistent state. Database recovery is triggered when a critical event occurs, such as a hardware error or application error.

KEY TERMS

atomicity, 417

atomic transaction property, 433

binary lock, 428

buffers, 434

checkpoints, 434

concurrency control, 420

consistency, 417

consistent database state, 414

database recovery, 433

database request, 414

deadlock, 430

deadly embrace, 430

deferred update, 434

deferred-write technique, 434

durability, 417

exclusive lock, 428

immediate update, 435

inconsistent retrievals, 422

isolation, 417

lock, 425

lock granularity, 425

   database-level lock, 425

   table-level lock, 426

   page-level lock, 426

   row-level lock, 427

   field-level lock, 428

lock manager, 425

lost updates, 420

monotonicity, 431

mutual exclusive rule, 429

optimistic approach, 433

page, 426

redundant transaction logs, 434

scheduler, 424

serializable schedule, 424

serializability, 417

shared lock, 428

time stamping, 431

transaction, 414

transaction log, 418

two-phase locking, 429

uncommitted data, 421

uniqueness, 431

wait/die, 432

wound/wait, 432

write-ahead-log protocol, 434

write-through technique, 435

ONLINE CONTENT

Answers to selected Review Questions and Problems for this chapter are contained in the Student Online Companion for this book.

REVIEW QUESTIONS

1. Explain the following statement: a transaction is a logical unit of work.

2. What is a consistent database state, and how is it achieved?

3. The DBMS does not guarantee that the semantic meaning of the transaction truly represents the real-world event. What are the possible consequences of that limitation? Give an example.

4. List and discuss the four transaction properties.

5. What is a transaction log, and what is its function?

6. What is a scheduler, what does it do, and why is its activity important to concurrency control?

7. What is a lock, and how, in general, does it work?

8. What is concurrency control, and what is its objective?

9. What is an exclusive lock, and under what circumstances is it granted?

10. What is a deadlock, and how can it be avoided? Discuss several strategies for dealing with deadlocks.

11. What are the three types of database critical events that can trigger the database recovery process? Give some examples for each one.

P R O B L E M S

1. Suppose you are a manufacturer of product ABC, which is composed of parts A, B, and C. Each time a new product ABC is created, it must be added to the product inventory, using the PROD_QOH in a table named PRODUCT. And each time the product is created, the parts inventory, using PART_QOH in a table named PART, must be reduced by one each of parts A, B, and C. The sample database contents are shown in the following tables.

TABLE
P10.1

TABLE NAME: PRODUCT	
PROD_CODE	PROD_QOH
ABC	1,205

TABLE NAME: PART	
PART_CODE	PART_QOH
A	567
B	98
C	549

Given that information, answer Questions a through e.

   a. How many database requests can you identify for an inventory update for both PRODUCT and PART?

   b. Using SQL, write each database request you identified in Step a.

   c. Write the complete transaction(s).

   d. Write the transaction log, using Table 10.1 as your template.

   e. Using the transaction log you created in Step d, trace its use in database recovery.

2. Describe the three most common concurrent transaction execution problems. Explain how concurrency control can be used to avoid those problems.

3. What DBMS component is responsible for concurrency control? How is this feature used to resolve conflicts?

4. Using a simple example, explain the use of binary and shared/exclusive locks in a DBMS.

5. Suppose your database system has failed. Describe the database recovery process and the use of deferred-write and write-through techniques.

ONLINE CONTENT

The **Ch10_ABC_Markets** database is located in the Student Online Companion for this book.

6. ABC Markets sell products to customers. The relational diagram shown in Figure P10.6 represents the main entities for ABC's database. Note the following important characteristics:

   • A customer may make many purchases, each one represented by an invoice.

      - The CUS_BALANCE is updated with each credit purchase or payment and represents the amount the customer owes.

      - The CUS_BALANCE is increased (+) with every credit purchase and decreased (–) with every customer payment.

      - The date of last purchase is updated with each new purchase made by the customer.

      - The date of last payment is updated with each new payment made by the customer.

FIGURE P10.6   The ABC Markets relational diagram

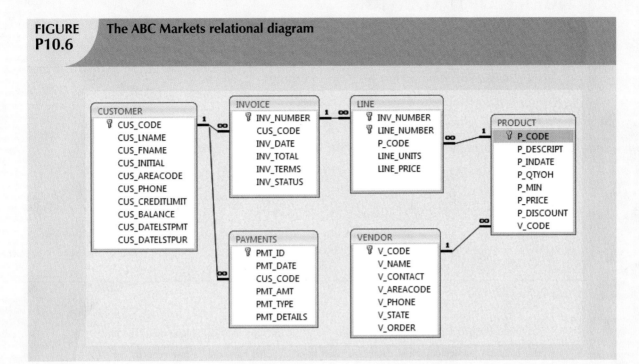

- An invoice represents a product purchase by a customer.

  - An INVOICE can have many invoice LINEs, one for each product purchased.

  - The INV_TOTAL represents the total cost of the invoice, including taxes.

  - The INV_TERMS can be "30," "60," or "90" (representing the number of days of credit) or "CASH," "CHECK," or "CC."

  - The invoice status can be "OPEN," "PAID," or "CANCEL."

- A product's quantity on hand (P_QTYOH) is updated (decreased) with each product sale.

- A customer may make many payments. The payment type (PMT_TYPE) can be one of the following:

  - "CASH" for cash payments.

  - "CHECK" for check payments.

  - "CC" for credit card payments.

- The payment details (PMT_DETAILS) are used to record data about check or credit card payments:

  - The bank, account number, and check number for check payments.

  - The issuer, credit card number, and expiration date for credit card payments.

Note: Not all entities and attributes are represented in this example. Use only the attributes indicated.

Using this database, write the SQL code to represent each of the following transactions. Use BEGIN TRANSACTION and COMMIT to group the SQL statements in logical transactions.

a. On May 11, 2008, customer 10010 makes a credit purchase (30 days) of one unit of product 11QER/31 with a unit price of $110.00; the tax rate is 8 percent. The invoice number is 10983, and this invoice has only one product line.

b. On June 3, 2008, customer 10010 makes a payment of $100 in cash. The payment ID is 3428.

c. Create a simple transaction log (using the format shown in Table 10.14) to represent the actions of the two previous transactions.

11

DATABASE PERFORMANCE TUNING AND QUERY OPTIMIZATION

ELEVEN

In this chapter, you will learn:

- Basic database performance-tuning concepts
- How a DBMS processes SQL queries
- About the importance of indexes in query processing
- About the types of decisions the query optimizer has to make
- Some common practices used to write efficient SQL code
- How to formulate queries and tune the DBMS for optimal performance

Database performance tuning is a critical topic, yet it usually receives minimal coverage in the database curriculum. Most databases used in classrooms have only a few records per table. As a result, the focus often is on making SQL queries perform an intended task, without considering the efficiency of the query process. In fact, even the most efficient query environment yields no visible performance improvements over the least efficient query environment when only 20 or 30 table rows (records) are queried. Unfortunately, that lack of attention to query efficiency can yield unacceptably slow results, when in the real world, queries are executed over tens of millions of records. In this chapter, you learn what it takes to create a more efficient query environment.

Preview

> **NOTE**
>
> Because this book focuses on databases, this chapter covers only those factors directly affecting *database* performance. Also, because performance-tuning techniques can be DBMS-specific, the material in this chapter might not be applicable under all circumstances, nor will it necessarily pertain to all DBMS types. This chapter is designed to build a foundation for the general understanding of database performance-tuning issues and to help you choose appropriate performance-tuning strategies. (For the most current information about tuning your database, consult the vendor's documentation.)

11.1 DATABASE PERFORMANCE-TUNING CONCEPTS

One of the main functions of a database system is to provide timely answers to end users. End users interact with the DBMS through the use of queries to generate information, using the following sequence:

1. The end-user (client-end) application generates a query.

2. The query is sent to the DBMS (server end).

3. The DBMS (server end) executes the query.

4. The DBMS sends the resulting data set to the end-user (client-end) application.

End users expect their queries to return results as quickly as possible. How do you know that the performance of a database is good? Good database performance is hard to evaluate. How do you know if a 1.06-second query response time is good enough? It's easier to identify bad database performance than good database performance—all it takes is end-user complaints about slow query results. Unfortunately, the same query might perform well one day and not so well two months later. Regardless of end-user perceptions, *the goal of database performance is to execute queries as fast as possible*. Therefore, database performance must be closely monitored and regularly tuned. **Database performance tuning** refers to a set of activities and procedures designed to reduce the response time of the database system—that is, to ensure that an end-user query is processed by the DBMS in the minimum amount of time.

The time required by a query to return a *result set* depends on many factors. Those factors tend to be wide-ranging and to vary from environment to environment and from vendor to vendor. The performance of a typical DBMS is constrained by three main factors: CPU processing power, available primary memory (RAM), and input/output (hard disk and network) throughput. Table 11.1 lists some system components and summarizes general guidelines for achieving better query performance.

TABLE 11.1   General Guidelines for Better System Performance

	SYSTEM RESOURCES	CLIENT	SERVER
Hardware	CPU	The fastest possible Dual-core CPU or higher	The fastest possible Multiple processors (Quad-core technology)
	RAM	The maximum possible	The maximum possible
	Hard Disk	Fast SATA/EIDE hard disk with sufficient free hard disk space	Multiple high-speed, high-capacity hard disks (SCSI / SATA / Firewire / Fibre Channel) in RAID configuration
	Network	High-speed connection	High-speed connection
Software	Operating System	Fine-tuned for best client application performance	Fine-tuned for best server application performance
	Network	Fine-tuned for best throughput	Fine-tuned for best throughput
	Application	Optimize SQL in client application	Optimize DBMS server for best performance

Naturally, the system will perform best when its hardware and software resources are optimized. However, in the real world, unlimited resources are not the norm; internal and external constraints always exist. Therefore, the system components should be optimized to obtain the best throughput possible with existing (and often limited) resources, which is why database performance tuning is important.

Fine-tuning the performance of a system requires a holistic approach. That is, *all* factors must be checked to ensure that each one operates at its optimum level and has sufficient resources to minimize the occurrence of bottlenecks. Because database design is such an important factor in determining the database system's performance efficiency, it is worth repeating this book's mantra:

Good database performance starts with good database design. *No amount of fine tuning will make a poorly designed database perform as well as a well-designed database.* This is particularly true in the case of redesigning existing databases, where the end user expects unrealistic performance gains from older databases.

What constitutes a good, efficient database design? From the performance tuning point of view, the database designer must ensure that the design makes use of the database features available in the DBMS that guarantee the integrity and optimal performance of the database. This chapter provides you with fundamental knowledge that will help you to optimize database performance by selecting the appropriate database server configuration, utilizing indexes, understanding table storage organization and data locations, and implementing the most efficient SQL query syntax.

11.1.1 PERFORMANCE TUNING: CLIENT AND SERVER

In general, database performance-tuning activities can be divided into those taking place on the client side and those taking place on the server side.

- On the client side, the objective is to generate a SQL query that returns the correct answer in the least amount of time, using the minimum amount of resources at the server end. The activities required to achieve that goal are commonly referred to as **SQL performance tuning**.

- On the server side, the DBMS environment must be properly configured to respond to clients' requests in the fastest way possible, while making optimum use of existing resources. The activities required to achieve that goal are commonly referred to as **DBMS performance tuning**.

ONLINE CONTENT

If you want to learn more about clients and servers, check **Appendix F, Client/Server Systems,** located in the Student Online Companion for this book.

Keep in mind that DBMS implementations are typically more complex than just a two-tier client/server configuration. However, even in multi-tier (client front-end, application middleware, and database server back-end) client/server environments, performance-tuning activities are frequently divided into subtasks to ensure the fastest possible response time between any two component points.

This chapter covers SQL performance-tuning practices on the client side and DBMS performance-tuning practices on the server side. But before you can start learning about the tuning processes, you must first learn more about the DBMS architectural components and processes and how those processes interact to respond to end-users requests.

11.1.2 DBMS ARCHITECTURE

The architecture of a DBMS is represented by the processes and structures (in memory and in permanent storage) used to manage a database. Such processes collaborate with one another to perform specific functions. Figure 11.1 illustrates the basic DBMS architecture.

FIGURE 11.1   Basic DBMS architecture

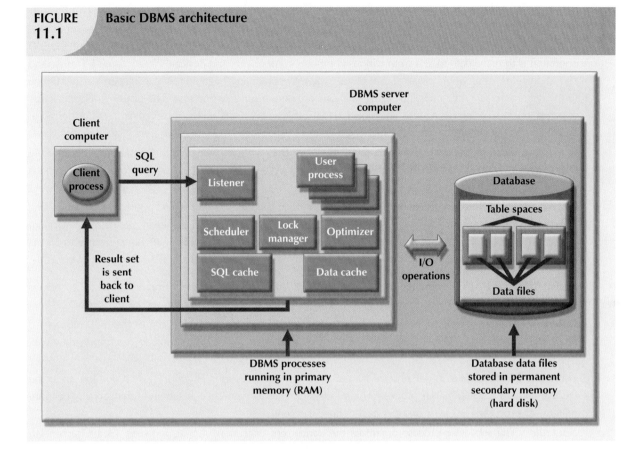

Note the following components and functions in Figure 11.1:

- All data in a database are stored in **data files**. A typical enterprise database is normally composed of several data files. A data file can contain rows from one single table, or it can contain rows from many different tables. A database administrator (DBA) determines the initial size of the data files that make up the database; however, as required, the data files can automatically expand in predefined increments known as **extends**. For example, if more space is required, the DBA can define that each new extend will be in 10 KB or 10 MB increments.

- Data files are generally grouped in file groups or table spaces. A **table space** or **file group** is a logical grouping of several data files that store data with similar characteristics. For example, you might have a *system* table space where the data dictionary table data are stored; a *user data* table space to store the user-created tables; an *index* table space to hold all indexes; and a *temporary* table space to do temporary sorts, grouping, and so on. Each time you create a new database, the DBMS automatically creates a minimum set of table spaces.

- The **data cache** or **buffer cache** is a shared, reserved memory area that stores the most recently accessed data blocks in RAM. The data cache is where the data read from the database data files are stored after the data have been read or before the data are written to the database data files. The data cache also caches system catalog data and the contents of the indexes.

- The **SQL cache** or **procedure cache** is a shared, reserved memory area that stores the most recently executed SQL statements or PL/SQL procedures, including triggers and functions. (To learn more about PL/SQL procedures, triggers, and SQL functions, study Chapter 8, Advanced SQL.) The SQL cache does not store the end-user written SQL. Rather, the SQL cache stores a "processed" version of the SQL that is ready for execution by the DBMS.

- To work with the data, the DBMS must retrieve the data from permanent storage (data files in which the data are stored) and place it in RAM (data cache).

- To move data from the permanent storage (data files) to the RAM (data cache), the DBMS issues I/O requests and waits for the replies. An **input/output (I/O) request** is a low-level (read or write) data access operation to and from computer devices, such as memory, hard disks, video, and printers. The purpose of the I/O operation is to move data to and from various computer components and devices. Note that an I/O disk read operation retrieves an entire physical disk block, generally containing multiple rows, from permanent storage to the data cache, even if you will be using only one attribute from only one row. The physical disk block size depends on the operating system and could be 4K, 8K, 16K, 32K, 64K, or even larger. Furthermore, depending on the circumstances, a DBMS might issue a single-block read request or a multiblock read request.

- Working with data in the data cache is many times faster than working with data in the data files because the DBMS doesn't have to wait for the hard disk to retrieve the data. This is because no hard disk I/O operations are needed to work within the data cache.

- The majority of performance-tuning activities focus on minimizing the number of I/O operations because using I/O operations is many times slower than reading data from the data cache. For example, as of this writing, RAM access times range from 5 to 70 ns (nanoseconds), while hard disk access times range from 5 to 15 ms (milliseconds). This means that hard disks are about six orders of magnitude (a million times) slower than RAM.[1]

Also illustrated in Figure 11.1 are some typical DBMS processes. Although the number of processes and their names vary from vendor to vendor, the functionality is similar. The following processes are represented in Figure 11.1:

- *Listener*. The listener process listens for clients' requests and handles the processing of the SQL requests to other DBMS processes. Once a request is received, the listener passes the request to the appropriate user process.

- *User*. The DBMS creates a user process to manage each client session. Therefore, when you log on to the DBMS, you are assigned a user process. This process handles all requests you submit to the server. There are many user processes—at least one per each logged-in client.

- *Scheduler*. The scheduler process organizes the concurrent execution of SQL requests. (See Chapter 10, Transaction Management and Concurrency Control.)

- *Lock manager*. This process manages all locks placed on database objects, including disk pages. (See Chapter 10.)

- *Optimizer*. The optimizer process analyzes SQL queries and finds the most efficient way to access the data. You will learn more about this process later in the chapter.

11.1.3 DATABASE STATISTICS

Another DBMS process that plays an important role in query optimization is gathering database statistics. The term **database statistics** refers to a number of measurements about database objects, such as number of processors used, processor speed, and temporary space available. Such statistics give a snapshot of database characteristics.

As you will learn later in this chapter, the DBMS uses these statistics to make critical decisions about improving query processing efficiency. Database statistics can be gathered manually by the DBA or automatically by the DBMS. For example, many DBMS vendors support the ANALYZE command in SQL to gather statistics. In addition, many vendors have their own routines to gather statistics. For example, IBM's DB2 uses the RUNSTATS procedure, while Microsoft's SQL Server uses the UPDATE STATISTICS procedure and provides the Auto-Update and Auto-Create Statistics options in its initialization parameters. A sample of measurements that the DBMS may gather about various database objects is shown in Table 11.2.

---

[1] Low Latency, Eliminating Application Jitters with Solaris, White Paper, May 2007, Sun Microsystems, http://www.sun.com/solutions/documents/white-papers/fn_lowlatency_solaris.pdf.

TABLE 11.2	Sample Database Statistics Measurements	
DATABASE OBJECT	**SAMPLE MEASUREMENTS**
Tables	Number of rows, number of disk blocks used, row length, number of columns in each row, number of distinct values in each column, maximum value in each column, minimum value in each column, and columns that have indexes
Indexes	Number and name of columns in the index key, number of key values in the index, number of distinct key values in the index key, histogram of key values in an index, and number of disk pages used by the index
Environment Resources	Logical and physical disk block size, location and size of data files, and number of extends per data file

If the object statistics exist, the DBMS will use them in query processing. Although some of the newer DBMSs (such as Oracle, SQL Server, and DB2) automatically gather statistics, others require the DBA to gather statistics manually. To generate the database object statistics manually, you could use the following syntax:

ANALYZE <TABLE/INDEX> object_name COMPUTE STATISTICS;

(In SQL Server, use UPDATE STATISTICS <object_name>, where the object_name refers to a table or a view.)

For example, to generate statistics for the VENDOR table, you would use the following command:

ANALYZE TABLE VENDOR COMPUTE STATISTICS;

(In SQL Server, use UPDATE STATISTICS VENDOR;.)

When you generate statistics for a table, all related indexes are also analyzed. However, you could generate statistics for a single index by using the following command:

ANALYZE INDEX VEND_NDX COMPUTE STATISTICS;

In the above example, VEND_NDX is the name of the index.

(In SQL Server, use UPDATE STATISTICS <table_name> <index_name>. For example: UPDATE STATISTICS VENDOR VEND_NDX;)

Database statistics are stored in the system catalog in specially designated tables. It is common to periodically regenerate the statistics for database objects, especially those database objects that are subject to frequent change. For example, if you are the owner of a video store and you have a video rental DBMS, your system will likely use a RENTAL table to store the daily video rentals. That RENTAL table (and its associated indexes) would be subject to constant inserts and updates as you record your daily rentals and returns. Therefore, the RENTAL table statistics you generated last week do not depict an accurate picture of the table as it exists today. The more current the statistics, the better the chances are for the DBMS to properly select the fastest way to execute a given query.

Now that you know the basic architecture of DBMS processes and memory structures, and the importance and timing of the database statistics gathered by the DBMS, you are ready to learn how the DBMS processes a SQL query request.

11.2 QUERY PROCESSING

What happens at the DBMS server end when the client's SQL statement is received? In simple terms, the DBMS processes a query in three phases:

1. *Parsing.* The DBMS parses the SQL query and chooses the most efficient access/execution plan.

2. *Execution.* The DBMS executes the SQL query using the chosen execution plan.

3. *Fetching.* The DBMS fetches the data and sends the result set back to the client.

The processing of SQL DDL statements (such as CREATE TABLE) is different from the processing required by DML statements. The difference is that a DDL statement actually updates the data dictionary tables or system catalog, while a DML statement (SELECT, INSERT, UPDATE, and DELETE) mostly manipulates end-user data. Figure 11.2 shows the general steps required for query processing. Each of the steps will be discussed in the following sections.

FIGURE 11.2   Query processing

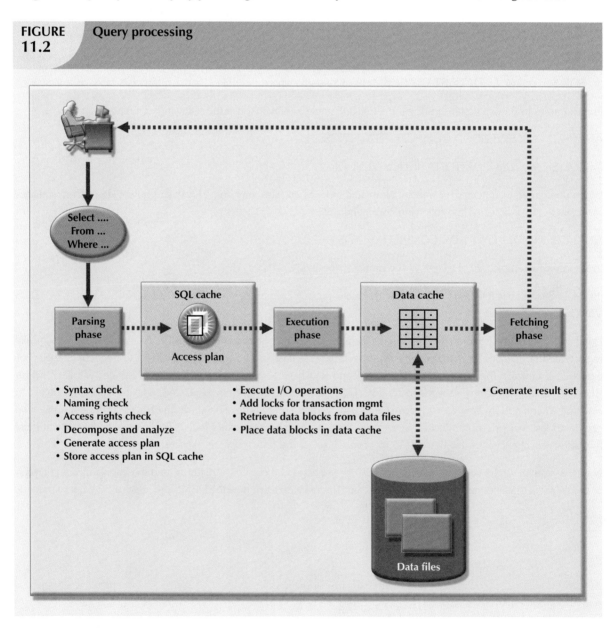

11.2.1 SQL PARSING PHASE

The optimization process includes breaking down—parsing—the query into smaller units and transforming the original SQL query into a slightly different version of the original SQL code, but one that is fully equivalent and more efficient. *Fully equivalent* means that the optimized query results are always the same as the original query. *More efficient* means that the optimized query will almost always execute faster than the original query. (Note that it *almost* always executes faster because, as explained earlier, many factors affect the performance of a database. Those factors include the network, the client computer's resources, and other queries running concurrently in the same database.) To determine the most efficient way to execute the query, the DBMS may use the database statistics you learned about earlier.

The SQL parsing activities are performed by the **query optimizer**, which analyzes the SQL query and finds the most efficient way to access the data. This process is the most time-consuming phase in query processing. Parsing a SQL query requires several steps, in which the SQL query is:

- Validated for syntax compliance.
- Validated against the data dictionary to ensure that tables and column names are correct.
- Validated against the data dictionary to ensure that the user has proper access rights.
- Analyzed and decomposed into more atomic components.
- Optimized through transformation into a fully equivalent but more efficient SQL query.
- Prepared for execution by determining the most efficient execution or access plan.

Once the SQL statement is transformed, the DBMS creates what is commonly known as an access or execution plan. An **access plan** is the result of parsing an SQL statement; it contains the series of steps a DBMS will use to execute the query and to return the result set in the most efficient way. First, the DBMS checks to see if an access plan already exists for the query in the SQL cache. If it does, the DBMS reuses the access plan to save time. If it doesn't, the optimizer evaluates various plans and makes decisions about what indexes to use and how to best perform join operations. The chosen access plan for the query is then placed in the SQL cache and made available for use and future reuse.

Access plans are DBMS-specific and translate the client's SQL query into the series of complex I/O operations required to read the data from the physical data files and generate the result set. Access plans are DBMS-specific; some commonly found I/O operations are illustrated in Table 11.3.

TABLE 11.3 **Sample DBMS Access Plan I/O Operations**

OPERATION	DESCRIPTION
Table Scan (Full)	Reads the entire table sequentially, from the first row to the last row, one row at a time (slowest)
Table Access (Row ID)	Reads a table row directly, using the row ID value (fastest)
Index Scan (Range)	Reads the index first to obtain the row IDs and then accesses the table rows directly (faster than a full table scan)
Index Access (Unique)	Used when a table has a unique index in a column
Nested Loop	Reads and compares a set of values to another set of values, using a nested loop style (slow)
Merge	Merges two data sets (slow)
Sort	Sorts a data set (slow)

Table 11.3 shows just a few database access I/O operations. (This illustration is based on an Oracle RDBMS.) However, Table 11.3 does illustrate the type of I/O operations that most DBMSs perform when accessing and manipulating data sets.

In Table 11.3, note that a table access using a row ID is the fastest method. A row ID is a unique identification for *every* row saved in permanent storage; it can be used to access the row directly. Conceptually, a row ID is similar to a parking slip you get when you park your car in an airport parking lot. The parking slip contains the section number and lot number. Using that information, you can go directly to your car without having to go through every section and lot.

11.2.2 SQL EXECUTION PHASE

In this phase, all I/O operations indicated in the access plan are executed. When the execution plan is run, the proper locks —if needed—are acquired for the data to be accessed, and the data are retrieved from the data files and placed in the DBMSs data cache. All transaction management commands are processed during the parsing and execution phases of query processing.

11.2.3 SQL FETCHING PHASE

After the parsing and execution phases are completed, all rows that match the specified condition(s) are retrieved, sorted, grouped, and/or aggregated (if required). During the fetching phase, the rows of the resulting query result set are returned to the client. The DBMS might use temporary table space to store temporary data. In this stage, the database server coordinates the movement of the result set rows from the server cache to the client cache. For example, a given query result set might contain 9,000 rows; the server would send the first 100 rows to the client and then wait for the client to request the next set of rows, until the entire result set is sent to the client.

11.2.4 QUERY PROCESSING BOTTLENECKS

The main objective of query processing is to execute a given query in the fastest way possible with the least amount of resources. As you have seen, the execution of a query requires the DBMS to break down the query into a series of interdependent I/O operations to be executed in a collaborative manner. The more complex a query is, the more complex the operations are, and the more likely it is that there will be bottlenecks. A **query processing bottleneck** is a delay introduced in the processing of an I/O operation that causes the overall system to slow down. In the same way, the more components a system has, the more interfacing among the components is required, and the more likely it is that there will be bottlenecks. Within a DBMS, there are five components that typically cause bottlenecks:

- *CPU.* The CPU processing power of the DBMS should match the system's expected work load. A high CPU utilization might indicate that the processor speed is too slow for the amount of work performed. However, heavy CPU utilization can be caused by other factors, such as a defective component, not enough RAM (the CPU spends too much time swapping memory blocks), a badly written device driver, or a rogue process. A CPU bottleneck will affect not only the DBMS but all processes running in the system.
- *RAM.* The DBMS allocates memory for specific usage, such as data cache and SQL cache. RAM must be shared among all running processes (operating system, DBMS, and all other running processes). If there is not enough RAM available, moving data among components that are competing for scarce RAM can create a bottleneck.
- *Hard disk.* Another common cause of bottlenecks is hard disk speed and data transfer rates. Current hard disk storage technology allows for greater storage capacity than in the past; however, hard disk space is used for more than just storing end user data. Current operating systems also use the hard disk for *virtual memory*, which refers to copying areas of RAM to the hard disk as needed to make room in RAM for more urgent tasks. Therefore, the greater the hard disk storage space is and the faster the data transfer rates are, the lesser likelihood of bottlenecks.
- *Network.* In a database environment, the database server and the clients are connected via a network. All networks have a limited amount of bandwidth that is shared among all clients. When many network nodes access the network at the same time, bottlenecks are likely.
- *Application code.* Not all bottlenecks are caused by limited hardware resources. One of the most common sources of bottlenecks is badly written application code. No amount of coding will make a poorly designed database perform better. We should also add: you can throw unlimited resources at a badly written application and it will still perform as a badly written application!

Learning how to avoid these bottlenecks and thus optimize database performance is the main focus of this chapter.

11.3 INDEXES AND QUERY OPTIMIZATION

Indexes are crucial in speeding up data access because they facilitate searching, sorting, and using aggregate functions and even join operations. The improvement in data access speed occurs because an index is an ordered set of values that contains the index key and pointers. The pointers are the row IDs for the actual table rows. Conceptually, a data index is similar to a book index. When you use a book index, you look up the word, similar to the index key, which is accompanied by the page number(s), similar to the pointer(s), which direct you to the appropriate page(s).

An index scan is more efficient than a full table scan because the index data are preordered and the amount of data is usually much smaller. Therefore, when performing searches, it is almost always better for the DBMS to use the index to access a table than to scan all rows in a table sequentially. For example, Figure 11.3 shows the index representation of a CUSTOMER table with 14,786 rows and the index STATE_NDX on the CUS_STATE attribute.

FIGURE 11.3   **Index representation for the CUSTOMER table**

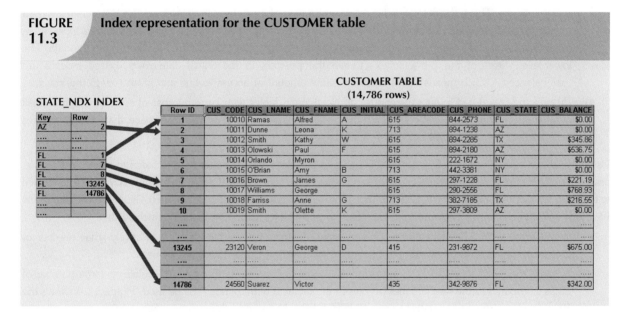

Suppose you submit the following query:

```
SELECT      CUS_NAME, CUS_STATE
FROM        CUSTOMER
WHERE       CUS_STATE = 'FL';
```

If there is no index, the DBMS will perform a full table scan, thus reading all 14,786 customer rows. Assuming that the index STATE_NDX is created (and ANALYZED), the DBMS will automatically use the index to locate the first customer with a state equal to 'FL' and then proceed to read all subsequent CUSTOMER rows, using the row IDs in the index as a guide. Assuming that only five rows meet the condition CUS_STATE = 'FL' then, there are 5 accesses to the index and 5 accesses to the data, for a total of 10 I/O accesses. The DBMS would save approximately 14,776 I/O requests for customer rows that do not meet the criteria. That's a lot of CPU cycles!

If indexes are so important, why not index every column in every table? It's not practical to do so. Indexing every column in every table taxes the DBMS too much in terms of index-maintenance processing, especially if the table has many attributes; has many rows; and/or requires many inserts, updates, and/or deletes.

One measure that determines the need for an index is the data *sparsity* of the column you want to index. **Data sparsity** refers to the number of different values a column could possibly have. For example, a STU_SEX column in a STUDENT table can have only two possible values, M or F; therefore that column is said to have low sparsity. In contrast, the STU_DOB column that stores the student date of birth can have many different date values; therefore, that column is said to have high sparsity. Knowing the sparsity helps you decide whether the use of an index is appropriate. For example, when you perform a search in a column with low sparsity, you are likely to read a high percentage of the table rows anyway; therefore, index processing might be unnecessary work. In Section 11.5, you learn how to determine when an index is recommended.

Most DBMSs implement indexes using one of the following data structures:

- *Hash indexes.* A hash algorithm is used to create a hash value from a key column. This value points to an entry in a hash table which in turn points to the actual location of the data row. This type of index is good for simple and fast lookup operations.

- *B-tree indexes.* This is the default and most common type of index used in databases. The B-tree index is used mainly in tables in which column values repeat a relative smaller number of times. The B-tree index is an ordered data structure organized as an upside down tree. The index tree is stored separate from the data. The lower-level leaves of the B-tree index contain the pointers to the actual data rows. B-tree indexes are "self-balanced," which means that it takes the same amount of access to find any given row in the index.

- *Bitmap indexes.* Used in data warehouse applications in tables with large number of rows in which a small number of column values repeat many times. Bitmap indexes tend to use less space than B-tree indexes because they use bits (instead of bytes) to store their data.

Using the above index characteristics, a database designer can determine the best type of index to use. For example, assume a CUSTOMER table with several thousand rows. The CUSTOMER table has two columns that are used extensively for query purposes: CUS_LNAME that represents a customer last name and REGION_CODE that could have one of four values (NE, NW, SW, and, SE). Based on this information, you could conclude that:

- Because the CUS_LNAME column contains many different values that repeat a relatively small number of times (compared to the total number of rows in the table), a B-tree index will be used.

- Because the REGION_CODE column contains fewer different values that repeat a relatively large number of times (compared to the total number of rows in the table), a bitmap index will be used. Figure 11.4 shows the B-tree and bitmap representations for a CUSTOMER table used in the previous discussion.

Current generation DBMSs are intelligent enough to determine the best type of index to use under certain circumstances (provided the DBMS has updated database statistics). Whatever the index chosen, the DBMS determines the best plan to execute a given query. The next section guides you through a simplified example of the type of choices that the query optimizer must perform.

FIGURE 11.4   B-tree and bitmap index representation

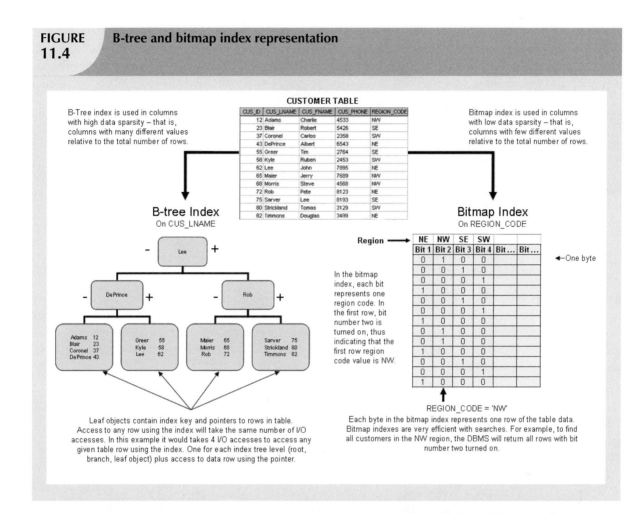

B-tree Index
On CUS_LNAME

Leaf objects contain index key and pointers to rows in table.
Access to any row using the index will take the same number of I/O
accesses. In this example it would takes 4 I/O accesses to access any
given table row using the index. One for each index tree level (root,
branch, leaf object) plus access to data row using the pointer.

Bitmap Index
On REGION_CODE

In the bitmap index, each bit represents one region code. In the first row, bit number two is turned on, thus indicating that the first row region code value is NW.

REGION_CODE = 'NW'

Each byte in the bitmap index represents one row of the table data.
Bitmap indexes are very efficient with searches. For example, to find
all customers in the NW region, the DBMS will return all rows with bit
number two turned on.

11.4 OPTIMIZER CHOICES

Query optimization is the central activity during the parsing phase in query processing. In this phase, the DBMS must choose what indexes to use, how to perform join operations, what table to use first, and so on. Each DBMS has its own algorithms for determining the most efficient way to access the data. The query optimizer can operate in one of two modes:

- A **rule-based optimizer** uses preset rules and points to determine the best approach to execute a query. The rules assign a "fixed cost" to each SQL operation; the costs are then added to yield the cost of the execution plan. For example, a full table scan has a set cost of 10, while a table access by row ID has a set cost of 3.

- A **cost-based optimizer** uses sophisticated algorithms based on the statistics about the objects being accessed to determine the best approach to execute a query. In this case, the optimizer process adds up the processing cost, the I/O costs, and the resource costs (RAM and temporary space) to come up with the total cost of a given execution plan.

The optimizer objective is to find alternative ways to execute a query—to evaluate the "cost" of each alternative and then to choose the one with the lowest cost. To understand the function of the query optimizer, let's use a simple example. Assume that you want to list all products provided by a vendor based in Florida. To acquire that information, you could write the following query:

```
SELECT     P_CODE, P_DESCRIPT, P_PRICE, V_NAME, V_STATE
FROM       PRODUCT, VENDOR
WHERE      PRODUCT.V_CODE = VENDOR.V_CODE
           AND VENDOR.V_STATE = 'FL';
```

Furthermore, let's assume that the database statistics indicate that:

- The PRODUCT table has 7,000 rows.
- The VENDOR table has 300 rows.
- Ten vendors are located in Florida.
- One thousand products come from vendors in Florida.

It's important to point out that only the first two items are available to the optimizer. The second two items are assumed to illustrate the choices that the optimizer must make. Armed with the information in the first two items, the optimizer would try to find the most efficient way to access the data. The primary factor in determining the most efficient access plan is the I/O cost. (Remember, the DBMS always tries to minimize I/O operations.) Table 11.4 shows two sample access plans for the previous query and their respective I/O costs.

TABLE 11.4   Comparing Access Plans and I/O Costs

PLAN	STEP	OPERATION	I/O OPERATIONS	I/O COST	RESULTING SET ROWS	TOTAL I/O COST
A	A1	Cartesian product (PRODUCT, VENDOR)	7,000 + 300	7,300	2,100,000	7,300
	A2	Select rows in A1 with matching vendor codes	2,100,000	2,100,000	7,000	2,107,300
	A3	Select rows in A2 with V_STATE = 'FL'	7,000	7,000	1,000	**2,114,300**
B	B1	Select rows in VENDOR with V_STATE = 'FL'	300	300	10	300
	B2	Cartesian Product (PRODUCT, B1)	7,000 + 10	7,010	70,000	7,310
	B3	Select rows in B2 with matching vendor codes	70,000	70,000	1,000	**77,310**

To make the example easier to understand, the I/O Operations and I/O Cost columns in Table 11.4 estimate only the number of I/O disk reads the DBMS must perform. For simplicity's sake, it is assumed that there are no indexes and that each row read has an I/O cost of 1. For example, in step A1, the DBMS must perform a Cartesian product of PRODUCT and VENDOR. To do that, the DBMS must read all rows from PRODUCT (7,000) and all rows from VENDOR (300), yielding a total of 7,300 I/O operations. The same computation is done in all steps. In Table 11.4, you can see how plan A has a total I/O cost that is almost 30 times higher than plan B. In this case, the optimizer will choose plan B to execute the SQL.

> **NOTE**
>
> Not all DBMSs optimize SQL queries the same way. As a matter of fact, Oracle parses queries differently than what is described in several sections in this chapter. Always read the documentation to examine the optimization requirements for your DBMS implementation.

Given the right conditions, some queries could be answered entirely by using only an index. For example, assume the PRODUCT table and the index P_QOH_NDX in the P_QOH attribute. Then a query such as SELECT MIN(P_QOH) FROM PRODUCT could be resolved by reading only the first entry in the P_QOH_NDX index, without the need to access any of the data blocks for the PRODUCT table. (Remember that the index defaults to ascending order.)

You learned in Section 11.3 that columns with low sparsity are not good candidates for index creation. However, there are cases where an index in a low sparsity column would be helpful. For example, assume that the EMPLOYEE table has 122,483 rows. If you want to find out how many female employees are in the company, you would write a query such as:

SELECT COUNT(EMP_SEX) FROM EMPLOYEE WHERE EMP_SEX = 'F';

If you do not have an index for the EMP_SEX column, the query would have to perform a full table scan to read all EMPLOYEE rows—and each full row includes attributes you do not need. However, if you have an index on EMP_SEX, the query could be answered by reading only the index data, without the need to access the employee data at all.

11.4.1 USING HINTS TO AFFECT OPTIMIZER CHOICES

Although the optimizer generally performs very well under most circumstances, in some instances the optimizer might not choose the best execution plan. Remember, the optimizer makes decisions based on the existing statistics. If the statistics are old, the optimizer might not do a good job in selecting the best execution plan. Even with current statistics, the optimizer choice might not be the most efficient one. There are some occasions when the end user would like to change the optimizer mode for the current SQL statement. In order to do that, you need to use hints. **Optimizer hints** are special instructions for the optimizer that are embedded inside the SQL command text. Table 11.5 summarizes a few of the most common optimizer hints used in standard SQL.

TABLE 11.5	Optimizer Hints
HINT	**USAGE**
ALL_ROWS	Instructs the optimizer to minimize the overall execution time, that is, to minimize the time it takes to return all rows in the query result set. This hint is generally used for batch mode processes. For example: SELECT /*+ ALL_ROWS */ * FROM PRODUCT WHERE P_QOH < 10;
FIRST_ROWS	Instructs the optimizer to minimize the time it takes to process the first set of rows, that is, to minimize the time it takes to return only the first set of rows in the query result set. This hint is generally used for interactive mode processes. For example: SELECT /*+ FIRST_ROWS */ * FROM PRODUCT WHERE P_QOH < 10;
INDEX(name)	Forces the optimizer to use the P_QOH_NDX index to process this query. For example: SELECT /*+ INDEX(P_QOH_NDX) */ * FROM PRODUCT WHERE P_QOH < 10;

Now that you are familiar with the way the DBMS processes SQL queries, let's turn our attention to some general SQL coding recommendations to facilitate the work of the query optimizer.

11.5 SQL PERFORMANCE TUNING

SQL performance tuning is evaluated from the client perspective. Therefore, the goal is to illustrate some common practices used to write efficient SQL code. A few words of caution are appropriate:

1. Most current-generation relational DBMSs perform automatic query optimization at the server end.

2. Most SQL performance optimization techniques are DBMS-specific, and therefore, are rarely portable, even across different versions of the same DBMS. Part of the reason for this behavior is the constant advancement in database technologies.

Does this mean that you should not worry about how a SQL query is written because the DBMS will always optimize it? No, because there is considerable room for improvement. (The DBMS uses *general* optimization techniques, rather than focusing on specific techniques dictated by the special circumstances of the query execution.) A poorly written SQL query can, *and usually will*, bring the database system to its knees from a performance point of view. The majority of current database performance problems are related to poorly written SQL code. Therefore, although a DBMS provides general optimizing services, a carefully written query almost always outperforms a poorly written one.

Although SQL data manipulation statements include many different commands (such as INSERT, UPDATE, DELETE, and SELECT), most recommendations in this section are related to the use of the SELECT statement, and in particular, the use of indexes and how to write conditional expressions.

11.5.1 INDEX SELECTIVITY

Indexes are the most important technique used in SQL performance optimization. The key is to know when an index is used. As a general rule, indexes are likely to be used:

- When an indexed column appears by itself in a search criteria of a WHERE or HAVING clause.
- When an indexed column appears by itself in a GROUP BY or ORDER BY clause.
- When a MAX or MIN function is applied to an indexed column.
- When the data sparsity on the indexed column is high.

Indexes are very useful when you want to select a small subset of rows from a large table based on a given condition. If an index exists for the column used in the selection, the DBMS may choose to use it. The objective is to create indexes with high selectivity. **Index selectivity** is a measure of how likely an index will be used in query processing. Here are some general guidelines for creating and using indexes:

- *Create indexes for each single attribute used in a WHERE, HAVING, ORDER BY, or GROUP BY clause.* If you create indexes in all single attributes *used in search conditions*, the DBMS will access the table using an index scan instead of a full table scan. For example, if you have an index for P_PRICE, the condition P_PRICE > 10.00 can be solved by accessing the index instead of sequentially scanning all table rows and evaluating P_PRICE for each row. Indexes are also used in join expressions, such as in CUSTOMER.CUS_CODE = INVOICE.CUS_CODE.

- *Do not use indexes in small tables or tables with low sparsity.* Remember, small tables and low-sparsity tables are not the same thing. A search condition in a table with low sparsity may return a high percentage of table rows anyway, making the index operation too costly and making the full table scan a viable option. Using the same logic, do not create indexes for tables with few rows and few attributes—*unless you must ensure the existence of unique values in a column.*

- *Declare primary and foreign keys so the optimizer can use the indexes in join operations.* All natural joins and old-style joins will benefit if you declare primary keys and foreign keys because the optimizer will use the available indexes at join time. (The declaration of a PK or FK will automatically create an index for the declared column.) Also, for the same reason, it is better to write joins using the SQL JOIN syntax. (See Chapter 8, Advanced SQL.)

- *Declare indexes in join columns other than PK or FK.* If you do join operations on columns other than the primary and foreign keys, you might be better off declaring indexes in those columns.

You cannot always use an index to improve performance. For example, using the data shown in Table 11.6 in the next section, the creation of an index for P_MIN will not help the search condition P_QOH > P_MIN * 1.10. The reason is because in some DBMSs, *indexes are ignored when you use functions in the table attributes.* However, major databases (such as Oracle, SQL Server, and DB2) now support function-based indexes. A **function-based index** is an index based on a specific SQL function or expression. For example, you could create an index on YEAR(INV_DATE). Function-based indexes are especially useful when dealing with derived attributes. For example, you could create an index on EMP_SALARY + EMP_COMMISSION.

How many indexes should you create? It bears repeating that you should not create an index for every column in a table. Too many indexes will slow down INSERT, UPDATE, and DELETE operations, especially if the table contains many thousands of rows. Furthermore, some query optimizers will choose only one index to be the driving index for a query, even if your query uses conditions in many different indexed columns. Which index does the optimizer use? If you use the cost-based optimizer, the answer will change with time as new rows are added or deleted from the tables. In any case, you should create indexes in all search columns and then let the optimizer choose. It's important to constantly evaluate the index usage—monitor, test, evaluate, and improve it if performance is not adequate.

11.5.2 CONDITIONAL EXPRESSIONS

A conditional expression is normally expressed within the WHERE or HAVING clauses of a SQL statement. Also known as conditional criteria, a conditional expression restricts the output of a query to only the rows matching the conditional criteria. Generally, the conditional criteria have the form shown in Table 11.6.

TABLE 11.6 Conditional Criteria

OPERAND1	CONDITIONAL OPERATOR	OPERAND2
P_PRICE	>	10.00
V_STATE	=	FL
V_CONTACT	LIKE	Smith%
P_QOH	>	P_MIN * 1.10

In Table 11.6, note that an operand can be:

- A simple column name such as P_PRICE or V_STATE.
- A literal or a constant such as the value 10.00 or the text 'FL'.
- An expression such as P_MIN * 1.10.

Most of the query optimization techniques mentioned next are designed to make the optimizer's work easier. Let's examine some common practices used to write efficient conditional expressions in SQL code.

- *Use simple columns or literals as operands in a conditional expression—avoid the use of conditional expressions with functions whenever possible.* Comparing the contents of a single column to a literal is faster than comparing to expressions. For example, P_PRICE > 10.00 is faster than P_QOH > P_MIN * 1.10 because the DBMS must evaluate the P_MIN * 1.10 expression first. The use of functions in expressions also adds to the total query execution time. For example, if your condition is UPPER (V_NAME) = 'JIM', try to use V_NAME = 'Jim' if all names in the V_NAME column are stored with proper capitalization.

- *Numeric field comparisons are faster than character, date, and NULL comparisons.* In search conditions, comparing a numeric attribute to a numeric literal is faster than comparing a character attribute to a character literal. In general, the CPU handles numeric comparisons (integer and decimal) faster than character and date comparisons. Because indexes do not store references to null values, NULL conditions involve additional processing, and therefore, tend to be the slowest of all conditional operands.

- *Equality comparisons are faster than inequality comparisons.* As a general rule, equality comparisons are processed faster than inequality comparisons. For example, P_PRICE = 10.00 is processed faster because the DBMS can do a direct search using the index in the column. If there are no exact matches, the condition is evaluated as false. However, if you use an inequality symbol (>, >=, <, <=), the DBMS must perform additional processing to complete the request. The reason is because there will almost always be more "greater than" or "less than" values than exactly "equal" values in the index. With the exception of NULL, the slowest of all comparison operators is LIKE with wildcard symbols, such as in V_CONTACT LIKE "%glo%". Also, using the "not equal" symbol (<>) yields slower searches, especially when the sparsity of the data is high, that is, when there are many more different values than there are equal values.

- *Whenever possible, transform conditional expressions to use literals.* For example, if your condition is P_PRICE – 10 = 7, change it to read P_PRICE = 17. Also, if you have a composite condition such as:

P_QOH < P_MIN AND P_MIN = P_REORDER AND P_QOH = 10

change it to read:

P_QOH = 10 AND P_MIN = P_REORDER AND P_MIN > 10

- *When using multiple conditional expressions, write the equality conditions first.* Note that this was done in the previous example. Remember, equality conditions are faster to process than inequality conditions. Although most RDBMSs will automatically do this for you, paying attention to this detail lightens the load for the query optimizer. The optimizer won't have to do what you have already done.

- *If you use multiple AND conditions, write the condition most likely to be false first.* If you use this technique, the DBMS will stop evaluating the rest of the conditions as soon as it finds a conditional expression that is evaluated to be false. Remember, for multiple AND conditions to be found true, all conditions must be evaluated as true. If one of the conditions evaluates to false, the whole set of conditions will be evaluated as false. If you use this technique, the DBMS won't waste time unnecessarily evaluating additional conditions. Naturally, the use of this technique implies an implicit knowledge of the sparsity of the data set. For example, look at the following condition list:

P_PRICE > 10 AND V_STATE = 'FL'

If you know that only a few vendors are located in Florida, you could rewrite this condition as:

V_STATE = 'FL' AND P_PRICE > 10

- *When using multiple OR conditions, put the condition most likely to be true first.* By doing this, the DBMS will stop evaluating the remaining conditions as soon as it finds a conditional expression that is evaluated to be true. Remember, for multiple OR conditions to evaluate to true, only one of the conditions must be evaluated to true.

NOTE

Oracle does not evaluate queries as described here. Instead, Oracle evaluates conditions from last to first.

- *Whenever possible, try to avoid the use of the NOT logical operator.* It is best to transform a SQL expression containing a NOT logical operator into an equivalent expression. For example:

NOT (P_PRICE > 10.00) can be written as P_PRICE <= 10.00.

Also, NOT (EMP_SEX = 'M') can be written as EMP_SEX = 'F'.

11.6 QUERY FORMULATION

Queries are usually written to answer questions. For example, if an end user gives you a sample output and tells you to match that output format, you must write the corresponding SQL. To get the job done, you must carefully evaluate what columns, tables, and computations are required to generate the desired output. And to do that, you must have a good understanding of the database environment and of the database that will be the focus of your SQL code.

This section focuses on SELECT queries because they are the queries you will find in most applications. To formulate a query, you would normally follow the steps outlined below.

1. *Identify what columns and computations are required*. The first step is to clearly determine what data values you want to return. Do you want to return just the names and addresses, or do you also want to include some computations? Remember that all columns in the SELECT statement should return single values.

   a. Do you need simple expressions? That is, do you need to multiply the price times the quantity on hand to generate the total inventory cost? You might need some single attribute functions such as DATE(), SYSDATE(), or ROUND().

   b. Do you need aggregate functions? If you need to compute the total sales by product, you should use a GROUP BY clause. In some cases, you might need to use a subquery.

   c. Determine the granularity of the raw data required for your output. Sometimes, you might need to summarize data that are not readily available on any table. In such cases, you might consider breaking the query into multiple subqueries and storing those subqueries as views. Then you could create a top-level query that joins those views and generates the final output.

2. *Identify the source tables*. Once you know what columns are required, you can determine the source tables used in the query. Some attributes appear in more than one table. In those cases, try to use the least number of tables in your query to minimize the number of join operations.

3. *Determine how to join the tables*. Once you know what tables you need in your query statement, you must properly identify how to join the tables. In most cases, you will use some type of natural join, but in some instances, you might need to use an outer join.

4. *Determine what selection criteria is needed*. Most queries involve some type of selection criteria. In this case, you must determine what operands and operators are needed in your criteria. Ensure that the data type and granularity of the data in the comparison criteria are correct.

   a. *Simple comparison*. In most cases, you will be comparing single values. For example, P_PRICE > 10.

   b. *Single value to multiple values*. If you are comparing a single value to multiple values, you might need to use an IN comparison operator. For example, V_STATE IN ('FL', 'TN', 'GA').

   c. *Nested comparisons*. In other cases, you might need to have some nested selection criteria involving subqueries. For example: P_PRICE > = (SELECT AVG(P_PRICE) FROM PRODUCT).

   d. *Grouped data selection*. On other occasions, the selection criteria might apply not to the raw data, but to the aggregate data. In those cases, you need to use the HAVING clause.

5. *Determine in what order to display the output*. Finally, the required output might be ordered by one or more columns. In those cases, you need to use the ORDER BY clause. Remember that the ORDER BY clause is one of the most resource-intensive operations for the DBMS.

11.7 DBMS PERFORMANCE TUNING

DBMS performance tuning includes global tasks such as managing the DBMS processes in primary memory (allocating memory for caching purposes) and managing the structures in physical storage (allocating space for the data files).

Fine-tuning the performance of the DBMS also includes applying several practices examined in the previous section. For example, the DBA must work with developers to ensure that the queries perform as expected—creating the indexes to speed up query response time and generating the database statistics required by cost-based optimizers.

DBMS performance tuning at the server end focuses on setting the parameters used for:

- *Data cache.* The data cache must be set large enough to permit as many data requests as possible to be serviced from the cache. Each DBMS has settings that control the size of the data cache; some DBMSs might require a restart. This cache is shared among all database users. The majority of primary memory resources will be allocated to the data cache.

- *SQL cache.* The SQL cache stores the most recently executed SQL statements (after the SQL statements have been parsed by the optimizer). Generally, if you have an application with multiple users accessing a database, the *same* query will likely be submitted by many different users. In those cases, the DBMS will parse the query only once and execute it many times, using the same access plan. In that way, the second and subsequent SQL requests for the same query are served from the SQL cache, skipping the parsing phase.

- *Sort cache.* The sort cache is used as a temporary storage area for ORDER BY or GROUP BY operations, as well as for index-creation functions.

- *Optimizer mode.* Most DBMSs operate in one of two optimization modes: cost-based or rule-based. Others automatically determine the optimization mode based on whether database statistics are available. For example, the DBA is responsible for generating the database statistics that are used by the cost-based optimizer. If the statistics are not available, the DBMS uses a rule-based optimizer.

Managing the physical storage details of the data files also plays an important role in DBMS performance tuning. Following are some general recommendations for the creation of databases.

- Use **RAID** (redundant array of independent disks) to provide balance between performance and fault tolerance. RAID systems use multiple disks to create virtual disks (storage volumes) formed by several individual disks. RAID systems provide performance improvement and fault tolerance. Table 11.7 shows the most common RAID configurations.

TABLE 11.7	Common RAID Configurations	
RAID LEVEL	**DESCRIPTION**	
0	The data blocks are spread over separate drives. Also known as striped array. Provides increased performance but no fault tolerance. (Fault tolerance means that in case of failure, data could be reconstructed and retrieved.) Requires a minimum of two drives.	
1	The same data blocks are written (duplicated) to separate drives. Also referred to as mirroring or duplexing. Provides increased read performance and fault tolerance via data redundancy. Requires a minimum of two drives.	
3	The data are striped across separate drives, and parity data are computed and stored in a dedicated drive. (Parity data are specially generated data that permit the reconstruction of corrupted or missing data.) Provides good read performance and fault tolerance via parity data. Requires a minimum of three drives.	
5	The data and the parity are striped across separate drives. Provides good read performance and fault tolerance via parity data. Requires a minimum of three drives.	

- Minimize disk contention. Use multiple, independent storage volumes with independent spindles (a spindle is a rotating disk) to minimize hard disk cycles. Remember, a database is composed of many table spaces, each with a particular function. In turn, each table space is composed of several data files in which the data are actually stored. A database should have at least the following table spaces:

  - *System table space.* This is used to store the data dictionary tables. It is the most frequently accessed table space and should be stored in its own volume.

  - *User data table space.* This is used to store end-user data. You should create as many user data table spaces and data files as are required to balance performance and usability. For example, you can create and assign a different user data table space for each application and/or for each distinct group of users; but not necessary for each user.

  - *Index table space.* This is used to store indexes. You can create and assign a different index table space for each application and/or for each group of users. The index table space data files should be stored on a storage volume that is separate from user data files or system data files.

  - *Temporary table space.* This is used as a temporary storage area for merge, sort, or set aggregate operations. You can create and assign a different temporary table space for each application and/or for each group of users.

  - *Rollback segment table space.* This is used for transaction-recovery purposes.

- Put high-usage tables in their own table spaces. By doing this, the database minimizes conflict with other tables.

- Assign separate data files in separate storage volumes for the indexes, system, and high-usage tables. This ensures that index operations will not conflict with end-user data or data dictionary table access operations. Another advantage of this approach is that you can use different disk block sizes in different volumes. For example, the data volume can use a 16K block size, while the index volume can use an 8K block size. Remember that the index record size is generally smaller, and by changing the block size you will be reducing contention and/or minimizing I/O operations. This is very important; many database administrators overlook indexes as a source of contention. By using separate storage volumes and different block sizes, the I/O operations on data and indexes will happen asynchronously (at different times), and more importantly, the likelihood of write operations blocking read operations is reduced (as page locks tend to lock less records).

- Take advantage of the various table storage organizations available in the database. For example, in Oracle consider the use of index organized tables (IOT); in SQL Server consider clustered index tables. An **index organized table** (or **clustered index table**) is a table that stores the end user data and the index data in consecutive locations on permanent storage. This type of storage organization provides a performance advantage to tables that are commonly accessed by a given index order. This is due to the fact that the index contains the index key as well as the data rows, and therefore the DBMS tends to perform fewer I/O operations.

- Partition tables based on usage. Some RDBMSs support horizontal partitioning of tables based on attributes. (See Chapter 12, Distributed Database Management Systems.) By doing so, a single SQL request could be processed by multiple data processors. Put the table partitions closest to where they are used the most.

- Use denormalized tables where appropriate. Another performance-improving technique involves taking a table from a higher normal form to a lower normal form—typically, from third to second normal form. This technique adds data duplication, but it minimizes join operations. (Denormalization was discussed in Chapter 5, Normalization of Database Tables.)

- Store computed and aggregate attributes in tables. In short, use derived attributes in your tables. For example, you might add the invoice subtotal, the amount of tax, and the total in the INVOICE table. Using derived attributes minimizes computations in queries and join operations.

11.8 QUERY OPTIMIZATION EXAMPLE

Now that you have learned the basis of query optimization, you are ready to test your new knowledge. Let's use a simple example to illustrate how the query optimizer works and how you can help it do its work. The example is based on the QOVENDOR and QOPRODUCT tables. Those tables are similar to the ones you used in previous chapters. However, the QO prefix is used for the table name to ensure that you do not overwrite previous tables.

ONLINE CONTENT

The databases and scripts used in this chapter can be found in the Student Online Companion for this book.

To perform this query optimization illustration, you will be using the Oracle SQL*Plus interface. Some preliminary work must be done before you can start testing query optimization. The following steps will guide you through this preliminary work:

1. Log in to Oracle SQL Plus using the username and password provided by your instructor.

2. Create a fresh set of tables using the QRYOPTDATA.SQL script file located on the Student Online Companion for this book. This step is necessary so Oracle has a new set of tables and the new tables contain no statistics. At the SQL> prompt, type:

    @path\ QRYOPTDATA.SQL

    where *path* is the location of the file in your computer.

3. Create the PLAN_TABLE. The PLAN_TABLE is a special table used by Oracle to store the access plan information for a given query. End users can then query the PLAN_TABLE to see how Oracle will execute the query. To create the PLAN_TABLE, run the UTLXPLAN.SQL script file located in the RDBMS\ADMIN folder of your Oracle RDBMS installation. The UTLXPLAN.SQL script file is also found in the Student Online Companion for this book. At the SQL prompt, type:

    @path\UTLXPLAN.SQL

You use the EXPLAIN PLAN command to store the execution plan of a SQL query in the PLAN_TABLE. Then, you would use the SELECT * FROM TABLE(DBMS_XPLAN.DISPLAY) command to display the access plan for a given SQL statement.

NOTE

Oracle 10g automatically defaults to cost-based optimization without giving you a choice. All previous versions of Oracle default to the Choose optimization mode, which implies that the DBMS will choose either rule-based or cost-based optimization, depending on the availability of table statistics.

To see the access plan used by the DBMS to execute your query, use the EXPLAIN PLAN and SELECT statements as shown in Figure 11.5. Note that the first SQL statement in Figure 11.5 generates the statistics for the QOVENDOR table. Also note that the initial access plan in Figure 11.5 uses a full table scan on the QOVENDOR table and that the cost of the plan is 4.

FIGURE 11.5   Initial explain plan

```
Oracle SQL*Plus                                                      _ |□| x|
File  Edit  Search  Options  Help
SQL> ANALYZE TABLE QOVENDOR COMPUTE STATISTICS;

Table analyzed.

SQL> EXPLAIN PLAN FOR SELECT * FROM QOVENDOR WHERE V_NAME LIKE 'B%' ORDER BY V_AREACODE;

Explained.

SQL> SELECT * FROM TABLE(DBMS_XPLAN.DISPLAY);

PLAN_TABLE_OUTPUT
-----------------------------------------------------------------------------
Plan hash value: 1837703589

-----------------------------------------------------------------------------
| Id  | Operation          | Name     | Rows  | Bytes | Cost (%CPU)| Time     |
-----------------------------------------------------------------------------
|   0 | SELECT STATEMENT   |          |     1 |    38 |     4  (25)| 00:00:01 |
|   1 |  SORT ORDER BY     |          |     1 |    38 |     4  (25)| 00:00:01 |
|*  2 |   TABLE ACCESS FULL| QOVENDOR |     1 |    38 |     3   (0)| 00:00:01 |
-----------------------------------------------------------------------------

Predicate Information (identified by operation id):

PLAN_TABLE_OUTPUT
-----------------------------------------------------------------------------
-------------------------------------------------------

   2 - filter("V_NAME" LIKE 'B%')

14 rows selected.

SQL> |
```

Let's now create an index on V_AREACODE (note that V_AREACODE is used in the ORDER BY clause) and see how that affects the access plan generated by the cost-based optimizer. The results are shown in Figure 11.6.

FIGURE 11.6   Explain plan after index on V_AREACODE

```
Oracle SQL*Plus
File  Edit  Search  Options  Help
SQL> CREATE INDEX QOV_NDX1 ON QOVENDOR(V_AREACODE);

Index created.

SQL> ANALYZE TABLE QOVENDOR COMPUTE STATISTICS;

Table analyzed.

SQL> EXPLAIN PLAN FOR SELECT * FROM QOVENDOR WHERE V_NAME LIKE 'B%' ORDER BY V_AREACODE;

Explained.

SQL> SELECT * FROM TABLE(DBMS_XPLAN.DISPLAY);

PLAN_TABLE_OUTPUT
--------------------------------------------------------------------------------
Plan hash value: 2305289760


---------------------------------------------------------------------------
| Id  | Operation                   | Name     | Rows | Bytes | Cost (%CPU)| Time     |
---------------------------------------------------------------------------
|   0 | SELECT STATEMENT            |          |    1 |    38 |    2   (0)| 00:00:01 |
|*  1 |  TABLE ACCESS BY INDEX ROWID| QOVENDOR |    1 |    38 |    2   (0)| 00:00:01 |
|   2 |   INDEX FULL SCAN           | QOV_NDX1 |   15 |       |    1   (0)| 00:00:01 |
---------------------------------------------------------------------------

Predicate Information (identified by operation id):
---------------------------------------------------

   1 - filter("V_NAME" LIKE 'B%')

14 rows selected.

SQL>
```

In Figure 11.6, note that the new access plan cuts the cost of executing the query by half! Also note that this new plan scans the QOV_NDX1 index and accesses the QOVENDOR rows, using the index row ID. (Remember that access by row ID is one of the fastest access methods.) In this case, the creation of the QOV_NDX1 index had a positive impact on overall query optimization results.

At other times, indexes do not necessarily help in query optimization. This is the case when you have indexes on small tables or when the query accesses a great percentage of table rows anyway. Let's see what happens when you create an index on V_NAME. The new access plan is shown in Figure 11.7. (Note that V_NAME is used on the WHERE clause as a conditional expression operand.)

FIGURE 11.7   **Explain plan after index on V_NAME**

```
Oracle SQL*Plus                                                          _□_|□|×|
File  Edit  Search  Options  Help
SQL> CREATE INDEX QOV_NDX2 ON QOVENDOR(V_NAME);

Index created.

SQL> ANALYZE TABLE QOVENDOR COMPUTE STATISTICS;

Table analyzed.

SQL> EXPLAIN PLAN FOR SELECT * FROM QOVENDOR WHERE V_NAME LIKE 'B%' ORDER BY V_AREACODE;

Explained.

SQL> SELECT * FROM TABLE(DBMS_XPLAN.DISPLAY);

PLAN_TABLE_OUTPUT
---------------------------------------------------------------------------------
Plan hash value: 2305289760

---------------------------------------------------------------------------------
| Id  | Operation                    | Name      | Rows | Bytes | Cost (%CPU)| Time     |
---------------------------------------------------------------------------------
|   0 | SELECT STATEMENT             |           |    1 |    38 |     2   (0)| 00:00:01 |
|*  1 |  TABLE ACCESS BY INDEX ROWID | QOVENDOR  |    1 |    38 |     2   (0)| 00:00:01 |
|   2 |   INDEX FULL SCAN            | QOV_NDX1  |   15 |       |     1   (0)| 00:00:01 |
---------------------------------------------------------------------------------

Predicate Information (identified by operation id):
---------------------------------------------------

   1 - filter("V_NAME" LIKE 'B%')

14 rows selected.

SQL>
```

As you can see in Figure 11.7, creation of the second index did not help the query optimization. However, there are occasions when an index might be used by the optimizer, but it is not executed because of the way in which the query is written. For example, Figure 11.8 shows the access plan for a different query using the V_NAME column.

FIGURE 11.8   Access plan using index on V_NAME

```
 Oracle SQL*Plus
File  Edit  Search  Options  Help
SQL> EXPLAIN PLAN FOR SELECT V_NAME, P_CODE FROM QOVENDOR V, QOPRODUCT P
  2                   WHERE  V.V_CODE = P.V_CODE AND V_NAME = 'ORDVA, Inc.';

Explained.

SQL> SELECT * FROM TABLE(DBMS_XPLAN.DISPLAY);

PLAN_TABLE_OUTPUT
--------------------------------------------------------------------------------
Plan hash value: 3956542569


--------------------------------------------------------------------------------
| Id | Operation                    | Name     | Rows | Bytes | Cost (%CPU)| Time     |
--------------------------------------------------------------------------------
|   0 | SELECT STATEMENT            |          |    2 |    74 |    6  (17)| 00:00:01 |
|*  1 |  HASH JOIN                  |          |    2 |    74 |    6  (17)| 00:00:01 |
|   2 |   TABLE ACCESS BY INDEX ROWID| QOVENDOR |    1 |    17 |    2   (0)| 00:00:01 |
|*  3 |    INDEX RANGE SCAN         | QOV_NDX2 |    1 |       |    1   (0)| 00:00:01 |
|   4 |   TABLE ACCESS FULL         | QOPRODUCT|   16 |   320 |    3   (0)| 00:00:01 |
--------------------------------------------------------------------------------


Predicate Information (identified by operation id):
---------------------------------------------------

   1 - access("V"."V_CODE"="P"."V_CODE")
   3 - access("V_NAME"='ORDVA, Inc.')

Note
-----
   - dynamic sampling used for this statement

21 rows selected.

SQL>
```

In Figure 11.8, note that the access plan for this new query uses the QOV_NDX2 index on the V_NAME column. What would happen if you wrote the same query, using the UPPER function on V_NAME? The results of that action are illustrated in Figure 11.9.

FIGURE 11.9   **Access plan using functions on indexed columns**

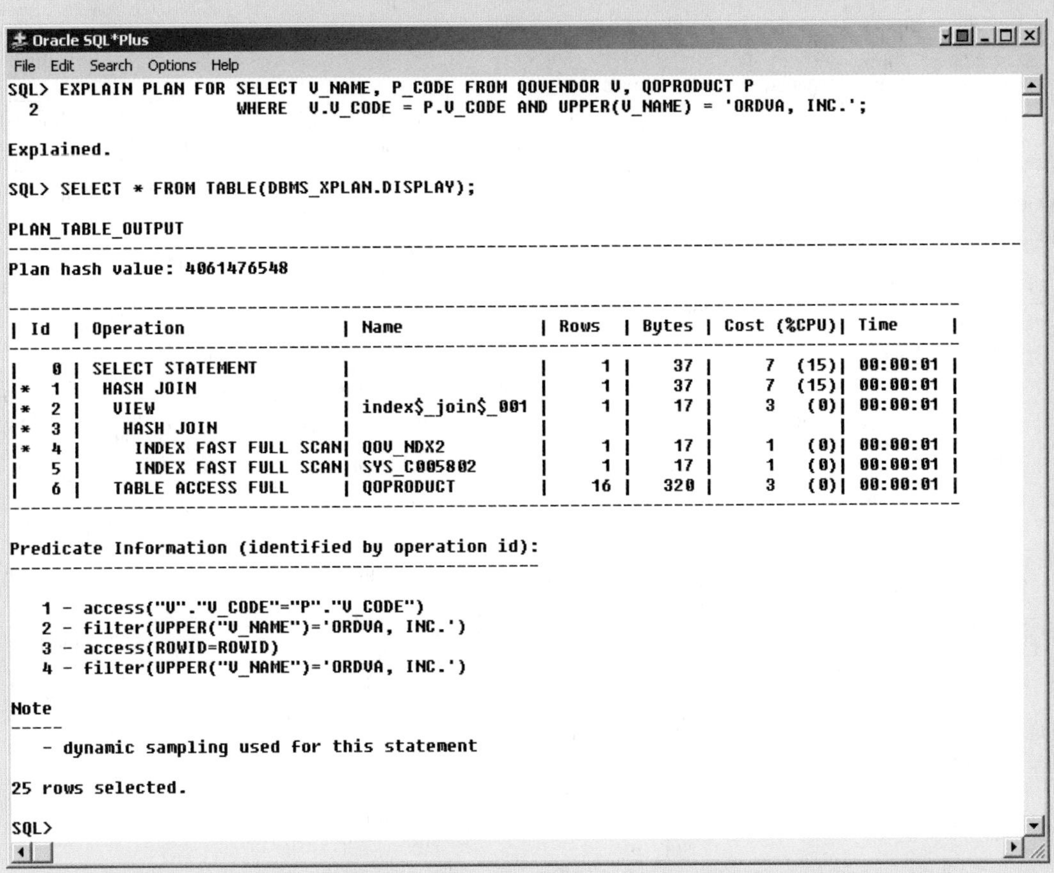

```
Oracle SQL*Plus                                                              _ |□| ×

File  Edit  Search  Options  Help

SQL> EXPLAIN PLAN FOR SELECT V_NAME, P_CODE FROM QOVENDOR V, QOPRODUCT P
  2                  WHERE   V.V_CODE = P.V_CODE AND UPPER(V_NAME) = 'ORDVA, INC.';

Explained.

SQL> SELECT * FROM TABLE(DBMS_XPLAN.DISPLAY);

PLAN_TABLE_OUTPUT
----------------------------------------------------------------------------------
Plan hash value: 4061476548


----------------------------------------------------------------------------------
| Id  | Operation              | Name           | Rows  | Bytes | Cost (%CPU)| Time     |
----------------------------------------------------------------------------------
|   0 | SELECT STATEMENT       |                |     1 |    37 |     7  (15)| 00:00:01 |
|*  1 |  HASH JOIN             |                |     1 |    37 |     7  (15)| 00:00:01 |
|*  2 |   VIEW                 | index$_join$_001 |   1 |    17 |     3   (0)| 00:00:01 |
|*  3 |    HASH JOIN           |                |       |       |            |          |
|*  4 |     INDEX FAST FULL SCAN| QOV_NDX2      |     1 |    17 |     1   (0)| 00:00:01 |
|   5 |     INDEX FAST FULL SCAN| SYS_C005802   |     1 |    17 |     1   (0)| 00:00:01 |
|   6 |   TABLE ACCESS FULL    | QOPRODUCT      |    16 |   320 |     3   (0)| 00:00:01 |
----------------------------------------------------------------------------------


Predicate Information (identified by operation id):
---------------------------------------------------

   1 - access("V"."V_CODE"="P"."V_CODE")
   2 - filter(UPPER("V_NAME")='ORDVA, INC.')
   3 - access(ROWID=ROWID)
   4 - filter(UPPER("V_NAME")='ORDVA, INC.')

Note
-----
   - dynamic sampling used for this statement

25 rows selected.

SQL>
```

As Figure 11.9 shows, the use of a function on an indexed column caused the DBMS to perform additional operations that increased the cost of the query. Note that the same query might produce different costs if your tables contain many more rows and if the index sparsity is different.

Let's now use the table QOPRODUCT to demonstrate how an index can help when aggregate function queries are being run. For example, Figure 11.10 shows the access plan for a SELECT statement using the MAX(P_PRICE) aggregate function. Note that this plan uses a full table scan with a total cost of 3.

FIGURE 11.10   First explain plan: aggregate function on a non-indexed column

```
Oracle SQL*Plus
File  Edit  Search  Options  Help
SQL> ANALYZE TABLE QOPRODUCT COMPUTE STATISTICS;

Table analyzed.

SQL> EXPLAIN PLAN FOR SELECT MAX(P_PRICE) FROM QOPRODUCT;

Explained.

SQL> SELECT * FROM TABLE(DBMS_XPLAN.DISPLAY);

PLAN_TABLE_OUTPUT
------------------------------------------------------------------------------
Plan hash value: 2544502246


-------------------------------------------------------------------------------
| Id | Operation          | Name      | Rows | Bytes | Cost (%CPU)| Time     |
-------------------------------------------------------------------------------
|  0 | SELECT STATEMENT   |           |    1 |     4 |     3  (0)| 00:00:01 |
|  1 |  SORT AGGREGATE    |           |    1 |     4 |           |          |
|  2 |   TABLE ACCESS FULL| QOPRODUCT |   16 |    64 |     3  (0)| 00:00:01 |
-------------------------------------------------------------------------------

9 rows selected.

SQL>
```

A cost of 3 is very low already, but could you improve it? Yes, you could improve the previous query performance by creating an index on P_PRICE. Figure 11.11 shows how the plan cost is reduced by two-thirds after the index is created and the QOPRODUCT table is analyzed. Also note that the second version of the access plan uses only the index QOP_NDX2 to answer the query; *the QOPRODUCT table* is *never accessed*.

FIGURE
11.11

Second explain plan: aggregate function on an indexed column

```
± Oracle SQL*Plus
File  Edit  Search  Options  Help
SQL> CREATE INDEX QOP_NDX2 ON QOPRODUCT(P_PRICE);

Index created.

SQL> ANALYZE TABLE QOPRODUCT COMPUTE STATISTICS;

Table analyzed.

SQL> EXPLAIN PLAN FOR SELECT MAX(P_PRICE) FROM QOPRODUCT;

Explained.

SQL> SELECT * FROM TABLE(DBMS_XPLAN.DISPLAY);

PLAN_TABLE_OUTPUT
--------------------------------------------------------------------------------
Plan hash value: 3423609809


------------------------------------------------------------------------
| Id  | Operation                  | Name     | Rows | Bytes | Cost (%CPU)| Time     |
------------------------------------------------------------------------
|   0 | SELECT STATEMENT           |          |    1 |     4 |    1   (0)| 00:00:01 |
|   1 |  SORT AGGREGATE            |          |    1 |     4 |           |          |
|   2 |   INDEX FULL SCAN (MIN/MAX)| QOP_NDX2 |   16 |    64 |    1   (0)| 00:00:01 |
------------------------------------------------------------------------

9 rows selected.

SQL>
```

Although the few examples in this section show how important proper index selection is for query optimization, you also saw examples in which index creation does not improve query performance. As a DBA, you should be aware that the main goal is to optimize overall database performance—not just for a single query, but for all requests and query types. Most database systems provide advanced graphical tools for performance monitoring and testing. For example, Figure 11.12 shows the graphical representation of the access plan using the Oracle 9*i* graphical tools. (Oracle 10g does not include this interface.)

FIGURE 11.12   Oracle 9*i* tools for query optimization

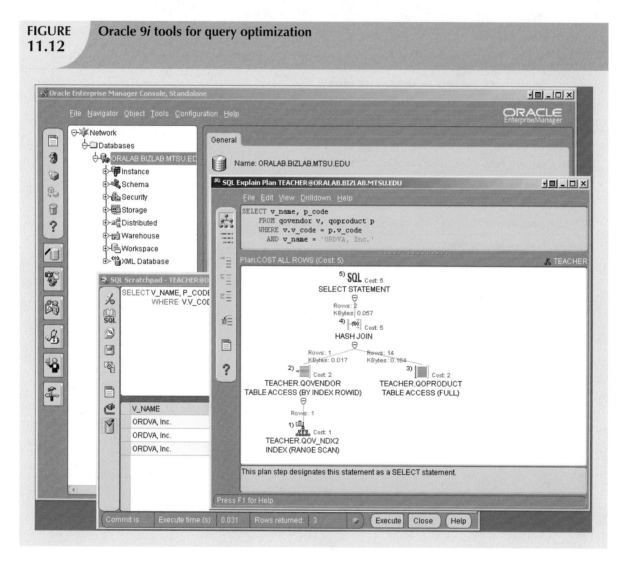

SUMMARY

▶ Database performance tuning refers to a set of activities and procedures designed to ensure that an end-user query is processed by the DBMS in the minimum amount of time.

▶ SQL performance tuning refers to the activities on the client side that are designed to generate SQL code that returns the correct answer in the least amount of time, using the minimum amount of resources at the server end.

▶ DBMS performance tuning refers to activities on the server side that are oriented to ensure that the DBMS is properly configured to respond to clients' requests in the fastest way possible while making optimum use of existing resources.

▶ The DBMS architecture is represented by the many processes and structures (in memory and in permanent storage) used to manage a database.

▶ Database statistics refers to a number of measurements gathered by the DBMS that describe a snapshot of the database objects' characteristics. The DBMS gathers statistics about objects such as tables, indexes, and available resources such as number of processors used, processor speed, and temporary space available. The DBMS uses the statistics to make critical decisions about improving the query processing efficiency.

▶ DBMSs process queries in three phases:

- *Parsing*. The DBMS parses the SQL query and chooses the most efficient access/execution plan.
- *Execution*. The DBMS executes the SQL query, using the chosen execution plan.
- *Fetching*. The DBMS fetches the data and sends the result set back to the client.

▶ Indexes are crucial in the process that speeds up data access. Indexes facilitate searching, sorting, and using aggregate functions and join operations. The improvement in data access speed occurs because an index is an ordered set of values that contains the index key and pointers. Data sparsity refers to the number of different values a column could possibly have. Indexes are recommended in high-sparsity columns used in search conditions.

▶ During query optimization, the DBMS must choose what indexes to use, how to perform join operations, which table to use first, and so on. Each DBMS has its own algorithms for determining the most efficient way to access the data. The two most common approaches are rule-based optimization and cost-based optimization.

- A rule-based optimizer uses preset rules and points to determine the best approach to execute a query. The rules assign a "fixed cost" to each SQL operation; the costs are then added to yield the cost of the execution plan.
- A cost-based optimizer uses sophisticated algorithms based on the statistics about the objects being accessed to determine the best approach to execute a query. In this case, the optimizer process adds up the processing cost, the I/O costs, and the resource costs (RAM and temporary space) to come up with the total cost of a given execution plan.

▶ Hints are used to change the optimizer mode for the current SQL statement. Hints are special instructions for the optimizer that are embedded inside the SQL command text.

▶ SQL performance tuning deals with writing queries that make good use of the statistics. In particular, queries should make good use of indexes. Indexes are very useful when you want to select a small subset of rows from a large table based on a condition. When an index exists for the column used in the selection, the DBMS may choose to use it. The objective is to create indexes with high selectivity. Index selectivity is a measure of how likely an index will be used in query processing. It is also important to write conditional statements using some common principles.

◗ Query formulation deals with how to translate business questions into specific SQL code to generate the required results. To do this, you must carefully evaluate what columns, tables, and computations are required to generate the desired output.

◗ DBMS performance tuning includes tasks such as managing the DBMS processes in primary memory (allocating memory for caching purposes) and managing the structures in physical storage (allocating space for the data files).

KEY TERMS

access plan, 449

cost-based optimizer, 453

database performance tuning, 443

database statistics, 446

data cache or buffer cache, 445

data files, 445

data sparsity, 452

DBMS performance tuning, 444

extends, 445

function-based index, 457

index organized table or cluster indexed table, 461

index selectivity, 456

input/output (I/O) request, 446

optimizer hints, 455

query optimizer, 449

query processing bottleneck, 450

RAID, 460

rule-based optimizer, 453

SQL cache or procedure cache, 445

SQL performance tuning, 444

table space or file group, 445

ONLINE CONTENT

Answers to selected Review Questions and Problems for this chapter are contained in the Student Online Companion for this book.

REVIEW QUESTIONS

1. What is SQL performance tuning?
2. What is database performance tuning?
3. What is the focus of most performance-tuning activities, and why does that focus exist?
4. What are database statistics, and why are they important?
5. How are database statistics obtained?
6. What database statistics measurements are typical of tables, indexes, and resources?
7. How is the processing of SQL DDL statements (such as CREATE TABLE) different from the processing required by DML statements?
8. In simple terms, the DBMS processes queries in three phases. What are those phases, and what is accomplished in each phase?
9. If indexes are so important, why not index every column in every table? (Include a brief discussion of the role played by data sparsity.)
10. What is the difference between a rule-based optimizer and a cost-based optimizer?
11. What are optimizer hints, and how are they used?
12. What are some general guidelines for creating and using indexes?
13. Most query optimization techniques are designed to make the optimizer's work easier. What factors should you keep in mind if you intend to write conditional expressions in SQL code?
14. What recommendations would you make for managing the data files in a DBMS with many tables and indexes?
15. What does RAID stand for, and what are some commonly used RAID levels?

Problems 1 and 2 are based on the following query:

```
SELECT      EMP_LNAME, EMP_FNAME, EMP_AREACODE, EMP_SEX
FROM        EMPLOYEE
WHERE       EMP_SEX = 'F' AND EMP_AREACODE = '615'
ORDER BY    EMP_LNAME, EMP_FNAME;
```

1. What is the likely data sparsity of the EMP_SEX column?

2. What indexes should you create? Write the required SQL commands.

3. Using Table 11.4 as an example, create two alternative access plans. Use the following assumptions:

    a. There are 8,000 employees.

    b. There are 4,150 female employees.

    c. There are 370 employees in area code 615.

    d. There are 190 female employees in area code 615.

Problems 4–6 are based on the following query:

```
SELECT      EMP_LNAME, EMP_FNAME, EMP_DOB, YEAR(EMP_DOB) AS YEAR
FROM        EMPLOYEE
WHERE       YEAR(EMP_DOB) = 1966;
```

4. What is the likely data sparsity of the EMP_DOB column?

5. Should you create an index on EMP_DOB? Why or why not?

6. What type of database I/O operations will likely be used by the query? (See Table 11.3.)

Problems 7–10 are based on the ER model shown in Figure P11.7 and on the query shown after the figure. Given the following query:

```
SELECT      P_CODE, P_PRICE
FROM        PRODUCT
WHERE       P_PRICE >= (SELECT AVG(P_PRICE) FROM PRODUCT);
```

7. Assuming that there are no table statistics, what type of optimization will the DBMS use?

8. What type of database I/O operations will likely be used by the query? (See Table 11.3.)

9. What is the likely data sparsity of the P_PRICE column?

10. Should you create an index? Why or why not?

Problems 11–14 are based on the following query:

```
SELECT      P_CODE, SUM(LINE_UNITS)
FROM        LINE
GROUP       BY P_CODE
HAVING      SUM(LINE_UNITS) > (SELECT MAX(LINE_UNITS) FROM LINE);
```

11. What is the likely data sparsity of the LINE_UNITS column?

12. Should you create an index? If so, what would the index column(s) be, and why would you create that index? If not, explain your reasoning.

13. Should you create an index on P_CODE? If so, write the SQL command to create that index. If not, explain your reasoning.

14. Write the command to create statistics for this table.

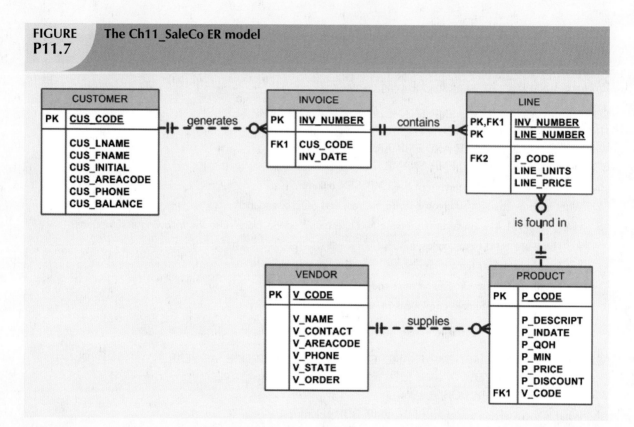

FIGURE P11.7   The Ch11_SaleCo ER model

Problems 15 and 16 are based on the following query:

```
SELECT      P_CODE, P_QOH*P_PRICE
FROM        PRODUCT
WHERE       P_QOH*P_PRICE > (SELECT AVG(P_QOH*P_PRICE) FROM PRODUCT)
```

15. What is the likely data sparsity of the P_QOH and P_PRICE columns?

16. Should you create an index, what would the index column(s) be, and why should you create that index?

Problems 17–21 are based on the following query:

```
SELECT      V_CODE, V_NAME, V_CONTACT, V_STATE
FROM        VENDOR
WHERE       V_STATE = 'TN'
ORDER BY    V_NAME;
```

17. What indexes should you create and why? Write the SQL command to create the indexes.

18. Assume that 10,000 vendors are distributed as shown in Table P11.18. What percentage of rows will be returned by the query?

19. What type of I/O database operations would most likely be used to execute that query?

20. Using Table 11.4 as an example, create two alternative access plans.

21. Assume that you have 10,000 different products stored in the PRODUCT table and that you are writing a Web-based interface to list all products with a quantity on hand (P_QOH) that is less than or equal to the minimum quantity, P_MIN. What optimizer hint would you use to ensure that your query returns the result set to the Web interface in the least time possible? Write the SQL code.

TABLE
P11.18

STATE	NUMBER OF VENDORS	STATE	NUMBER OF VENDORS
AK	15	MS	47
AL	55	NC	358
AZ	100	NH	25
CA	3244	NJ	645
CO	345	NV	16
FL	995	OH	821
GA	75	OK	62
HI	68	PA	425
IL	89	RI	12
IN	12	SC	65
KS	19	SD	74
KY	45	TN	113
LA	29	TX	589
MD	208	UT	36
MI	745	VA	375
MO	35	WA	258

Problems 22–24 are based on the following query:

```
SELECT      P_CODE, P_DESCRIPT, P_PRICE, P.V_CODE, V_STATE
FROM        PRODUCT P, VENDOR V
WHERE       P.V_CODE = V.V_CODE
            AND V_STATE = 'NY'
            AND V_AREACODE = '212'
ORDER BY    P_PRICE;
```

22. What indexes would you recommend?

23. Write the commands required to create the indexes you recommended in Problem 22.

24. Write the command(s) used to generate the statistics for the PRODUCT and VENDOR tables.

Problems 25 and 26 are based on the following query:

```
SELECT      P_CODE, P_DESCRIPT, P_QOH, P_PRICE, V_CODE
FROM        PRODUCT
WHERE       V_CODE = '21344'
ORDER BY    P_CODE;
```

25. What index would you recommend, and what command would you use?

26. How should you rewrite the query to ensure that it uses the index you created in your solution to Problem 25?

Problems 27 and 28 are based on the following query:

```
SELECT      P_CODE, P_DESCRIPT, P_QOH, P_PRICE, V_CODE
FROM        PRODUCT
WHERE       P_QOH < P_MIN
            AND P_MIN = P_REORDER
            AND P_REORDER = 50
ORDER BY    P_QOH;
```

27. Use the recommendations given in Section 11.5.2 to rewrite the query to produce the required results more efficiently.

28. What indexes would you recommend? Write the commands to create those indexes.

Problems 29–32 are based on the following query:

```
SELECT      CUS_CODE, MAX(LINE_UNITS*LINE_PRICE)
FROM        CUSTOMER NATURAL JOIN INVOICE NATURAL JOIN LINE
WHERE       CUS_AREACODE = '615'
GROUP BY    CUS_CODE;
```

29. Assuming that you generate 15,000 invoices per month, what recommendation would you give the designer about the use of derived attributes?

30. Assuming that you follow the recommendations you gave in Problem 29, how would you rewrite the query?

31. What indexes would you recommend for the query you wrote in Problem 30, and what SQL commands would you use?

32. How would you rewrite the query to ensure that the index you created in Problem 31 is used?

DISTRIBUTED DATABASE MANAGEMENT SYSTEMS

In this chapter, you will learn:

- What a distributed database management system (DDBMS) is and what its components are
- How database implementation is affected by different levels of data and process distribution
- How transactions are managed in a distributed database environment
- How database design is affected by the distributed database environment

Preview

In this chapter, you learn that a single database can be divided into several fragments. The fragments can be stored on different computers within a network. Processing, too, can be dispersed among several different network sites, or nodes. The multisite database forms the core of the distributed database system.

The growth of distributed database systems has been fostered by the dispersion of business operations across the country and the world, along with the rapid pace of technological change that has made local and wide area networks practical and more reliable. The network-based distributed database system is very flexible: it can serve the needs of a small business operating two stores in the same town while at the same time meeting the needs of a global business.

Although a distributed database system requires a more sophisticated DBMS, the end user should not be burdened by increased operational complexity. That is, the greater complexity of a distributed database system should be transparent to the end user.

The distributed database management system (DDBMS) treats a distributed database as a single logical database; therefore, the basic design concepts you learned in earlier chapters apply. However, although the end user need not be aware of the distributed database's special characteristics, the distribution of data among different sites in a computer network clearly adds to a system's complexity. For example, the design of a distributed database must consider the location of the data and the partitioning of the data into database fragments. You examine such issues in this chapter.

12.1 THE EVOLUTION OF DISTRIBUTED DATABASE MANAGEMENT SYSTEMS

A **distributed database management system (DDBMS)** governs the storage and processing of logically related data over interconnected computer systems in which both data and processing functions are distributed among several sites. To understand how and why the DDBMS is different from the DBMS, it is useful to briefly examine the changes in the business environment that set the stage for the development of the DDBMS.

During the 1970s, corporations implemented centralized database management systems to meet their structured information needs. Structured information is usually presented as regularly issued formal reports in a standard format. Such information, generated by procedural programming languages, is created by specialists in response to precisely channeled requests. Thus, structured information needs are well served by centralized systems.

The use of a centralized database required that corporate data be stored in a single central site, usually a mainframe computer. Data access was provided through dumb terminals. The centralized approach, illustrated in Figure 12.1, worked well to fill the structured information needs of corporations, but it fell short when quickly moving events required faster response times and equally quick access to information. The slow progression from information request to approval, to specialist, to user simply did not serve decision makers well in a dynamic environment. What was needed was quick, unstructured access to databases, using ad hoc queries to generate on-the-spot information.

| FIGURE 12.1 | Centralized database management system |

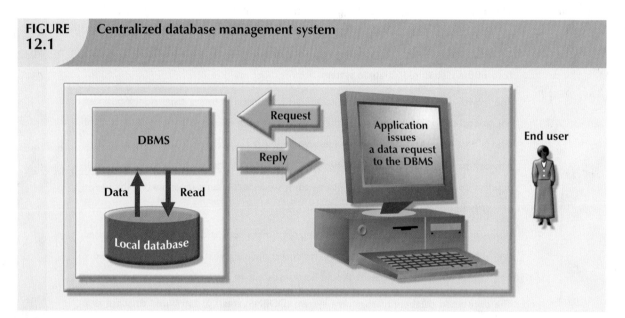

Database management systems based on the relational model could provide the environment in which unstructured information needs would be met by employing ad hoc queries. End users would be given the ability to access data when needed. Unfortunately, the early relational model implementations did not yet deliver acceptable throughput when compared to the well-established hierarchical or network database models.

The last two decades gave birth to a series of crucial social and technological changes that affected database development and design. Among those changes were:

- Business operations became decentralized.
- Competition increased at the global level.
- Customer demands and market needs favored a decentralized management style.

- Rapid technological change created low-cost computers with mainframe-like power, impressive multifunction handheld portable wireless devices with cellular phone and data services, and increasingly complex and fast networks to connect them. As a consequence, corporations have increasingly adopted advanced network technologies as the platform for their computerized solutions.

- The large number of applications based on DBMSs and the need to protect investments in centralized DBMS software made the notion of data sharing attractive. Data realms are converging in the digital world more and more. As a result, single applications manage multiple different types of data (voice, video, music, images, etc.), and such data are accessed from multiple geographically dispersed locations.

Those factors created a dynamic business environment in which companies had to respond quickly to competitive and technological pressures. As large business units restructured to form leaner and meaner, quickly reacting, dispersed operations, two database requirements became obvious:

- *Rapid ad hoc data access* became crucial in the quick-response decision-making environment.

- *The decentralization of management structures* based on the decentralization of business units made decentralized multiple-access and multiple-location databases a necessity.

During recent years, the factors just described became even more firmly entrenched. However, the way those factors were addressed was strongly influenced by:

- *The growing acceptance of the Internet as the platform for data access and distribution.* The World Wide Web (WWW or just the Web) is, in effect, the *repository* for distributed data.

- *The wireless revolution.* The widespread use of wireless digital devices, such as personal digital assistants (PDAs) like Palm and BlackBerry and multipurpose "smart phones" like the iPhone, has created high demand for data access. Such devices access data from geographically dispersed locations and require varied data exchanges in multiple formats (data, voice, video, music, pictures, etc.) Although distributed data access does not necessarily imply distributed databases; performance and failure tolerance requirements often make use of data replication techniques similar to the ones found in distributed databases.

- *The accelerated growth of companies providing "application as a service" type of services.* This new type of service provides remote application services to companies wanting to outsource their application development, maintenance, and operations. The company data is generally stored on central servers and is not necessarily distributed. Just as with wireless data access, this type of service may not require fully distributed data functionality; however, other factors such as performance and failure tolerance often require the use of data replication techniques similar to the ones found in distributed databases.

- *The increased focus on data analysis that led to data mining and data warehousing.* Although a data warehouse is not usually a distributed database, it does rely on techniques such as data replication and distributed queries that facilitate data extraction and integration. (Data warehouse design, implementation, and use are discussed in Chapter 13, Business Intelligence and Data Warehouses.)

ONLINE CONTENT

To learn more about the Internet's impact on data access and distribution, see **Appendix I** in the Student Online Companion, **Databases in Electronic Commerce.**

At this point, the long-term impact of the Internet and the wireless revolution on *distributed* database design and management is still unclear. Perhaps the success of the Internet and wireless technologies will foster the use of distributed databases as bandwidth becomes a more troublesome bottleneck. Perhaps the resolution of bandwidth problems will simply confirm the centralized database standard. In any case, distributed databases exist today and many distributed database operating concepts and components are likely to find a place in future database developments.

The decentralized database is especially desirable because centralized database management is subject to problems such as:

- *Performance degradation* due to a growing number of remote locations over greater distances.

- *High costs* associated with maintaining and operating large central (mainframe) database systems.

- *Reliability problems* created by dependence on a central site (single point of failure syndrome) and the need for data replication.

- *Scalability problems* associated with the physical limits imposed by a single location (power, temperature conditioning, and power consumption.)

- *Organizational rigidity* imposed by the database might not support the flexibility and agility required by modern global organizations.

The dynamic business environment and the centralized database's shortcomings spawned a demand for applications based on accessing data from different sources at multiple locations. Such a multiple-source/multiple-location database environment is best managed by a distributed database management system (DDBMS).

12.2 DDBMS ADVANTAGES AND DISADVANTAGES

Distributed database management systems deliver several advantages over traditional systems. At the same time, they are subject to some problems. Table 12.1 summarizes the advantages and disadvantages associated with a DDBMS.

TABLE 12.1   **Distributed DBMS Advantages and Disadvantages**

ADVANTAGES	DISADVANTAGES
• *Data are located near the greatest demand site.* The data in a distributed database system are dispersed to match business requirements. • *Faster data access.* End users often work with only a locally stored subset of the company's data. • *Faster data processing.* A distributed database system spreads out the systems workload by processing data at several sites. • *Growth facilitation.* New sites can be added to the network without affecting the operations of other sites. • *Improved communications.* Because local sites are smaller and located closer to customers, local sites foster better communication among departments and between customers and company staff. • *Reduced operating costs.* It is more cost-effective to add workstations to a network than to update a mainframe system. Development work is done more cheaply and more quickly on low-cost PCs than on mainframes. • *User-friendly interface.* PCs and workstations are usually equipped with an easy-to-use graphical user interface (GUI). The GUI simplifies training and use for end users. • *Less danger of a single-point failure.* When one of the computers fails, the workload is picked up by other workstations. Data are also distributed at multiple sites. • *Processor independence.* The end user is able to access any available copy of the data, and an end user's request is processed by any processor at the data location.	• *Complexity of management and control.* Applications must recognize data location, and they must be able to stitch together data from various sites. Database administrators must have the ability to coordinate database activities to prevent database degradation due to data anomalies. • *Technological difficulty.* Data integrity, transaction management, concurrency control, security, backup, recovery, query optimization, access path selection, and so on, must all be addressed and resolved. • *Security.* The probability of security lapses increases when data are located at multiple sites. The responsibility of data management will be shared by different people at several sites. • *Lack of standards.* There are no standard communication protocols *at the database level.* (Although TCP/IP is the de facto standard at the network level, there is no standard at the application level.) For example, different database vendors employ different—and often incompatible—techniques to manage the distribution of data and processing in a DDBMS environment. • *Increased storage and infrastructure requirements.* Multiple copies of data are required at different sites, thus requiring additional disk storage space. • *Increased training cost.* Training costs are generally higher in a distributed model than they would be in a centralized model, sometimes even to the extent of offsetting operational and hardware savings. • *Costs.* Distributed databases require duplicated infrastructure to operate (physical location, environment, personnel, software, licensing, etc.)

Distributed databases are used successfully but have a long way to go before they will yield the full flexibility and power of which they are theoretically capable. The inherently complex distributed data environment increases the urgency for standard protocols governing transaction management, concurrency control, security, backup, recovery, query optimization, access path selection, and so on. Such issues must be addressed and resolved before DDBMS technology is widely embraced.

The remainder of this chapter explores the basic components and concepts of the distributed database. Because the distributed database is usually based on the relational database model, relational terminology is used to explain the basic concepts and components of a distributed database.

12.3 DISTRIBUTED PROCESSING AND DISTRIBUTED DATABASES

In **distributed processing**, a database's logical processing is shared among two or more physically independent sites that are connected through a network. For example, the data input/output (I/O), data selection, and data validation might be performed on one computer, and a report based on that data might be created on another computer.

A basic distributed processing environment is illustrated in Figure 12.2, which shows that a distributed processing system shares the database processing chores among three sites connected through a communications network. Although the database resides at only one site (Miami), each site can access the data and update the database. The database is located on Computer A, a network computer known as the *database server*.

FIGURE 12.2   Distributed processing environment

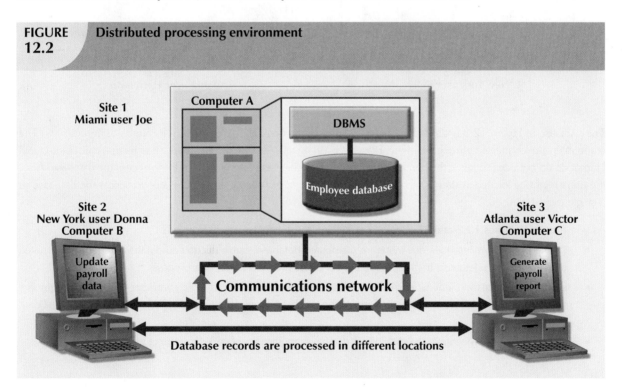

Site 1
Miami user Joe

Computer A

DBMS

Employee database

Site 2
New York user Donna
Computer B

Update payroll data

Communications network

Site 3
Atlanta user Victor
Computer C

Generate payroll report

Database records are processed in different locations

A **distributed database**, on the other hand, stores a logically related database over two or more physically independent sites. The sites are connected via a computer network. In contrast, the distributed processing system uses only a single-site database but shares the processing chores among several sites. In a distributed database system, a database is composed of several parts known as **database fragments**. The database fragments are located at different sites and can be replicated among various sites. Each database fragment is, in turn, managed by its local database process. An example of a distributed database environment is shown in Figure 12.3.

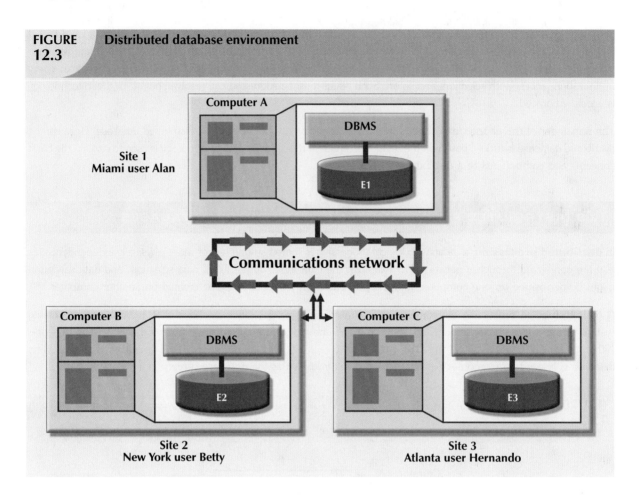

FIGURE 12.3   Distributed database environment

The database in Figure 12.3 is divided into three database fragments (E1, E2, and E3) located at different sites. The computers are connected through a network system. In a fully distributed database, the users Alan, Betty, and Hernando do not need to know the name or location of each database fragment in order to access the database. Also, the users might be located at sites other than Miami, New York, or Atlanta, and still be able to access the database as a single logical unit.

As you examine Figures 12.2 and 12.3, you should keep the following points in mind:

- Distributed processing does not require a distributed database, but a distributed database requires distributed processing (each database fragment is managed by its own local database process).
- Distributed processing may be based on a single database located on a single computer. For the management of distributed data to occur, copies or parts of the database processing functions must be distributed to all data storage sites.
- Both distributed processing and distributed databases require a network to connect all components.

12.4 CHARACTERISTICS OF DISTRIBUTED DATABASE MANAGEMENT SYSTEMS

A DDBMS governs the storage and processing of logically related data over interconnected computer systems in which both data and processing functions are distributed among several sites. A DBMS must have at least the following functions to be classified as distributed:

- *Application interface* to interact with the end user, application programs, and other DBMSs within the distributed database.

- *Validation* to analyze data requests for syntax correctness.

- *Transformation* to decompose complex requests into atomic data request components.

- *Query optimization* to find the best access strategy. (Which database fragments must be accessed by the query, and how must data updates, if any, be synchronized?)

- *Mapping* to determine the data location of local and remote fragments.

- *I/O interface* to read or write data from or to permanent local storage.

- *Formatting* to prepare the data for presentation to the end user or to an application program.

- *Security* to provide data privacy at both local and remote databases.

- *Backup and recovery* to ensure the availability and recoverability of the database in case of a failure.

- *DB administration features* for the database administrator.

- *Concurrency control* to manage simultaneous data access and to ensure data consistency across database fragments in the DDBMS.

- *Transaction management* to ensure that the data moves from one consistent state to another. This activity includes the synchronization of local and remote transactions as well as transactions across multiple distributed segments.

A fully distributed database management system must perform all of the functions of a centralized DBMS, as follows:

1. Receive an application's (or an end user's) request.

2. Validate, analyze, and decompose the request. The request might include mathematical and/or logical operations such as the following: Select all customers with a balance greater than $1,000. The request might require data from only a single table, or it might require access to several tables.

3. Map the request's logical-to-physical data components.

4. Decompose the request into several disk I/O operations.

5. Search for, locate, read, and validate the data.

6. Ensure database consistency, security, and integrity.

7. Validate the data for the conditions, if any, specified by the request.

8. Present the selected data in the required format.

In addition, a distributed DBMS must handle all necessary functions imposed by the distribution of data and processing. And it must perform those additional functions *transparently* to the end user. The DDBMS's transparent data access features are illustrated in Figure 12.4.

The single logical database in Figure 12.4 consists of two database fragments, A1 and A2, located at sites 1 and 2, respectively. Mary can query the database as if it were a local database; so can Tom. Both users "see" only one logical database and *do not need to know the names of the fragments*. In fact, the end users do not even need to know that the database is divided into fragments, *nor do they need to know where the fragments are located*.

To better understand the different types of distributed database scenarios, let's first define the distributed database system's components.

FIGURE 12.4   A fully distributed database management system

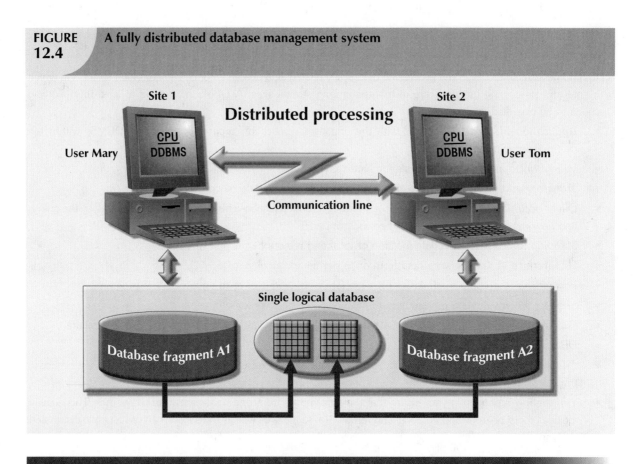

12.5 DDBMS COMPONENTS

The DDBMS must include at least the following components:

- *Computer workstations* (sites or nodes) that form the network system. The distributed database system must be independent of the computer system hardware.

- *Network hardware and software* components that reside in each workstation. The network components allow all sites to interact and exchange data. Because the components—computers, operating systems, network hardware, and so on—are likely to be supplied by different vendors, it is best to ensure that distributed database functions can be run on multiple platforms.

- *Communications media* that carry the data from one workstation to another. The DDBMS must be communications-media-independent; that is, it must be able to support several types of communications media.

- The **transaction processor** (**TP**), which is the software component found in each computer that requests data. The transaction processor receives and processes the application's data requests (remote and local). The TP is also known as the **application processor** (**AP**) or the **transaction manager** (**TM**).

- The **data processor** (**DP**), which is the software component residing on each computer that stores and retrieves data located at the site. The DP is also known as the **data manager** (**DM**). A data processor may even be a centralized DBMS.

Figure 12.5 illustrates the placement of the components and the interaction among them. The communication among TPs and DPs shown in Figure 12.5 is made possible through a specific set of rules, or *protocols*, used by the DDBMS.

FIGURE 12.5 | Distributed database system management components

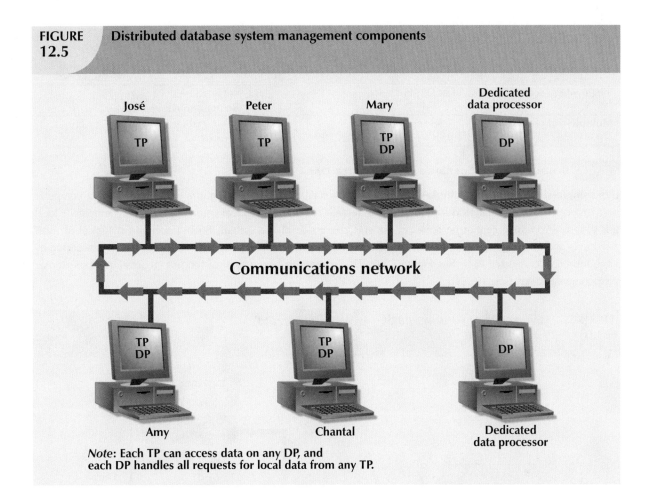

Note: Each TP can access data on any DP, and each DP handles all requests for local data from any TP.

The protocols determine how the distributed database system will:

- Interface with the network to transport data and commands between data processors (DPs) and transaction processors (TPs).

- Synchronize all data received from DPs (TP side) and route retrieved data to the appropriate TPs (DP side).

- Ensure common database functions in a distributed system. Such functions include security, concurrency control, backup, and recovery.

DPs and TPs can be added to the system without affecting the operation of the other components. A TP and a DP can reside on the same computer, allowing the end user to access local as well as remote data transparently. In theory, a DP can be an independent centralized DBMS with proper interfaces to support remote access from other independent DBMSs in the network.

12.6 LEVELS OF DATA AND PROCESS DISTRIBUTION

Current database systems can be classified on the basis of how process distribution and data distribution are supported. For example, a DBMS may store data in a single site (centralized DB) or in multiple sites (distributed DB) and may support data processing at a single site or at multiple sites. Table 12.2 uses a simple matrix to classify database systems according to data and process distribution. These types of processes are discussed in the sections that follow.

TABLE 12.2	Database Systems: Levels of Data and Process Distribution	
	SINGLE-SITE DATA	MULTIPLE-SITE DATA
Single-site process	Host DBMS	Not applicable (Requires multiple processes)
Multiple-site process	File server Client/server DBMS (LAN DBMS)	Fully distributed Client/server DDBMS

12.6.1 SINGLE-SITE PROCESSING, SINGLE-SITE DATA (SPSD)

In the **single-site processing**, **single-site data** (**SPSD**) scenario, all processing is done on a single host computer (single-processor server, multiprocessor server, mainframe system) and all data are stored on the host computer's local disk system. Processing cannot be done on the end user's side of the system. Such a scenario is typical of most mainframe and midrange server computer DBMSs. The DBMS is located on the host computer, which is accessed by dumb terminals connected to it. See Figure 12.6. This scenario is also typical of the first generation of single-user microcomputer databases.

FIGURE 12.6	Single-site processing, single-site data (centralized)

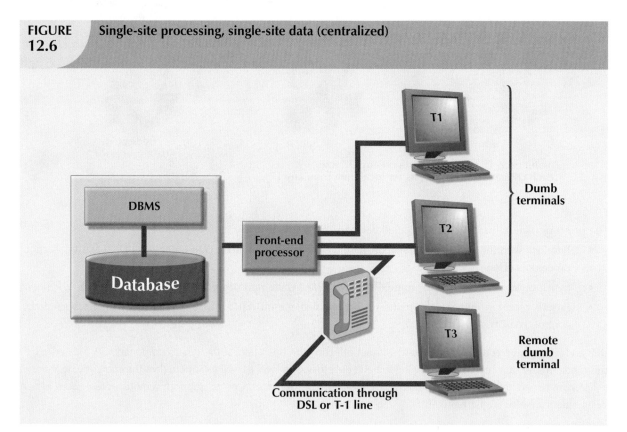

Using Figure 12.6 as an example, you can see that the functions of the TP and the DP are embedded within the DBMS located on a single computer. The DBMS usually runs under a time-sharing, multitasking operating system, which allows several processes to run concurrently on a host computer accessing a single DP. All data storage and data processing are handled by a single host computer.

12.6.2 MULTIPLE-SITE PROCESSING, SINGLE-SITE DATA (MPSD)

Under the **multiple-site processing, single-site data (MPSD)** scenario, multiple processes run on different computers sharing a single data repository. Typically, the MPSD scenario requires a network file server running conventional applications that are accessed through a network. Many multiuser accounting applications running under a personal computer network fit such a description. (See Figure 12.7.)

FIGURE 12.7   Multiple-site processing, single-site data

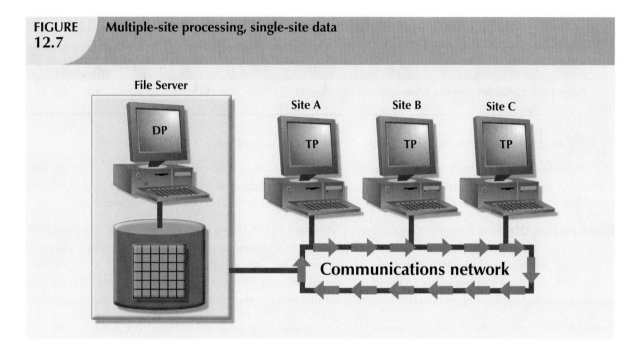

As you examine Figure 12.7, note that:

- The TP on each workstation acts only as a redirector to route all network data requests to the file server.
- The end user sees the file server as just another hard disk. Because only the data storage input/output (I/O) is handled by the file server's computer, the MPSD offers limited capabilities for distributed processing.
- The end user must make a direct reference to the file server in order to access remote data. All record- and file-locking activities are done at the end-user location.
- All data selection, search, and update functions take place at the workstation, thus requiring that entire files travel through the network for processing at the workstation. Such a requirement increases network traffic, slows response time, and increases communication costs.

The inefficiency of the last condition can be illustrated easily. For example, suppose the file server computer stores a CUSTOMER table containing 10,000 data rows, 50 of which have balances greater than $1,000. Suppose site A issues the following SQL query:

```
SELECT      *
FROM        CUSTOMER
WHERE       CUS_BALANCE > 1000;
```

All 10,000 CUSTOMER rows must travel through the network to be evaluated at site A.

A variation of the multiple-site processing, single-site data approach is known as client/server architecture. **Client/ server architecture** is similar to that of the network file server *except that all database processing is done at the server site, thus reducing network traffic.* Although both the network file server and the client/server systems perform multiple-site processing, the latter's processing is distributed. Note that the network file server approach requires the database to be located at a single site. In contrast, the client/server architecture is capable of supporting data at multiple sites.

ONLINE CONTENT

Appendix F, Client/Server Systems, is located in the Student Online Companion for this book.

12.6.3 MULTIPLE-SITE PROCESSING, MULTIPLE-SITE DATA (MPMD)

The **multiple-site processing, multiple-site data (MPMD)** scenario describes a fully distributed DBMS with support for multiple data processors and transaction processors at multiple sites. Depending on the level of support for various types of centralized DBMSs, DDBMSs are classified as either homogeneous or heterogeneous.

Homogeneous DDBMSs integrate only one type of centralized DBMS over a network. Thus, the same DBMS will be running on different server platforms (single processor server, multi-processor server, server farms, or server blades). In contrast, **heterogeneous DDBMSs** integrate different types of centralized DBMSs over a network. See Figure 12.8. A **fully heterogeneous DDBMS** will support different DBMSs that may even support different data models (relational, hierarchical, or network) running under different computer systems, such as mainframes and PCs.

Some DDBMS implementations support several platforms, operating systems, and networks and allow remote data access to another DBMS. However, such DDBMSs still are subject to certain restrictions. For example:

- Remote access is provided on a read-only basis and does not support write privileges.
- Restrictions are placed on the number of remote tables that may be accessed in a single transaction.
- Restrictions are placed on the number of distinct databases that may be accessed.
- Restrictions are placed on the database model that may be accessed. Thus, access may be provided to relational databases but not to network or hierarchical databases.

The preceding list of restrictions is by no means exhaustive. The DDBMS technology continues to change rapidly, and new features are added frequently. Managing data at multiple sites leads to a number of issues that must be addressed and understood. The next section will examine several key features of distributed database management systems.

FIGURE 12.8	Heterogeneous distributed database scenario

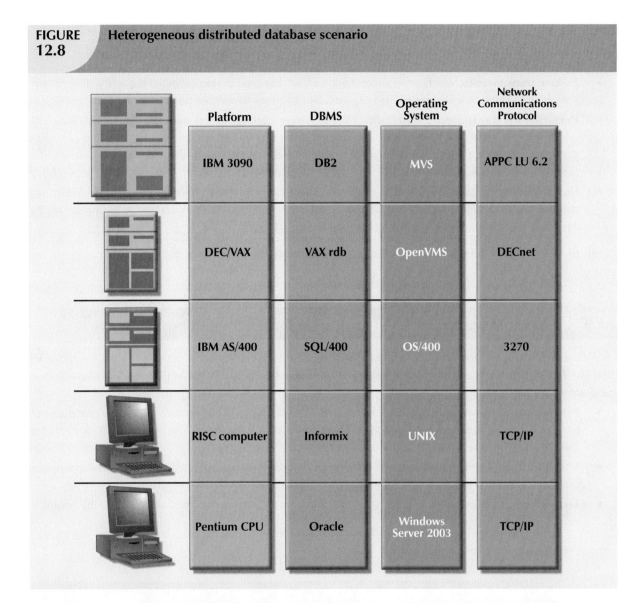

Platform	DBMS	Operating System	Network Communications Protocol
IBM 3090	DB2	MVS	APPC LU 6.2
DEC/VAX	VAX rdb	OpenVMS	DECnet
IBM AS/400	SQL/400	OS/400	3270
RISC computer	Informix	UNIX	TCP/IP
Pentium CPU	Oracle	Windows Server 2003	TCP/IP

12.7 DISTRIBUTED DATABASE TRANSPARENCY FEATURES

A distributed database system requires functional characteristics that can be grouped and described as transparency features. DDBMS transparency features have the common property of allowing the end user to feel like the database's only user. In other words, the user believes that (s)he is working with a centralized DBMS; all complexities of a distributed database are hidden, or transparent, to the user.

The DDBMS transparency features are:

- **Distribution transparency**, which allows a distributed database to be treated as a single logical database. If a DDBMS exhibits distribution transparency, the user does not need to know:
  - That the data are partitioned—meaning the table's rows and columns are split vertically or horizontally and stored among multiple sites.
  - That the data can be replicated at several sites.
  - The data location.

- **Transaction transparency**, which allows a transaction to update data at more than one network site. Transaction transparency ensures that the transaction will be either entirely completed or aborted, thus maintaining database integrity.

- **Failure transparency**, which ensures that the system will continue to operate in the event of a node failure. Functions that were lost because of the failure will be picked up by another network node.

- **Performance transparency**, which allows the system to perform as if it were a centralized DBMS. The system will not suffer any performance degradation due to its use on a network or due to the network's platform differences. Performance transparency also ensures that the system will find the most cost-effective path to access remote data.

- **Heterogeneity transparency**, which allows the integration of several different local DBMSs (relational, network, and hierarchical) under a common, or global, schema. The DDBMS is responsible for translating the data requests from the global schema to the local DBMS schema.

Distribution, transaction, and performance transparency features will be examined in greater detail in the next few sections.

12.8 DISTRIBUTION TRANSPARENCY

Distribution transparency allows a physically dispersed database to be managed as though it were a centralized database. The level of transparency supported by the DDBMS varies from system to system. Three levels of distribution transparency are recognized:

- **Fragmentation transparency** is the highest level of transparency. The end user or programmer does not need to know that a database is partitioned. Therefore, neither fragment names nor fragment locations are specified prior to data access.

- **Location transparency** exists when the end user or programmer must specify the database fragment names but does not need to specify where those fragments are located.

- **Local mapping transparency** exists when the end user or programmer must specify both the fragment names and their locations.

Transparency features are summarized in Table 12.3.

TABLE 12.3	A Summary of Transparency Features		
IF THE SQL STATEMENT REQUIRES:			
FRAGMENT NAME?	LOCATION NAME?	THEN THE DBMS SUPPORTS	LEVEL OF DISTRIBUTON TRANSPARENCY
Yes	Yes	Local mapping	Low
Yes	No	Location transparency	Medium
No	No	Fragmentation transparency	High

As you examine Table 12.3, you might ask why there is no reference to a situation in which the fragment name is "No" and the location name is "Yes." The reason for not including that scenario is simple: you cannot have a location name that fails to reference an existing fragment. (If you don't need to specify a fragment name, its location is clearly irrelevant.)

To illustrate the use of various transparency levels, suppose you have an EMPLOYEE table containing the attributes EMP_NAME, EMP_DOB, EMP_ADDRESS, EMP_DEPARTMENT, and EMP_SALARY. The EMPLOYEE data are distributed over three different locations: New York, Atlanta, and Miami. The table is divided by location; that is, New

York employee data are stored in fragment E1, Atlanta employee data are stored in fragment E2, and Miami employee data are stored in fragment E3. See Figure 12.9.

FIGURE 12.9   **Fragment locations**

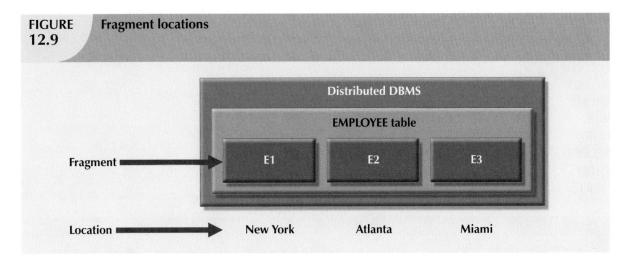

Now suppose the end user wants to list all employees with a date of birth prior to January 1, 1960. To focus on the transparency issues, also suppose the EMPLOYEE table is fragmented and each fragment is unique. The **unique fragment** condition indicates that each row is unique, regardless of the fragment in which it is located. Finally, assume that no portion of the database is replicated at any other site on the network.

Depending on the level of distribution transparency support, you may examine three query cases.

Case 1: The Database Supports Fragmentation Transparency

The query conforms to a nondistributed database query format; that is, it does not specify fragment names or locations. The query reads:

```
SELECT      *
FROM        EMPLOYEE
WHERE       EMP_DOB < '01-JAN-1960';
```

Case 2: The Database Supports Location Transparency

Fragment names must be specified in the query, but fragment location is not specified. The query reads:

```
SELECT      *
FROM        E1
WHERE       EMP_DOB < '01-JAN-1960';
UNION
SELECT      *
FROM        E2
WHERE       EMP_DOB < '01-JAN-1960';
UNION
SELECT      *
FROM        E3
WHERE       EMP_DOB < '01-JAN-1960';
```

Case 3: The Database Supports Local Mapping Transparency

Both the fragment name and location must be specified in the query. Using pseudo-SQL:

```
SELECT       *
FROM         El NODE NY
WHERE        EMP_DOB < '01-JAN-1960';
UNION
SELECT       *
FROM         E2 NODE ATL
WHERE        EMP_DOB < '01-JAN-1960';
UNION
SELECT       *
FROM         E3 NODE MIA
WHERE        EMP_DOB < '01-JAN-1960';
```

> **NOTE**
>
> NODE indicates the location of the database fragment. NODE is used for illustration purposes and is not part of the standard SQL syntax.

As you examine the preceding query formats, you can see how distribution transparency affects the way end users and programmers interact with the database.

Distribution transparency is supported by a **distributed data dictionary (DDD)**, or a **distributed data catalog (DDC)**. The DDC contains the description of the entire database as seen by the database administrator. The database description, known as the **distributed global schema**, is the common database schema used by local TPs to translate user requests into subqueries (remote requests) that will be processed by different DPs. The DDC is itself distributed, and it is replicated at the network nodes. Therefore, the DDC must maintain consistency through updating at all sites.

Keep in mind that some of the current DDBMS implementations impose limitations on the level of transparency support. For instance, you might be able to distribute a database, but not a table, across multiple sites. Such a condition indicates that the DDBMS supports location transparency but not fragmentation transparency.

12.9 TRANSACTION TRANSPARENCY

Transaction transparency is a DDBMS property that ensures that database transactions will maintain the distributed database's integrity and consistency. Remember that a DDBMS database transaction can update data stored in many different computers connected in a network. Transaction transparency ensures that the transaction will be completed only when all database sites involved in the transaction complete their part of the transaction.

Distributed database systems require complex mechanisms to manage transactions and to ensure the database's consistency and integrity. To understand how the transactions are managed, you should know the basic concepts governing remote requests, remote transactions, distributed transactions, and distributed requests.

12.9.1 DISTRIBUTED REQUESTS AND DISTRIBUTED TRANSACTIONS[1]

Whether or not a transaction is distributed, it is formed by one or more database requests. The basic difference between a nondistributed transaction and a distributed transaction is that the latter can update or request data from

---

[1]The details of distributed requests and transactions were originally described in David McGoveran and Colin White, "Clarifying Client/Server," *DBMS* 3(12), November 1990, pp. 78–89.

several different remote sites on a network. To better illustrate the distributed transaction concepts, let's begin by establishing the difference between remote and distributed transactions, using the BEGIN WORK and COMMIT WORK transaction format. Assume the existence of location transparency to avoid having to specify the data location.

A **remote request**, illustrated in Figure 12.10, lets a single SQL statement access the data that are to be processed by a single remote database processor. In other words, the SQL statement (or request) can reference data at only one remote site.

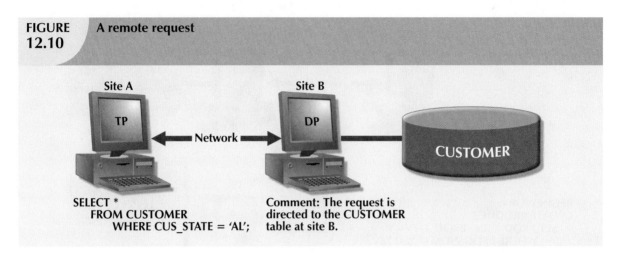

FIGURE 12.10   A remote request

Site A
TP

Network

Site B
DP

CUSTOMER

SELECT *
    FROM CUSTOMER
        WHERE CUS_STATE = 'AL';

Comment: The request is directed to the CUSTOMER table at site B.

Similarly, a **remote transaction**, composed of several requests, accesses data at a single remote site. A remote transaction is illustrated in Figure 12.11.

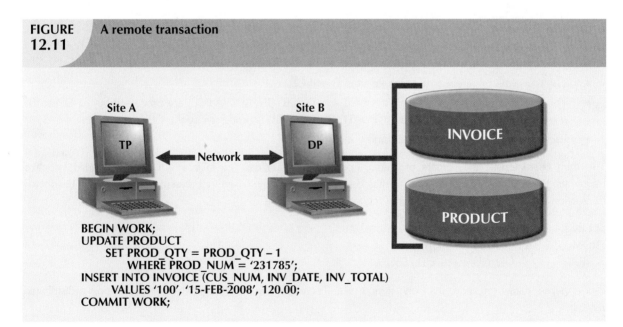

FIGURE 12.11   A remote transaction

Site A
TP

Network

Site B
DP

INVOICE

PRODUCT

BEGIN WORK;
UPDATE PRODUCT
    SET PROD_QTY = PROD_QTY – 1
        WHERE PROD_NUM = '231785';
INSERT INTO INVOICE (CUS_NUM, INV_DATE, INV_TOTAL)
    VALUES '100', '15-FEB-2008', 120.00;
COMMIT WORK;

As you examine Figure 12.11, note the following remote transaction features:

- The transaction updates the PRODUCT and INVOICE tables (located at site B).
- The remote transaction is sent to and executed at the remote site B.
- The transaction can reference only one remote DP.
- Each SQL statement (or request) can reference only one (the same) remote DP at a time, and the entire transaction can reference and be executed at only one remote DP.

A **distributed transaction** allows a transaction to reference several different local or remote DP sites. Although each single request can reference only one local or remote DP site, the transaction as a whole can reference multiple DP sites because each request can reference a different site. The distributed transaction process is illustrated in Figure 12.12.

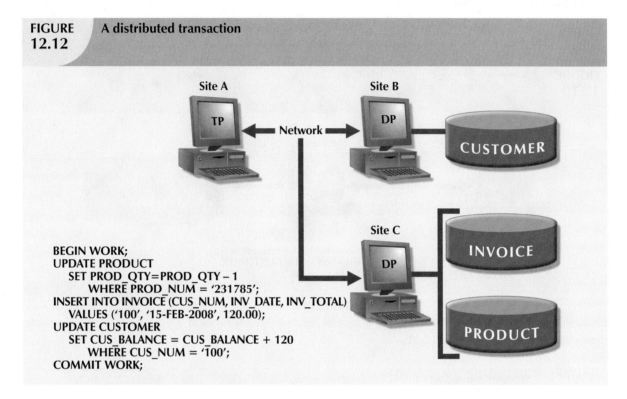

FIGURE 12.12 A distributed transaction

Note the following features in Figure 12.12:

- The transaction references two remote sites (B and C).
- The first two requests (UPDATE PRODUCT and INSERT INTO INVOICE) are processed by the DP at the remote site C, and the last request (UPDATE CUSTOMER) is processed by the DP at the remote site B.
- Each request can access only one remote site at a time.

The third characteristic may create problems. For example, suppose the table PRODUCT is divided into two fragments, PRODl and PROD2, located at sites B and C, respectively. Given that scenario, the preceding distributed transaction cannot be executed because the request:

```
SELECT      *
FROM        PRODUCT
WHERE       PROD_NUM = &'231785';
```

cannot access data from more than one remote site. Therefore, the DBMS must be able to support a distributed request.

A **distributed request** lets a single SQL statement reference data located at several different local or remote DP sites. Because each request (SQL statement) can access data from more than one local or remote DP site, a transaction can access several sites. The ability to execute a distributed request provides fully distributed database processing capabilities because of the ability to:

- Partition a database table into several fragments.
- Reference one or more of those fragments with only one request. In other words, there is fragmentation transparency.

The location and partition of the data should be transparent to the end user. Figure 12.13 illustrates a distributed request. As you examine Figure 12.13, note that the transaction uses a single SELECT statement to reference two tables, CUSTOMER and INVOICE. The two tables are located at two different sites, B and C.

FIGURE
12.13

A distributed request

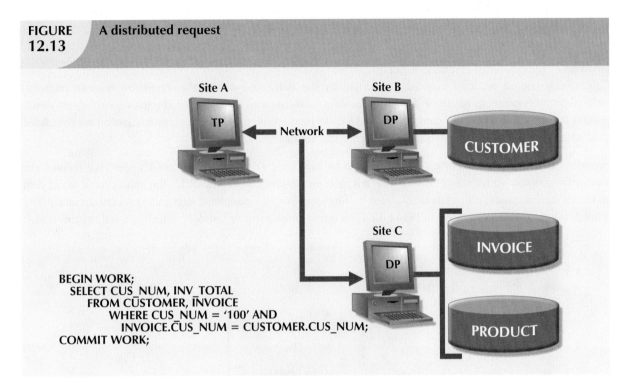

The distributed request feature also allows a single request to reference a physically partitioned table. For example, suppose a CUSTOMER table is divided into two fragments, C1 and C2, located at sites B and C, respectively. Further suppose the end user wants to obtain a list of all customers whose balances exceed $250. The request is illustrated in Figure 12.14. Full fragmentation transparency support is provided only by a DDBMS that supports distributed requests.

FIGURE
12.14

Another distributed request

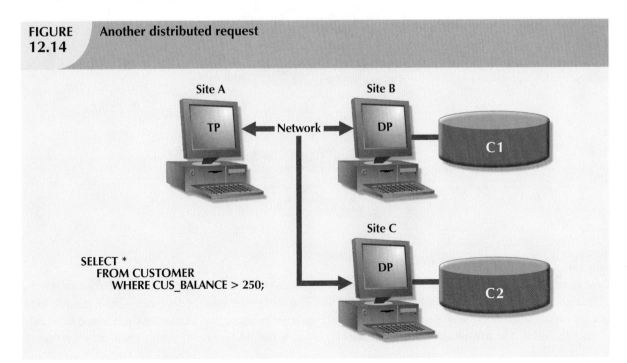

Understanding the different types of database requests in distributed database systems helps you address the transaction transparency issue more effectively. Transaction transparency ensures that distributed transactions are treated as centralized transactions, ensuring the serializability of transactions. (Review Chapter 10, Transaction Management and Concurrency Control, if necessary.) That is, the execution of concurrent transactions, whether or not they are distributed, will take the database from one consistent state to another.

12.9.2 DISTRIBUTED CONCURRENCY CONTROL

Concurrency control becomes especially important in the distributed database environment because multisite, multiple-process operations are more likely to create data inconsistencies and deadlocked transactions than single-site systems are. For example, the TP component of a DDBMS must ensure that all parts of the transaction are completed at all sites before a final COMMIT is issued to record the transaction.

Suppose each transaction operation was committed by each local DP, but one of the DPs could not commit the transaction's results. Such a scenario would yield the problems illustrated in Figure 12.15: the transaction(s) would yield an inconsistent database, with its inevitable integrity problems, because committed data cannot be uncommitted! The solution for the problem illustrated in Figure 12.15 is a *two-phase commit protocol*, which you will explore next.

FIGURE 12.15   **The effect of a premature COMMIT**

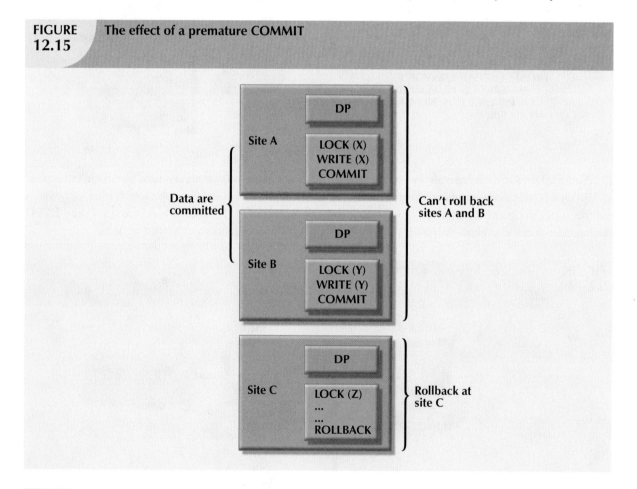

12.9.3 TWO-PHASE COMMIT PROTOCOL

Centralized databases require only one DP. All database operations take place at only one site, and the consequences of database operations are immediately known to the DBMS. In contrast, distributed databases make it possible for a transaction to access data at several sites. A final COMMIT must not be issued until all sites have committed their parts of the transaction. The **two-phase commit protocol** guarantees that if a portion of a transaction operation cannot be committed, all changes made at the other sites participating in the transaction will be undone to maintain a consistent database state.

Each DP maintains its own transaction log. The two-phase commit protocol requires that the transaction entry log for each DP be written before the database fragment is actually updated. (See Chapter 10.) Therefore, the two-phase commit protocol requires a DO-UNDO-REDO protocol and a write-ahead protocol.

The **DO-UNDO-REDO protocol** is used by the DP to roll back and/or roll forward transactions with the help of the system's transaction log entries. The DO-UNDO-REDO protocol defines three types of operations:

- DO performs the operation and records the "before" and "after" values in the transaction log.
- UNDO reverses an operation, using the log entries written by the DO portion of the sequence.
- REDO redoes an operation, using the log entries written by the DO portion of the sequence.

To ensure that the DO, UNDO, and REDO operations can survive a system crash while they are being executed, a write-ahead protocol is used. The **write-ahead protocol** forces the log entry to be written to permanent storage before the actual operation takes place.

The two-phase commit protocol defines the operations between two types of nodes: the **coordinator** and one or more **subordinates**, or *cohorts*. The participating nodes agree on a coordinator. Generally, the coordinator role is assigned to the node that initiates the transaction. However, different systems implement various, more sophisticated election methods. The protocol is implemented in two phases:

Phase 1: Preparation

The coordinator sends a PREPARE TO COMMIT message to all subordinates.

1. The subordinates receive the message; write the transaction log, using the write-ahead protocol; and send an acknowledgment (YES/PREPARED TO COMMIT or NO/NOT PREPARED) message to the coordinator.
2. The coordinator makes sure that all nodes are ready to commit, or it aborts the action.

If all nodes are PREPARED TO COMMIT, the transaction goes to phase 2. If one or more nodes reply NO or NOT PREPARED, the coordinator broadcasts an ABORT message to all subordinates.

Phase 2: The Final COMMIT

1. The coordinator broadcasts a COMMIT message to all subordinates and waits for the replies.
2. Each subordinate receives the COMMIT message, and then updates the database using the DO protocol.
3. The subordinates reply with a COMMITTED or NOT COMMITTED message to the coordinator.

If one or more subordinates did not commit, the coordinator sends an ABORT message, thereby forcing them to UNDO all changes.

The objective of the two-phase commit is to ensure that each node commits its part of the transaction; otherwise, the transaction is aborted. If one of the nodes fails to commit, the information necessary to recover the database is in the transaction log, and the database can be recovered with the DO-UNDO-REDO protocol. (Remember that the log information was updated using the write-ahead protocol.)

12.10 PERFORMANCE TRANSPARENCY AND QUERY OPTIMIZATION

One of the most important functions of a database is its ability to make data available. Because all data reside at a single site in a centralized database, the DBMS must evaluate every data request and find the most efficient way to access the local data. In contrast, the DDBMS makes it possible to partition a database into several fragments, thereby rendering the query translation more complicated, because the DDBMS must decide which fragment of the database to access. In addition, the data may also be replicated at several different sites. The data replication makes the access problem even more complex, because the database must decide which copy of the data to access. The DDBMS uses query optimization techniques to deal with such problems and to ensure acceptable database performance.

The objective of a query optimization routine is to minimize the total cost associated with the execution of a request. The costs associated with a request are a function of the:

- Access time (I/O) cost involved in accessing the physical data stored on disk.
- Communication cost associated with the transmission of data among nodes in distributed database systems.
- CPU time cost associated with the processing overhead of managing distributed transactions.

Although costs are often classified as either communication or processing costs, it is difficult to separate the two. Not all query optimization algorithms use the same parameters, and all algorithms do not assign the same weight to each parameter. For example, some algorithms minimize total time; others minimize the communication time; and still others do not factor in the CPU time, considering it insignificant relative to other cost sources.

NOTE

Chapter 11, Database Performance Tuning and Query Optimization, provides additional details about query optimization.

To evaluate query optimization, keep in mind that the TP must receive data from the DP, synchronize it, assemble the answer, and present it to the end user or an application. Although that process is standard, you should consider that a particular query may be executed at any one of several different sites. The response time associated with remote sites cannot be easily predetermined because some nodes are able to finish their part of the query in less time than others.

One of the most important characteristics of query optimization in distributed database systems is that it must provide distribution transparency as well as *replica* transparency. (Distribution transparency was explained earlier in this chapter.) **Replica transparency** refers to the DDBMS's ability to hide the existence of multiple copies of data from the user.

Most of the algorithms proposed for query optimization are based on two principles:

- The selection of the optimum execution order.
- The selection of sites to be accessed to minimize communication costs.

Within those two principles, a query optimization algorithm can be evaluated on the basis of its *operation mode* or the *timing of its optimization*.

Operation modes can be classified as manual or automatic. **Automatic query optimization** means that the DDBMS finds the most cost-effective access path without user intervention. **Manual query optimization** requires that the optimization be selected and scheduled by the end user or programmer. Automatic query optimization is clearly more desirable from the end user's point of view, but the cost of such convenience is the increased overhead that it imposes on the DDBMS.

Query optimization algorithms can also be classified according to when the optimization is done. Within this timing classification, query optimization algorithms can be classified as static or dynamic.

- **Static query optimization** takes place at compilation time. In other words, the best optimization strategy is selected when the query is compiled by the DBMS. This approach is common when SQL statements are embedded in procedural programming languages such as C# or Visual Basic .NET. When the program is submitted to the DBMS for compilation, it creates the plan necessary to access the database. When the program is executed, the DBMS uses that plan to access the database.

- **Dynamic query optimization** takes place at execution time. Database access strategy is defined when the program is executed. Therefore, access strategy is dynamically determined by the DBMS at run time, using the most up-to-date information about the database. Although dynamic query optimization is efficient, its cost is measured by run-time processing overhead. The best strategy is determined every time the query is executed; this could happen several times in the same program.

Finally, query optimization techniques can be classified according to the type of information that is used to optimize the query. For example, queries may be based on statistically based or rule-based algorithms.

- A **statistically based query optimization algorithm** uses statistical information about the database. The statistics provide information about database characteristics such as size, number of records, average access time, number of requests serviced, and number of users with access rights. These statistics are then used by the DBMS to determine the best access strategy.

- The statistical information is managed by the DDBMS and is generated in one of two different modes: dynamic or manual. In the **dynamic statistical generation mode**, the DDBMS automatically evaluates and updates the statistics after each access. In the **manual statistical generation mode**, the statistics must be updated periodically through a user-selected utility such as IBM's RUNSTAT command used by DB2 DBMSs.

- A **rule-based query optimization algorithm** is based on a set of user-defined rules to determine the best query access strategy. The rules are entered by the end user or database administrator, and they typically are very general in nature.

12.11 DISTRIBUTED DATABASE DESIGN

Whether the database is centralized or distributed, the design principles and concepts described in Chapter 3, The Relational Database Model; Chapter 4, Entity Relationship Modeling; and Chapter 5, Normalization of Database Tables, are still applicable. However, the design of a distributed database introduces three new issues:

- How to partition the database into fragments.
- Which fragments to replicate.
- Where to locate those fragments and replicas.

Data fragmentation and data replication deal with the first two issues, and data allocation deals with the third issue.

12.11.1 DATA FRAGMENTATION

Data fragmentation allows you to break a single object into two or more segments or fragments. The object might be a user's database, a system database, or a table. Each fragment can be stored at any site over a computer network. Information about data fragmentation is stored in the distributed data catalog (DDC), from which it is accessed by the TP to process user requests.

Data fragmentation strategies, as discussed here, are based at the table level and consist of dividing a table into logical fragments. You will explore three types of data fragmentation strategies: horizontal, vertical, and mixed. (Keep in mind that a fragmented table can always be re-created from its fragmented parts by a combination of unions and joins.)

- **Horizontal fragmentation** refers to the division of a relation into subsets (fragments) of tuples (rows). Each fragment is stored at a different node, and each fragment has unique rows. However, the unique rows all have the same attributes (columns). In short, each fragment represents the equivalent of a SELECT statement, with the WHERE clause on a single attribute.

- **Vertical fragmentation** refers to the division of a relation into attribute (column) subsets. Each subset (fragment) is stored at a different node, and each fragment has unique columns—with the exception of the key column, which is common to all fragments. This is the equivalent of the PROJECT statement in SQL.

- **Mixed fragmentation** refers to a combination of horizontal and vertical strategies. In other words, a table may be divided into several horizontal subsets (rows), each one having a subset of the attributes (columns).

To illustrate the fragmentation strategies, let's use the CUSTOMER table for the XYZ Company, depicted in Figure 12.16. The table contains the attributes CUS_NUM, CUS_NAME, CUS_ADDRESS, CUS_STATE, CUS_LIMIT, CUS_BAL, CUS_RATING, and CUS_DUE.

ONLINE CONTENT

The databases used to illustrate the material in this chapter are found in the Student Online Companion for this book.

FIGURE 12.16 **A sample CUSTOMER table**

Table name: CUSTOMER

CUS_NUM	CUS_NAME	CUS_ADDRESS	CUS_STATE	CUS_LIMIT	CUS_BAL	CUS_RATING	CUS_DUE
10	Sinex, Inc.	12 Main St.	TN	3500.00	2700.00	3	1245.00
11	Martin Corp.	321 Sunset Blvd.	FL	6000.00	1200.00	1	0.00
12	Mynux Corp.	910 Eagle St.	TN	4000.00	3500.00	3	3400.00
13	BTBC, Inc.	Rue du Monde	FL	6000.00	5890.00	3	1090.00
14	Victory, Inc.	123 Maple St.	FL	1200.00	550.00	1	0.00
15	NBCC Corp.	909 High Ave.	GA	2000.00	350.00	2	50.00

Horizontal Fragmentation

Suppose XYZ Company's corporate management requires information about its customers in all three states, but company locations in each state (TN, FL, and GA) require data regarding local customers only. Based on such requirements, you decide to distribute the data by state. Therefore, you define the horizontal fragments to conform to the structure shown in Table 12.4.

TABLE 12.4 **Horizontal Fragmentation of the Customer Table by State**

FRAGMENT NAME	LOCATION	CONDITION	NODE NAME	CUSTOMER NUMBERS	NUMBER OF ROWS
CUST_H1	Tennessee	CUS_STATE = 'TN'	NAS	10, 12	2
CUST_H2	Georgia	CUS_STATE = 'GA'	ATL	15	1
CUST_H3	Florida	CUS_STATE = 'FL'	TAM	11, 13, 14	3

Each horizontal fragment may have a different number of rows, but each fragment *must* have the same attributes. The resulting fragments yield the three tables depicted in Figure 12.17.

Vertical Fragmentation

You may also divide the CUSTOMER relation into vertical fragments that are composed of a collection of attributes. For example, suppose the company is divided into two departments: the service department and the collections department. Each department is located in a separate building, and each has an interest in only a few of the CUSTOMER table's attributes. In this case, the fragments are defined as shown in Table 12.5.

FIGURE 12.17 — Table fragments in three locations

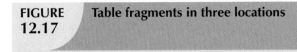

Table name: CUST_H1   **Location: Tennessee**   **Node: NAS**

CUS_NUM	CUS_NAME	CUS_ADDRESS	CUS_STATE	CUS_LIMIT	CUS_BAL	CUS_RATING	CUS_DUE
10	Sinex, Inc.	12 Main St.	TN	3500.00	2700.00	3	1245.00
12	Mynux Corp.	910 Eagle St.	TN	4000.00	3500.00	3	3400.00

Table name: CUST_H2   **Location: Georgia**   **Node: ATL**

CUS_NUM	CUS_NAME	CUS_ADDRESS	CUS_STATE	CUS_LIMIT	CUS_BAL	CUS_RATING	CUS_DUE
15	NBCC Corp.	909 High Ave.	GA	2000.00	350.00	2	50.00

Table name: CUST_H3   **Location: Florida**   **Node: TAM**

CUS_NUM	CUS_NAME	CUS_ADDRESS	CUS_STATE	CUS_LIMIT	CUS_BAL	CUS_RATING	CUS_DUE
11	Martin Corp.	321 Sunset Blvd.	FL	6000.00	1200.00	1	0.00
13	BTBC, Inc.	Rue du Monde	FL	6000.00	5890.00	3	1090.00
14	Victory, Inc.	123 Maple St.	FL	1200.00	550.00	1	0.00

TABLE 12.5 — Vertical Fragmentation of the Customer Table

FRAGMENT NAME	LOCATION	NODE NAME	ATTRIBUTE NAMES
CUST_V1	Service Bldg.	SVC	CUS_NUM, CUS_NAME, CUS_ADDRESS, CUS_STATE
CUST_V2	Collection Bldg.	ARC	CUS_NUM, CUS_LIMIT, CUS_BAL, CUS_RATING, CUS_DUE

Each vertical fragment must have the same number of rows, but the inclusion of the different attributes depends on the key column. The vertical fragmentation results are displayed in Figure 12.18. Note that the key attribute (CUS_NUM) is common to both fragments CUST_V1 and CUST_V2.

FIGURE 12.18 — Vertically fragmented table contents

Table name: CUST_V1   **Location: Service Building**   **Node: SVC**

CUS_NUM	CUS_NAME	CUS_ADDRESS	CUS_STATE
10	Sinex, Inc.	12 Main St.	TN
11	Martin Corp.	321 Sunset Blvd.	FL
12	Mynux Corp.	910 Eagle St.	TN
13	BTBC, Inc.	Rue du Monde	FL
14	Victory, Inc.	123 Maple St.	FL
15	NBCC Corp.	909 High Ave.	GA

Table name: CUST_V2   **Location: Collection Building**   **Node: ARC**

CUS_NUM	CUS_LIMIT	CUS_BAL	CUS_RATING	CUS_DUE
10	3500.00	2700.00	3	1245.00
11	6000.00	1200.00	1	0.00
12	4000.00	3500.00	3	3400.00
13	6000.00	5890.00	3	1090.00
14	1200.00	550.00	1	0.00
15	2000.00	350.00	2	50.00

Mixed Fragmentation

The XYZ Company's structure requires that the CUSTOMER data be fragmented horizontally to accommodate the various company locations; within the locations, the data must be fragmented vertically to accommodate the two departments (service and collection). In short, the CUSTOMER table requires mixed fragmentation.

Mixed fragmentation requires a two-step procedure. First, horizontal fragmentation is introduced for each site based on the location within a state (CUS_STATE). The horizontal fragmentation yields the subsets of customer tuples (horizontal fragments) that are located at each site. Because the departments are located in different buildings, vertical fragmentation is used within each horizontal fragment to divide the attributes, thus meeting each department's information needs at each subsite. Mixed fragmentation yields the results displayed in Table 12.6.

TABLE 12.6	Mixed Fragmentation of the Customer Table				
FRAGMENT NAME	LOCATION	HORIZONTAL CRITERIA	NODE NAME	RESULTING ROWS AT SITE	VERTICAL CRITERIA ATTRIBUTES AT EACH FRAGMENT
CUST_M1	TN-Service	CUS_STATE = 'TN'	NAS-S	10, 12	CUS_NUM, CUS_NAME CUS_ADDRESS, CUS_STATE
CUST_M2	TN-Collection	CUS_STATE = 'TN'	NAS-C	10, 12	CUS_NUM, CUS_LIMIT, CUS_BAL, CUS_RATING, CUS_DUE
CUST_M3	GA-Service	CUS_STATE = 'GA'	ATL-S	15	CUS_NUM, CUS_NAME CUS_ADDRESS, CUS_STATE
CUST_M4	GA-Collection	CUS_STATE = 'GA'	ATL-C	15	CUS_NUM, CUS_LIMIT, CUS_BAL, CUS_RATING, CUS_DUE
CUST_M5	FL-Service	CUS_STATE = 'FL'	TAM-S	11, 13, 14	CUS_NUM, CUS_NAME CUS_ADDRESS, CUS_STATE
CUST_M6	FL-Collection	CUS_STATE = 'FL'	TAM-C	11, 13, 14	CUS_NUM, CUS_LIMIT, CUS_BAL, CUS_RATING, CUS_DUE

Each fragment displayed in Table 12.6 contains customer data by state and, within each state, by department location, to fit each department's data requirements. The tables corresponding to the fragments listed in Table 12.6 are shown in Figure 12.19.

FIGURE 12.19   Table contents after the mixed fragmentation process

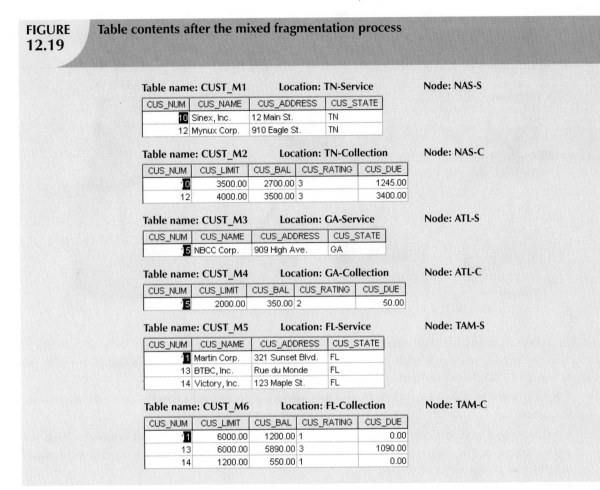

Table name: CUST_M1   Location: TN-Service   Node: NAS-S

CUS_NUM	CUS_NAME	CUS_ADDRESS	CUS_STATE
10	Sinex, Inc.	12 Main St.	TN
12	Mynux Corp.	910 Eagle St.	TN

Table name: CUST_M2   Location: TN-Collection   Node: NAS-C

CUS_NUM	CUS_LIMIT	CUS_BAL	CUS_RATING	CUS_DUE
10	3500.00	2700.00	3	1245.00
12	4000.00	3500.00	3	3400.00

Table name: CUST_M3   Location: GA-Service   Node: ATL-S

CUS_NUM	CUS_NAME	CUS_ADDRESS	CUS_STATE
15	NBCC Corp.	909 High Ave.	GA

Table name: CUST_M4   Location: GA-Collection   Node: ATL-C

CUS_NUM	CUS_LIMIT	CUS_BAL	CUS_RATING	CUS_DUE
15	2000.00	350.00	2	50.00

Table name: CUST_M5   Location: FL-Service   Node: TAM-S

CUS_NUM	CUS_NAME	CUS_ADDRESS	CUS_STATE
11	Martin Corp.	321 Sunset Blvd.	FL
13	BTBC, Inc.	Rue du Monde	FL
14	Victory, Inc.	123 Maple St.	FL

Table name: CUST_M6   Location: FL-Collection   Node: TAM-C

CUS_NUM	CUS_LIMIT	CUS_BAL	CUS_RATING	CUS_DUE
11	6000.00	1200.00	1	0.00
13	6000.00	5890.00	3	1090.00
14	1200.00	550.00	1	0.00

12.11.2 DATA REPLICATION

Data replication refers to the storage of data copies at multiple sites served by a computer network. Fragment copies can be stored at several sites to serve specific information requirements. Because the existence of fragment copies can enhance data availability and response time, data copies can help to reduce communication and total query costs.

Suppose database A is divided into two fragments, A1 and A2. Within a replicated distributed database, the scenario depicted in Figure 12.20 is possible: fragment A1 is stored at sites S1 and S2, while fragment A2 is stored at sites S2 and S3.

Replicated data are subject to the mutual consistency rule. The **mutual consistency rule** requires that all copies of data fragments be identical. Therefore, to maintain data consistency among the replicas, the DDBMS must ensure that a database update is performed at all sites where replicas exist.

Although replication has some benefits (such as improved data availability, better load distribution, improved data failure-tolerance, and reduced query costs), it also imposes additional DDBMS processing overhead— because each data copy must be maintained by the system. Furthermore, because the data are replicated at another site, there are

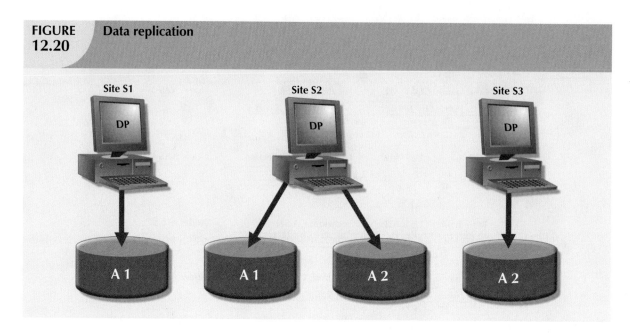

FIGURE 12.20   Data replication

associated storage costs and increased transaction times (as data must be updated at several sites concurrently to comply with the mutual consistency rule). To illustrate the replica overhead imposed on a DDBMS, consider the processes that the DDBMS must perform to use the database.

- If the database is fragmented, the DDBMS must decompose a query into *subqueries* to access the appropriate fragments.

- If the database is replicated, the DDBMS must decide which copy to access. A READ operation selects the *nearest copy* to satisfy the transaction. A WRITE operation requires that *all copies* be selected and updated to satisfy the mutual consistency rule.

- The TP sends a data request to each selected DP for execution.

- The DP receives and executes each request and sends the data back to the TP.

- The TP assembles the DP responses.

The problem becomes more complex when you consider additional factors such as network topology and communication throughputs.

Three replication scenarios exist: a database can be fully replicated, partially replicated, or unreplicated.

- A **fully replicated database** stores multiple copies of *each* database fragment at multiple sites. In this case, all database fragments are replicated. A fully replicated database can be impractical due to the amount of overhead it imposes on the system.

- A **partially replicated database** stores multiple copies of *some* database fragments at multiple sites. Most DDBMSs are able to handle the partially replicated database well.

- An **unreplicated database** stores each database fragment at a single site. Therefore, there are no duplicate database fragments.

Several factors influence the decision to use data replication:

- Database size. The amount of data replicated will have an impact on the storage requirements and also on the data transmission costs. Replicating large amounts of data requires a window of time and higher network bandwidth that could affect other applications.

- Usage frequency. The frequency of data usage determines how frequently the data needs to be updated. Frequently used data needs to be updated more often, for example, than large data sets that are used only every quarter.
- Costs, including those for performance, software overhead, and management associated with synchronizing transactions and their components vs. fault-tolerance benefits that are associated with replicated data.

When the usage frequency of remotely located data is high and the database is large, data replication can reduce the cost of data requests. Data replication information is stored in the distributed data catalog (DDC), whose contents are used by the TP to decide which copy of a database fragment to access. The data replication makes it possible to restore lost data.

12.11.3 DATA ALLOCATION

Data allocation describes the process of deciding where to locate data. Data allocation strategies are as follows:

- With **centralized data allocation**, the entire database is stored at one site.
- With **partitioned data allocation**, the database is divided into two or more disjointed parts (fragments) and stored at two or more sites.
- With **replicated data allocation**, copies of one or more database fragments are stored at several sites.

Data distribution over a computer network is achieved through data partition, through data replication, or through a combination of both. Data allocation is closely related to the way a database is divided or fragmented. Most data allocation studies focus on one issue: *which* data to locate *where*.

Data allocation algorithms take into consideration a variety of factors, including:

- Performance and data availability goals.
- Size, number of rows, and number of relations that an entity maintains with other entities.
- Types of transactions to be applied to the database and the attributes accessed by each of those transactions.
- Disconnected operation for mobile users. In some cases, the design might consider the use of loosely disconnected fragments for mobile users, particularly for read-only data that does not require frequent updates and for which the replica update windows (the amount of time available to perform a certain data processing task that cannot be executed concurrently with other tasks) may be longer.

Some algorithms include external data, such as network topology or network throughput. No optimal or universally accepted algorithm exists yet, and very few algorithms have been implemented to date.

12.12 CLIENT/SERVER VS. DDBMS

Because the trend toward distributed databases is firmly established, many database vendors have used the "client/server" label to indicate distributed database capability. However, distributed databases do not always accurately reflect the characteristics implied by the client/server label.

Client/server architecture refers to the way in which computers interact to form a system. The client/server architecture features a *user* of resources, or a client, and a *provider* of resources, or a server. The client/server architecture can be used to implement a DBMS in which the client is the TP and the server is the DP.

Client/server interactions in a DDBMS are carefully scripted. The client (TP) interacts with the end user and sends a request to the server (DP). The server receives, schedules, and executes the request, *selecting only those records that are needed by the client*. The server then sends the data to the client *only* when the client requests the data.

Client/server applications offer several advantages.

- Client/server solutions tend to be less expensive than alternate minicomputer or mainframe solutions in terms of startup infrastructure requirements.

- Client/server solutions allow the end user to use the microcomputer's GUI, thereby improving functionality and simplicity. In particular, using the ubiquitous Web browser in conjunction with Java and .NET frameworks provides a familiar end-user interface.

- More people in the job market have PC skills than mainframe skills. The majority of new generation students are learning Java and .NET programming skills.

- The PC is well established in the workplace. In addition, the increased use of the Internet as a business channel, coupled with security advances (SSL, Virtual Private Networks, multifactor authentication, etc.) provide a more reliable and secure platform for business transactions.

- Numerous data analysis and query tools exist to facilitate interaction with many of the DBMSs that are available in the PC market.

- There is a considerable cost advantage to offloading applications development from the mainframe to powerful PCs.

Client/server applications are also subject to some disadvantages.

- The client/server architecture creates a more complex environment in which different platforms (LANs, operating systems, and so on) are often difficult to manage.

- An increase in the number of users and processing sites often paves the way for security problems.

- The client/server environment makes it possible to spread data access to a much wider circle of users. Such an environment increases the demand for people with a broad knowledge of computers and software applications. The burden of training increases the cost of maintaining the environment.

ONLINE CONTENT

Refer to **Appendix F, Client/Server Systems,** for complete coverage of client/server computing concepts, components, and managerial implications.

12.13 C. J. DATE'S TWELVE COMMANDMENTS FOR DISTRIBUTED DATABASES

The notion of distributed databases has been around for at least 20 years. With the rise of relational databases, most vendors implemented their own versions of distributed databases, generally highlighting their respective product's strengths. To make the comparison of distributed databases easier, C. J. Date formulated twelve "commandments" or basic principles of distributed databases.[2] Although no current DDBMS conforms to all of them, they constitute a useful target. The twelve rules are as follows:

1. *Local site independence.* Each local site can act as an independent, autonomous, centralized DBMS. Each site is responsible for security, concurrency control, backup, and recovery.

2. *Central site independence.* No site in the network relies on a central site or any other site. All sites have the same capabilities.

3. *Failure independence.* The system is not affected by node failures. The system is in continuous operation even in the case of a node failure or an expansion of the network.

4. *Location transparency.* The user does not need to know the location of data in order to retrieve those data.

---

[2] Date, C. J. "Twelve Rules for a Distributed Database," *Computer World*, June 8, 1987, 2(23) pp. 77–81.

5. *Fragmentation transparency.* Data fragmentation is transparent to the user, who sees only one logical database. The user does not need to know the name of the database fragments in order to retrieve them.

6. *Replication transparency.* The user sees only one logical database. The DDBMS transparently selects the database fragment to access. To the user, the DDBMS manages all fragments transparently.

7. *Distributed query processing.* A distributed query may be executed at several different DP sites. Query optimization is performed transparently by the DDBMS.

8. *Distributed transaction processing.* A transaction may update data at several different sites, and the transaction is executed transparently.

9. *Hardware independence.* The system must run on any hardware platform.

10. *Operating system independence.* The system must run on any operating system platform.

11. *Network independence.* The system must run on any network platform.

12. *Database independence.* The system must support any vendor's database product.

Summary

▮ A distributed database stores logically related data in two or more physically independent sites connected via a computer network. The database is divided into fragments, which can be horizontal (a set of rows) or vertical (a set of attributes). Each fragment can be allocated to a different network node.

▮ Distributed processing is the division of logical database processing among two or more network nodes. Distributed databases require distributed processing. A distributed database management system (DDBMS) governs the processing and storage of logically related data through interconnected computer systems.

▮ The main components of a DDBMS are the transaction processor (TP) and the data processor (DP). The transaction processor component is the software that resides on each computer node that requests data. The data processor component is the software that resides on each computer that stores and retrieves data.

▮ Current database systems can be classified by the extent to which they support processing and data distribution. Three major categories are used to classify distributed database systems: (1) single-site processing, single-site data (SPSD); (2) multiple-site processing, single-site data (MPSD); and (3) multiple-site processing, multiple-site data (MPMD).

▮ A homogeneous distributed database system integrates only one particular type of DBMS over a computer network. A heterogeneous distributed database system integrates several different types of DBMSs over a computer network.

▮ DDBMS characteristics are best described as a set of transparencies: distribution, transaction, failure, heterogeneity, and performance. All transparencies share the common objective of making the distributed database behave as though it were a centralized database system; that is, the end user sees the data as part of a single logical centralized database and is unaware of the system's complexities.

▮ A transaction is formed by one or more database requests. An undistributed transaction updates or requests data from a single site. A distributed transaction can update or request data from multiple sites.

▮ Distributed concurrency control is required in a network of distributed databases. A two-phase COMMIT protocol is used to ensure that all parts of a transaction are completed.

▮ A distributed DBMS evaluates every data request to find the optimum access path in a distributed database. The DDBMS must optimize the query to reduce access, communications, and CPU costs associated with the query.

▮ The design of a distributed database must consider the fragmentation and replication of data. The designer must also decide how to allocate each fragment or replica to obtain better overall response time and to ensure data availability to the end user.

▮ A database can be replicated over several different sites on a computer network. The replication of the database fragments has the objective of improving data availability, thus decreasing access time. A database can be partially, fully, or not replicated. Data allocation strategies are designed to determine the location of the database fragments or replicas.

▮ Database vendors often label software as client/server database products. The client/server architecture label refers to the way in which two computers interact over a computer network to form a system.

KEY TERMS

application processor (AP), 484

automatic query optimization, 498

client/server architecture, 488

coordinator, 497

data allocation, 505
   centralized, 505
   partitioned, 505
   replicated, 505

database fragments, 481

data fragmentation, 499
   horizontal, 499
   mixed, 499
   vertical, 499

data manager (DM), 484

data processor (DP), 484

data replication, 503

distributed database, 481

distributed database management
   system (DDBMS), 478

distributed data catalog (DDC), 492

distributed data dictionary
   (DDD), 492

distributed global schema, 492

distributed processing, 481

distributed request, 494

distributed transaction, 494

distribution transparency, 489

DO-UNDO-REDO protocol, 497

dynamic query optimization, 499

dynamic statistical generation
   mode, 499

failure transparency, 490

fragmentation transparency, 490

fully heterogeneous DDBMS, 488

fully replicated database, 504

heterogeneity transparency, 490

heterogeneous DDBMS, 488

homogeneous DDBMS, 488

local mapping transparency, 490

location transparency, 490

manual query optimization, 498

manual statistical generation
   mode, 499

multiple-site processing, multiple-site
   data (MPMD), 488

multiple-site processing, single-site
   data (MPSD), 487

mutual consistency rule, 503

partially replicated database, 504

performance transparency, 490

remote request, 493

remote transaction, 493

replica transparency, 498

rule-based query optimization
   algorithm, 499

single-site processing, single-site
   data (SPSD), 487

static query optimization, 498

statistically based query optimization
   algorithm, 499

subordinates, 497

transaction manager (TM), 484

transaction processor (TP), 484

transaction transparency, 490

two-phase commit protocol, 496

unique fragment, 491

unreplicated database, 504

write-ahead protocol, 497

ONLINE CONTENT

Answers to selected Review Questions and Problems for this chapter are contained in the Student Online Companion for this book.

REVIEW QUESTIONS

1. Describe the evolution from centralized DBMSs to distributed DBMSs.

2. List and discuss some of the factors that influenced the evolution of the DDBMS.

3. What are the advantages of the DDBMS?

4. What are the disadvantages of the DDBMS?

5. Explain the difference between a distributed database and distributed processing.

6. What is a fully distributed database management system?

7. What are the components of a DDBMS?

8. List and explain the transparency features of a DDBMS.

9. Define and explain the different types of distribution transparency.

10. Describe the different types of database requests and transactions.

11. Explain the need for the two-phase commit protocol. Then describe the two phases.

12. What is the objective of query optimization functions?

13. To which transparency feature are the query optimization functions related?

14. What are the different types of query optimization algorithms?

15. Describe the three data fragmentation strategies. Give some examples of each.

16. What is data replication, and what are the three replication strategies?

17. Explain the difference between distributed databases and client/server architecture.

PROBLEMS

The first problem is based on the DDBMS scenario in Figure P12.1.

FIGURE P12.1   **The DDBMS scenario for Problem 1**

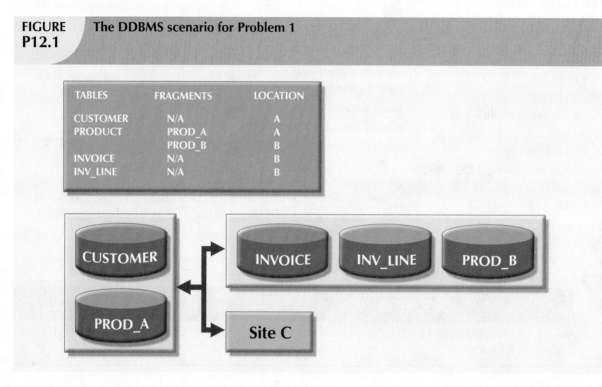

TABLES	FRAGMENTS	LOCATION
CUSTOMER	N/A	A
PRODUCT	PROD_A	A
	PROD_B	B
INVOICE	N/A	B
INV_LINE	N/A	B

1. Specify the minimum type(s) of operation(s) the database must support (remote request, remote transaction, distributed transaction, or distributed request) to perform the following operations:

    At site C

    a. SELECT   *
       FROM   CUSTOMER;

    b. SELECT   *
       FROM   INVOICE
       WHERE   INV_TOT > 1000;

    c. SELECT   *
       FROM   PRODUCT
       WHERE   PROD_ QOH < 10;

d. BEGIN WORK;

UPDATE	CUSTOMER
SET	CUS_BAL = CUS_BAL + 100
WHERE	CUS_NUM = '10936';
INSERT	INTO INVOICE(INV_NUM, CUS_NUM, INV_DATE, INV_TOTAL)
	VALUES ('986391', '10936', '15-FEB-2008', 100);
INSERT	INTO LINE(INV_NUM, PROD_NUM, LINE_PRICE)
	VALUES('986391', '1023', 100);
UPDATE	PRODUCT
SET	PROD_QOH = PROD_ QOH –1
WHERE	PROD_NUM = '1023'; COMMIT WORK;

e. BEGIN WORK;

INSERT	INTO CUSTOMER(CUS_NUM, CUS_NAME, CUS_ADDRESS, CUS_BAL)
	VALUES ('34210', 'Victor Ephanor', '123 Main St.', 0.00);
INSERTINTO INVOICE(INV_NUM, CUS_NUM, INV_DATE, INV_TOTAL)	
	VALUES ('986434', '34210', '10-AUG-2007', 2.00);

COMMIT WORK;

At site A

f.
SELECT	CUS_NUM,CUS_NAME,INV_TOTAL
FROM	CUSTOMER, INVOICE
WHERE	CUSTOMER.CUS_NUM = INVOICE.CUS_NUM;

g.
SELECT	*
FROM	INVOICE
WHERE	INV_TOTAL > 1000;

h.
SELECT	*
FROM	PRODUCT
WHERE	PROD_QOH < 10;

At site B

i.
SELECT	*
FROM	CUSTOMER;

j.
SELECT	CUS_NAME, INV_TOTAL
FROM	CUSTOMER, INVOICE
WHERE	INV_TOTAL > 1000
	AND CUSTOMER.CUS_NUM = INVOICE.CUS_NUM;

k.
SELECT	*
FROM	PRODUCT
WHERE	PROD_QOH < 10;

2. The following data structure and constraints exist for a magazine publishing company:

   a. The company publishes one regional magazine in each region: Florida (FL), South Carolina (SC), Georgia (GA), and Tennessee (TN).

   b. The company has 300,000 customers (subscribers) distributed throughout the four states listed in Part a.

   c. On the first of each month, an annual subscription INVOICE is printed and sent to each customer whose subscription is due for renewal. The INVOICE entity contains a REGION attribute to indicate the state (FL, SC, GA, TN) in which the customer resides:

   CUSTOMER (CUS_NUM, CUS_NAME, CUS_ADDRESS, CUS_CITY, CUS_ZIP, CUS_SUBSDATE)
   INVOICE (INV_NUM, INV_REGION, CUS_NUM, INV_DATE, INV_TOTAL)

   The company's management is aware of the problems associated with centralized management and has decided to decentralize management of the subscriptions into the company's four regional subsidiaries. Each subscription site will handle its own customer and invoice data. The management at company headquarters, however, will have access to customer and invoice data to generate annual reports and to issue ad hoc queries such as:

   - List all current customers by region.
   - List all new customers by region.
   - Report all invoices by customer and by region.

   Given those requirements, how must you partition the database?

3. Given the scenario and the requirements in Question 2, answer the following questions:

   a. What recommendations will you make regarding the type and characteristics of the required database system?

   b. What type of data fragmentation is needed for each table?

   c. What criteria must be used to partition each database?

   d. Design the database fragments. Show an example with node names, location, fragment names, attribute names, and demonstration data.

   e. What type of distributed database operations must be supported at each remote site?

   f. What type of distributed database operations must be supported at the headquarters site?

BUSINESS INTELLIGENCE AND DATA WAREHOUSES

In this chapter, you will learn:

- How business intelligence is a comprehensive framework to support business decision making
- How operational data and decision support data differ
- What a data warehouse is, how to prepare data for one, and how to implement one
- What star schemas are and how they are constructed
- What data mining is and what role it plays in decision support
- About online analytical processing (OLAP)
- How SQL extensions are used to support OLAP-type data manipulations

Preview

Data are crucial raw material in this information age, and data storage and management have become the focus of database design and implementation. Ultimately, the reason for collecting, storing, and managing data is to generate information that becomes the basis for rational decision making. Decision support systems (DSSs) were originally developed to facilitate the decision-making process. However, as the complexity and range of information requirements increased, so did the difficulty of extracting all the necessary information from the data structures typically found in an operational database. Therefore, a new data storage facility, called a data warehouse, was developed. The data warehouse extracts or obtains its data from operational databases as well as from external sources, providing a more comprehensive data pool.

In parallel with data warehouses, new ways to analyze and present decision support data were developed. Online analytical processing (OLAP) provides advanced data analysis and presentation tools (including multidimensional data analysis). Data mining employs advanced statistical tools to analyze the wealth of data now available through data warehouses and other sources and to identify possible relationships and anomalies.

Business intelligence (BI) is the collection of best practices and software tools developed to support business decision making in this age of globalization, emerging markets, rapid change, and increasing regulation. BI encompasses tools and techniques such as data warehouses and OLAP, with a more comprehensive focus on integrating them from a company-wide perspective.

This chapter explores the main concepts and components of business intelligence and decision support systems that gather, generate, and present information for business decision makers, focusing especially on the use of data warehouses.

13.1 THE NEED FOR DATA ANALYSIS

Organizations tend to grow and prosper as they gain a better understanding of their environment. Most managers want to be able to track daily transactions to evaluate how the business is performing. By tapping into the operational database, management can develop strategies to meet organizational goals. In addition, data analysis can provide information about short-term tactical evaluations and strategies such as these: Are our sales promotions working? What market percentage are we controlling? Are we attracting new customers? Tactical and strategic decisions are also shaped by constant pressure from external and internal forces, including globalization, the cultural and legal environment, and (perhaps most importantly) technology.

Given the many and varied competitive pressures, managers are always looking for a competitive advantage through product development and maintenance, service, market positioning, sales promotion, and so on. Managers understand that the business climate is dynamic, and thus, mandates their prompt reaction to change in order to remain competitive. In addition, the modern business climate requires managers to approach increasingly complex problems that involve a rapidly growing number of internal and external variables. It should also come as no surprise that interest is growing in creating support systems dedicated to facilitating quick decision making in a complex environment.

Different managerial levels require different decision support needs. For example, transaction-processing systems, based on operational databases, are tailored to serve the information needs of people who deal with short-term inventory, accounts payable, and purchasing. Middle-level managers, general managers, vice presidents, and presidents focus on strategic and tactical decision making. Those managers require detailed information designed to help them make decisions in a complex data and analysis environment.

Companies and software vendors addressed these multilevel decision support needs by creating independent applications to fit the needs of particular areas (finance, customer management, human resources, product support, etc.). Applications were also tailored to different industry sectors such as education, retail, health care, or financial. This approach worked well for some time, but changes in the business world (globalization, expanding markets, mergers and acquisitions, increased regulation, and more) called for new ways of integrating and managing data across levels and sectors. This more comprehensive and integrated decision support framework within organizations became known as business intelligence.

13.2 BUSINESS INTELLIGENCE

Business intelligence (BI)[1] is a term used to describe a comprehensive, cohesive, and integrated set of tools and processes used to capture, collect, integrate, store, and analyze data with the purpose of generating and presenting information used to support business decision making. As the names implies, BI is about creating intelligence about a business. This intelligence is based on learning and understanding the facts about a business environment. BI is a framework that allows a business to transform data into information, information into knowledge, and knowledge into wisdom. BI has the potential to positively affect a company's culture by creating "business wisdom" and distributing it to all users in an organization. This business wisdom empowers users to make sound business decisions based on the accumulated

---

[1] In 1989, while working at Gartner Inc., Howard Dresner popularized "BI" as an umbrella term to describe a set of concepts and methods to improve business decision making by using fact-based support systems. Source: http://www.computerworld.com/action/article.do?command=viewArticleBasic&articleId=266298

knowledge of the business as reflected on recorded facts (historic operational data). Table 13.1 gives some real-world examples of companies that have implemented BI tools (data warehouse, data mart, OLAP, and/or data mining tools) and shows how the use of such tools benefited the companies.

TABLE 13.1 **Solving Business Problems and Adding Value with BI Tools**

COMPANY	PROBLEM	BENEFIT
MOEN Manufacturer of bathroom and kitchen fixtures and supplies Source: Cognos Corp. *www.cognos.com*	• Information generation very limited and time-consuming. • How to extract data using a 3GL known by only five people. • Response time unacceptable for managers' decision-making purposes.	• Provided quick answers to ad hoc questions for decision making. • Provided access to data for decision-making purposes. • Received in-depth view of product performance and customer margins.
NASDAQ Largest U.S. electronic stock market trading organization Source: Oracle *www.oracle.com*	• Inability to provide real-time ad hoc query and standard reporting to executives, business analysts, and other users. • Excessive storage costs for many terabytes of data.	• Reduced storage cost by moving to a multitier storage solution. • Implemented new data warehouse center with support for ad hoc query and reporting and near real-time data access for end users.
Sega of America, Inc. Interactive entertainment systems and video games Source: Oracle Corp. *www.oracle.com*	• Needed a way to rapidly analyze a great amount of data. • Needed to track advertising, coupons, and rebates associated with effects of pricing changes. • Used to do it with Excel spreadsheets, leading to human-caused errors.	• Eliminated data-entry errors. • Identified successful marketing strategies to dominate interactive entertainment niches. • Used product analysis to identify better markets/product offerings.
Owens and Minor, Inc. Medical and surgical supply distributor Source: *CFO Magazine* *www.cfomagazine.com*	• Lost its largest customer, which represented 10% of its annual revenue ($360 million). • Stock plunged 23%. • Cumbersome process to get information out of antiquated mainframe system.	• Increased earnings per share in just five months. • Gained more business, thanks to opening the data warehouse to its clients. • Managers gained quick access to data for decision-making purposes.
Amazon.com Leading online retailer Source: *PC Week Online* *whitepapers.zdnet.com/whitepaper.aspx?docid=241748*	• Difficulty in managing a very rapidly growing data environment. • Existing data warehouse solution not capable of supporting extremely rapid growth. • Needed more flexible and reliable data warehouse solution to protect its investment in data and infrastructure.	• Implemented new data warehouse with superior scalability and performance. • Improved business intelligence. • Improved management of product flow through the entire supply chain. • Improved customer experience.

BI is a comprehensive endeavor because it encompasses all business processes within an organization. *Business processes* are the central units of operation in a business. Implementing BI in an organization involves capturing not only business data (internal and external) but also the metadata, or knowledge about the data. In practice, BI is a complex proposition that requires a deep understanding and alignment of the business processes, the internal and external data, and the information needs of users at all levels in an organization.

BI is not a product by itself, but a framework of concepts, practices, tools, and technologies that help a business better understand its core capabilities, provide snapshots of the company situation, and identify key opportunities to create competitive advantage. In practice, BI provides a well-orchestrated framework for the management of data that works across all levels of the organization. BI involves the following general steps:

1. Collecting and storing operational data

2. Aggregating the operational data into decision support data

3. Analyzing decision support data to generate information

4. Presenting such information to the end user to support business decisions

5. Making business decisions, which in turn generate more data that is collected, stored, etc. (restarting the process)

6. Monitoring results to evaluate outcomes of the business decisions (providing more data to be collected, stored, etc.)

To implement all these steps, BI uses varied components and technologies. In the following sections, you will learn about the basic BI architecture and implementations.

13.3 BUSINESS INTELLIGENCE ARCHITECTURE

BI covers a range of technologies and applications to manage the entire data life cycle from acquisition to storage, transformation, integration, analysis, monitoring, presentation, and archiving. BI functionality ranges from simple data gathering and extraction to very complex data analysis and presentation applications. There is no single BI architecture; instead, it ranges from highly integrated applications from a single vendor to a loosely integrated, multivendor environment. However, there are some general types of functionality that all BI implementations share.

Like any critical business IT infrastructure, the BI architecture is composed of data, people, processes, technology, and the management of such components. Figure 13.1 depicts how all those components fit together within the BI framework.

Remember that the main focus of BI is to gather, integrate, and store business data for the purpose of creating information. As depicted in Figure 13.1, BI integrates people and processes using technology in order to add value to the business. Such value is derived from how end users use such information in their daily activities, and in particular, their daily business decision making. Also note that the BI technology components are varied. This chapter will explain those components in greater detail in the following sections.

The focus of traditional information systems was on operational automation and reporting; in contrast, BI tools focus on the strategic and tactical use of information. In order to achieve this goal, BI recognizes that technology alone is not enough. Therefore, BI uses an arrangement of best management practices to manage data as a corporate asset. One of the most recent developments in this area is the use of master data management techniques. **Master data management (MDM)** is a collection of concepts, techniques, and processes for the proper identification, definition, and management of data elements within an organization. MDM's main goal is to provide a comprehensive and consistent definition of all data within an organization. MDM ensures that all company resources (people, procedures, and IT systems) that operate over data have uniform and consistent views of the company's data.

FIGURE 13.1 Business intelligence framework

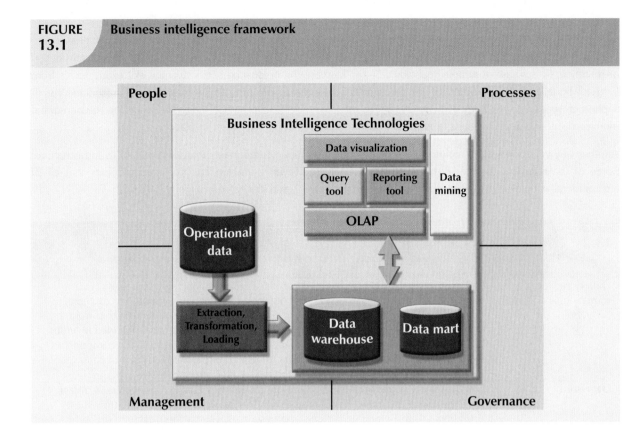

An added benefit of this meticulous approach to data management and decision making is that it provides a framework for business governance. **Governance** is a method or process of government. In this case, BI provides a method for controlling and monitoring business health and for consistent decision making. Furthermore, having such governance creates accountability for business decisions. In the present age of business flux, accountability is increasingly important. Had governance been as pivotal to business operations a few years back, crises precipitated by the likes of Enron, WorldCom, and Arthur Andersen might have been avoided.

Monitoring a business's health is crucial to understanding where the company is and where it is headed. In order to do this, BI makes extensive use of a special type of metrics known as key performance indicators. **Key performance indicators (KPI)** are quantifiable measurements (numeric or scale based) that assess the company's effectiveness or success in reaching its strategic and operational goals. There are many different KPI used by different industries. Some examples of KPI are:

- *General.* Year-to-year measurements of profit by line of business, same store sales, product turnovers, product recalls, sales by promotion, sales by employee, etc.
- *Finance.* Earnings per share, profit margin, revenue per employee, percentage of sales to account receivables, assets to sales, etc.
- *Human resources.* Applicants to job openings, employee turnover, employee longevity, etc.
- *Education.* Graduation rates, number of incoming freshmen, student retention rates, etc.

KPIs are determined after the main strategic, tactical, and operational goals for a business are defined. To tie the KPI to the strategic master plan of an organization, a KPI will be compared to a desired goal within a specific time frame. For example, if you are in an academic environment, you might be interested in ways to measure student satisfaction or retention. In this case, a sample goal would be to "Increase the graduating senior average exit exam grades from 9 to 12 by fall, 2010." Another sample KPI would be: "Increase the returning student rate of freshman year to sophomore year from 60% to 75% by 2012." In this case, such performance indicators would be measured and monitored on a year-to-year basis, and plans to achieve such goals would be set in place.

Another way to understand BI architecture is by describing the basic components that form part of its infrastructure. Some of the components have overlapping functionality; however, there are four basic components that all BI environments should provide. These are described in Table 13.2 and illustrated in Figure 13.2.

TABLE 13.2	Basic BI Architectural Components
COMPONENT	**DESCRIPTION**
Data extraction, transformation, and loading (ETL) tools	This component is in charge of collecting, filtering, integrating, and aggregating operational data to be saved into a data store optimized for decision support. For example, to determine the relative market share by selected product lines, you require data from competitors' products. Such data can be located in external databases provided by industry groups or by companies that market the data. As the name implies, this component extracts the data, filters the extracted data to select the relevant records, and packages the data in the right format to be added to the data store component.
Data store	The data store is optimized for decision support and is generally represented by a data warehouse or a data mart. The data store contains business data extracted from the operational database and from external data sources. The business data are stored in structures that are optimized for data analysis and query speed. The external data sources provide data that cannot be found within the company but that are relevant to the business, such as stock prices, market indicators, marketing information (such as demographics), and competitors' data.
Data query and analysis tools	This component performs data retrieval, data analysis, and data mining tasks using the data in the data store and business data analysis models. This component is used by the data analyst to create the queries that access the database. Depending on the implementation, the query tool accesses either the operational database, or more commonly, the data store. This tool advises the user on which data to select and how to build a reliable business data model. This component is generally represented in the form of an OLAP tool.
Data presentation and visualization tools	This component is in charge of presenting the data to the end user in a variety of ways. This component is used by the data analyst to organize and present the data. This tool helps the end user select the most appropriate presentation format, such as summary report, map, pie or bar graph, or mixed graphs. The query tool and the presentation tool are the front end to the BI environment.

FIGURE 13.2   Business intelligence components

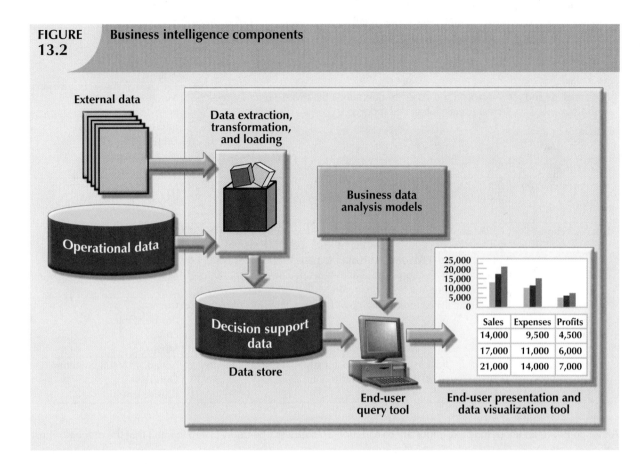

Each BI component shown in Table 13.2 has generated a fast-growing market for specialized tools. And thanks to the advancement of client/server technologies, those components can interact with other components to form a truly open architecture. As a matter of fact, you can integrate multiple tools from different vendors into a single BI framework. Table 13.3 shows a sample of common BI tools and vendors.

TABLE 13.3   Sample of Business Intelligence Tools

TOOL	DESCRIPTION	SAMPLE VENDORS
Decision support systems	A **decision support system (DSS)** is an arrangement of computerized tools used to assist managerial decision making within a business. *Decision support* systems were the precursors of modern BI systems. A DSS typically has a much narrower focus and reach than a BI solution.	SAP Teradata IBM Proclarity
Dashboards and business activity monitoring	**Dashboards** use Web-based technologies to present key business performance indicators or information in a single integrated view, generally using graphics in a clear, concise, and easy to understand manner.	*Salesforce* VisualCalc Cognos BusinessObjects Information Builders Actuate
Portals	**Portals** provide a unified, single point of entry for information distribution. Portals are a Web-based technology that uses a Web browser to integrate data from multiple sources into a single Web page. Many different types of BI functionality can be accessed through a portal.	Oracle Portal Actuate Microsoft

TABLE 13.3	Sample of Business Intelligence Tools (continued)	
TOOL	**DESCRIPTION**	**SAMPLE VENDORS**
Data analysis and reporting tools	Advanced tools used to query multiple diverse data sources to create single integrated reports.	Mircrosoft Reporting Services Information Builders Eclipse BIRT MicroStrategy SAS WebReportStudio
Data mining tools	Tools that provide advanced statistical analysis to uncover problems and opportunities hidden within business data.	MicroStrategy Intelligence Server MS Analytics Services
Data warehouses	The data warehouse is the foundation on which a BI infrastructure is built. Data is captured from the OLTP system and placed on the DW on near-real time basis. BI provides companywide integration of data and the capability to respond to business issues in a timely manner.	Microsoft Oracle IBM MicroStrategy
OLAP tools	Online analytical processing provides multidimensional data analysis.	Cognos BusinessObjects Oracle Microsoft
Data visualization	Tools that provide advanced visual analysis and techniques to enhance understanding of business data.	Advanced Visual Systems Dundas iDashboards

Although BI has an unquestionably important role in modern business operations, keep in mind that the *manager* must initiate the decision support process by asking the appropriate questions. The BI environment exists to support the manager; it does *not* replace the management function. If the manager fails to ask the appropriate questions, problems will not be identified and solved, and opportunities will be missed. In spite of the very powerful BI presence, the human component is still at the center of business technology.

NOTE

> Although the term BI includes a variety of components and tools, this chapter focuses on its data warehouse component.

13.4 DECISION SUPPORT DATA

Although BI is used at strategic and tactical managerial levels within organizations, *its effectiveness depends on the quality of data gathered at the operational level*. Yet operational data are seldom well suited to the decision support tasks. The differences between operational data and decision support data are examined in the next section.

13.4.1 OPERATIONAL DATA VS. DECISION SUPPORT DATA

Operational data and decision support data serve different purposes. Therefore, it is not surprising to learn that their formats and structures differ.

Most operational data are stored in a relational database in which the structures (tables) tend to be highly normalized. Operational data storage is optimized to support transactions that represent daily operations. For example, each time an item is sold, it must be accounted for. Customer data, inventory data, and so on, are in a frequent update mode. To provide effective update performance, operational systems store data in many tables, each with a minimum number of fields. Thus, a simple sales transaction might be represented by five or more different tables (for example, invoice,

invoice line, discount, store, and department). Although such an arrangement is excellent in an operational database, it is not efficient for query processing. For example, to extract a simple invoice, you would have to join several tables. Whereas operational data are useful for capturing daily business transactions, decision support data give tactical and strategic business meaning to the operational data. From the data analyst's point of view, decision support data differ from operational data in three main areas: time span, granularity, and dimensionality.

- *Time span.* Operational data cover a short time frame. In contrast, decision support data tend to cover a longer time frame. Managers are seldom interested in a specific sales invoice to customer X; rather, they tend to focus on sales generated during the last month, the last year, or the last five years.

- *Granularity (level of aggregation).* Decision support data must be presented at different levels of aggregation, from highly summarized to near-atomic. For example, if managers must analyze sales by region, they must be able to access data showing the sales by region, by city within the region, by store within the city within the region, and so on. In that case, summarized data to compare the regions is required, but also data in a structure that enables a manager to **drill down**, or decompose, the data into more atomic components (that is, finer-grained data at lower levels of aggregation). In contrast, when you **roll up** the data, you are aggregating the data to a higher level.

- *Dimensionality.* Operational data focus on representing individual transactions rather than on the effects of the transactions over time. In contrast, data analysts tend to include many data dimensions and are interested in how the data relate over those dimensions. For example, an analyst might want to know how product X fared relative to product Z during the past six months by region, state, city, store, and customer. In that case, both place and time are part of the picture.

Figure 13.3 shows how decision support data can be examined from multiple dimensions (such as product, region, and year), using a variety of filters to produce each dimension. The ability to analyze, extract, and present information in meaningful ways is one of the differences between decision support data and transaction-at-a-time operational data.

From the designer's point of view, the differences between operational and decision support data are as follows:

- Operational data represent transactions as they happen in real time. Decision support data are a snapshot of the operational data at a given point in time. Therefore, decision support data are historic, representing a time slice of the operational data.

- Operational and decision support data are different in terms of transaction *type* and transaction *volume*. Whereas operational data are characterized by update transactions, decision support data are mainly characterized by query (read-only) transactions. Decision support data also require *periodic* updates to load new data that are summarized from the operational data. Finally, the concurrent transaction volume in operational data tends to be very high when compared with the low-to-medium levels found in decision support data.

- Operational data are commonly stored in many tables, and the stored data represent the information about a given transaction only. Decision support data are generally stored in a few tables that store data derived from the operational data. The decision support data do not include the details of each operational transaction. Instead, decision support data represent transaction *summaries*; therefore, the decision support database stores data that are integrated, aggregated, and summarized for decision support purposes.

- The degree to which decision support data are summarized is very high when contrasted with operational data. Therefore, you will see a great deal of derived data in decision support databases. For example, rather than storing all 10,000 sales transactions for a given store on a given day, the decision support database might simply store the total number of units sold and the total sales dollars generated during that day. Decision support data might be collected to monitor such aggregates as total sales for each store or for each product. The purpose of the summaries is simple: they are to be used to establish and evaluate sales trends, product sales comparisons, and so on, that serve decision needs. (How well are items selling? Should this product be discontinued? Has the advertising been effective as measured by increased sales?)

- The data models that govern operational data and decision support data are different. The operational database's frequent and rapid data updates make data anomalies a potentially devastating problem. Therefore,

FIGURE 13.3 Transforming operational data into decision support data

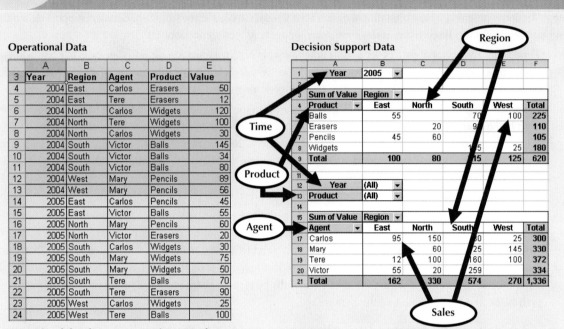

Operational Data

	A	B	C	D	E
3	Year	Region	Agent	Product	Value
4	2004	East	Carlos	Erasers	50
5	2004	East	Tere	Erasers	12
6	2004	North	Carlos	Widgets	120
7	2004	North	Tere	Widgets	100
8	2004	North	Carlos	Widgets	30
9	2004	South	Victor	Balls	145
10	2004	South	Victor	Balls	34
11	2004	South	Victor	Balls	80
12	2004	West	Mary	Pencils	89
13	2004	West	Mary	Pencils	56
14	2005	East	Carlos	Pencils	45
15	2005	East	Victor	Balls	55
16	2005	North	Mary	Pencils	60
17	2005	North	Victor	Erasers	20
18	2005	South	Carlos	Widgets	30
19	2005	South	Mary	Widgets	75
20	2005	South	Mary	Widgets	50
21	2005	South	Tere	Balls	70
22	2005	South	Tere	Erasers	90
23	2005	West	Carlos	Widgets	25
24	2005	West	Tere	Balls	100

Decision Support Data

	A	B	C	D	E	F
1	Year	2005				
3	Sum of Value	Region				
4	Product	East	North	South	West	Total
	Balls	55		70	100	225
	Erasers		20	90		110
7	Pencils	45	60			105
8	Widgets			155	25	180
9	Total	100	80	315	125	620

	A	B	C	D	E	F
12	Year	(All)				
13	Product	(All)				
15	Sum of Value	Region				
	Agent	East	North	South	West	Total
17	Carlos	95	150	30	25	300
18	Mary		60	125	145	330
19	Tere	12	100	160	100	372
20	Victor	55	20	259		334
21	Total	162	330	574	270	1,336

Operational data have a narrow time span, low granularity, and single focus. Such data are usually presented in tabular format, in which each row represents a single transaction. This format often makes it difficult to derive useful information.

Decision support system (DSS) data focus on a broader time span, tend to have high levels of granularity, and can be examined in multiple dimensions. For example, note these possible aggregations:
- Sales by product, region, agent, etc.
- Sales for all years or only a few selected years.
- Sales for all products or only a few selected products.

ONLINE CONTENT

The operational data in Figure 13.3 are found in the Student Online Companion for this book. The decision support data in Figure 13.3 shows the output for the solution to Problem 2 at the end of this chapter.

the data requirements in a typical relational transaction (operational) system generally require normalized structures that yield many tables, each of which contains the minimum number of attributes. In contrast, the decision support database is not subject to such transaction updates, and the focus is on querying capability. Therefore, decision support databases tend to be non-normalized and include few tables, each of which contains a large number of attributes.

- Query activity (frequency and complexity) in the operational database tends to be low to allow additional processing cycles for the more crucial update transactions. Therefore, queries against operational data typically are narrow in scope, low in complexity, and speed-critical. In contrast, decision support data exist for the sole purpose of serving query requirements. Queries against decision support data typically are broad in scope, high in complexity, and less speed-critical.

- Finally, decision support data are characterized by very large amounts of data. The large data volume is the result of two factors. First, data are stored in non-normalized structures that are likely to display many data redundancies and duplications. Second, the same data can be categorized in many different ways to represent different snapshots. For example, sales data might be stored in relation to product, store, customer, region, and manager.

Table 13.4 summarizes the differences between operational and decision support data from the database designer's point of view.

TABLE 13.4

Contrasting Operational and Decision Support Data Characteristics

CHARACTERISTIC	OPERATIONAL DATA	DECISION SUPPORT DATA
Data currency	Current operations Real-time data	Historic data Snapshot of company data Time component (week/month/year)
Granularity	Atomic-detailed data	Summarized data
Summarization level	Low; some aggregate yields	High; many aggregation levels
Data model	Highly normalized Mostly relational DBMS	Non-normalized Complex structures Some relational, but mostly multidimensional DBMS
Transaction type	Mostly updates	Mostly query
Transaction volumes	High update volumes	Periodic loads and summary calculations
Transaction speed	Updates are critical	Retrievals are critical
Query activity	Low to medium	High
Query scope	Narrow range	Broad range
Query complexity	Simple to medium	Very complex
Data volumes	Hundreds of megabytes, up to gigabytes	Hundreds of gigabytes, up to terabytes

The many differences between operational data and decision support data are good indicators of the requirements of the decision support database, described in the next section.

13.4.2 DECISION SUPPORT DATABASE REQUIREMENTS

A decision support database is a specialized DBMS tailored to provide fast answers to complex queries. There are four main requirements for a decision support database: the database schema, data extraction and loading, the end-user analytical interface, and database size.

TABLE 13.5

Ten-Year Sales History for a Single-Department, in Millions of Dollars

YEAR	SALES
1998	8,227
1999	9,109
2000	10,104
2001	11,553
2002	10,018
2003	11,875
2004	12,699
2005	14,875
2006	16,301
2007	19,986

Database Schema

The decision support database schema must support complex (non-normalized) data representations. As noted earlier, the decision support database must contain data that are aggregated and summarized. In addition to meeting those requirements, the queries must be able to extract multidimensional time slices. If you are using an RDBMS, the conditions suggest using non-normalized and even duplicated data. To see why this must be true, take a look at the 10-year sales history for a single store containing a single department. At this point, the data are fully normalized within the single table, as shown in Table 13.5.

This structure works well when you have only one store with only one department. However, it is very unlikely that such a simple environment has much need for a decision support database. One would suppose that a decision support database becomes a factor when dealing with more than one store, each of which has more than one department. To support all of the decision support requirements, the database must contain data for all of the stores and all of their departments—and the database must be able to support multidimensional queries that track sales by stores, by departments, and over time. For simplicity, suppose there are only two stores (A and B) and two departments (1 and 2) within each store. Let's also change the time dimension to include yearly data. Table 13.6 shows the sales figures under the specified conditions. Only 1998, 2002, and 2007 are shown; ellipses (...) are used to indicate that data values were omitted. You can see in Table 13.6 that the number of rows and attributes already multiplies quickly and that the table exhibits multiple redundancies.

TABLE 13.6	Yearly Sales Summaries, Two Stores and Two Departments per Store, in Millions of Dollars		
YEAR	STORE	DEPARTMENT	SALES
1998	A	1	1,985
1998	A	2	2,401
1998	B	1	1,879
1998	B	2	1,962
...	...	...	...
2002	A	1	3,912
2002	A	2	4,158
2002	B	1	3,426
2002	B	2	1,203
...	...	...	...
2007	A	1	7,683
2007	A	2	6,912
2007	B	1	3,768
2007	B	2	1,623

Now suppose that the company has 10 departments per store and 20 stores nationwide. And suppose you want to access *yearly* sales summaries. Now you are dealing with 200 rows and 12 monthly sales attributes per row. (Actually, there are 13 attributes per row if you add each store's sales total for each year.)

The decision support database schema must also be optimized for query (read-only) retrievals. To optimize query speed, the DBMS must support features such as bitmap indexes and data partitioning to increase search speed. In addition, the DBMS query optimizer must be enhanced to support the non-normalized and complex structures found in decision support databases.

Data Extraction and Filtering

The decision support database is created largely by extracting data from the operational database and by importing additional data from external sources. Thus, the DBMS must support advanced data extraction and data filtering tools. To minimize the impact on the operational database, the data extraction capabilities should allow batch and scheduled data extraction. The data extraction capabilities should also support different data sources: flat files and hierarchical, network, and relational databases, as well as multiple vendors. Data filtering capabilities must include the ability to check for inconsistent data or data validation rules. Finally, to filter and integrate the operational data into the decision support database, the DBMS must support advanced data integration, aggregation, and classification.

Using data from multiple external sources also usually means having to solve data-formatting conflicts. For example, data such as Social Security numbers and dates can occur in different formats; measurements can be based on different scales, and the same data elements can have different names. In short, data must be filtered and purified to ensure that only the pertinent decision support data are stored in the database and that they are stored in a standard format.

End-User Analytical Interface

The decision support DBMS must support advanced data modeling and data presentation tools. Using those tools makes it easy for data analysts to define the nature and extent of business problems. Once the problems have been defined, the decision support DBMS must generate the necessary queries to retrieve the appropriate data from the decision support database. If necessary, the query results may then be evaluated with data analysis tools supported by the decision support DBMS. Because queries yield crucial information for decision makers, the queries must be optimized for speedy processing. The end-user analytical interface is one of the most critical DBMS components. When properly implemented, an analytical interface permits the user to navigate through the data to simplify and accelerate the decision-making process.

Database Size

Decision support databases tend to be very large; gigabyte and terabyte ranges are not unusual. For example, in 2005, Wal-Mart, the world's largest company, had 260 terabytes of data in its data warehouses. As mentioned earlier, the decision support database typically contains redundant and duplicated data to improve data retrieval and simplify information generation. Therefore, the DBMS must be capable of supporting **very large databases (VLDBs)**. To support a VLDB adequately, the DBMS might be required to use advanced hardware, such as multiple disk arrays, and even more importantly, to support multiple-processor technologies, such as a symmetric multiprocessor (SMP) or a massively parallel processor (MPP).

The complex information requirements and the ever-growing demand for sophisticated data analysis sparked the creation of a new type of data repository. This repository contains data in formats that facilitate data extraction, data analysis, and decision making. This data repository is known as a data warehouse and has become the foundation for a new generation of decision support systems.

13.5 THE DATA WAREHOUSE

Bill Inmon, the acknowledged "father" of the **data warehouse**, defines the term as "an *integrated, subject-oriented, time-variant, nonvolatile* collection of data (italics added for emphasis) that provides support for decision making."[2] To understand that definition, let's take a more detailed look at its components.

- *Integrated.* The data warehouse is a centralized, consolidated database that integrates data derived from the entire organization and from multiple sources with diverse formats. Data integration implies that all business entities, data elements, data characteristics, and business metrics *are described in the same way throughout the enterprise.* Although this requirement sounds logical, you would be amazed to discover how many different measurements for "sales performance" can exist within an organization; the same scenario holds true for any other business element. For instance, the status of an order might be indicated with text labels such as "open," "received," "cancelled," and "closed" in one department and as "1," "2," "3," and "4" in another department. A student's status might be defined as "freshman," "sophomore," "junior," or "senior" in the accounting department and as "FR," "SO," "JR," or "SR" in the computer information systems department. To avoid the potential format tangle, the data in the data warehouse must conform to a common format acceptable throughout the organization. This integration can be time-consuming, but once accomplished, it enhances decision making and helps managers better understand the company's operations. This understanding can be translated into recognition of strategic business opportunities.
- *Subject-oriented.* Data warehouse data are arranged and optimized to provide answers to questions coming from diverse functional areas within a company. Data warehouse data are organized and summarized by topic, such as sales, marketing, finance, distribution, and transportation. For each topic, the data warehouse contains

---

[2] Inmon, Bill and Chuck Kelley. "The Twelve Rules of Data Warehouse for a Client/Server World," *Data Management Review*, 4(5), May 1994, pp. 6–16.

specific subjects of interest—products, customers, departments, regions, promotions, and so on. This form of data organization is quite different from the more functional or process-oriented organization of typical transaction systems. For example, an invoicing system designer concentrates on designing normalized data structures (relational tables) to support the business process by storing invoice components in two tables: INVOICE and INVLINE. In contrast, the data warehouse has a *subject* orientation. Data warehouse designers focus specifically on the data rather than on the processes that modify the data. (After all, data warehouse data are not subject to numerous real-time data updates!) Therefore, instead of storing an invoice, the data warehouse stores its "sales by product" and "sales by customer" components because decision support activities require the retrieval of sales summaries by product or customer.

- *Time-variant.* In contrast to operational data, which focus on current transactions, warehouse data represent the flow of data through time. The data warehouse can even contain projected data generated through statistical and other models. It is also time-variant in the sense that once data are periodically uploaded to the data warehouse, all time-dependent aggregations are recomputed. For example, when data for previous weekly sales are uploaded to the data warehouse, the weekly, monthly, yearly, and other time-dependent aggregates for products, customers, stores, and other variables are also updated. Because data in a data warehouse constitute a snapshot of the company history as measured by its variables, the time component is crucial. The data warehouse contains a time ID that is used to generate summaries and aggregations by week, month, quarter, year, and so on. Once the data enter the data warehouse, the time ID assigned to the data cannot be changed.

- *Nonvolatile.* Once data enter the data warehouse, they are never removed. Because the data in the warehouse represent the company's history, the operational data, representing the near-term history, are always added to it. Because data are never deleted and new data are continually added, the data warehouse is always growing. That's why the DBMS must be able to support multigigabyte and even multiterabyte databases, operating on multiprocessor hardware. Table 13.7 summarizes the differences between data warehouses and operational databases.

TABLE 13.7	Characteristics of Data Warehouse Data and Operational Database Data	
CHARACTERISTIC	OPERATIONAL DATABASE DATA	DATA WAREHOUSE DATA
Integrated	Similar data can have different representations or meanings. For example, Social Security numbers may be stored as ###-##-#### or as #########, and a given condition may be labeled as T/F or 0/1 or Y/N. A sales value may be shown in thousands or in millions.	Provide a unified view of all data elements with a common definition and representation for all business units.
Subject-oriented	Data are stored with a functional, or process, orientation. For example, data may be stored for invoices, payments, and credit amounts.	Data are stored with a subject orientation that facilitates multiple views of the data and facilitates decision making. For example, sales may be recorded by product, by division, by manager, or by region.
Time-variant	Data are recorded as current transactions. For example, the sales data may be the sale of a product on a given date, such as $342.78 on 12-MAY-2008.	Data are recorded with a historical perspective in mind. Therefore, a time dimension is added to facilitate data analysis and various time comparisons.
Nonvolatile	Data updates are frequent and common. For example, an inventory amount changes with each sale. Therefore, the data environment is fluid.	Data cannot be changed. Data are added only periodically from historical systems. Once the data are properly stored, no changes are allowed. Therefore, the data environment is relatively static.

In summary, the data warehouse is usually a read-only database optimized for data analysis and query processing. Typically, data are extracted from various sources and are then transformed and integrated—in other words, passed through a data filter—before being loaded into the data warehouse. Users access the data warehouse via front-end tools and/or end-user application software to extract the data in usable form. Figure 13.4 illustrates how a data warehouse is created from the data contained in an operational database.

FIGURE 13.4   Creating a data warehouse

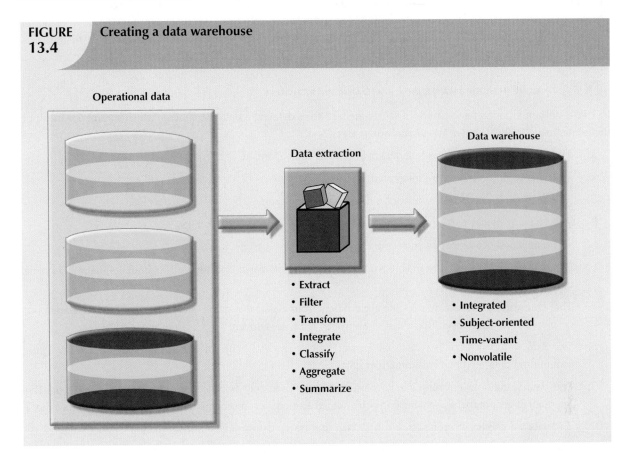

Although the centralized and integrated data warehouse can be a very attractive proposition that yields many benefits, managers may be reluctant to embrace this strategy. Creating a data warehouse requires time, money, and considerable managerial effort. Therefore, it is not surprising that many companies begin their foray into data warehousing by focusing on more manageable data sets that are targeted to meet the special needs of small groups within the organization. These smaller data stores are called data marts. A **data mart** is a small, single-subject data warehouse subset that provides decision support to a small group of people. In addition, a data mart could also be created from data extracted from a larger data warehouse with the specific function to support faster data access to a target group or function. That is, data marts and data warehouses can coexist within a business intelligence environment.

Some organizations choose to implement data marts not only because of the lower cost and shorter implementation time, but also because of the current technological advances and inevitable "people issues" that make data marts attractive. Powerful computers can provide a customized decision support system to small groups in ways that might not be possible with a centralized system. Also, a company's culture may predispose its employees to resist major changes, but they might quickly embrace relatively minor changes that lead to demonstrably improved decision support. In addition, people at different organizational levels are likely to require data with different summarization, aggregation, and presentation formats. Data marts can serve as a test vehicle for companies exploring the potential benefits of data warehouses. By migrating gradually from data marts to data warehouses, a specific department's decision support needs can be addressed within a reasonable time frame (six months to one year), as compared to the longer time frame usually required to implement a data warehouse (one to three years). Information technology (IT) departments also benefit from this approach because their personnel have the opportunity to learn the issues and develop the skills required to create a data warehouse.

The only difference between a data mart and a data warehouse is the size and scope of the problem being solved. Therefore, the problem definitions and data requirements are essentially the same for both. To be useful, the data warehouse must conform to uniform structures and formats to avoid data conflicts and to support decision making. In fact, before a decision support database can be considered a true data warehouse, it must conform to the rules described in the next section.

13.5.1 TWELVE RULES THAT DEFINE A DATA WAREHOUSE

In 1994, William H. Inmon and Chuck Kelley created 12 rules defining a data warehouse, which summarize many of the points made in this chapter about data warehouses.[3]

1. The data warehouse and operational environments are separated.

2. The data warehouse data are integrated.

3. The data warehouse contains historical data over a long time.

4. The data warehouse data are snapshot data captured at a given point in time.

5. The data warehouse data are subject oriented.

6. The data warehouse data are mainly read-only with periodic batch updates from operational data. No online updates are allowed.

7. The data warehouse development life cycle differs from classical systems development. The data warehouse development is data-driven; the classical approach is process-driven.

8. The data warehouse contains data with several levels of detail: current detail data, old detail data, lightly summarized data, and highly summarized data.

9. The data warehouse environment is characterized by read-only transactions to very large data sets. The operational environment is characterized by numerous update transactions to a few data entities at a time.

10. The data warehouse environment has a system that traces data sources, transformations, and storage.

11. The data warehouse's metadata are a critical component of this environment. The metadata identify and define all data elements. The metadata provide the source, transformation, integration, storage, usage, relationships, and history of each data element.

12. The data warehouse contains a chargeback mechanism for resource usage that enforces optimal use of the data by end users.

Note how those 12 rules capture the complete data warehouse life cycle—from its introduction as an entity separate from the operational data store to its components, functionality, and management processes. The next section illustrates the historical progression of decision support architectural styles. This discussion will help you understand how the data store components evolved to produce the data warehouse.

13.5.2 DECISION SUPPORT ARCHITECTURAL STYLES

Several decision support database architectural styles are available. These architectures provide advanced decision support features, and some are capable of providing access to multidimensional data analysis. Table 13.8 summarizes the main architectural styles that you are likely to encounter in the decision support database environment.

---

[3] Inmon, Bill and Chuck Kelley. "The Twelve Rules of Data Warehouse for a Client/Server World," *Data Management Review*, 4 (5), May 1994, pp. 6–16.

TABLE 13.8 Decision Support Architectural Styles

SYSTEM TYPE	SOURCE DATA	DATA EXTRACTION/ INTEGRATION PROCESS	DECISION SUPPORT DATA STORE	END-USER QUERY TOOL	END USER PRESENTATION TOOL
Traditional mainframe-based online transaction processing (OLTP)	Operational data	None Reports, reads, and summarizes data directly from operational data	None Temporary files used for reporting purposes	Very basic Predefined reporting formats Basic sorting, totaling, and averaging	Very basic Menu-driven, predefined reports, text and numbers only
Managerial information system (MIS) with third-generation language (3GL)	Operational data	Basic extraction and aggregation Reads, filters, and summarizes operational data into intermediate data store	Lightly aggregated data in RDBMS	Same as above, in addition to some ad hoc reporting using SQL	Same as above, in addition to some ad hoc columnar report definitions
First-generation departmental DSS	Operational data	Data extraction and integration process to populate a DSS data store; is run periodically	First DSS database generation Usually RDBMS	Query tool with some analytical capabilities and predefined reports	Advanced presentation tools with plotting and graphics capabilities
First-generation enterprise data warehouse using RDBMS	Operational data External data (census data)	Advanced data extraction and integration tools Features include access to diverse data sources, transformations, filters, aggregations, classifications, scheduling, and conflict resolution	Data warehouse integrated decision support database to support the entire organization Uses RDBMS technology optimized for query purposes Star schema model	Same as above, in addition to support for more advanced queries and analytical functions with extensions	Same as above, in addition to additional multidimensional presentation tools with drill-down capabilities
Second-generation data warehouse using multidimensional database management system (MDBMS)	Operational data External data (Industry group data)	Same as above	Data warehouse stores data by using MDBMS technology based on data structures; referred to as cubes with multiple dimensions	Same as above, but uses different query interface to access MDBMS (proprietary)	Same as above, but uses cubes and multidimensional matrixes Limited in terms of cube size

You might be tempted to think that the data warehouse is just a big summarized database. The previous discussion indicates that a good data warehouse is much more than that. A complete data warehouse architecture includes support for a decision support data store, a data extraction and integration filter, and a specialized presentation interface. In the next section you will learn more about a common decision support architectural style known as Online Analytical Processing (OLAP).

13.6 ONLINE ANALYTICAL PROCESSING

The need for more intensive decision support prompted the introduction of a new generation of tools. Those new tools, called **online analytical processing** (**OLAP**), create an advanced data analysis environment that supports decision making, business modeling, and operations research. OLAP systems share four main characteristics:

- They use multidimensional data analysis techniques.
- They provide advanced database support.
- They provide easy-to-use end-user interfaces.
- They support client/server architecture.

Let's examine each of those characteristics.

13.6.1 MULTIDIMENSIONAL DATA ANALYSIS TECHNIQUES

The most distinct characteristic of modern OLAP tools is their capacity for multidimensional analysis. In multidimensional analysis, data are processed and viewed as part of a multidimensional structure. This type of data analysis is particularly attractive to business decision makers because they tend to view business data as data that are related to other business data.

To better understand this view, let's examine how, as a business data analyst, you might investigate sales figures. In this case, you are probably interested in the sales figures as they relate to other business variables such as customers and time. In other words, customers and time are viewed as different dimensions of sales. Figure 13.5 illustrates how the operational (one-dimensional) view differs from the multidimensional view of sales.

Note in Figure 13.5 that the tabular (operational) view of sales data is not well suited to decision support, because the relationship between INVOICE and LINE does not provide a business perspective of the sales data. On the other hand, the end user's view of sales data *from a business perspective* is more closely represented by the multidimensional view of sales than by the tabular view of separate tables. Note also that the multidimensional view allows end users to consolidate or aggregate data at different levels: total sales figures by customers and by date. Finally, the multidimensional view of data allows a business data analyst to easily switch business perspectives (dimensions) from sales by customer to sales by division, by region, and so on.

Multidimensional data analysis techniques are augmented by the following functions:

- *Advanced data presentation functions.* 3-D graphics, pivot tables, crosstabs, data rotation, and three-dimensional cubes. Such facilities are compatible with desktop spreadsheets, statistical packages, and query and report packages.
- *Advanced data aggregation, consolidation, and classification functions.* These allow the data analyst to create multiple data aggregation levels, slice and dice data (see Section 13.6.3), and drill down and roll up data across different dimensions and aggregation levels. For example, aggregating data across the time dimension (by week, month, quarter, and year) allows the data analyst to drill down and roll up across time dimensions.
- *Advanced computational functions.* These include business-oriented variables (market share, period comparisons, sales margins, product margins, and percentage changes), financial and accounting ratios (profitability, overhead, cost allocations, and returns), and statistical and forecasting functions. These functions are provided automatically, and the end user does not need to redefine their components each time they are accessed.
- *Advanced data modeling functions.* These provide support for what-if scenarios, variable assessment, variable contributions to outcome, linear programming, and other modeling tools.

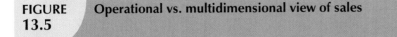

FIGURE 13.5   Operational vs. multidimensional view of sales

Database name: Ch13_Text

Table name: DW_INVOICE

INV_NUM	INV_DATE	CUS_NAME	INV_TOTAL
2034	15-May-08	Dartonik	1400.00
2035	15-May-08	Summer Lake	1200.00
2036	16-May-08	Dartonik	1350.00
2037	16-May-08	Summer lake	3100.00
2038	16-May-08	Trydon	400.00

Table name: DW_LINE

INV_NUM	LINE_NUM	PROD_DESCRIPTION	LINE_PRICE	LINE_QUANTITY	LINE_AMOUNT
2034	1	Optical Mouse	45.00	20	900.00
2034	2	Wireless RF remote and laser pointer	50.00	10	500.00
2035	1	Everlast Hard Drive, 60 GB	200.00	6	1200.00
2036	1	Optical Mouse	45.00	30	1350.00
2037	1	Optical Mouse	45.00	10	450.00
2037	2	Roadster 56KB Ext. Modem	120.00	5	600.00
2037	3	Everlast Hard Drive, 60 GB	205.00	10	2050.00
2038	1	NoTech Speaker Set	50.00	8	400.00

Multidimensional View of Sales

Customer Dimension	Time Dimension		Totals
	15-May-08	16-May-08	
Dartonik	$1,400.00	$1,350.00	$2,750.00
Summer Lake	$1,800.00	$3,100.00	$4,900.00
Trydon		$400.00	$400.00
Totals	$3,200.00	$4,850.00	$8,050.00

Sales are located in the intersection of a customer row and time column

Aggregations are provided for both dimensions

Because many analysis and presentation functions are common to desktop spreadsheet packages, most OLAP vendors have closely integrated their systems with spreadsheets such as Microsoft Excel and IBM Lotus 1-2-3. Using the features available in graphical end-user interfaces such as Windows, the OLAP menu option simply becomes another option within the spreadsheet menu bar, as shown in Figure 13.6. This seamless integration is an advantage for OLAP systems and for spreadsheet vendors because end users gain access to advanced data analysis features by using familiar programs and interfaces. Therefore, additional training and development costs are minimized.

13.6.2 ADVANCED DATABASE SUPPORT

To deliver efficient decision support, OLAP tools must have advanced data access features. Such features include:

- Access to many different kinds of DBMSs, flat files, and internal and external data sources.
- Access to aggregated data warehouse data as well as to the detail data found in operational databases.
- Advanced data navigation features such as drill-down and roll-up.
- Rapid and consistent query response times.

FIGURE 13.6   Integration of OLAP with a spreadsheet program

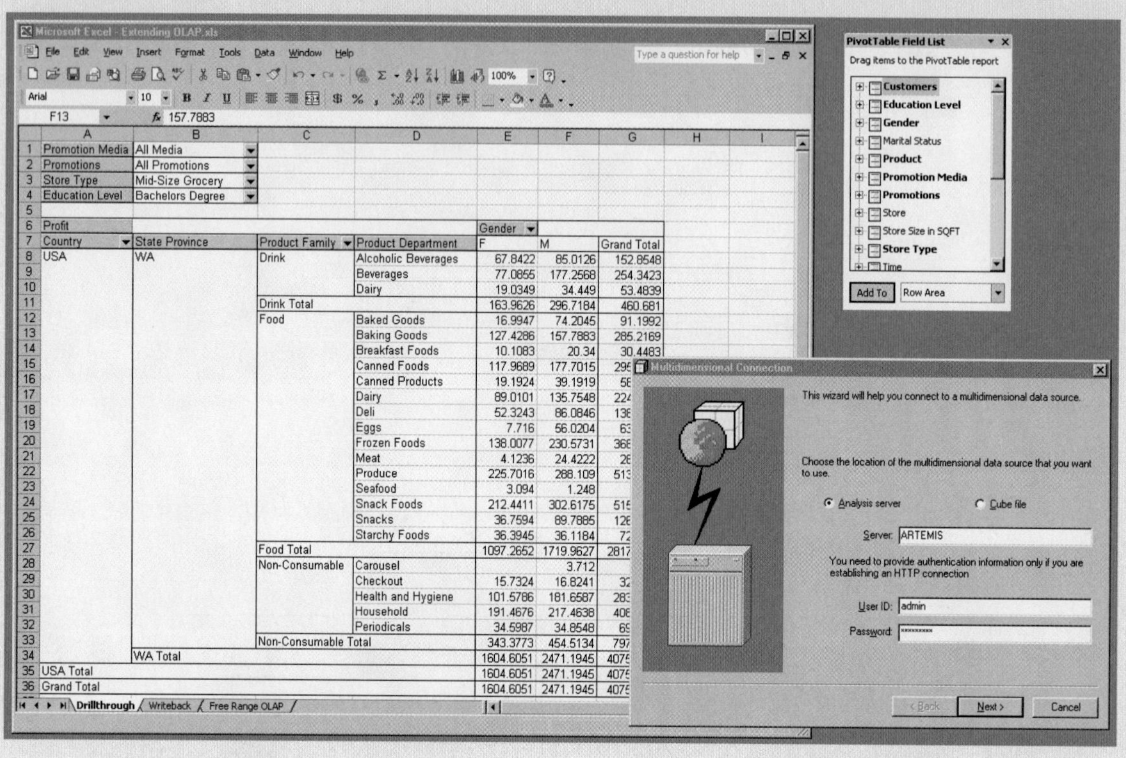

- The ability to map end-user requests, expressed in either business or model terms, to the appropriate data source and then to the proper data access language (usually SQL). The query code must be optimized to match the data source, regardless of whether the source is operational or data warehouse data.

- Support for very large databases. As already explained, the data warehouse can easily and quickly grow to multiple gigabytes and even terabytes.

To provide a seamless interface, OLAP tools map the data elements from the data warehouse and from the operational database to their own data dictionaries. These metadata are used to translate end-user data analysis requests into the proper (optimized) query codes, which are then directed to the appropriate data source(s).

13.6.3 EASY-TO-USE END-USER INTERFACE

Advanced OLAP features become more useful when access to them is kept simple. OLAP tool vendors learned this lesson early and have equipped their sophisticated data extraction and analysis tools with easy-to-use graphical interfaces. Many of the interface features are "borrowed" from previous generations of data analysis tools that are already familiar to end users. This familiarity makes OLAP easily accepted and readily used.

13.6.4 CLIENT/SERVER ARCHITECTURE

Client/server architecture provides a framework within which new systems can be designed, developed, and implemented. The client/server environment enables an OLAP system to be divided into several components that define its architecture. Those components can then be placed on the same computer, or they can be distributed among several computers. Thus, OLAP is designed to meet ease-of-use requirements while keeping the system flexible.

ONLINE CONTENT

If necessary, review the coverage in **Appendix F, Client/Server Systems** in the Student Online Companion for this book, which provides an in-depth look at client/server system architecture and principles.

13.6.5 OLAP ARCHITECTURE

OLAP operational characteristics can be divided into three main modules:

- Graphical user interface (GUI).
- Analytical processing logic.
- Data-processing logic.

In the client/server environment, those three OLAP modules make the defining features of OLAP possible: multidimensional data analysis, advanced database support, and an easy-to-use interface. Figure 13.7 illustrates OLAP's client/server components and attributes.

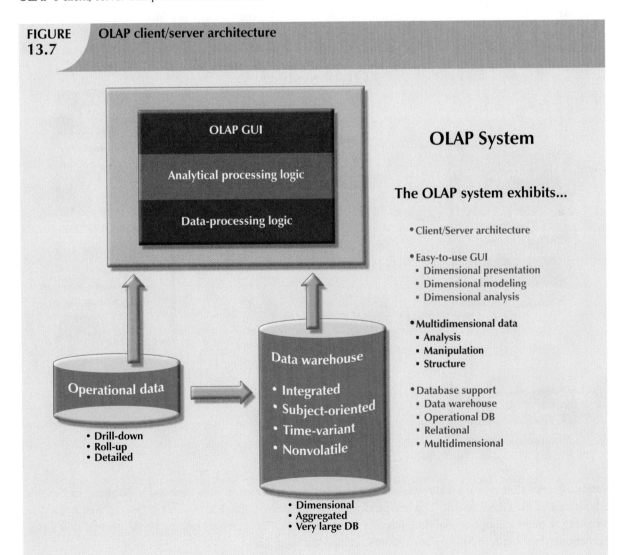

FIGURE 13.7 OLAP client/server architecture

As Figure 13.7 illustrates, OLAP systems are designed to use both operational and data warehouse data. Figure 13.7 shows the OLAP system components located on a single computer, but this single-user scenario is only one of many. In fact, one problem with the installation shown here is that each data analyst must have a powerful computer to store the OLAP system and perform all data processing locally. In addition, each analyst uses a separate copy of the data. Therefore, the data copies must be synchronized to ensure that analysts are working with the same data. In other words, each end user must have his/her own "private" copy (extract) of the data and programs, thus returning to the *islands of information* problems discussed in Chapter 1, Database Systems. This approach does not provide the benefits of a single business image shared among all users.

A more common and practical architecture is one in which the OLAP GUI runs on client workstations, while the OLAP engine, or server, composed of the OLAP analytical processing logic and OLAP data-processing logic, runs on a shared computer. In that case, the OLAP server will be a front end to the data warehouse's decision support data. This front end or middle layer (because it sits between the data warehouse and the end-user GUI) accepts and processes the data-processing requests generated by the many end-user analytical tools. The end-user GUI might be a custom-made program or, more likely, a plug-in module that is integrated with Lotus 1-2-3, Microsoft Excel, or a third-party data analysis and query tool. Figure 13.8 illustrates such an arrangement.

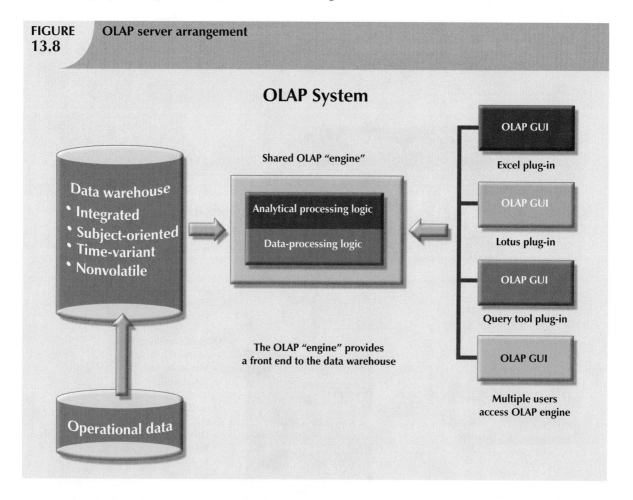

FIGURE 13.8   OLAP server arrangement

Note in Figure 13.8 that the data warehouse is created and maintained by a process or software tool that is independent of the OLAP system. This independent software performs the data extraction, filtering, and integration necessary to transform operational data into data warehouse data. This scenario reflects the fact that in most cases, the data warehousing and data analysis activities are handled separately.

At this point, you might ask why you need a data warehouse if OLAP provides the necessary multidimensional data analysis of operational data. The answer lies in the definition of OLAP. OLAP is defined as an "advanced data analysis environment that supports decision making, business modeling, and research activities." The keyword here is *environment*, which includes client/server technology. Environment is defined as "surroundings or atmosphere." And an atmosphere surrounds a nucleus. *In this case, the nucleus is composed of all business activities within an organization as represented by the operational data.* Just as there are several layers within the atmosphere, there are several layers of data processing, each outer layer representing a more aggregated data analysis. The fact is that an OLAP system might access both data storage types (operational or data warehouse) or only one; it depends on the vendor's implementation of the product selected. In any case, multidimensional data analysis requires some type of multidimensional data representation, which is normally provided by the OLAP engine.

In most implementations, the data warehouse and OLAP are interrelated, complementary environments. While the data warehouse holds integrated, subject-oriented, time-variant, and nonvolatile decision support data, the OLAP system provides the front end through which end users access and analyze such data. Yet an OLAP system can also directly access operational data, transforming it and storing it in a multidimensional structure. In other words, the OLAP system can provide a multidimensional data store component, as shown in Figure 13.9.

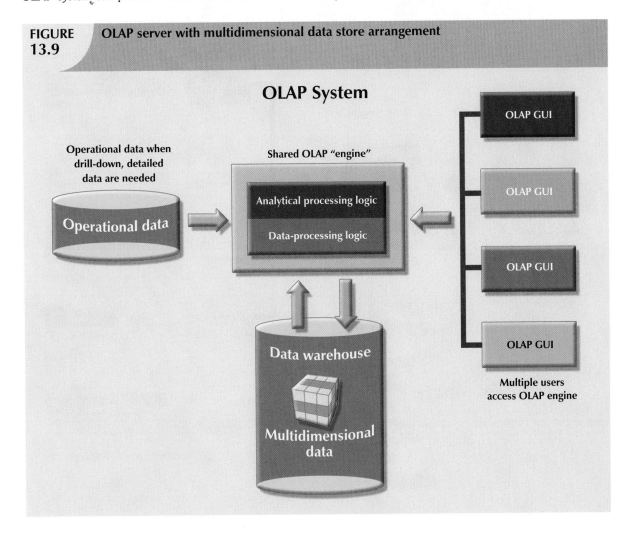

FIGURE 13.9   **OLAP server with multidimensional data store arrangement**

Figure 13.9 represents a scenario in which the OLAP engine extracts data from an operational database and then stores it in a multidimensional structure for further data analysis. The extraction process follows the same conventions used with data warehouses. Therefore, the OLAP provides a mini data-warehouse component that looks remarkably like the data mart mentioned in previous sections. In this scenario, the OLAP engine has to perform all of the data extraction, filtering, integration, classification, and aggregation functions that the data warehouse normally provides. In fact, when properly implemented, the data warehouse performs all data preparation functions instead of letting OLAP perform those chores; as a result, there is no duplication of functions. Better yet, the data warehouse handles the data component more efficiently than OLAP does; so you can appreciate the benefits of having a central data warehouse serve as the large enterprise decision support database.

To provide better performance, some OLAP systems merge the data warehouse and data mart approaches by storing small extracts of the data warehouse at end-user workstations. The objective is to increase the speed of data access and data visualization (the graphic representations of data trends and characteristics). The logic behind that approach is the assumption that most end users usually work with fairly small, stable data warehouse data subsets. For example, a sales analyst is most likely to work with sales data, whereas a customer representative is likely to work with customer data. Figure 13.10 illustrates that scenario.

FIGURE 13.10   OLAP server with local mini data marts

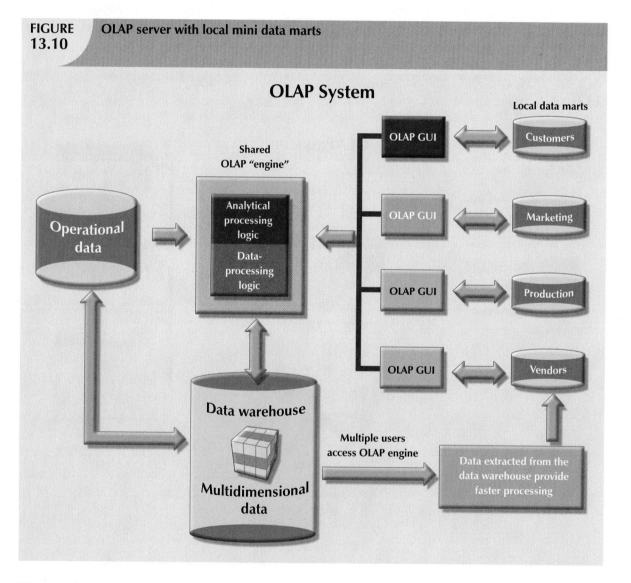

Whatever the arrangement of the OLAP components, one thing is certain: multidimensional data must be used. But how are multidimensional data best stored and managed? OLAP proponents are sharply divided. Some favor the use of relational databases to store the multidimensional data; others argue for the superiority of specialized multidimensional databases to store multidimensional data. The basic characteristics of each approach are examined next.

RELATIONAL OLAP

Relational online analytical processing (ROLAP) provides OLAP functionality by using relational databases and familiar relational query tools to store and analyze multidimensional data. That approach builds on existing relational technologies and represents a natural extension to all of the companies that already use relational database management systems within their organizations. ROLAP adds the following extensions to traditional RDBMS technology:

- Multidimensional data schema support within the RDBMS.
- Data access language and query performance optimized for multidimensional data.
- Support for very large databases (VLDBs).

Multidimensional Data Schema Support within the RDBMS

Relational technology uses normalized tables to store data. The reliance on normalization as the design methodology for relational databases is seen as a stumbling block to its use in OLAP systems. Normalization divides business entities into smaller pieces to produce the normalized tables. For example, sales data components might be stored in four or five different tables. The reason for using normalized tables is to reduce redundancies, thereby eliminating data anomalies, and to facilitate data updates. Unfortunately, for decision support purposes, it is easier to understand data when they are seen with respect to other data. (See the example in Figure 13.5.) Given that view of the data environment, this book has stressed that decision support data tend to be non-normalized, duplicated, and pre-aggregated. Those characteristics seem to preclude the use of standard relational design techniques and RDBMSs as the foundation for multidimensional data.

Fortunately for those heavily invested in relational technology, ROLAP uses a special design technique to enable RDBMS technology to support multidimensional data representations. This special design technique is known as a star schema, which is covered in detail in Section 13.7.

The star schema is designed to optimize data query operations rather than data update operations. Naturally, changing the data design foundation means that the tools used to access such data will have to change. End users who are familiar with the traditional relational query tools will discover that those tools do not work efficiently with the new star schema. However, ROLAP saves the day by adding support for the star schema when familiar query tools are used. ROLAP provides advanced data analysis functions and improves query optimization and data visualization methods.

Data Access Language and Query Performance Optimized for Multidimensional Data

Another criticism of relational databases is that SQL is not suited for performing advanced data analysis. Most decision support data requests require the use of multiple-pass SQL queries or multiple nested SQL statements. To answer this criticism, ROLAP extends SQL so that it can differentiate between access requirements for data warehouse data (based on the star schema) and operational data (normalized tables). In that way, a ROLAP system is able to generate the SQL code required to access the star schema data.

Query performance is also improved because the query optimizer is modified to identify the SQL code's intended query targets. For example, if the query target is the data warehouse, the optimizer passes the requests to the data warehouse. However, if the end user performs drill-down queries against operational data, the query optimizer identifies that operation and properly optimizes the SQL requests before passing them through to the operational DBMS.

Another source of improved query performance is the use of advanced indexing techniques such as bitmapped indexes within relational databases. As the name suggests, a bitmapped index is based on 0 and 1 bits to represent a given condition. For example, if the REGION attribute in Figure 13.3 has only four outcomes—North, South, East, and West—those outcomes may be represented as shown in Table 13.9. (Only the first 10 rows from Figure 13.3 are represented in Table 13.9. The "1" represents "bit on," and the "0" represents "bit off." For example, to represent a row with a REGION attribute = "East," only the "East" bit would be on. Note that each row must be represented in the index table.)

TABLE 13.9 Bitmap Representation of Region Values

NORTH	SOUTH	EAST	WEST
0	0	1	0
0	0	1	0
1	0	0	0
1	0	0	0
1	0	0	0
0	1	0	0
0	1	0	0
0	1	0	0
0	0	0	1
0	0	0	1

Note that the index in Table 13.9 takes a minimum amount of space. Therefore, bitmapped indexes are more efficient at handling large amounts of data than are the indexes typically found in many relational databases. But do keep in mind that bitmapped indexes are primarily used in situations where the number of possible values for an attribute (in other words, the attribute domain) is fairly small. For example, REGION has only four outcomes in this example. Marital status—married, single, widowed, divorced—would be another good bitmapped index candidate, as would gender—M or F.

ROLAP tools are mainly client/server products in which the end-user interface, the analytical processing, and the data processing take place on different computers. Figure 13.11 shows the interaction of the client/server ROLAP components.

FIGURE 13.11 Typical ROLAP client/server architecture

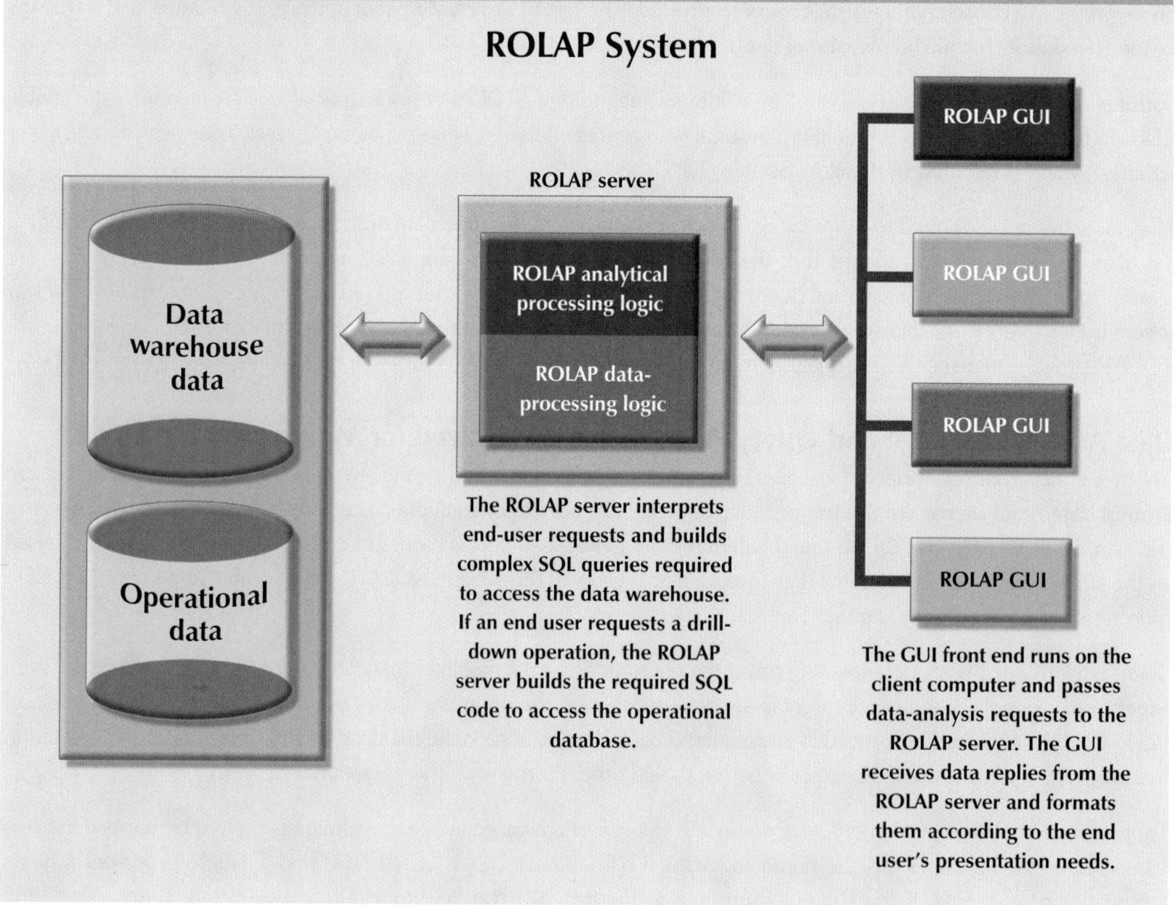
ROLAP System
Data warehouse data
Operational data
ROLAP server
ROLAP analytical processing logic
ROLAP data-processing logic
The ROLAP server interprets end-user requests and builds complex SQL queries required to access the data warehouse. If an end user requests a drill-down operation, the ROLAP server builds the required SQL code to access the operational database.
ROLAP GUI
ROLAP GUI
ROLAP GUI
ROLAP GUI
The GUI front end runs on the client computer and passes data-analysis requests to the ROLAP server. The GUI receives data replies from the ROLAP server and formats them according to the end user's presentation needs.

Support for Very Large Databases

Recall that support for VLDBs is a requirement for decision support databases. Therefore, when the relational database is used in a decision support role, it also must be able to store very large amounts of data. Both the storage capability and the process of loading data into the database are crucial. Therefore, the RDBMS must have the proper tools to import, integrate, and populate the data warehouse with data. Decision support data are normally loaded in bulk (batch) mode from the operational data. However, batch operations require that both the source and the destination databases be reserved (locked). The speed of the data-loading operations is important, especially when you realize that most operational systems run 24 hours a day, 7 days a week, 52 weeks a year. Therefore, the window of opportunity for maintenance and batch loading is open only briefly, typically during slack periods.

With an open client/server architecture, ROLAP provides advanced decision support capabilities that are scalable to the entire enterprise. Clearly, ROLAP is a logical choice for companies that already use relational databases for their operational data. Given the size of the relational database market, it is hardly surprising that most current RDBMS vendors have extended their products to support data warehouses.

13.6.7 MULTIDIMENSIONAL OLAP

Multidimensional online analytical processing (**MOLAP**) extends OLAP functionality to **multidimensional database management systems** (**MDBMSs**). (An MDBMS uses special proprietary techniques to store data in matrix-like *n*-dimensional arrays.) MOLAP's premise is that multidimensional databases are best suited to manage, store, and analyze multidimensional data. Most of the proprietary techniques used in MDBMSs are derived from engineering fields such as computer-aided design/computer-aided manufacturing (CAD/CAM) and geographic information systems (GIS).

Conceptually, MDBMS end users visualize the stored data as a three-dimensional cube known as a **data cube**. The location of each data value in the data cube is a function of the x-, y-, and z-axes in a three-dimensional space. The x-, y-, and z-axes represent the dimensions of the data value. The data cubes can grow to *n* number of dimensions, thus becoming *hypercubes*. Data cubes are created by extracting data from the operational databases or from the data warehouse. One important characteristic of data cubes is that they are static; that is, they are not subject to change and must be created before they can be used. Data cubes cannot be created by ad hoc queries. Instead, you query pre-created cubes with defined axes; for example, a cube for sales will have the product, location, and time dimensions, and you can query only those dimensions. Therefore, the data cube creation process is critical and requires in-depth front-end design work. The front-end design work may be well justified because MOLAP databases are known to be much faster than their ROLAP counterparts, especially when dealing with small to medium data sets. To speed data access, data cubes are normally held in memory in what is called the **cube cache**. (A data cube is only a window to a predefined subset of data in the database. A *data cube* and a *database* are not the same thing.) Because MOLAP also benefits from a client/server infrastructure, the cube cache can be located at the MOLAP server, at the MOLAP client, or in both locations. Figure 13.12 shows the basic MOLAP architecture.

Because the data cube is predefined with a set number of dimensions, the addition of a new dimension requires that the entire data cube be re-created. This re-creation process is time consuming. Therefore, when data cubes are created too often, the MDBMS loses some of its speed advantage over the relational database. And although MDBMSs have performance advantages over relational databases, the MDBMS is best suited to small and medium data sets. Scalability is somewhat limited because the size of the data cube is restricted to avoid lengthy data access times caused by having less work space (memory) available for the operating system and the application programs. In addition, the MDBMS makes use of proprietary data storage techniques that, in turn, require proprietary data access methods using a multidimensional query language.

Multidimensional data analysis is also affected by how the database system handles sparsity. **Sparsity** is a measurement of the density of the data held in the data cube and is computed by dividing the total number of actual values in the cube by the total number of cells in the cube. Because the data cube's dimensions are predefined, not all cells are populated. In other words, some cells are empty. Returning to the sales example, there may be many products

FIGURE 13.12   MOLAP client/server architecture

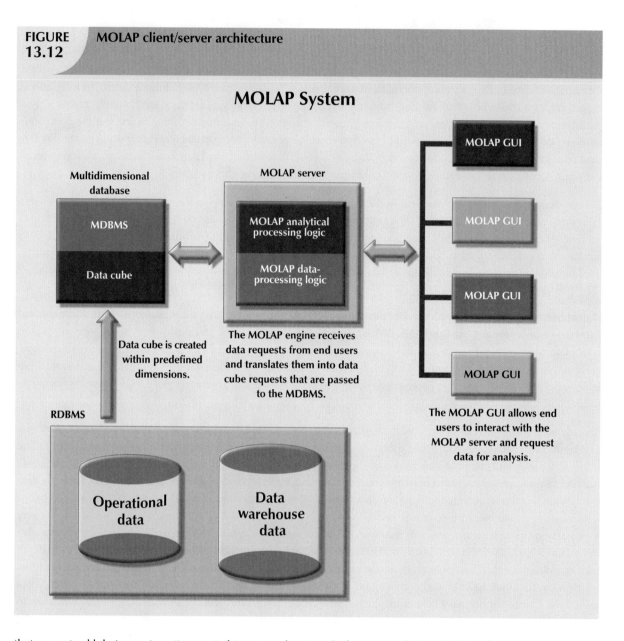

that are not sold during a given time period in a given location. In fact, you will often find that fewer than 50 percent of the data cube's cells are populated. In any case, multidimensional databases must handle sparsity effectively to reduce processing overhead and resource requirements.

Relational proponents also argue that using proprietary solutions makes it difficult to integrate the MDBMS with other data sources and tools used within the enterprise. Although it takes a substantial investment of time and effort to integrate the new technology and the existing information systems architecture, MOLAP may be a good solution for those situations in which small- to medium-sized databases are the norm and application software speed is critical.

13.6.8 RELATIONAL VS. MULTIDIMENSIONAL OLAP

Table 13.10 summarizes some OLAP and MOLAP pros and cons. Keep in mind, too, that the selection of one or the other often depends on the evaluator's vantage point. For example, a proper evaluation of OLAP must include price, supported hardware platforms, compatibility with the existing DBMS, programming requirements, performance, and availability of administrative tools. The summary in Table 13.10 provides a useful starting point for comparison.

TABLE 13.10	Relational vs. Multidimensional OLAP	
CHARACTERISTIC	ROLAP	MOLAP
Schema	Uses star schema Additional dimensions can be added dynamically	Uses data cubes Additional dimensions require re-creation of the data cube
Database size	Medium to large	Small to medium
Architecture	Client/server Standards-based Open	Client/server Proprietary
Access	Supports ad hoc requests Unlimited dimensions	Limited to predefined dimensions
Resources	High	Very high
Flexibility	High	Low
Scalability	High	Low
Speed	Good with small data sets; average for medium to large data sets	Faster for small to medium data sets; average for large data sets

ROLAP and MOLAP vendors are working toward the integration of their respective solutions within a unified decision support framework. Many OLAP products are able to handle tabular and multidimensional data with the same ease. For example, if you are using Excel OLAP functionality, as shown earlier in Figure 13.6, you can access relational OLAP data in a SQL server as well as cube (multidimensional data) in the local computer. In the meantime, relational databases successfully use the star schema design to handle multidimensional data, and their market share makes it unlikely that their popularity will fade anytime soon.

13.7 STAR SCHEMAS

The **star schema** is a data modeling technique used to map multidimensional decision support data into a relational database. In effect, the star schema creates the near equivalent of a multidimensional database schema from the existing relational database. The star schema was developed because existing relational modeling techniques, ER, and normalization did not yield a database structure that served advanced data analysis requirements well.

Star schemas yield an easily implemented model for multidimensional data analysis while still preserving the relational structures on which the operational database is built. The basic star schema has four components: facts, dimensions, attributes, and attribute hierarchies.

13.7.1 FACTS

Facts are numeric measurements (values) that represent a specific business aspect or activity. For example, sales figures are numeric measurements that represent product and/or service sales. Facts commonly used in business data analysis are units, costs, prices, and revenues. Facts are normally stored in a fact table that is the center of the star schema. The **fact table** contains facts that are linked through their dimensions, which are explained in the next section.

Facts can also be computed or derived at run time. Such computed or derived facts are sometimes called **metrics** to differentiate them from stored facts. The fact table is updated periodically (daily, weekly, monthly, and so on) with data from operational databases.

13.7.2 DIMENSIONS

Dimensions are qualifying characteristics that provide additional perspectives to a given fact. Recall that dimensions are of interest because *decision support data are almost always viewed in relation to other data*. For instance, sales might be compared by product from region to region and from one time period to the next. The kind of problem typically addressed by a BI system might be to make a comparison of the sales of unit X by region for the first quarters of 1998 through 2007. In that example, sales have product, location, and time dimensions. In effect, dimensions are the magnifying glass through which you study the facts. Such dimensions are normally stored in **dimension tables**. Figure 13.13 depicts a star schema for sales with product, location, and time dimensions.

FIGURE 13.13	Simple star schema

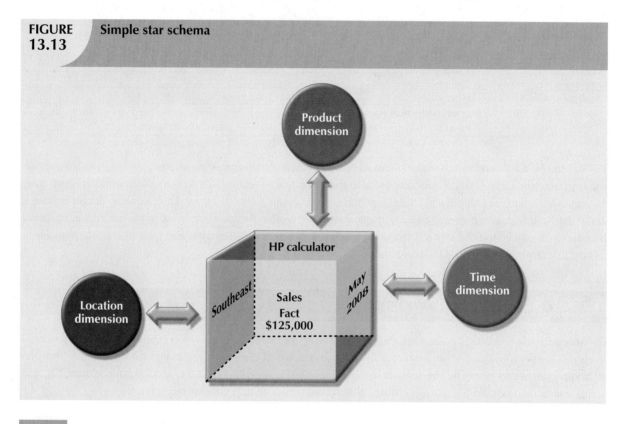

13.7.3 ATTRIBUTES

Each dimension table contains attributes. Attributes are often used to search, filter, or classify facts. *Dimensions provide descriptive characteristics about the facts through their attributes.* Therefore, the data warehouse designer must define common business attributes that will be used by the data analyst to narrow a search, group information, or describe dimensions. Using a sales example, some possible attributes for each dimension are illustrated in Table 13.11.

TABLE 13.11	Possible Attributes for Sales Dimensions	
DIMENSION NAME	DESCRIPTION	POSSIBLE ATTRIBUTES
Location	Anything that provides a description of the location. For example, Nashville, Store 101, South Region, and TN	Region, state, city, store, and so on
Product	Anything that provides a description of the product sold. For example, hair care product, shampoo, Natural Essence brand, 5.5-oz. bottle, and blue liquid	Product type, product ID, brand, package, presentation, color, size, and so on
Time	Anything that provides a time frame for the sales fact. For example, the year 2008, the month of July, the date 07/29/2008, and the time 4:46 p.m.	Year, quarter, month, week, day, time of day, and so on

These product, location, and time dimensions add a business perspective to the sales facts. The data analyst can now group the sales figures for a given product, in a given region, and at a given time. The star schema, through its facts and dimensions, can provide the data in the required format when the data are needed. And it can do so without imposing the burden of the additional and unnecessary data (such as order number, purchase order number, and status) that commonly exist in operational databases.

Conceptually, the sales example's multidimensional data model is best represented by a three-dimensional cube. Of course, this does not imply that there is a limit on the number of dimensions that can be associated to a fact table. There is no mathematical limit to the number of dimensions used. However, using a three-dimensional model makes it easy to visualize the problem. In this three-dimensional example, the multidimensional data analysis terminology, the cube illustrated in Figure 13.14 represents a view of sales dimensioned by product, location, and time.

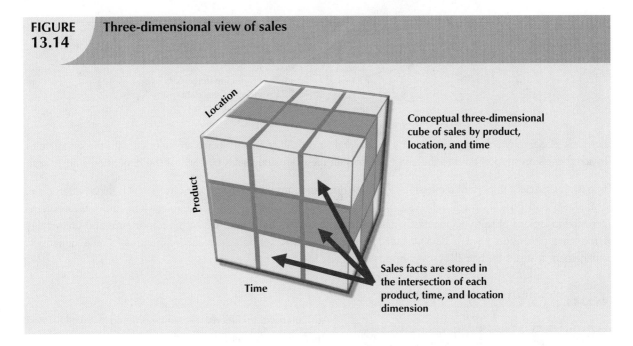

FIGURE 13.14   Three-dimensional view of sales

Conceptual three-dimensional cube of sales by product, location, and time

Sales facts are stored in the intersection of each product, time, and location dimension

Note that each sales value stored in the cube in Figure 13.14 is associated with the location, product, and time dimensions. However, keep in mind that this cube is only a *conceptual* representation of multidimensional data, and it does not show how the data are physically stored in a data warehouse. A ROLAP engine stores data in an RDBMS and uses its own data analysis logic and the end-user GUI to perform multidimensional analysis. A MOLAP system stores data in an MDBMS, using proprietary matrix and array technology to simulate this multidimensional cube.

Whatever the underlying database technology, one of the main features of multidimensional analysis is its ability to focus on specific "slices" of the cube. For example, the product manager may be interested in examining the sales of a product while the store manager is interested in examining the sales made by a particular store. In multidimensional terms, the ability to focus on slices of the cube to perform a more detailed analysis is known as **slice and dice**. Figure 13.15 illustrates the slice-and-dice concept. As you look at Figure 13.15, note that each cut across the cube yields a slice. Intersecting slices produce small cubes that constitute the "dice" part of the "slice-and-dice" operation.

To slice and dice, it must be possible to identify each slice of the cube. That is done by using the values of each attribute in a given dimension. For example, to use the location dimension, you might need to define a STORE_ID attribute in order to focus on a particular store.

Given the requirement for attribute values in a slice-and-dice environment, let's reexamine Table 13.11. Note that each attribute adds an additional perspective to the sales facts, thus setting the stage for finding new ways to search, classify, and possibly aggregate information. For example, the location dimension adds a geographic perspective of where the

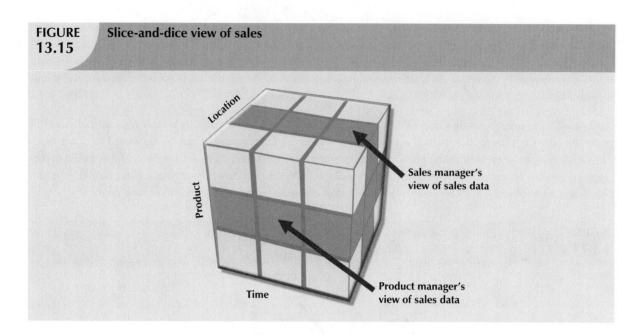

FIGURE 13.15   Slice-and-dice view of sales

Sales manager's view of sales data

Product manager's view of sales data

sales took place: in which region, state, city, store, and so on. All of the attributes are selected with the objective of providing decision support data to the end users so that they can study sales by each of the dimension's attributes.

Time is an especially important dimension. The time dimension provides a framework from which sales patterns can be analyzed and possibly predicted. Also, the time dimension plays an important role when the data analyst is interested in looking at sales aggregates by quarter, month, week, and so on. Given the importance and universality of the time dimension from a data analysis perspective, many vendors have added automatic time dimension management features to their data warehousing products.

13.7.4   ATTRIBUTE HIERARCHIES

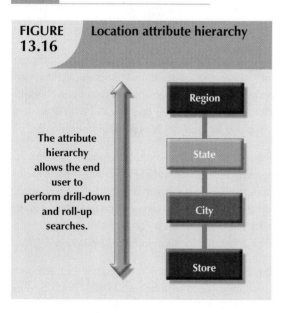

FIGURE 13.16   Location attribute hierarchy

Region

State

City

Store

The attribute hierarchy allows the end user to perform drill-down and roll-up searches.

Attributes within dimensions can be ordered in a well-defined attribute hierarchy. The **attribute hierarchy** provides a top-down data organization that is used for two main purposes: aggregation and drill-down/roll-up data analysis. For example, Figure 13.16 shows how the location dimension attributes can be organized in a hierarchy by region, state, city, and store.

The attribute hierarchy provides the capability to perform drill-down and roll-up searches in a data warehouse. For example, suppose a data analyst looks at the answers to the query, How does the 2007 month-to-date sales performance compare to the 2008 month-to-date sales performance? The data analyst spots a sharp sales decline for March 2008. The data analyst might decide to drill down inside the month of March to see how sales by regions compared to the previous year. By doing that, the analyst can determine whether the low March sales were reflected in all regions or in only a particular region. This type of drill-down operation can even be extended until the data analyst identifies the store that is performing below the norm.

The March sales scenario is possible because the attribute hierarchy allows the data warehouse and OLAP systems to have a defined path that will identify how data are to be decomposed and aggregated for drill-down and roll-up operations. It is not necessary for all attributes to be part of an attribute hierarchy; some attributes exist merely to provide narrative descriptions of the dimensions. But keep in mind that the attributes from different dimensions can be grouped to form a hierarchy. For example, after you drill down from city to store, you might want to drill down using the product dimension so the manager can identify slow products in the store. The product dimension can be based on the product group (dairy, meat, and so on) or on the product brand (Brand A, Brand B, and so on).

Figure 13.17 illustrates a scenario in which the data analyst studies sales facts, using the product, time, and location dimensions. In this example, the product dimension is set to "All products," meaning that the data analyst will see all products on the y-axis. The time dimension (x-axis) is set to "Quarter," meaning that the data are aggregated by quarters (for example, total sales of products A, B, and C in Q1, Q2, Q3, and Q4). Finally, the location dimension is initially set to "Region," thus ensuring that each cell contains the total regional sales for a given product in a given quarter.

FIGURE 13.17   **Attribute hierarchies in multidimensional analysis**

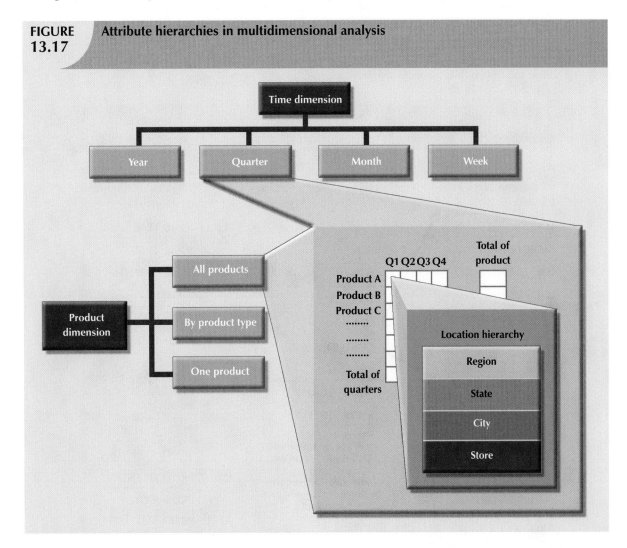

The simple data analysis scenario illustrated in Figure 13.17 provides the data analyst with three different information paths. On the product dimension (the y-axis), the data analyst can request to see all products, products grouped by type, or just one product. On the time dimension (the x-axis), the data analyst can request time-variant data at different levels of aggregation: year, quarter, month, or week. Each sales value initially shows the total sales, by region, of each product. When a GUI is used, clicking on the region cell enables the data analyst to drill down to see sales by states within the region. Clicking again on one of the state values yields the sales for each city in the state, and so forth.

As the preceding examples illustrate, attribute hierarchies determine how the data in the data warehouse are extracted and presented. The attribute hierarchy information is stored in the DBMS's data dictionary and is used by the OLAP tool to access the data warehouse properly. Once such access is ensured, query tools must be closely integrated with the data warehouse's metadata and they must support powerful analytical capabilities.

13.7.5 STAR SCHEMA REPRESENTATION

Facts and dimensions are normally represented by physical tables in the data warehouse database. The fact table is related to each dimension table in a many-to-one (M:1) relationship. In other words, many fact rows are related to each dimension row. Using the sales example, you can conclude that each product appears many times in the SALES fact table.

Fact and dimension tables are related by foreign keys and are subject to the familiar primary key/foreign key constraints. The primary key on the "1" side, the dimension table, is stored as part of the primary key on the "many" side, the fact table. *Because the fact table is related to many dimension tables, the primary key of the fact table is a composite primary key.* Figure 13.18 illustrates the relationships among the sales fact table and the product, location, and time dimension tables. To show you how easily the star schema can be expanded, a customer dimension has been added to the mix. Adding the customer dimension merely required including the CUST_ID in the SALES fact table and adding the CUSTOMER table to the database.

FIGURE 13.18   Star schema for SALES

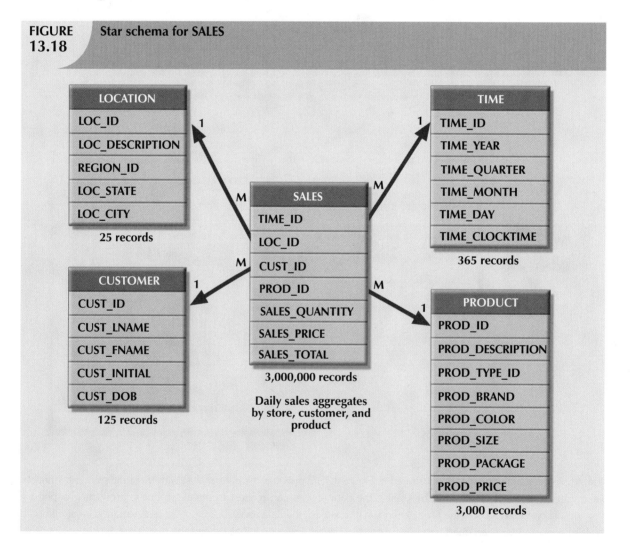

The composite primary key for the SALES fact table is composed of TIME_ID, LOC_ID, CUST_ID, and PROD_ID. Each record in the SALES fact table is uniquely identified by the combination of values for each of the fact table's foreign keys. *By default, the fact table's primary key is always formed by combining the foreign keys pointing to the dimension tables to which they are related.* In this case, each sales record represents each product sold to a specific customer, at a specific time, and in a specific location. In this schema, the TIME dimension table represents daily periods, so the SALES fact table represents daily sales aggregates by product and by customer. Because fact tables contain the actual values used in the decision support process, those values are repeated many times in the fact tables. Therefore, the fact tables are always the largest tables in the star schema. Because the dimension tables contain only nonrepetitive information (all unique salespersons, all unique products, and so on), the dimension tables are always smaller than the fact tables.

In a typical star schema, each dimension record is related to thousands of fact records. For example, "widget" appears only once in the product dimension, but it has thousands of corresponding records in the SALES fact table. That characteristic of the star schema facilitates data retrieval functions because most of the time the data analyst will look at the facts through the dimension's attributes. Therefore, a data warehouse DBMS that is optimized for decision support first searches the smaller dimension tables before accessing the larger fact tables.

Data warehouses usually have many fact tables. Each fact table is designed to answer specific decision support questions. For example, suppose you develop a new interest in orders while maintaining your original interest in sales. In that scenario, you should maintain an ORDERS fact table and a SALES fact table in the same data warehouse. If orders are considered to be an organization's key interest, the ORDERS fact table should be the center of a star schema that might have vendor, product, and time dimensions. In that case, an interest in vendors yields a new vendor dimension, represented by a new VENDOR table in the database. The product dimension is represented by the same product table used in the initial sales star schema. However, given the interest in orders as well as sales, the time dimension now requires special attention. If the orders department uses the same time periods as the sales department, time can be represented by the same time table. If different time periods are used, you must create another table, perhaps named ORDER_TIME, to represent the time periods used by the orders department. In Figure 13.19, the orders star schema shares the product, vendor, and time dimensions.

Multiple fact tables also can be created for performance and semantic reasons. The following section explains several performance-enhancing techniques that can be used within the star schema.

FIGURE 13.19   Orders star schema

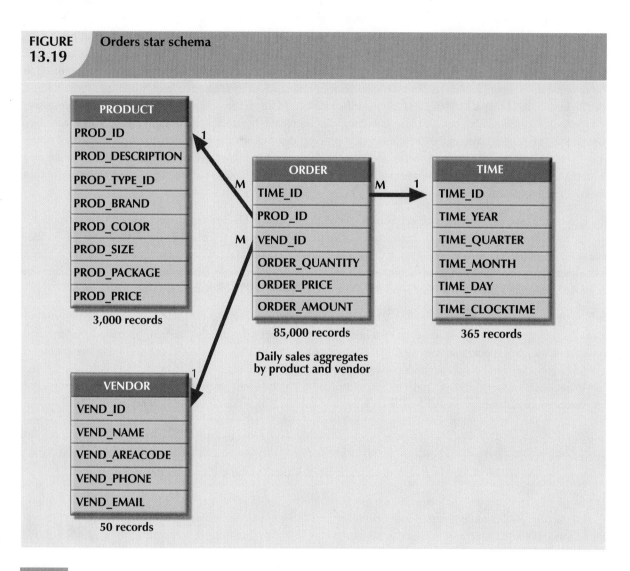

PRODUCT

PROD_ID
PROD_DESCRIPTION
PROD_TYPE_ID
PROD_BRAND
PROD_COLOR
PROD_SIZE
PROD_PACKAGE
PROD_PRICE

3,000 records

ORDER

TIME_ID
PROD_ID
VEND_ID
ORDER_QUANTITY
ORDER_PRICE
ORDER_AMOUNT

85,000 records

Daily sales aggregates
by product and vendor

TIME

TIME_ID
TIME_YEAR
TIME_QUARTER
TIME_MONTH
TIME_DAY
TIME_CLOCKTIME

365 records

VENDOR

VEND_ID
VEND_NAME
VEND_AREACODE
VEND_PHONE
VEND_EMAIL

50 records

13.7.6 PERFORMANCE-IMPROVING TECHNIQUES FOR THE STAR SCHEMA

The creation of a database that provides fast and accurate answers to data analysis queries is the data warehouse design's prime objective. Therefore, performance-enhancement actions might target query speed through the facilitation of SQL code as well as through better semantic representation of business dimensions. Four techniques are often used to optimize data warehouse design:

- Normalizing dimensional tables.
- Maintaining multiple fact tables to represent different aggregation levels.
- Denormalizing fact tables.
- Partitioning and replicating tables.

Normalizing Dimensional Tables

Dimensional tables are normalized to achieve semantic simplicity and facilitate end-user navigation through the dimensions. For example, if the location dimension table contains transitive dependencies among region, state, and city, you can revise those relationships to the 3NF (third normal form), as shown in Figure 13.20. (If necessary, review normalization techniques in Chapter 5, Normalization of Database Tables.) The star schema shown in Figure 13.20 is known as a **snowflake schema**, which is a type of star schema in which the dimension tables can have their own dimension tables. The snowflake schema is usually the result of normalizing dimension tables.

FIGURE 13.20   Normalized dimension tables

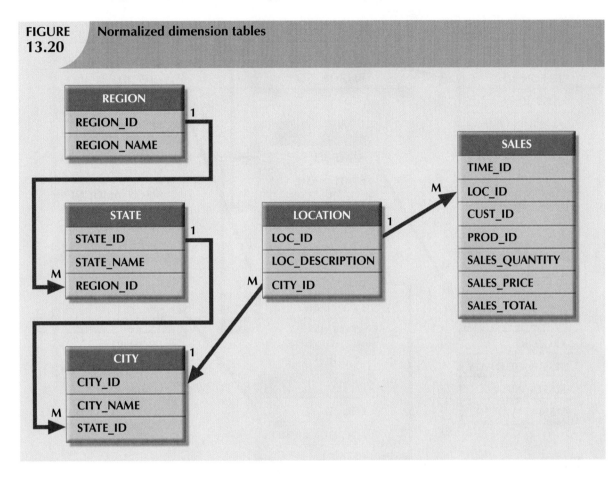

By normalizing the dimension tables, you simplify the data-filtering operations related to the dimensions. In this example, the region, state, city, and location contain very few records compared to the SALES fact table. Only the location table is directly related to the sales fact table.

> **NOTE**
>
> Although using the dimension tables shown in Figure 13.20 gains structural simplicity, there is a price to pay for that simplicity. For example, if you want to aggregate the data by region, you must use a four-table join, thus increasing the complexity of the SQL statements. The star schema in Figure 13.18 uses a LOCATION dimension table that greatly facilitates data retrieval by eliminating multiple join operations. This is yet another example of the trade-offs that designers must consider.

Maintaining Multiple Fact Tables that Represent Different Aggregation Levels

You can also speed up query operations by creating and maintaining multiple fact tables related to each level of aggregation (region, state, and city) in the location dimension. These aggregate tables are precomputed at the data-loading phase rather than at run time. The purpose of this technique is to save processor cycles at run time, thereby speeding up data analysis. An end-user query tool optimized for decision analysis then properly accesses the summarized fact tables instead of computing the values by accessing a lower level of detail fact table. This technique is illustrated in Figure 13.21, which adds aggregate fact tables for region, state, and city to the initial sales example.

FIGURE 13.21   Multiple fact tables

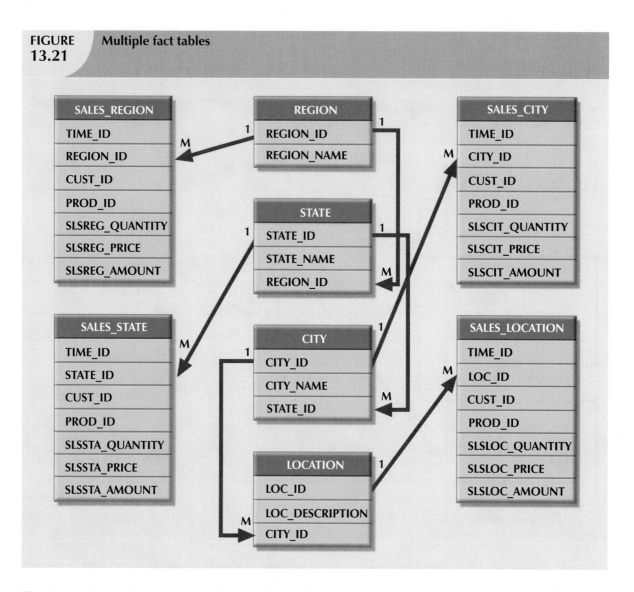

The data warehouse designer must identify which levels of aggregation to precompute and store in the database. These multiple aggregate fact tables are updated during each load cycle in batch mode. And because the objective is to minimize access and processing time, according to the expected frequency of use and the processing time required to calculate a given aggregation level at run time, the data warehouse designer must select which aggregation fact tables to create.

Denormalizing Fact Tables

Denormalizing fact tables improves data access performance and saves data storage space. The latter objective, however, is becoming less of an issue. Data storage costs decrease almost daily, and DBMS limitations that restrict database and table size limits, record size limits, and the maximum number of records in a single table have far more negative effects than raw storage space costs.

Denormalization improves performance by using a single record to store data that normally take many records. For example, to compute the total sales for all products in all regions, you might have to access the region sales aggregates and summarize all of the records in this table. If you have 300,000 product sales, you could be summarizing at least 300,000 rows. Although this might not be a very taxing operation for a DBMS, a comparison of, say, 10 years' worth of previous sales begins to bog down the system. In such cases, it is useful to have special aggregate tables that are denormalized. For example, a YEAR_TOTALS table might contain the following fields: YEAR_ID, MONTH_1, MONTH_2 ... MONTH_12, and each year's total. Such tables can easily be used to serve as a basis for year-to-year comparisons at the top month level, the quarter level, or the year level. Here again, design criteria, such as frequency of use and performance requirements, are evaluated against the possible overload placed on the DBMS to manage the denormalized relations.

Partitioning and Replicating Tables

Because table partitioning and replication were covered in detail in Chapter 12, Distributed Database Management Systems, those techniques are discussed here only as they specifically relate to the data warehouse. Table partitioning and replication are particularly important when a BI system is implemented in dispersed geographic areas. **Partitioning** splits a table into subsets of rows or columns and places the subsets close to the client computer to improve data access time. **Replication** makes a copy of a table and places it in a different location, also to improve access time.

No matter which performance-enhancement scheme is used, time is the most common dimension used in business data analysis. Therefore, it is very common to have one fact table for each level of aggregation defined within the time dimension. For example, in the sales example, you might have five aggregate sales fact tables: daily, weekly, monthly, quarterly, and yearly. Those fact tables must have an implicit or explicit periodicity defined. **Periodicity**, usually expressed as current year only, previous years, or all years, provides information about the time span of the data stored in the table.

At the end of each year, daily sales for the current year are moved to another table that contains previous years' daily sales only. This table actually contains all sales records from the beginning of operations, with the exception of the current year. The data in the current year and previous years' tables thus represent the complete sales history of the company. The previous years' sales table can be replicated at several locations to avoid remote access to the historic sales data, which can cause slow response time. The possible size of this table is enough to intimidate all but the bravest of query optimizers. Here is one case in which denormalization would be of value!

13.8 IMPLEMENTING A DATA WAREHOUSE

Organization-wide information system development is subject to many constraints. Some of the constraints are based on available funding. Others are a function of management's view of the role played by an IS department and of the extent and depth of the information requirements. Add the constraints imposed by corporate culture, and you understand why no single formula can describe perfect data warehouse development. Therefore, rather than proposing a single data warehouse design and implementation methodology, this section identifies a few factors that appear to be common to data warehousing.

13.8.1 THE DATA WAREHOUSE AS AN ACTIVE DECISION SUPPORT FRAMEWORK

Perhaps the first thing to remember is that a data warehouse is not a static database. Instead, it is a dynamic framework for decision support that is, almost by definition, always a work in progress. Because it is the foundation of a modern BI environment, the design and implementation of the data warehouse means that you are involved in the design and implementation of a complete database-system-development infrastructure for company-wide decision support. Although it is easy to focus on the data warehouse database as the BI central data repository, you must remember that the decision support infrastructure includes hardware, software, people, and procedures, as well as data. The argument that the data warehouse is the only *critical* BI success component is as misleading as the argument that a human being needs only a heart or a brain to function. The data warehouse is a critical component of a modern BI environment, but it is certainly not the only critical component. Therefore, its design and implementation must be examined in light of the entire infrastructure.

13.8.2 A COMPANY-WIDE EFFORT THAT REQUIRES USER INVOLVEMENT

Designing a data warehouse means being given an opportunity to help develop an integrated data model that captures the data that are considered to be essential to the organization, from both end-user and business perspectives. Data warehouse data cross departmental lines and geographical boundaries. Because the data warehouse represents an attempt to model all of the organization's data, you are likely to discover that organizational components (divisions, departments, support groups, and so on) often have conflicting goals, and it certainly will be easy to find data inconsistencies and damaging redundancies. Information is power, and the control of its sources and uses is likely to trigger turf battles, end-user resistance, and power struggles at all levels. Building the perfect data warehouse is not just a matter of knowing how to create a star schema; it requires managerial skills to deal with conflict resolution, mediation, and arbitration. In short, the designer must:

- Involve end users in the process.
- Secure end users' commitment from the beginning.
- Solicit continuous end-user feedback.
- Manage end-user expectations.
- Establish procedures for conflict resolution.

13.8.3 SATISFY THE TRILOGY: DATA, ANALYSIS, AND USERS

Great managerial skills are not, of course, solely sufficient. The technical aspects of the data warehouse must be addressed as well. The old adage of input-process-output repeats itself here. The data warehouse designer must satisfy:

- Data integration and loading criteria.
- Data analysis capabilities with acceptable query performance.
- End-user data analysis needs.

The foremost technical concern in implementing a data warehouse is to provide end-user decision support with advanced data analysis capabilities—at the right moment, in the right format, with the right data, and at the right cost.

13.8.4 APPLY DATABASE DESIGN PROCEDURES

You learned about the database life cycle and the database design process in Chapter 9, Database Design, so perhaps it is wise to review the traditional database design procedures. These design procedures must then be adapted to fit the data warehouse requirements. If you remember that the data warehouse derives its data from operational databases, you will understand why a solid foundation in operational database design is important. (It's difficult to produce good data warehouse data when the operational database data are corrupted.) Figure 13.22 depicts a simplified process for implementing the data warehouse.

As noted, developing a data warehouse is a company-wide effort that requires many resources: human, financial, and technical. Providing company-wide decision support requires a sound architecture based on a mix of people skills, technology, and managerial procedures that is often difficult to find and implement. For example:

- The sheer and often mind-boggling quantity of decision support data is likely to require the latest hardware and software—that is, advanced computers with multiple processors, advanced database systems, and large-capacity storage units. In the not-too-distant past, those requirements usually prompted the use of a mainframe-based system. Today's client/server technology offers many other choices to implement a data warehouse.
- Very detailed procedures are necessary to orchestrate the flow of data from the operational databases to the data warehouse. Data flow control includes data extraction, validation, and integration.
- To implement and support the data warehouse architecture, you also need people with advanced database design, software integration, and management skills.

FIGURE 13.22   Data warehouse design and implementation road map

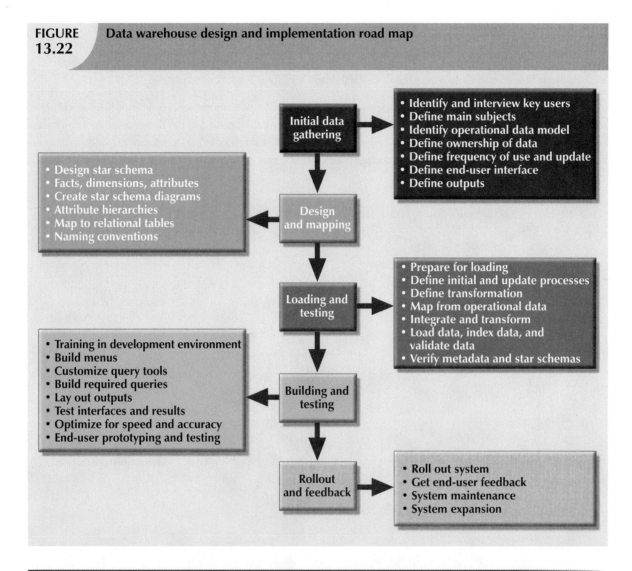

13.9 DATA MINING

The purpose of data analysis is to discover previously unknown data characteristics, relationships, dependencies, or trends. Such discoveries then become part of the information framework on which decisions are built. *A typical data analysis tool relies on the end users to define the problem, select the data, and initiate the appropriate data analyses to generate the information that helps model and solve problems that the end users uncover.* In other words, the end user reacts to an external stimulus—the discovery of the problem itself. If the end user fails to detect a problem, no action is taken. Given that limitation, some current BI environments now support various types of automated alerts. The alerts are software agents that constantly monitor certain parameters, such as sales indicators and inventory levels, and then perform specified actions (send e-mail or alert messages, run programs, and so on) when such parameters reach predefined levels.

In contrast to the traditional (reactive) BI tools, data mining is *proactive*. Instead of having the end user define the problem, select the data, and select the tools to analyze the data, *data-mining tools automatically search the data for anomalies and possible relationships, thereby identifying problems that have not yet been identified by the end user.* In other words, **data mining** refers to the activities that analyze the data, uncover problems or opportunities hidden in the data relationships, form computer models based on their findings, and then use the models to predict business behavior—requiring minimal end-user intervention. Therefore, the end user is able to use the system's findings to gain knowledge that might yield competitive advantages. Data mining describes a new breed of specialized decision support tools that automate data analysis. In short, data-mining tools *initiate* analyses to create knowledge. Such knowledge can be used to address any number of business problems. For example, banks and credit card companies use knowledge-based analysis to detect fraud, thereby decreasing fraudulent transactions.

To put data mining in perspective, look at the pyramid in Figure 13.23, which represents how knowledge is extracted from data. *Data* form the pyramid base and represent what most organizations collect in their operational databases. The second level contains *information* that represents the purified and processed data. Information forms the basis for decision making and business understanding. *Knowledge* is found at the pyramid's apex and represents highly specialized information.

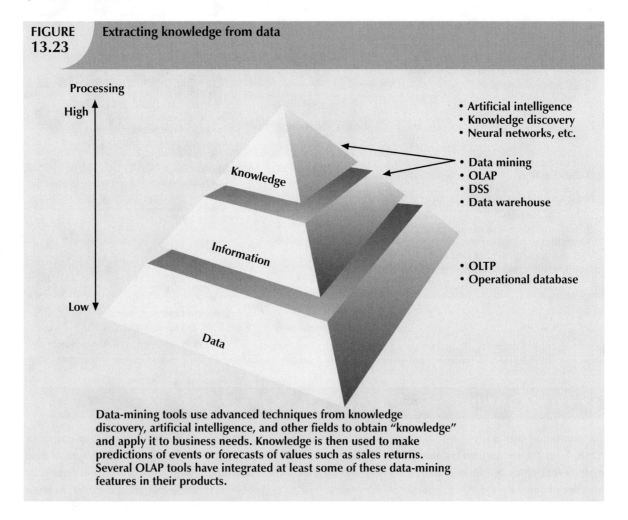

FIGURE 13.23 Extracting knowledge from data

Data-mining tools use advanced techniques from knowledge discovery, artificial intelligence, and other fields to obtain "knowledge" and apply it to business needs. Knowledge is then used to make predictions of events or forecasts of values such as sales returns. Several OLAP tools have integrated at least some of these data-mining features in their products.

It is difficult to provide a precise list of characteristics of data-mining tools. For one thing, the current generation of data-mining tools contains many design and application variations to fit data-mining requirements. Additionally, the many variations exist because there are no established standards that govern the creation of data-mining tools. Each data-mining tool seems to be governed by a different approach and focus, thus generating families of data-mining tools that focus on market niches such as marketing, retailing, finance, healthcare, investments, insurance, and banking. Within a given niche, data-mining tools can use certain algorithms, and those algorithms can be implemented in different ways and/or applied over different data.

In spite of the lack of precise standards, data mining is subject to four general phases:

1. Data preparation.
2. Data analysis and classification.
3. Knowledge acquisition.
4. Prognosis.

In the *data preparation phase*, the main data sets to be used by the data mining operation are identified and cleansed of any data impurities. Because the data in the data warehouse are already integrated and filtered, the data warehouse usually is the target set for data mining operations.

The *data analysis and classification phase* studies the data to identify common data characteristics or patterns. During this phase, the data-mining tool applies specific algorithms to find:

*   Data groupings, classifications, clusters, or sequences.
*   Data dependencies, links, or relationships.
*   Data patterns, trends, and deviations.

The *knowledge acquisition phase* uses the results of the data analysis and classification phase. During the knowledge acquisition phase, the data-mining tool (with possible intervention by the end user) selects the appropriate modeling or knowledge acquisition algorithms. The most common algorithms used in data mining are based on neural networks, decision trees, rules induction, genetic algorithms, classification and regression trees, memory-based reasoning, and nearest neighbor and data visualization. A data-mining tool may use many of these algorithms in any combination to generate a computer model that reflects the behavior of the target data set.

Although many data-mining tools stop at the knowledge-acquisition phase, others continue to the *prognosis phase*. In that phase, the data mining findings are used to predict future behavior and forecast business outcomes. Examples of data mining findings can be:

*   Sixty-five percent of customers who did not use a particular credit card in the last six months are 88 percent likely to cancel that account.
*   Eighty-two percent of customers who bought a 27-inch or larger TV are 90 percent likely to buy an entertainment center within the next four weeks.
*   If age < 30 and income <= 25,000 and credit rating < 3 and credit amount > 25,000, then the minimum loan term is 10 years.

The complete set of findings can be represented in a decision tree, a neural net, a forecasting model, or a visual presentation interface that is used to project future events or results. For example, the prognosis phase might project the likely outcome of a new product rollout or a new marketing promotion. Figure 13.24 illustrates the different phases of the data mining techniques.

Because data mining technology is still in its infancy, some of the data mining findings might fall outside the boundaries of what business managers expect. For example, a data-mining tool might find a close relationship between a customer's favorite brand of soda and the brand of tires on the customer's car. Clearly, that relationship might not be held in high regard among sales managers. (In regression analysis, those relationships are commonly described by the label "idiot correlation.") Fortunately, data mining usually yields more meaningful results. In fact, data mining has proved to be very helpful in finding practical relationships among data that help define customer buying patterns, improve product development and acceptance, reduce healthcare fraud, analyze stock markets, and so on.

Ideally, you can expect the development of databases that not only store data and various statistics about data usage, but also have the ability to learn about and extract knowledge from the stored data. Such database management systems, also known as inductive or intelligent databases, are the focus of intense research in many laboratories. Although those databases have yet to lay claim to substantial commercial market penetration, both "add-on" and DBMS-integrated data mining tools have proliferated in the data warehousing database market.

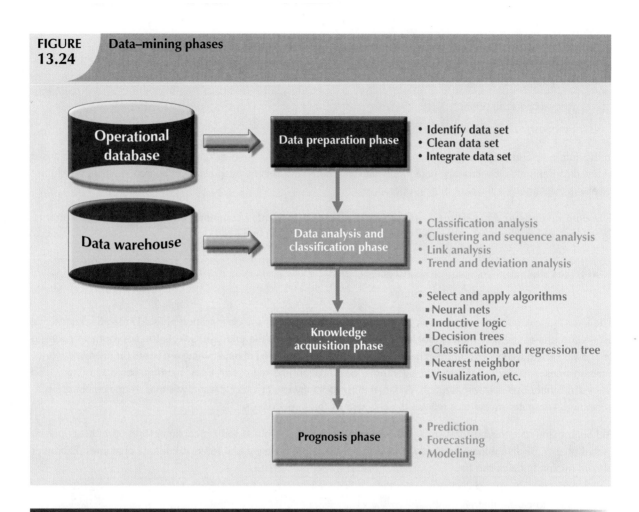

FIGURE 13.24   Data–mining phases

- Identify data set
- Clean data set
- Integrate data set

Data preparation phase

- Classification analysis
- Clustering and sequence analysis
- Link analysis
- Trend and deviation analysis

Data analysis and classification phase

- Select and apply algorithms
  - Neural nets
  - Inductive logic
  - Decision trees
  - Classification and regression tree
  - Nearest neighbor
  - Visualization, etc.

Knowledge acquisition phase

- Prediction
- Forecasting
- Modeling

Prognosis phase

13.10 SQL EXTENSIONS FOR OLAP

The proliferation of OLAP tools has fostered the development of SQL extensions to support multidimensional data analysis. Most SQL innovations are the result of vendor-centric product enhancements. However, many of the innovations have made their way into standard SQL. This section introduces some of the new SQL extensions that have been created to support OLAP-type data manipulations.

The SaleCo snowflake schema shown in Figure 13.25 will be used to demonstrate the use of the SQL extensions. Note that this snowflake schema has a central DWSALESFACT fact table and three dimension tables: DWCUSTOMER, DWPRODUCT, and DWTIME. The central fact table represents daily sales by product and customer. However, as you examine the star schema shown in Figure 13.25 more carefully, you will see that the DWCUSTOMER and DWPRODUCT dimension tables have their own dimension tables: DWREGION and DWVENDOR.

Keep in mind that a database is at the core of all data warehouses. Therefore, all SQL commands (such as CREATE, INSERT, UPDATE, DELETE, and SELECT) will work in the data warehouse as expected. However, most queries you run in a data warehouse tend to include a lot of data groupings and aggregations over multiple columns. That's why this section introduces two extensions to the GROUP BY clause that are particularly useful: ROLLUP and CUBE. In addition, you will learn about using materialized views to store preaggregated rows in the database.

FIGURE 13.25	SaleCo snowflake schema

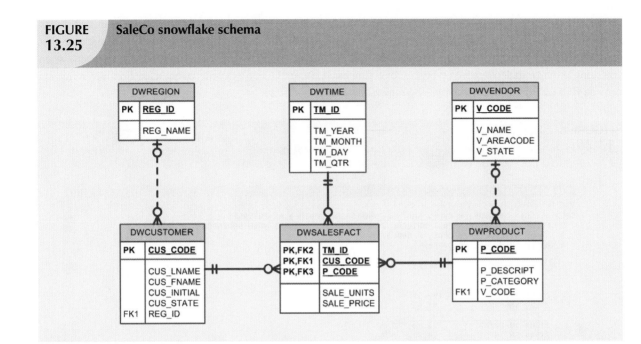

ONLINE CONTENT

The script files used to populate the database and run the SQL commands are available in the Student Online Companion.

NOTE

This section uses the Oracle RDBMS to demonstrate the use of SQL extensions to support OLAP functionality. If you use a different DBMS, consult the documentation to verify whether the vendor supports similar functionality and what the proper syntax is for your DBMS.

13.10.1 THE ROLLUP EXTENSION

The ROLLUP extension is used with the GROUP BY clause to generate aggregates by different dimensions. As you know, the GROUP BY clause will generate only one aggregate for each new value combination of attributes listed in the GROUP BY clause. The ROLLUP extension goes one step further; it enables you to get a subtotal for each column listed except for the last one, which gets a grand total instead. The syntax of the GROUP BY ROLLUP is as follows:

```
SELECT      column1, column2 [, …], aggregate_function(expression)
FROM        table1 [,table2, …]
[WHERE      condition]
GROUP BY    ROLLUP (column1, column2 [, …])
[HAVING     condition]
[ORDER BY   column1 [, column2, …]]
```

The order of the column list within the GROUP BY ROLLUP is very important. The last column in the list will generate a grand total. All other columns will generate subtotals. For example, Figure 13.26 shows the use of the ROLLUP extension to generate subtotals by vendor and product.

Note that Figure 13.26 shows the subtotals by vendor code and a grand total for all product codes. Contrast that with the normal GROUP BY clause that will generate only the subtotals for each vendor and product combination rather than the subtotals *by vendor* and the grand total for *all products*. The ROLLUP extension is particularly useful when you want to obtain multiple nested subtotals for a dimension hierarchy. For example, within a location hierarchy, you can use ROLLUP to generate subtotals by region, state, city, and store.

FIGURE 13.26   **ROLLUP extension**

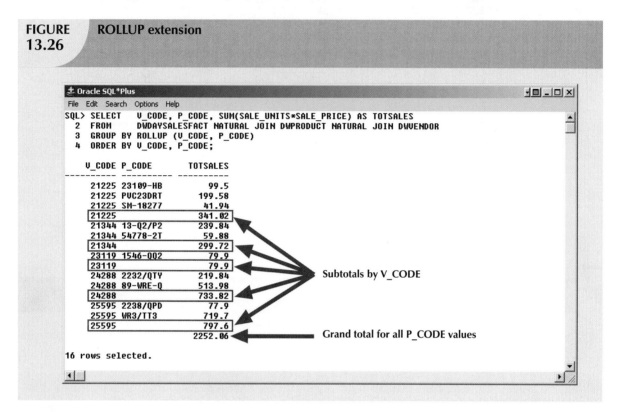

13.10.2 THE CUBE EXTENSION

The CUBE extension is also used with the GROUP BY clause to generate aggregates by the listed columns, including the last one. The CUBE extension will enable you to get a subtotal for each column listed in the expression, in addition to a grand total for the last column listed. The syntax of the GROUP BY CUBE is as follows:

```
SELECT      column1 [, column2, ...], aggregate_function(expression)
FROM        table1 [,table2, ...]
[WHERE      condition]
GROUP BY    CUBE (column1, column2 [, ...])
[HAVING     condition]
[ORDER BY   column1 [, column2, ...]]
```

For example, Figure 13.27 shows the use of the CUBE extension to compute the sales subtotals by month and by product, as well as a grand total.

In Figure 13.27, note that the CUBE extension generates the subtotals for each combination of month and product, in addition to subtotals by month and by product, as well as a grand total. The CUBE extension is particularly useful when you want to compute all possible subtotals within groupings based on multiple dimensions. Cross-tabulations are especially good candidates for application of the CUBE extension.

FIGURE 13.27   CUBE extension

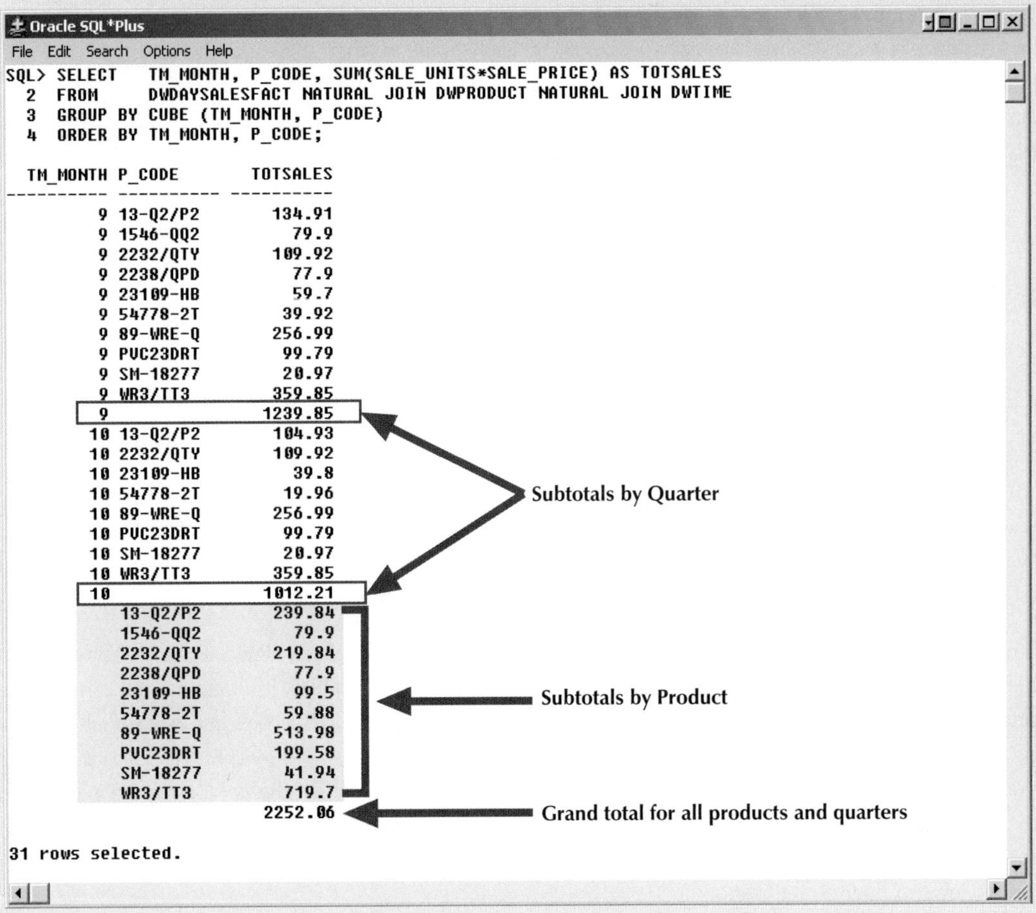

```
± Oracle SQL*Plus                                                  ⌐□ _□ ×
File  Edit  Search  Options  Help
SQL> SELECT   TM_MONTH, P_CODE, SUM(SALE_UNITS*SALE_PRICE) AS TOTSALES
  2  FROM     DWDAYSALESFACT NATURAL JOIN DWPRODUCT NATURAL JOIN DWTIME
  3  GROUP BY CUBE (TM_MONTH, P_CODE)
  4  ORDER BY TM_MONTH, P_CODE;

  TM_MONTH P_CODE       TOTSALES
---------- ---------- ----------
         9 13-Q2/P2      134.91
         9 1546-QQ2       79.9
         9 2232/QTY      109.92
         9 2238/QPD       77.9
         9 23109-HB       59.7
         9 54778-2T       39.92
         9 89-WRE-Q      256.99
         9 PVC23DRT       99.79
         9 SM-18277       20.97
         9 WR3/TT3       359.85
         9             1239.85                ←    Subtotals by Quarter
        10 13-Q2/P2      104.93
        10 2232/QTY      109.92
        10 23109-HB       39.8
        10 54778-2T       19.96
        10 89-WRE-Q      256.99
        10 PVC23DRT       99.79
        10 SM-18277       20.97
        10 WR3/TT3       359.85
        10            1012.21                ←
           13-Q2/P2      239.84
           1546-QQ2       79.9
           2232/QTY      219.84
           2238/QPD       77.9
           23109-HB       99.5
           54778-2T       59.88             ←    Subtotals by Product
           89-WRE-Q      513.98
           PVC23DRT      199.58
           SM-18277       41.94
           WR3/TT3       719.7
                        2252.06             ←    Grand total for all products and quarters

31 rows selected.
```

13.10.3 MATERIALIZED VIEWS

The data warehouse normally contains fact tables that store specific measurements of interest to an organization. Such measurements are organized by different dimensions. The vast majority of OLAP business analysis of "everyday activities" is based on comparisons of data that are aggregated at different levels, such as totals by vendor, by product, and by store.

Because businesses normally use a predefined set of summaries for benchmarking, it is reasonable to predefine such summaries for future use by creating summary fact tables. (See Section 13.5.6 for a discussion of additional performance-improving techniques.) However, creating multiple summary fact tables that use GROUP BY queries with multiple table joins could become a resource-intensive operation. In addition, data warehouses must also be able to maintain up-to-date summarized data at all times. So what happens with the summary fact tables after new sales data have been added to the base fact tables? Under normal circumstances, the summary fact tables are re-created. This operation requires that the SQL code be run again to re-create all summary rows, even when only a few rows needed updating. Clearly, this is a time-consuming process.

To save query processing time, most database vendors have implemented additional "functionality" to manage aggregate summaries more efficiently. This new functionality resembles the standard SQL views for which the SQL code is predefined in the database. However, the added functionality difference is that the views also store the preaggregated rows, something like a summary table. For example, Microsoft SQL Server provides indexed views, while Oracle provides materialized views. This section explains the use of materialized views.

A **materialized view** is a dynamic table that not only contains the SQL query command to generate the rows, but also stores the actual rows. The materialized view is created the first time the query is run and the summary rows are stored in the table. The materialized view rows are automatically updated when the base tables are updated. That way, the data warehouse administrator will create the view but will not have to worry about updating the view. The use of materialized views is totally transparent to the end user. The OLAP end user can create OLAP queries, using the standard fact tables, and the DBMS query optimization feature will automatically use the materialized views if those views provide better performance.

The basic syntax for the materialized view is:

```
CREATE MATERIALIZED VIEW view_name
BUILD {IMMEDIATE | DEFERRED}
REFRESH {[FAST | COMPLETE | FORCE]} ON COMMIT
[ENABLE QUERY REWRITE]
AS select_query;
```

The BUILD clause indicates when the materialized view rows are actually populated. IMMEDIATE indicates that the materialized view rows are populated right after the command is entered. DEFERRED indicates that the materialized view rows will be populated at a later time. Until then, the materialized view is in an "unusable" state. The DBMS provides a special routine that an administrator runs to populate materialized views.

The REFRESH clause lets you indicate when and how to update the materialized view when new rows are added to the base tables. FAST indicates that whenever a change is made in the base tables, the materialized view updates only the affected rows. COMPLETE indicates that a complete update will be made for all rows in the materialized view when the select query on which the view is based is rerun. FORCE indicates that the DBMS will first try to do a FAST update; otherwise, it will do a COMPLETE update. The ON COMMIT clause indicates that the updates to the materialized view will take place as part of the commit process of the underlying DML statement, that is, as part of the commit of the DML transaction that updated the base tables. The ENABLE QUERY REWRITE option allows the DBMS to use the materialized views in query optimization.

To create materialized views, you must have specified privileges and you must complete specified prerequisite steps. As always, you must defer to the DBMS documentation for the latest updates. In the case of Oracle, you must create materialized view logs on the base tables of the materialized view. Figure 13.28 shows the steps required to create the MONTH_SALES_MV materialized view in the Oracle RDBMS.

FIGURE 13.28   **Creating a materialized view**

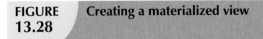

```
± Oracle SQL*Plus                                                    ◄□ _□ x
File  Edit  Search  Options  Help
SQL> CREATE MATERIALIZED VIEW LOG ON DWTIME
  2   WITH ROWID, SEQUENCE INCLUDING NEW VALUES;

Materialized view log created.

SQL> CREATE MATERIALIZED VIEW LOG ON DWDAYSALESFACT
  2   WITH ROWID, SEQUENCE INCLUDING NEW VALUES;

Materialized view log created.

SQL> CREATE MATERIALIZED VIEW SALES_MONTH_MV
  2      BUILD IMMEDIATE
  3      REFRESH FORCE ON COMMIT
  4      ENABLE QUERY REWRITE
  5      AS SELECT TM_YEAR, TM_MONTH, P_CODE,
  6             SUM(SALE_UNITS), SUM(SALE_PRICE*SALE_UNITS) AS SUM_SALES
  7         FROM DWTIME T, DWDAYSALESFACT S
  8         WHERE  S.TM_ID = T.TM_ID
  9         GROUP BY  TM_YEAR, TM_MONTH, P_CODE;

Materialized view created.

SQL> SELECT * FROM SALES_MONTH_MV;

  TM_YEAR   TM_MONTH P_CODE    SUM(SALE_UNITS)  SUM_SALES
---------- ---------- -------- ---------------- ----------
      2005          9 WR3/TT3                 3     359.85
      2005          9 13-Q2/P2                9     134.91
      2005          9 1546-QQ2                2       79.9
      2005          9 2232/QTY                1     109.92
      2005          9 2238/QPD                2       77.9
      2005          9 23109-HB                6       59.7
      2005          9 54778-2T                8      39.92
      2005          9 89-WRE-Q                1     256.99
      2005          9 PVC23DRT               17      99.79
      2005          9 SM-18277                3      20.97
      2005         10 WR3/TT3                 3     359.85
      2005         10 13-Q2/P2                7     104.93
      2005         10 2232/QTY                1     109.92
      2005         10 23109-HB                4       39.8
      2005         10 54778-2T                4      19.96
      2005         10 89-WRE-Q                1     256.99
      2005         10 PVC23DRT               17      99.79
      2005         10 SM-18277                3      20.97

18 rows selected.

SQL> COMMIT;

Commit complete.
```

The materialized view in Figure 13.28 computes the monthly total units sold and the total sales aggregates by product. The SALES_MONTH_MV materialized view is configured to automatically update after each change in the base tables. Note that the last row of SALES_MONTH_MV indicates that during October, the sales of product 'SM-18277' are three units, for a total of $20.97. Figure 13.29 shows the effects of an update to the DWDAYSALESFACT base table.

FIGURE 13.29   Refreshing a materialized view

```
Oracle SQL*Plus
File  Edit  Search  Options  Help
SQL> INSERT INTO DWDAYSALESFACT VALUES (207,10017,'SM-18277',1,6.99);

1 row created.

SQL>
SQL> COMMIT;

Commit complete.

SQL>
SQL> SELECT * FROM SALES_MONTH_MV;

   TM_YEAR   TM_MONTH P_CODE     SUM(SALE_UNITS)  SUM_SALES
---------- ---------- ---------- --------------- ----------
      2005          9 WR3/TT3                  3     359.85
      2005          9 13-Q2/P2                 9     134.91
      2005          9 1546-QQ2                 2       79.9
      2005          9 2232/QTY                 1     109.92
      2005          9 2238/QPD                 2       77.9
      2005          9 23109-HB                 6       59.7
      2005          9 54778-2T                 8      39.92
      2005          9 89-WRE-Q                 1     256.99
      2005          9 PVC23DRT                17      99.79
      2005          9 SM-18277                 3      20.97
      2005         10 WR3/TT3                  3     359.85
      2005         10 13-Q2/P2                 7     104.93
      2005         10 2232/QTY                 1     109.92
      2005         10 23109-HB                 4       39.8
      2005         10 54778-2T                 4      19.96
      2005         10 89-WRE-Q                 1     256.99
      2005         10 PVC23DRT                17      99.79
      2005         10 SM-18277                 4      27.96

18 rows selected.
```

Figure 13.29 shows how the materialized view was automatically updated after the insertion of a new row in the DWDAYSALESFACT table. Note that the last row of the SALES_MONTH_MV now shows that in October, the sales of product 'SM-18277' are four units, for a total of $27.96.

Although all of the examples in this section focus on SQL extensions to support OLAP reporting in an Oracle DBMS, you have seen just a small fraction of the many business intelligence features currently provided by most DBMS vendors. For example, most vendors provide rich graphical user interfaces to manipulate, analyze, and present the data in multiple formats. Figure 13.30 shows two sample screens, one for Oracle and one for Microsoft OLAP products.

FIGURE 13.30   **Sample OLAP applications**

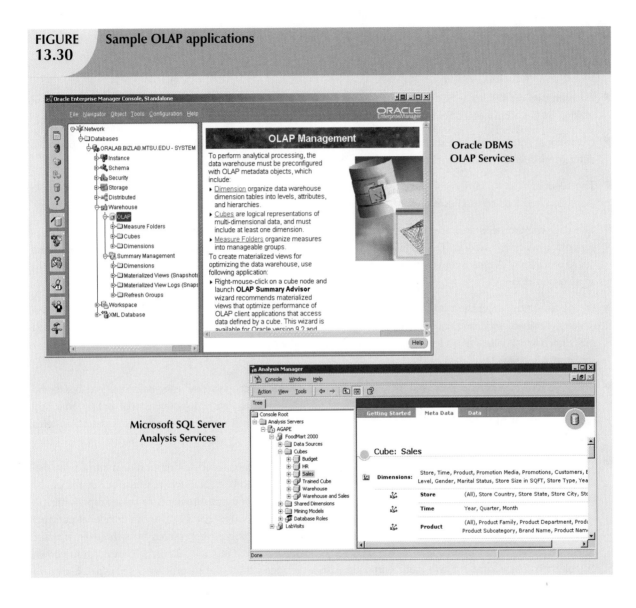

SUMMARY

- Business intelligence (BI) is a term used to describe a comprehensive, cohesive, and integrated set of applications used to capture, collect, integrate, store, and analyze data with the purpose of generating and presenting information used to support business decision making.

- BI covers a range of technologies and applications to manage the entire data life cycle from acquisition to storage, transformation, integration, analysis, monitoring, presentation, and archiving. BI functionality ranges from simple data gathering and extraction to very complex data analysis and presentation.

- Decision support systems (DSS) refers to an arrangement of computerized tools used to assist managerial decision making within a business. DSS were the original precursor of current generation BI systems.

- Operational data are not well-suited for decision support. From the end-user point of view, decision support data differ from operational data in three main areas: time span, granularity, and dimensionality.

- The requirements for a decision support DBMS are divided into four main categories: database schema, data extraction and loading, end-user analytical interface, and database size requirements.

- The data warehouse is an integrated, subject-oriented, time-variant, nonvolatile collection of data that provides support for decision making. The data warehouse is usually a read-only database optimized for data analysis and query processing. A data mart is a small, single-subject data warehouse subset that provides decision support to a small group of people.

- Online analytical processing (OLAP) refers to an advanced data analysis environment that supports decision making, business modeling, and operations research. OLAP systems have four main characteristics: use of multidimensional data analysis techniques, advanced database support, easy-to-use end-user interfaces, and client/server architecture.

- Relational online analytical processing (ROLAP) provides OLAP functionality by using relational databases and familiar relational query tools to store and analyze multidimensional data. Multidimensional online analytical processing (MOLAP) provides OLAP functionality by using multidimensional database management systems (MDBMSs) to store and analyze multidimensional data.

- The star schema is a data-modeling technique used to map multidimensional decision support data into a relational database with the purpose of performing advanced data analysis. The basic star schema has four components: facts, dimensions, attributes, and attribute hierarchies. Facts are numeric measurements or values representing a specific business aspect or activity. Dimensions are general qualifying categories that provide additional perspectives to a given fact. Conceptually, the multidimensional data model is best represented by a three-dimensional cube. Attributes can be ordered in well-defined attribute hierarchies. The attribute hierarchy provides a top-down organization that is used for two main purposes: to permit aggregation and to provide drill-down/roll-up data analysis.

- Four techniques are generally used to optimize data warehouse design: normalizing dimensional tables, maintaining multiple fact tables representing different aggregation levels, denormalizing fact tables, and partitioning and replicating tables.

- Data mining automates the analysis of operational data with the intention of finding previously unknown data characteristics, relationships, dependencies, and/or trends. The data mining process has four phases: data preparation, data analysis and classification, knowledge acquisition, and prognosis.

- SQL has been enhanced with extensions that support OLAP-type processing and data generation.

KEY TERMS

attribute hierarchy, 544

cube cache, 539

dashboard, 519

data cube, 539

data mart, 527

data mining, 553

data store, 518

data warehouse, 525

decision support system (DSS), 519

dimensions, 542

dimension tables, 542

drill down, 521

facts, 541

fact table, 541

governance, 517

key performance indicators (KPI), 517

master data management (MDM), 516

materialized view, 560

metrics, 541

multidimensional database management system (MDBMS), 539

multidimensional online analytical processing (MOLAP), 539

online analytical processing (OLAP), 530

partitioning, 551

periodicity, 551

relational online analytical processing (ROLAP), 537

replication, 551

roll up, 521

slice and dice, 543

snowflake schema, 549

sparsity, 539

star schema, 541

very large databases (VLDBs), 525

ONLINE CONTENT

Answers to selected Review Questions and Problems for this chapter are contained in the Student Online Companion for this book.

REVIEW QUESTIONS

1. What is business intelligence?

2. Describe the BI framework.

3. What are decision support systems, and what role do they play in the business environment?

4. Explain how the main components of the BI architecture interact to form a system.

5. What are the most relevant differences between operational and decision support data?

6. What is a data warehouse, and what are its main characteristics?

7. Give three examples of problems likely to be encountered when operational data are integrated into the data warehouse.

Use the following scenario to answer Questions 8–14.

While working as a database analyst for a national sales organization, you are asked to be part of its data warehouse project team.

8. Prepare a high-level summary of the main requirements for evaluating DBMS products for data warehousing.

9. Your data warehousing project group is debating whether to prototype a data warehouse before its implementation. The project group members are especially concerned about the need to acquire some data warehousing skills before implementing the enterprise-wide data warehouse. What would you recommend? Explain your recommendations.

10. Suppose you are selling the data warehouse idea to your users. How would you define multidimensional data analysis for them? How would you explain its advantages to them?

11. Before making a commitment, the data warehousing project group has invited you to provide an OLAP overview. The group's members are particularly concerned about the OLAP client/server architecture requirements and how OLAP will fit the existing environment. Your job is to explain to them the main OLAP client/server components and architectures.

12. One of your vendors recommends using an MDBMS. How would you explain this recommendation to your project leader?

13. The project group is ready to make a final decision, choosing between ROLAP and MOLAP. What should be the basis for this decision? Why?

14. The data warehouse project is in the design phase. Explain to your fellow designers how you would use a star schema in the design.

15. Briefly discuss the decision support architectural styles and their evolution. What major technologies influenced this evolution?

16. What is OLAP, and what are its main characteristics?

17. Explain ROLAP and give the reasons you would recommend its use in the relational database environment.

18. Explain the use of facts, dimensions, and attributes in the star schema.

19. Explain multidimensional cubes and describe how the slice-and-dice technique fits into this model.

20. In the star schema context, what are attribute hierarchies and aggregation levels, and what is their purpose?

21. Discuss the most common performance improvement techniques used in star schemas.

22. Explain some of the most important issues in data warehouse implementation.

23. What is data mining, and how does it differ from traditional decision support tools?

24. How does data mining work? Discuss the different phases in the data mining process.

PROBLEMS

ONLINE CONTENT

The databases used for this problem set are found in the Student Online Companion for this book. These databases are stored in Microsoft Access 2000 format. The databases, named **Ch13_P1.mdb**, **Ch13_P3. mdb**, and **Ch13_P4.mdb**, contain the data for Problems 1, 3, and 4, respectively. The data for Problem 2 are stored in Microsoft Excel format in the Student Online Companion for this book. The spreadsheet filename is **Ch13_P2.xls**.

1. The university computer lab's director keeps track of lab usage, measured by the number of students using the lab. This particular function is important for budgeting purposes. The computer lab director assigns you the task of developing a data warehouse in which to keep track of the lab usage statistics. The main requirements for this database are to:

   • Show the total number of users by different time periods.

   • Show usage numbers by time period, by major, and by student classification.

   • Compare usage for different majors and different semesters.

Use the **Ch13_P1.mdb** database, which includes the following tables:

- USELOG contains the student lab access data.
- STUDENT is a dimension table containing student data.

Given the three bulleted requirements and using the **Ch13_P1.mdb** data, complete Problems 1a–1g.

    a. Define the main facts to be analyzed. (*Hint*: These facts become the source for the design of the fact table.)

    b. Define and describe the appropriate dimensions. (*Hint*: These dimensions become the source for the design of the dimension tables.)

    c. Draw the lab usage star schema, using the fact and dimension structures you defined in Problems 1a and 1b.

    d. Define the attributes for each of the dimensions in Problem 1b.

    e. Recommend the appropriate attribute hierarchies.

    f. Implement your data warehouse design, using the star schema you created in Problem 1c and the attributes you defined in Problem 1d.

    g. Create the reports that will meet the requirements listed in this problem's introduction.

2. Ms. Victoria Ephanor manages a small product distribution company. Because the business is growing fast, Ms. Ephanor recognizes that it is time to manage the vast information pool to help guide the accelerating growth. Ms. Ephanor, who is familiar with spreadsheet software, currently employs a small sales force of four people. She asks you to develop a data warehouse application prototype that will enable her to study sales figures by year, region, salesperson, and product. (This prototype is to be used as the basis for a future data warehouse database.)

Using the data supplied in the **Ch13_P2.xls** file, complete the following seven problems:

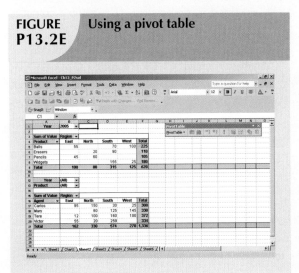

FIGURE P13.2E   **Using a pivot table**

    a. Identify the appropriate fact table components.

    b. Identify the appropriate dimension tables.

    c. Draw a star schema diagram for this data warehouse.

    d. Identify the attributes for the dimension tables that will be required to solve this problem.

    e. Using a Microsoft Excel spreadsheet (or any other spreadsheet capable of producing pivot tables), generate a pivot table to show the sales by product and by region. The end user must be able to specify the display of sales for any given year. (The sample output is shown in the first pivot table in Figure P13.2E.)

    f. Using Problem 2e as your base, add a second pivot table (see Figure P13.2E) to show the sales by salesperson and by region. The end user must be able to specify sales for a given year or for all years and for a given product or for all products.

FIGURE P13.2G   **3-D bar graph showing the relationships among agent, product, and region**

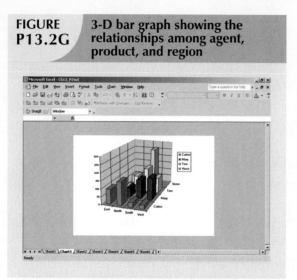

FIGURE P13.3   **Crosstab report: orders by product and department**

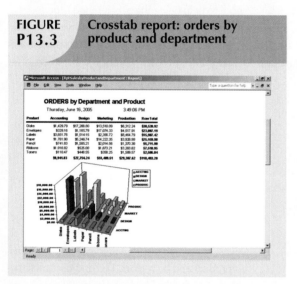

g. Create a 3-D bar graph to show sales by sales-person, by product, and by region. (See the sample output in Figure P13.2G.)

3. Mr. David Suker, the inventory manager for a marketing research company, is interested in studying the use of supplies within the different company departments. Mr. Suker has heard that his friend, Ms. Ephanor, has developed a small spreadsheet-based data warehouse model (see Problem 2) that she uses to analyze sales data. Mr. Suker is interested in developing a small data warehouse model like Ms. Ephanor's so he can analyze orders by department and by product. He will use Microsoft Access as the data warehouse DBMS and Microsoft Excel as the analysis tool.

a. Develop the order star schema.

b. Identify the appropriate dimensions attributes.

c. Identify the attribute hierarchies required to support the model.

d. Develop a crosstab report (in Microsoft Access), using a 3-D bar graph to show orders by product and by department. (The sample output is shown in Figure P13.3.)

4. ROBCOR, whose sample data are contained in the database named **Ch13_P4.mdb**, provides "on-demand" aviation charters, using a mix of different aircraft and aircraft types. Because ROBCOR has grown rapidly, its owner has hired you to be its first database manager. (The company's database, developed by an outside consulting team, already has a charter database in place to help manage all of its operations.) Your first critical assignment is to develop a decision support system to analyze the charter data. (Review Problems 30–34 in Chapter 3, The Relational Database Model, in which the operations have been described.) The charter operations manager wants to be able to analyze charter data such as cost, hours flown, fuel used, and revenue. She would also like to be able to drill down by pilot, type of airplane, and time periods.

Given those requirements, complete the following:

a. Create a star schema for the charter data.

b. Define the dimensions and attributes for the charter operation's star schema.

c. Define the necessary attribute hierarchies.

d. Implement the data warehouse design, using the design components you developed in Problems 4a–4c.

e. Generate the reports that will illustrate that your data warehouse meets the specified information requirements.

Using the data provided in the SaleCo snowflake schema in Figure 13.25, solve the following problems.

ONLINE CONTENT

The script files used to populate the database are available in the Student Online Companion. The script files assume an Oracle RDBMS. If you use a different DBMS, consult the documentation to verify whether the vendor supports similar functionality and what the proper syntax is for your DBMS.

5. What is the SQL command to list the total sales by customer and by product, with subtotals by customer and a grand total for all product sales? (*Hint*: Use the ROLLUP command.)

6. What is the SQL command to list the total sales by customer, month, and product, with subtotals by customer and by month and a grand total for all product sales? (*Hint*: Use the ROLLUP command.)

7. What is the SQL command to list the total sales by region and customer, with subtotals by region and a grand total for all sales? (*Hint*: Use the ROLLUP command.)

8. What is the SQL command to list the total sales by month and product category, with subtotals by month and a grand total for all sales? (*Hint*: Use the ROLLUP command.)

9. What is the SQL command to list the number of product sales (number of rows) and total sales by month, with subtotals by month and a grand total for all sales? (*Hint*: Use the ROLLUP command.)

10. What is the SQL command to list the number of product sales (number of rows) and total sales by month and product category, with subtotals by month and product category and a grand total for all sales? (*Hint*: Use the ROLLUP command.)

11. What is the SQL command to list the number of product sales (number of rows) and total sales by month, product category, and product, with subtotals by month and product category and a grand total for all sales? (*Hint*: Use the ROLLUP command.)

12. Using the answer to Problem 10 as your base, what command would you need to generate the same output but with subtotals in all columns? (*Hint*: Use the CUBE command.)

PART

V

DATABASES AND THE INTERNET

DATABASE CONNECTIVITY AND WEB TECHNOLOGIES	14

CASIO UPGRADES CUSTOMER WEB EXPERIENCE

Business Vignette

A global leader, Casio Computer Co., Ltd., has been developing consumer electronics since 1957. While the company creates high tech devices such as LCD TVs, digital cameras, handheld computers, and other communications devices, its Web site was behind the times. The site was written in HTML only, with little content and with an e-commerce page that was managed by a third party.

"It was an OK site, but not the kind of breakthrough experience we were hoping to offer our customers," says Casio's Internet services manager Michael McCormick.

The company wanted to upgrade the look and feel and add new functionality, including better merchandising capabilities and a more comprehensive shopping cart. The company chose Macromedia's ColdFusion MX running on a Linux platform. (Macromedia recently merged with Adobe and the product is now Adobe ColdFusion.) A 15-person team including designers, programmers, and testers spent five months creating the new site. ColdFusion's tag-based language, its reusability of code modules, and its debugging tools expedited this development. ColdFusion's open architecture Web application server also facilitated integration with the company's enterprise system. Now, users can browse through thousands of products, make purchases easily, and track their orders.

Casio also enjoys much greater administrative functionality, accessing inventory figures, sales reports, membership information, and order process and fulfillment data through the site. In addition, the company now can manage the site's content itself, rather than turning to its Web development partner, Pipeline Interactive, each time it needs to update one of its 50,000 screens of content.

Yet, the most critical feature of the new site is its ability to cross-sell and upsell. "We can cross-sell a printer when someone buys a digital camera," explains McCormick. "Or we can suggest additional ink cartridges when someone buys a printer. If a particular SKU isn't in stock, [we] suggest substitute products that are similar—perhaps a different color."

The result is that Casio's e-commerce sales have doubled since the site was launched. The site boasts more than 700,000 registered users with more than one million page views per day.

14

DATABASE CONNECTIVITY AND WEB TECHNOLOGIES

In this chapter, you will learn:

- About the various database connectivity technologies
- How Web-to-database middleware is used to integrate databases with the Internet
- About Web browser plug-ins and extensions
- What services are provided by Web application servers
- What Extensible Markup Language (XML) is and why it is important for Web database development

As you know, a database is a central repository for critical business data. Such data can be generated through traditional business applications or via newer business channels such as the Web, a phone connection, a wireless PDA, or a smart phone. To be useful universally, the data must be available to all business users. Those users need access to the data via many avenues: a spreadsheet, a user-developed Visual Basic application, a Web front end, Microsoft Access forms and reports, and so on. In this chapter, you learn about the architectures used by applications to connect to databases.

The Internet has changed how organizations of all types operate. For example, buying goods and services via the Internet has become commonplace. In today's environment, interconnectivity occurs not only between an application and the database, but also between applications interchanging messages and data. Extensible Markup Language (XML) provides a standard way of exchanging unstructured and structured data between applications.

Given the growing relationship between the Web and databases, database professionals must know how to create, use, and manage Web interfaces to those databases. This chapter examines the basics of Web database technologies.

Preview

14.1 DATABASE CONNECTIVITY

The term *database connectivity* refers to the mechanisms through which application programs connect and communicate with data repositories. Database connectivity software is also known as **database middleware** because it interfaces between the application program and the database. The data repository, also known as the *data source*, represents the data management application (that is, an Oracle RDBMS, SQL Server DBMS, or IBM DBMS) that will be used to store the data generated by the application program. Ideally, a data source or data repository could be located anywhere and hold any type of data. For example, the data source could be a relational database, a hierarchical database, a spreadsheet, or a text data file.

The need for standard database connectivity interfaces cannot be overstated. Just as SQL has become the de facto data manipulation language, there is a need for a standard database connectivity interface that will enable applications to connect to data repositories. There are many different ways to achieve database connectivity. This section will cover only the following interfaces:

- Native SQL connectivity (vendor provided).
- Microsoft's Open Database Connectivity (ODBC).
- Data Access Objects (DAO) and Remote Data Objects (RDO).
- Microsoft's Object Linking and Embedding for Database (OLE-DB).
- Microsoft's ActiveX Data Objects (ADO.NET).
- Sun's Java Database Connectivity (JDBC).

You should not be surprised to learn that most interfaces you are likely to encounter are Microsoft offerings. After all, client applications connect to databases, and the majority of those applications run on computers that are powered by some version of Microsoft Windows. The data connectivity interfaces illustrated here are dominant players in the market, and more importantly, they enjoy the support of the majority of database vendors. In fact, ODBC, OLE-DB, and ADO.NET form the backbone of Microsoft's **Universal Data Access** (**UDA**) architecture, a collection of technologies used to access any type of data source and manage the data through a common interface. As you will see, Microsoft's database connectivity interfaces have evolved over time: each interface builds on top of the other, thus providing enhanced functionality, features, flexibility, and support.

14.1.1 NATIVE SQL CONNECTIVITY

Most DBMS vendors provide their own methods for connecting to their databases. Native SQL connectivity refers to the connection interface that is provided by the database vendor and that is unique to that vendor. The best example of that type of native interface is the Oracle RDBMS. To connect a client application to an Oracle database, you must install and configure the Oracle's SQL*Net interface in the client computer. Figure 14.1 shows the configuration of Oracle SQL*Net interface on the client computer.

Native database connectivity interfaces are optimized for "their" DBMS, and those interfaces support access to most, if not all, of the database features. However, maintaining multiple native interfaces for different databases can become a burden for the programmer. Therefore, the need for "universal" database connectivity arises. Usually, the native database connectivity interface provided by the vendor is not the only way to connect to a database; most current DBMS products support other database connectivity standards, the most common being ODBC.

14.1.2 ODBC, DAO, AND RDO

Developed in early 1990s, **Open Database Connectivity** (**ODBC**) is Microsoft's implementation of a superset of the SQL Access Group **Call Level Interface** (**CLI**) standard for database access. ODBC is probably the most widely supported database connectivity interface. ODBC allows any Windows application to access relational data sources, using SQL via a standard **application programming interface** (**API**). The Webopedia online dictionary

FIGURE 14.1   ORACLE native connectivity

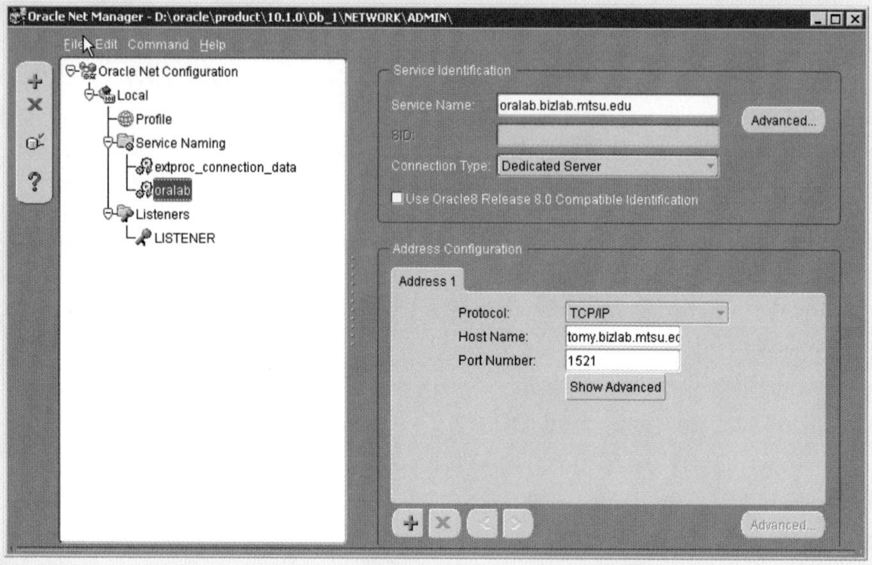

(*www.webopedia.com*) defines an API as "a set of routines, protocols, and tools for building software applications." A good API makes it easy to develop a program by providing all of the building blocks; the programmer puts the blocks together. Most operating environments, such as Microsoft Windows, provide an API so programmers can write applications consistent with the operating environment. Although APIs are designed for programmers, they are ultimately good for users because they guarantee that all programs using a common API will have similar interfaces. That makes it easy for users to learn new programs.

ODBC was the first widely adopted database middleware standard, and it enjoyed rapid adoption in Windows applications. As programming languages evolved, ODBC did not provide significant functionality beyond the ability to execute SQL to manipulate relational style data. Therefore, programmers needed a better way to access data. To answer that need, Microsoft developed two other data access interfaces:

- **Data Access Objects (DAO)** is an object-oriented API used to access MS Access, MS FoxPro, and dBase databases (using the Jet data engine) from Visual Basic programs. DAO provided an optimized interface that exposed to programmers the functionality of the Jet data engine (on which the MS Access database is based). The DAO interface can also be used to access other relational style data sources.

- **Remote Data Objects (RDO)** is a higher-level object-oriented application interface used to access remote database servers. RDO uses the lower-level DAO and ODBC for direct access to databases. RDO was optimized to deal with server-based databases, such as MS SQL Server, Oracle, and DB2.

Figure 14.2 illustrates how Windows applications can use ODBC, DAO, and RDO to access local and remote relational data sources.

As you can tell by examining Figure 14.2, client applications can use ODBC to access relational data sources. However, the DAO and RDO object interfaces provide more functionality. DAO and RDO make use of the underlying ODBC data services. ODBC, DAO, and RDO are implemented as shared code that is dynamically linked to the Windows operating environment through **dynamic-link libraries (DLLs)** which are stored as files with the .dll extension. Running as a DLL, the code speeds up load and run times.

FIGURE 14.2   **Using ODBC, DAO, and RDO to access databases**

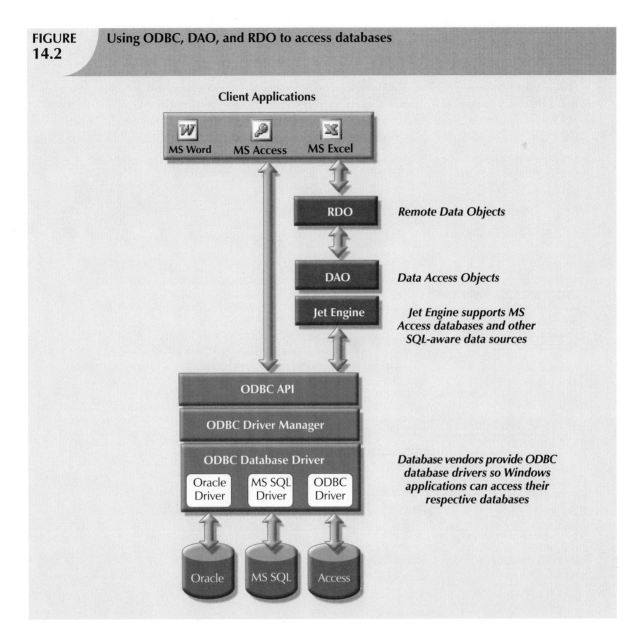

The basic ODBC architecture has three main components:

- A high-level *ODBC API* through which application programs access ODBC functionality.
- A *driver manager* that is in charge of managing all database connections.
- An *ODBC driver* that communicates directly to the DBMS.

Defining a data source is the first step in using ODBC. To define a data source, you must create a **data source name (DSN)** for the data source. To create a DSN you need to provide:

- *An ODBC driver.* You must identify the driver to use to connect to the data source. The ODBC driver is normally provided by the database vendor, although Microsoft provides several drivers that connect to most common databases. For example, if you are using an Oracle DBMS, you will select the Oracle ODBC driver provided by Oracle, or if desired, the Microsoft-provided ODBC driver for Oracle.

- A *DSN name*. This is a unique name by which the data source will be known to ODBC, and therefore, to applications. ODBC offers two types of data sources: user and system. *User data sources* are available only to the user. *System data sources* are available to all users, including operating system services.

- *ODBC driver parameters*. Most ODBC drivers require specific parameters in order to establish a connection to the database. For example, if you are using an MS Access database, you must point to the location of the MS Access (.mdb) file, and if necessary, provide a username and password. If you are using a DBMS server, you must provide the server name, the database name, the username, and the password needed to connect to the database. Figure 14.3 shows the ODBC screens required to create a System ODBC data source for an Oracle DBMS. Note that some ODBC drivers use the native driver provided by the DBMS vendor.

FIGURE 14.3 Configuring an Oracle ODBC data source

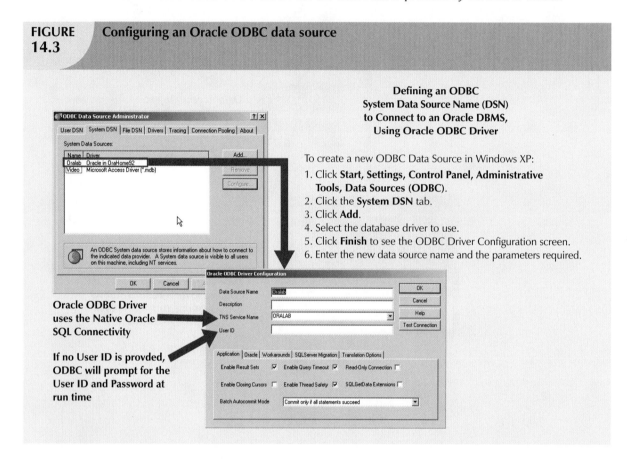

Defining an ODBC System Data Source Name (DSN) to Connect to an Oracle DBMS, Using Oracle ODBC Driver

To create a new ODBC Data Source in Windows XP:

1. Click **Start, Settings, Control Panel, Administrative Tools, Data Sources (ODBC)**.
2. Click the **System DSN** tab.
3. Click **Add**.
4. Select the database driver to use.
5. Click **Finish** to see the ODBC Driver Configuration screen.
6. Enter the new data source name and the parameters required.

Oracle ODBC Driver uses the Native Oracle SQL Connectivity

If no User ID is provded, ODBC will prompt for the User ID and Password at run time

Once the ODBC data source is defined, application programmers can write to the ODBC API by issuing specific commands and providing the required parameters. The ODBC Driver Manager will properly route the calls to the appropriate data source. The ODBC API standard defines three levels of compliance: Core, Level-1, and Level-2, which provide increasing levels of functionality. For example, Level-1 might provide support for most SQL DDL and DML statements, including subqueries and aggregate functions, but no support for procedural SQL or cursors. The database vendors can choose which level to support. However, to interact with ODBC, the database vendor must implement all of the features indicated in that ODBC API support level.

Figure 14.4 shows how you could use MS Excel to retrieve data from an Oracle RDBMS, using ODBC. Because much of the functionality provided by these interfaces is oriented to accessing relational data sources, the use of the interfaces was limited when they were used with other data source types. With the advent of object-oriented programming languages, it has become more important to provide access to other nonrelational data sources.

FIGURE 14.4   MS EXCEL uses ODBC to connect to an Oracle database

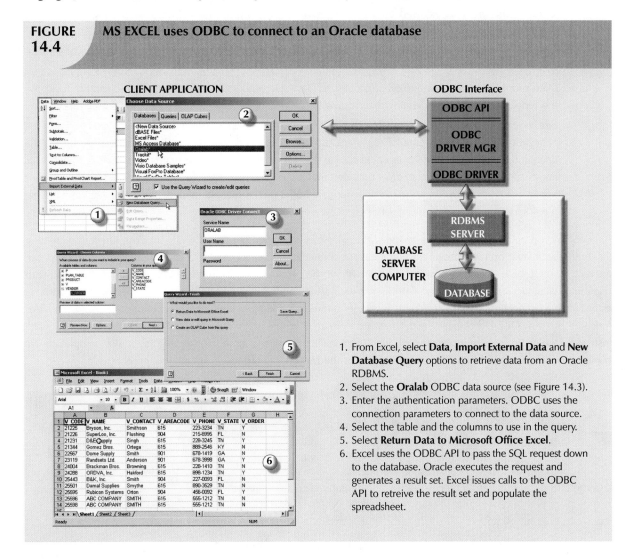

1. From Excel, select **Data**, **Import External Data** and **New Database Query** options to retrieve data from an Oracle RDBMS.
2. Select the **Oralab** ODBC data source (see Figure 14.3).
3. Enter the authentication parameters. ODBC uses the connection parameters to connect to the data source.
4. Select the table and the columns to use in the query.
5. Select **Return Data to Microsoft Office Excel**.
6. Excel uses the ODBC API to pass the SQL request down to the database. Oracle executes the request and generates a result set. Excel issues calls to the ODBC API to retreive the result set and populate the spreadsheet.

14.1.3 OLE-DB

Although ODBC, DAO, and RDO were widely used, they did not provide support for nonrelational data. To answer that need and to simplify data connectivity, Microsoft developed **Object Linking and Embedding for Database (OLE-DB)**. Based on Microsoft's Component Object Model (COM), OLE-DB is database middleware that adds object-oriented functionality for access to relational and nonrelational data. OLE-DB was the first part of Microsoft's strategy to provide a unified object-oriented framework for the development of next-generation applications.

OLE-DB is composed of a series of COM objects that provide low-level database connectivity for applications. Because OLE-DB is based on COM, the objects contain data and methods, also known as the interface. The OLE-DB model is better understood when you divide its functionality into two types of objects:

- *Consumers* are objects (applications or processes) that request and use data. The data consumers request data by invoking the methods exposed by the data provider objects (public interface) and passing the required parameters.

- *Providers* are objects that manage the connection with a data source and provide data to the consumers. Providers are divided into two categories: data providers and service providers.

  - *Data providers* provide data to other processes. Database vendors create data provider objects that expose the functionality of the underlying data source (relational, object-oriented, text, and so on).

  - *Service providers* provide additional functionality to consumers. The service provider is located between the data provider and the consumer. The service provider requests data from the data provider, transforms the data, and then provides the transformed data to the data consumer. In other words, the service provider acts like a data consumer of the data provider and as a data provider for the data consumer (end-user application). For example, a service provider could offer cursor management services, transaction management services, query processing services, and indexing services.

As a common practice, many vendors provide OLE-DB objects to augment their ODBC support, effectively creating a shared object layer on top of their existing database connectivity (ODBC or native) through which applications can interact. The OLE-DB objects expose functionality about the database; for example, there are objects that deal with relational data, hierarchical data, and flat-file text data. Additionally, the objects implement specific tasks, such as establishing a connection, executing a query, invoking a stored procedure, defining a transaction, or invoking an OLAP function. By using OLE-DB objects, the database vendor can choose what functionality to implement in a modular way, instead of being forced to include all of the functionality all of the time. Table 14.1 shows a sample of the object-oriented classes used by OLE-DB and some of the methods (interfaces) exposed by the objects.

TABLE 14.1   Sample OLE-DB Classes and Interfaces

OBJECT CLASS	USAGE	SAMPLE INTERFACES
Session	Used to create an OLE-DB session between a data consumer application and a data provider.	IGetDataSource ISessionProperties
Command	Used to process commands to manipulate a data provider's data. Generally, the command object will create RowSet objects to hold the data returned by a data provider.	ICommandPrepare ICommandProperties
RowSet	Used to hold the result set returned by a relational style database or a database that supports SQL. Represents a collection of rows in a tabular format.	IRowsetInfo IRowsetFind IRowsetScroll

OLE-DB provided additional capabilities for the applications accessing the data. However, it did not provide support for scripting languages, especially the ones used for Web development, such as Active Server Pages (ASP) and ActiveX. (A **script** is written in a programming language that is not compiled, but is interpreted and executed at run time.) To provide that support, Microsoft developed a new object framework called **ActiveX Data Objects** (**ADO**), which provides a high-level application-oriented interface to interact with OLE-DB, DAO, and RDO. ADO provides a unified interface to access data from any programming language that uses the underlying OLE-DB objects. Figure 14.5 illustrates the ADO/OLE-DB architecture, showing how it interacts with ODBC and native connectivity options.

FIGURE 14.5   OLE-DB architecture

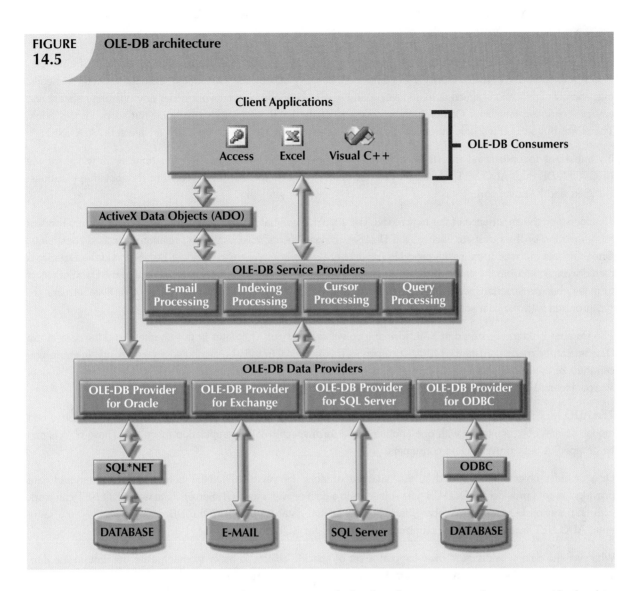

ADO introduced a simpler object model that was composed of only a few interacting objects to provide the data manipulation services required by the applications. Sample objects in ADO are shown in Table 14.2.

TABLE 14.2   Sample ADO Objects

OBJECT CLASS	USAGE
Connection	Used to set up and establish a connection with a data source. ADO will connect to any OLE-DB data source. The data source can be of any type.
Command	Used to execute commands against a specific connection (data source).
Recordset	Contains the data generated by the execution of a command. It will also contain any new data to be written to the data source. The Recordset can be disconnected from the data source.
Fields	Contains a collection of Field descriptions for each column in the Recordset.

Although the ADO model is a tremendous improvement over the OLE-DB model, Microsoft is actively encouraging programmers to use its new data access framework, ADO.NET.

14.1.4 ADO.NET

Based on ADO, **ADO.NET** is the data access component of Microsoft's .NET application development framework. The **Microsoft .NET framework** is a component-based platform for developing distributed, heterogeneous, interoperable applications aimed at manipulating any type of data over any network under any operating system and any programming language. Comprehensive coverage of the .NET framework is beyond the scope of this book. Therefore, this section will only introduce the basic data access component of the .NET architecture, ADO.NET.

It's important to understand that the .NET framework extends and enhances the functionality provided by the ADO/OLE-DB duo. ADO.NET introduced two new features critical for the development of distributed applications: DataSets and XML support.

To understand the importance of this new model, you should know that a **DataSet** is a disconnected memory-resident representation of the database. That is, the DataSet contains tables, columns, rows, relationships, and constraints. Once the data are read from a data provider, the data are placed on a memory-resident DataSet, and the DataSet is then disconnected from the data provider. The data consumer application interacts with the data in the DataSet object to make changes (inserts, updates, and deletes) in the DataSet. Once the processing is done, the DataSet data are synchronized with the data source and the changes are made permanent.

The DataSet is internally stored in XML format (you will learn about XML later in this chapter), and the data in the DataSet can be made persistent as XML documents. This is critical in today's distributed environments. In short, you can think of the DataSet as an XML-based, in-memory database that represents the persistent data stored in the data source. Figure 14.6 illustrates the main components of the ADO.NET object model.

The ADO.NET framework consolidates all data access functionality under one integrated object model. In this object model, several objects interact with one another to perform specific data manipulation functions. Those objects can be grouped as data providers and consumers.

Data provider objects are provided by the database vendors. However, ADO.NET comes with two standard data providers: a data provider for OLE-DB data sources and a data provider for SQL Server. That way ADO.NET can work with any previously supported database, including an ODBC database with an OLE-DB data provider. At the same time, ADO.NET includes a highly optimized data provider for SQL Server.

Whatever the data provider is, it must support a set of specific objects in order to manipulate the data in the data source. Some of those objects are shown in Figure 14.6. A brief description of the objects follows.

- *Connection.* The Connection object defines the data source used, the name of the server, the database, and so on. This object enables the client application to open and close a connection to a database.
- *Command.* The Command object represents a database command to be executed within a specified database connection. This object contains the actual SQL code or a stored procedure call to be run by the database. When a SELECT statement is executed, the Command object returns a set of rows and columns.
- *DataReader.* The DataReader object is a specialized object that creates a read-only session with the database to retrieve data sequentially (forward only) in a very fast manner.
- *DataAdapter.* The DataAdapter object is in charge of managing a DataSet object. This is the most specialized object in the ADO.NET framework. The DataAdapter object contains the following objects that aid in managing the data in the DataSet: SelectCommand, InsertCommand, UpdateCommand, and DeleteCommand. The DataAdapter object uses those objects to populate and synchronize the data in the DataSet with the permanent data source data.
- *DataSet.* The DataSet object is the in-memory representation of the data in the database. This object contains two main objects. The DataTableCollection object contains a collection of DataTable objects that make up the "in-memory" database, and the DataRelationCollection object contains a collection of objects describing the data relationships and ways to associate one row in a table to the related row in another table.

FIGURE 14.6   ADO.NET framework

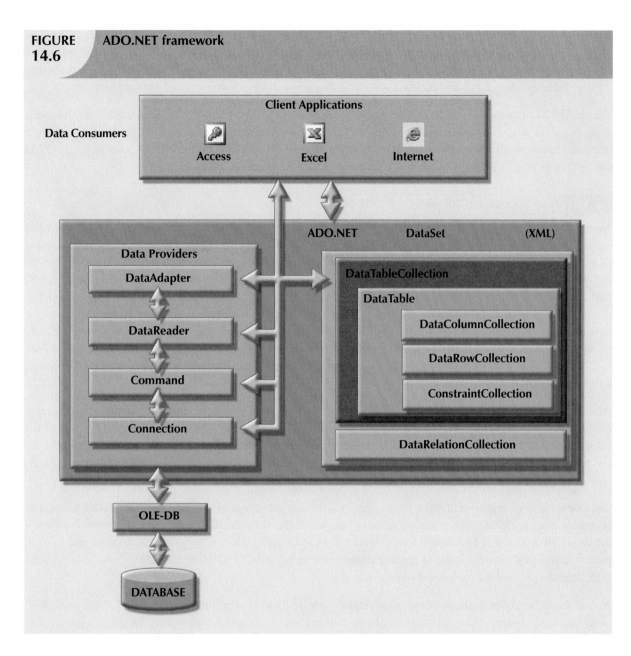

- *DataTable.* The DataTable object represents the data in tabular format. This object has one very important property: PrimaryKey, which allows the enforcement of entity integrity. In turn, the DataTable object is composed of three main objects:
  - *DataColumnCollection* contains one or more column descriptions. Each column description has properties such as column name, data type, nulls allowed, maximum value, and minimum value.
  - *DataRowCollection* contains zero rows, one row, or more than one row with data as described in the DataColumnCollection.
  - *ConstraintCollection* contains the definition of the constraints for the table. Two types of constraints are supported: ForeignKeyConstraint and UniqueConstraint.

As you can see, a DataSet is, in fact, a simple database with tables, rows, and constraints. Even more important, the DataSet doesn't require a permanent connection to the data source. The DataAdapter uses the SelectCommand object to populate the DataSet from a data source. However, once the DataSet is populated, it is completely independent of the data source, which is why it's called "disconnected."

Additionally, DataTable objects in a DataSet can come from different data sources. This means that you could have an EMPLOYEE table in an Oracle database and a SALES table in a SQL Server database. You could then create a DataSet that relates both tables as though they were located in the same database. In short, the DataSet object paves the way for truly heterogeneous distributed database support within applications.

The ADO.NET framework is optimized to work in disconnected environments. In a disconnected environment, applications exchange messages in request/reply format. The most common example of a disconnected system is the Internet. Modern applications rely on the Internet as the network platform and on the Web browser as the graphical user interface. In the next section, you will learn details about how Internet databases work.

14.1.5 JAVA DATABASE CONNECTIVITY (JDBC)

Java is an object-oriented programming language developed by Sun Microsystems that runs on top of Web browser software. Java is one of the most common programming languages for Web development. Sun Microsystems created Java as a "write once, run anywhere" environment. That means that a programmer can write a Java application once and then without any modification, run the application in multiple environments (Microsoft Windows, Apple OS X, IBM AIX, etc.). The cross-platform capabilities of Java are based on its portable architecture. Java code is normally stored in pre-processed chunks known as applets that run on a virtual machine environment in the host operating system. This environment has well-defined boundaries and all interactivity with the host operating system is closely monitored. Sun provides Java runtime environments for most operating systems (from computers to hand-held devices to TV set-top boxes.) Another advantage of using Java is its "on-demand" architecture. When a Java application loads, it can dynamically download all its modules or required components via the Internet.

When Java applications want to access data outside the Java runtime environment, they use pre-defined application programming interfaces. **Java Database Connectivity (JDBC)** is an application programming interface that allows a Java program to interact with a wide range of data sources (relational databases, tabular data sources, spreadsheets, and text files). JDBC allows a Java program to establish a connection with a data source, prepare and send the SQL code to the database server, and process the result set.

One of the main advantages of JDBC is that it allows a company to leverage its existing investment in technology and personnel training. JDBC allows programmers to use their SQL skills to manipulate the data in the company's databases. As a matter of fact, JDBC allows direct access to a database server or access via database middleware. Furthermore, JDBC provides a way to connect to databases through an ODBC driver. Figure 14.7 illustrates the basic JDBC architecture and the various database access styles.

As you see in Figure 14.7, the database access architecture in JDBC is very similar to the ODBC/OLE/ADO.NET architecture. All database access middleware shares similar components and functionality. One advantage of JDBC over other middleware is that it requires no configuration on the client side. The JDBC driver is automatically downloaded and installed as part of the Java applet download. Because Java is a Web-based technology, applications can connect to a database directly using a simple URL. Once the URL is invoked, the Java architecture comes into place, the necessary applets are downloaded to the client (including the JDBC database driver and all configuration information), and then the applets are executed securely in the client's runtime environment.

Every day, more and more companies are investing resources in developing and expanding their Web presence and finding ways to do more business on the Internet. Such business will generate increasing amounts of data that will be stored in databases. Java and the .NET framework are part of the trend toward increasing reliance on the Internet as a critical business resource. In fact, it has been said that the Internet will become the development platform of the future. In the next section you will learn more about Internet databases and how they are used.

FIGURE
14.7   JDBC architecture

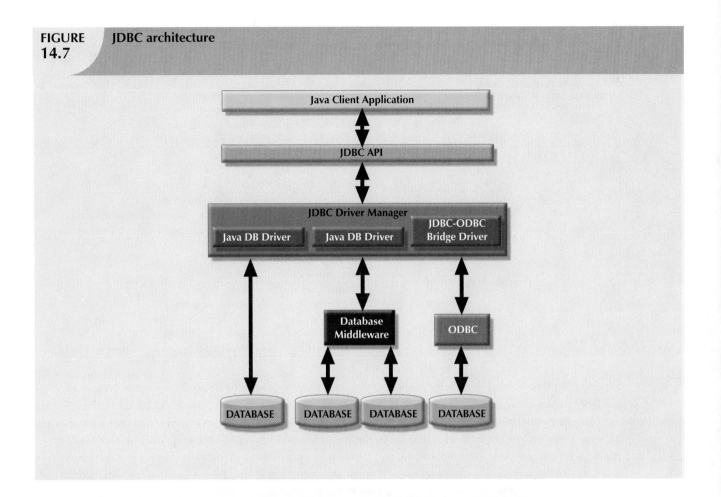

14.2 INTERNET DATABASES

Millions of people all over the world use computers and Web browser software to access the Internet, connecting to databases over the Web. Web database connectivity opens the door to new innovative services that:

- Permit rapid responses to competitive pressures by bringing new services and products to market quickly.
- Increase customer satisfaction through the creation of Web-based support services.
- Yield fast and effective information dissemination through universal access from across the street or across the globe.

Given those advantages, many organizations rely on their IS departments to create universal data access architectures based on Internet standards. Table 14.3 shows a sample of Internet technology characteristics and the benefits they provide.

TABLE 14.3	Characteristics and Benefits of Internet Technologies
INTERNET CHARACTERISTIC	**BENEFIT**
Hardware and software independence	Savings in equipment/software acquisition Ability to run on most existing equipment Platform independence and portability No need for multiple platform development
Common and simple user interface	Reduced training time and cost Reduced end-user support cost No need for multiple platform development
Location independence	Global access through Internet infrastructure Reduced requirements (and costs!) for dedicated connections
Rapid development at manageable costs	Availability of multiple development tools Plug-and-play development tools (open standards) More interactive development Reduced development times Relatively inexpensive tools Free client access tools (Web browsers) Low entry costs. Frequent availability of free Web servers Reduced costs of maintaining private networks Distributed processing and scalability, using multiple servers

In the current business and global information environment, it's easy to see why many database professionals consider the DBMS connection to the Internet to be a critical element in IS development. As you will learn in the following sections, database application development—and, in particular, the creation and management of user interfaces and database connectivity—are profoundly affected by the Web. However, having a Web-based database interface does not negate the database design and implementation issues that were addressed in the previous chapters. In the final analysis, whether you make a purchase by going online or by standing in line, the system-level transaction details are essentially the same, and they require the same basic database structures and relationships. If any immediate lesson is to be learned, it is this: *The effects of bad database design, implementation, and management are multiplied in an environment in which transactions might be measured in hundreds of thousands per day, rather than in hundreds per day.*

The Internet is rapidly changing the way information is generated, accessed, and distributed. At the core of this change is the Web's ability to access data in databases (local and remote), the simplicity of the interface, and cross-platform (heterogeneous) functionality. The Web has helped create a new information dissemination standard.

The following sections examine how Web-to-database middleware enables end users to interact with databases over the Web.

14.2.1   WEB-TO-DATABASE MIDDLEWARE: SERVER-SIDE EXTENSIONS

In general, the Web server is the main hub through which all Internet services are accessed. For example, when an end user uses a Web browser to dynamically query a database, the client browser requests a Web page. When the Web server receives the page request, it looks for the page on the hard disk; when it finds the page (for example, a stock quote, product catalog information, or an airfare listing), the server sends it back to the client.

ONLINE CONTENT

Client/server systems are covered in detail in **Appendix F, Client/Server Systems,** located in the Student Online Companion for this book.

Dynamic Web pages are at the heart of current generation Web sites. In this database-query scenario, the Web server generates the Web page contents before it sends the page to the client Web browser. The only problem with the preceding query scenario is that the Web server must include the database query result on the page *before* it sends that page back to the client. Unfortunately, neither the Web browser nor the Web server knows how to connect to and read data from the database. Therefore, to support this type of request (database query), the Web server's capability must be extended so it can understand and process database requests. This job is done through a server-side extension.

A **server-side extension** is a program that interacts directly with the Web server to handle specific types of requests. In the preceding database query example, the server-side extension program retrieves the data from databases and passes the retrieved data to the Web server, which, in turn, sends the data to the client's browser for display purposes. The server-side extension makes it possible to retrieve and present the query results, but what's more important is that *it provides its services to the Web server in a way that is totally transparent to the client browser*. In short, the server-side extension adds significant functionality to the Web server, and therefore, to the Internet.

A database server-side extension program is also known as **Web-to-database middleware**. Figure 14.8 shows the interaction between the browser, the Web server, and the Web-to-database middleware.

Trace the Web-to-database middleware actions in Figure 14.8:

1. The client browser sends a page request to the Web server.

2. The Web server receives and validates the request. In this case, the server will pass the request to the Web-to-database middleware for processing. Generally, the requested page contains some type of scripting language to enable the database interaction.

3. The Web-to-database middleware reads, validates, and executes the script. In this case, it connects to the database and passes the query using the database connectivity layer.

4. The database server executes the query and passes the result back to the Web-to-database middleware.

5. The Web-to-database middleware compiles the result set, dynamically generates an HTML-formatted page that includes the data retrieved from the database, and sends it to the Web server.

6. The Web server returns the just-created HTML page, which now includes the query result, to the client browser.

7. The client browser displays the page on the local computer.

The interaction between the Web server and the Web-to-database middleware is crucial to the development of a successful Internet database implementation. Therefore, the middleware must be well integrated with the other Internet services and the components that are involved in its use. For example, when installing Web-to-database middleware, the middleware must verify the type of Web server being used and install itself to match that Web server's requirements. In addition, how well the Web server and the Web-to-database service interact will depend on the Web server interfaces that are supported by the Web server.

FIGURE 14.8   Web-to-database middleware

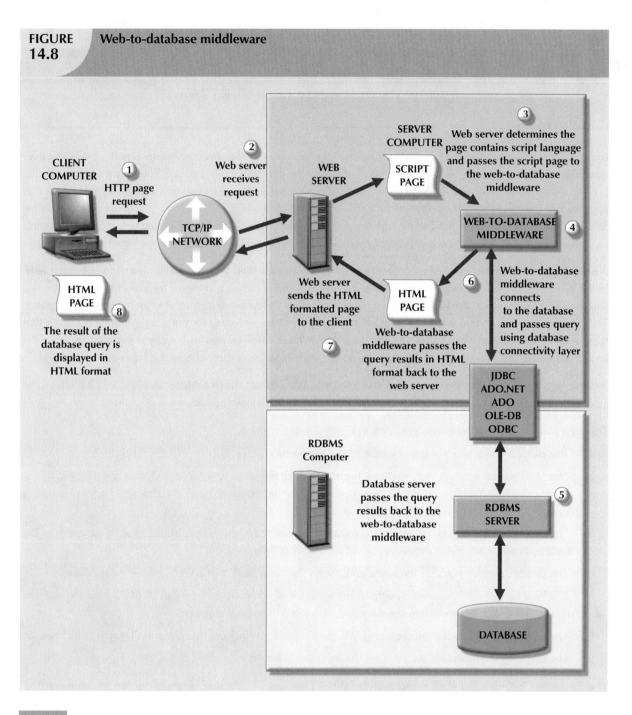

14.2.2 WEB SERVER INTERFACES

Extending Web server functionality implies that the Web server and the Web-to-database middleware will properly communicate with each other. (Database professionals often use the word *interoperate* to indicate that each party can respond to the communications of the other. This book's use of *communicate* assumes interoperation.) If a Web server is to communicate successfully with an external program, both programs must use a standard way to exchange messages and to respond to requests. A Web server interface defines how a Web server communicates with external programs. Currently, there are two well-defined Web server interfaces:

- Common Gateway Interface (CGI).
- Application programming interface (API).

The **Common Gateway Interface** (**CGI**) uses script files that perform specific functions based on the client's parameters that are passed to the Web server. The script file is a small program containing commands written in a programming language—usually Perl, C++, or Visual Basic. The script file's contents can be used to connect to the database and to retrieve data from it, using the parameters passed by the Web server. Next, the script converts the retrieved data to HTML format and passes the data to the Web server, which sends the HTML-formatted page to the client.

The main disadvantage of using CGI scripts is that the script file is an external program that is individually executed for each user request. That scenario decreases system performance. For example, if you have 200 concurrent requests, the script is loaded 200 *different* times, which takes significant CPU and memory resources away from the Web server. The language and method used to create the script also can affect system performance. For example, performance is degraded by using an interpreted language or by writing the script inefficiently.

An application programming interface (API) is a newer Web server interface standard that is more efficient and faster than a CGI script. APIs are more efficient because they are implemented as shared code or as dynamic-link libraries (DLLs). That means the API is treated as part of the Web server program that is dynamically invoked when needed.

APIs are faster than CGI scripts because the code resides in memory, so there is no need to run an external program for each request. Instead, the same API serves all requests. Another advantage is that an API can use a shared connection to the database instead of creating a new one every time, as is the case with CGI scripts.

Although APIs are more efficient in handling requests, they have some disadvantages. Because the APIs share the same memory space as the Web server, an API error can bring down the server. The other disadvantage is that APIs are specific to the Web server and to the operating system.

At the time of this writing, there are four well-established Web server APIs:

- Netscape API (NSAPI) for Netscape servers.
- Internet Server API (ISAPI) for Microsoft Windows Web servers.
- WebSite API (WSAPI) for O'Reilly Web servers.
- JDBC to provide database connectivity for Java applications.

The various types of Web interfaces are illustrated in Figure 14.9.

Regardless of the type of Web server interface used, the Web-to-database middleware program must be able to connect with the database. That connection can be accomplished in one of two ways:

- Use the native SQL access middleware provided by the vendor. For example, you can use SQL*Net if you are using Oracle.
- Use the services of general database connectivity standards such as Open Database Connectivity (ODBC), Object Linking and Embedding for Database (OLE-DB), ActiveX Data Objects (ADO), the ActiveX Data Objects for .NET (ADO.NET) interface, or JDBC for Java connectivity.

14.2.3 THE WEB BROWSER

The Web browser is the application software in the client computer, such as Microsoft Internet Explorer, Apple Safari, or Mozilla Firefox, that lets end users navigate (browse) the Web. Each time the end user clicks a hyperlink, the browser generates an HTTP GET page request that is sent to the designated Web server, using the TCP/IP Internet protocol.

The Web browser's job is to *interpret* the HTML code that it receives from the Web server and to present the various page components in a standard formatted way. Unfortunately, the browser's interpretation and presentation capabilities are not sufficient to develop Web-based applications. That is because the Web is a **stateless system**— which means that at any given time, a Web server does not know the status of any of the clients communicating with it. That is, there is no open communication line between the server and each client accessing it, which, of course, is impractical in a *worldwide* Web! Instead, client and server computers interact in very short "conversations" that follow

FIGURE 14.9   Web server CGI and API interfaces

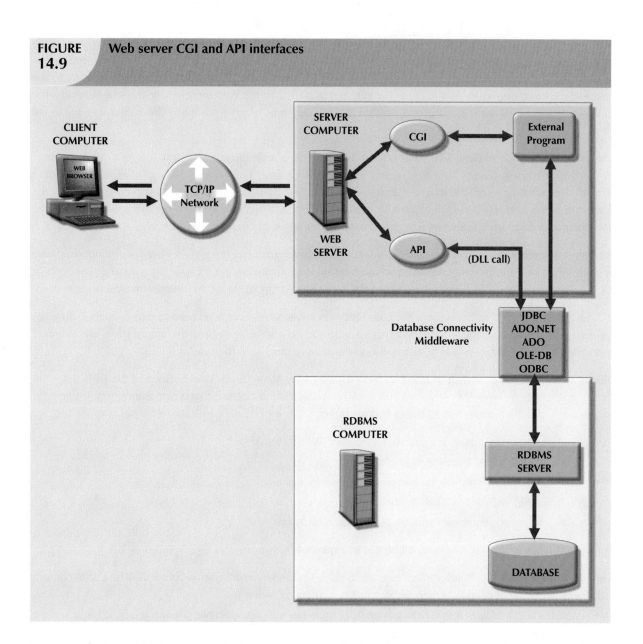

the request-reply model. For example, the browser is concerned only with the *current* page, so there is no way for the second page to know what was done in the first page. The only time the client and server computers communicate is when the client requests a page—when the user clicks a link—and the server sends the requested page to the client. Once the client receives the page and its components, the client/server communication is ended. Therefore, although you may be browsing a page and *think* that the communication is open, you are actually just browsing the HTML document stored in the local cache (temporary directory) of your browser. The server does not have any idea what the end user is doing with the document, what data is entered in a form, what option is selected, and so on. On the Web, if you want to act on a client's selection, you need to jump to a new page (go back to the Web server), therefore losing track of whatever was done before!

A Web browser's function is to display a page on the client computer. The browser—through its use of HTML—does not have computational abilities beyond formatting output text and accepting form field inputs. Even when the browser accepts form field data, there is no way to perform immediate data entry validation. Therefore, to perform such crucial processing in the client, the Web defers to other Web programming languages such as Java, JavaScript, and VBScript. The browser resembles a dumb terminal that displays only data and can perform only rudimentary processing such as accepting form data inputs. To improve capabilities on the client side of the Web browser, you must use plug-ins and other client-side extensions. On the server side, Web application servers provide the necessary processing power.

14.2.4 CLIENT-SIDE EXTENSIONS

Client-side extensions add functionality to the Web browser. Although client-side extensions are available in various forms, the most commonly encountered extensions are:

- Plug-ins.
- Java and JavaScript.
- ActiveX and VBScript.

A **plug-in** is an external application that is automatically invoked by the browser when needed. Because it is an *external* application, the plug-in is operating-system specific. The plug-in is associated with a data object—generally using the file extension—to allow the Web server to properly handle data that are not originally supported. For example, if one of the page components is a PDF document, the Web server will receive the data, recognize it as a "portable document format" object, and launch Adobe Acrobat Reader to present the document on the client computer.

As noted earlier, Java runs on top of the Web browser software. Java applications are compiled and stored in the Web server. (In many respects, Java resembles C++.) Calls to Java routines are embedded inside the HTML page. When the browser finds this call, it downloads the Java classes (code) from the Web server and runs that code in the client computer. Java's main advantage is that it enables application developers to develop their applications once and run them in many environments. (For developing Web applications, interoperability is a very important issue. Unfortunately, different client browsers are not 100 percent interoperable, thus limiting portability.)

JavaScript is a scripting language (one that enables the running of a series of commands or macros) that allows Web authors to design interactive sites. Because JavaScript is simpler to generate than Java, it is easier to learn. JavaScript code is embedded in the Web pages. It is downloaded with the Web page and is activated when a specific event takes place—such as a mouse click on an object or a page being loaded from the server into memory.

ActiveX is Microsoft's alternative to Java. ActiveX is a specification for writing programs that will run inside the Microsoft client browser (Internet Explorer). Because ActiveX is oriented mainly to Windows applications, it has low portability. ActiveX extends the Web browser by adding "controls" to Web pages. (Examples of such controls are drop-down lists, a slider, a calendar, and a calculator.) Those controls, downloaded from the Web server when needed, let you manipulate data inside the browser. ActiveX controls can be created in several programming languages; C++ and Visual Basic are most commonly used. Microsoft's .NET framework allows for wider interoperability of ActiveX-based applications (such as ADO.NET) across multiple operating environments.

VBScript is another Microsoft product that is used to extend browser functionality. VBScript is derived from Microsoft Visual Basic. Like JavaScript, VBScript code is embedded inside an HTML page and is activated by triggering events such as clicking a link.

From the developer's point of view, using routines that permit data validation on the client side is an absolute necessity. For example, when data are entered on a Web form and no data validation is done on the client side, the entire data set must be sent to the Web server. That scenario requires the server to perform all data validation, thus wasting valuable CPU processing cycles. Therefore, client-side data input validation is one of the most basic requirements for Web applications. Most of the data validation routines are done in Java, JavaScript, ActiveX, or VBScript.

14.2.5 WEB APPLICATION SERVERS

A **Web application server** is a middleware application that expands the functionality of Web servers by linking them to a wide range of services, such as databases, directory systems, and search engines. The Web application server also provides a consistent run-time environment for Web applications.

Web application servers can be used to:

- Connect to and query a database from a Web page.
- Present database data in a Web page, using various formats.
- Create dynamic Web search pages.
- Create Web pages to insert, update, and delete database data.
- Enforce referential integrity in the application program logic.
- Use simple and nested queries and programming logic to represent business rules.

Web application servers provide features such as:

- An integrated development environment with session management and support for persistent application variables.
- Security and authentication of users through user IDs and passwords.
- Computational languages to represent and store business logic in the application server.
- Automatic generation of HTML pages integrated with Java, JavaScript, VBScript, ASP, and so on.
- Performance and fault-tolerant features.
- Database access with transaction management capabilities.
- Access to multiple services, such as file transfers (FTP), database connectivity, e-mail, and directory services.

As of this writing, popular Web application servers include ColdFusion by Adobe, Oracle Application Server by Oracle, WebLogic by BEA Systems, NetDynamics by Sun Microsystems, Fusion by NetObjects, Visual Studio.NET by Microsoft, and WebObjects by Apple. All Web application servers offer the ability to connect Web servers to multiple data sources and other services. They vary in terms of the range of available features, robustness, scalability, ease of use, compatibility with other Web and database tools, and extent of the development environment.

ONLINE CONTENT

To see and try a particular Web-to-database interface in action, consult **Appendix J, Web Database Development with ColdFusion,** in the Student Online Companion for this book. This appendix steps you through the process of creating and using a simple Web-to-database interface, and gives more detailed information on developing Web databases with Adobe ColdFusion middleware.

Current-generation systems involve more than just the development of Web-enabled database applications. They also require applications capable of intercommunicating with each other and with other systems not based on the Web. Clearly, systems must be able to exchange data in a standard-based format. That's the role of XML.

14.3 EXTENSIBLE MARKUP LANGUAGE (XML)

The Internet has brought about new technologies that facilitate the exchange of business data among business partners and consumers. Companies are using the Internet to create new types of systems that integrate their data to increase efficiency and reduce costs. Electronic commerce (e-commerce) enables all types of organizations to market and sell products and services to a global market of millions of users. E-commerce transactions—the sale of products or services—can take place between businesses (business-to-business, or B2B) or between a business and a consumer (business-to-consumer, or B2C).

Most e-commerce transactions take place between businesses. Because B2B e-commerce integrates business processes among companies, it requires the transfer of business information among different business entities. But the way in which businesses represent, identify, and use data tends to differ substantially from company to company. (Is a *product code* the same thing as an *item ID*?)

Until recently, the expectation was that a purchase order traveling over the Web would be in the form of an HTML document. The HTML Web page displayed on the Web browser would include formatting tags as well as the order details. HTML **tags** describe how something *looks* on the Web page, such as bold type or heading style, and often come in pairs to start and end formatting features. For example, the following HTML tags would put the words FOR SALE in bold in the Arial font:

<font face=Arial>FOR SALE</font>

If an application wants to get the order data from the Web page, there is no easy way to extract the order details (such as the order number, the date, the customer number, the item, the quantity, the price, or payment details) from an HTML document. The HTML document can only describe how to display the order in a Web browser; it does not permit the manipulation of the order's data elements, that is, date, shipping information, payment details, product information, and so on. To solve that problem, a new markup language, known as Extensible Markup Language, or XML, was developed.

Extensible Markup Language (XML) is a metalanguage used to represent and manipulate data elements. XML is designed to facilitate the exchange of structured documents, such as orders and invoices, over the Internet. The World Wide Web Consortium (W3C)[1] published the first XML 1.0 standard definition in 1998. That standard sets the stage for giving XML the real-world appeal of being a true vendor-independent platform. Therefore, it is not surprising that XML has rapidly become the data exchange standard for e-commerce applications.

The XML metalanguage allows the definition of new tags, such as <ProdPrice>, to describe the data elements used in an XML document. This ability to *extend* the language explains the *X* in XML; the language is said to be *extensible*. XML is derived from the Standard Generalized Markup Language (SGML), an international standard for the publication and distribution of highly complex technical documents. For example, documents used by the aviation industry and the military services are too complex and unwieldy for the Web. Just like HTML, which was also derived from SGML, an XML document is a text file. However, it has a few very important additional characteristics, as follows:

- XML allows the definition of new tags to describe data elements, such as <ProductId>.
- XML is case sensitive: <ProductID> is not the same as <Productid>.
  - XML tags must be well formed; that is, each opening tag has a corresponding closing tag. For example, the product identification would require the format <ProductId>2345-AA</ProductId>.
  - XML tags must be properly nested. For example, a properly nested XML tag might look like this: <Product><ProductId>2345-AA</ProductId></Product>.
- You can use the <-- and --> symbols to enter comments in the XML document.
- The *XML* and *xml* prefixes are reserved for XML tags only.

XML is *not* a new version or replacement for HTML. XML is concerned with the description and representation of the data, rather than the way the data are displayed. XML provides the semantics that facilitate the sharing, exchange, and manipulation of structured documents over organizational boundaries. In short, XML and HTML perform complementary, rather than overlapping, functions. Extensible Hypertext Markup Language (XHTML) is the next generation of HTML based on the XML framework. The XHTML specification expands the HTML standard to include XML features. Although more powerful than HTML, XHTML requires very strict adherence to syntax requirements.

---

[1]You can visit the W3C Web page, located at *www.w3.org*, to get additional information about the efforts that were made to develop the XML standard.

As an illustration of the use of XML for data exchange purposes, consider a B2B example in which Company A uses XML to exchange product data with Company B over the Internet. Figure 14.10 shows the contents of the ProductList.xml document.

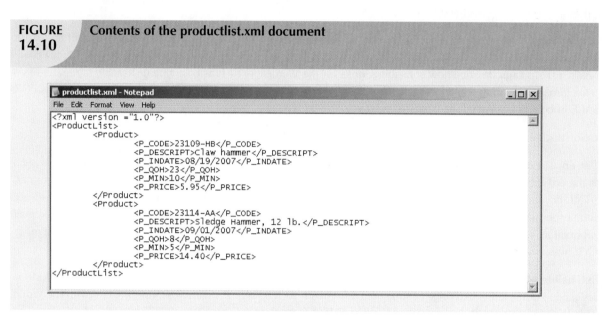

FIGURE 14.10   **Contents of the productlist.xml document**

```
productlist.xml - Notepad
File  Edit  Format  View  Help
<?xml version ="1.0"?>
<ProductList>
        <Product>
                <P_CODE>23109-HB</P_CODE>
                <P_DESCRIPT>Claw hammer</P_DESCRIPT>
                <P_INDATE>08/19/2007</P_INDATE>
                <P_QOH>23</P_QOH>
                <P_MIN>10</P_MIN>
                <P_PRICE>5.95</P_PRICE>
        </Product>
        <Product>
                <P_CODE>23114-AA</P_CODE>
                <P_DESCRIPT>Sledge Hammer, 12 lb.</P_DESCRIPT>
                <P_INDATE>09/01/2007</P_INDATE>
                <P_QOH>8</P_QOH>
                <P_MIN>5</P_MIN>
                <P_PRICE>14.40</P_PRICE>
        </Product>
</ProductList>
```

The XML example shown in Figure 14.10 illustrates several important XML features, as follows:

- The first line represents the XML document declaration, and it is mandatory.

- Every XML document has a *root element*. In the example, the second line declares the ProductList root element.

- The root element contains *child elements* or sub-elements. In the example, line 3 declares Product as a child element of ProductList.

- Each element can contain *sub-elements*. For example, each Product element is composed of several child elements, represented by P_CODE, P_DESCRIPT, P_INDATE, P_QOH, P_MIN, and P_PRICE.

- The XML document reflects a hierarchical tree structure where elements are related in a parent-child relationship; each parent element can have many children elements. For example, the root element is ProductList. Product is the child element of ProductList. Product has six child elements: P_CODE, P_DESCRIPT, P_INDATE, P_QOH, P_MIN, and P_PRICE.

Once Company B receives the ProductList.xml document, it can process the document—assuming it understands the tags created by Company A. The meaning of the XML tags in the example shown in Figure 14.10 is fairly self-evident, but there is no easy way to validate the data or to check whether the data are complete. For example, you could encounter a P_INDATE value of "25/14/2007"—but is that value correct? And what happens if Company B expects a Vendor element as well? How can companies share data descriptions about their business data elements? The next section will show how document type definitions and XML schemas are used to address those concerns.

14.3.1   DOCUMENT TYPE DEFINITIONS (DTD) AND XML SCHEMAS

B2B solutions require a high degree of business integration between companies. Companies that use B2B transactions must have a way to understand and validate each other's tags. One way to accomplish that task is through the use of Document Type Definitions. A **Document Type Definition (DTD)** is a file with a .dtd extension that describes XML elements—in effect, a DTD file provides the composition of the database's logical model and defines the syntax rules or valid tags for each type of XML document. (The DTD component is similar to having a public data dictionary for business data.) Companies that intend to engage in e-commerce business transactions must develop and share DTDs. Figure 14.11 shows the productlist.dtd document for the productlist.xml document shown earlier in Figure 14.10.

FIGURE 14.11   Contents of the productlist.dtd document

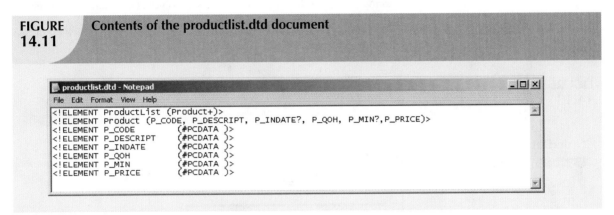

In Figure 14.11, note that the productlist.dtd file provides definitions of the elements in the productlist.xml document. In particular, note that:

- The first line declares the ProductList root element.
- The ProductList root element has one child, the Product element.
- The plus "+" symbol indicates that Product occurs one or more times within ProductList.
- An asterisk "*" would mean that the child element occurs zero or more times.
- A question mark "?" would mean that the child element is optional.
- The second line describes the Product element.
- The question mark "?" after the P_INDATE and P_MIN indicates that they are optional elements.
- The third through eighth lines show that the Product element has six child elements.
- The #PCDATA keyword represents the actual text data.

To be able to use a DTD file to define elements within an XML document, the DTD must be referenced from within that XML document. Figure 14.12 shows the productlistv2.xml document that includes the reference to the productlist.dtd in the second line.

FIGURE 14.12   Contents of the productlistv2.xml document

```
productlistv2.xml - Notepad
File  Edit  Format  View  Help
<?xml version ="1.0"?>
<!DOCTYPE ProductList SYSTEM "ProductList.dtd">
<ProductList>
        <Product>
                <P_CODE>23109-HB</P_CODE>
                <P_DESCRIPT>Claw hammer</P_DESCRIPT>
                <P_QOH>23</P_QOH>
                <P_PRICE>5.95</P_PRICE>
        </Product>
        <Product>
                <P_CODE>23114-AA</P_CODE>
                <P_DESCRIPT>Sledge hammer, 12 lb.</P_DESCRIPT>
                <P_QOH>8</P_QOH>
                <P_MIN>5</P_MIN>
                <P_PRICE>14.40</P_PRICE>
        </Product>
</ProductList>
```

In Figure 14.12, note that the P_INDATE and P_MIN do not appear in all Product definitions because they were declared to be optional elements. The DTD can be referenced by many XML documents of the same type. For example, if Company A routinely exchanges product data with Company B, it will need to create the DTD only once. All subsequent XML documents will refer to the DTD, and Company B will be able to verify the data being received.

To further demonstrate the use of XML and DTD for e-commerce business data exchanges, assume the case of two companies exchanging order data. Figure 14.13 shows the DTD and XML documents for that scenario.

FIGURE 14.13   **DTD and XML documents for order data**

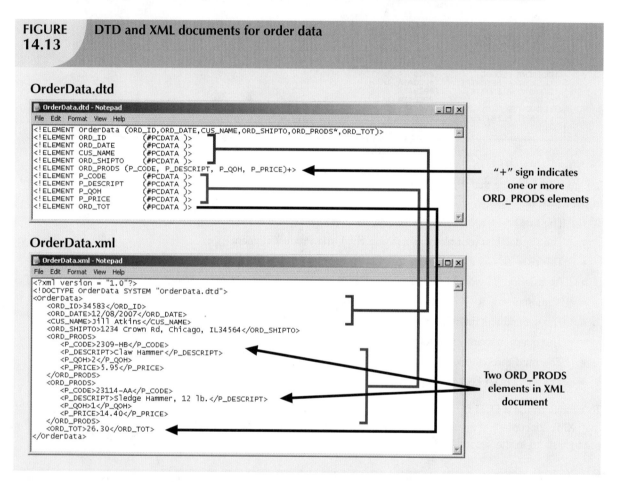

Although the use of DTDs is a great improvement for data sharing over the Web, a DTD provides only descriptive information for understanding how the elements—root, parent, child, mandatory, or optional—relate to one another. A DTD provides limited additional semantic value, such as data type support or data validation rules. That information is very important for database administrators who are in charge of large e-commerce databases. To solve the DTD problem, the W3C published an XML Schema standard in May 2001 to provide a better way to describe XML data.

The **XML schema** is an advanced data definition language that is used to describe the structure (elements, data types, relationship types, ranges, and default values) of XML data documents. One of the main advantages of an XML schema is that it more closely maps to database terminology and features. For example, an XML schema will be able to define common database types such as date, integer or decimal, minimum and maximum values, list of valid values, and required elements. Using the XML schema, a company would be able to validate the data for values that may be out of range, incorrect dates, valid values, and so on. For example, a university application must be able to specify that a GPA value be between zero and 4.0, and it must be able to detect an invalid birth date such as "14/13/1987." (There is no 14th month.) Many vendors are adopting this new standard and are supplying tools to translate DTD documents into XML Schema Definition (XSD) documents. It is widely expected that XML schemas will replace DTD as the method to describe XML data.

Unlike a DTD document, which uses a unique syntax, an **XML schema definition (XSD)** file uses a syntax that resembles an XML document. Figure 14.14 shows the XSD document for the OrderData XML document.

FIGURE 14.14   **The XML schema document for the order data**

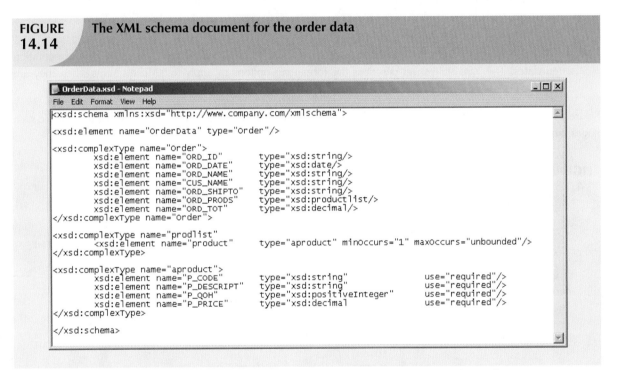

The code shown in Figure 14.14 is a simplified version of the XML schema document. As you can see, the XML schema syntax is similar to the XML document syntax. In addition, the XML schema introduces additional semantic information for the OrderData XML document, such as string, date, and decimal data types; required elements; and minimum and maximum cardinalities for the data elements.

14.3.2 XML PRESENTATION

One of the main benefits of XML is that it separates data structure from its presentation and processing. By separating data and presentation, you are able to present the same data in different ways—which is similar to having views in SQL. But what mechanisms are used to present data?

The Extensible Style Language (XSL) specification provides the mechanism to display XML data. *XSL* is used to define the rules by which XML data are formatted and displayed. The XSL specification is divided in two parts: Extensible Style Language Transformations (XSLT) and XSL style sheets.

- *Extensible Style Language Transformations (XSLT)* describe the general mechanism that is used to extract and process data from one XML document and enable its transformation within another document. Using XSLT, you can extract data from an XML document and convert it into a text file, an HTML Web page, or a Web page that is formatted for a mobile device. What the user sees in those cases is actually a view (or HTML representation) of the actual XML data. XSLT can also be used to extract certain elements from an XML document, such as the product codes and product prices, to create a product catalog. XSLT can even be used to transform one XML document into another XML document.

- *XSL style sheets* define the presentation rules applied to XML elements—something like presentation templates. The XSL style sheet describes the formatting options to apply to XML elements when they are displayed on a browser, cellular phone display, PDA screen, and so on.

Figure 14.15 illustrates the framework used by the various components to translate XML documents into viewable Web pages, an XML document, or some other document.

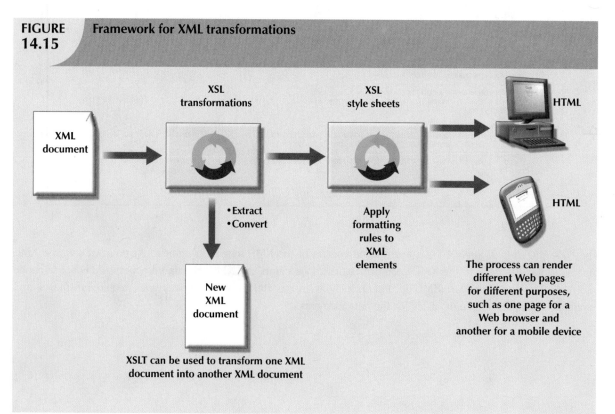

FIGURE 14.15   Framework for XML transformations

To display the XML document with Microsoft Internet Explorer (MSIE) 5.0 or later, enter the URL of the XML document in the browser's address bar. Figure 14.16 is based on the productlist.xml document created earlier. As you examine Figure 14.16, note that MSIE shows the XML data in a color-coded, collapsible, treelike structure. (Actually, this is the MSIE default style sheet that is used to render XML documents.)

FIGURE 14.16   **Displaying XML documents**

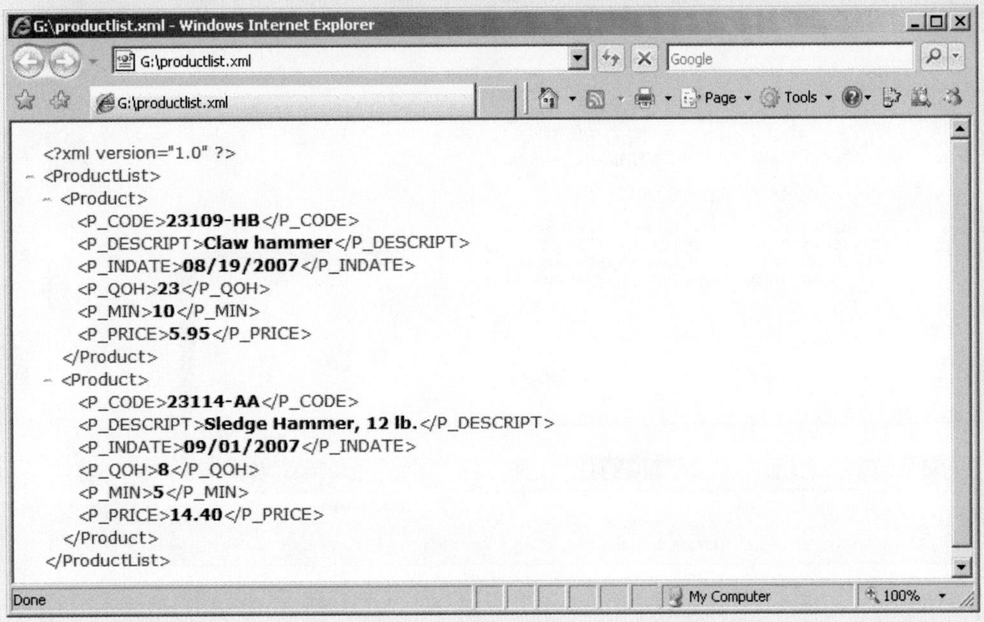

```
<?xml version="1.0" ?>
- <ProductList>
  - <Product>
      <P_CODE>23109-HB</P_CODE>
      <P_DESCRIPT>Claw hammer</P_DESCRIPT>
      <P_INDATE>08/19/2007</P_INDATE>
      <P_QOH>23</P_QOH>
      <P_MIN>10</P_MIN>
      <P_PRICE>5.95</P_PRICE>
    </Product>
  - <Product>
      <P_CODE>23114-AA</P_CODE>
      <P_DESCRIPT>Sledge Hammer, 12 lb.</P_DESCRIPT>
      <P_INDATE>09/01/2007</P_INDATE>
      <P_QOH>8</P_QOH>
      <P_MIN>5</P_MIN>
      <P_PRICE>14.40</P_PRICE>
    </Product>
  </ProductList>
```

Internet Explorer also provides *data binding* of XML data to HTML documents. Figure 14.17 shows the HTML code that is used to bind an XML document to an HTML table. The example uses the <xml> tag to include the XML data in the HTML document to later bind it to the HTML table. This example works in MSIE 5.0 or later.

FIGURE 14.17   XML data binding

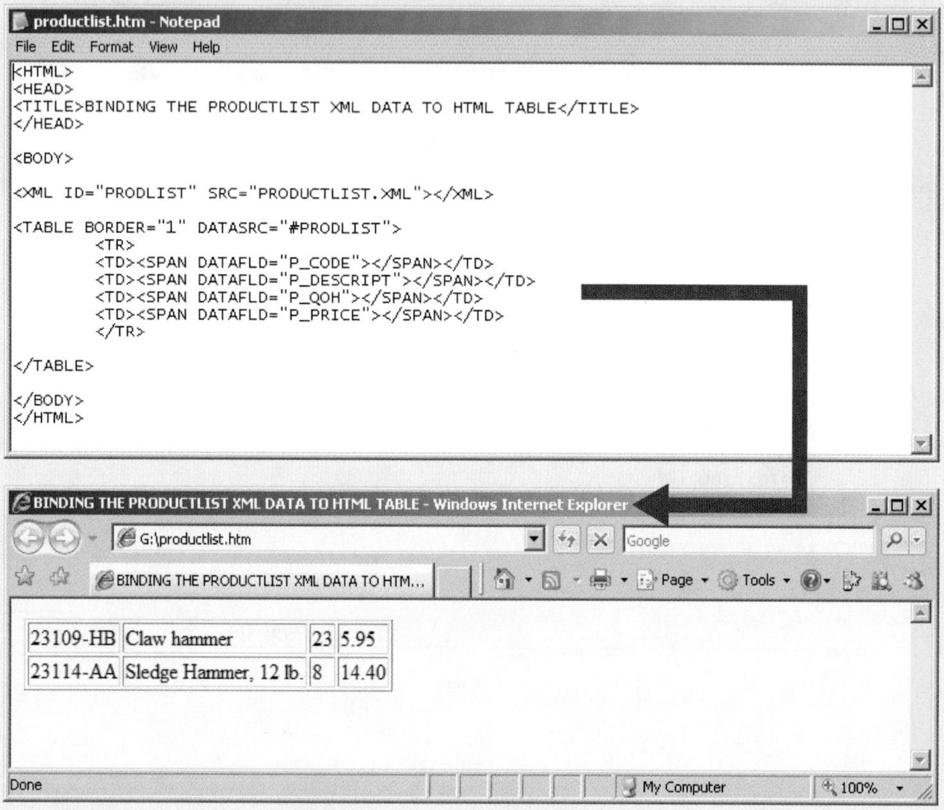

14.3.3 XML APPLICATIONS

Now that you have some idea what XML is, the next question is, how can you use it? What kinds of applications lend themselves particularly well to XML? This section will list some of the uses of XML. Keep in mind that the future use of XML is limited only by the imagination and creativity of the developers, designers, and programmers.

- *B2B exchanges.* As noted earlier, XML enables the exchange of B2B data, providing the standard for all organizations that need to exchange data with partners, competitors, the government, or customers. In particular, XML is positioned to replace EDI as the standard for the automation of the supply chain because it is less expensive and more flexible.

- *Legacy systems integration.* XML provides the "glue" to integrate legacy system data with modern e-commerce Web systems. Web and XML technologies could be used to inject some new life in "old but trusted" legacy applications. Another example is the use of XML to import transaction data from multiple operational databases to a data warehouse database.

- *Web page development.* XML provides several features that make it a good fit for certain Web development scenarios. For example, Web portals with large amounts of personalized data can use XML to pull data from multiple external sources (such as news, weather, and stocks) and apply different presentation rules to format pages on desktop computers as well as mobile devices.

- *Database support.* Databases are at the heart of *e-commerce* applications. A DBMS that supports XML exchanges will be able to integrate with external systems (Web, mobile data, legacy systems, and so on) and thus enable the creation of new types of systems. These databases can import or export data in XML format or generate XML documents from SQL queries while still storing the data, using their native data model format. Alternatively, a DBMS can also support an XML data type to store XML data in its native format. The implications of these capabilities are far-reaching—you would even be able to store a hierarchical-like tree structure inside a relational structure. Of course, such activities would also require that the query language be extended to support queries on XML data.

- *Database meta-dictionaries.* XML can also be used to create meta-dictionaries, or vocabularies, for databases. These meta-dictionaries can be used by applications that need to access other external data sources. (Until now, each time an application wanted to exchange data with another application, a new interface had to be built for that purpose.) DBMS vendors can publish meta-dictionaries to facilitate data exchanges and the creation of data views from multiple applications—hierarchical, relational, object-oriented, object-relational, or extended-relational. The meta-dictionaries would all use a common language regardless of the DBMS type. The development of industry-specific meta-dictionaries is expected. These meta-dictionaries would enable the development of complex B2B interactions, such as those likely to be found in the aviation, automotive, and pharmaceutical industries. Also likely are application-specific initiatives that would create XML meta-dictionaries for data warehousing, system management, and complex statistical applications. Even the United Nations and a not-for-profit standards-promoting organization named Oasis are working on a new specification called *ebXML* that will create a standard XML vocabulary for e-business. Other examples of meta-dictionaries are HR-XML for the human resources industry; the metadata encoding and transmission standard (METS) from the Library of Congress; the clinical accounting information (CLAIM) data exchange standard for patient data exchange in electronic medical record systems; and the extensible business reporting language (XBRL) standard for exchanging business and financial information.

- *XML databases.*[2] Given the huge number of expected XML-based data exchanges, businesses are already looking for ways to better manage and utilize the data. Currently, many different products are on the market to address this problem. The approaches range from simple middleware XML software, to object databases with XML interfaces, to full XML database engines and servers. The current generation of relational databases is tuned for the storage of normalized rows—that is, manipulating one row of data at a time. Because business data do not always conform to such a requirement, XML databases provide for the storage of data in complex relationships. For example, an XML database would be well suited to store the contents of a book. (The book's structure would dictate its database structure: a book typically consists of chapters, sections, paragraphs, figures, charts, footnotes, endnotes, and so on.) Examples of XML databases are Oracle, IBM DB2, MS SQL Server, Ipedo XML Database (*www.ipedo.com*), Tamino from Software AG (*www.softwareag.com), and* the open source dbXML from *http://sourceforge.net/projects/dbxml-core*.

- *XML services.* Many companies are already working on the development of a new breed of services based on XML and Web technologies. These services promise to break down the interoperability barriers among systems and companies alike. XML provides the infrastructure that facilitates heterogeneous systems to work together across the desk, the street, and the world. Services would use XML and other Internet technologies to publish their interfaces. Other services, wanting to interact with existing services, would locate them and learn their vocabulary (service request and replies) to establish a "conversation."

---

[2] For a comprehensive analysis of XML database products, see "XML Database Products" by Ronald Bourret at *www.rpbourret.com.*

S U M M A R Y

- Database connectivity refers to the mechanisms through which application programs connect and communicate with data repositories. Database connectivity software is also known as *database middleware*. The data repository is also known as the *data source* because it represents the data management application (that is, an Oracle RDBMS, SQL Server DBMS, or IBM DBMS) that will be used to store the data generated by the application program.

- Microsoft database connectivity interfaces are dominant players in the market and enjoy the support of most database vendors. In fact, ODBC, OLE-DB, and ADO.NET form the backbone of Microsoft's Universal Data Access (UDA) architecture. UDA is a collection of technologies used to access any type of data source and manage any type of data, using a common interface.

- Native database connectivity refers to the connection interface that is provided by the database vendor and is unique to that vendor. Open Database Connectivity (ODBC) is Microsoft's implementation of a superset of the SQL Access Group Call Level Interface (CLI) standard for database access. ODBC is probably the most widely supported database connectivity interface. ODBC allows any Windows application to access relational data sources, using standard SQL. Data Access Objects (DAO) is an object-oriented API used to access MS Access, MS FoxPro, and dBase databases (using the Jet data engine) from Visual Basic programs. Remote Data Objects (RDO) is a higher-level object-oriented application interface used to access remote database servers. RDO uses the lower-level DAO and ODBC for direct access to databases. RDO was optimized to deal with server-based databases, such as MS SQL Server and Oracle.

- Based on Microsoft's Component Object Model (COM), Object Linking and Embedding for Database (OLE-DB) is a database middleware developed with the goal of adding object-oriented functionality for access to relational and nonrelational data. ActiveX Data Objects (ADO) provides a high-level application-oriented interface to interact with OLE-DB, DAO, and RDO. Based on ADO, ADO.NET is the data access component of Microsoft's .NET application development framework, a component-based platform for developing distributed, heterogeneous, interoperable applications aimed at manipulating any type of data over any network under any operating system and any programming language. Java Database Connectivity (JDBC) is the standard way to interface Java applications with data sources (relational, tabular, and text files).

- Database access through the Web is achieved through middleware. To improve capabilities on the client side of the Web browser, you must use plug-ins and other client-side extensions such as Java and Javascript, or ActiveX and VBScript. On the server side, Web application servers are middleware that expands the functionality of Web servers by linking them to a wide range of services, such as databases, directory systems, and search engines.

- Extensible Markup Language (XML) facilitates the exchange of B2B and other data over the Internet. XML provides the semantics that facilitates the exchange, sharing, and manipulation of structured documents across organizational boundaries. XML produces the description and the representation of data, thus setting the stage for data manipulation in ways that were not possible before XML. XML documents can be validated through the use of Document Type Definition (DTD) documents and XML Schema Definition (XSD) documents. The use of DTD, XML schemas, and XML documents permits a greater level of integration among diverse systems than was possible before this technology was made available.

ORECK REVISES DISASTER RECOVERY PLAN AFTER KATRINA

Business Vignette

Because companies design disaster recovery plans during normal business operations, holes in the plan often become apparent only during crises. Oreck Corporation, the vacuum manufacturer, had a decent disaster recovery plan. The company, headquartered in New Orleans, had arranged with IBM to host its AS/400-based applications in a data center in Boulder, CO. The staff in New Orleans would relocate to Long Beach, Mississippi, where the company had a large manufacturing and distribution center. On Sunday, August 28, two days before Hurricane Katrina hit New Orleans, the IT staff put the plan into operation.

"It took 12 hours to make two backups," remembers Michael Evanson, Oreck's vice president of IT. As roads out of the city were closing or clogged, his staff was still backing up the AS/400 data. Finally, on Monday morning, CEO Tom Oreck grabbed the tapes and his family, and they flew a private plane to Houston, where he overnighted the tapes to Boulder. By the time Mr. Oreck arrived in Houston, however, a major oversight in their plan had become apparent.

"We had two facilities, and we assumed that at least one would survive the hurricane. They're about 80 miles apart, but Katrina went right through the middle of the two," Evanson explains. The hurricane had flooded the Long Beach factory, and many of its 900 employees had evacuated the area. The company's Intel-based data which were needed for its facilities elsewhere had been left behind — in New Orleans. Like most backup and recovery programs, Oreck's plan did not cover all components of its IT system, and in establishing its priorities, it had failed to prepare for a disaster that could take down both its headquarters and its backup center.

The company scrambled to recover quickly. It established temporary headquarters in Dallas, located workers by setting up an 800 number, bought RVs for its homeless employees and generators for the plant, and brought in contractors to repair physical damage. Two weeks later, the Long Beach facility was up and running.

Today, the company has revised its plan to provide for better access to and protection of the hardware, software, and data it needs in the event of an emergency. Oreck now runs backups every two hours instead of every eight hours. The company tests its data recovery and contingency plans regularly. Recently, the company opened a new plant further away in Cookeville, Tennessee.

"The plans that we had in place before Katrina, which I think served us well, have been highly modified," says Oreck. "Honestly, the plan hadn't been updated in a long time," says Oreck. "We've obviously changed our view on that."

15

DATABASE ADMINISTRATION AND SECURITY

FIFTEEN

In this chapter, you will learn:

- That data are a valuable business asset requiring careful management
- How a database plays a critical role in an organization
- That the introduction of a DBMS has important technological, managerial, and cultural organizational consequences
- What the database administrator's managerial and technical roles are
- About data security, database security, and the information security framework
- About several database administration tools and strategies
- How various database administration technical tasks are performed with Oracle

Preview

This chapter shows you the basis for a successful database administration strategy. Such a strategy requires that data be considered important and valuable resources to be treated and managed as corporate assets.

The chapter explores how a database fits within an organization, what the data views and requirements are at various management levels, and how the DBMS supports those views and requirements. Database administration must be fully understood and accepted within an organization before a sound data administration strategy can be implemented. In this chapter, you learn about important data management issues by looking at the managerial and technical roles of the database administrator (DBA). This chapter also explores database security issues, such as the confidentiality, integrity, and availability of data. In our information-based society, one of the key aspects of data management is to ensure that the data are protected against intentional or unintentional access by unauthorized personnel. It is also essential to ensure that the data are available when and where needed, even in the face of natural disaster or hardware failure, and to maintain the integrity of the data in the database.

The technical aspects of database administration are augmented by a discussion of database administration tools and the corporate-wide data architectural framework. The managerial aspects of database administration are explained by showing you how the database administration function fits within classical organizational structures. Because Oracle is the current leader in mid- to high-level corporate database markets, you learn how a DBA performs some typical database management functions in Oracle.

15.1 DATA AS A CORPORATE ASSET

In Chapter 1, Database Systems, you learned that data are the raw material from which information is produced. Therefore, it is not surprising that in today's information-driven environment, data are a valuable asset that requires careful management.

To assess data's monetary value, take a look at what's stored in a company database: data about customers, suppliers, inventory, operations, and so on. How many opportunities are lost if the data are lost? What is the actual cost of data loss? For example, an accounting firm whose entire database is lost would incur significant direct and indirect costs. The accounting firm's problems would be magnified if the data loss occurred during tax season. Data loss puts any company in a difficult position. The company might be unable to handle daily operations effectively, it might be faced with the loss of customers who require quick and efficient service, and it might lose the opportunity to gain new customers.

Data are a valuable *resource* that can translate into *information*. If the information is accurate and timely, it is likely to trigger actions that enhance the company's competitive position and generate wealth. In effect, an organization is subject to a *data-information-decision cycle*; that is, the data *user* applies intelligence to *data* to produce *information* that is the basis of *knowledge* used in *decision making* by the user. This cycle is illustrated in Figure 15.1.

FIGURE 15.1   The data-information-decision-making cycle

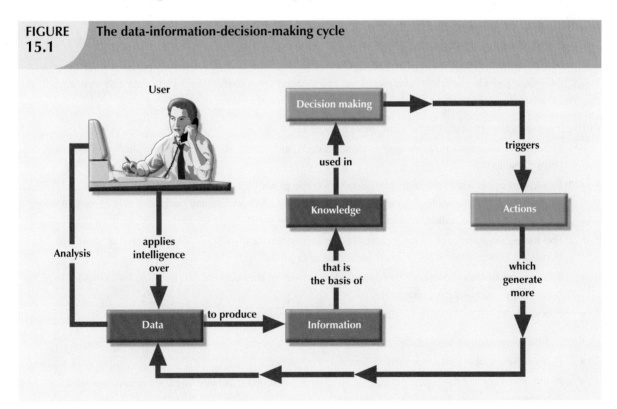

Note in Figure 15.1 that the decisions made by high-level managers trigger actions within the organization's lower levels. Such actions produce additional data to be used for monitoring company performance. In turn, the additional data must be recycled within the data/information/decision framework. Thus, data form the basis for decision making, strategic planning, control, and operations monitoring.

A critical success factor of an organization is efficient asset management. To manage data as a corporate asset, managers must understand the value of information—that is, processed data. In fact, there are companies (for example, those that provide credit reports) whose only product is information and whose success is solely a function of information management.

15.2 THE NEED FOR AND ROLE OF A DATABASE IN AN ORGANIZATION

Data are used by different people in different departments for different reasons. Therefore, data management must address the concept of shared data. Chapter 1 showed how the need for data sharing made the DBMS almost inevitable. Used properly, the DBMS facilitates:

- *Interpretation* and *presentation* of data in useful formats by transforming raw data into information.
- *Distribution* of data and information to the right people at the right time.
- Data *preservation* and *monitoring* the data usage for adequate periods of time.
- *Control* over data duplication and use, both internally and externally.

Whatever the type of organization, the database's predominant role is *to support managerial decision making at all levels in the organization while preserving data privacy and security.*

An organization's managerial structure might be divided into three levels: top, middle, and operational. Top-level management makes strategic decisions, middle management makes tactical decisions, and operational management makes daily operational decisions. Operational decisions are short term and affect only daily operations; for example, deciding to change the price of a product to clear it from inventory. Tactical decisions involve a longer time frame and affect larger-scale operations; for example, changing the price of a product in response to competitive pressures. Strategic decisions are those that affect the long-term well-being of the company or even its survival; for example, changing pricing strategy across product lines to capture market share.

The DBMS must provide tools that give each level of management a useful view of the data and that support the required level of decision making. The following activities are typical of each management level.

At the *top management* level, the database must be able to:

- Provide the information necessary for strategic decision making, strategic planning, policy formulation, and goals definition.
- Provide access to external and internal data to identify growth opportunities and to chart the direction of such growth. (Direction refers to the nature of the operations: Will a company become a service organization, a manufacturing organization, or some combination of the two?)
- Provide a framework for defining and enforcing organizational policies. (Remember that such polices are translated into business rules at lower levels in the organization.)
- Improve the likelihood of a positive return on investment for the company by searching for new ways to reduce costs and/or by boosting productivity.
- Provide feedback to monitor whether the company is achieving its goals.

At the *middle management* level, the database must be able to:

- Deliver the data necessary for tactical decisions and planning.
- Monitor and control the allocation and use of company resources and evaluate the performance of the various departments.
- Provide a framework for enforcing and ensuring the security and privacy of the data in the database. **Security** means protecting the data against accidental or intentional use by unauthorized users. **Privacy** deals with the rights of individuals and the organization to determine the "who, what, when, where, and how" of data usage.

At the *operational management* level, the database must be able to:

- Represent and support the company operations as closely as possible. The data model must be flexible enough to incorporate all required present and expected data.

- Produce query results within specified performance levels. Keep in mind that the performance requirements increase for lower levels of management and operations. Thus, the database must support fast responses to a greater number of transactions at the operational management level.

- Enhance the company's short-term operational ability by providing timely information for customer support and for application development and computer operations.

A general objective for any database is to provide a seamless flow of information throughout the company.

The company's database is also known as the corporate or enterprise database. The **enterprise database** might be defined as "the company's data representation that provides support for all present and expected future operations." Most of today's successful organizations depend on the enterprise database to provide support for all of their operations—from design to implementation, from sales to services, and from daily decision making to strategic planning.

15.3 INTRODUCTION OF A DATABASE: SPECIAL CONSIDERATIONS

Having a computerized database management system does not guarantee that the data will be properly used to provide the best solutions required by managers. A DBMS is a tool for managing data; like any tool, it must be used effectively to produce the desired results. Consider this analogy: in the hands of a carpenter, a hammer can help produce furniture; in the hands of a child, it might do damage. The solution to company problems is not the mere existence of a computer system or its database, but, rather, its effective management and use.

The introduction of a DBMS represents a big change and challenge; throughout the organization, the DBMS is likely to have a profound impact, which might be positive or negative depending on how it is administered. For example, one key consideration is adapting the DBMS to the organization rather than forcing the organization to adapt to the DBMS. The main issue should be the organization's needs rather than the DBMS's technical capabilities. However, the introduction of a DBMS cannot be accomplished without affecting the organization. The flood of new DBMS-generated information has a profound effect on the way the organization functions, and therefore, on its corporate culture.

The introduction of a DBMS into an organization has been described as a process that includes three important aspects:[1]

- *Technological:* DBMS software and hardware.
- *Managerial:* Administrative functions.
- *Cultural:* Corporate resistance to change.

The *technological* aspect includes selecting, installing, configuring, and monitoring the DBMS to make sure that it efficiently handles data storage, access, and security. The person or people in charge of addressing the technological aspect of the DBMS installation must have the technical skills necessary to provide or secure adequate support for the various users of the DBMS: programmers, managers, and end users. Therefore, database administration staffing is a key technological consideration in the DBMS introduction. The selected personnel must exhibit the right mix of technical and managerial skills to provide a smooth transition to the new shared-data environment.

The *managerial* aspect of the DBMS introduction should not be taken lightly. A high-quality DBMS does not guarantee a high-quality information system, just as having the best race car does not guarantee winning a race.

The introduction of a DBMS into an organization requires careful planning to create an appropriate organizational structure to accommodate the person or people responsible for administering the DBMS. The organizational structure must also be subject to well-developed monitoring and controlling functions. The administrative personnel must have excellent interpersonal and communications skills combined with broad organizational and business understanding.

---

[1] Murray, John P. "The Managerial and Cultural Issues of a DBMS," *370/390 Database Management* 1(8), September 1991, pp. 32–33.

Top management must be committed to the new system and must define and support the data administration functions, goals, and roles within the organization.

The *cultural* impact of the introduction of a database system must be assessed carefully. The DBMS's existence is likely to have an effect on people, functions, and interactions. For example, additional personnel might be added, new roles might be allocated to existing personnel, and employee performance might be evaluated using new standards.

A cultural impact is likely because the database approach creates a more controlled and structured information flow. Department managers who are used to handling their own data must surrender their subjective ownership to the data administration function and must share their data with the rest of the company. Application programmers must learn and follow new design and development standards. Managers might be faced with what they consider to be an information overload and might require some time to adjust to the new environment.

When the new database comes online, people might be reluctant to use the information provided by the system and might question its value or accuracy. (Many will be surprised and possibly chagrined to discover that the information does not fit their preconceived notions and strongly held beliefs.) The database administration department must be prepared to open its doors to end users, listen to their concerns, act on those concerns when possible, and educate end users about the system's uses and benefits.

15.4 THE EVOLUTION OF THE DATABASE ADMINISTRATION FUNCTION

Data administration has its roots in the old, decentralized world of the file system. The cost of data and managerial duplication in such file systems gave rise to a centralized data administration function known as the electronic data processing (EDP) or data processing (DP) department. The DP department's task was to pool all computer resources to support all departments *at the operational level*. The DP administration function was given the authority to manage all existing company file systems as well as resolve data and managerial conflicts created by the duplication and/or misuse of data.

The advent of the DBMS and its shared view of data produced a new level of data management sophistication and led the DP department to evolve into an **information systems (IS) department**. The responsibilities of the IS department were broadened to include:

- A *service* function to provide end users with active data management support.
- A *production* function to provide end users with specific solutions for their information needs through integrated application or management information systems.

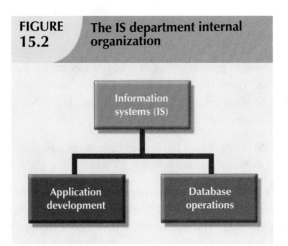

FIGURE 15.2 The IS department internal organization

The functional orientation of the IS department was reflected in its internal organizational structure. IS departments typically were structured as shown in Figure 15.2. As the demand for application development grew, the IS application development segment was subdivided by the type of supported system: accounting, inventory, marketing, and so on. However, this development meant that the database administration responsibilities were divided. The application development segment was in charge of gathering database requirements and logical database design, whereas the database operations segment took charge of implementing, monitoring, and controlling the DBMS operations.

As the number of database applications grew, data management became an increasingly complex job, thus leading to the development of the database administration function.

The person responsible for the control of the centralized and shared database became known as the **database administrator (DBA)**.

The size and role of the DBA function varies from company to company, as does its placement within a company's organizational structure. On the organization chart, the DBA function might be defined as either a staff or line position. Placing the DBA function in a staff position often creates a consulting environment in which the DBA is able to devise the data administration strategy but does not have the authority to enforce it or to resolve possible conflicts.[2] The DBA function in a line position has both the responsibility and the authority to plan, define, implement, and enforce the policies, standards, and procedures used in the data administration activity. The two possible DBA function placements are illustrated in Figure 15.3.

FIGURE 15.3   The placement of the DBA function

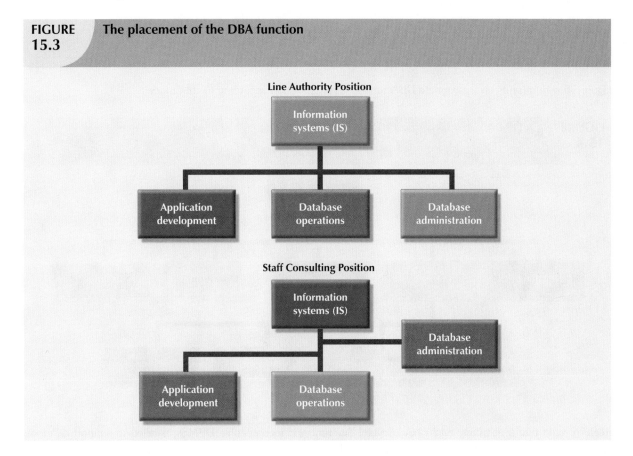

There is no standard for how the DBA function fits in an organization's structure. In part, that is because the DBA function itself is probably the most dynamic of any organization's functions. In fact, the fast-paced changes in DBMS technology dictate changing organizational styles. For example:

- The development of distributed databases can force an organization to decentralize the data administration function further. The distributed database requires the system DBA to define and delegate the responsibilities of each local DBA, thus imposing new and more complex *coordinating* activities on the system DBA.

- The growing use of Internet-accessible data and the growing number of data warehousing applications are likely to add to the DBA's data modeling and design activities, thus expanding and diversifying the DBA's job.

- The increasing sophistication and power of microcomputer-based DBMS packages provide an easy platform for the development of user-friendly, cost-effective, and efficient solutions to the needs of specific departments.

[2]For a historical perspective on the development of the DBA function and a broader coverage of its organizational placement alternatives, refer to Jay-Louise Weldon's classic *Data Base Administration* (New York, Plenum Press, 1981). Although you might think that the book's publication date renders it obsolete, a surprising number of its topics are returning to the current operational database scene.

But such an environment also invites data duplication, not to mention the problems created by people who lack the technical qualifications to produce good database designs. In short, the new microcomputer environment requires the DBA to develop a new set of technical and managerial skills.

It is common practice to define the DBA function by dividing the DBA operations according to the Database Life Cycle (DBLC) phases. If that approach is used, the DBA function requires personnel to cover the following activities:

- Database planning, including the definition of standards, procedures, and enforcement.
- Database requirements gathering and conceptual design.
- Database logical and transaction design.
- Database physical design and implementation.
- Database testing and debugging.
- Database operations and maintenance, including installation, conversion, and migration.
- Database training and support.

Figure 15.4 represents an appropriate DBA functional organization according to that model.

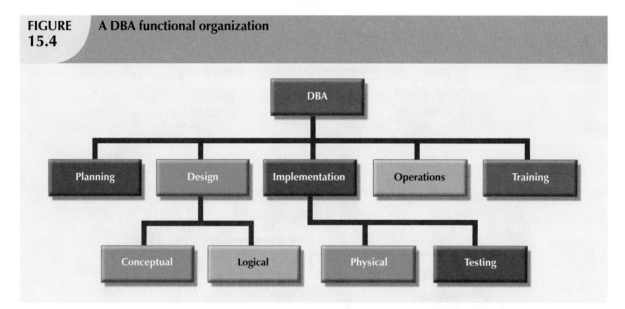

FIGURE 15.4   A DBA functional organization

Keep in mind that a company might have several different and incompatible DBMSs installed to support different operations. For example, it is not uncommon to find corporations with a hierarchical DBMS to support the daily transactions at the operational level and a relational database to support middle and top management's ad hoc information needs. There may also be a variety of microcomputer DBMSs installed in the different departments. In such an environment, the company might have one DBA assigned for each DBMS. The general coordinator of all DBAs is sometimes known as the **systems administrator**; that position is illustrated in Figure 15.5.

There is a growing trend toward specialization in the data management function. For example, the organization charts used by some of the larger corporations make a distinction between a DBA and the **data administrator (DA)**. The DA, also known as the **information resource manager (IRM)**, usually reports directly to top management and is given a higher degree of responsibility and authority than the DBA, although the two roles overlap some.

The DA is responsible for controlling the overall corporate data resources, both computerized and manual. Thus, the DA's job description covers a larger area of operations than that of the DBA because the DA is in charge of controlling not only the computerized data, but also the data outside the scope of the DBMS. The placement of the DBA within the expanded organizational structure may vary from company to company. Depending on the structure's components, the DBA might report to the DA, the IRM, the IS manager, or directly to the company's CEO.

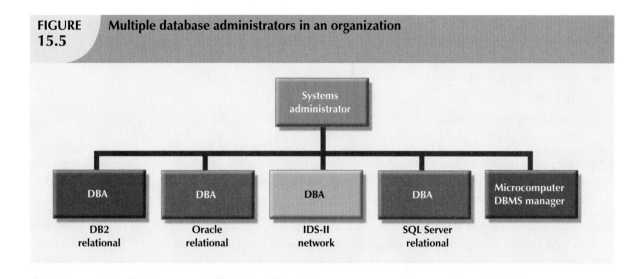

FIGURE 15.5   Multiple database administrators in an organization

15.5 THE DATABASE ENVIRONMENT'S HUMAN COMPONENT

A substantial portion of this book is devoted to relational database design and implementation and to examining DBMS features and characteristics. Thus far the book has focused on the very important technical aspects of the database. However, there is another important side of the database coin: even the most carefully crafted database system cannot operate without the human component. So in this section, you will explore how people perform the data administration activities that make a good database design useful.

Effective data administration requires both technical and managerial skills. For example, the DA's job typically has a strong managerial orientation with company-wide scope. In contrast, the DBA's job tends to be more technically oriented and has a narrower DBMS-specific scope. However, the DBA, too, must have a considerable store of people skills. After all, both the DA and the DBA perform "people" functions common to all departments in an organization. For example, both the DA and DBA direct and control personnel staffing and training within their respective departments.

Table 15.1 contrasts the general characteristics of both positions by summarizing the typical DA and DBA activities. All activities flowing from the characteristics shown in Table 15.1 are invested in the DBA if the organization does not employ both a DA and a DBA.

TABLE 15.1   Contrasting DA and DBA Activities and Characteristics

DATA ADMINISTRATOR (DA)	DATABASE ADMINISTRATOR (DBA)
Does strategic planning	Controls and supervises
Sets long-term goals	Executes plans to reach goals
Sets policies and standards	Enforces policies and procedures Enforces programming standards
Is broad in scope	Is narrow in scope
Focuses on the long term	Focuses on the short term (daily operations)
Has a managerial orientation	Has a technical orientation
Is DBMS-independent	Is DBMS-specific

Note that the DA is responsible for providing a global and comprehensive administrative strategy for all of the organization's data. In other words, the DA's plans must consider the entire data spectrum. Thus, the DA is responsible for the consolidation and consistency of both manual and computerized data.

The DA also must set data administration goals. Those goals are defined by issues such as:

- Data "sharability" and time availability.
- Data consistency and integrity.
- Data security and privacy.
- Data quality standards.
- Extent and type of data use.

Naturally, that list can be expanded to fit the organization's specific data needs. Regardless of how data management is conducted—and despite the fact that much authority is invested in the DA or DBA to define and control the way company data are used—the DA and DBA do not own the data. Instead, DA and DBA functions are defined to emphasize that data are a shared company asset.

The preceding discussion should not lead you to believe that there are universally accepted DA and DBA administrative standards. As a matter of fact, the style, duties, organizational placement, and internal structure of both functions vary from company to company. For example, many companies distribute DA duties between the DBA and the manager of information systems. For simplicity and to avoid confusion, the label DBA is used here as a general title that encompasses all appropriate data administration functions. Having made that point, let's move on to the DBA's role as an arbitrator between data and users.

The arbitration of interactions between the two most important assets of any organization, people and data, places the DBA in the dynamic environment portrayed in Figure 15.6.

FIGURE 15.6   A summary of DBA activities

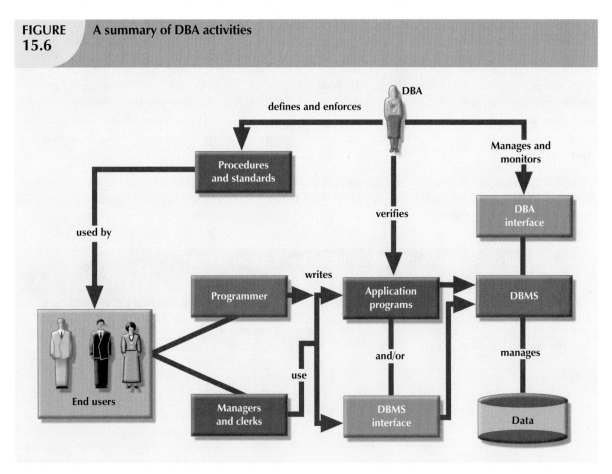

As you examine Figure 15.6, note that the DBA is the focal point for data/user interaction. The DBA defines and enforces the procedures and standards to be used by programmers and end users during their work with the DBMS. The DBA also verifies that programmer and end-user access meets the required quality and security standards.

Database users might be classified by the:

- Type of decision-making support required (operational, tactical, or strategic).
- Degree of computer knowledge (novice, proficient, or expert).
- Frequency of access (casual, periodic, or frequent).

Those classifications are not exclusive and usually overlap. For example, an operational user can be an expert with casual database access. Nevertheless, a typical top-level manager might be a strategic novice user with periodic database access. On the other hand, a database application programmer is an operational expert and frequent database user. Thus, each organization employs people whose levels of database expertise span an entire spectrum. The DBA must be able to interact with all of those people, understand their different needs, answer questions at all levels of the expertise scale, and communicate effectively.

The DBA activities portrayed in Figure 15.6 suggest the need for a diverse mix of skills. In large companies, such skills are likely to be distributed among several people who work within the DBA function. In small companies, the skills might be the domain of just one individual. The skills can be divided into two categories—managerial and technical—as summarized in Table 15.2.

TABLE 15.2   Desired DBA Skills

MANAGERIAL	TECHNICAL
Broad business understanding	Broad data-processing background
Coordination skills	Systems development life cycle knowledge

Analytical skills	Structured methodologies:
	    Data flow diagrams
	    Structure charts
	    Programming languages

Conflict resolution skills	Database life cycle knowledge

Communications skills (oral and written)	Database modeling and design skills
	    Conceptual
	    Logical
	    Physical

Negotiation skills	Operational skills: database implementation, data dictionary management, security, and so on
Experience: 10 years in a large DP department	

As you examine Table 15.2, keep in mind that the DBA must perform two distinct roles. The DBA's managerial role is focused on personnel management and on interactions with the end-user community. The DBA's technical role involves the use of the DBMS—database design, development, and implementation—as well as the production, development, and use of application programs. The DBA's managerial and technical roles will be examined in greater detail in the following sections.

15.5.1   THE DBA'S MANAGERIAL ROLE

As a manager, the DBA must concentrate on the control and planning dimensions of database administration. Therefore, the DBA is responsible for:

- Coordinating, monitoring, and allocating database administration resources: people and data.
- Defining goals and formulating strategic plans for the database administration function.

More specifically, the DBA's responsibilities are shown in Table 15.3.

TABLE 15.3	DBA Activities and Services	
DBA ACTIVITY		**DBA SERVICE**
Planning		End-user support
Organizing		Policies, procedures, and standards
Testing	of	Data security, privacy, and integrity
Monitoring		Data backup and recovery
Delivering		Data distribution and use

Table 15.3 illustrates that the DBA is generally responsible for planning, organizing, testing, monitoring, and delivering quite a few services. Those services might be performed by the DBA or, more likely, by the DBA's personnel. Let's examine the services in greater detail.

End-User Support

The DBA interacts with the end user by providing data and information support services to the organization's departments. Because end users usually have dissimilar computer backgrounds, end-user support services include:

- *Gathering user requirements.* The DBA must work within the end-user community to help gather the data required to identify and describe the end users' problems. The DBA's communications skills are very important at this stage because the DBA works closely with people who have varying computer backgrounds and communication styles. The gathering of user requirements requires the DBA to develop a precise understanding of the users' views and needs and to identify present and future information needs.

- *Building end-user confidence.* Finding adequate solutions to end users' problems increases end-user trust and confidence in the DBA function. The DBA function is also to educate the end-user in the services provided and how those services enhance data stewardship and data security.

- *Resolving conflicts and problems.* Finding solutions to end users' problems in one department might trigger conflicts with other departments. End users are typically concerned with their own specific data needs rather than with those of others, and they are not likely to consider how their data affect other departments within the organization. When data/information conflicts arise, the DBA function has the authority and responsibility to resolve them.

- *Finding solutions to information needs.* The ability and authority to resolve data conflicts enables the DBA to develop solutions that will properly fit within the existing data management framework. The DBA's primary objective is to provide solutions to the end users' information needs. Given the growing importance of the Internet, those solutions are likely to require the development and management of Web servers to interface with the databases. In fact, the explosive growth of e-commerce requires the use of *dynamic* interfaces to facilitate interactive product queries and product sales.

- *Ensuring quality and integrity of data and applications.* Once the right solution has been found, it must be properly implemented and used. Therefore, the DBA must work with both application programmers and end users to teach them the database standards and procedures required for data quality, access, and manipulation. The DBA must also make sure that the database transactions do not adversely affect the quality of the data. Likewise, certifying the quality of the application programs that access the database is a crucial DBA function. Special attention must be given to the DBMS Internet interfaces because those interfaces are prone to security issues.

- *Managing the training and support of DBMS users.* One of the most time-consuming DBA activities is teaching end users how to use the database properly. The DBA must ensure that all users accessing the database have a basic understanding of the functions and use of the DBMS software. The DBA coordinates and monitors all activities concerning end-user education.

Policies, Procedures, and Standards

A prime component of a successful data administration strategy is the continuous enforcement of the policies, procedures, and standards for correct data creation, usage, distribution, and deletion within the database. The DBA must define, document, and communicate the policies, procedures, and standards before they can be enforced. Basically:

- **Policies** are general statements of direction or action that communicate and support DBA goals.
- **Standards** describe the minimum requirements of a given DBA activity; they are more detailed and specific than policies. In effect, standards are rules that are used to evaluate the quality of the activity. For example, standards define the structure of application programs and the naming conventions programmers must use.
- **Procedures** are written instructions that describe a series of steps to be followed during the performance of a given activity. Procedures must be developed within existing working conditions, and they must support and enhance that environment.

To illustrate the distinctions among policies, standards, and procedures, look at the following examples:

Policies

> All users must have passwords.
>
> Passwords must be changed every six months.

Standards

> A password must have a minimum of five characters.
>
> A password must have a maximum of 12 characters.
>
> Social Security numbers, names, and birth dates cannot be used as passwords.

Procedures

> To create a password, (1) the end user sends to the DBA a written request for the creation of an account; (2) the DBA approves the request and forwards it to the computer operator; (3) the computer operator creates the account, assigns a temporary password, and sends the account information to the end user; (4) a copy of the account information is sent to the DBA; and (5) the user changes the temporary password to a permanent one.

Standards and procedures defined by the DBA are used by all end users who want to benefit from the database. Standards and procedures must complement each other and must constitute an extension of data administration policies. Procedures must facilitate the work of end users and the DBA. The DBA must define, communicate, and enforce procedures that cover areas such as:

- *End-user database requirements gathering.* What documentation is required? What forms must be used?
- *Database design and modeling.* What database design methodology is to be used (normalization or object-oriented methodology)? What tools are to be used (CASE tools, data dictionaries, UML or ER diagrams)?
- *Documentation and naming conventions.* What documentation must be used in the definition of all data elements, sets, and programs that access the database?
- *Design, coding, and testing of database application programs.* The DBA must define the standards for application program coding, documentation, and testing. The DBA standards and procedures are given to the application programmers, and the DBA must enforce those standards.
- *Database software selection.* The selection of the DBMS package and any other software related to the database must be properly managed. For example, the DBA might require that software be properly interfaced with existing software, that it have the features needed by the organization, and that it provide a positive return on investment. In today's Internet environment, the DBA must also work with Web administrators to implement efficient and secure Web-to-database connectivity.

- *Database security and integrity.* The DBA must define the policies governing security and integrity. Database security is especially crucial. Security standards must be clearly defined and strictly enforced. Security procedures must be designed to handle a multitude of security scenarios to ensure that security problems are minimized. Although no system can ever be completely secure, security procedures must be designed to meet critical standards. The growing use of Internet interfaces to databases opens the door to new security threats that are far more complex and difficult to manage than those encountered with more traditional internally generated and controlled interfaces. Therefore, the DBA must work closely with Internet security specialists to ensure that the databases are properly protected from attacks launched inadvertently or deliberately.

- *Database backup and recovery.* Database backup and recovery procedures must include the information necessary to guarantee proper execution and management of the backups.

- *Database maintenance and operation.* The DBMS's daily operations must be clearly documented. Operators must keep job logs, and they must write operator instructions and notes. Such notes are helpful in pinpointing the causes and solutions of problems. Operational procedures must also include precise information concerning backup and recovery procedures.

- *End-user training.* A full-featured training program must be established within the organization, and procedures governing the training must be clearly specified. The objective is to indicate clearly who does what, when, and how. Each end user must be aware of the type and extent of the available training methodology.

Procedures and standards must be revised at least annually to keep them up to date and to ensure that the organization can adapt quickly to changes in the work environment. Naturally, the introduction of new DBMS software, the discovery of security or integrity violations, the reorganization of the company, and similar changes require revision of the procedures and standards.

Data Security, Privacy, and Integrity

The security, privacy, and integrity of the data in the database are of great concern to DBAs who manage current DBMS installations. Technology has pointed the way to greater productivity through information management. Technology has also resulted in the distribution of data across multiple sites, thus making it more difficult to maintain data control, security, and integrity. The multiple-site data configuration has made it imperative that the DBA use the security and integrity mechanisms provided by the DBMS to enforce the database administration policies defined in the previous section. In addition, DBAs must team up with Internet security experts to build security mechanisms to safeguard data from possible attacks or unauthorized access. Section 15.6 covers security issues in more detail.

Data Backup and Recovery

When data are not readily available, companies face potentially ruinous losses. Therefore, data backup and recovery procedures are critical in all database installations. The DBA also must ensure that the data in the database can be fully recovered in case of physical data loss or loss of database integrity.

Data loss can be partial or total. A partial loss is caused by a physical loss of part of the database or when part of the database has lost integrity. A total loss might mean that the database continues to exist but its integrity is entirely lost or that the entire database is physically lost. In any case, backup and recovery procedures are the cheapest database insurance you can buy.

The management of database security, integrity, backup, and recovery is so critical that many DBA departments have created a position called the **database security officer (DSO)**. The DSO's sole job is to ensure database security and integrity. In large organizations, the DSO's activities are often classified as *disaster management*.

Disaster management includes all of the DBA activities designed to secure data availability following a physical disaster or a database integrity failure. Disaster management includes all planning, organizing, and testing of database contingency plans and recovery procedures. The backup and recovery measures must include at least:

- *Periodic data and applications backups.* Some DBMSs include tools to ensure backup and recovery of the data in the database. The DBA should use those tools to render the backup and recovery tasks automatic. Products such as IBM's DB2 allow the creation of different backup types: full, incremental, and concurrent. A **full backup**, also known as a **database dump**, produces a complete copy of the entire database. An **incremental backup** produces a backup of all data since the last backup date; a **concurrent backup** takes place while the user is working on the database.

- *Proper backup identification.* Backups must be clearly identified through detailed descriptions and date information, thus enabling the DBA to ensure that the correct backups are used to recover the database. The most common backup medium is tape; the storage and labeling of tapes must be done diligently by the computer operators, and the DBA must keep track of tape currency and location. However, organizations that are large enough to hire a DBA do not typically use CDs and DVDs for enterprise backup. Other emerging backup solutions include optical and disk-based backup devices. Such backup solutions include online storage based on Network Attached Storage (NAS) and Storage Area Networks (SAN). Enterprise backup solutions use a layered backup approach in which the data are first backed up to fast disk media for intermediate storage and fast restoration. Later, the data is transferred to tape for archival storage.

- *Convenient and safe backup storage.* There must be multiple backups of the same data, and each backup copy must be stored in a different location. The storage locations must include sites inside and outside the organization. (Keeping different backups in the same place defeats the purpose of having multiple backups in the first place.) The storage locations must be properly prepared and may include fire-safe and quakeproof vaults, as well as humidity and temperature controls. The DBA must establish a policy to respond to two questions: (1) Where are the backups to be stored? (2) How long are backups to be stored?

- *Physical protection of both hardware and software.* Protection might include the use of closed installations with restricted access, as well as preparation of the computer sites to provide air conditioning, backup power, and fire protection. Physical protection also includes the provision of a backup computer and DBMS to be used in case of emergency. For example, when Hurricane Katrina hit the U.S. Gulf Coast in 2005, New Orleans suffered almost total destruction of its communications infrastructure. Many organizations and educational institutions did not have adequate disaster recovery plans for such an extreme level of service interruption (see the Part VI Business Vignette at the beginning of this chapter for one example).[3]

- *Personal access control to the software of a database installation.* Multilevel passwords and privileges and hardware and software challenge/response tokens can be used to properly identify authorized users of resources.

- *Insurance coverage for the data in the database.* The DBA or security officer must secure an insurance policy to provide financial protection in the event of a database failure. The insurance might be expensive, but it is less expensive than the disaster created by massive data loss.

Two additional points are worth making.

- Data recovery and contingency plans must be thoroughly tested and evaluated, and they must be practiced frequently. So-called fire drills are not to be disparaged, and they require top-level management's support and enforcement.

- A backup and recovery program is not likely to cover all components of an information system. Therefore, it is appropriate to establish priorities concerning the nature and extent of the data recovery process.

[3]See "AAUP Responds to Katrina's Impact on New Orleans Universities," http://www.aaup.org/AAUP/pubsres/academe/2006/MA/AW/kat.htm.

Data Distribution and Use

Data are useful only when they reach the right users in a timely fashion. The DBA is responsible for ensuring that the data are distributed to the right people, at the right time, and in the right format. The DBA's data distribution and use tasks can become very time-consuming, especially when the data delivery capacity is based on a typical applications programming environment, where users depend on programmers to deliver the programs to access the data in the database. Although the Internet and its intranet and extranet extensions have opened databases to corporate users, their use has also created a new set of challenges for the DBA.

Current data distribution philosophy makes it easy for *authorized* end users to access the database. One way to accomplish that task is to facilitate the use of a new generation of more sophisticated query tools and the new Internet Web front ends. They enable the DBA to educate end users to produce the required information without being dependent on applications programmers. Naturally, the DBA must ensure that all users adhere to appropriate standards and procedures.

This data-sharing philosophy is common today, and it is likely that it will become more common as database technology marches on. Such an environment is more flexible for the end user. Clearly, enabling end users to become relatively self-sufficient in the acquisition and use of data can lead to more efficient use of data in the decision process. Yet this "data democracy" can also produce some troublesome side effects. Letting end users micromanage their data subsets could inadvertently sever the connection between those users and the data administration function. The DBA's job under those circumstances might become sufficiently complicated to compromise the efficiency of the data administration function. Data duplication might flourish again without checks at the organizational level to ensure the uniqueness of data elements. Thus, end users who do not completely understand the nature and sources of data might make improper use of the data elements.

15.5.2 THE DBA'S TECHNICAL ROLE

The DBA's technical role requires a broad understanding of DBMS functions, configuration, programming languages, data modeling and design methodologies, and so on. For example, the DBA's technical activities include the selection, installation, operation, maintenance, and upgrading of the DBMS and utility software, as well as the design, development, implementation, and maintenance of the application programs that interact with the database.

Many of the DBA's technical activities are a logical extension of the DBA's managerial activities. For example, the DBA deals with database security and integrity, backup and recovery, and training and support. Thus, the DBA's dual role might be conceptualized as a capsule whose technical core is covered by a clear managerial shell.

The technical aspects of the DBA's job are rooted in the following areas of operation:

- Evaluating, selecting, and installing the DBMS and related utilities.
- Designing and implementing databases and applications.
- Testing and evaluating databases and applications.
- Operating the DBMS, utilities, and applications.
- Training and supporting users.
- Maintaining the DBMS, utilities, and applications.

The following sections will explore the details of those operational areas.

Evaluating, Selecting, and Installing the DBMS and Utilities

One of the DBA's first and most important technical responsibilities is selecting the database management system, utility software, and supporting hardware to be used in the organization. Therefore, the DBA must develop and execute a plan for evaluating and selecting the DBMS, utilities, and hardware. That plan must be based primarily on the organization's needs rather than on specific software and hardware features. The DBA must recognize that the search is for solutions to problems rather than for a computer or DBMS software. Put simply, a DBMS is a management tool and not a technological toy.

The first and most important step of the evaluation and acquisition plan is to determine company needs. To establish a clear picture of those needs, the DBA must make sure that the entire end-user community, including top- and mid-level managers, is involved in the process. Once the needs are identified, the objectives of the data administration function can be clearly established and the DBMS features and selection criteria can be defined.

To match DBMS capability to the organization's needs, the DBA would be wise to develop a checklist of desired DBMS features. That DBMS checklist should address at least these issues:

- *DBMS model.* Are the company's needs better served by a relational, object-oriented, or object/relational DBMS? If a data warehouse application is required, should a relational or multidimensional DBMS be used? Does the DBMS support star schemas?

- *DBMS storage capacity.* What maximum disk and database size is required? How many disk packages must be supported? How many tape units are needed? What are other storage needs?

- *Application development support.* Which programming languages are supported? What application development tools (database schema design, data dictionary, performance monitoring, and screen and menu painters) are available? Are end-user query tools provided? Does the DBMS provide Web front-end access?

- *Security and integrity.* Does the DBMS support referential and entity integrity rules, access rights, and so on? Does the DBMS support the use of audit trails to spot errors and security violations? Can the audit trail size be modified?

- *Backup and recovery.* Does the DBMS provide some automated backup and recovery tools? Does the DBMS support tape, optical disc, or network-based backups? Does the DBMS automatically back up the transaction logs?

- *Concurrency control.* Does the DBMS support multiple users? What levels of isolation (table, page, row) does the DBMS offer? How much manual coding is needed in the application programs?

- *Performance.* How many transactions per second does the DBMS support? Are additional transaction processors needed?

- *Database administration tools.* Does the DBMS offer some type of DBA management interface? What type of information does the DBA interface provide? Does the DBMS provide alerts to the DBA when errors or security violations occur?

- *Interoperability and data distribution.* Can the DBMS work with other DBMS types in the same environment? What coexistence or interoperability level is achieved? Does the DBMS support READ and WRITE operations to and from other DBMS packages? Does the DBMS support a client/server architecture?

- *Portability and standards.* Can the DBMS run on different operating systems and platforms? Can the DBMS run on mainframes, midrange computers, and personal computers? Can the DBMS applications run without modification on all platforms? What national and industry standards does the DBMS follow?

- *Hardware.* What hardware does the DBMS require?

- *Data dictionary.* Does the DBMS have a data dictionary? If so, what information is kept in it? Does the DBMS interface with any data dictionary tool? Does the DBMS support any CASE tools?

- *Vendor training and support.* Does the vendor offer in-house training? What type and level of support does the vendor provide? Is the DBMS documentation easy to read and helpful? What is the vendor's upgrade policy?

- *Available third-party tools.* What additional tools are offered by third-party vendors (query tools, data dictionary, access management and control, storage allocation management tools)?

- *Cost.* What costs are involved in the acquisition of the software and hardware? How many additional personnel are required, and what level of expertise is required of them? What are the recurring costs? What is the expected payback period?

Pros and cons of several alternative solutions must be evaluated during the selection process. Available alternatives are often restricted because software must be compatible with the organization's existing computer system. Remember that a DBMS is just part of a solution; it requires support from collateral hardware, application software, and utility programs. For example, the DBMS's use is likely to be constrained by the available CPU(s), front-end processor(s), auxiliary storage devices, data communication devices, the operating system, a transaction processor system, and so on. The costs associated with the hardware and software components must be included in the estimations.

The selection process must also consider the site's preparation costs. For example, the DBA must include both one-time and recurring expenditures involved in the preparation and maintenance of the computer room installations.

The DBA must supervise the installation of all software and hardware designated to support the data administration strategy; must have a thorough understanding of the components being installed; and must be familiar with the installation, configuration, and startup procedures of such components. The installation procedures include details such as the location of backup and transaction log files, network configuration information, and physical storage details.

Keep in mind that installation and configuration details are DBMS-dependent. Therefore, such details cannot be addressed in this book. Consult the installation and configuration sections of your system's DBMS administration guide for those details.

Designing and Implementing Databases and Applications

The DBA function also provides data modeling and design services to end-users. Such services are often coordinated with an application development group within the data-processing department. Therefore, one of the primary activities of a DBA is to determine and enforce standards and procedures to be used. Once the appropriate standards and procedures framework are in place, the DBA must ensure that the database modeling and design activities are performed within the framework. The DBA then provides the necessary assistance and support during the design of the database at the conceptual, logical, and physical levels. (Remember that the conceptual design is both DBMS- and hardware-independent, the logical design is DBMS-dependent and hardware-independent, and the physical design is both DBMS- and hardware-dependent.)

The DBA function usually requires that several people be dedicated to database modeling and design activities. Those people might be grouped according to the organizational areas covered by the application. For example, database modeling and design personnel may be assigned to production systems, financial and managerial systems, or executive and decision support systems. The DBA schedules the design jobs to coordinate the data design and modeling activities. That coordination may require reassignment of available resources based on externally determined priorities.

The DBA also works with applications programmers to ensure the quality and integrity of database design and transactions. Such support services include reviewing the database application design to ensure that transactions are:

- *Correct:* The transactions mirror real-world events.
- *Efficient:* The transactions do not overload the DBMS.
- *Compliant.* Complies with integrity rules and standards.

These activities require personnel with broad database design and programming skills.

The implementation of the applications requires the implementation of the physical database. Therefore, the DBA must provide assistance and oversight during the physical design, including storage space determination and creation, data loading, conversion, and database migration services. The DBA's implementation tasks also include the generation, compilation, and storage of the application's access plan. An **access plan** is a set of instructions generated at application completion time that predetermines how the application will access the database at run time. To be able to create and validate the access plan, the user must have the required rights to access the database (see Chapter 11, Database Performance Tuning and Query Optimization).

Before an application comes online, the DBA must develop, test, and implement the operational procedures required by the new system. Such operational procedures include utilizing training, security, and backup and recovery plans, as well as assigning responsibility for database control and maintenance. Finally, the DBA must authorize application users to access the database from which the applications draw the required data.

The addition of a new database might require the fine-tuning and/or reconfiguring of the DBMS. Remember that the DBMS assists all applications by managing the shared corporate data repository. Therefore, when data structures are added or modified, the DBMS might require the assignment of additional resources to service the new and original users with equal efficiency (see Chapter 11, Database Performance Tuning and Query Optimization).

Testing and Evaluating Databases and Applications

The DBA must also provide testing and evaluation services for all of the database and end-user applications. Those services are the logical extension of the design, development, and implementation services described in the preceding section. Clearly, testing procedures and standards must already be in place before any application program can be approved for use in the company.

Although testing and evaluation services are closely related to database design and implementation services, they usually are maintained independently. The reason for the separation is that application programmers and designers often are too close to the problem being studied to detect errors and omissions.

Testing usually starts with the loading of the testbed database. That database contains test data for the applications, and its purpose is to check the data definition and integrity rules of the database and application programs.

The testing and evaluation of a database application cover all aspects of the system—from the simple collection and creation of data to its use and retirement. The evaluation process covers:

- Technical aspects of both the applications and the database. Backup and recovery, security and integrity, use of SQL, and application performance must be evaluated.
- Evaluation of the written documentation to ensure that the documentation and procedures are accurate and easy to follow.
- Observance of standards for naming, documenting, and coding.
- Data duplication conflicts with existing data.
- The enforcement of all data validation rules.

Following the thorough testing of all applications, the database, and the procedures, the system is declared operational and can be made available to end users.

Operating the DBMS, Utilities, and Applications

DBMS operations can be divided into four main areas:

- System support.
- Performance monitoring and tuning.
- Backup and recovery.
- Security auditing and monitoring.

System support activities cover all tasks directly related to the day-to-day operations of the DBMS and its applications. These activities include filling out job logs, changing tape, and verifying the status of computer hardware, disk packages, and emergency power sources. System-related activities include periodic, occasional tasks such as running special programs and resource configurations for new and/or upgraded versions of database applications.

Performance monitoring and tuning require much of the DBA's attention and time. These activities are designed to ensure that the DBMS, utilities, and applications maintain satisfactory performance levels. To carry out the performance monitoring and tuning tasks, the DBA must:

- Establish DBMS performance goals.
- Monitor the DBMS to evaluate whether the performance objectives are being met.
- Isolate the problem and find solutions (if performance objectives are not met).
- Implement the selected performance solutions.

DBMSs often include performance-monitoring tools that allow the DBA to query database usage information. Performance-monitoring tools are also available from many different sources: DBMS utilities are provided by third-party vendors, or they might be included in operating system utilities or transaction processor facilities. Most of the performance-monitoring tools allow the DBA to focus on selected system bottlenecks. The most common bottlenecks in DBMS performance tuning are related to the use of indexes, query-optimization algorithms, and management of storage resources.

Because improper index selection can have a deleterious effect on system performance, most DBMS installations adhere to a carefully defined index creation and usage plan. Such a plan is especially important in a relational database environment.

To produce satisfactory performance, the DBA is likely to spend much time trying to educate programmers and end users on the proper use of SQL statements. Typically, DBMS programmers' manuals and administration manuals contain useful performance guidelines and examples that demonstrate the proper use of SQL statements, both in the command-line mode and within application programs. Because relational systems do not give the user an index choice within a query, the DBMS makes the index selection for the user. Therefore, the DBA should create indexes that can be used to improve system performance. (For examples of database performance tuning, see Chapter 11, Database Performance Tuning and Query Optimization.)

Query-optimization routines are usually integrated into the DBMS package, thereby allowing few tuning options. Query-optimization routines are oriented to improve concurrent access to the database. Several database packages let the DBA specify parameters for determining the desired level of concurrency. Concurrency is also affected by the types of locks used by the DBMS and requested by the applications. Because the concurrency issue is important to the efficient operation of the system, the DBA must be familiar with the factors that influence concurrency. (See Chapter 10, Transaction Management and Concurrency Control, for more information on that subject.)

During DBMS performance tuning, the DBA must also consider available storage resources in terms of both primary and secondary memory. The allocation of storage resources is determined when the DBMS is configured. Storage configuration parameters can be used to determine:

- The number of databases that may be opened concurrently.
- The number of application programs or users supported concurrently.
- The amount of primary memory (buffer pool size) assigned to each database and each database process.
- The size and location of the log files. (Remember that these files are used to recover the database. The log files can be located in a separate volume to reduce the disk's head movement and to increase performance.)

Performance-monitoring issues are DBMS-specific. Therefore, the DBA must become familiar with the DBMS manuals to learn the technical details involved in the performance-monitoring task (see Chapter 11).

Because data loss is likely to be devastating to the organization, *backup and recovery activities* are of primary concern during the DBMS operation. The DBA must establish a schedule for backing up database and log files at appropriate intervals. Backup frequency is dependent on the application type and on the relative importance of the data. All critical system components—the database, the database applications, and the transaction logs—must be backed up periodically.

Most DBMS packages include utilities that schedule automated database backups, be they full or incremental. Although incremental backups are faster than full backups, an incremental backup requires the existence of a periodic full backup to be useful for recovery purposes.

Database recovery after a media or systems failure requires application of the transaction log to the correct database copy. The DBA must plan, implement, test, and enforce a "bulletproof" backup and recovery procedure.

Security auditing and monitoring assumes the appropriate assignment of access rights and the proper use of access privileges by programmers and end users. The technical aspects of security auditing and monitoring involve creating users, assigning access rights, using SQL commands to grant and revoke access rights to users and database objects, and creating audit trails to discover security violations or attempted violations. The DBA must periodically generate an audit trail report to determine whether there have been actual or attempted security violations—and, if so, from what locations, and if possible, by whom.

Training and Supporting Users

Training people to use the DBMS and its tools is included in the DBA's technical activities. In addition, the DBA provides or secures technical training in the use of the DBMS and its utilities for the applications programmers. Applications programmer training covers the use of the DBMS tools as well as the procedures and standards required for database programming.

Unscheduled, on-demand technical support for end users and programmers is also included in the DBA's activities. A technical troubleshooting procedure can be developed to facilitate such support. The technical procedure might include the development of a technical database used to find solutions to common technical problems.

Part of the support is provided by interaction with DBMS vendors. Establishing good relationships with software suppliers is one way to ensure that the company has a good external support source. Vendors are the source for up-to-date information concerning new products and personnel retraining. Good vendor–company relations also are likely to give organizations an edge in determining the future direction of database development.

Maintaining the DBMS, Utilities, and Applications

The maintenance activities of the DBA are an extension of the operational activities. Maintenance activities are dedicated to the preservation of the DBMS environment.

Periodic DBMS maintenance includes management of the physical or secondary storage devices. One of the most common maintenance activities is reorganizing the physical location of data in the database. (That is usually done as part of the DBMS fine-tuning activities.) The reorganization of a database might be designed to allocate contiguous disk-page locations to the DBMS to increase performance. The reorganization process also might free the space allocated to deleted data, thus providing more disk space for new data.

Maintenance activities also include upgrading the DBMS and utility software. The upgrade might require the installation of a new version of the DBMS software or an Internet front-end tool. Or it might create an additional DBMS gateway to allow access to a host DBMS running on a different host computer. DBMS gateway services are very common in distributed DBMS applications running in a client/server environment. Also, new-generation databases include features such as spatial data support, data warehousing and star query support, and support for Java programming interfaces for Internet access (see Chapter 14, Database Connectivity and Web Technologies).

Quite often companies are faced with the need to exchange data in dissimilar formats or between databases. The maintenance efforts of the DBA include migration and conversion services for data in incompatible formats or for different DBMS software. Such conditions are common when the system is upgraded from one version to another or when the existing DBMS is replaced by an entirely new DBMS. Database conversion services also include downloading data from the host DBMS (mainframe-based) to an end user's personal computer to allow that user to perform a variety of activities—spreadsheet analysis, charting, statistical modeling, and so on. Migration and conversion services can be done at the logical level (DBMS- or software-specific) or at the physical level (storage media or operating-system-specific). Current generation DBMSs support XML as a standard format for data exchange among database systems and applications (see Chapter 14).

15.6 SECURITY

Security refers to activities and measures to ensure the confidentiality, integrity, and availability of an information system and its main asset, data.[4] It is important to understand that securing data requires a comprehensive, company-wide approach. That is, you cannot secure data if you do not secure all the processes and systems around it. Indeed, securing data entails securing the overall information system architecture, including hardware systems, software applications, the network and its devices, people (internal and external users), procedures, and the data itself. To understand the scope of data security, let's discuss each of the three security goals in more detail:

- **Confidentiality** deals with ensuring that data is protected against unauthorized access, and if the data are accessed by an authorized user, that the data are used only for an authorized purpose. In other words, confidentiality entails safeguarding data against disclosure of any information that would violate the privacy rights of a person or organization. Data must be evaluated and classified according to the level of confidentiality: highly restricted (very few people have access), confidential (only certain groups have access), and unrestricted (can be accessed by all users). The data security officer spends a great amount of time ensuring that the organization is in compliance with the desired levels of confidentiality. **Compliance** refers to activities undertaken to meet data privacy and security reporting guidelines. These reporting guidelines are either part of internal procedures or are imposed by external regulatory agencies such as the federal government. Examples of U.S. legislation enacted with the purpose of ensuring data privacy and confidentiality include the Health Insurance Portability and Accountability Act (HIPAA), Gramm-Leach-Bliley Act (GLBA), and Sarbanes-Oxley Act (SOX).[5]

- **Integrity**, within the data security framework, is concerned with keeping data consistent, free of errors or anomalies. Integrity focuses on maintaining the data free of inconsistencies and anomalies (see Chapter 1, Database Systems, to review the concepts of data inconsistencies and data anomalies). The DBMS plays a pivotal role in ensuring the integrity of the data in the database. However, from the security point of view, integrity deals not only with the data in the database, but also with ensuring that organizational processes, users, and usage patterns maintain such integrity. For example, a work-at-home employee using the Internet to access product costing could be considered an acceptable use; however, security standards might require the employee to use a secure connection and follow strict procedures to manage the data at home (shredding printed reports, using encryption to copy data to the local hard drive, etc.). Maintaining the integrity of the data is a process that starts with data collection and continues with data storage, processing, usage, and archival (see Chapter 13, Business Intelligence and Data Warehouses). The rationale behind integrity is to treat data as the most valuable asset in the organization and therefore to ensure that rigorous data validation is carried out at all levels within the organization.

- **Availability** refers to the accessibility of data whenever required by authorized users and for authorized purposes. To ensure data availability, the entire system (not only the data component) must be protected from service degradation or interruption caused by any source (internal or external). Service interruptions could be very costly for companies and users alike--recall the JetBlue[6] case in the Part V Business Vignette of this book, and, more recently the case of SKYPE, the voice over IP (VoIP) telephone service provider who suffered a 48-hour worldwide service interruption.[7] System availability is an important goal of security.

---

[4]The National Security Telecommunications and Information Systems Security Committee (NSTISSC) defines the CIA framework. See *http://www.nsa.gov/snac/wireless/I332-005R-2005.pdf*.

[5]To find additional information about these various laws, please visit *http://library.uis.edu/findinfo/govinfo/federal/law.html*.

[6]"JetBlue's C.E.O. Is 'Mortified' After Fliers Are Stranded," Jeff Baily, February 19, 2007, New York Times, *http://www.nytimes.com/2007/02/19/business/19jetblue.html?ex=1189051200&en=d63f3b54a602bf0d&ei=5070*.

[7]"Skype protection is limited," Andrew Garcia, *eWeek*, p. 59, August 27, 2007.

15.6.1 SECURITY POLICIES

Normally, the tasks of securing the system and its main asset, the data, are performed by the database security officer and the database administrator(s), who work together to establish a cohesive data security strategy. Such security strategy begins with defining a sound and comprehensive security policy. A **security policy** is a collection of standards, policies and procedures created to guarantee the security of a system and ensure auditing and compliance. The security audit process starts by identifying the security vulnerabilities in the organization's information system infrastructure and identifying measures to protect the system and data against those vulnerabilities.

15.6.2 SECURITY VULNERABILITIES

A **security vulnerability** is a weakness in a system component that could be exploited to allow unauthorized access or cause service disruptions. The nature of such vulnerabilities could be of multiple types: technical (such as a flaw in the operating system or Web browser), managerial (for example, not educating users about critical security issues), cultural (hiding passwords under the keyboard or not shredding confidential reports), procedural (not requiring complex passwords or not checking user IDs), and so on. Whatever the case, when a security vulnerability is left unchecked, it could become a security threat. A **security threat** is an imminent security violation that could occur at any time due to unchecked security vulnerability.

A **security breach** occurs when a security threat is exploited to negatively affect the integrity, confidentiality, or availability of the system. Security breaches can yield a database whose integrity is either preserved or corrupted:

- *Preserved:* Action is required to avoid the repetition of similar security problems, but data recovery may not be necessary. As a matter of fact, most security violations are produced by unauthorized and unnoticed access for information purposes, but such snooping does not disrupt the database.

- *Corrupted:* Action is required to avoid the repetition of similar security problems, and the database must be recovered to a consistent state. Corrupting security breaches include database access by computer viruses and by hackers whose actions are intended to destroy or alter data.

Table 15.4 illustrates some security vulnerabilities that systems components are exposed to and some measures typically taken to protect against them.

TABLE 15. 4	Sample Security Vulnerabilities and Related Measures	
SYSTEM COMPONENT	**SECURITY VULNERABILITY**	**SECURITY MEASURES**
People	• User sets a blank password. • Password is short or includes birth date. • User leaves office door open all the time. • User leaves payroll information on screen for long periods of time.	• Enforce complex password policies. • Use multilevel authentication. • Use security screens and screen savers. • Educate users about sensitive data. • Install security cameras. • Use automatic door locks.
Workstations and Servers	• User copies data to flash drive. • Workstation is used by multiple users. • Power failure crashes computer. • Unauthorized personnel can use computer. • Sensitive data stored in laptop computer. • Data lost due to stolen hard disk/laptop. • Natural disasters—earthquake, flood, etc.	• Use group policies to restrict use of flash drives. • Assign user access rights to workstations. • Install Uninterrupted Power Supplies (UPS). • Add security lock devices to computers. • Implement a "kill" switch for stolen laptops. • Create and test data backup and recovery plans. • Protect system against natural disasters—use co-location strategies.

TABLE 15.4	Sample Security Vulnerabilities and Related Measures (continued)	
SYSTEM COMPONENT	**SECURITY VULNERABILITY**	**SECURITY MEASURES**
Operating System	• Buffer overflow attacks. • Virus attacks. • Root kits and worm attacks. • Denial of service attacks. • Trojan horses. • Spyware applications. • Password crackers.	• Apply OS security patches and updates. • Apply application server patches. • Install antivirus and antispyware software. • Enforce audit trails on the computers. • Perform periodic system backups. • Install only authorized applications. • Use group policies to prevent unauthorized installs.
Applications	• Application bugs—buffer overflow. • SQL injection, session hijacking, etc. • Application vulnerabilities—cross site scripting, nonvalidated inputs. • E-mail attacks: spamming, phishing, etc. • Social engineering e-mails.	• Test application programs extensively. • Built safeguards in code. • Do extensive vulnerability testing in applications. • Install spam filter/antivirus for e-mail system. • Use secure coding techniques (see *www.owasp.org*). • Educate users about social engineering attacks.
Network	• IP spoofing. • Packet sniffers. • Hacker attacks. • Clear passwords on network.	• Install firewalls. • Virtual Private Networks (VPN). • Intrusion Detection Systems (IDS). • Network Access Control (NAC). • Network activity monitoring.
Data	• Data shares are open to all users. • Data can be accessed remotely. • Data can be deleted from shared resource.	• Implement file system security. • Implement share access security. • Use access permission. • Encrypt data at the file system or database level.

15.6.3 DATABASE SECURITY

Database security refers to the use of the DBMS features and other related measures to comply with the security requirements of the organization. From the DBA's point of view, security measures should be implemented to protect the DBMS against service degradation and the database against loss, corruption, or mishandling. In short, the DBA should secure the DBMS from the point of installation through operation and maintenance.

> **NOTE**
>
> James Martin provides an excellent enumeration and description of the desirable attributes of a database security strategy that remains relevant today (James Martin, *Managing the Database Environment*, Englewood Cliffs, NJ: Prentice-Hall, 1977). Martin's security strategy is based on the seven essentials of database security and may be summarized as one in which:
>
> | Data are | Users are |
> |---|---|
> | Protected | Identifiable |
> | Reconstructable | Authorized |
> | Auditable | Monitored |
> | Tamperproof | |

To protect the DBMS against service degradation there are certain minimum recommended security safeguards. For example: change default system passwords, change default installation paths, apply the latest patches, secure installation folders with proper access rights, make sure only required services are running, set up auditing logs, set up session logging, and require session encryption. Furthermore, the DBA should work closely with the network administrator to implement network security to protect the DBMS and all services running on the network. In current organizations, one of the most critical components in the information architecture is the network.

Protecting the data in the database is a function of authorization management. **Authorization management** defines procedures to protect and guarantee database security and integrity. Those procedures include, but are not limited to, user access management, view definition, DBMS access control, and DBMS usage monitoring.

- *User access management.* This function is designed to limit access to the database and likely includes at least the following procedures:
  - *Define each user to the database.* This is achieved at the operating system level and at the DBMS level. At the operating system level, the DBA can request the creation of a logon user ID that allows the end user to log on to the computer system. At the DBMS level, the DBA can either create a different user ID or employ the same user ID to authorize the end user to access the DBMS.
  - *Assign passwords to each user.* This, too, can be done at both operating system and DBMS levels. The database passwords can be assigned with predetermined expiration dates. The use of expiration dates enables the DBA to screen end users periodically and to remind users to change their passwords periodically, thus making unauthorized access less probable.
  - *Define user groups.* Classifying users into user groups according to common access needs facilitates the DBA's job of controlling and managing the access privileges of individual users. Also, the DBA can use database roles and resource limits to minimize the impact of rogue users in the system (see Section 15.9.6 for more information about these topics).
  - *Assign access privileges.* The DBA assigns access privileges or access rights to specific users to access specified databases. An access privilege describes the type of authorized access. For example, access rights may be limited to read-only, or the authorized access might include READ, WRITE, and DELETE privileges. Access privileges in relational databases are assigned through SQL GRANT and REVOKE commands.
  - *Control physical access.* Physical security can prevent unauthorized users from directly accessing the DBMS installation and facilities. Some common physical security practices found in large database installations include secured entrances, password-protected workstations, electronic personnel badges, closed-circuit video, voice recognition, and biometric technology.
- *View definition.* The DBA must define data views to protect and control the scope of the data that are accessible to an authorized user. The DBMS must provide the tools that allow the definition of views that are composed of one or more tables and the assignment of access rights to a user or a group of users. The SQL command CREATE VIEW is used in relational databases to define views. Oracle DBMS offers Virtual Private Database (VPD), which allows the DBA to create customized views of the data for multiple different users. With this feature, the DBA could restrict a regular user querying a payroll database to see only the rows and columns necessary, while the department manager would see only the rows and columns pertinent to that department.
- *DBMS access control.* Database access can be controlled by placing limits on the use of DBMS query and reporting tools. The DBA must make sure that those tools are used properly and only by authorized personnel.
- *DBMS usage monitoring.* The DBA must also audit the use of the data in the database. Several DBMS packages contain features that allow the creation of an **audit log**, which automatically records a brief description of the database operations performed by all users. Such audit trails enable the DBA to pinpoint access violations. The audit trails can be tailored to record all database accesses or just failed database accesses.

The integrity of a database could be lost because of external factors beyond the DBA's control. For example, the database might be damaged or destroyed by an explosion, a fire, or an earthquake. Whatever the reason, the specter of database corruption or destruction makes backup and recovery procedures crucial to any DBA.

15.7 DATABASE ADMINISTRATION TOOLS

The importance of the data dictionary as a prime DBA tool cannot be overstated. This section will examine the data dictionary as a data administration tool, as well as the DBA's use of computer-aided software engineering (CASE) tools to support database analysis and design.

15.7.1 THE DATA DICTIONARY

In Chapter 1, a *data dictionary* was defined as "a DBMS component that stores the definition of data characteristics and relationships." You may recall that such "data about data" are called *metadata*. The DBMS data dictionary provides the DBMS with its self-describing characteristic. In effect, the data dictionary resembles an x-ray of the company's entire data set, and it is a crucial element in data administration.

Two main types of data dictionaries exist: *integrated* and *standalone*. An integrated data dictionary is included with the DBMS. For example, all relational DBMSs include a built-in data dictionary or system catalog that is frequently accessed and updated by the RDBMS. Other DBMSs, especially older types, do not have a built-in data dictionary; instead, the DBA may use third-party *standalone data dictionary* systems.

Data dictionaries can also be classified as *active* or *passive*. An **active data dictionary** is automatically updated by the DBMS with every database access, thereby keeping its access information up to date. A **passive data dictionary** is not updated automatically and usually requires running a batch process. Data dictionary access information is normally used by the DBMS for query optimization purposes.

The data dictionary's main function is to store the description of all objects that interact with the database. Integrated data dictionaries tend to limit their metadata to the data managed by the DBMS. Standalone data dictionary systems are usually more flexible and allow the DBA to describe and manage all of the organization's data, whether or not they are computerized. Whatever the data dictionary's format, its existence provides database designers and end users with a much-improved ability to communicate. In addition, the data dictionary is the tool that helps the DBA resolve data conflicts.

Although there is no standard format for the information stored in the data dictionary, several features are common. For example, the data dictionary typically stores descriptions of all:

- *Data elements that are defined in all tables of all databases.* Specifically, the data dictionary stores the names, data types, display format, internal storage format, and validation rules. The data dictionary tells where an element is used, by whom it is used, and so on.
- *Tables defined in all databases.* For example, the data dictionary is likely to store the name of the table creator, the date of creation, access authorizations, and the number of columns.
- *Indexes defined for each database table.* For each index, the DBMS stores at least the index name, the attributes used, the location, specific index characteristics, and the creation date.
- *Defined databases.* This includes who created each database, when the database was created, where the database is located, who the DBA is, and so on.
- *End users and administrators of the database.*
- *Programs that access the database.* This includes screen formats, report formats, application programs, and SQL queries.
- *Access authorizations for all users of all databases.*
- *Relationships among data elements.* This includes which elements are involved, whether the relationships are mandatory or optional, and what the connectivity and cardinality requirements are.

If the data dictionary can be organized to include data external to the DBMS itself, it becomes an especially flexible tool for more general corporate resource management. The management of such an extensive data dictionary thus makes it possible to manage the use and allocation of all of the organization's information, regardless of whether the information has its roots in the database data. That is why some managers consider the data dictionary to be a key element of information resource management. And that is also why the data dictionary might be described as the *information resource dictionary*.

The metadata stored in the data dictionary are often the basis for monitoring database use and for assigning access rights to the database users. The information stored in the data dictionary is usually based on a relational table format, thus enabling the DBA to query the database with SQL commands. For example, SQL commands can be used to extract information about the users of a specific table or about the access rights of a particular user. In the following example, the IBM DB2 system catalog tables will be used as the basis for several examples of how a data dictionary is used to derive information.

SYSTABLES stores one row for each table or view.
SYSCOLUMNS stores one row for each column of each table or view.
SYSTABAUTH stores one row for each authorization given to a user for a table or view in a database.

Examples of Data Dictionary Usage
Example 1

List the names and creation dates of all tables created by the user JONESVI in the current database.

```
SELECT      NAME, CTIME
FROM        SYSTABLES
WHERE       CREATOR = 'JONESVI';
```

Example 2

List the names of the columns for all tables created by JONESVI in the current database.

```
SELECT      NAME
FROM        SYSCOLUMNS
WHERE       TBCREATOR = 'JONESVI';
```

Example 3

List the names of all tables for which the user JONESVI has DELETE authorization.

```
SELECT      TTNAME
FROM        SYSTABAUTH
WHERE       GRANTEE = 'JONESVI' AND DELETEAUTH = 'Y';
```

Example 4

List the names of all users who have some type of authority over the INVENTORY table.

```
SELECT      DISTINCT GRANTEE
FROM        SYSTABAUTH
WHERE       TTNAME = 'INVENTORY';
```

Example 5

List the user and table names for all users who can alter the database structure for any table in the database.

```
SELECT      GRANTEE, TTNAME
FROM        SYSTABAUTH
WHERE       ALTERAUTH = 'Y'
ORDER BY    GRANTEE, TTNAME;
```

As you can see in the preceding examples, the data dictionary can be a tool for monitoring the security of data in the database by checking the assignment of data access privileges. Although the preceding examples targeted database tables and users, information about the application programs that access the database can also be drawn from the data dictionary.

The DBA can use the data dictionary to support data analysis and design. For example, the DBA can create a report that lists all data elements to be used in a particular application; a list of all users who access a particular program; a report that checks for data redundancies, duplications, and the use of homonyms and synonyms; and a number of other reports that describe data users, data access, and data structure. The data dictionary can also be used to ensure that applications programmers have met all of the naming standards for the data elements in the database and that the data validation rules are correct. Thus, the data dictionary can be used to support a wide range of data administration activities and to facilitate the design and implementation of information systems. Integrated data dictionaries are also essential to the use of computer-aided software engineering tools.

15.7.2 CASE TOOLS

CASE is the acronym for **computer-aided systems engineering**. A CASE tool provides an automated framework for the Systems Development Life Cycle (SDLC). CASE uses structured methodologies and powerful graphical interfaces. Because they automate many tedious system design and implementation activities, CASE tools play an increasingly important role in information systems development.

CASE tools are usually classified according to the extent of support they provide for the SDLC. For example, **front-end CASE tools** provide support for the planning, analysis, and design phases; **back-end CASE tools** provide support for the coding and implementation phases. The benefits associated with CASE tools include:

- A reduction in development time and costs.
- Automation of the SDLC.
- Standardization of systems development methodologies.
- Easier maintenance of application systems developed with CASE tools.

One of the CASE tools' most important components is an extensive data dictionary, which keeps track of all objects created by the systems designer. For example, the CASE data dictionary stores data flow diagrams, structure charts, descriptions of all external and internal entities, data stores, data items, report formats, and screen formats. A CASE data dictionary also describes the relationships among the components of the system.

Several CASE tools provide interfaces that interact with the DBMS. Those interfaces allow the CASE tool to store its data dictionary information by using the DBMS. Such CASE/DBMS interaction demonstrates the interdependence that exists between systems development and database development, and it helps create a fully integrated development environment.

In a CASE development environment, the database and application designers use the CASE tool to store the description of the database schema, data elements, application processes, screens, reports, and other data relevant to the development process. The CASE tool integrates all systems development information in a common repository, which can be checked by the DBA for consistency and accuracy.

As an additional benefit, a CASE environment tends to improve the extent and quality of communication among the DBA, the application designers, and the end users. The DBA can interact with the CASE tool to check the definition of the data schema for the application, the observance of naming conventions, the duplication of data elements, the validation rules for the data elements, and a host of other developmental and managerial variables. When the CASE tool indicates conflicts, rule(s) violations, and inconsistencies, it facilitates making corrections. Better yet, a correction is transported by the CASE tool to cascade its effects throughout the applications environment, thus greatly simplifying the job of the DBA and the application designer.

A typical CASE tool provides five components:

- Graphics designed to produce structured diagrams such as data flow diagrams, ER diagrams, class diagrams, and object diagrams.

- Screen painters and report generators to produce the information system's input/output formats (for example, the end-user interface).

- An integrated repository for storing and cross-referencing the system design data. This repository includes a comprehensive data dictionary.

- An analysis segment to provide a fully automated check on system consistency, syntax, and completeness.

- A program documentation generator.

Figure 15.7 illustrates how Microsoft Visio Professional can be used to produce an ER diagram.

FIGURE 15.7   An example of a CASE tool: Visio Professional

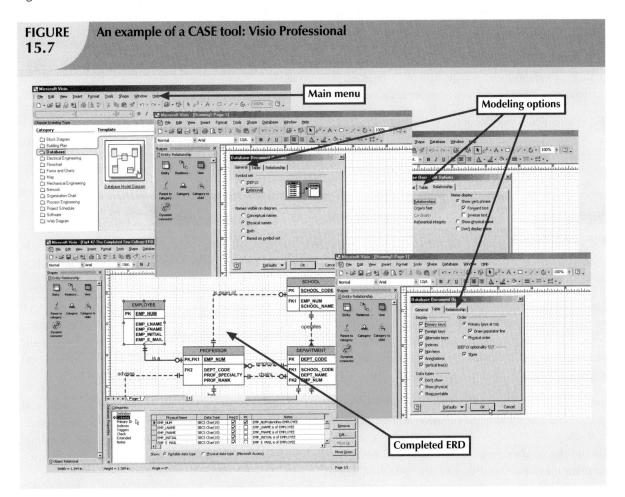

One CASE tool, ERwin Data Modeler by Computer Associates, produces fully documented ER diagrams that can be displayed at different abstraction levels. In addition, ERwin can produce detailed relational designs. The user specifies the attributes and primary keys for each entity and describes the relations. ERwin then assigns foreign keys based on the specified relationships among the entities. Changes in primary keys are always updated automatically throughout the system. Table 15.5 shows a short list of the many available CASE tool vendors.

TABLE 15.5	CASE Tools	
COMPANY	**PRODUCT**	**WEB SITE**
Casewise	Corporate Modeler Suite	www.casewise.com
Computer Associates	ERwin	www3.ca.com/Solutions/Product.asp?ID=260
Embarcadero Technologies	ER/Studio	www.embarcadero.com/products/erstudio
Microsoft	Visio	office.microsoft.com/en-us/FX010857981033.aspx
Oracle	Designer	www.oracle.com/technology/products/designer
Telelogic	System Architect	www.telelogic.com/products/system_architect/sa
Sybase	Power Designer	www.sybase.com/products/developmentintegration/powerdesigner
Visible	Visible Analyst	www.visible.com/Products/Analyst

Major relational DBMS vendors, such as Oracle, now provide fully integrated CASE tools for their own DBMS software as well as for RDBMSs supplied by other vendors. For example, Oracle's CASE tools can be used with IBM's DB2, SQL/DS, and Microsoft's SQL Server to produce fully documented database designs. Some vendors even take nonrelational DBMSs, develop their schemas, and produce the equivalent relational designs automatically.

There is no doubt that CASE has enhanced the database designer's and the applications programmer's efficiency. But no matter how sophisticated the CASE tool, its users must be well versed in conceptual design ideas. In the hands of database novices, CASE tools simply produce impressive-looking but bad designs.

15.8 DEVELOPING A DATA ADMINISTRATION STRATEGY

For a company to succeed, its activities must be committed to its main objectives or mission. Therefore, regardless of a company's size, a critical step for any organization is to ensure that its information system supports its strategic plans for each of its business areas.

The database administration strategy must not conflict with the information systems plans. After all, the information systems plans are derived from a detailed analysis of the company's goals, its condition or situation, and its business needs. Several methodologies are available to ensure the compatibility of data administration and information systems plans and to guide the strategic plan development. The most commonly used methodology is known as information engineering.

Information engineering (IE) allows for the translation of the company's strategic goals into the data and applications that will help the company achieve those goals. IE focuses on the description of the corporate data instead of the processes. The IE rationale is simple: business data types tend to remain fairly stable. In contrast, processes change often and thus require the frequent modification of existing systems. By placing the emphasis on data, IE helps decrease the impact on systems when processes change.

The output of the IE process is an **information systems architecture (ISA)** that serves as the basis for planning, development, and control of future information systems. Figure 15.8 shows the forces that affect ISA development.

Implementing IE methodologies in an organization is a costly process that involves planning, a commitment of resources, management liability, well-defined objectives, identification of critical factors, and control. An ISA provides a framework that includes the use of computerized, automated, and integrated tools such as a DBMS and CASE tools.

FIGURE 15.8   Forces affecting the development of the ISA

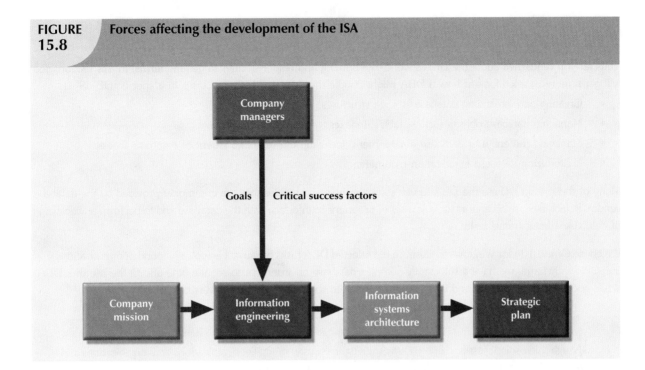

The success of the overall information systems strategy, and therefore, of the data administration strategy depends on several critical success factors. Understanding the critical success factors helps the DBA develop a successful corporate data administration strategy. Critical success factors include managerial, technological, and corporate culture issues, such as:

- *Management commitment.* Top-level management commitment is necessary to enforce the use of standards, procedures, planning, and controls. The example must be set at the top.
- *Thorough company situation analysis.* The current situation of the corporate data administration must be analyzed to understand the company's position and to have a clear vision of what must be done. For example, how are database analysis, design, documentation, implementation, standards, codification, and other issues handled? Needs and problems should be identified first; then prioritized.
- *End-user involvement.* End-user involvement is another aspect critical to the success of the data administration strategy. What is the degree of organizational change involved? Successful organizational change requires that people be able to adapt to the change. Users should be given an open communication channel to upper-level management to ensure success of the implementation. Good communication is key to the overall process.
- *Defined standards.* Analysts and programmers must be familiar with appropriate methodologies, procedures, and standards. If analysts and programmers lack familiarity, they might need to be trained in the use of the procedures and standards.
- *Training.* The vendor must train the DBA personnel in the use of the DBMS and other tools. End users must be trained to use the tools, standards, and procedures to obtain and demonstrate the maximum benefit, thereby increasing end-user confidence. Key personnel should be trained first so they can train others.
- *A small pilot project.* A small project is recommended to ensure that the DBMS will work in the company, that the output is what was expected, and that the personnel have been trained properly.

That list of factors is not and cannot be comprehensive. Nevertheless, it does provide the initial framework for the development of a successful strategy. Remember that no matter how comprehensive the list of success factors is, it must be based on the notion that development and implementation of a successful data administration strategy are tightly integrated with the overall information systems planning activity of the organization.

15.9 THE DBA AT WORK: USING ORACLE FOR DATABASE ADMINISTRATION

Thus far you've learned about the DBA's work environment and responsibilities in general terms. In this section, you will get a more detailed look at how a DBA might handle the following technical tasks in a specific DBMS:

- Creating and expanding database storage structures.
- Managing database objects such as tables, indexes, triggers, and procedures.
- Managing the end-user database environment, including the type and extent of database access.
- Customizing database initialization parameters.

Many of those tasks require the DBA to use software tools and utilities that are commonly provided by the database vendor. In fact, all DBMS vendors provide a set of programs to interface with the database and to perform a wide range of database administrative tasks.

We chose Oracle 10g for Windows to illustrate the selected DBA tasks because it is typically found in organizations that are sufficiently large and have a sufficiently complex database environment to require (and afford) the use of a DBA, it has good market presence, and it is also often found in small colleges and universities.

> **NOTE**
>
> Although Microsoft Access is a superb DBMS, it is typically used in smaller organizations or in organizations and departments with relatively simple data environments. Access yields a superior database prototyping environment, and given its easy-to-use GUI tools, rapid front-end application development is a snap. Also, Access is one of the components in the MS Office suite, thus making end-user applications integration relatively simple and seamless. Finally, Access does provide some important database administration tools. However, an Access-based database environment does not typically require the services of a DBA. Therefore, MS Access does not fit this section's mission.

Keep in mind that most of the tasks described in this section are encountered by DBAs regardless of their DBMS or their operating system. However, the *execution* of those tasks tends to be specific to the DBMS and the operating system. Therefore, if you use IBM DB2 Universal Database or Microsoft SQL Server, you must adapt the procedures shown here to your DBMS. And because these examples run under the Windows operating system, if you use some other OS, you must adapt the procedures shown in this section to your OS.

This section will not serve as a database administration manual. Instead, it will offer a brief introduction to the way some typical DBA tasks would be performed in Oracle. Before learning how to use Oracle to accomplish specific database administration tasks, you should become familiar with the tools Oracle offers for database administration and with the procedures for logging on, which will be discussed in the next two sections.

15.9.1 ORACLE DATABASE ADMINISTRATION TOOLS

All database vendors supply a set of database administration tools. In Oracle, you perform most DBA tasks via the Oracle Enterprise Manager interface. See Figure 15.9.

In Figure 15.9, note that it shows the status of the current database. (This section uses the ORALAB database.) In the following sections, you will examine the tasks most commonly encountered by a DBA.

FIGURE 15.9   The Oracle Enterprise Manager interface

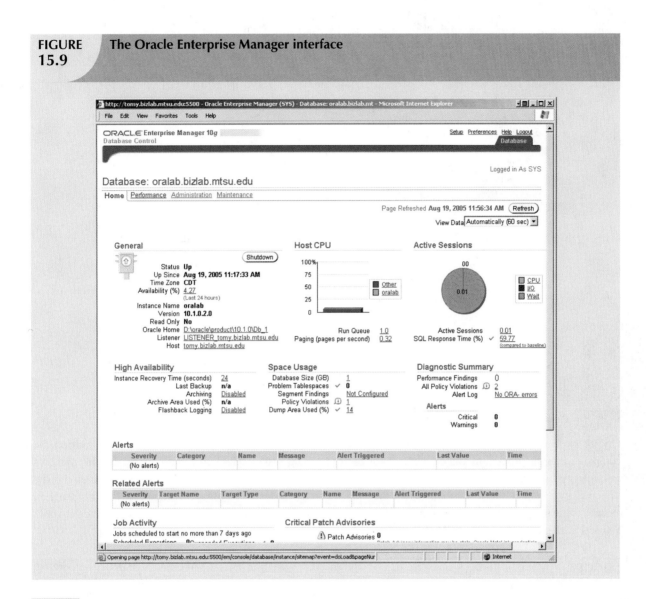

15.9.2 THE DEFAULT LOGIN

To perform any administrative task, you must connect to the database, using a username with administrative (DBA) privileges. By default, Oracle automatically creates SYSTEM and SYS user IDs that have administrative privileges with every new database you create. You can define the preferred credentials for each database by clicking on the **Preferences** link at the top of the page, then click on **Preferred Credentials**. Finally, choose your target username under **Set Credentials**. Figure 15.10 shows the Edit Local Preferred Credentials page that defines the user ID (SYS) used to log on to the ORALAB database.

Keep in mind that usernames and passwords are database-specific. Therefore, each database can have different usernames and passwords. One of the first things you must do is change the password for the SYSTEM and SYS users. Immediately after doing that, you can start defining your users and assigning them database privileges.

FIGURE 15.10   The Oracle Edit Local Preferred Credentials page

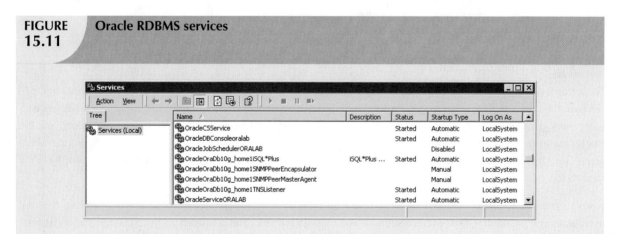

15.9.3 ENSURING AN AUTOMATIC RDBMS START

One of the basic DBA tasks is to ensure that your database access is automatically started when you turn on the computer. Startup procedures will be different for each operating system. Because Oracle is used for this section's examples, you would need to identify the required services to ensure automatic database startup. (A *service* is the Windows system name for a special program that runs automatically as part of the operating system. This program ensures the availability of required services to the system and to end users on the local computer or over the network.) Figure 15.11 shows the required Oracle services that are started automatically when Windows starts up.

FIGURE 15.11   Oracle RDBMS services

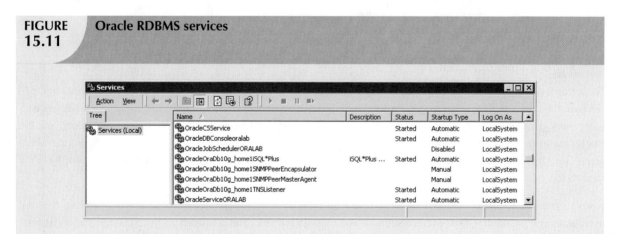

As you examine Figure 15.11, note the following Oracle services:

- *OracleOraDb10g_home1TNSListener* is the process that "listens to" and processes the end-user connection requests over the network. For example, when a SQL connection request such as "connect userid/ password@ORALAB" is sent over the network, the listener service will take the request, validate it, and establish the connection.

- *OracleServiceORALAB* refers to the Oracle processes running in memory that are associated with the ORALAB database instance. You can think of a **database instance** as a separate location in memory that is reserved to run your database. Because you can have several databases (and, therefore, several instances) running in memory at the same time, you need to identify each database instance uniquely, using a different suffix for each one.

15.9.4 CREATING TABLESPACES AND DATAFILES

Each DBMS manages data storage differently. In this example, the Oracle RDBMS will be used to illustrate how the database manages data storage at the logical and the physical levels. In Oracle:

A database is *logically* composed of one or more tablespaces. A **tablespace** is a logical storage space. Tablespaces are used primarily to group related data logically.

The tablespace data are *physically* stored in one or more datafiles. A **datafile** physically stores the database's data. Each datafile is associated with one and only one tablespace, but each datafile can reside in a different directory on the hard disk or even on one or more different hard disks.

Given the preceding description of tablespaces and datafiles, you can conclude that a database has a one-to-many relationship with tablespaces and that a tablespace has a one-to-many relationship with datafiles. This set of 1:M hierarchical relationships isolates the end user from any physical details of the data storage. However, *the DBA must be aware of these details in order to properly manage the database.*

To perform database storage management tasks such as creating and managing tablespaces and datafiles, the DBA uses Enterprise Manager, Administration, Storage option. See Figure 15.12.

When the DBA creates a database, Oracle automatically creates the tablespaces and datafiles shown in Figure 15.12. A few of them are described here.

- The *SYSTEM* tablespace is used to store the data dictionary data.
- The *USERS* tablespace is used to store the table data created by the end users.
- The *TEMP* tablespace is used to store the temporary tables and indexes created during the execution of SQL statements. For example, temporary tables are created when your SQL statement contains an ORDER BY, GROUP BY, or HAVING clause.
- The *UNDOTBS1* tablespace is used to store database transaction recovery information. If for any reason a transaction must be rolled back (usually to preserve database integrity), the UNDOTBS1 tablespace is used to store the undo information.

FIGURE 15.12	The Oracle Storage Manager

Using the Storage Manager, the DBA can:

- Create additional tablespaces to organize the data in the database. Therefore, if you have a database with several hundred users, you can create several user tablespaces to segment the data storage for different types of users. For example, you might create a teacher tablespace and a student tablespace.

- Create additional tablespaces to organize the various subsystems that exist within the database. For example, you might create different tablespaces for human resources data, payroll data, accounting data, and manufacturing data. Figure 15.13 shows the page used to create a new tablespace called ROBCOR to hold the tables used in this book. This tablespace will be stored in the datafile named D:\ORACLE\PRODUCT\ 10.1.0\ORADATA\ORALAB\ROBCOR.DBF and its initial size is 5 megabytes. Note in Figure 15.13 that the tablespace will be put online immediately so it is available to users for data storage purposes. Note also the "Show SQL" button at the top of the page. You can use this button to see the SQL code generated by Oracle to create the tablespace. (Actually, all DBA tasks can also be accomplished through the direct use of SQL commands. In fact, some die-hard DBAs prefer writing their own SQL code rather than using the "easy-way-out" GUI.)

- Expand the tablespace storage capacity by creating additional datafiles. Remember that the datafiles can be stored in the same directory or on different hard disks to increase access performance. For example, you could increase storage and access performance to the USERS tablespace by creating a new datafile in a different drive.

FIGURE 15.13   Creating a new tablespace

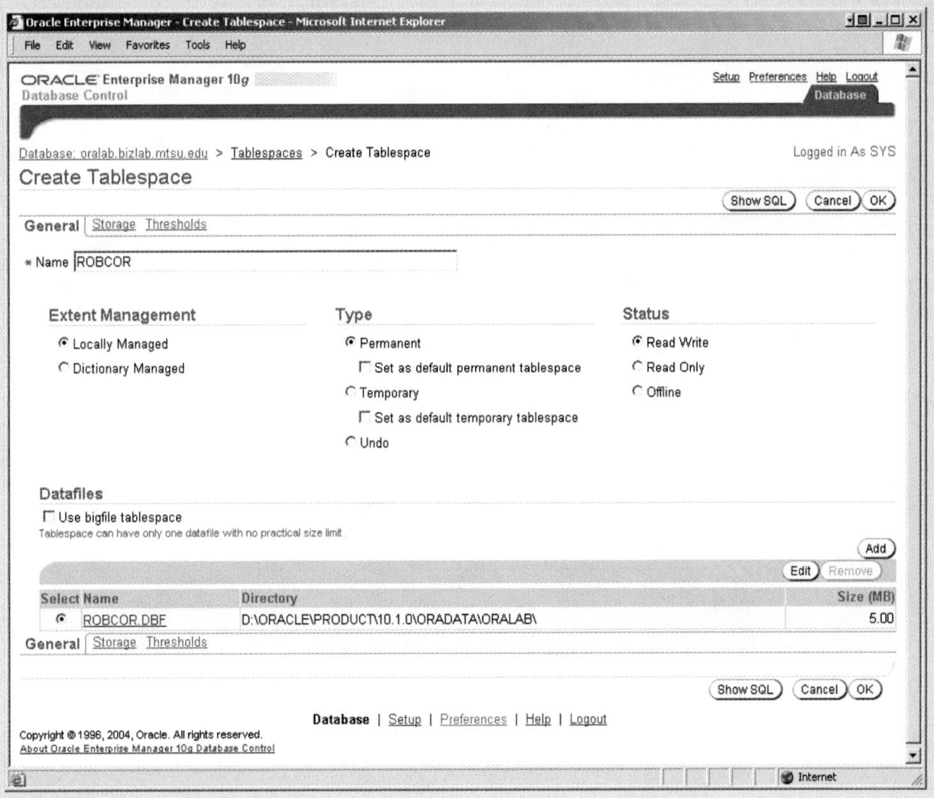

15.9.5 MANAGING THE DATABASE OBJECTS: TABLES, VIEWS, TRIGGERS, AND PROCEDURES

Another important aspect of managing a database is monitoring the database objects that were created in the database. The Oracle Enterprise Manager gives the DBA a graphical user interface to create, edit, view, and delete database objects in the database. A **database object** is basically any object created by end users; for example, tables, views, indexes, stored procedures, and triggers. Figure 15.14 shows some of the different types of objects listed in the Oracle Schema Manager.

An Oracle **schema** is a logical section of the database that belongs to a given user, and that schema is identified by the username. For example, if the user named SYSTEM creates a VENDOR table, the table will belong to the SYSTEM schema. Oracle prefixes the table name with the username. Therefore, the SYSTEM's VENDOR table name will be named SYSTEM.VENDOR by Oracle. Similarly, if the user PEROB creates a VENDOR table, that table will be created in the PEROB schema and will be named PEROB.VENDOR.

Within the schema, users can create their own tables and other objects. The database can contain as many different schemas as there are users. Because users see only their own object(s), each user might gain the impression that there are no other users of the database.

Normally, users are authorized to access only the objects that belong to their own schemas. Users could, of course, give other users access to their data by changing access rights. In fact, all users with DBA authorization have access to all objects in all schemas in the database.

As you can see in Figure 15.14, the Schema Manager presents an organized view of all of the objects in the database schema. With this program, the DBA can create, edit, view, and delete tables, indexes, views, functions, triggers, procedures, and other specialized objects.

FIGURE 15.14   The Oracle Schema Manager

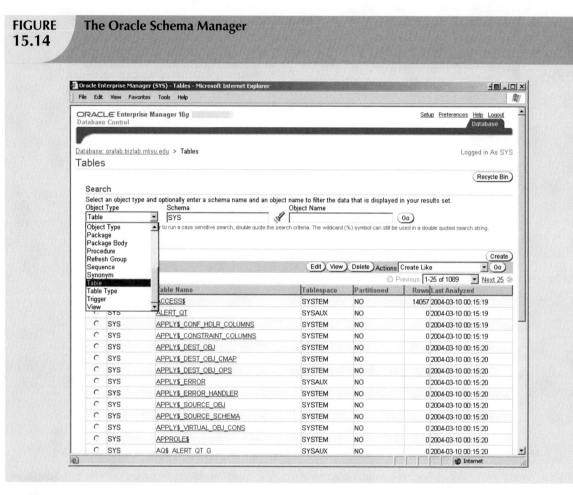

15.9.6   MANAGING USERS AND ESTABLISHING SECURITY

One of the most common database administration activities is creating and managing database users. (Actually, the creation of user IDs is just the first component of any well-planned database security function. As was indicated earlier in this chapter, database security is one of the most important database administration tasks.)

The Security section of the Oracle Enterprise Manager's Administration page enables the DBA to create users, roles, and profiles.

- A **user** is a uniquely identifiable object that allows a given person to log on to the database. The DBA assigns privileges for accessing the objects in the database. Within the privilege assignment, the DBA may specify a set of limits that define how many of the database's resources the user can use.

- A **role** is a named collection of database access privileges that authorize a user to connect to the database and use the database system resources. Examples of roles are as follows:
  - *CONNECT* allows a user to connect to the database and create and modify tables, views, and other data-related objects.

- RESOURCE allows a user to create triggers, procedures, and other data management objects.
- DBA gives the user database administration privileges.
- A **profile** is a named collection of settings that control how much of the database resource a given user can use. (If you consider the possibility that a runaway query could cause the database to lock up or to stop responding to the user's commands, you'll understand why it is important to limit access to the database resource.) By specifying profiles, the DBA can limit how much storage space a user can use, how long a user can be connected, how much idle time may be used before the user is disconnected, and so on. In an ideal world, all users would have unlimited access to all resources at all times, but in the real world, such access is neither possible nor desirable.

Figure 15.15 shows the Oracle Enterprise Manager Administration page. From here, the DBA can manage the database and create security objects (users, roles, and profiles).

FIGURE 15.15

The Oracle Enterprise Manager Administration page

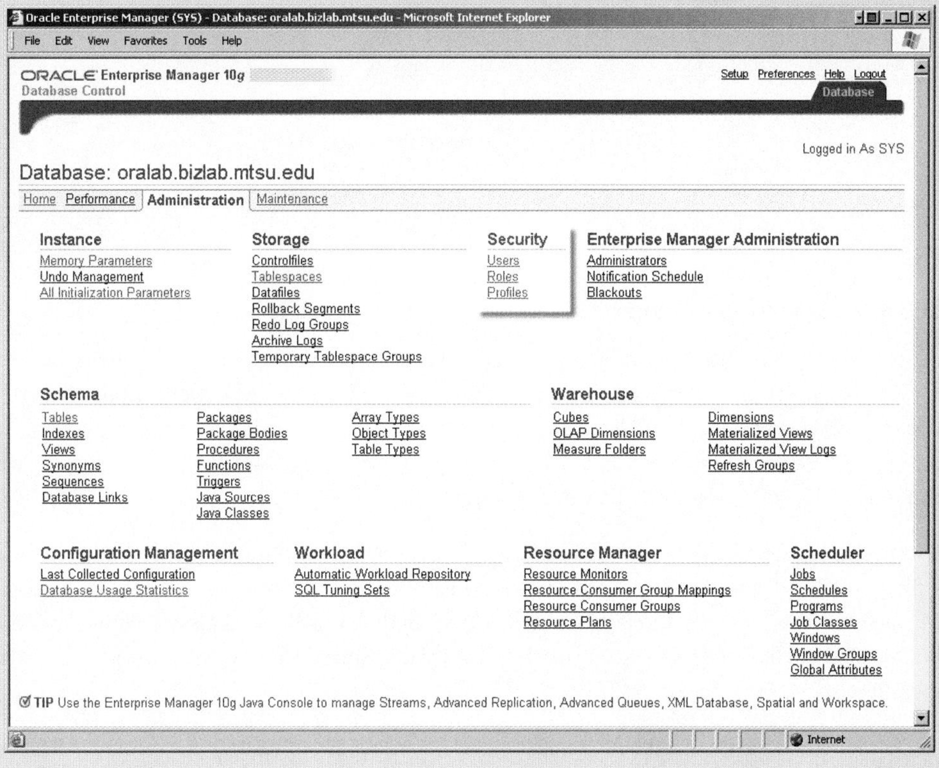

To create a new user, the DBA uses the Create User page, shown in Figure 15.16.

The Create User page contains many links; the most important ones are as follows:
- The General link allows the DBA to assign the name, profile, and password to the new user. Also in this page, the DBA defines the default tablespace used to store table data and the temporary tablespace for temporary data.
- The Roles link allows the DBA to assign the roles for a user.
- The Object Privileges link is used by the DBA to assign specific access rights to other database objects.
- The Quotas link allows the DBA to specify the maximum amount of storage that the user can have in each assigned tablespace.

> **FIGURE 15.16**   The Create User page

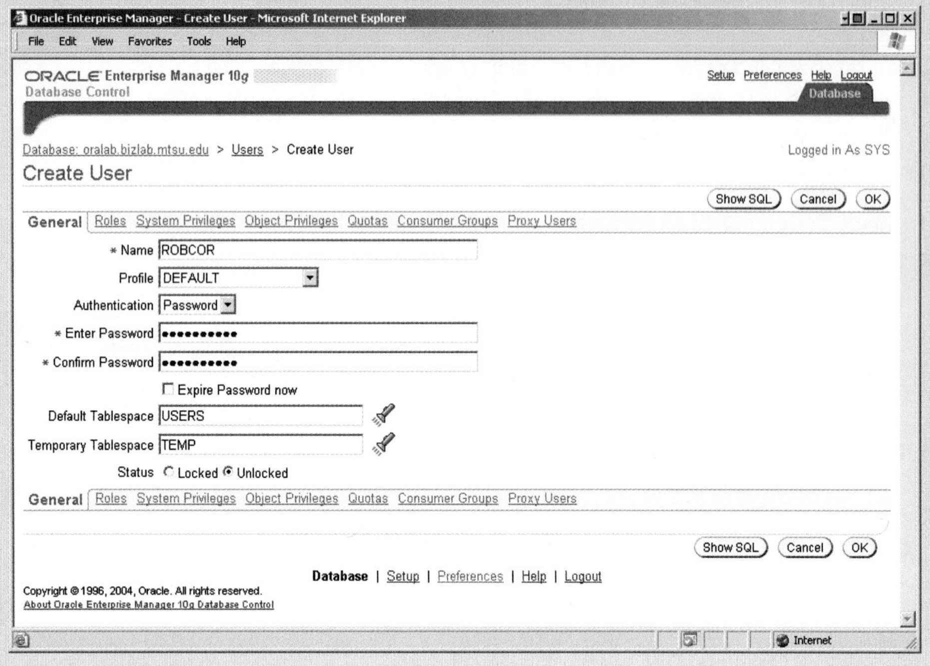

15.9.7   CUSTOMIZING THE DATABASE INITIALIZATION PARAMETERS

Fine-tuning a database is another important DBA task. This task usually requires the modification of database configuration parameters, some of which can be changed in real time, using SQL commands. Others require the database to be shut down and restarted. Also, some parameters may affect only the database instance, while others affect the entire RDBMS and all instances running. So it is very important that the DBA become familiar with database configuration parameters, especially those that affect performance.

Each database has an associated database initialization file that stores its run-time configuration parameters. The initialization file is read at instance startup and is used to set the working environment for the database. Oracle's Enterprise Manager allows the DBA to start up, shut down, and view/edit the database configuration parameters (stored in the initialization file) of a database instance. The Oracle Enterprise Manager interface provides a GUI to modify that text file, shown in Figure 15.17.

One of the important functions provided by the initialization parameters is to reserve the resources that must be used by the database at run time. One of those resources is the primary memory to be reserved for database caching. Such caching is used to fine-tune database performance. For example, the "db_cache_size" parameter sets the amount of memory reserved for database caching. This parameter should be set to a value that is large enough to support all concurrent transactions.

Once you modify the initialization parameters, you may be required to restart the database. As you have seen in this brief section, the DBA is responsible for a wide range of tasks. The quality and completeness of the administration tools available to the DBA go a long way toward making the DBA job easier. Even so, the DBA must become familiar with the tools and technical details of the RDBMS to perform the DBA tasks properly and efficiently.

FIGURE 15.17	The Oracle Enterprise Manager – Initialization Parameters page

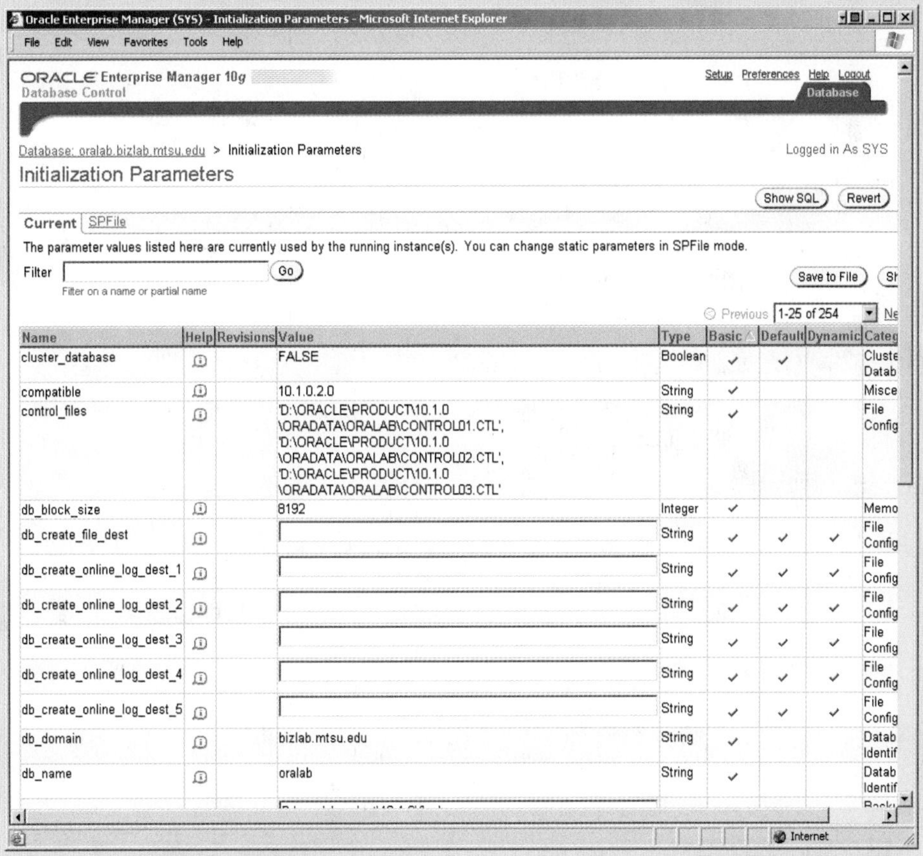

15.9.8 CREATING A NEW DATABASE

Although the general database creation format tends to be generic, its execution tends to be DBMS-specific. The leading RDBMS vendors offer the DBA the option to create databases manually, using SQL commands or using a GUI-based process. Which option is selected depends on the DBA's sense of control and style.

Using the Oracle Database Configuration Assistant, it is simple to create a database. The DBA uses a wizard interface to answer a series of questions to establish the parameters for the database to be created. Figures 15.18 through 15.30 show you how to create a database with the help of the Oracle Database Configuration Assistant.

FIGURE
15.18

Creating a new database with the Database Configuration Assistant

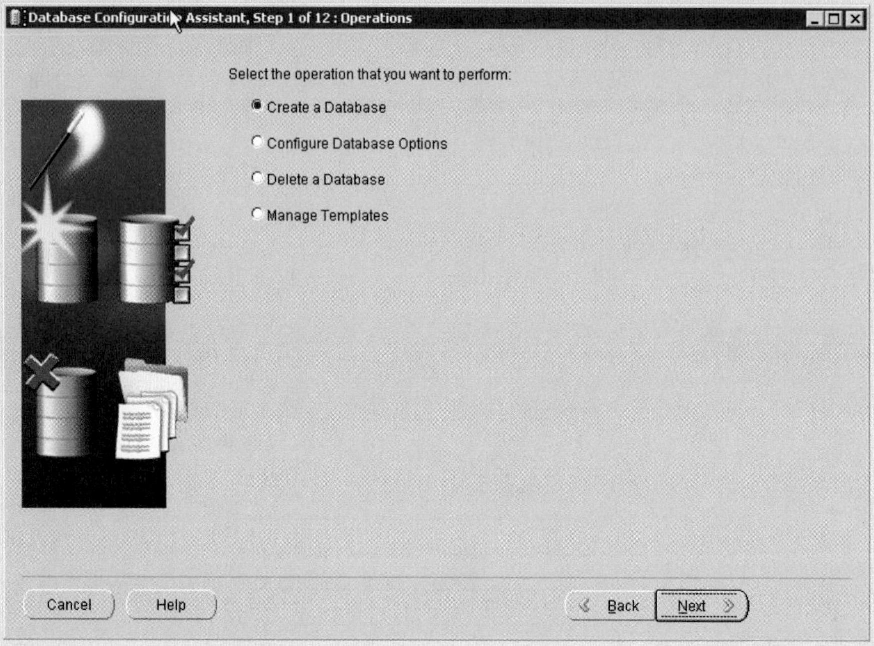

FIGURE
15.19

Selecting the new database template

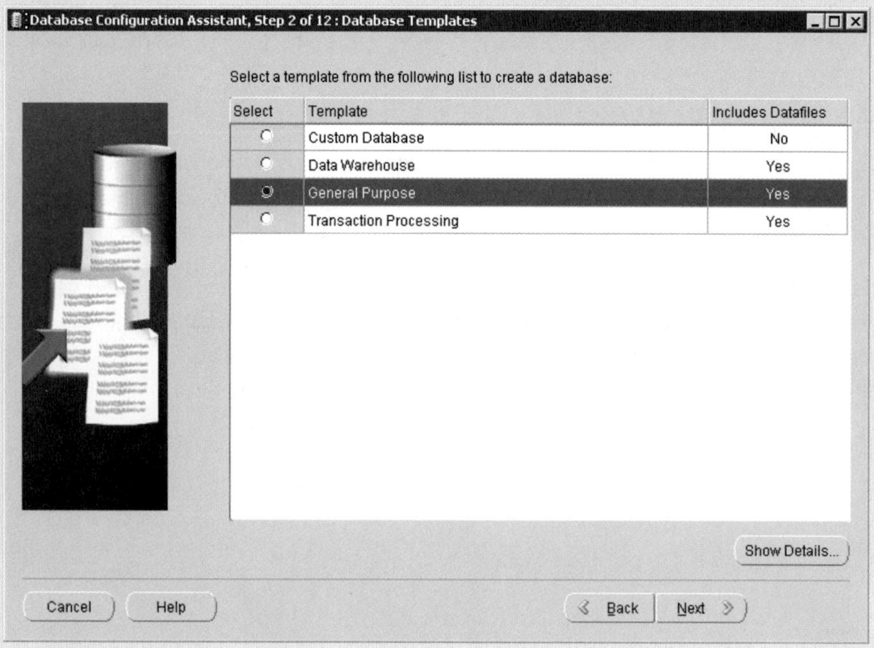

**FIGURE
15.20**   **Naming the database**

**FIGURE
15.21**   **Selecting management options**

FIGURE 15.22 Specifying database credentials

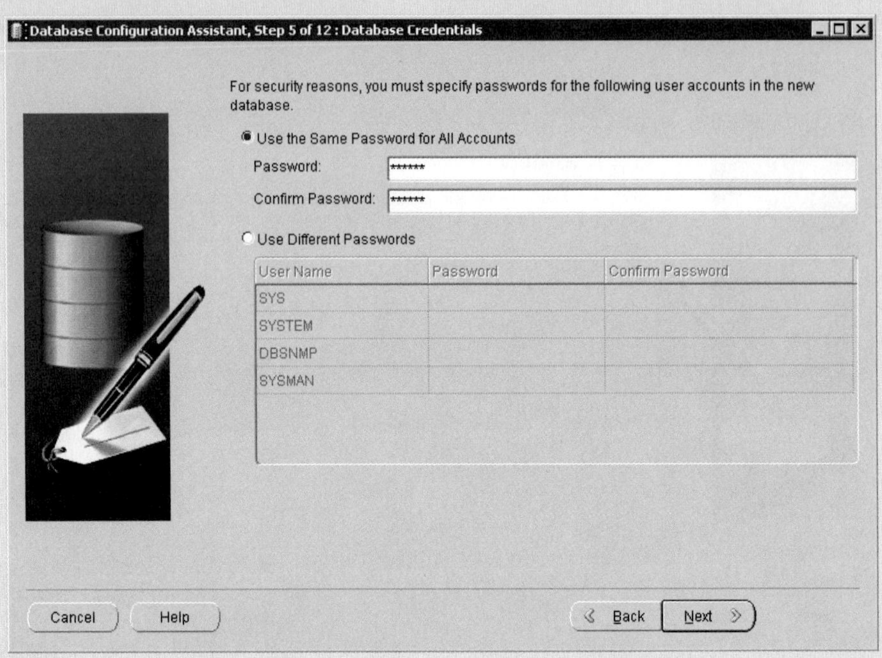

FIGURE 15.23 Selecting storage options

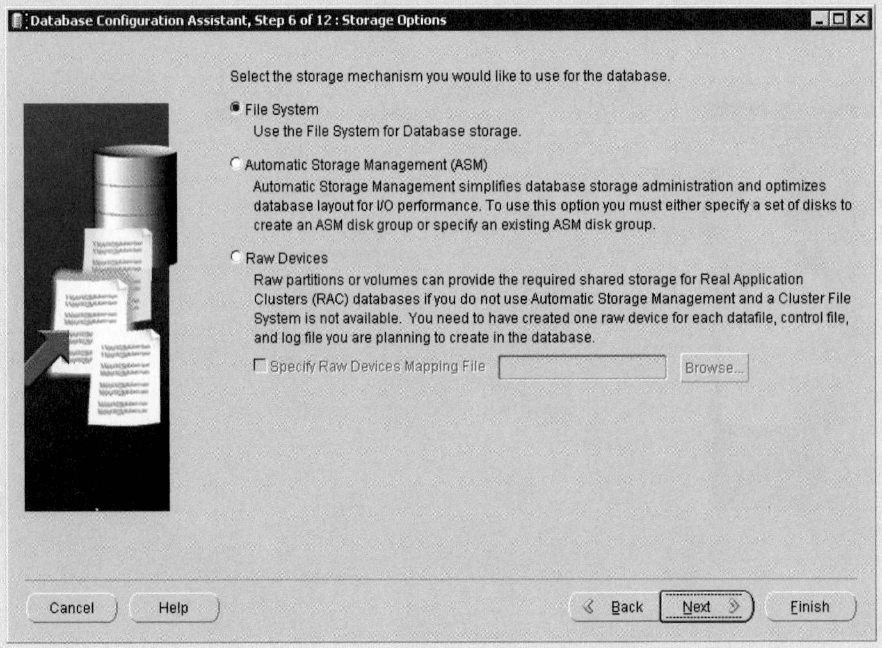

**FIGURE
15.24**   **Specifying database file locations**

Database Configuration Assistant, Step 7 of 12 : Database File Locations

Specify locations for the Database files to be created:

● Use Database File Locations from Template

○ Use Common Location for All Database Files

Database Files Location: [   ] Browse...

○ Use Oracle-Managed Files

Database Area: [   ] Browse...

Multiplex Redo Logs and Control Files...

ⓘ If you want to specify different locations for any database files, pick either of the above options and use the Storage page to specify each location.

File Location Variables...

Cancel   Help   Back   Next   Finish

**FIGURE
15.25**   **Specifying database recovery configuration**

Database Configuration Assistant, Step 8 of 12 : Recovery Configuration

Choose the recovery options for the database:

☑ Specify Flash Recovery Area

This is used as the default for all backup and recovery operations, and is also required for automatic backup using Enterprise Manager. Oracle recommends that the database files and recovery files be located on physically different disks for data protection and performance.

Flash Recovery Area: [{ORACLE_BASE}\flash_recovery_] Browse...

Flash Recovery Area Size: [2048] M Bytes

☐ Enable Archiving   Edit Archive Mode Parameters...

File Location Variables...

Cancel   Help   Back   Next   Finish

FIGURE 15.26   Selecting database sample content

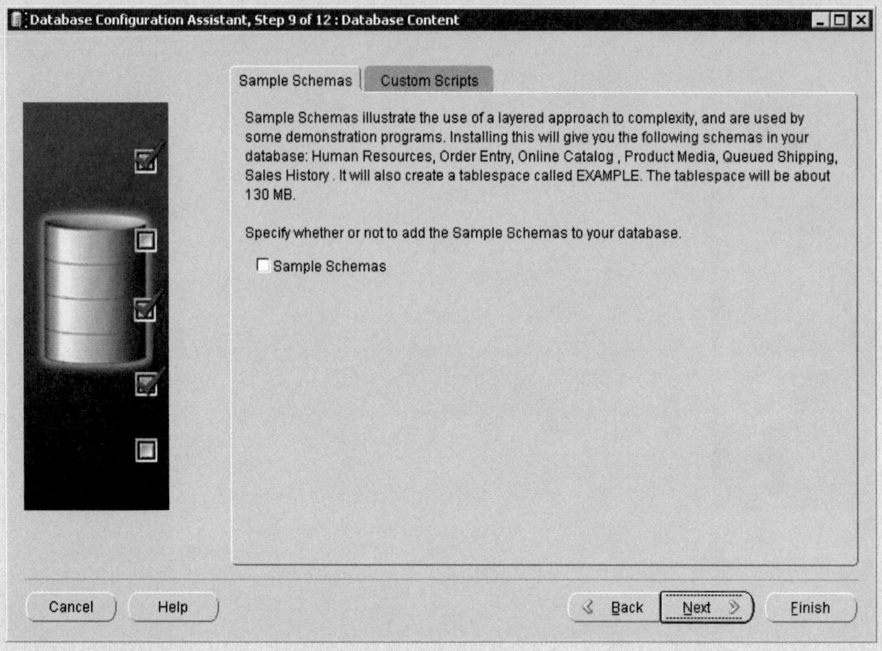

FIGURE 15.27   Selecting database initialization parameters

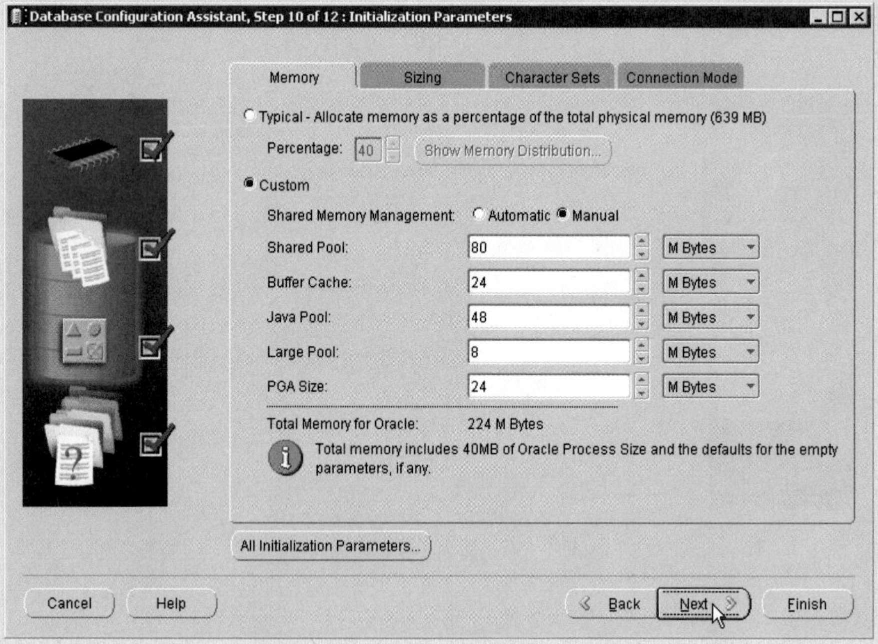

FIGURE
15.28

Confirming database storage parameters

Database Configuration Assistant, Step 11 of 12 : Database Storage

- Storage
  - Controlfile
  - Datafiles
  - Redo Log Groups

Database Storage

From the **Database Storage** page, you can specify storage parameters for the database creation. This page displays a tree listing and summary view (multi-column lists) to allow you to change and view the following objects:

- Control files
- Tablespaces
- Datafiles
- Rollback Segments
- Redo Log Groups

From any object type folder, click **Create** to create a new object. To delete an object, select the specific object from within the object type folder and click **Delete**.

Important: If you select a database template including data files, you will not be able to add or remove data files, tablespaces, or rollback segments. Selecting this type of template allows you to change the following:

- Destination of the datafiles
- Control files or log groups.

Create   Delete

File Location Variables...

Cancel   Help   Back   Next   Finish

FIGURE
15.29

Confirming database creation options

Database Configuration Assistant, Step 12 of 12 : Creation Options

Select the database creation options:

☑ Create Database

☐ Save as a Database Template

Name: ROBCOR

Description:

Cancel   Help   Back   Next   Finish

FIGURE 15.30 Database creation progress

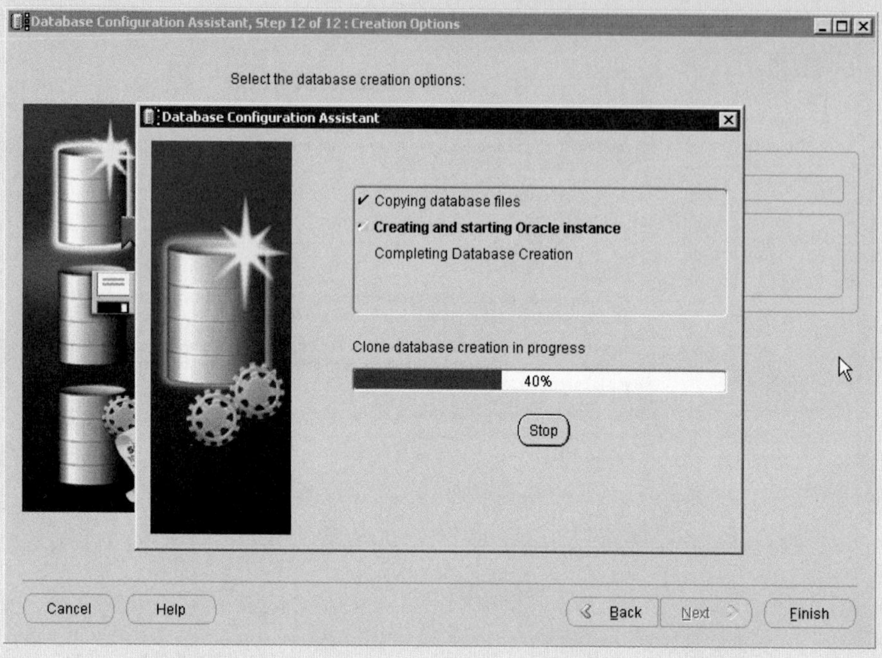

Finally, after confirming all of the database options selected in Figures 15.18 through 15.29, the database creation process starts. This process creates the database structure, including the necessary data dictionary tables, the administrator user accounts, and other supporting processes required by the DBMS to manage the database. Figure 15.30 shows the rate at which the database creation process proceeds.

One of the disadvantages of using this graphical interface to create databases is that there are no records in the form of SQL scripts to document the steps. Even given the GUI's help, the database creation process requires a solid understanding of the database's underlying structures and components.

SUMMARY

◢ Data management is a critical activity for any organization. Data must be treated as a corporate asset. The value of a data set is measured by the utility of the information derived from it. Good data management is likely to produce good information, which is the basis for better decision making.

◢ The DBMS is the most commonly used electronic tool for corporate data management. The DBMS supports strategic, tactical, and operational decision making at all levels of the organization. The company data that are managed by the DBMS are stored in the corporate or enterprise database.

◢ The introduction of a DBMS into an organization is a very delicate job. In addition to managing the technical details of DBMS introduction, the impact of the DBMS on the organization's managerial and cultural framework must be carefully examined.

◢ Development of the data administration function is based on the evolution from departmental data processing to the more centralized electronic data processing (EDP) department to the more formal "data as a corporate asset" information systems (IS) department. Typical file systems were characterized by applications that tended to behave as distinct "islands of information." As applications began to share a common data repository, the need for centralized data management to control such data became clear.

◢ The database administrator (DBA) is responsible for managing the corporate database. The internal organization of the database administration function varies from company to company. Although no standard exists, it is common practice to divide DBA operations according to the database life cycle phases. Some companies have created a position with a broader data management mandate to manage computerized and other data within the organization. This broader data management activity is handled by the data administrator (DA).

◢ The DA and the DBA functions tend to overlap. Generally speaking, the DA is more managerially oriented than the more technically-oriented DBA. Compared to the DBA function, the DA function is DBMS-independent, with a broader and longer-term focus. However, when the organization chart does not include a DA position, the DBA executes all of the DA's functions. Because the DBA has both technical and managerial responsibilities, the DBA must have a diverse mix of skills.

◢ The managerial services of the DBA function include at least: supporting the end-user community; defining and enforcing policies, procedures, and standards for the database function; ensuring data security, privacy, and integrity; providing data backup and recovery services; and monitoring the distribution and use of the data in the database.

◢ The technical role requires the DBA to be involved in at least these activities: evaluating, selecting, and installing the DBMS; designing and implementing databases and applications; testing and evaluating databases and applications; operating the DBMS, utilities, and applications; training and supporting users; and maintaining the DBMS, utilities, and applications.

◢ Security refers to activities and measures to ensure the confidentiality, integrity, and availability of an information system and its main asset, data. A security policy is a collection of standards, policies, and practices created to guarantee the security of a system and ensure auditing and compliance.

◢ A security vulnerability is weakness in a system component that could be exploited to allow unauthorized access or service disruption. A security threat is an imminent security violation caused by an unchecked security vulnerability. Security vulnerabilities exist in all components of an information system: people, hardware, software, network, procedures, and data. Therefore, it is critical to have robust database security. Database security refers to the use of DBMS features and related measures to comply with the security requirements of the organization.

▼ The development of the data administration strategy is closely related to the company's mission and objectives. Therefore, the development of an organization's strategic plan corresponds to that of data administration, requiring a detailed analysis of company goals, situation, and business needs. To guide the development of this overall plan, an integrating methodology is required. The most commonly used integrating methodology is known as information engineering (IE).

▼ To help translate strategic plans into operational plans, the DBA has access to an arsenal of database administration tools. These tools include the data dictionary and computer-aided software engineering (CASE) tools.

KEY TERMS

access plan, 622

active data dictionary, 630

audit log, 629

authorization management, 629

back-end CASE tools, 632

CASE (computer-aided systems engineering), 632

concurrent backup, 619

data administrator (DA), 612

database administrator (DBA), 611

database instance (Oracle), 639

database object (Oracle), 641

database security, 628

database security officer (DSO), 618

datafile (Oracle), 639

disaster management, 619

enterprise database, 609

front-end CASE tools, 632

full backup (database dump), 619

incremental backup, 619

information engineering (IE), 634

information resource dictionary, 631

information resource manager (IRM), 612

information systems architecture (ISA), 634

information systems (IS) department, 610

passive data dictionary, 630

policies, 617

privacy, 608

procedures, 617

profile (Oracle), 643

role (Oracle), 642

schema (Oracle), 641

security, 627

security vulnerability, 627

security threat, 627

security breach, 627

standards, 0

systems administrator , 612

tablespace (Oracle), 639

user (Oracle), 642

ONLINE CONTENT

Answers to selected Review Questions for this chapter are contained in the Student Online Companion for this book.

REVIEW QUESTIONS

1. Explain the difference between data and information. Give some examples of raw data and information.

2. Explain the interactions among end user, data, information, and decision making. Draw a diagram and explain the interactions.

3. Suppose you are a DBA staff member. What data dimensions would you describe to top-level managers to obtain their support for the data administration function?

4. How and why did database management systems become the organizational data management standard? Discuss some advantages of the database approach over the file-system approach.

5. Using a single sentence, explain the role of databases in organizations. Then explain your answer.

6. Define *security* and *privacy*. How are those two concepts related?

7. Describe and contrast the information needs at the strategic, tactical, and operational levels in an organization. Use examples to explain your answer.

8. What special considerations must you take into account when contemplating the introduction of a DBMS into an organization?

9. Describe the DBA's responsibilities.

10. How can the DBA function be placed within the organization chart? What effect(s) will that placement have on the DBA function?

11. Why and how are new technological advances in computers and databases changing the DBA's role?

12. Explain the DBA department's internal organization, based on the DBLC approach.

13. Explain and contrast the differences and similarities between the DBA and DA.

14. Explain how the DBA plays an arbitration role between an organization's two main assets. Draw a diagram to facilitate your explanation.

15. Describe and characterize the skills desired for a DBA.

16. What are the DBA's managerial roles? Describe the managerial activities and services provided by the DBA.

17. What DBA activities are used to support the end-user community?

18. Explain the DBA's managerial role in the definition and enforcement of policies, procedures, and standards.

19. Protecting data security, privacy, and integrity are important database functions. What activities are required in the DBA's managerial role of enforcing those functions?

20. Discuss the importance and characteristics of database backup and recovery procedures. Then describe the actions that must be detailed in backup and recovery plans.

21. Assume that your company assigned you the responsibility of selecting the corporate DBMS. Develop a checklist for the technical and other aspects involved in the selection process.

22. Describe the activities that are typically associated with the design and implementation services of the DBA technical function. What technical skills are desirable in the DBA's personnel?

23. Why are testing and evaluation of the database and applications not done by the same people who are responsible for design and implementation? What minimum standards must be met during the testing and evaluation process?

24. Identify some bottlenecks in DBMS performance. Then propose some solutions used in DBMS performance tuning.

25. What are typical activities involved in the maintenance of the DBMS, utilities, and applications? Would you consider application performance tuning to be part of the maintenance activities? Explain your answer.

26. How do you normally define security? How is your definition of security similar to or different from the definition of database security in this chapter?

27. What are the levels of data confidentiality?

28. What are security vulnerabilities? What is a security threat? Give some examples of security vulnerabilities that exist in different IS components.

29. Define the concept of a data dictionary. Discuss the different types of data dictionaries. If you were to manage an organization's entire data set, what characteristics would you look for in the data dictionary?

30. Using SQL statements, give some examples of how you would use the data dictionary to monitor the security of the database.

> **NOTE**
>
> If you use IBM DB2, the names of the main tables are SYSTABLES, SYSCOLUMNS, and SYSTABAUTH.

31. What characteristics do a CASE tool and a DBMS have in common? How can those characteristics be used to enhance the data administration function?

32. Briefly explain the concepts of information engineering (IE) and information systems architecture (ISA). How do those concepts affect the data administration strategy?

33. Identify and explain some of the critical success factors in the development and implementation of a successful data administration strategy.

34. What is the tool used by Oracle to create users?

35. In Oracle, what is a tablespace?

36. In Oracle, what is a database role?

37. In Oracle, what is a datafile? How does it differ from a file systems file?

38. In Oracle, what is a database profile?

39. In Oracle, what is a database schema?

40. In Oracle, what role is required to create triggers and procedures?

A

access plan—A set of instructions, generated at application compilation time, that is created and managed by a DBMS. The access plan predetermines the way an application's query will access the database at run time.

active data dictionary—A data dictionary that is automatically updated by the database management system every time the database is accessed, thereby keeping its information current. See also *data dictionary*.

ActiveX—Microsoft's alternative to Java. A specification for writing programs that will run inside the Microsoft client browser (Internet Explorer). Oriented mainly to Windows applications, it is not portable. It adds "controls" such as drop-down windows and calendars to Web pages.

ActiveX Data Objects (ADO)—A Microsoft object framework that provides a high-level application-oriented interface to interact with OLE-DB, DAO, and RDO. ADO provides a unified interface to access data from any programming language that uses the underlying OLE-DB objects.

ad hoc query—A "spur-of-the-moment" question.

ADO.NET—The data access component of Microsoft's .NET application development framework. The Microsoft .NET framework is a component-based platform for developing distributed, heterogeneous, and interoperable applications aimed at manipulating any type of data over any network under any operating system and programming language.

alias—An alternative name given to a column or table in any SQL statement.

ALTER TABLE—The SQL command used to make changes to table structure. Followed by a keyword (ADD or MODIFY), it adds a column or changes column characteristics.

American National Standards Institute (ANSI)—The group that accepted the DBTG recommendations and augmented database standards in 1975 through its SPARC committee.

AND—The SQL logical operator used to link multiple conditional expressions in a WHERE or HAVING clause. It requires that all conditional expressions evaluate to true.

anonymous PL/SQL block—A PL/SQL block that has not been given a specific name.

application processor—See *transaction processor (TP)*.

application programming interface (API)—Software through which programmers interact with middleware. Allows the use of generic SQL code, thereby allowing client processes to be database server-independent.

associative entity—See *composite entity*.

atomic attribute—An attribute that cannot be further subdivided to produce meaningful components. For example, a person's last name attribute cannot be meaningfully subdivided into other name components; therefore, the last name attribute is atomic.

atomicity—See *atomic transaction property*.

atomic transaction property—A property of transactions that states that all parts of a transaction must be treated as a single logical unit of work in which all operations must be completed (committed) to produce a consistent database.

attribute—A characteristic of an entity or object. An attribute has a name and a data type.

attribute domain—See *domain*.

attribute hierarchy—Provides a top-down data organization that is used for two main purposes: aggregation and drill-down/roll-up data analysis.

audit log—A database management system security feature that automatically records a brief description of the database operations performed by all users.

authentication—The process through which a DBMS verifies that only registered users are able to access the database.

authorization management—Defines procedures to protect and guarantee database security and integrity. Such procedures include: user access management, view definition, DBMS access control, and DBMS usage monitoring.

automatic query optimization—A method by which a DBMS takes care of finding the most efficient access path for the execution of a query.

AVG—A SQL aggregate function that outputs the mean average for the specified column or expression.

B

back-end CASE tools—A computer-aided software tool that has been classified as "back end" because it provides support for the coding and implementation phases of the SDLC. In comparison, front-end case tools provide support for the planning, analysis, and design phases.

base table—The table on which a view is based.

batch update routine—A routine that pools transactions into a single "batch" to update a master table in a single operation.

BETWEEN—In SQL, a special comparison operator used to check whether a value is within a range of specified values.

binary lock—A lock that has only two states: *locked* (1) and *unlocked* (0). If a data item is locked by a transaction, no other transaction can use that data item. See also *lock*.

binary relationship—An ER term used to describe an association (relationship) between two entities. Example: PROFESSOR teaches COURSE.

Boolean algebra—A branch of mathematics that deals with the use of the logical operators OR, AND, and NOT.

bottom-up design—A design philosophy that begins by identifying individual design components and then aggregates those components into larger units. In database design, it is a process that begins by defining attributes and then groups those attributes into entities. Compare to *top-down design*.

boundaries—The external limits to which any proposed system is subjected. These include budget, personnel, and existing hardware and software.

Boyce-Codd normal form (BCNF)—A special form of third normal form (3NF) in which every determinant is a candidate key. A table that is in BCNF must be in 3NF. See also *determinant*.

bridge entity—See *composite entity*.

buffer—See *buffer cache*.

buffer cache—A shared, reserved memory area that stores the most recently accessed data blocks in RAM. Also called *data cache*. Used to take advantage of a computer's fast primary memory compared to the slower secondary memory, thereby minimizing the number of input/output (I/O) operations between the primary and secondary memories. Also called *data cache*.

business rule—Narrative descriptions of a policy, procedure, or principle within an organization. Examples: A pilot cannot be on duty for more than 10 hours during a 24-hour period. A professor may teach up to four classes during any one semester.

C

Call Level Interface (CLI)—A standard developed by the SQL Access Group for database access.

candidate key—See *key*.

cardinality—Assigns a specific value to connectivity. Expresses the range (minimum to maximum) of allowed entity occurrences associated with a single occurrence of the related entity.

cascading order sequence—Refers to a nested ordering sequence for a set of rows. For example, a list in which all last names are alphabetically ordered and, within the last names, all first names are ordered represents a cascading sequence.

CASE—See *computer-assisted software engineering (CASE)*.

centralized database—A database located at a single site.

centralized design—A process in which a single conceptual design is modeled to match an organization's database requirements. Typically used when a data component consists of a relatively small number of objects and procedures. Compare to *decentralized design*.

checkpoint—In transaction management, an operation in which the database management system writes all of its updated buffers to disk.

Chen notation—See *entity relationship (ER) model*.

class—A collection of like objects with shared structure (attributes) and behavior (methods). A class encapsulates an object's data representation and a method's implementation. Classes are organized in a class hierarchy.

class diagram—Used to represent data and their relationships in UML object modeling system notation.

class hierarchy—The organization of classes in a hierarchical tree where each "parent" class is a *superclass* and each "child" class is a *subclass*. See also *inheritance*.

client/server architecture—Refers to the arrangement of hardware and software components to form a system composed of clients, servers, and middleware. The client/server architecture features a user of resources, or a client, and a provider of resources, or a server.

client-side extensions—These extensions add functionality to a Web browser. Although available in various forms, the most commonly encountered extensions are plug-ins, Java, JavaScript, ActiveX, and VBScript.

closure—A property of relational operators that permits the use of relational algebra operators on existing tables (relations) to produce new relations.

cluster organized table—See *index organized table*.

cohesivity—The strength of the relationships between a module's components. Module cohesivity must be high.

COMMIT—The SQL command that permanently saves data changes to a database.

Common Gateway Interface (CGI)—A Web server interface standard that uses script files to perform specific functions based on a client's parameters.

completeness constraint—A constraint that specifies whether each entity supertype occurrence must also be a member of at least one subtype. The completeness constraint can be partial or total. Partial completeness means that not every supertype occurrence is a member of a subtype; that is, there may be some supertype occurrences that are not members of any subtype. Total completeness means that every supertype occurrence must be a member of at least one subtype.

composite attribute—An attribute that can be further subdivided to yield additional attributes. For example, a phone number (615-898-2368) may be divided into an area code (615), an exchange number (898), and a four-digit code (2368). Compare to *simple attribute*.

composite entity—An entity designed to transform an M:N relationship into two 1:M relationships. The composite entity's primary key comprises at least the primary keys of the entities that it connects. Also known as a *bridge entity*. See also *linking table*.

composite identifier—In ER modeling, a key composed of more than one attribute.

composite key—A multiple-attribute key.

computer-assisted software engineering (CASE)—Tools used to automate part or all of the Systems Development Life Cycle.

conceptual design—A process that uses data modeling techniques to create a model of a database structure that represents the real-world objects in the most realistic way possible. Both software- and hardware-independent.

conceptual model—The output of the conceptual design process. The conceptual model provides a global view of an entire database. Describes the main data objects, avoiding details.

conceptual schema—A representation of the conceptual model, usually expressed graphically. See also *conceptual model*.

concurrency control—A DBMS feature that is used to coordinate the simultaneous execution of transactions in a multiprocessing database system while preserving data integrity.

concurrent backup—A backup that takes place while one or more users are working on a database.

Conference on Data Systems Languages (CODASYL)—A group originally formed to help standardize COBOL; its DBTG subgroup helped to develop database standards in the early 1970s.

connectivity—Describes the classification of the relationship between entities. Classifications include 1:1, 1:M, and M:N.

consistency—A database condition in which all data integrity constraints are satisfied. To ensure consistency of a database, every transaction must begin with the database in a known consistent state. If the database is not in a consistent state, the transaction will yield an inconsistent database that violates its integrity and business rules.

consistent database state—A database state in which all data integrity constraints are satisfied.

constraint—A restriction placed on data. Constraints are normally expressed in the form of rules. Example: "A student's GPA must be between 0.00 and 4.00." Constraints are important because they help to ensure data integrity.

coordinator—The transaction processor (TP) node that coordinates the execution of a two-phase COMMIT in a DDBMS. See also *data processor (DP)*, *transaction processor (TP)*, and *two-phase commit protocol*.

correlated subquery—A subquery that executes once for each row in the outer query.

cost-based optimizer—A query optimizer technique that uses an algorithm based on statistics about the objects being accessed, that is, number of rows, indexes available, indexes sparsity, and so on.

COUNT—A SQL aggregate function that outputs the number of rows containing not null values for a given column or expression, sometimes used in conjunction with the DISTINCT clause.

CREATE INDEX—A SQL command that creates indexes on the basis of any selected attribute or attributes.

CREATE TABLE—A SQL command used to create a table's structures, using the characteristics and attributes given.

CREATE VIEW—A SQL command that creates a logical, "virtual" table based on stored end-user tables. The view can be treated as a real table.

cross join—A join that performs a relational product (also known as the Cartesian product) of two tables.

Crow's Foot notation—A representation of the entity relationship diagram using a three-pronged symbol to represent the "many" sides of the relationship.

cube cache—In multidimensional OLAP, refers to the shared, reserved memory area where data cubes are held. Using the cube cache assists in speeding up data access.

cursor—A special construct used in procedural SQL to hold the data rows returned by a SQL query. A cursor may be thought of as a reserved area of memory in which the output of the query is stored, like an array holding columns and rows. Cursors are held in a reserved memory area in the DBMS server, not in the client computer.

D

dashboard—In business intelligence, refers to a Web-based system that presents key business performance indicators or information in a single, integrated view. Generally uses graphics in a clear, concise, and easily understood manner.

data—Raw facts, that is, facts that have not yet been processed to reveal their meaning to the end user.

Data Access Objects (DAO)—An object-oriented API (application programming interface) used to access MS Access, MS FoxPro, and dBase databases (using the Jet data engine) from Visual Basic programs. DAO provides an optimized interface that exposes the functionality of the Jet data engine (on which MS Access database is based) to programmers. The DAO interface can also be used to access other relational style data sources.

data administrator (DA)—The person responsible for managing the entire data resource, whether computerized or not. The DA has broader authority and responsibility than the database administrator (DBA). Also known as an *information resource manager (IRM)*.

data allocation—In a distributed DBMS, describes the process of deciding where to locate data fragments.

data anomaly—A data abnormality that exists when inconsistent changes to a database have been made. For example, an employee moves, but the address change is corrected in only one file and not across all files in the database.

data cache—A shared, reserved memory area that stores the most recently accessed data blocks in RAM. Also called *buffer cache*.

data cube—Refers to the multidimensional data structure used to store and manipulate data in a multidimensional DBMS. The location of each data value in the data cube is based on the x-, y-, and z-axes of the cube. Data cubes are static (must be created before they are used), so they cannot be created by an ad hoc query.

data definition language (DDL)—The language that allows a database administrator to define the database structure, schema, and subschema.

data dependence—A data condition in which the data representation and manipulation are dependent on the physical data storage characteristics.

data dictionary—A DBMS component that stores metadata—data about data. Thus, the data dictionary contains the data definition as well as its characteristics and relationships. A data dictionary may also include data that are external to the DBMS. Also known as an *information resource dictionary*. See also *active data dictionary*, *metadata*, and *passive data dictionary*.

Data Encryption Standard (DES)—The most widely used standard for private-key encryption. DES is used by the U.S. government.

data extraction—A process used to extract and validate data taken from an operational database and external data sources prior to their placement in a data warehouse.

database—A shared, integrated computer structure that houses a collection of related data. A database contains two types of data: end-user data (raw facts) and metadata. The metadata consist of data about data, that is, the data characteristics and relationships.

database administrator (DBA)—The person responsible for planning, organizing, controlling, and monitoring the centralized and shared corporate database. The DBA is the general manager of the database administration department.

database design—The process that yields the description of the database structure. The database design process determines the database components. Database design is the second phase of the database life cycle.

database development—A term used to describe the process of database design and implementation.

database fragments—Subsets of a distributed database. Although the fragments may be stored at different sites within a computer network, the set of all fragments is treated as a single database. See also *horizontal fragmentation* and *vertical fragmentation*.

database instance—In an Oracle DBMS, refers to the collection of processes and data structures used to manage a specific database.

database-level lock—A type of lock that restricts database access to only the owner of the lock. It allows only one user at a time to access the database. Successful for batch processes, but unsuitable for online multiuser DBMSs.

database life cycle (DBLC)—Traces the history of a database within an information system. Divided into six phases: initial study, design, implementation and loading, testing and evaluation, operation and maintenance, and evolution.

database management system (DBMS)—Refers to the collection of programs that manages the database structure and controls access to the data stored in the database.

database middleware—Database connectivity software through which application programs connect and communicate with data repositories.

database object—Any object in a database, such as a table, a view, an index, a stored procedure, or a trigger.

database performance tuning—A set of activities and procedures designed to reduce the response time of a database system, that is, to ensure that an end-user query is processed by the DBMS in the minimum amount of time.

database recovery—The process of restoring a database to a previous consistent state.

database request—The equivalent of a single SQL statement in an application program or a transaction.

database security—The use of DBMS features and other related measures to comply with the security requirements of an organization.

database security officer (DSO)—Person responsible for the security, integrity, backup, and recovery of the database.

database statistics—In query optimization, refers to measurements about database objects, such as the number of rows in a table, number of disk blocks used, maximum and average row length, number of columns in each row, number of distinct values in each column, etc. Such statistics give a snapshot of database characteristics.

database system—An organization of components that defines and regulates the collection, storage, management, and use of data in a database environment.

database task group (DBTG)—A CODASYL committee that helped develop database standards in the early 1970s. See also *Conference on Data Systems Languages (CODASYL)*.

datafile—See *data files*.

data files—A named physical storage space that stores a database's data. It can reside in a different directory on a hard disk or on one or more different hard disks. All data in a database are stored in data files. A typical enterprise database is normally composed of several data files. A data file can contain rows from one table, or it can contain rows from many different tables.

data filtering—See *data extraction*.

data fragmentation—A characteristic of a DDBMS that allows a single object to be broken into two or more segments or fragments. The object might be a user's database, a system database, or a table. Each fragment can be stored at any site over a computer network.

data inconsistency—A condition in which different versions of the same data yield different (inconsistent) results.

data independence—A condition that exists when data access is unaffected by changes in the physical data storage characteristics.

data integrity—In a relational database, refers to a condition in which the data in the database is in compliance with all entity and referential integrity constraints.

data management—A process that focuses on data collection, storage, and retrieval. Common data management functions include addition, deletion, modification, and listing.

data manager (DM)—See *data processing (DP) manager*.

data manipulation language (DML)—The language (set of commands) that allows an end user to manipulate the data in the database (SELECT, INSERT, UPDATE, DELETE, COMMIT, and ROLLBACK).

data mart—A small, single-subject data warehouse subset that provides decision support to a small group of people.

data mining—A process that employs automated tools to analyze data in a data warehouse and other sources and to proactively identify possible relationships and anomalies.

data model—A representation, usually graphic, of a complex "real-world" data structure. Data models are used in the database design phase of the database life cycle.

data processing (DP) manager—A DP specialist who evolved into a department supervisor. Roles include managing the technical and human resources, supervising the senior programmers, and troubleshooting the program. Also known as a *data manager (DM)*.

data processor (DP)—The software component residing on a computer that stores and retrieves data through a DDBMS. The DP is responsible for managing the local data in the computer and coordinating access to that data. See also *transaction processor (TP)*.

data redundancy—A condition that exists when a data environment contains redundant (unnecessarily duplicated) data.

data replication—The storage of duplicated database fragments at multiple sites on a DDBMS. Duplication of the fragments is transparent to the end user. Used to provide fault tolerance and performance enhancements.

DataSet—In ADO.NET, refers to a disconnected memory-resident representation of the database. That is, the DataSet contains tables, columns, rows, relationships, and constraints.

data source name (DSN)—Identifies and defines an ODBC data source.

data sparsity—A column distribution of values or the number of different values a column could have.

data store—The component of the decision support system that acts as a database for storage of business data and business model data. The data in the data store has already been extracted and filtered from the external and operational data and will be stored for access by the end-user query tool for the business data model.

data warehouse—Bill Inmon, the acknowledged "father of the data warehouse," defines the term as "an integrated, subject-oriented, time-variant, nonvolatile collection of data that provides support for decision making."

DBMS performance tuning—Refers to the activities required to ensure that clients' requests are responded to in the fastest way possible, while making optimum use of existing resources.

deadlock—A condition that exists when two or more transactions wait indefinitely for each other to release the lock on a previously locked data item. Also called *deadly embrace*. See also *lock*.

deadly embrace—See *deadlock*.

decentralized design—A process in which conceptual design is used to model subsets of an organization's database requirements. After verification of the views, processes, and constraints, the subsets are then aggregated into a complete design. Such modular designs are typical of complex systems in which the data component consists of a relatively large number of objects and procedures. Compare to *centralized design*.

decision support system (DSS)—An arrangement of computerized tools used to assist managerial decision making within a business.

deferred update—In transaction management, refers to a condition when transaction operations do not immediately update a physical database. Also called *deferred write technique*.

deferred write technique—See *deferred update*.

DELETE—A SQL command that allows specific data rows to be deleted from a table.

denormalization—A process by which a table is changed from a higher level normal form to a lower level normal form. Usually done to increase processing speed. Potentially yields data anomalies.

dependency diagram—A representation of all data dependencies (primary key, partial, or transitive) within a table.

derived attribute—An attribute that does not physically exist within the entity and is derived via an algorithm. Example: Age = current date − birth date.

description of operations—A document that provides a precise, detailed, up-to-date, and thoroughly reviewed description of the activities that define an organization's operating environment.

design trap—Occurs when a relationship is improperly or incompletely identified and, therefore, is represented in a way that is not consistent with the real world. The most common design trap is known as a *fan trap*.

desktop database—A single-user database that runs on a personal computer.

determinant—Any attribute in a specific row whose value directly determines other values in that row. See also *Boyce-Codd normal form (BCNF)*.

determination—The role of a key. In the context of a database table, the statement "A determines B" indicates that knowing the value of attribute A means that (determine) the value of attribute B can be looked up (determined).

differential backup—A level of database backup in which only the last modifications to the database (when compared with a previous full backup copy) are copied.

dimensions—In a star schema design, refers to qualifying characteristics that provide additional perspectives to a given fact.

dimension tables—In a data warehouse, used to search, filter, or classify facts within a star schema. The fact table is in a one-to-many relationship with dimension tables.

disaster management—The set of DBA activities dedicated to securing data availability following a physical disaster or a database integrity failure.

disjoint subtype (nonoverlapping subtype)—In a specialization hierarchy, refers to a unique and nonoverlapping subtype entity set.

diskpage—In permanent storage, the equivalent of a disk block, which can be describe as a directly addressable section of a disk. A diskpage has a fixed size, such as 4K, 8K, or 16K.

DISTINCT—A SQL clause designed to produce a list of only those values that are different from one another.

distributed database—A logically related database that is stored over two or more physically independent sites.

distributed database management system (DDBMS)—A DBMS that supports a database distributed across several different sites; governs the storage and processing of logically related data over interconnected computer systems in which both data and processing functions are distributed among several sites.

distributed data catalog (DDC)—A data dictionary that contains the description (fragment names, locations) of a distributed database. Also known as a *distributed data dictionary (DDD)*.

distributed data dictionary (DDD)—See *distributed data catalog*.

distributed global schema—The database schema description of a distributed database as seen by the database administrator.

distributed processing—The activity of sharing (dividing) the logical processing of a database over two or more sites connected by a network.

distributed request—A database request that allows a single SQL statement to access data in several remote DPs in a distributed database.

distributed transaction—A database transaction that accesses data in several remote DPs in a distributed database.

distribution transparency—A DDBMS feature that allows a distributed database to appear to the end-user as though it were a single logical database.

document type definition (DTD)—A file with a .dtd filename extension that describes XML elements; in effect, a DTD file provides the description of a document's composition and defines the syntax rules or valid tags for each type of XML document.

domain—In data modeling, refers to the construct used to organize and describe an attribute's set of possible values.

DO-UNDO-REDO protocol—Used by a DP to roll back and/or roll forward transactions with the help of a system's transaction log entries.

drill down—To decompose data into more atomic components, that is, data at lower levels of aggregation. Used primarily in a decision support system to focus on specific geographic areas, business types, and so on. See also *roll up*.

DROP—A SQL command used to delete database objects such as tables, views, indexes, and users.

durability—The transaction property indicating the permanence of a database's consistent state. Transactions that have been completed will not be lost in the event of a system failure if the database has proper durability.

dynamic-link libraries (DLLs)—Shared code libraries that are treated as part of the operating system or server process so they can be dynamically invoked at run time.

dynamic query optimization—Refers to the process of determining the SQL access strategy at run time, using the most up-to-date information about the database. Contrast with *static query optimization*.

dynamic SQL—A term used to describe an environment in which the SQL statement is not known in advance, but instead is generated at run time. In a dynamic SQL environment, a program can generate the SQL statements at run time that are required to respond to ad hoc queries.

dynamic statistical generation mode—In a DBMS, the capability to automatically evaluate and update the database access statistics after each data access.

dynamic-link libraries (DLLs)—Shared code modules that are treated as part of the operating system or server process so they can be dynamically invoked at run time.

E

EER diagram (EERD)—Refers to the entity-relationship diagram resulting from the application of extended entity relationship concepts that provide additional semantic content in the ER model.

embedded SQL—A term used to refer to SQL statements that are contained within an application programming language such as COBOL, C++, ASP, Java, and ColdFusion.

end-user presentation tool—Used by a data analyst to organize and present selected data compiled by the end-user query tool.

end-user query tool—Used by a data analyst to create the queries that access the specific desired information from the data store.

enterprise database—The overall company data representation, which provides support for present and expected future needs.

entity—Something about which someone wants to store data; typically a person, a place, a thing, a concept, or an event. See also *attribute*.

entity cluster—A "virtual" entity type used to represent multiple entities and relationships in the ERD. An entity cluster is formed by combining multiple interrelated entities into a single abstract entity object. An entity cluster is considered "virtual" or "abstract" in the sense that it is not actually an entity in the final ERD.

entity instance—A term used in ER modeling to refer to a specific table row. Also known as an *entity occurrence*.

entity integrity—The property of a relational table that guarantees that each entity has a unique value in a primary key and that there are no null values in the primary key.

entity occurrence—See *entity instance*.

entity relationship diagram (ERD)—A diagram that depicts an entity relationship model's entities, attributes, and relations.

entity relationship (ER) model—A data model developed by P. Chen in 1975. It describes relationships (1:1, 1:M, and M:N) among entities at the conceptual level with the help of ER diagrams.

entity set—In a relational model, refers to a grouping of related entities.

entity subtype—In a generalization/specialization hierarchy, refers to a subset of an entity supertype where the entity supertype contains the common characteristics and the entity subtypes contain the unique characteristics of each entity subtype.

entity supertype—In a generalization/specialization hierarchy, refers to a generic entity type that contains the common characteristics of entity subtypes.

equijoin—A join operator that links tables based on an equality condition that compares specified columns of the tables.

exclusive lock—A lock that is reserved by a transaction. An exclusive lock is issued when a transaction requests permission to write (update) a data item and no locks are previously held on that data item by any other transaction. An exclusive lock does not allow any other transactions to access the database. See also *shared lock*.

existence-dependent—A property of an entity whose existence depends on one or more other entities. In an existence-dependent environment, the existence-independent table must be created and loaded first because the existence-dependent key cannot reference a table that does not yet exist.

existence-independent—An entity that can exist apart from one or more related entities. It must be created first when referencing an existence-dependent table to it.

EXISTS—In SQL, a comparison operator used to check whether a subquery returns any rows.

explicit cursor—In procedural SQL, a cursor created to hold the output of a SQL statement that may return two or more rows (but could return zero rows or only one row).

extended entity relationship model (EERM)—Sometimes referred to as the enhanced entity relationship model; the result of adding more semantic constructs (entity supertypes, entity subtypes, and entity clustering) to the original entity relationship (ER) model.

extended relational data model (ERDM)—A model that includes the object-oriented model's best features in an inherently simpler relational database structural environment. See EERM.

extends—In a DBMS environment, refers to the data files' ability to automatically expand in size, using predefined increments.

Extensible Markup Language (XML)—A metalanguage used to represent and manipulate data elements. Unlike other markup languages, XML permits the manipulation of a document's data elements. XML is designed to facilitate the exchange of structured documents such as orders and invoices over the Internet.

external model—The application programmer's view of the data environment. Given its business-unit focus, an external model works with a data subset of the global database schema.

external schema—The specific representation of an external view, that is, the end user's view of the data environment.

F

facts—In a data warehouse, refers to the measurements (values) that represent a specific business aspect or activity. For example, sales figures are numeric measurements that represent product and/or service sales. Facts commonly used in business data analysis are units, costs, prices, and revenues.

fact table—In a data warehouse, refers to the star schema center table containing facts that are linked and classified through their common dimensions. A fact table is in a one-to-many relationship with each associated dimension table.

failure transparency—A DDBMS feature that allows continuous operation of a DDBMS, even in the event of a failure in one of the nodes of the network.

fan trap—A design trap that occurs when one entity is in two 1:M relationships to other entities, thus producing an association among the other entities that is not expressed in the model.

field—A character or group of characters (alphabetic or numeric) that defines a characteristic of a person, place, or thing. For example, a person's Social Security number, address, phone number, and bank balance all constitute fields.

field-level lock—Allows concurrent transactions to access the same row as long as they require the use of different fields (attributes) within that row. Yields the most flexible multiuser data access but requires a high level of computer overhead.

file—A named collection of related records.

file group—See *table space*.

first normal form (1NF)—The first stage in the normalization process. It describes a relation depicted in tabular format, with no repeating groups and with a primary key identified. All nonkey attributes in the relation are dependent on the primary key.

flags—Special codes implemented by designers to trigger a required response, to alert end users to specified conditions, or to encode values. Flags may be used to prevent nulls by bringing attention to the absence of a value in a table.

foreign key—See *key*.

fourth normal form (4NF)—A table is in 4NF when it is in 3NF and contains no multiple independent sets of multivalued dependencies.

fragmentation transparency—A DDBMS feature that allows a system to treat a distributed database as a single database even though the database is divided into two or more fragments.

front-end CASE tools—A computer-aided software tool that has been classified as "front end" because it provides support for the planning, analysis, and design phases of the SDLC. In comparison, back- end case tools provide support for the coding and implementation phases.

full backup (database dump)—A complete copy of an entire database saved and periodically updated in a separate memory location. Ensures a full recovery of all data in the event of a physical disaster or a database integrity failure.

full functional dependence—A condition in which an attribute is functionally dependent on a composite key but not on any subset of that composite key.

fully heterogeneous distributed database system (fully heterogeneous DDBMS)—Integrates different types of database management systems (hierarchical, network, and relational) over a network. It supports different database management systems that may even support different data models running under different computer systems, such as mainframes, minicomputers, and microcomputers. See also *heterogeneous DDBMS* and *homogeneous DDBMS*.

fully replicated database—In a DDBMS, refers to the distributed database that stores multiple copies of each database fragment at multiple sites. See also *partially replicated database*.

function-based index—A type of index based on a specific SQL function or expression.

functional dependence—Within a relation R, an attribute B is functionally dependent on an attribute A if and only if a given value of the attribute A determines exactly one value of the attribute B. The relationship "B is dependent on A" is equivalent to "A determines B" and is written as A→B.

G

granularity—Refers to the level of detail represented by the values stored in a table's row. Data stored at their lowest level of granularity are said to be *atomic data*.

governance—In business intelligence, the methods for controlling and monitoring business health and promoting consistent decision making.

GROUP BY—A SQL clause used to create frequency distributions when combined with any of the aggregate functions in a SELECT statement.

H

hardware independence—Means that a model does not depend on the hardware used in the implementation of the model. Therefore, changes in the hardware will have no effect on the database design at the conceptual level.

HAVING—A restriction placed on the GROUP BY clause output. The HAVING clause is applied to the output of a GROUP BY operation to restrict the selected rows.

heterogeneity transparency—A DDBMS feature that allows a system to integrate several different centralized DBMSs into one logical DDBMS.

665

heterogeneous DDBMS—Integrates different types of centralized database management systems over a network. See also *fully heterogeneous distributed database system (fully heterogeneous DDBMS)* and *homogeneous DDBMS*.

hierarchical model—No longer a major player in the current database market; important to know, however, because the basic concepts and characteristics form the basis for subsequent database development. This model is based on an "upside-down" tree structure in which each record is called a segment. The top record is the root segment. Each segment has a 1:M relationship to the segment directly below it.

homogeneous DDBMS—Integrates only one particular type of centralized database management system over a network. See also *heterogeneous DDBMS* and *fully heterogeneous distributed database system (fully heterogeneous DDBMS)*.

homonym—Indicates the use of the same name to label different attributes; generally should be avoided. Some relational software automatically checks for homo-nyms and either alerts the user to their existence or automatically makes the appropriate adjustments. See also *synonym*.

horizontal fragmentation—The distributed database design process that breaks up a table into subsets consisting of unique rows. See also *database fragments* and *vertical fragmentation*.

host language—A term used to describe any language that contains embedded SQL statements.

I

identifiers—The ERM uses identifiers to uniquely identify each entity instance. In the relational model, such identifiers are mapped to primary keys in tables.

identifying relationship—A relationship that exists when the related entities are existence-dependent. Also called a *strong relationship* or *strong identifying relationship* because the dependent entity's primary key contains the primary key of the parent entity.

immediate update—When a database is immediately updated by transaction operations during the transaction's execution, even before the transaction reaches its commit point.

implicit cursor—A cursor that is automatically created in procedural SQL when the SQL statement returns only one value.

IN—In SQL, a comparison operator used to check whether a value is among a list of specified values.

inconsistent retrievals—A concurrency control problem that arises when a transaction calculating summary (aggregate) functions over a set of data—while other transactions are updating the data—yields erroneous results.

incremental backup—A process that makes a backup only of data that has changed in the database since the last backup (incremental or full).

index—An ordered array composed of index key values and row ID values (pointers). Indexes are generally used to speed up and facilitate data retrieval. Also known as an *index key*.

index key—See *index*.

index organized table—In a DBMS, a type of table storage organization that stores the end user data and the index data in consecutive locations on permanent storage. Also known as *clustered index table*.

index selectivity—A measure of how likely an index will be used in query processing.

information—The result of processing raw data to reveal its meaning. Information consists of transformed data and facilitates decision making.

information engineering (IE)—A methodology that translates a company's strategic goals into data and applications that will help the company achieve its goals.

information resource dictionary—See *data dictionary*.

information resource manager (IRM)—See *data administrator (DA)*.

information system (IS)—A system that provides for data collection, storage, and retrieval; facilitates the transformation of data into information and the management of both data and information. An information system is composed of hardware, software (DMBS and applications), the database(s), people, and procedures.

information systems architecture (ISA)—The output of the information engineering (IE) process that serves as the basis for planning, developing, and controlling future information systems. (IE allows for the translation of a company's strategic goals into the data and applications that will help the company achieve those goals. IE focuses on the description of the corporate data instead of the processes.)

information systems (IS) department—An evolution of the data-processing department when responsibilities were broadened to include service and production functions.

inheritance—In the object-oriented data model, the ability of an object to inherit the data structure and methods of the classes above it in the class hierarchy. See also *class hierarchy*.

inner join—A join operation in which only rows that meet a given criteria are selected. The join criteria can be an equality condition (natural join or equijoin) or an inequality condition (theta join). Inner join is the most commonly used type of join. Contrast with *outer join*.

inner query—A query that is embedded (or nested) inside another query. Also known as a *nested query* or a *subquery*.

input/output (IO) request—A low-level (read or write) data access operation to/from computer devices (such as memory, hard disks, video, and printers).

INSERT—A SQL command that allows the insertion of data rows into a table, one row at a time or multiple rows at a time, using a subquery.

internal model—In database modeling, refers to a level of data abstraction that adapts the conceptual model to a specific DBMS model for implementation.

internal schema—Depicts a specific representation of an internal model, using the database constructs supported by the chosen database. (The internal model is the representation of a database as "seen" by the DBMS. In other words, the internal model requires a designer to match the conceptual model's characteristics and constraints to those of the selected implementation model.)

IS NULL—In SQL, a comparison operator used to check whether an attribute has a value.

islands of information—A term used in the old-style file system environment to refer to independent, often duplicated, and inconsistent data pools created and managed by different organizational departments.

isolation—A property of a database transaction that guarantees that a data item used by one transaction is not available to other transactions until the first transaction ends.

iterative process—A process based on repetition of steps and procedures.

J

Java—An object-oriented programming language developed by Sun Microsystems that runs on top of the Web browser software. Java applications are compiled and stored in the Web server. Java's main advantage is its ability to let application developers develop their applications once and run them in many environments.

Java Database Connectivity (JDBC)—An application programming interface that allows a Java program to interact with a wide range of data sources (relational databases, tabular data sources, spreadsheets, and text files).

JavaScript—A scripting language (one that enables the running of a series of commands or macros) developed by Netscape that allows Web authors to design interactive Web sites. JavaScript code is embedded in Web pages. This JavaScript is downloaded with the page and is activated when a specific event takes place, such as a mouse click on an object.

join columns—A term used to refer to the columns that join two tables. The join columns generally share similar values.

K

key—An entity identifier based on the concept of functional dependence; may be classified as follows: *Superkey:* An attribute (or combination of attributes) that uniquely identifies each entity in a table. *Candidate key:* A minimal superkey, that is, one that does not contain a subset of attributes that is itself a superkey. *Primary key (PK):* A candidate key selected as a unique entity identifier. *Secondary key:* A key that is used strictly for data retrieval purposes. For example, a customer is not likely to know his or her customer number (primary key), but the combination of last name, first name, middle initial, and telephone number is likely to make a match to the appropriate table row. *Foreign key:* An attribute (or combination of attributes) in one table whose values must match the primary key in another table or whose values must be null.

key attribute(s)—The attribute(s) that form(s) a primary key. See also *prime attribute*.

key performance indicators (KPI)—In business intelligence, refers to quantifiable measurements (numeric or scale-based) that assess a company's effectiveness or success in reaching strategic and operational goals. Examples of KPI are product turnovers, sales by promotion, sales by employee, earnings per share, etc.

knowledge—The body of information and facts about a specific subject. Knowledge implies familiarity, awareness, and understanding of information as it applies to an environment. A key characteristic of knowledge is that "new" knowledge can be derived from "old" knowledge.

L

left outer join—In a pair of tables to be joined, a left outer join yields all of the rows in the left table, including those that have no matching values in the other table. For example, a left outer join of Customer with Agent will yield all of the Customer rows, including the ones that do not have a matching Agent row. See also *outer join* and *right outer join*.

LIKE—In SQL, a comparison operator used to check whether a attribute's text value matches a specified string pattern.

linking table—In the relational model, a table that implements a M:M relationship. See also *composite entity*.

local mapping transparency—A property of a DDBMS in which access to the data requires the end user to know both the name and the location of the fragments in order to access the database. See also *location transparency*.

location transparency—The property of a DDBMS in which access to the data requires that only the name of the database fragments be known. (Fragment locations need not be known.) See also *local mapping transparency*.

lock—A device that is employed to guarantee unique use of a data item to a particular transaction operation, thereby preventing other transactions from using that data item. A transaction requires a lock prior to data access, and that lock is released (unlocked) after the operation's execution to enable other transactions to lock the data item for their use.

lock granularity—Indicates the level of lock use. Lock-ing can take place at the following levels: database, table, page, row, and field (attribute).

lock manager—A DBMS component that is responsible for assigning and releasing locks.

logical data format—The way in which a human being views data.

logical design—A stage in the design phase that matches the conceptual design to the requirements of the selected DBMS and is, therefore, software-dependent. It is used to translate the conceptual design into the internal model for a selected database management system, such as DB2, SQL Server, Oracle, IMS, Informix, Access, and Ingress.

logical independence—A condition that exists when the internal model can be changed without affecting the conceptual model. (The internal model is hardware-independent because it is unaffected by the choice of computer on which the software is installed. Therefore, a change in storage devices or even a change in operating systems will not affect the internal model.)

lost updates—A concurrency control problem in which data updates are lost during the concurrent execution of transactions.

M

---

mandatory participation—A term used to describe a relationship in which one entity occurrence must have a corresponding occurrence in another entity. Example: EMPLOYEE works in DIVISION. (A person cannot be an employee if he or she is not assigned to a company's division.)

manual query optimization—An operation mode that requires the end user or programmer to define the access path for the execution of a query.

manual statistical generation mode—One mode of generating statistical data access information used for query optimization. In this mode, the DBA must periodically run a routine to generate the data access statistics; for example, running the RUNSTAT command in an IBM DB2 database.

many-to-many (M:N or M:M) relationships—One of three types of relationships (associations among two or more entities) in which one occurrence of an entity is associated with many occurrences of a related entity and one occurrence of the related entity is associated with many occurrences of the first entity.

master data management (MDM)— In business intelligence, a collection of concepts, techniques, and processes for the proper identification, definition, and management of data elements within an organization.

materialized view—A dynamic table that not only contains the SQL query command to generate the rows, but also stores the actual rows. The materialized view is created the first time the query is run and the summary rows are stored in the table. The materialized view rows are automatically updated when the base tables are updated.

MAX—A SQL aggregate function that yields the maximum attribute value encountered in a given column.

metadata—Data about data, that is, data concerning data characteristics and relationships. See also *data dictionary*.

method—In the object-oriented data model, a named set of instructions to perform an action. Methods represent real-world actions. Methods are invoked through messages.

metrics—In a data warehouse, numeric facts that measure a business characteristic of interest to the end user.

Microsoft .NET framework—A component-based platform for the development of distributed, heterogeneous, interoperable applications aimed at manipulating any type of data over any network under any operating system and any programming language.

MIN—A SQL aggregate function that yields the minimum attribute value encountered in a given column.

minimal data rule—Defined as "All that is needed is there, and all that is there is needed." In other words, all data elements required by database transactions must be defined in the model, and all data elements defined in the model must be used by at least one database transaction.

mixed fragmentation—Regarding data fragmentation, refers to a combination of horizontal and vertical strategies, meaning a table may be divided into several rows, each row having a subset of the attributes (columns).

module—(1) A design segment that can be implemented as an autonomous unit, sometimes linked to produce a system. (2) An information system component that handles a specific function, such as inventory, orders, or payroll.

module coupling—A description of the extent to which modules are independent of one another.

monotonicity—Ensures that time stamp values always increase. (The time stamping approach to scheduling concurrent transactions assigns a global, unique time stamp to each transaction. The time stamp value produces an explicit order in which transactions are submitted to the DBMS.)

multidimensional database management system (MDBMS)—A database management system that uses proprietary techniques to store data in matrixlike arrays of *n*-dimensions, known as cubes.

multidimensional online analytical processing (MOLAP)—Extends online analytical processing functionality to multidimensional database management systems.

multiple-site processing, multiple-site data (MPMD)—A scenario describing a fully distributed database management system with support for multiple DPs and transaction processors at multiple sites.

multiple-site processing, single-site data (MPSD)—A scenario in which multiple processes run on different computers sharing a single data repository.

multiuser database—A database that supports multiple concurrent users.

multivalued attribute—An attribute that can have many values for a single entity occurrence. For example, an EMP_DEGREE attribute might store the string "BBA, MBA, PHD" to indicate three different degrees held.

mutual consistency rule—A data replication rule requiring that all copies of data fragments be identical.

mutual exclusive rule—A condition in which only one transaction at a time can own an exclusive lock on the same object.

N

natural join—A relational operation that links tables by selecting only the rows with common values in their common attribute(s).

natural key (natural identifier)—A real-world, generally accepted identifier used to identify real-world objects. As its name implies, a natural key is familiar to end users and forms part of their day-to-day business vocabulary.

nested query—In SQL, refers to a query that is embedded in another query. See *subquery*.

network model—A data model standard created by the CODASYL Data Base Task Group in the late 1960s. It represented data as a collection of record types and relationships as predefined sets with an owner record type and a member record type in a 1:M relationship.

non-identifying relationship—A relationship that occurs when the primary key of the dependent (many side) entity does not contain the primary key of the related parent entity. Also known as a *weak relationship*.

nonkey attribute—See *nonprime attribute*.

nonprime attribute—An attribute that is not part of a key.

normalization—A process that assigns attributes to entities in such a way that data redundancies are reduced or eliminated.

NOT—A SQL logical operator that negates a given predicate.

null—In SQL, refers to the absence of an attribute value. Note: A null is not a blank.

O

object—An abstract representation of a real-world entity that has a unique identity, embedded properties, and the ability to interact with other objects and with itself.

Object Linking and Embedding for Database (OLE-DB)—Based on Microsoft's Component Object Model (COM), OLE-DB is database middleware that adds object-oriented functionality for accessing relational and nonrelational data. OLE-DB was the first part of Microsoft's strategy to provide a unified object-oriented framework for the development of next- generation applications.

object-oriented database management system (OODBMS)—Data management software used to manage data found within an object-oriented database model.

object-oriented data model (OODM)—A data model whose basic modeling structure is an object.

object-oriented programming (OOP)—An alternative to conventional programming methods based on object-oriented concepts. It reduces programming time and lines of code and increases programmers' productivity.

object/relational database management system (O/RDBMS)—A DBMS based on the extended relational model (ERDM). The ERDM, championed by many relational database researchers, constitutes the relational model's response to the OODM. This model includes many of the object-oriented model's best features within an inherently simpler relational database structural environment.

one-to-many (1:M) relationship—One of three types of relationships (associations among two or more entities) that are used by data models. In a 1:M relationship, one entity instance is associated with many instances of the related entity.

one-to-one (1:1) relationship—One of three types of relationships (associations among two or more entities) that are used by data models. In a 1:1 relationship, one entity instance is associated with only one instance of the related entity.

online analytical processing (OLAP)—Decision support system (DSS) tools that use multidimensional data analysis techniques. OLAP creates an advanced data analysis environment that supports decision making, business modeling, and operations research activities.

Open Database Connectivity (ODBC)—Database middleware developed by Microsoft to provide a database access API to Windows applications.

operational database—A database that is designed primarily to support a company's day-to-day operations. Also known as a *transactional database* or *production database*.

optimistic approach—In transaction management, refers to a concurrency control technique based on the assumption that the majority of database operations do not conflict.

optimizer hints—Special instructions for the query optimizer that are embedded inside the SQL command text.

optional attribute—In ER modeling, refers to an attribute that does not require a value, therefore it can be left empty.

optional participation—In ER modeling, refers to a condition where one entity occurrence does not require a corresponding entity occurrence in a particular relationship.

OR—The SQL logical operator used to link multiple conditional expressions in a WHERE or HAVING clause. It requires that only one of the conditional expressions be true.

ORDER BY—A SQL clause useful for ordering the output of a SELECT query (for example, in ascending or descending order).

outer join—A relational-algebra JOIN operation that produces a table in which all unmatched pairs are retained; unmatched values in the related table are left null. Contrast with *inner join*. See also *left outer join* and *right outer join*.

overlapping—In a specialization hierarchy, describes a condition where each entity instance (row) of the supertype can appear in more than one subtype.

P

page—See *diskpage*.

page-level lock—In this type of lock, the database management system will lock an entire diskpage, or section of a disk. A diskpage can contain data for one or more rows and from one or more tables.

partial completeness—In a generalization hierarchy, means that not *every* supertype occurrence is a member of a subtype; that is, there may be some supertype occurrences that are not members of any subtype.

partial dependency—In normalization, a condition in which an attribute is dependent on only a portion (subset) of the primary key.

partially replicated database—A distributed database in which copies of only some database fragments are stored at multiple sites. See also *fully replicated database*.

participants—An ER term used to label the entities that participate in a relationship. Example: PROFESSOR teaches CLASS. (The *teaches* relationship is based on the participants PROFESSOR and CLASS.)

partitioning—The process of splitting a table into subsets of rows or columns.

passive data dictionary—A DBMS data dictionary that requires an end-user-initiated command to update its data access statistics. See also *data dictionary*.

performance transparency—A DDBMS feature that allows a system to perform as though it were a centralized DBMS (no degradation of response times).

performance tuning—Activities that make a database perform more efficiently in terms of storage and access speed.

periodicity—Usually expressed as current year only, previous years, or all years; provides information about the time span of data stored in a table.

persistent stored module (PSM)—A block of code (containing standard SQL statements and procedural extensions) that is stored and executed at the DBMS server.

personalization—Customization of a Web page for individual users.

physical data format—The way in which a computer "sees" (stores) data.

physical design—A stage of database design that maps the data storage and access characteristics of a database. Since these characteristics are a function of the types of devices supported by the hardware, the data access methods supported by the system (and the selected DBMS) physical design is both hardware- and software-dependent. See also *physical model*.

physical independence—A condition that exists when the physical model can be changed without affecting the internal model.

physical model—A model in which the physical characteristics (location, path, and format) are described for the data. Both hardware- and software-dependent. See also *physical design*.

plug-in—In the World Wide Web (WWW), a client-side, external application that is automatically invoked by the browser when it is needed to manage specific types of data.

policies—General statements of direction that are used to manage company operations through the communication and support of the organization's objectives.

predicate logic—Used extensively in mathematics, provides a framework in which an assertion (statement of fact) can be verified as either true or false. For example, suppose that a student with a student ID of 12345678 is named Melissa Sanduski. That assertion can easily be demonstrated to be true or false.

primary key (PK)—In the relational model, an identifier composed of one or more attributes that uniquely identifies a row. See also *key*.

prime attribute—A key attribute, that is, an attribute that is part of a key or is the whole key. See also *key attribute*.

privacy—Control of data usage dealing with the rights of individuals and organizations to determine the "who, what, when, where, and how" of data access.

procedural SQL (PL/SQL)—A type of SQL that allows the use of procedural code and SQL statements that are stored in a database as a single callable object that can be invoked by name.

procedure cache—A shared, reserved memory area that stores the most recently executed SQL statements or PL/SQL procedures (including triggers and functions). Also called *SQL cache*.

procedures—Series of steps to be followed during the performance of a given activity or process.

production database—The main database designed to keep track of the day-to-day operations of a company. See also *transactional database*.

profile—In Oracle, a named collection of settings that controls how much of the database resource a given user can use.

Q

query—A question or task asked by an end user of a database in the form of SQL code. A specific request for data manipulation issued by the end user or the application to the DBMS.

query language—A nonprocedural language that is used by a DBMS to manipulate its data. An example of a query language is SQL.

query optimizer—A DBMS process that analyzes SQL queries and finds the most efficient way to access the data. The query optimizer generates the access or execution plan for the query.

query processing bottleneck—In query optimization, a delay introduced in the processing of an I/O operation that causes the overall system to slow down.

query result set—The collection of data rows that are returned by a query.

R

RAID—An acronym that means Redundant Array of Independent Disks. RAID is used to provide balance between performance and fault tolerance. RAID systems use multiple disks to create virtual disks (storage volumes) formed by several individual disks. RAID systems provide performance improvement and fault tolerance.

record—A collection of related (logically connected) fields.

recursive query—A nested query that joins a table to itself. For example, a recursive query joins the EMPLOYEE table to itself.

recursive relationship—A relationship that is found within a single entity type. For example, an EMPLOYEE is married to an EMPLOYEE or a PART is a component of another PART.

redundant transaction logs—Most database management systems keep several copies of the transaction log to ensure that the physical failure of a disk will not impair the DBMS's ability to recover data.

referential integrity—A condition by which a dependent table's foreign key must have either a null entry or a matching entry in the related table. Even though an attribute may not have a *corresponding* attribute, it is impossible to have an invalid entry.

relation—In a relational database model, an entity set. Relations are implemented as tables. Relations (tables) are related to each other through the sharing of a common entity characteristic (value in a column).

relational algebra—A set of mathematical principles that form the basis of the manipulation of relational table contents; composed of eight main functions: SELECT, PROJECT, JOIN, INTERSECT, UNION, DIFFERENCE, PRODUCT, and DIVIDE.

relational database management system (RDBMS)—A collection of programs that manages a relational database. The RDBMS software translates a user's logical requests (queries) into commands that physically locate and retrieve the requested data. A good RDBMS also creates and maintains a data dictionary (system catalog) to help provide data security, data integrity, concurrent access, easy access, and system administration to the data in the database through a query language (SQL) and application programs.

relational diagram—A graphical representation of a relational database's entities, the attributes within those entities, and the relationships among those entities.

relational model—Developed by E. F. Codd (of IBM) in 1970, it represents a major breakthrough for users and designers because of its conceptual simplicity. The relational model, based on mathematical set theory, represents data as independent relations. Each relation (table) is conceptually represented as a matrix of intersecting rows and columns. The relations are related to each other through the sharing of common entity characteristics (values in columns).

relational online analytical processing (ROLAP)—Provides online analytical processing functionality by using relational databases and familiar relational query tools to store and analyze multidimensional data.

relational schema—The description of the organization of a relational database as seen by the database administrator.

relationship—An association between entities.

relationship degree—Indicates the number of entities or participants associated with a relationship. A relationship degree can be unary, binary, ternary, or higher level.

Remote Data Objects (RDO)—A higher-level object-oriented application interface used to access remote database servers. RDO uses the lower-level DAO and ODBC for direct access to databases. RDO was optimized to deal with server-based databases such as MS SQL Server, Oracle, and DB2.

remote request—A DDBMS feature that allows a single SQL statement to access data in a single remote DP. See also *remote transaction*.

remote transaction—A DDBMS feature that allows a transaction (formed by several requests) to access data in a single remote DP. See also *remote request*.

repeating group—In a relation, a characteristic describing a group of multiple entries of the same type that exist for a single key attribute occurrence. For example, a car can have multiple colors (top, interior, bottom, trim, and so on).

replicated data allocation—A data allocation strategy by which copies of one or more database fragments are stored at several sites.

replica transparency—Refers to the DDBMS's ability to hide the existence of multiple copies of data from the user.

replication—The process of creating and managing duplicate versions of a database. Used to place copies in different locations and to improve access time and fault tolerance.

required attribute—In ER modeling, refers to an attribute that must have a value. In other words, it cannot be left empty.

reserved words—Words that are used by a system and that cannot be used for any other purpose. For example, in Oracle SQL , the word INITIAL cannot be used to name tables or columns.

right outer join—In a pair of tables to be joined, a right outer join yields all of the rows in the right table, including the ones with no matching values in the other table. For example, a right outer join of CUSTOMER with AGENT will yield all of the agent rows, including the ones that do not have a matching CUSTOMER row. See also *left outer join* and *outer join*.

role—In Oracle, a named collection of database access privileges that authorize a user to connect to a database and use the database system resources.

ROLLBACK—A SQL command that restores the database table contents to their original condition (the condition that existed after the last COMMIT statement).

roll up—In SQL, an OLAP extension used with the GROUP BY clause to aggregate data by different dimensions. (Rolling up the data is the exact opposite of drilling down the data.) See also *drill down*.

row-level lock—A comparatively less restrictive database lock where the DBMS allows concurrent transactions to access different rows of the same table, even when the rows are located on the same page.

row-level trigger—A trigger that is executed once for each row affected by the triggering SQL statement. A row-level trigger requires the use of the FOR EACH ROW keywords in the trigger declaration.

rule-based optimizer—A query optimization mode based on the rule-based query optimization algorithm.

rule-based query optimization algorithm—A query optimization technique that uses a set of preset rules and points to determine the best approach to executing a query.

rules of precedence—Basic algebraic rules that specify the order in which operations are performed, such as conditions within parentheses being executed first. For example, in the equation $2 + (3 \times 5)$, the multiplication portion is calculated first, making the correct answer 17.

S

scheduler—The DBMS component that is responsible for establishing the order in which concurrent transaction operations are executed. The scheduler *interleaves* the execution of database operations in a specific order (sequence) to ensure *serializability*.

schema—A logical grouping of database objects (tables, indexes, views, queries, etc.) that are related to each other. Usually, a schema belongs to a single user or application.

scope—That part of a system that defines the extent of the design, according to operational requirements.

script—A programming language that is not compiled, but rather is interpreted and executed at run time.

search services—Business-enabling Web services that allow Web sites to perform searches on their contents.

secondary key—A key that is used strictly for data retrieval purposes. For example, a customer is not likely to know his or her customer number (primary key), but the combination of last name, first name, middle initial, and telephone number is likely to make a match to the appropriate table row. See also *key*.

second normal form (2NF)—The second stage in the normalization process in which a relation is in 1NF and there are no partial dependencies (dependencies in only part of the primary key).

security—Refers to activities and measures to ensure the confidentiality, integrity and availability of an information system and its main asset, data.

security breach—An event that occurs when a security threat is exploited to negatively affect the integrity, confidentiality, or availability of the system.

security threat—An imminent security violation that could occur at any time due to unchecked security vulnerabilities.

security vulnerability—A weakness in a system's component that could be exploited to allow unauthorized access or cause service disruptions.

segment—In the hierarchical data model, the equivalent of a file system's record type.

SELECT—A SQL command that yields the values of all rows or a subset of rows in a table. The SELECT statement is used to retrieve data from tables.

semantic data model—The first of a series of data models that more closely represented the real world, modeling both data and their relationships in a single structure known as an object. The SDM, published in 1981, was developed by M. Hammer and D. McLeod.

semistructured data—Data that have already been processed to some extent.

serializable schedule—In transaction management, a schedule of transaction operations in which the interleaved execution of the transactions yields the same result as if the transactions were executed in serial order.

serializability—A transaction property that ensures that the selected order of transaction operations creates a final database state that would have been produced if the transactions had been executed in a serial fashion.

server-side extension—A program that interacts directly with the server process to handle specific types of requests. They add significant functionality to Web servers and to intranets.

set theory—A mathematical science component that deals with sets, or groups of things, and is used as the basis for data manipulation in the relational model.

shared lock—A lock that is issued when a transaction requests permission to read data from a database and no exclusive locks are held on that data by another transaction. A shared lock allows other read-only transactions to access the database. See also *exclusive lock*.

simple attribute—An attribute that cannot be subdivided into meaningful components. Compare to *composite attribute*.

single-site processing, single-site data (SPSD)—A scenario in which all processing is done on a single CPU or host computer (mainframe, minicomputer, or PC) and all data are stored on the host computer's local disk.

single-user database—A database that supports only one user at a time.

single-valued attribute—An attribute that can have only one value.

slice and dice—Multidimensional jargon meaning the ability to cut slices off of the data cube (drill down or drill up) to perform a more detailed analysis.

snowflake schema—A type of star schema in which the dimension tables can have their own dimension tables. The snowflake schema is usually the result of normalizing dimension tables.

software independence—A property of any model or application that does not depend on the software used to implement it.

sparsity—In multidimensional data analysis, a measurement of the density of the data held in the data cube.

specialization hierarchy—A hierarchy that is based on the top-down process of identifying lower-level, more specific entity subtypes from a higher-level entity supertype. Specialization is based on grouping unique characteristics and relationships of the subtypes.

SQL cache—A shared, reserved memory area that stores the most recently executed SQL statements or PL/SQL procedures (including triggers and functions). Also called *procedure cache*.

SQL performance tuning—Activities oriented toward generating a SQL query that returns the correct answer in the least amount of time, using the minimum amount of resources at the server end.

standards—A detailed and specific set of instructions that describes the minimum requirements for a given activity. Standards are used to evaluate the quality of the output.

star schema—A data modeling technique used to map multidimensional decision support data into a relational database. The star schema represents data, using a central table known as a fact table, in a 1:M relationship with one or more dimension tables.

stateless system—Describes the fact that at any given time, a Web server does not know the status of any of the clients communicating with it. The Web does not reserve memory to maintain an open communications "state" between the client and the server.

statement-level trigger—A SQL trigger that is assumed if the FOR EACH ROW keywords are omitted. This type of trigger is executed once, before or after the triggering statement completes, and is the default case.

static query optimization—A query optimization mode in which the access path to a database is predetermined at compilation time. Contrast with *dynamic query optimization*.

static SQL—A style of embedded SQL in which the SQL statements do not change while the application is running.

statistically based query optimization algorithm—A query optimization technique that uses statistical information about a database. These statistics are then used by the DBMS to determine the best access strategy.

stored function—A named group of procedural and SQL statements that returns a value, indicated by a RETURN statement in its program code.

stored procedure—(1) A named collection of procedural and SQL statements. (2) Business logic stored on a server in the form of SQL code or some other DBMS-specific procedural language.

strong (identifying) relationship—When two entities are existence-dependent; from a database design perspective, this exists whenever the primary key of the related entity contains the primary key of the parent entity.

structural dependence—A data characteristic that exists when a change in the database schema affects data access, thus requiring changes in all access programs.

structural independence—A data characteristic that exists when changes in the database schema do not affect data access.

structured data—Unstructured data that have been formatted (structured) to facilitate storage, use, and information generation.

Structured Query Language (SQL)—A powerful and flexible relational database language composed of commands that enable users to create database and table structures, perform various types of data manipulation and data administration, and query the database to extract useful information.

subordinate—In a DDBMS, a DP node that participates in a distributed transaction, using the two-phase COMMIT protocol.

subquery—A query that is embedded (or nested) inside another query. Also known as a *nested query* or an *inner query*.

subschema—In the network model, the portion of the database "seen" by the application programs that produce the desired information from the data contained within the database.

subtype (entity set)—An entity (set) that contains the unique characteristics (attributes) of an entity whose general characteristics are found in another, more broadly defined entity known as a supertype. In a generalization hierarchy, it is any entity that is found below a parent entity. Example: The subtype PILOT of the supertype EMPLOYEE.

subtype discriminator—The attribute in the supertype entity that determines to which entity subtype each supertype occurrence is related.

SUM—A SQL aggregate function that yields the sum of all values for a given column or expression.

superkey—See *key*.

supertype (entity set)—An entity (set) that contains the general (commonly shared) characteristics of an entity (see *subtype*). If the entity set can include characteristics that are not common to all entities within the set, the supertype becomes the parent to one or more subtypes in a generalization hierarchy.

surrogate key—A system-assigned primary key, generally numeric and auto-incremented.

synonym—The use of different names to identify the same object, such as an entity, an attribute, or a relationship; should generally be avoided. See also *homonym*.

system catalog—A detailed system data dictionary that describes all objects in a database.

systems administrator—The person responsible for coordinating the activities of the data processing function.

systems analysis—The process that establishes the need for and the extent of an information system.

systems development—The process of creating an information system.

Systems Development Life Cycle (SDLC)—The cycle that traces the history (life cycle) of an information system. The SDLC provides the big picture within which the database design and application development can be mapped out and evaluated.

T

table—A (conceptual) matrix composed of intersecting rows (entities) and columns (attributes) that represents an entity set in the relational model. Also called a relation.

table-level lock—A locking scheme that allows only one transaction at a time to access a table. A table-level lock locks an entire table, preventing access to any row by transaction T2 while transaction T1 is using the table.

table space—In a DBMS, a logical storage space used to group related data. Also known as *file group*.

tag—In markup languages such as HTML and XML, a command inserted in a document to specify how the document should be formatted. Tags are used in server-side markup languages and interpreted by a Web browser for presenting data.

ternary relationship—An ER term used to describe an association (relationship) between three entities. Example: A CONTRIBUTOR contributes money to a FUND from which a RECIPIENT receives money.

theta join—A join operator that links tables, using an inequality comparison operator (<, >, <=, >=) in the join condition.

third normal form (3NF)—A table is in 3NF when it is in 2NF and no nonkey attribute is functionally depen-dent on another nonkey attribute; that is, it cannot include transitive dependencies.

time stamping—In transaction management, a technique used in scheduling concurrent transactions that assigns a global unique time stamp to each transaction.

time-variant data—Data whose values are a function of time. For example, time variant data can be seen at work when the history of all administrative appointments (date of appointment and date of termination) are tracked.

top-down design—A design philosophy that begins by defining the main (macro) structures of a system and then moves to define the smaller units within those structures. In database design, it is a process that first identifies entities and then defines the attributes within the entities. Compare to *bottom-up design*.

total completeness—In a generalization/specialization hierarchy, a condition in which every supertype occurrence must be a member of at least one subtype.

transaction—A sequence of database operations (one or more database requests) that accesses the database. A transaction is a logical unit of work; that is, it must be *entirely* completed or aborted—no intermediate ending states are accepted. All transactions must have the following properties: (1) *Atomicity* requires that, unless all operations (parts) of a transaction are completed, the transaction be aborted. A transaction is treated as a single, indivisible logical unit of work. (2) *Consistency* indicates the permanence of the database consistent state. Once a transaction is completed, the database reaches a consistent state. (3) *Isolation* assures that the data used during the execution of a transaction cannot be used by a second transaction until the first one is completed. (4) *Durability* assures that once transaction changes are done, they cannot be undone or lost, even in the event of a system failure.

transactional database—A database designed to keep track of the day-to-day transactions of an organization. See also *production database*.

transaction log—A feature used by the DBMS to keep track of all transaction operations that update the database. The information stored in this log is used by the DBMS for recovery purposes.

transaction log backup—Backs up only the transaction log operations that are not reflected in a previous backup copy of the database.

transaction manager (TM)—See *transaction processor (TP)*.

transaction processor (TP)—In a DDBMS, the software component on each computer that requests data. The TP is responsible for the execution and coordination of all databases issued by a local application that access data on any DP. Also called *transaction manager (TM)*. See also *data processor (DP)*.

transaction transparency—A DDBMS property that ensures that database transactions will maintain the distributed database's integrity and consistency. They ensure that a transaction will be completed only when all database sites involved in the transaction complete their part of the transaction.

transitive dependency—A condition in which an attribute is dependent on another attribute that is not part of the primary key.

trigger—A procedural SQL code that is automatically invoked by the relational database management system upon the occurrence of a data manipulation event.

tuple—In the relational model, a table row.

two-phase commit protocol—In a DDBMS, an algorithm used to ensure atomicity of transactions and database consistency as well as integrity in distributed transactions.

two-phase locking—A set of rules that governs the way transactions acquire and relinquish locks. Two-phase locking guarantees serializability, but it does not prevent deadlocks. The two-phase locking protocol is divided into two phases: (1) A *growing phase* occurs when the transaction acquires all of the locks that it needs without unlocking any *existing* data locks. Once all locks have been acquired, the transaction is in its *locked* point. (2) A *shrinking phase* occurs when the transaction releases all locks and cannot obtain a new lock.

U

unary relationship—An ER term used to describe an association *within* an entity. Example: A COURSE is a prerequisite to another COURSE.

uncommitted data—When trying to achieve concurrency control, uncommitted data causes data integrity and consistency problems. It occurs when two transactions are executed concurrently and the first transaction is rolled back after the second transaction has already accessed the uncommitted data, thus violating the isolation property of transactions.

Unified Modeling Language (UML)—A language based on object-oriented concepts that provides tools such as diagrams and symbols used to graphically model a system.

union-compatible—Two or more tables are union-compatible when they share the same column names and the columns have compatible data types or domains.

unique fragment—In a DDBMS, a condition indicating that each row is unique, regardless of which fragment it is located in.

unique index—An index in which the index key can have only one pointer value (row) associated with it.

uniqueness—In concurrency control, a property of time stamping that ensures that no equal time stamp values can exist.

Universal Data Access (UDA)—Within the Microsoft application framework, a collection of technologies used to access any type of data source and to manage the data through a common interface.

unreplicated database—A distributed database in which each database fragment is stored at a single site.

unstructured data—Data that exist in their original (raw) state; that is in the format in which they were collected.

updatable view—A view that can be used to update attributes in base tables that are used in the view.

UPDATE—A SQL command that allows attribute values to be changed in one or more rows of a table.

user—In a system, a uniquely identifiable object that allows a given person or process to log on to the database.

V

VBScript—A client-side extension in the form of a Microsoft language product used to extend a browser's functionality; derived from Visual Basic.

vertical fragmentation—In distributed database design, the process that breaks up a table into fragments consisting of a subset of columns from the original table. Fragments must share a common primary key. See also *database fragments* and *horizontal fragmentation*.

very large databases (VLDBs)—As the name implies, databases that contain huge amounts of data—gigabyte, terabyte, and petabytes ranges are not unusual.

view—A virtual table based on a SELECT query.

W

wait/die—A concurrency control scheme that says that if the transaction requesting the lock in the older, it waits for the younger transaction to complete and release the locks. Otherwise, the newer transaction dies and it is rescheduled.

weak entity—An entity that displays existence dependence and inherits the primary key of its parent entity. Example: A DEPENDENT requires the existence of an EMPLOYEE.

weak relationship—A relationship that exists when the PK of the related entity does not contain a PK component of the parent entity. Also known as a *non-identifying relationship*.

Web application server—A middleware application that expands the functionality of Web servers by linking them to a wide range of services, such as databases, directory systems, and search engines.

Web-to-database middleware—A database server-side extension program that retrieves data from databases and passes it on to the Web server, which sends it to the client's browser for display purposes.

wildcard character—A symbol that can be used as a general substitute for one or more characters in an SQL LIKE clause condition. The wildcard characters used in SQL are the _ and % symbols.

workgroup database—A multiuser database that supports a relatively small number of users (usually fewer than 50) or that is used for a specific department in an organization.

wound/wait—A concurrency control scheme that says that if the transaction requesting the lock in the older, it preempts the younger transaction and reschedules it. Otherwise, the newer transaction waits until the older transaction finishes.

write-ahead-log protocol—In concurrency control, a process that ensures that transaction logs are always written to permanent storage before any database data are actually updated. Also called write-ahead protocol.

write-ahead protocol—See *write-ahead-log protocol*.

write-through technique—In concurrency control, a process that ensures that a database is immediately updated by transaction operations during the transaction's execution, even before the transaction reaches its commit point.

X

XML—See *Extensible Markup Language (XML)*.

XML database—A database system that stores and manages semistructured XML data.

XML schema—An advanced data definition language that is used to describe the structure (elements, data types, relationship types, ranges, and default values) of XML data documents. One of the main advantages of an XML schema is that it more closely maps to database terminology and features. For example, an XML schema will be able to define common database types such as date, integer or decimal, minimum and maximum values, list of valid values, and required elements. Using the XML schema, a company could validate the data for values that may be out of range, incorrect dates, valid values, and so on.

XML schema definition (XSD)—A file containing the description of an XML document.

XSL (Extensible Style Language)—A specification used to define the rules by which XML data are formatted and displayed. The XSL specification is divided into two parts: Extensible Style Language Transformations (XSLT) and XSL style sheets.

XSL style sheets—Similar to presentation templates, define the presentation rules applied to XML elements. The XSL style sheet describes the formatting options to apply to XML elements when they are displayed on a browser, cellular phone display, PDA screen, and so on.

XSLT (Extensible Style Language Transformations)—A term that describes the general mechanism used to extract and process data from one XML document and to enable its transformation within another document.